AF654063

About Pearson

Pearson is the world's learning company, with presence across 70 countries worldwide. Our unique insights and world-class expertise comes from a long history of working closely with renowned teachers, authors and thought leaders, as a result of which, we have emerged as the preferred choice for millions of teachers and learners across the world.

We believe learning opens up opportunities, creates fulfilling careers and hence better lives. We hence collaborate with the best of minds to deliver you class-leading products, spread across the Higher Education and K12 spectrum.

Superior learning experience and improved outcomes are at the heart of everything we do. This product is the result of one such effort.

Your feedback plays a critical role in the evolution of our products and you can contact us – reachus@pearson.com. We look forward to it.

STATISTICS FOR THE LIFE SCIENCES

Fifth Edition

Myra L. Samuels
Purdue University

Jeffrey A. Witmer
Oberlin College

Andrew A. Schaffner
California Polytechnic State University, San Luis Obispo

Pearson

Original edition, entitled *Statistics for the Life Sciences,* 5th Edition, by Samuels, Myra L.; Witmer, Jeffrey A.; Schaffner, Andrew, published by Pearson Education, Limited. Copyright © 2016 by Pearson Education, Inc.

Indian Subcontinent Reprint

ISBN 978-93-530-6847-9

First Impression

Published by Pearson India Education Services Pvt. Ltd, CIN: U72200TN2005PTC057128.

Head Office: 15th Floor, Tower-B, World Trade Tower, Plot No. 1, Block-C, Sector 16, Noida 201 301, Uttar Pradesh, India.
Registered Office: 4th Floor, Software Block, Elnet Software City, TS 140, Block 2 & 9, Rajiv Gandhi Salai, Taramani, Chennai 600 113, Tamil Nadu, India.
Fax: 080-30461003, Phone: 080-30461060
Website: in.pearson.com; Email: companysecretary.india@pearson.com

Printed in India at Sai Printo Pack Pvt Ltd

CONTENTS

*Indicates optional chapters

PREFACE

Statistics for the Life Sciences is an introductory text in statistics, specifically addressed to students specializing in the life sciences. Its primary aims are (1) to show students how statistical reasoning is used in biological, medical, and agricultural research; (2) to enable students to confidently carry out simple statistical analyses and to interpret the results; and (3) to raise students' awareness of basic statistical issues such as randomization, confounding, and the role of independent replication.

Style and Approach

The style of *Statistics for the Life Sciences* is informal and uses only minimal mathematical notation. There are no prerequisites except elementary algebra; anyone who can read a biology or chemistry textbook can read this text. It is suitable for use by graduate or undergraduate students in biology, agronomy, medical and health sciences, nutrition, pharmacy, animal science, physical education, forestry, and other life sciences.

Use of Real Data Real examples are more interesting and often more enlightening than artificial ones. *Statistics for the Life Sciences* includes hundreds of examples and exercises that use real data, representing a wide variety of research in the life sciences. Each example has been chosen to illustrate a particular statistical issue. The exercises have been designed to reduce computational effort and focus students' attention on concepts and interpretations.

Emphasis on Ideas The text emphasizes statistical ideas rather than computations or mathematical formulations. Probability theory is included only to support statistical concepts. The text stresses interpretation throughout the discussion of descriptive and inferential statistics. By means of salient examples, we show why it is important that an analysis be appropriate for the research question to be answered, for the statistical design of the study, and for the nature of the underlying distributions. We help the student avoid the common blunder of confusing statistical nonsignificance with practical insignificance and encourage the student to use confidence intervals to assess the magnitude of an effect. The student is led to recognize the impact on real research of design concepts such as random sampling, randomization, efficiency, and the control of extraneous variation by blocking or adjustment. Numerous exercises amplify and reinforce the student's grasp of these ideas.

The Role of Technology The analysis of research data is usually carried out with the aid of a computer. Computer-generated graphs are shown at several places in the text. However, in studying statistics it is desirable for the student to gain experience working directly with data, using paper and pencil and a hand-held calculator, as well as a computer. This experience will help the student appreciate the nature and purpose of the statistical computations. The student is thus prepared to make intelligent use of the computer—to give it appropriate instructions and properly interpret the output. Accordingly, most of the exercises in this text are intended for hand calculation.

Organization

This text is organized to permit coverage in one semester of the maximum number of important statistical ideas, including power, multiple inference, and the basic principles of design. By including or excluding optional sections, the instructor can also use the text for a one-quarter course or a two-quarter course. It is suitable for a terminal course or for the first course of a sequence.

The following is a brief outline of the text.

Unit I: Data and Distributions

Chapter 1: Introduction. The nature and impact of variability in biological data. The hazards of observational studies, in contrast with experiments. Random sampling.

Chapter 2: Description of distributions. Frequency distributions, descriptive statistics, the concept of population versus sample.

Chapters 3, 4, and 5: Theoretical preparation. Probability, binomial and normal distributions, sampling distributions.

Unit II: Inference for Means

Chapter 6: Confidence intervals for a single mean and for a difference in means.

Chapter 7: Hypothesis testing, with emphasis on the *t* test. The randomization test, the Wilcoxon-Mann-Whitney test.

Chapter 8: Inference for paired samples. Confidence interval, *t* test, sign test, and Wilcoxon signed-rank test.

Unit III: Inference for Categorical Data

Chapter 9: Inference for a single proportion. Confidence intervals and the chi-square goodness-of-fit test.

Chapter 10: Relationships in categorical data. Conditional probability, contingency tables. Optional sections cover Fisher's exact test, McNemar's test, and odds ratios.

Unit IV: Modeling Relationships

Chapter 11: Analysis of variance. One-way layout, multiple comparison procedures, one-way blocked ANOVA, two-way ANOVA. Contrasts and multiple comparisons are included in optional sections.

Chapter 12: Correlation and regression. Descriptive and inferential aspects of correlation and simple linear regression and the relationship between them.

Chapter 13: A summary of inference methods.

Most sections within each chapter conclude with section-specific exercises. Chapters and units conclude with supplementary exercises that provide opportunities for students to practice integrating the breadth of methods presented within the chapter or across the entire unit.

Changes to the Fifth Edition

- Chapters are grouped by unit, and feature Unit Highlights with reflections, summaries, and additional examples and exercises at the end of each unit that often require connecting ideas from multiple chapters.
- We added material on randomization-based inference to introduce or motivate most inference procedures presented in this text. There are now presentations of randomization methods at the beginnings of Chapters 7, 8, 10, 11, and 12.
- New exercises have been added throughout the text. Many exercises from the previous edition that involved calculation and reading tables have been updated to exercises that require interpretation of computer output.
- We replaced many older examples throughout the text with examples from current research from a variety life science disciplines.

Instructor Supplements

Instructor's Solutions Manual (downloadable) (ISBN-13: 978-1-292-10183-5; ISBN-10: 1-292-10183-0) Solutions to all exercises are available as a downloadable manual from Pearson Education's online catalog at www.pearsoned.co.in/Samuels. Careful attention has been paid to ensure that all methods of solution and notation are consistent with those used in the core text.

PowerPoint Slides (downloadable) (ISBN-13: 978-1-292-10184-2; ISBN-10: 1-292-10184-9) Selected figures and tables from throughout the textbook are available as downloadable PowerPoint slides for use in creating custom PowerPoint lecture presentations. These slides are available for download at www.pearsoned.co.in/Samuels.

Acknowledgments for the Fifth Edition

The fifth edition of *Statistics for the Life Science* retains the style and spirit of the writing of Myra Samuels. Prior to her tragic death from cancer, Myra wrote the first edition of the text, based on her experience both as a teacher of statistics and as a statistical consultant. We hope that the book retains her vision.

Many researchers have contributed sets of data to the text, which have enriched the text considerably. We have benefited from countless conversations over the years with David Moore, Dick Scheaffer, Murray Clayton, Alan Agresti, Don Bentley, George Cobb, and many others who have our thanks.

We are grateful for the sound editorial guidance and encouragement of Katherine Roz. We are also grateful for adopters of the earlier editions, particularly Robert Wolf and Jeff May, whose suggestions led to improvements in the current edition. Finally, we express our gratitude to the reviewers of this edition:

Jeffrey Schmidt (University of Wisconsin-Parkside), Liansheng Tang (George Mason University), Tim Hanson (University of South Carolina), Mohammed Kazemi (University of North Carolina–Charlotte), Kyoungmi Kim (University of California, Davis), and Leslie Hendrix (University of South Carolina)

Special Thanks

To Merrilee, for her steadfast support.
JAW

To Michelle, for her patience and encouragement, and for my sons, Ganden and Tashi, for their curiosity and interest in learning something new every day.
AAS

Pearson wishes to thank and acknowledge the following people for their work on this Edition:

Contributor

C. V. Vinay, *JSS Academy of Technical Education*

Dilip Nath, *Gauhati University*

Reviewers

D. V. Chandrashekhar, *Vivekananda Institute of Technology*

Sunil Jacob John, *National Institute of Technology Calicut*

D. V. Jayalakshmamma, *Vemana Institute of Technology*

INTRODUCTION

Chapter 1

OBJECTIVES

In this chapter we will look at a series of examples of areas in the life sciences in which statistics is used, with the goal of understanding the scope of the field of statistics. We will also

- explain how experiments differ from observational studies.
- discuss the concepts of placebo effect, blinding, and confounding.
- discuss the role of random sampling in statistics.

1.1 Statistics and the Life Sciences

Researchers in the life sciences carry out investigations in various settings: in the clinic, in the laboratory, in the greenhouse, in the field. Generally, the resulting data exhibit some *variability*. For instance, patients given the same drug respond somewhat differently; cell cultures prepared identically develop somewhat differently; adjacent plots of genetically identical wheat plants yield somewhat different amounts of grain. Often the degree of variability is substantial even when experimental conditions are held as constant as possible.

The challenge to the life scientist is to discern the patterns that may be more or less obscured by the variability of responses in living systems. The scientist must try to distinguish the "signal" from the "noise."

Statistics is the science of understanding data and of making decisions in the face of variability and uncertainty. The discipline of statistics has evolved in response to the needs of scientists and others whose data exhibit variability. The concepts and methods of statistics enable the investigator to describe variability and to plan research so as to take variability into account (i.e., to make the "signal" strong in comparison to the background "noise" in data that are collected). Statistical methods are used to analyze data so as to extract the maximum information and also to quantify the reliability of that information.

We begin with some examples that illustrate the degree of variability found in biological data and the ways in which variability poses a challenge to the biological researcher. We will briefly consider examples that illustrate some of the statistical issues that arise in life sciences research and indicate where in this book the issues are addressed.

The first two examples provide a contrast between an experiment that showed no variability and another that showed considerable variability.

Example 1.1.1

Vaccine for Anthrax Anthrax is a serious disease of sheep and cattle. In 1881, Louis Pasteur conducted a famous experiment to demonstrate the effect of his vaccine against anthrax. A group of 24 sheep were vaccinated; another group of 24 unvaccinated sheep served as controls. Then, all 48 animals were inoculated with a virulent culture of anthrax bacillus. Table 1.1.1 shows the results.[1] The data of Table 1.1.1 show no variability; all the vaccinated animals survived and all the unvaccinated animals died. ■

Table 1.1.1 Response of sheep to anthrax

	Treatment	
Response	Vaccinated	Not vaccinated
Died of anthrax	0	24
Survived	24	0
Total	24	24
Percent survival	100%	0%

Example 1.1.2

Bacteria and Cancer To study the effect of bacteria on tumor development, researchers used a strain of mice with a naturally high incidence of liver tumors. One group of mice were maintained entirely germ free, while another group were exposed to the intestinal bacteria *Escherichia coli*. The incidence of liver tumors is shown in Table 1.1.2.[2]

Table 1.1.2 Incidence of liver tumors in mice

	Treatment	
Response	*E. coli*	Germ free
Liver tumors	8	19
No liver tumors	5	30
Total	13	49
Percent with liver tumors	62%	39%

In contrast to Table 1.1.1, the data of Table 1.1.2 show variability; mice given the same treatment did not all respond the same way. Because of this variability, the results in Table 1.1.2 are equivocal; the data suggest that exposure to *E. coli* increases the risk of liver tumors, but the possibility remains that the observed difference in percentages (62% versus 39%) might reflect only chance variation rather than an effect of *E. coli*. If the experiment were replicated with different animals, the percentages might change substantially.

One way to explore what might happen if the experiment were replicated is to simulate the experiment, which could be done as follows. Take 62 cards and write "liver tumors" on 27 (= 8 + 19) of them and "no liver tumors" on the other 35 (= 5 + 30). Shuffle the cards and randomly deal 13 cards into one stack (to correspond to the *E. coli* mice) and 49 cards into a second stack. Next, count the number of cards in the "*E. coli* stack" that have the words "liver tumors" on them—to correspond to mice exposed to *E. coli* who develop liver tumors—and record whether this number is greater than or equal to 8. This process represents distributing 27 cases of liver tumors to two groups of mice (*E. coli* and germ free) randomly, with *E. coli* mice no more likely, nor any less likely, than germ-free mice to end up with liver tumors.

If we repeat this process many times (say, 10,000 times, with the aid of a computer in place of a physical deck of cards), it turns out that roughly 12% of the time we get 8 or more *E. coli* mice with liver tumors. Since something that happens 12% of the time is not terribly surprising, Table 1.1.2 does not provide significant evidence that exposure to *E. coli* increases the incidence of liver tumors. ■

In Chapter 10 we will discuss statistical techniques for evaluating data such as those in Tables 1.1.1 and 1.1.2. Of course, in some experiments variability is minimal and the message in the data stands out clearly without any special statistical analysis. It is worth noting, however, that absence of variability is itself an experimental result that must be justified by sufficient data. For instance, because Pasteur's anthrax data (Table 1.1.1) show no variability at all, it is intuitively plausible to conclude that the data provide "solid" evidence for the efficacy of the vaccination. But note that this conclusion involves a judgment; consider how much *less* "solid" the evidence would be if Pasteur had included only 3 animals in each group, rather than 24. Statistical analyses can be used to make such a judgment, that is, to determine if the variability is indeed negligible. Thus, a statistical view can be helpful even in the absence of variability.

The next two examples illustrate additional questions that a statistical approach can help to answer.

Example 1.1.3

Flooding and ATP In an experiment on root metabolism, a plant physiologist grew birch tree seedlings in the greenhouse. He flooded four seedlings with water for one day and kept four others as controls. He then harvested the seedlings and analyzed the roots for adenosine triphosphate (ATP). The measured amounts of ATP (nmoles per mg tissue) are given in Table 1.1.3 and displayed in Figure 1.1.1.[3]

Table 1.1.3 ATP concentration in birch tree roots (nmol/mg)

Flooded	Control
1.45	1.70
1.19	2.04
1.05	1.49
1.07	1.91

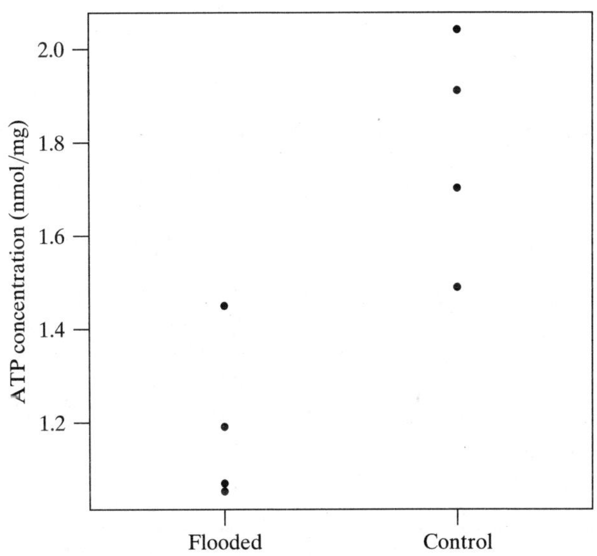

Figure 1.1.1 ATP concentration in birch tree roots

The data of Table 1.1.3 raise several questions: How should one summarize the ATP values in each experimental condition? How much information do the data provide about the effect of flooding? How confident can one be that the reduced ATP in the flooded group is really a response to flooding rather than just random variation? What size experiment would be required in order to firmly corroborate the apparent effect seen in these data? ■

Chapters 2, 6, and 7 address questions like those posed in Example 1.1.3. One question that we can address here is whether the data in Table 1.1.3 are consistent with the claim that flooding has no effect on ATP concentration, or instead provide significant evidence that flooding affects ATP concentrations. If the claim of no effect is true, then should we be surprised to see that all four of the flooded observations are smaller than each of the control observations? Might this happen by chance alone? If we wrote each of the numbers 1.05, 1.07, 1.19, 1.45, 1.49, 1.91, 1.70, and 2.04 on cards, shuffled the eight cards, and randomly dealt them into two piles, what is the chance that the four smallest numbers would end up in one pile and the four largest numbers in the other pile? It turns out that we could expect this to happen 1 time in 35 random shufflings, so "chance alone" would only create the kind of imbalance seen in Figure 1.1.1 about 2.9% of the time (since $1/35 = 0.029$). Thus, we have some evidence that flooding has an effect on ATP concentration. We will develop this idea more fully in Chapter 7.

Example 1.1.4

MAO and Schizophrenia Monoamine oxidase (MAO) is an enzyme that is thought to play a role in the regulation of behavior. To see whether different categories of patients with schizophrenia have different levels of MAO activity, researchers collected blood specimens from 42 patients and measured the MAO activity in the platelets. The results are given in Table 1.1.4 and displayed in Figure 1.1.2. (Values are expressed as nmol benzylaldehyde product per 10^8 platelets per hour.[4]) Note that it is much easier to get a feeling for the data by looking at the graph (Figure 1.1.2) than it is to read through the data in the table. The use of graphical displays of data is a very important part of data analysis. ■

Table 1.1.4 MAO activity in patients with schizophrenia

Diagnosis	MAO activity				
I:	6.8	4.1	7.3	14.2	18.8
Chronic	9.9	7.4	11.9	5.2	7.8
undifferentiated schizophrenia	7.8	8.7	12.7	14.5	10.7
(18 patients)	8.4	9.7	10.6		
II:	7.8	4.4	11.4	3.1	4.3
Undifferentiated	10.1	1.5	7.4	5.2	10.0
with paranoid features	3.7	5.5	8.5	7.7	6.8
(16 patients)	3.1				
III:	6.4	10.8	1.1	2.9	4.5
Paranoid	5.8	9.4	6.8		
schizophrenia (8 patients)					

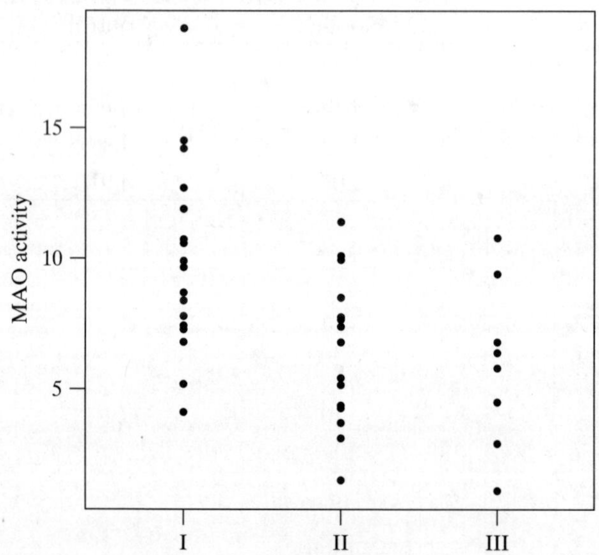

Figure 1.1.2 MAO activity in patients with schizophrenia

To analyze the MAO data, one would naturally want to make comparisons among the three groups of patients, to describe the reliability of those comparisons, and to characterize the variability within the groups. To go beyond the data to a biological interpretation, one must also consider more subtle issues, such as the following: How were the patients selected? Were they chosen from a common hospital population, or were the three groups obtained at different times or places? Were precautions taken so that the person measuring the MAO was unaware of the patient's diagnosis? Did the investigators consider various ways of subdividing the patients before choosing the particular diagnostic categories used in Table 1.1.4? At first glance, these questions may seem irrelevant—can we not let the measurements speak for themselves? We will see, however, that the proper interpretation of data always requires careful consideration of how the data were obtained.

Sections 1.2 and 1.3. as well as Chapters 2 and 8, include discussions of selection of experimental subjects and of guarding against unconscious investigator bias. In Chapter 11 we will show how sifting through a data set in search of patterns can lead to serious misinterpretations and we will give guidelines for avoiding the pitfalls in such searches.

The next example shows how the effects of variability can distort the results of an experiment and how this distortion can be minimized by careful design of the experiment.

Example 1.1.5 **Food Choice by Insect Larvae** The clover root curculio, *Sitona hispidulus*, is a root-feeding pest of alfalfa. An entomologist conducted an experiment to study food choice by *Sitona* larvae. She wished to investigate whether larvae would preferentially choose alfalfa roots that were nodulated (their natural state) over roots whose nodulation had been suppressed. Larvae were released in a dish where both nodulated and nonnodulated roots were available. After 24 hours, the investigator counted the larvae that had clearly made a choice between root types. The results are shown in Table 1.1.5.[5]

The data in Table 1.1.5 appear to suggest rather strongly that *Sitona* larvae prefer nodulated roots. But our description of the experiment has obscured an important point—we have not stated how the roots were arranged. To see the relevance of the arrangement, suppose the experimenter had used only one dish, placing all the nodulated roots on one side of the dish and all the nonnodulated roots on the other side, as shown in Figure 1.1.3(a), and had then released 120 larvae in the center of the dish. This experimental arrangement would be seriously deficient, because the data of Table 1.1.5 would then permit several competing interpretations—for instance, (a) perhaps the larvae really do prefer nodulated roots; or (b) perhaps the two sides of the dish were at slightly different temperatures and the larvae were responding to temperature rather than nodulation; or (c) perhaps one larva chose the nodulated roots just by chance and the other larvae followed its trail. Because of these possibilities the experimental arrangement shown in Figure 1.1.3(a) can yield only weak information about larval food preference.

Table 1.1.5 Food choice by *Sitona* larvae

Choice	Number of larvae
Chose nodulated roots	46
Chose nonnodulated roots	12
Other (no choice, died, lost)	62
Total	120

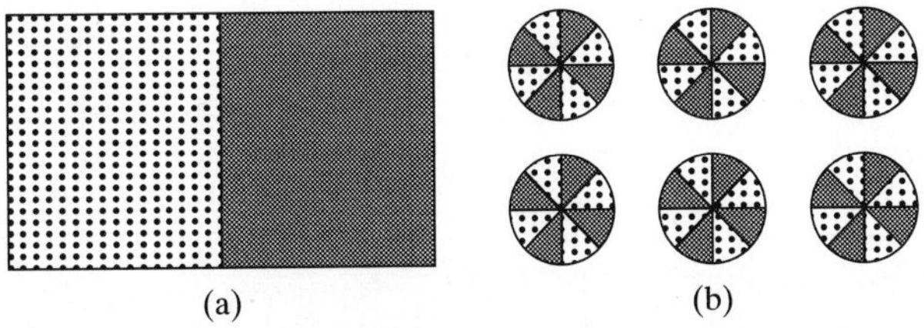

Figure 1.1.3 Possible arrangements of food choice experiment. The dark-shaded areas contain nodulated roots and the light-shaded areas contain nonnodulated roots.
(a) A poor arrangement.
(b) A good arrangement.

The experiment was actually arranged as in Figure 1.1.3(b), using six dishes with nodulated and nonnodulated roots arranged in a symmetric pattern. Twenty larvae were released into the center of each dish. This arrangement avoids the pitfalls of the arrangement in Figure 1.1.3(a). Because of the alternating regions of nodulated and nonnodulated roots, any fluctuation in environmental conditions (such as temperature) would tend to affect the two root types equally. By using several dishes, the experimenter has generated data that can be interpreted even if the larvae do tend to follow each other. To analyze the experiment properly, we would need to know the results in each dish; the condensed summary in Table 1.1.5 is not adequate. ■

In Chapter 11 we will describe various ways of arranging experimental material in space and time so as to yield the most informative experiment, as well as how to analyze the data to extract as much information as possible and yet resist the temptation to overinterpret patterns that may represent only random variation.

The following example is a study of the relationship between two measured quantities.

Example 1.1.6 **Body Size and Energy Expenditure** How much food does a person need? To investigate the dependence of nutritional requirements on body size, researchers used underwater weighing techniques to determine the fat-free body mass for each of seven men. They also measured the total 24-hour energy expenditure during conditions of quiet sedentary activity; this was repeated twice for each subject. The results are shown in Table 1.1.6 and plotted in Figure 1.1.4.[6]

Table 1.1.6 Fat-free mass and energy expenditure

Subject	Fat-free mass (kg)	24-hour energy expenditure (kcal)	
1	49.3	1,851	1,936
2	59.3	2,209	1,891
3	68.3	2,283	2,423
4	48.1	1,885	1,791
5	57.6	1,929	1,967
6	78.1	2,490	2,567
7	76.1	2,484	2,653

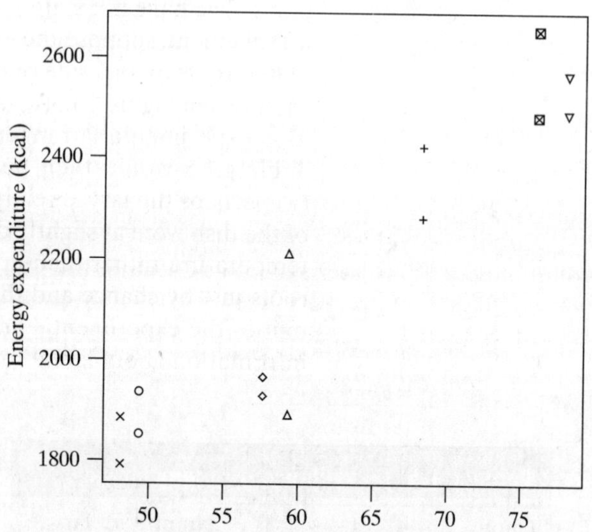

Figure 1.1.4 Fat-free mass and energy expenditure in seven men. Each man is represented by a different symbol.

A primary goal in the analysis of these data would be to describe the relationship between fat-free mass and energy expenditure—to characterize not only the overall trend of the relationship, but also the degree of scatter or variability in the relationship. (Note also that, to analyze the data, one needs to decide how to handle the duplicate observations on each subject.) ■

The focus of Example 1.1.6 is on the relationship between two variables: fat-free mass and energy expenditure. Chapter 12 deals with methods for describing such relationships, and also for quantifying the reliability of the descriptions.

A LOOK AHEAD

Where appropriate, statisticians make use of the computer as a tool in data analysis; computer-generated output and statistical graphics appear throughout this book. The computer is a powerful tool, but it must be used with caution. Using the computer to perform calculations allows us to concentrate on concepts. The danger when using a computer in statistics is that we will jump straight to the calculations without looking closely at the data and asking the right questions about the data. Our goal is to analyze, understand, and interpret data—which are numbers *in a specific context*—not just to perform calculations.

In order to understand a data set it is necessary to know how and why the data were collected. In addition to considering the most widely used methods in statistical inference, we will consider issues in data collection and experimental design. Together, these topics should provide the reader with the background needed to read the scientific literature and to design and analyze simple research projects.

The preceding examples illustrate the kind of data to be considered in this book. In fact, each of the examples will reappear as an exercise or example in an appropriate chapter. As the examples show, research in the life sciences is usually concerned with the comparison of two or more groups of observations, or with the relationship between two or more variables. We will begin our study of statistics by focusing on a simpler situation—observations of a *single* variable for a *single* group. Many of the basic ideas of statistics will be introduced in this oversimplified context. Two-group comparisons and more complicated analyses will then be discussed in Chapter 7 and later chapters.

1.2 Types of Evidence

Researchers gather information and make inferences about the state of nature in a variety of settings. Much of statistics deals with the *analysis* of data, but statistical considerations often play a key role in the planning and *design* of a scientific investigation. We begin with examples of the three major kinds of evidence that one encounters.

Example 1.2.1

Lightning and Deafness On 15 July 1911, 65-year-old Mrs. Jane Decker was struck by lightning while in her house. She had been deaf since birth, but after being struck, she recovered her hearing, which led to a headline in the *New York Times*, "Lightning Cures Deafness."[7] Is this compelling evidence that lightning is a cure for deafness? Could this event have been a coincidence? Are there other explanations for her cure? ■

The evidence discussed in Example 1.2.1 is **anecdotal evidence**. An anecdote is a short story or an example of an interesting event, in this case, of lightning curing deafness. The accumulation of anecdotes often leads to conjecture and to scientific investigation, but it is predictable pattern, not anecdote, that establishes a scientific theory.

Example 1.2.2 **Sexual Orientation** Some research has suggested that there is a genetic basis for sexual orientation. One such study involved measuring the midsagittal area of the anterior commissure (AC) of the brain for 30 homosexual men, 30 heterosexual men, and 30 heterosexual women. The researchers found that the AC tends to be larger in heterosexual women than in heterosexual men and that it is even larger in homosexual men. These data are summarized in Table 1.2.1 and are shown graphically in Figure 1.2.1.

Table 1.2.1 Midsagittal area of the anterior commissure (mm^2)

Group	Average midsagittal area (mm^2) of the anterior commissure
Homosexual men	14.20
Heterosexual men	10.61
Heterosexual women	12.03

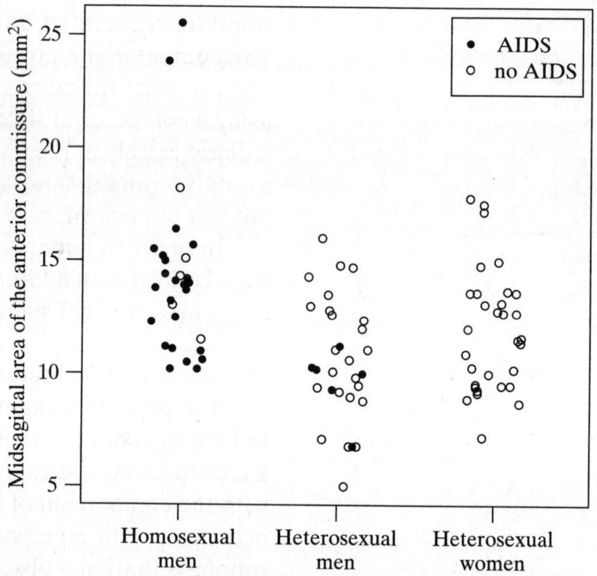

Figure 1.2.1 Midsagittal area of the anterior commissure (mm^2)

The data suggest that the size of the AC in homosexual men is more like that of heterosexual women than that of heterosexual men. When analyzing these data, we should take into account two things. (1) The measurements for two of the homosexual men were much larger than any of the other measurements; sometimes one or two such outliers can have a big impact on the conclusions of a study. (2) Twenty-four of the 30 homosexual men had AIDS, as opposed to 6 of the 30 heterosexual men; if AIDS affects the size of the anterior commissure, then this factor could account for some of the difference between the two groups of men.[8] ■

Example 1.2.2 presents an **observational study**. In an observational study the researcher systematically collects data from subjects, but only as an observer and not as someone who is manipulating conditions. By systematically examining all the data that arise in observational studies, one can guard against selectively viewing and reporting only evidence that supports a previous view. However, observational studies can be misleading due to *confounding variables*. In Example 1.2.2 we noted that having AIDS may affect the size of the anterior commissure. We would say that the effect of AIDS is confounded with the effect of sexual orientation in this study.

Note that the *context* in which the data arose is of central importance in statistics. This is quite clear in Example 1.2.2. The numbers themselves can be used to compute averages or to make graphs, like Figure 1.2.1, but if we are to understand what the data have to say, we must have an understanding of the context in which they arose. This context tells us to be on the alert for the effects that other factors, such as the impact of AIDS, may have on the size of the anterior commissure. Data analysis without reference to context is meaningless.

Example 1.2.3

Health and Marriage A study conducted in Finland found that people who were married at midlife were less likely to develop cognitive impairment (particularly Alzheimer's disease) later in life.[9] However, from an observational study such as this we don't know whether marriage *prevents* later problems or whether persons who are likely to develop cognitive problems are less likely to get married. ■

Example 1.2.4

Toxicity in Dogs Before new drugs are given to human subjects, it is common practice to first test them in dogs or other animals. In part of one study, a new investigational drug was given to eight male and eight female dogs at doses of 8 mg/kg and 25 mg/kg. Within each sex, the two doses were assigned at random to the eight dogs. Many "endpoints" were measured, such as cholesterol, sodium, glucose, and so on, from blood samples, in order to screen for toxicity problems in the dogs before starting studies on humans. One endpoint was alkaline phosphatase level (or APL, measured in U/l). The data are shown in Table 1.2.2 and plotted in Figure 1.2.2.[10]

Table 1.2.2 Alkaline phosphatase level (U/l)

Dose (mg/kg)	Male	Female
8	171	150
	154	127
	104	152
	143	105
Average	**143**	**133.5**
25	80	101
	149	113
	138	161
	131	197
Average	**124.5**	**143**

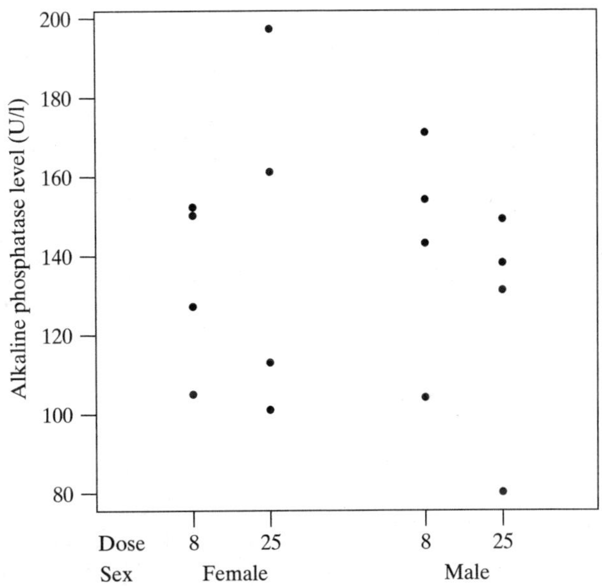

Figure 1.2.2 Alkaline phosphatase level in dogs

The design of this experiment allows for the investigation of the interaction between two factors: sex of the dog and dose. These factors interacted in the following sense: For females, the effect of increasing the dose from 8 to 25 mg/kg was positive, although small (the average APL increased from 133.5 to 143 U/l), but for males the effect of increasing the dose from 8 to 25 mg/kg was negative (the average APL dropped from 143 to 124.5 U/l). Techniques for studying such interactions will be considered in Chapter 11. ■

Example 1.2.4 presents an **experiment**, in that the researchers imposed the conditions—in this case, doses of a drug—on the subjects (the dogs). By randomly assigning treatments (drug doses) to subjects (dogs), we can get around the problem of confounding that complicates observational studies and limits the conclusions that we can reach from them. Randomized experiments are considered the "gold standard" in scientific investigation, but they can also be plagued by difficulties.

Often human subjects in experiments are given a **placebo**—an inert substance, such as a sugar pill. It is well known that people often exhibit a *placebo response*; that is, they tend to respond favorably to *any* treatment, even if it is only inert. This psychological effect can be quite powerful. Research has shown that placebos are effective for roughly one-third of people who are in pain; that is, one-third of pain sufferers report their pain ending after being giving a "painkiller" that is, in fact, an inert pill. For diseases such as bronchial asthma, angina pectoris (recurrent chest pain caused by decreased blood flow to the heart), and ulcers, the use of placebos has been shown to produce clinically beneficial results in over 60% of patients.[11] Of course, if a placebo control is used, then the subjects must not be told which group they are in—the group getting the active treatment or the group getting the placebo.

Example 1.2.5

Autism Autism is a serious condition in which children withdraw from normal social interactions and sometimes engage in aggressive or repetitive behavior. In 1997, an autistic child responded remarkably well to the digestive enzyme secretin. This led to an experiment (a "clinical trial") in which secretin was compared to a placebo. In this experiment, children who were given secretin improved considerably. However, the children given the placebo also improved considerably. There was no statistically significant difference between the two groups. Thus, the favorable response in the secretin group was considered to be only a "placebo response," meaning, unfortunately, that secretin was not found to be beneficial (beyond inducing a positive response associated simply with taking a substance as part of an experiment).[12] ■

The word *placebo* means "I shall please." The word *nocebo* ("I shall harm") is sometimes used to describe adverse reactions to perceived, but nonexistent, risks. The following example illustrates the strength that psychological effects can have.

Example 1.2.6

Bronchial Asthma A group of patients suffering from bronchial asthma were given a substance that they were told was a chest-constricting chemical. After being given this substance, several of the patients experienced bronchial spasms. However, during part of the experiment, the patients were given a substance that they were told would alleviate their symptoms. In this case, bronchial spasms were prevented. In reality, the second substance was identical to the first substance: Both were distilled water. It appears that it was the power of suggestion that brought on the bronchial spasms; the same power of suggestion prevented spasms.[13] ■

Similar to placebo treatment is *sham* treatment, which can be used on animals as well as humans. An example of sham treatment is injecting control animals with an inert substance such as saline. In some studies of surgical treatments, control animals (even, occasionally, humans) are given a "mock" surgery.

Example 1.2.7

Renal Denervation A surgical procedure called "renal denervation" was developed to help people with hypertension who do not respond to medication. An early study suggested that renal denervation (which uses radiotherapy to destroy some nerves in arteries feeding the kidney) reduces blood pressure. In that experiment, patients who received surgery had an average improvement in systolic blood pressure of 33 mmHg more than did control patients who received no surgery. Later an experiment was conducted in which patients were randomly assigned to one of two groups. Patients in the treatment group received the renal denervation surgery. Patients in the control group received a sham operation in which a catheter was inserted, as in the real operation, but 20 minutes later the catheter was removed *without* radiotherapy being used. These patients had no way of knowing that their operation was a sham. The rates of improvement in the two groups of patients were nearly identical.[14] ■

BLINDING

In experiments on humans, particularly those that involve the use of placebos, **blinding** is often used. This means that the treatment assignment is kept secret from the experimental subject. The purpose of blinding the subject is to minimize the extent to which his or her expectations influence the results of the experiment. If subjects exhibit a psychological reaction to getting a medication, that placebo response will tend to balance out between the two groups so that any difference between the groups can be attributed to the effect of the active treatment.

In many experiments the persons who evaluate the responses of the subjects are also kept blind; that is, during the experiment they are kept ignorant of the treatment assignment. Consider, for instance, the following:

> In a study to compare two treatments for lung cancer, a radiologist reads X-rays to evaluate each patient's progress. The X-ray films are coded so that the radiologist cannot tell which treatment each patient received.
>
> Mice are fed one of three diets; the effects on their liver are assayed by a research assistant who does not know which diet each mouse received.

Of course, *someone* needs to keep track of which subject is in which group, but that person should not be the one who measures the response variable. The most obvious reason for blinding the person making the evaluations is to reduce the possibility of subjective bias influencing the observation process itself: Someone who *expects* or *wants* certain results may unconsciously influence those results. Such bias can enter even apparently "objective" measurements through subtle variation in dissection techniques, titration procedures, and so on.

In medical studies of human beings, blinding often serves additional purposes. For one thing, a patient must be asked whether he or she consents to participate in a medical study. Suppose the physician who asks the question already knows which treatment the patient will receive. By discouraging certain patients and encouraging others, the physician can (consciously or unconsciously) create noncomparable treatment groups. The effect of such biased assignment can be surprisingly large, and it has been noted that it generally favors the "new" or "experimental" treatment.[15] Another reason for blinding in medical studies is that a physician may (consciously or unconsciously) provide more psychological encouragement, or even better care, to the patients who are receiving the treatment that the physician regards as superior.

An experiment in which both the subjects and the persons making the evaluations of the response are blinded is called a **double-blind** experiment. The first mammary artery ligation experiment described in Example 1.2.7 was conducted as a double-blind experiment.

THE NEED FOR CONTROL GROUPS

Example 1.2.8 **Clofibrate** An experiment was conducted in which subjects were given the drug clofibrate, which was intended to lower cholesterol and reduce the chance of death from coronary disease. The researchers noted that many of the subjects did not take all the medication that the experimental protocol called for them to take. They calculated the percentage of the prescribed capsules that each subject took and divided the subjects into two groups according to whether or not the subjects took at least 80% of the capsules they were given. Table 1.2.3 shows that the 5-year mortality rate for those who took at least 80% of their capsules was much lower than the corresponding rate for subjects who took fewer than 80% of the capsules. On the surface, this suggests that taking the medication lowers the chance of death. However, there was a placebo control group in the experiment and many of the placebo subjects took fewer than 80% of their capsules. The mortality rates for the two placebo groups—those who adhered to the protocol and those who did not—are quite similar to the rates for the clofibrate groups.

Table 1.2.3 Mortality rates for the clofibrate experiment

	Clofibrate		Placebo	
Adherence	n	5-year mortality	n	5-year mortality
≥80%	708	15.0%	1813	15.1%
<80%	357	24.6%	882	28.2%

The clofibrate experiment seems to indicate that there are two kinds of subjects: those who adhere to the protocol and those who do not. The first group had a much lower mortality rate than the second group. This might be due simply to better health habits among people who show stronger adherence to a scientific protocol for 5 years than among people who only adhere weakly, if at all. A further conclusion from the experiment is that clofibrate does not appear to be any more effective than placebo in reducing the death rate. Were it not for the presence of the placebo control group, the researchers might well have drawn the wrong conclusion from the study and attributed the lower death rate among strong adherers to clofibrate itself, rather than to other confounded effects that make the strong adherers different from the nonadherers.[16] ■

Example 1.2.9 **The Common Cold** Many years ago, investigators invited university students who believed themselves to be particularly susceptible to the common cold to be part of an experiment. Volunteers were randomly assigned to either the treatment group, in which case they took capsules of an experimental vaccine, or to the control group, in which case they were told that they were taking a vaccine, but in fact were given a placebo—capsules that looked like the vaccine capsules but that contained lactose in place of the vaccine.[17] As shown in Table 1.2.4, both groups reported having dramatically fewer colds during the study than they had had in the previous year. The average number of colds per person dropped 70% in the treatment group. This would have been startling evidence that the vaccine had an effect, except that the corresponding drop in the control group was 69%. ■

Table 1.2.4 Number of colds in cold-vaccine experiment

	Vaccine	Placebo
n	201	203
Average number of colds		
Previous year (from memory)	5.6	5.2
Current year	1.7	1.6
% reduction	70%	69%

We can attribute much of the large drop in colds in Example 1.2.9 to the placebo effect. However, another statistical concern is **panel bias**, which is bias attributable to the study having influenced the behavior of the subjects—that is, people who know they are being studied often change their behavior. The students in this study reported from memory the number of colds they had suffered in the previous year. The fact that they were part of a study might have influenced their behavior so that they were less likely to catch a cold during the study. Being in a study might also have affected the way in which they defined having a cold—during the study, they were "instructed to report to the health service whenever a cold developed"—so that some illness may have gone unreported during the study. (How sick do you have to be before you classify yourself as having a cold?)

Example 1.2.10

Diet and Cancer Prevention A diet that is high in fruits and vegetables may yield many health benefits, but how can we be sure? During the 1990s, the medical community believed that such a diet would reduce the risk of cancer. This belief was based on comparisons from **case-control studies**. In such studies patients with cancer were matched with "control subjects"—persons of the same age, race, sex, and so on—who did not have cancer; then the diets of the two groups were compared, and it was found that the control patients ate more fruits and vegetables than did the cancer patients. This would seem to indicate that cancer rates go down as consumption of fruits and vegetables goes up. The use of case-control studies is quite sensible because it allows researchers to make comparisons (e.g., of diets, etc.) while taking into consideration important characteristics such as age.

Nonetheless, a case-control study is not perfect. Not all people agree to be interviewed and to complete health information surveys, and these individuals thus might be excluded from a case-control study. People who agree to be interviewed about their health are generally more healthy than those who decline to participate. In addition to eating more fruits and vegetables than the average person, they are also less likely to smoke and more likely to exercise.[18] Thus, even though case-control studies took into consideration age, race, and other characteristics, they overstated the benefits of fruits and vegetables. The observed benefits are likely also the result of other healthy lifestyle factors.* Drawing a cause–effect conclusion that fruit and vegetable consumption protects against cancer is dangerous. ■

HISTORICAL CONTROLS

Researchers may be particularly reluctant to use randomized allocation in medical experiments on human beings. Suppose, for instance, that researchers want to evaluate a promising new treatment for a certain illness. It can be argued that it would be unethical to withhold the treatment from any patients, and that therefore all current patients should receive the new treatment. But then who would serve as a control group? One possibility is to use historical controls—that is, previous patients with the same illness who were treated with another therapy. One difficulty with historical controls is that there is often a tendency for later patients to show a better response—even to the same therapy—than earlier patients with the same diagnosis. This tendency has been confirmed, for instance, by comparing experiments conducted at the same medical centers in different years.[19] One major reason for the tendency is that the overall characteristics of the patient population may change with time. For

*A more informative kind of study is a prospective study or cohort study in which people with varying diets are followed over time to see how many of them develop cancer; however, such a study can be difficult to carry out.

instance, because diagnostic techniques tend to improve, patients with a given diagnosis (say, breast cancer) in 2001 may have a better chance of recovery (even with the same treatment) than those with the same diagnosis in 1991 because they were diagnosed earlier in the course of the disease. This is one reason that patients diagnosed with kidney cancer in 1995 had a 61% chance of surviving for at least 5 years but those with the same diagnosis in 2005 had a 75% 5-year survival rate.[20]

Medical researchers do not agree on the validity and value of historical controls. The following example illustrates the importance of this controversial issue.

Example 1.2.11

Coronary Artery Disease Disease of the coronary arteries is often treated by surgery (such as bypass surgery), but it can also be treated with drugs only. Many studies have attempted to evaluate the effectiveness of surgical treatment for this common disease. In a review of 29 of these studies, each study was classified as to whether it used randomized controls or historical controls; the conclusions of the 29 studies are summarized in Table 1.2.5.[21]

Table 1.2.5 Coronary artery disease studies

	Conclusion about effectiveness of surgery		
Type of controls	Effective	Not effective	Total number of studies
Randomized	1	7	8
Historical	16	5	21

It would appear from Table 1.2.5 that enthusiasm for surgery is much more common among researchers who use historical controls than among those who use randomized controls. ■

Example 1.2.12

Healthcare Trials A medical intervention, such as a new surgical procedure or drug, will often be used at one time in a nonrandomized clinical trial and at another time in a clinical trial of patients with the same condition who are assigned to groups randomly. Nonrandomized trials, which include the use of historical controls, tend to overstate the effectiveness of interventions. One analysis of many pairs of studies found that the nonrandomized trial showed a larger intervention effect than the corresponding randomized trial 22 times out of 26 comparisons; see Table 1.2.6.[22] Researchers concluded that overestimates of effectiveness are "due to poorer prognosis in non-randomly selected control groups compared with randomly selected control groups."[23] That is, if you give a new drug to relatively healthy patients and compare them to very sick patients taking the standard drug, the new drug is going to look better than it really is.

Even when randomization is used, trials may or may not be run double-blind. A review of 250 controlled trials found that trials that were not run double-blind produced significantly larger estimates of treatment effects than did trials that were double-blind.[24] ■

Table 1.2.6 Randomized versus nonrandomized trials

	Larger estimate of effect of the (common) intervention		
	Not randomized	Randomized	Total
Number of studies	22	4	26

Proponents of the use of historical controls argue that statistical adjustment can provide meaningful comparison between a current group of patients and a group of historical controls; for instance, if the current patients are younger than the historical controls, then the data can be analyzed in a way that adjusts, or corrects, for the effect of age. Critics reply that such adjustment may be grossly inadequate.

The concept of historical controls is not limited to medical studies. The issue arises whenever a researcher compares current data with past data. Whether the data are from the lab, the field, or the clinic, the researcher must confront the question: Can the past and current results be meaningfully compared? One should always at least ask whether the experimental material, and/or the environmental conditions, may have changed enough over time to distort the comparison.

Exercises 1.2.1–1.2.10

1.2.1 Fluoridation of drinking water has long been a controversial issue in the United States. One of the first communities to add fluoride to their water was Newburgh, New York. In March 1944, a plan was announced to begin to add fluoride to the Newburgh water supply on April 1 of that year. During the month of April, citizens of Newburgh complained of digestive problems, which were attributed to the fluoridation of the water. However, there had been a delay in the installation of the fluoridation equipment so that fluoridation did not begin until May 2.[25] Explain how the placebo effect/nocebo effect is related to this example.

1.2.2 Olestra is a no-calorie, no-fat additive that is used in the production of some potato chips. After the Food and Drug Administration approved the use of olestra, some consumers complained that olestra caused stomach cramps and diarrhea. A randomized, double-blind experiment was conducted in which some subjects were given bags of potato chips made with olestra and other subjects were given ordinary potato chips. In the olestra group, 38% of the subjects reported having gastrointestinal symptoms. However, in the group given regular potato chips the corresponding percentage was 37%. (The two percentages are not statistically significantly different.)[26] Explain how the placebo effect/nocebo effect is related to this example. Also explain why it was important for this experiment to be double-blind.

1.2.3 (Hypothetical) In a study of acupuncture, patients with headaches are randomly divided into two groups. One group is given acupuncture and the other group is given aspirin. The acupuncturist evaluates the effectiveness of the acupuncture and compares it to the results from the aspirin group. Explain how lack of blinding biases the experiment in favor of acupuncture.

1.2.4 Randomized, controlled experiments have found that vitamin C is not effective in treating terminal cancer patients.[27] However, a 1976 research paper reported that terminal cancer patients given vitamin C survived much longer than did historical controls. The patients treated with vitamin C were selected by surgeons from a group of cancer patients in a hospital.[28] Explain how this experiment was biased in favor of vitamin C.

1.2.5 On 3 November 2009, the blog lifehacker.com contained a posting by an individual with chronic toenail fungus. He remarked that after many years of suffering and trying all sorts of cures, he resorted to sanding his toenail as thin as he could tolerate, followed by daily application of vinegar and hydrogen-peroxide-soaked bandaids on his toenail. He repeated the vinegar peroxide bandaging for 100 days. After this time his nail grew out and the fungus was gone. Using the language of statistics, what kind of evidence is this? Is this convincing evidence that this procedure is an effective cure of toenail fungus?

1.2.6 For each of the following cases [(a) (b)],

(I) state whether the study should be observational or experimental.

(II) state whether the study should be run blind, double-blind, or neither. If the study should be run blind or double-blind, who should be blinded?

(a) An investigation of whether taking aspirin reduces one's chance of having a heart attack.

(b) An investigation of whether babies born into poor families (family income below $25,000) are more likely to weigh less than 5.5 pounds at birth than babies born into wealthy families (family income above $65,000).

1.2.7 For each of the following cases [(a) and (b)],

(I) state whether the study should be observational or experimental.

(II) state whether the study should be run blind, double-blind, or neither. If the study should be run blind or double-blind, who should be blinded?

(a) An investigation of whether the size of the midsagittal plane of the anterior commissure (a part of the brain) of a man is related to the sexual orientation of the man.

(b) An investigation of whether drinking more than 1 liter of water per day helps with weight loss for people who are trying to lose weight.

1.2.8 (Hypothetical) In order to assess the effectiveness of a new fertilizer, researchers applied the fertilizer to the tomato plants on the west side of a garden but did not fertilize the plants on the east side of the garden. They later measured the weights of the tomatoes produced by each plant and found that the fertilized plants grew larger tomatoes than did the nonfertilized plants. They concluded that the fertilizer works.

(a) Was this an experiment or an observational study? Why?

(b) This study is seriously flawed. Use the language of statistics to explain the flaw and how this affects the validity of the conclusion reached by the researchers.

(c) Could this study have used the concept of blinding (i.e., does the word "blind" apply to this study)? If so, how? Could it have been double-blind? If so, how?

1.2.9 Reseachers studied 1,718 persons over age 65 living in North Carolina. They found that those who attended religious services regularly were more likely to have strong immune systems (as determined by the blood levels of the protein interleukin-6) than those who didn't.[29] Does this mean that attending religious services improves one's health? Why or why not?

1.2.10 Researchers studied 300,818 golfers in Sweden and found that the "standardized mortality ratios" for golfers, adjusting for age, sex, and socioeconomic status, were lower than for nongolfers, meaning that golfers tend to live longer.[30] Does this mean that playing golf improves one's health? Why or why not?

1.3 Random Sampling

In order to address research questions with data, we first must consider how those data are to be gathered. How we gather our data has tremendous implications on our choice of analysis methods and even on the validity of our studies. In this section we will examine some common types of data-gathering methods with special emphasis on the **simple random sample**.

SAMPLES AND POPULATIONS

Before gathering data, we first consider the scope of our study by identifying the **population**. The population consists of all subjects/animals/specimens/plants, and so on, of interest. The following are all examples of populations:

- All birch tree seedlings in Florida
- All raccoons in Montaña de Oro State Park
- All people with schizophrenia in the United States
- All 100-ml water specimens in Chorro Creek

Typically we are unable to observe the entire population; therefore, we must be content with gathering data from a subset of the population, a **sample** of size n. From this sample we make inferences about the population as a whole (see Figure 1.3.1). The following are all examples of samples:

- A selection of eight ($n = 8$) Florida birch seedlings grown in a greenhouse.

Figure 1.3.1 Sampling from a population

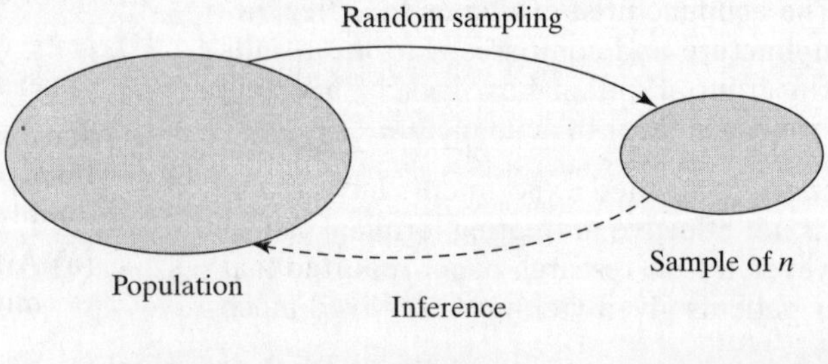

- Thirteen ($n = 13$) raccoons captured in traps at the Montaña de Oro campground.
- Forty-two ($n = 42$) patients with schizophrenia who respond to an advertisement in a U.S. newspaper.
- Ten ($n = 10$) 100-ml vials of water collected one day at 10 locations along Chorro Creek.

Remark There is some potential for confusion between the statistical meaning of the term *sample* and the sense in which this word is sometimes used in biology. If a biologist draws blood from 20 people and measures the glucose concentration in each, she might say she has 20 samples of blood. However, the statistician says she has *one* sample of 20 glucose measurements; the sample size is $n = 20$. In the interest of clarity, throughout this book we will use the term *specimen* where a biologist might prefer *sample.* So we would speak of glucose measurements on a sample of 20 specimens of blood.

Ideally our sample will be a representative subset of the population; however, unless we are careful, we may end up obtaining a **biased** sample. A biased sample systematically overestimates or systematically underestimates a characteristic of the population. For example, consider the raccoons from the sample described previously that are captured in traps at a campground. These raccoons may systematically differ from the population; they may be larger (from having ample access to food from dumpsters and campers), less timid (from being around people who feed them), and may be even longer lived than the general population of raccoons in the entire park.

One method to ensure that samples will be (in the long run) representative of the population is to use random sampling.

DEFINITION OF A SIMPLE RANDOM SAMPLE

Informally, the process of obtaining a simple random sample can be visualized in terms of labeled tickets, such as those used in a lottery or raffle. Suppose that each member of the population (e.g., raccoon, patient, plant) is represented by one ticket, and that the tickets are placed in a large box and thoroughly mixed. Then n tickets are drawn from the box by a blindfolded assistant, with new mixing after each ticket is removed. These n tickets constitute the sample. (Equivalently, we may visualize that n assistants reach in the box simultaneously, each assistant drawing one ticket.)

More abstractly, we may define random sampling as follows.

A Simple Random Sample

A *simple random sample* of n items is a sample in which (a) every member of the population has the same chance of being included in the sample, and (b) the members of the sample are chosen independently of each other. [Requirement (b) means that the chance of a given member of the population being chosen does not depend on which other members are chosen.]*

*Technically, requirement (b) is that every pair of members of the population has the same chance of being selected for the sample, every group of 3 members of the population has the same chance of being selected for the sample, and so on. In contrast to this, suppose we had a population with 30 persons in it and we wrote the names of 3 persons on each of 10 tickets. We could then choose one ticket in order to get a sample of size $n = 3$, but this would not be a simple random sample, since the pair (1,2) could end up in the sample but the pair (1,4) could not. Here the selections of members of the sample are not independent of each other. (This kind of sampling is known as "cluster sampling," with 10 clusters of size 3.) If the population is infinite, then the technical definition that all subsets of a given size are equally likely to be selected as part of the sample is equivalent to the requirement that the members of the sample are chosen independently.

Simple random sampling can be thought of in other, equivalent, ways. We may envision the sample members being chosen one at a time from the population; under simple random sampling, at each stage of the drawing, every remaining member of the population is equally likely to be the next one chosen. Another view is to consider the totality of possible samples of size n. If all possible samples are equally likely to be obtained, then the process gives a simple random sample.

EMPLOYING RANDOMNESS

When conducting statistical investigations, we will need to make use of randomness. As previously discussed, we obtain simple random samples randomly—every member of the population has the same chance of being selected. In Chapter 7 we shall discuss experiments in which we wish to compare the effects of different treatments on members of a sample. To conduct these experiments we will have to assign the treatments to subjects randomly—so that every subject has the same chance of receiving treatment A as they do treatment B.

Unfortunately, as a practical matter, humans are not very capable of mentally employing randomness. We are unable to eliminate unconscious bias that often leads us to systematically exclude or include certain individuals in our sample (or at least decrease or increase the chance of choosing certain individuals). For this reason, we must use external resources for selecting individuals when we want a random sample: mechanical devices such as dice, coins, and lottery tickets; electronic devices that produce random digits such as computers and calculators; or tables of random digits such as Table 1 in the back of this book. Although straightforward, using mechanical devices such as tickets in a box is impractical, so we will focus on the use of random digits for sample selection.

HOW TO CHOOSE A RANDOM SAMPLE

The following is a simple procedure for choosing a random sample of n items from a finite population of items.

(a) Create the **sampling frame**: a list of all members of the population with unique identification numbers for each member. All identification numbers must have the same number of digits; for instance, if the population contains 75 items, the identification numbers could be 01, 02, . . . , 75.

(b) Read numbers from Table 1, a calculator, or computer. Reject any numbers that do not correspond to any population member. (For example, if the population has 75 items that have been assigned identification numbers 01, 02, . . . , 75, then skip over the numbers 76, 77, . . . , 99, and 00.) Continue until n numbers have been acquired. (Ignore any repeated occurrence of the same number.)

(c) The population members with the chosen identification numbers constitute the sample.

The following example illustrates this procedure.

Example 1.3.1

Suppose we are to choose a random sample of size 6 from a population of 75 members. Label the population members 01, 02, . . . , 75. Use Table 1, a calculator, or a computer to generate a string of random digits.* For example, our calculator might produce the following string:

8 3 8 7 1 7 9 4 0 1 6 2 5 3 4 5 9 7 5 3 9 8 2 2

*Most calculators generate random numbers expressed as decimal numbers between 0 and 1; to convert these to random digits, simply ignore the leading zero and decimal and read the digits that follow the decimal. To generate a long string of random digits, simply call the random number function on the calculator repeatedly.

As we examine two-digit pairs of numbers, we ignore numbers greater than 75 as well as any pairs that identify a previously chosen individual.

~~83~~ ~~87~~ 17 ~~94~~ 01 62 53 45 ~~97~~ ~~53~~ ~~98~~ 22

Thus, the population members with the following identification numbers will constitute the sample: 17, 01, 62, 53, 45, 22. ■

Remark In calling the digits in Table 1 or your calculator or computer *random* digits, we are using the term *random* loosely. Strictly speaking, random digits are digits produced by a random *process*—for example, tossing a 10-sided die. The digits in Table 1 or in your calculator or computer are actually *pseudorandom* digits; they are generated by a deterministic (although possibly very complex) process that is designed to produce sequences of digits that mimic randomly generated sequences.

Remark If the population is large, then computer software can be quite helpful in generating a sample. If you need a random sample of size 15 from a population with 2,500 members, have the computer (or calculator) generate 15 random numbers between 1 and 2,500. (If there are duplicates in the set of 15, then go back and get more random numbers.)

PRACTICAL CONCERNS WHEN RANDOM SAMPLING

In many cases, obtaining a proper simple random sample is difficult or impossible. For example, to obtain a random sample of raccoons from Montaña de Oro State Park, one would first have to create the sampling frame, which provides a unique number for each raccoon in the park. Then, after generating the list of random numbers to identify our sample, one would have to capture those particular raccoons. This is likely an impossible task.

In practice, when it is possible to obtain a proper random sample, one should. When a proper random sample is impractical, it is important to take all precautions to ensure that the subjects in the study may be viewed *as if* they were obtained by random sampling from some population. That is, the sample should be comprised of individuals that all have the same chance of being selected from the population, and the individuals should be chosen independently. To do this, the first step is to define the population. The next step is to scrutinize the procedure by which the observational units are selected and to ask: Could the *observations* have been chosen at random? With the raccoon example, this might mean that we first define the population of raccoons by creating a sharp geographic boundary based on raccoon habitat and place traps at randomly chosen locations within the population habitat using a variety of baits and trap sizes. (We could use random numbers to generate latitude and longitude coordinates within the population habitat.) Although still less than ideal (some raccoons might be trap shy, and baby raccoons may not enter the traps at all), this is certainly better than simply capturing raccoons at one nonrandomly chosen atypical location (e.g., the campground) within the park. Presumably, the vast majority of raccoons now have the same chance of being trapped (i.e., equally likely to be selected), and capturing one raccoon has little or no bearing on the capture of any other (i.e., they can be considered to be independently chosen). Thus, it seems reasonable to treat the observations as if they were chosen at random.

NONSIMPLE RANDOM SAMPLING METHODS

There are other kinds of sampling that are random in a sense, but that are not simple. Two common nonsimple random sampling techniques are the **random cluster sample** and **stratified random sample**. To illustrate the concept of a cluster sample, consider a modification to the lottery method of generating a simple random sample. With cluster sampling, rather than assigning a unique ticket (or ID number) for each member of the population, IDs are assigned to entire groups of individuals. As tickets are drawn from the box, entire groups of individuals are selected for the sample as in the following example and Figure 1.3.2.

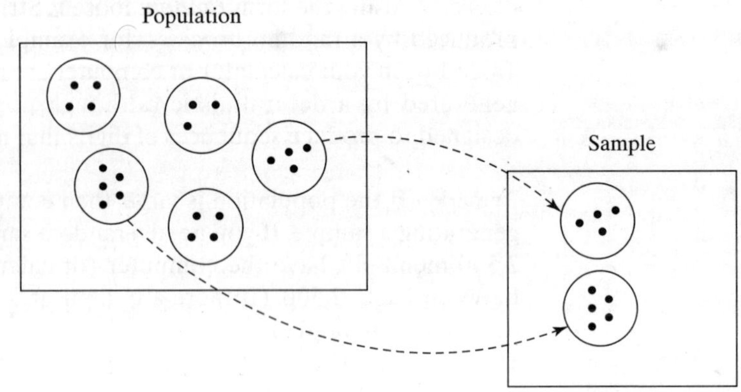

Figure 1.3.2 Random cluster sampling. The dots represent individuals within the population that are grouped into clusters (circles). Individuals in entire clusters are sampled from the population to form the sample.

Example 1.3.2 **La Graciosa Thistle** The La Graciosa thistle (*Cirsium loncholepis*) is an endangered plant native to the Guadalupe Dunes on the central coast of California. In a seed germination study, 30 plants were randomly chosen from the population of plants in the Guadalupe Dunes and all seeds from the 30 plants were harvested. The seeds form a cluster sample from the population of all La Graciosa thistle seeds in Guadalupe while the individual plants were used to identify the clusters.[31] ■

A stratified random sample is chosen by first dividing the population into **strata**—homogeneous collections of individuals. Then, many simple random samples are taken—one within each stratum—and combined to comprise the sample (see Figure 1.3.3). The following is an example of a stratified random sample.

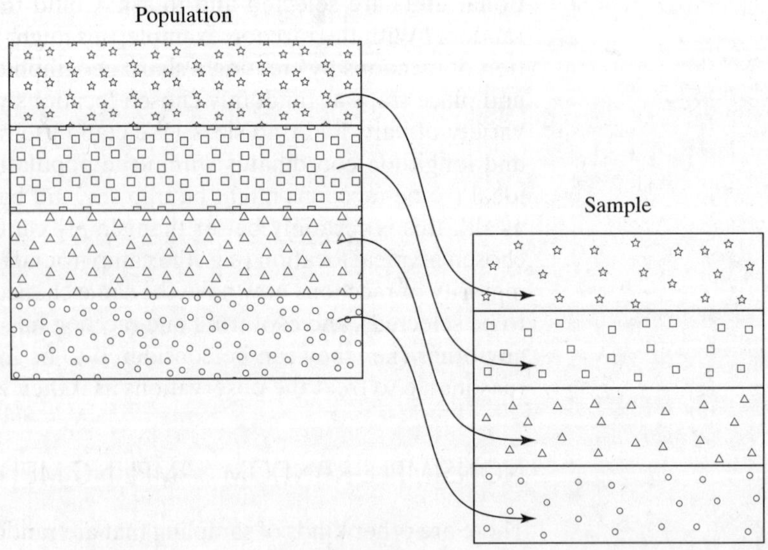

Figure 1.3.3 Stratified random sampling. The dots represent individuals within the population that are grouped into strata. Individuals from each stratum are randomly sampled and combined to form the sample.

Example 1.3.3

Sand Crabs In a study of parasitism of sand crabs (*Emerita analoga*), researchers obtained a stratified random sample of crabs by dividing a beach into 5-meter strips parallel to the water's edge. These strips were chosen as the strata because crab parasite loads may differ systematically based on the distance to the water's edge, thus making the parasite load for crabs within each stratum more similar than loads across strata. The first stratum was the 5-meter strip of beach just under the water's edge parallel to the shoreline. The second stratum was the 5-meter strip of beach just above the shoreline, followed by the third and fourth strata—the next two 5-meter strips above the shoreline. Within each strata, 25 crabs were randomly sampled, yielding a total sample size of 100 crabs.[32] ■

The majority of statistical methods discussed in this textbook will assume we are working with data gathered from a simple random sample. A sample chosen by simple random sampling is often called a *random sample.* But note that it is actually the *process* of sampling rather than the sample itself that is defined as random; randomness is not a property of the particular sample that happens to be chosen.

SAMPLING ERROR

How can we provide a rationale for inference from a limited sample to a much larger population? The approach of statistical theory is to refer to an idealized model of the sample–population relationship. In this model, which is called the **random sampling model**, the sample is chosen from the population by random sampling. The model is represented schematically in Figure 1.3.1.

The random sampling model is useful because it provides a basis for answering the question, How representative (of the population) is a sample likely to be? The model can be used to determine how much an inference might be influenced by chance, or "luck of the draw." More explicitly, a randomly chosen sample will usually not exactly resemble the population from which it was drawn. The discrepancy between the sample and the population is called **chance error due to sampling** or **sampling error**. We will see in later chapters how statistical theory derived from the random sampling model enables us to set limits on the likely amount of error due to sampling in an experiment. The quantification of such error is a major contribution that statistical theory has made to scientific thinking.

Because our samples are chosen randomly, there will always be sampling error present. If we sample nonrandomly, however, we may exacerbate the sampling error in unpredictable ways such as by introducing **sampling bias**, which is a systematic tendency for some individuals of the population to be selected more readily than others. The following two examples illustrate sampling bias.

Example 1.3.4

Lengths of Fish A biologist plans to study the distribution of body length in a certain population of fish in the Chesapeake Bay. The sample will be collected using a fishing net. Smaller fish can more easily slip through the holes in the net. Thus, smaller fish are less likely to be caught than larger ones, so the sampling procedure is biased. ■

Example 1.3.5

Sizes of Nerve Cells A neuroanatomist plans to measure the sizes of individual nerve cells in cat brain tissue. In examining a tissue specimen, the investigator must decide which of the hundreds of cells in the specimen should be selected for measurement. Some of the nerve cells are incomplete because the microtome cut through them when the tissue was sectioned. If the size measurement can be made only on complete cells, a bias arises because the smaller cells had a greater chance of being missed by the microtome blade. ■

When the sampling procedure is biased, the sample may not accurately represent the population, because it is systematically distorted. For instance, in Example 1.3.4 smaller fish will tend to be underrepresented in the sample, so the length of the fish in the sample will tend to be larger than those in the population.

The following example illustrates a kind of nonrandomness that is different from bias.

Example 1.3.6 **Sucrose in Beet Roots** An agronomist plans to sample beet roots from a field in order to measure their sucrose content. Suppose she were to take all her specimens from a randomly selected small area of the field. This sampling procedure would not be biased but would tend to produce *too homogeneous* a sample, because environmental variation across the field would not be reflected in the sample. ■

Example 1.3.6 illustrates an important principle that is sometimes overlooked in the analysis of data: In order to check applicability of the random sampling model, one needs to ask not only whether the sampling procedure might be biased, but also whether the sampling procedure will adequately reflect the variability inherent in the population. Faulty information about variability can distort scientific conclusions just as seriously as bias can.

We now consider some examples where the random sampling model might reasonably be applied.

Example 1.3.7 **Fungus Resistance in Corn** A certain variety of corn is resistant to fungus disease. To study the inheritance of this resistance, an agronomist crossed the resistant variety with a nonresistant variety and measured the degree of resistance in the progeny plants. The actual progeny in the experiment can be regarded as a random sample from a conceptual population of all *potential* progeny of that particular cross. ■

When the purpose of a study is to *compare* two or more experimental conditions, a very narrow definition of the population may be satisfactory, as illustrated in the next example.

Example 1.3.8 **Nitrite Metabolism** To study the conversion of nitrite to nitrate in the blood, researchers injected four New Zealand White rabbits with a solution of radioactively labeled nitrite molecules. Ten minutes after injection, they measured for each rabbit the percentage of the nitrite that had been converted to nitrate.[33] Although the four animals were not literally chosen at random from a specified population, it might be reasonable, nevertheless, to view the measurements of nitrite metabolism as a random sample from similar measurements made on all New Zealand White rabbits. (This formulation assumes that age and sex are irrelevant to nitrite metabolism.) ■

Example 1.3.9 **Treatment of Ulcerative Colitis** A medical team conducted a study of two therapies, A and B, for treatment of ulcerative colitis. All the patients in the study were referral patients in a clinic in a large city. Each patient was observed for satisfactory "response" to therapy. In applying the random sampling model, the researchers might want to make an inference to the population of all ulcerative colitis patients in urban referral clinics. First, consider inference about the actual probabilities of response; such an inference would be valid if the probability of response to each therapy is the same at all urban referral clinics. However, this assumption might be somewhat questionable, and the investigators might believe that the population should be defined very narrowly—for instance, as "the type of ulcerative colitis patients who are referred to this clinic." Even such a narrow population can be of interest in a comparative study. For instance, if treatment A is better than treatment B for the narrow population, it might be reasonable to infer that A would be better than B for a broader population (even if the actual response probabilities might be different in the broader population). In fact, it might even be argued that the broad population should include all ulcerative colitis patients, not merely those in urban referral clinics. ■

It often happens in research that, for practical reasons, the population actually studied is narrower than the population that is of real interest. In order to apply the kind of rationale illustrated in Example 1.3.9, one must argue that the results in the narrowly defined population (or, at least, some aspects of those results) can meaningfully be extrapolated to the population of interest. This extrapolation is not a *statistical* inference; it must be defended on biological, not statistical, grounds.

In Section 2.8 we will say more about the connection between samples and populations as we further develop the concept of statistical inference.

NONSAMPLING ERRORS

In addition to sampling errors, other concerns can arise in statistical studies. A **nonsampling error** is an error that is not caused by the sampling method; that is, a nonsampling error is one that would have arisen even if the researcher had a census of the entire population. For example, the way in which questions are worded can greatly influence how people answer them, as Example 1.3.10 shows.

Example 1.3.10

Abortion Funding In 1991, the U.S. Supreme Court made a controversial ruling upholding a ban on abortion counseling in federally financed family-planning clinics. Shortly after the ruling, a sample of 1,000 people were asked, "As you may know, the U.S. Supreme Court recently ruled that the federal government is not required to use taxpayer funds for family planning programs to perform, counsel, or refer for abortion as a method of family planning. In general, do you favor or oppose this ruling?" In the sample, 48% favored the ruling, 48% were opposed, and 4% had no opinion.

A separate opinion poll conducted at nearly the same time, but by a different polling organization, asked over 1,200 people, "Do you favor or oppose that Supreme Court decision preventing clinic doctors and medical personnel from discussing abortion in family-planning clinics that receive federal funds?" In this sample, 33% favored the decision and 65% opposed it.[34] The difference in the percentages favoring the opinion is too large to be attributed to chance error in the sampling. It seems that the way in which the question was worded had a strong impact on the respondents. ■

Another type of nonsampling error is **nonresponse bias**, which is bias caused by persons not responding to some of the questions in a survey or not returning a written survey. It is common to have only one-third of those receiving a survey in the mail complete the survey and return it to the researchers. (We consider the people receiving the survey to be part of the sample, even if some of them don't complete the entire survey, or even return the survey at all.) If the people who respond are unlike those who choose not to respond—and this is often the case, since people with strong feelings about an issue tend to complete a questionnaire, while others will ignore it—then the data collected will not accurately represent the population.

Example 1.3.11 **HIV Testing** A sample of 949 men were asked if they would submit to an HIV test of their blood. Of the 782 who agreed to be tested, 8 (1.02%) were found to be HIV positive. However, some of the men refused to be tested. The health researchers conducting the study had access to serum specimens that had been taken earlier from these 167 men and found that 9 of them (5.4%) were HIV positive.[35] Thus, those who refused to be tested were much more likely to have HIV than those who agreed to be tested. An estimate of the HIV rate based only on persons who agree to be tested is likely to substantially underestimate the true prevalence. ■

There are other cases in which an experimenter is faced with the vexing problem of **missing data**—that is, observations that were planned but could not be made. In addition to nonresponse, this can arise because experimental animals or plants die, because equipment malfunctions, or because human subjects fail to return for a follow-up observation.

A common approach to the problem of missing data is to simply use the remaining data and ignore the fact that some observations are missing. This approach is temptingly simple but must be used with extreme caution, because comparisons based on the remaining data may be seriously biased. For instance, if observations on some experimental mice are missing because the mice died of causes related to the treatment they received, it is obviously not valid to simply compare the mice that survived. As another example, if patients drop out of a medical study because they think their treatment is not working, then analysis of the remaining patients could produce a greatly distorted picture.

Naturally, it is best to make every effort to avoid missing data. But if data are missing, it is crucial that the possible reasons for the omissions be considered in interpreting and reporting the results.

Data can also be misleading if there is bias in how the data are collected. People have difficulty remembering the dates on which events happen and they tend to give unreliable answers if asked a question such as "How many times per week do you exercise?" They may also be biased as they make observations, as the following example shows.

Example 1.3.12 **Sugar and Hyperactivity** Mothers who thought that their young sons were "sugar sensitive" were randomly divided into two groups. Those in the first group were told that their sons had been given a large dose of sugar, whereas those in the second group were told that their sons had been given a placebo. In fact, all the boys had been given the placebo. Nonetheless, the mothers in the first group rated their sons to be much more hyperactive during a 25-minute study period than did the mothers in the second group.[36] Neutral measurements found that boys in the first group were actually a bit *less* active than those in the second group. Numerous other studies have failed to find a link between sugar consumption and activity in children, despite the widespread belief that sugar causes hyperactive behavior. It seems that the expectations that these mothers had colored their observations.[37] ■

Exercises 1.3.1–1.3.7

1.3.1 In each of the following studies, identify which sampling technique best describes the way the data were collected (or could be treated as if they were collected): simple random sampling, random cluster sampling, or stratified random sampling. For cluster samples identify the clusters, and for stratified samples identify the strata.

(a) All 257 leukemia patients from three randomly chosen pediatric clinics in the United States were enrolled in a clinical trial for a new drug.

(b) A total of twelve 10-g soil specimens were collected from random locations on a farm to study physical and chemical soil profiles.

(c) In a pollution study three 100-ml air specimens were collected at each of four specific altitudes (100 m, 500 m, 1000 m, 2000 m) for a total of twelve 100-ml specimens.

(d) A total of 20 individual grapes were picked, one from each of 20 random vines in a vineyard, to evaluate readiness for harvest.

(e) Twenty-four dogs (eight randomly chosen small breed, eight randomly chosen medium breed, and eight randomly chosen large breed) were enrolled in an experiment to evaluate a new training program.

1.3.2 For each of the following studies, identify the source(s) of sampling bias and describe (i) how it might affect the study conclusions and (ii) how you might alter the sampling method to avoid the bias.

(a) Eight hundred volunteers were recruited from nightclubs to enroll in an experiment to evaluate a new treatment for social anxiety.

(b) In a water pollution study, water specimens were collected from a stream on 15 rainy days.

(c) To study the size (radius) distribution of scrub oaks (shrubby oak trees), 20 oak trees were selected by using random latitude/longitude coordinates. If the random coordinate fell within the canopy of a tree, the tree was selected; if not, another random location was generated.

1.3.3 For each of the following studies, identify the source(s) of sampling bias and describe (i) how it might affect the study conclusions and (ii) how you might alter the sampling method to avoid the bias.

(a) To study the size distribution of rock cod (*Epinephelus puscus*) off the coast of southeastern Australia, scientists recorded the lengths and weights for all cod captured by a commercial fishing vessel on one day (using standard hook-and-line fishing methods).

(b) A nutritionist is interested in the eating habits of college students and observes what each student who enters a dining hall between 8:00 A.M. and 8:30 A.M. chooses for breakfast on a Monday morning.

(c) To study how fast an experimental painkiller relieves headache pain residents of a nursing home who complain of headaches are given the painkiller and are later asked how quickly their headaches subsided.

1.3.4 (A fun activity) Write the digits 1, 2, 3, 4 in order on an index card. Bring this card to a busy place (e.g., dining hall, library, university union) and ask at least 30 people to look at the card and select one of the digits at random in their head. Record their responses.

(a) If people can think "randomly," about what fraction of the people should respond with the digit 1? 2? 3? 4?

(b) What fraction of those surveyed responded with the digit 1? 2? 3? 4?

(c) Do the results suggest anything about people's ability to choose randomly?

1.3.5 Consider a population consisting of 600 individuals with unique IDs: 001, 002, . . . , 600. Use the following string of random digits to select a simple random sample of 5 individuals. List the IDs of the individuals selected for your sample.

728121876442121593787803547216596851

1.3.6 (Sampling exercise) Refer to the collection of 100 ellipses shown in the accompanying figure, which can be thought of as representing a natural population of the mythical organism *C. ellipticus*. The ellipses have been given identification numbers 00, 01, . . . , 99 for convenience in sampling. Certain individuals of *C. ellipticus* are mutants and have two tail bristles.

(a) Use your *judgment* to choose a sample of size 10 from the population that you think is representative of the entire population. Note the number of mutants in the sample.

(b) Use *random digits* (from Table 1 or your calculator or computer) to choose a random sample of size 10 from the population and note the number of mutants in the sample.

1.3.7 (Sampling exercise) Refer to the collection of 100 ellipses.

(a) Use random digits (from Table 1 or your calculator or computer) to choose a random sample of size 5 from the population and note the number of mutants in the sample.

(b) Repeat part (a) nine more times, for a total of 10 samples. (Some of the 10 samples may overlap.)

To facilitate pooling of results from the entire class, report your results in the following format:

Number of mutants	Nonmutants	Frequency (no. of samples)
0	5	
1	4	
2	3	
3	2	
4	1	
5	0	
		Total: 10

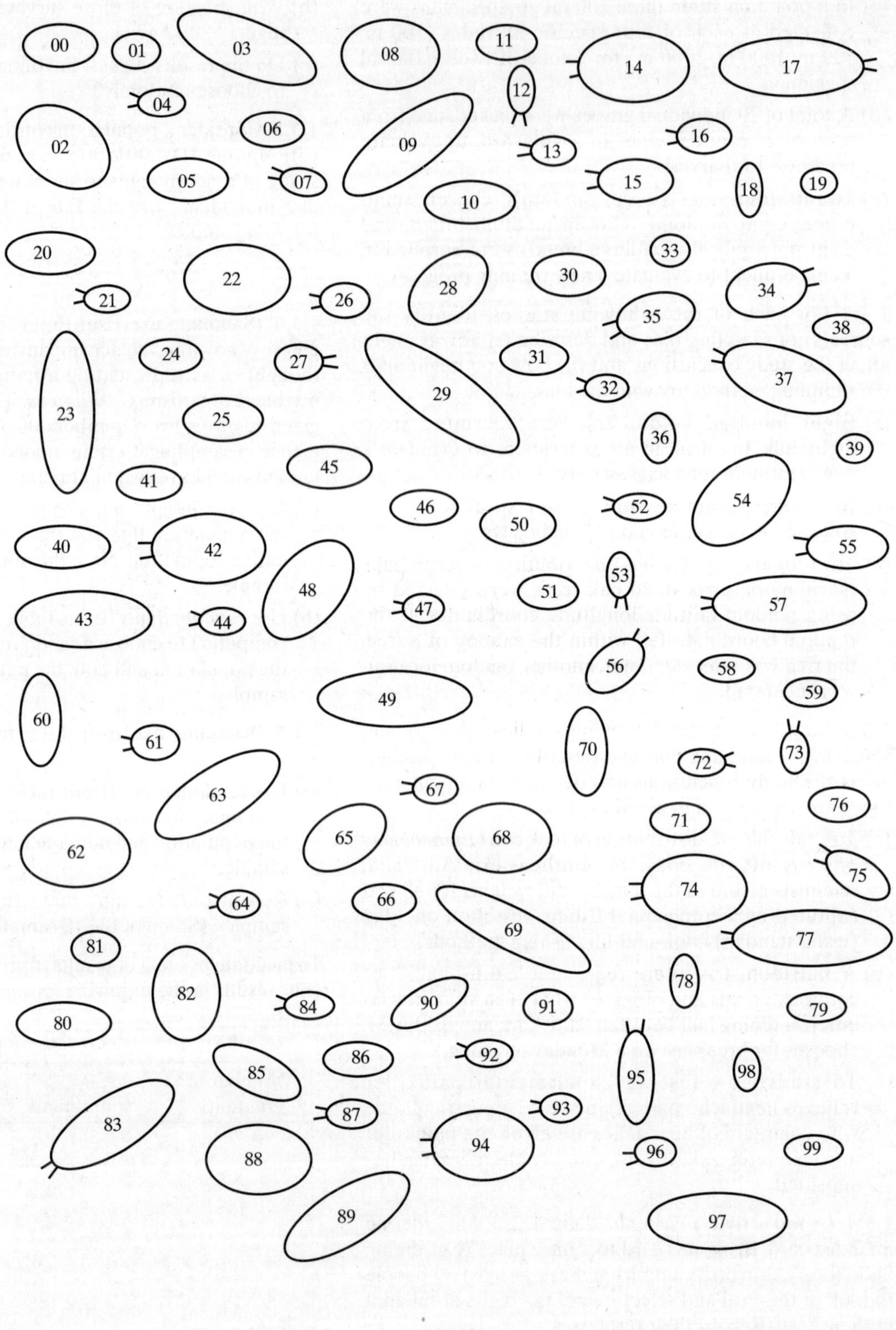
00 01 02 03 04 05 06 07 08 09
10 11 12 13 14 15 16 17 18 19
20 21 22 23 24 25 26 27 28 29
30 31 32 33 34 35 36 37 38 39
40 41 42 43 44 45 46 47 48 49
50 51 52 53 54 55 56 57 58 59
60 61 62 63 64 65 66 67 68 69
70 71 72 73 74 75 76 77 78 79
80 81 82 83 84 85 86 87 88 89
90 91 92 93 94 95 96 97 98 99

DESCRIPTION OF SAMPLES AND POPULATIONS

Chapter 2

OBJECTIVES

In this chapter we will study how to describe data. In particular, we will

- show how frequency distributions are used to make bar charts and histograms.
- compare the mean and median as measures of center.
- demonstrate how to construct and read a variety of graphics including dotplots, boxplots, and scatterplots.
- compare several measures of variability with emphasis on the standard deviation.
- examine how transformations of variables affect distributions.
- consider the relationship between populations and samples.

2.1 Introduction

Statistics is the science of analyzing and learning from data. In this section we introduce some terminology and notation for dealing with data.

VARIABLES

We begin with the concept of a **variable**. A variable is a characteristic of a person or a thing that can be assigned a number or a category. For example, blood type (A, B, AB, O) and age are two variables we might measure on a person.

Blood type is an example of a **categorical variable***: A categorical variable is a variable that records which of several categories a person or thing is in. Examples of categorical variables are

Blood type of a person: A, B, AB, O
Sex of a fish: male, female
Color of a flower: red, pink, white
Shape of a seed: wrinkled, smooth

Age is an example of a **numeric variable**, that is, a variable that records the amount of something. A **continuous variable** is a numeric variable that is measured on a continuous scale. Examples of continuous variables are

Weight of a baby
Cholesterol concentration in a blood specimen
Optical density of a solution

A variable such as weight is continuous because, in principle, two weights can be arbitrarily close together. Some types of numeric variables are not continuous but fall on a discrete scale, with spaces between the possible values. A **discrete variable** is a numeric variable for which we can list the possible values. For example, the number of eggs in a bird's nest is a discrete variable because only the values $0, 1, 2, 3, \ldots,$ are possible. Other examples of discrete variables are

Number of bacteria colonies in a petri dish
Number of cancerous lymph nodes detected in a patient
Length of a DNA segment in basepairs

*For some categorical variables, the categories can be arrayed in a meaningful rank order. Such a variable is said to be **ordinal**. For example, the response of a patient to therapy might be none, partial, or complete.

The distinction between continuous and discrete variables is not a rigid one. After all, physical measurements are always rounded off. We may measure the weight of a steer to the nearest kilogram, of a rat to the nearest gram, or of an insect to the nearest milligram. The scale of the actual measurements is always discrete, strictly speaking. The continuous scale can be thought of as an approximation to the actual scale of measurement.

OBSERVATIONAL UNITS

When we collect a sample of n persons or things and measure one or more variables on them, we call these persons or things **observational units** or cases. The following are some examples of samples.

Sample	Variable	Observational unit
150 babies born in a certain hospital	Birthweight (kg)	A baby
73 *Cecropia* moths caught in a trap	Sex	A moth
81 plants that are a progeny of a single parental cross	Flower color	A plant
Bacterial colonies in each of six petri dishes	Number of colonies	A petri dish

NOTATION FOR VARIABLES AND OBSERVATIONS

We will adopt a notational convention to distinguish between a variable and an observed value of that variable. We will denote variables by uppercase letters such as Y. We will denote the observations themselves (that is, the data) by lowercase letters such as y. Thus, we distinguish, for example, between $Y =$ birthweight (the variable) and $y = 7.9$ lb (the observation). This distinction will be helpful in explaining some fundamental ideas concerning variability.

Exercises 2.1.1–2.1.5

For each of the following settings in Exercises 2.1.1–2.1.5, (i) identify the variable(s) in the study, (ii) for each variable tell the type of variable (e.g., categorical and ordinal, discrete, etc.), (iii) identify the observational unit (the thing sampled), and (iv) determine the sample size.

2.1.1

(a) A paleontologist measured the width (in mm) of the last upper molar in 36 specimens of the extinct mammal *Acropithecus rigidus*.

(b) The birthweight, date of birth, and the mother's race were recorded for each of 65 babies.

2.1.2

(a) A physician measured the height and weight of each of 37 children.

(b) During a blood drive, a blood bank offered to check the cholesterol of anyone who donated blood. A total of 129 persons donated blood. For each of them, the blood type and cholesterol levels were recorded.

2.1.3

(a) A biologist measured the number of leaves on each of 25 plants.

(b) A physician recorded the number of seizures that each of 20 patients with severe epilepsy had during an eight-week period.

2.1.4

(a) A conservationist recorded the weather (clear, partly cloudy, cloudy, rainy) and number of cars parked at noon at a trailhead on each of 18 days.

(b) An enologist measured the pH and residual sugar content (g/l) of seven barrels of wine.

2.1.5

(a) A biologist measured the body mass (g) and sex of each of 123 blue jays.

(b) A biologist measured the lifespan (in days), the thorax length (in mm), and the percent of time spent sleeping for each of 125 fruit flies.

2.2 Frequency Distributions

A first step toward understanding a set of data on a given variable is to explore the data and describe the data in summary form. In this chapter we discuss three mutually complementary aspects of data description: frequency distributions, measures of center, and measures of dispersion. These tell us about the shape, center, and spread of the data.

A **frequency distribution** is simply a display of the **frequency**, or number of occurrences, of each value in the data set. The information can be presented in tabular form or, more vividly, with a graph. A **bar chart** is a graph of categorical data showing the number of observations in each category. Here are two examples of frequency distributions for categorical data.

Example 2.2.1

Color of Poinsettias Poinsettias can be red, pink, or white. In one investigation of the hereditary mechanism controlling the color, 182 progeny of a certain parental cross were categorized by color.[1] The bar graph in Figure 2.2.1 is a visual display of the results given in Table 2.2.1. ■

Figure 2.2.1 Bar chart of color of 182 poinsettias

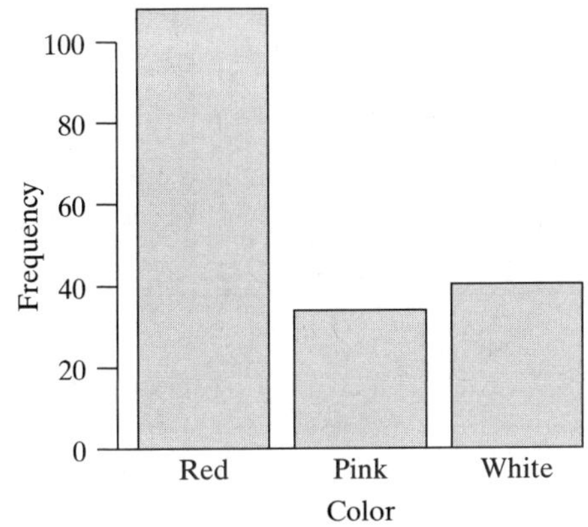

Table 2.2.1 Color of 182 poinsettias

Color	Frequency (number of plants)
Red	108
Pink	34
White	40
Total	182

Example 2.2.2

School Bags and Neck Pain Physiologists in Australia were concerned that carrying a school bag loaded with heavy books was a cause of neck pain in adolescents, so they asked a sample of 585 teenage girls how often they get neck pain when carrying their school bag (never, almost never, sometimes, often, always). A summary of the results reported to them is given in Table 2.2.2 and displayed as a bar graph in Figure 2.2.2(a).[2] As the variable incidence is an ordinal categorical variable, our tables and graphs should respect the natural ordering. Figure 2.2.2(b) shows the same data but with the categories in alphabetical order (a default setting for much software), which obscures the information in the data. ■

Table 2.2.2 Neck pain associated with carrying a school bag

Incidence	Frequency (number of girls)
Never	179
Almost never	159
Sometimes	173
Often	64
Always	10
Total	585

Figure 2.2.2 (a) Bar chart of incidence of neck pain reported by 585 adolescents; (b) the same data but with the categories in alphabetical order

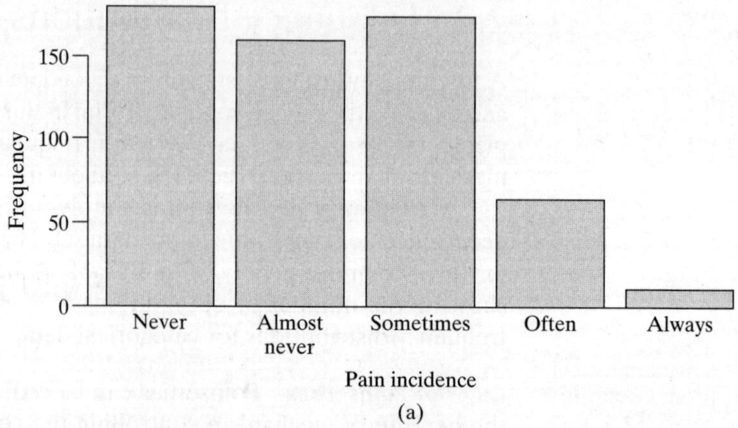

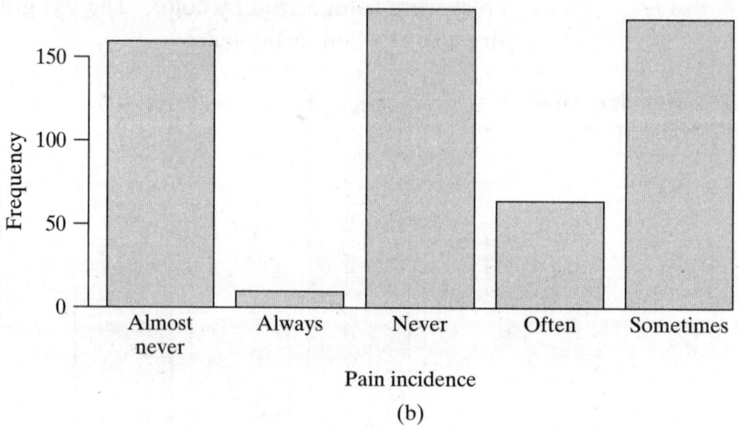

A **dotplot** is a simple graph that can be used to show the distribution of a numeric variable when the sample size is small. To make a dotplot, we draw a number line covering the range of the data and then put a dot above the number line for each observation, as the following example shows.

Example 2.2.3

Infant Mortality Table 2.2.3 shows the infant mortality rate (infant deaths per 1,000 live births) in each of seven countries in South Asia, as of 2013.[3] The distribution is shown in Figure 2.2.3. ■

Table 2.2.3 Infant mortality in seven South Asian countries

Country	Infant mortality rate (deaths per 1,000 live births)
Bangladesh	47.3
Bhutan	40.0
India	44.6
Maldives	25.5
Nepal	41.8
Pakistan	59.4
Sri Lanka	9.2

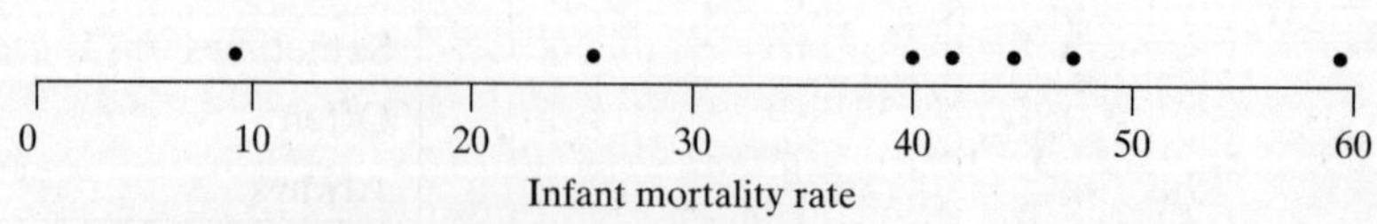

Figure 2.2.3 Dotplot of infant mortality in seven South Asian countries

When two or more observations take on the same value, we stack the dots in a dotplot on top of each other. This gives an effect similar to the effect of the bars in a bar chart. If we create bars in place of the stacks of dots, we then have a **histogram**. A histogram is like a bar chart, except that a histogram displays a numeric variable, which means that there is a natural order and scale for the variable. In a bar chart the amount of space between the bars (if any) is arbitrary, since the data being displayed are categorical. In a histogram the scale of the variable determines the placement of the bars. The following example shows a dotplot and a histogram for a frequency distribution.

Example 2.2.4 **Litter Size of Sows** A group of thirty-six 2-year-old sows of the same breed ($\frac{3}{4}$ Duroc, $\frac{1}{4}$ Yorkshire) were bred to Yorkshire boars. The number of piglets surviving to 21 days of age was recorded for each sow.[4] The results are given in Table 2.2.4 and displayed as a dotplot in Figure 2.2.4 and as a histogram in Figure 2.2.5. ■

Table 2.2.4 Number of surviving piglets of 36 sows

Number of piglets	Frequency (number of sows)
5	1
6	0
7	2
8	3
9	3
10	9
11	8
12	5
13	3
14	2
Total	36

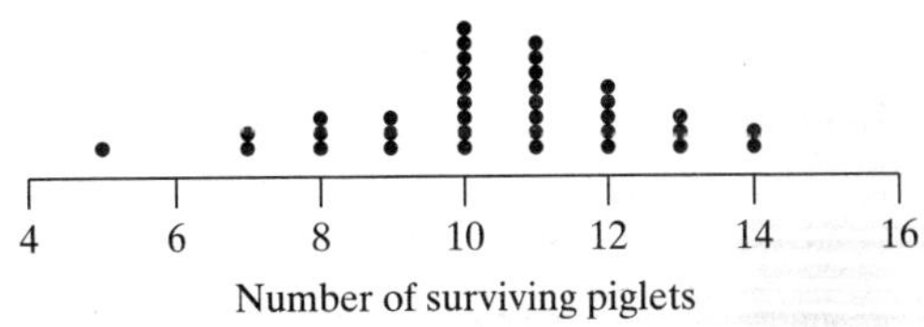

Figure 2.2.4 Dotplot of number of surviving piglets of 36 sows

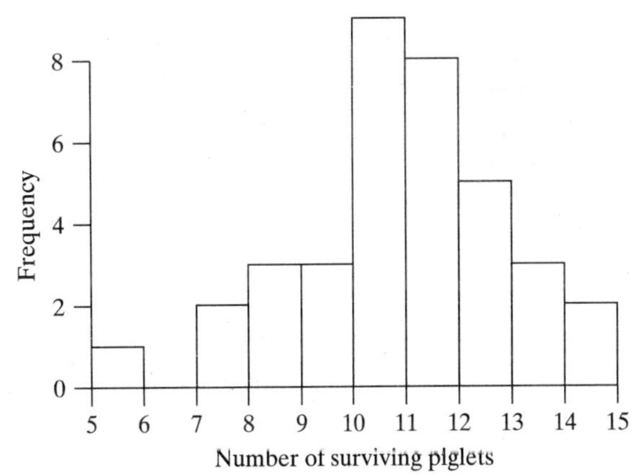

Figure 2.2.5 Histogram of number of surviving piglets of 36 sows

RELATIVE FREQUENCY

The frequency scale is often replaced by a **relative frequency** scale:

$$\text{Relative frequency} = \frac{\text{Frequency}}{n}$$

The relative frequency scale is useful if several data sets of different sizes (n's) are to be displayed together for comparison. As another option, a relative frequency can be expressed as a percentage frequency. The shape of the display is not affected by the choice of frequency scale, as the following example shows.

Example 2.2.5 **Color of Poinsettias** The poinsettia color distribution of Example 2.2.1 is expressed as frequency, relative frequency, and percent frequency in Table 2.2.5 and Figure 2.2.6. ■

Table 2.2.5 Color of 182 poinsettias

Color	Frequency	Relative frequency	Percent frequency
Red	108	.59	59
Pink	34	.19	19
White	40	.22	22
Total	182	1.00	100

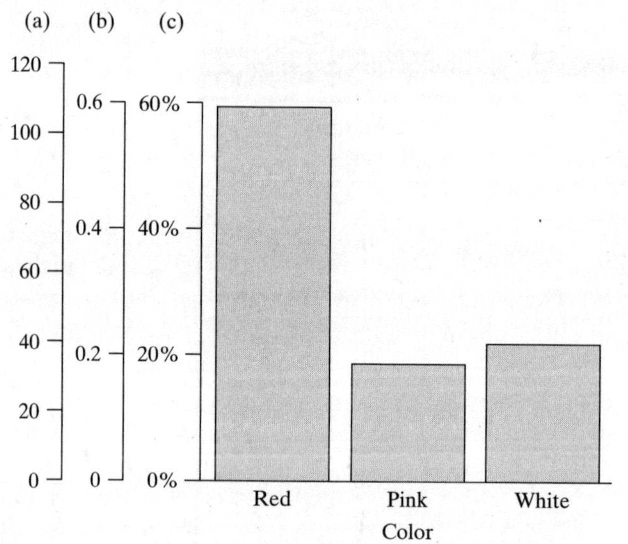

Figure 2.2.6 Bar chart of poinsettia colors on three scales:
(a) Frequency
(b) Relative frequency
(c) Percent frequency

GROUPED FREQUENCY DISTRIBUTIONS

In the preceding examples, simple ungrouped frequency distributions provided concise summaries of the data. For many data sets, it is necessary to group the data in order to condense the information adequately. (This is usually the case with continuous variables.) The following example shows a grouped frequency distribution.

Example 2.2.6 **Serum CK** Creatine phosphokinase (CK) is an enzyme related to muscle and brain function. As part of a study to determine the natural variation in CK concentration, blood was drawn from 36 male volunteers. Their serum concentrations of CK (measured in U/l) are given in Table 2.2.6.[5] Table 2.2.7 shows these data grouped into **classes**. For instance, the frequency of the class [20,40) (all values in the interval $20 \leq y < 40$) is 1, which means that one CK value fell in this range. The grouped frequency distribution is displayed as a histogram in Figure 2.2.7. ■

Table 2.2.6 Serum CK values for 36 men

121	82	100	151	68	58
95	145	64	201	101	163
84	57	139	60	78	94
119	104	110	113	118	203
62	83	67	93	92	110
25	123	70	48	95	42

Table 2.2.7 Frequency distribution of serum CK values for 36 men

Serum CK (U/l)	Frequency (number of men)
[20,40)	1
[40,60)	4
[60,80)	7
[80,100)	8
[100,120)	8
[120,140)	3
[140,160)	2
[160,180)	1
[180,200)	0
[200,220)	2
Total	36

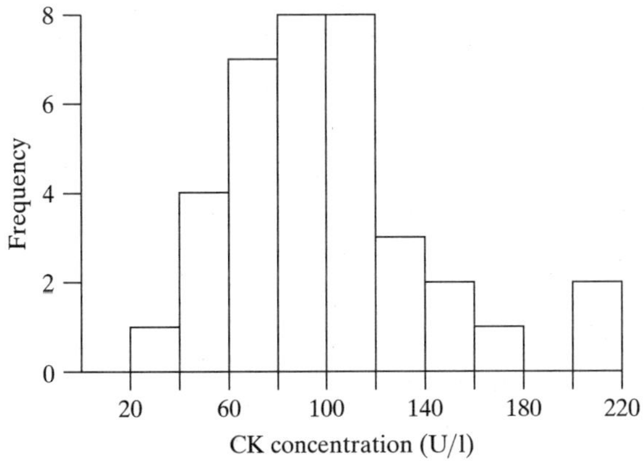

Figure 2.2.7 Histogram of serum CK concentrations for 36 men

A grouped frequency distribution should display the essential features of the data. For instance, the histogram of Figure 2.2.7 shows that the average CK value is about 100 U/l, with the majority of the values falling between 60 and 140 U/l. In addition, the histogram shows the *shape* of the distribution. Note that the CK values are piled up around a central peak, or **mode**. On either side of this mode, the frequencies decline and ultimately form the **tails** of the distribution. These shape features are labeled in Figure 2.2.8. The CK distribution is not symmetric but is a bit **skewed to the right**, which means that the right tail is more stretched out than the left.*

Figure 2.2.8 Shape features of the CK distribution

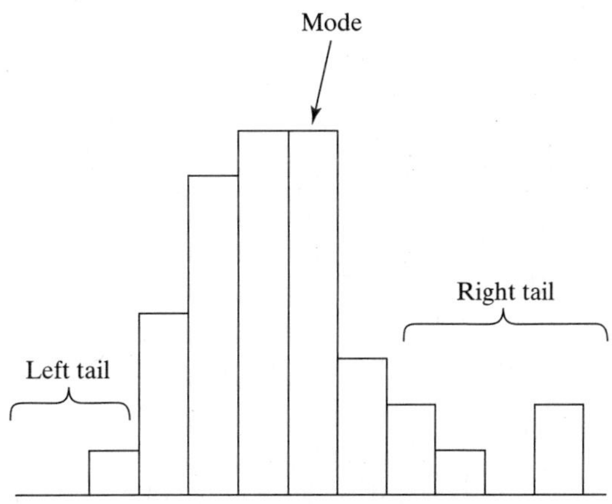

*To help remember which tail of a skewed distribution is the longer tail, think of skew as stretch. Which side of the distribution is more stretched away from the center? A distribution that is skewed to the right is one in which the right tail stretches out more than the left.

When making a histogram, we need to decide how many classes to have and how wide the classes should be. If we use computer software to generate a histogram, the program will choose the number of classes and the class width for us, but most software allows the user to change the number of classes and to specify the class width. If a data set is large and is quite spread out, it is a good idea to look at more than one histogram of the data, as is done in Example 2.2.7.

Example 2.2.7

Heights of Students A sample of 510 college students were asked how tall they were. Note that they were not measured; rather, they just reported their heights.[6] Figure 2.2.9 shows the distribution of the self-reported values, using 7 classes and a class width of 3 (inches). By using only 7 classes, the distribution appears to be reasonably symmetric, with a single peak around 66 inches.

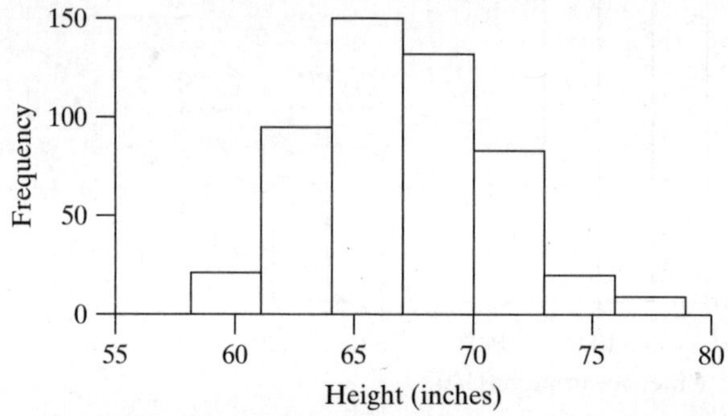

Figure 2.2.9 Heights of students, using 7 classes (class width = 3)

Figure 2.2.10 shows the height data, but in a histogram that uses 18 classes and a class width of 1.1. This view of the data shows two modes—one for women and one for men.

Figure 2.2.11 shows the height data again, this time using 37 classes, each of width 0.5. Using such a large number of classes makes the distribution look jagged. In this case, we see an alternating pattern between classes with lots of observations and classes with few observations. In the middle of the distribution we see that there were many students who reported a height of 63 inches, few who reported a height of 63.5 inches, many who reported a height of 64 inches, and so on. It seems that most students round off to the nearest inch! ■

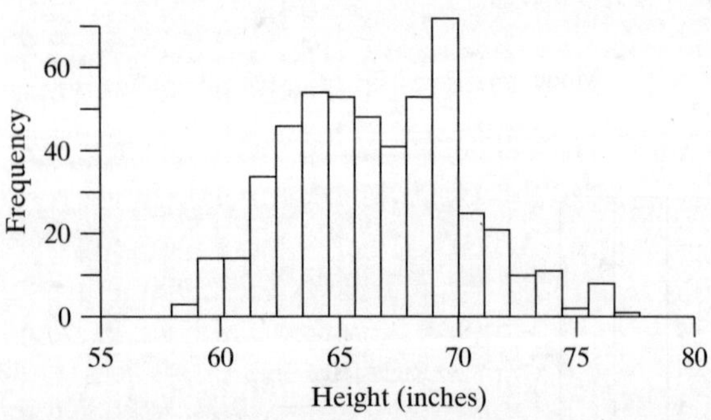

Figure 2.2.10 Heights of students, using 18 classes (class width = 1.1)

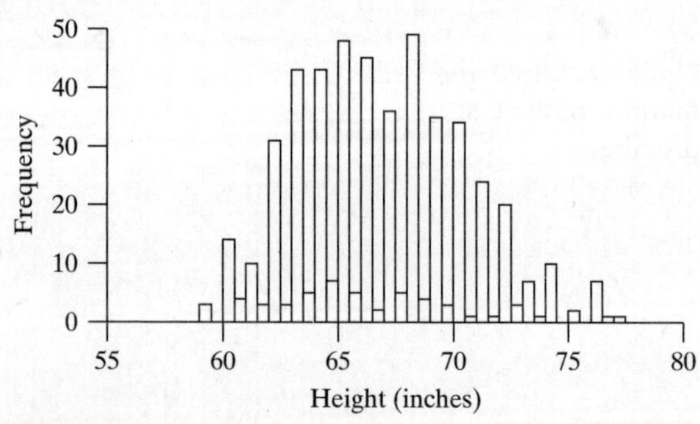

Figure 2.2.11 Heights of students, using 37 classes (class width = 0.5)

INTERPRETING AREAS IN A HISTOGRAM

A histogram can be looked at in two ways. The tops of the bars sketch out the shape of the distribution. But the *areas* within the bars also have a meaning. The area of each bar is proportional to the corresponding frequency. Consequently, the area of one or several bars can be interpreted as expressing the number of observations in the classes represented by the bars. For example, Figure 2.2.12 shows a histogram of the CK distribution of Example 2.2.6. The shaded area is 42% of the total area in all the bars. Accordingly, 42% of the CK values are in the corresponding classes; that is, 15 of 36 or 42% of the values are between 60 U/I and 100 U/l.*

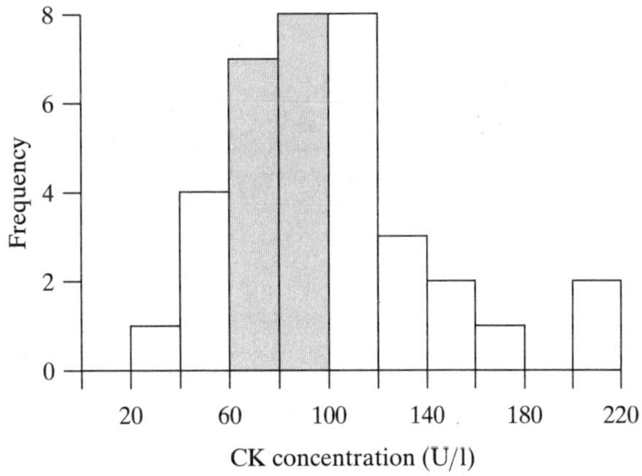

Figure 2.2.12 Histogram of CK distribution. The shaded area is 42% of the total area and represents 42% of the observations.

The area interpretation of histograms is a simple but important idea. In our later work with distributions we will find the idea to be indispensable.

SHAPES OF DISTRIBUTIONS

When discussing a set of data, we want to describe the shape, center, and spread of the distribution. In this section we concentrate on the shapes of frequency distributions and illustrate some of the diversity of distributions encountered in the life sciences. The shape of a distribution can be indicated by a smooth curve that approximates the histogram, as shown in Figure 2.2.13.

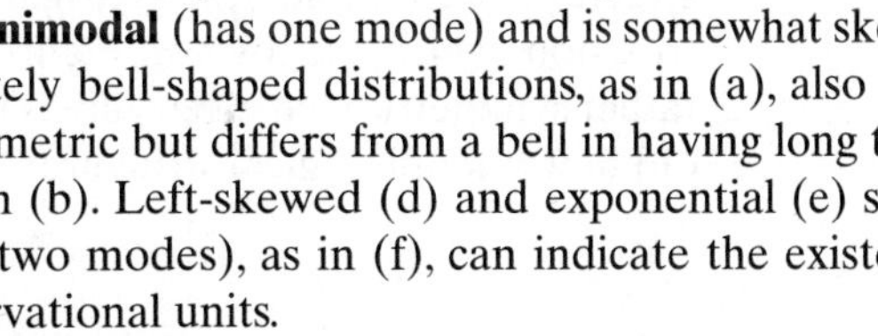

Figure 2.2.13 Approximation of a histogram by a smooth curve

Some distributional shapes are shown in Figure 2.2.14. A common shape for biological data is **unimodal** (has one mode) and is somewhat skewed to the right, as in (c). Approximately bell-shaped distributions, as in (a), also occur. Sometimes a distribution is symmetric but differs from a bell in having long tails; an exaggerated version is shown in (b). Left-skewed (d) and exponential (e) shapes are less common. **Bimodality** (two modes), as in (f), can indicate the existence of two distinct subgroups of observational units.

Notice that the shape characteristics we are emphasizing, such as number of modes and degree of symmetry, are *scale free;* that is, they are not affected by the arbitrary choices of vertical and horizontal scale in plotting the distribution. By contrast, a characteristic such as whether the distribution appears short and fat, or tall and skinny, is affected by how the distribution is plotted and so is not an inherent feature of the biological variable.

*Strictly speaking, between 60 U/l and 99 U/l, inclusive.

The following three examples illustrate biological frequency distributions with various shapes. In the first example, the shape provides evidence that the distribution is in fact biological rather than nonbiological.

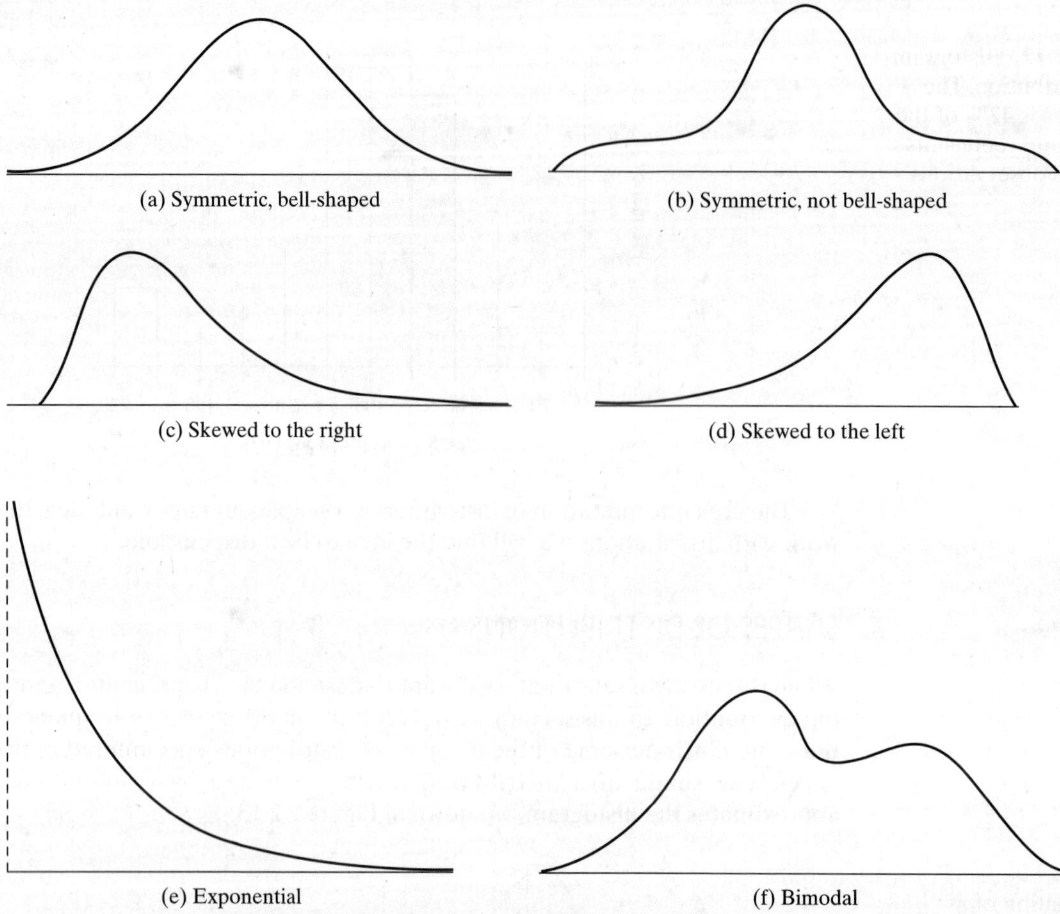

Figure 2.2.14 Shapes of distributions

Example 2.2.8

Microfossils In 1977, paleontologists discovered microscopic fossil structures, resembling algae, in rocks 3.5 billion years old. A central question was whether these structures were biological in origin. One line of argument focused on their size distribution, which is shown in Figure 2.2.15. This distribution, with its unimodal and rather symmetric shape, resembles that of known microbial populations, but not that of known nonbiological structures.[7] ■

Figure 2.2.15 Sizes of microfossils

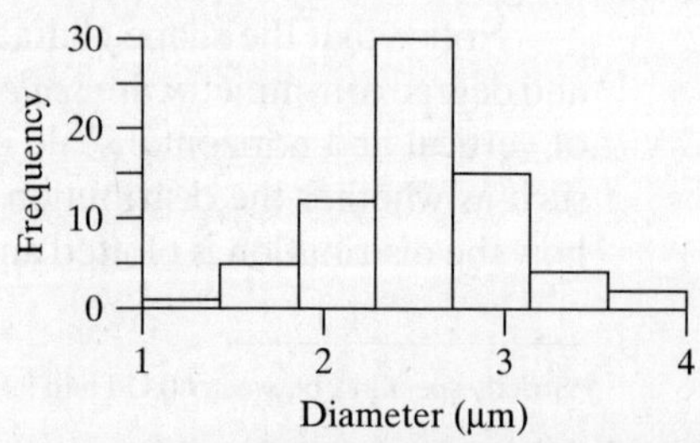

Example 2.2.9

Cell Firing Times A neurobiologist observed discharges from rat muscle cells grown in culture together with nerve cells. The time intervals between 308 successive discharges were distributed as shown in Figure 2.2.16. Note the exponential shape of the distribution.[8] ■

Figure 2.2.16 Time intervals between electrical discharges in rat muscle cells

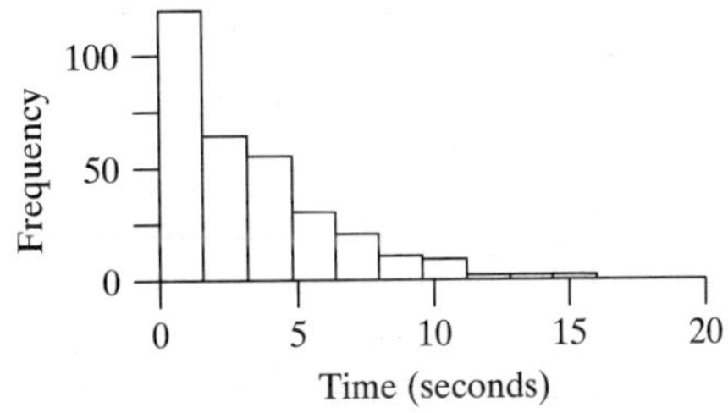

Example 2.2.10

Brain Weight In 1888, P. Topinard published data on the brain weights of hundreds of French men and women. The data for males and females are shown in Figure 2.2.17(a) and (b). The male distribution is fairly symmetric and bell shaped; the female distribution is somewhat skewed to the right. Part (c) of the figure shows the brain weight distribution for males and females combined. This combined distribution is slightly bimodal.[9] ■

Figure 2.2.17 Brain weights

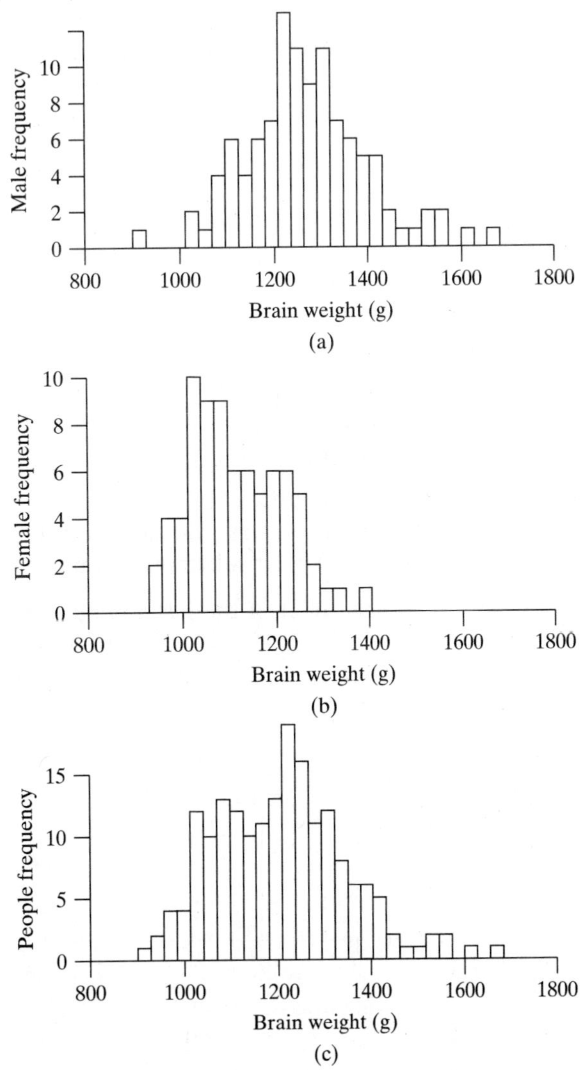

SOURCES OF VARIATION

In interpreting biological data, it is helpful to be aware of sources of variability. The variation among observations in a data set often reflects the combined effects of several underlying factors. The following two examples illustrate such situations.

Example 2.2.11 **Weights of Seeds** In a classic experiment to distinguish environmental from genetic influence, a geneticist weighed seeds of the princess bean *Phaseolus vulgaris.* Figure 2.2.18 shows the weight distributions of (a) 5,494 seeds from a commercial seed lot, and (b) 712 seeds from a highly inbred line that was derived from a single seed from the original lot. The variability in (a) is due to both environmental and genetic factors; in (b), because the plants are nearly genetically identical, the variation in weights is due largely to environmental influence.[10] Thus, there is less variability in the inbred line. ■

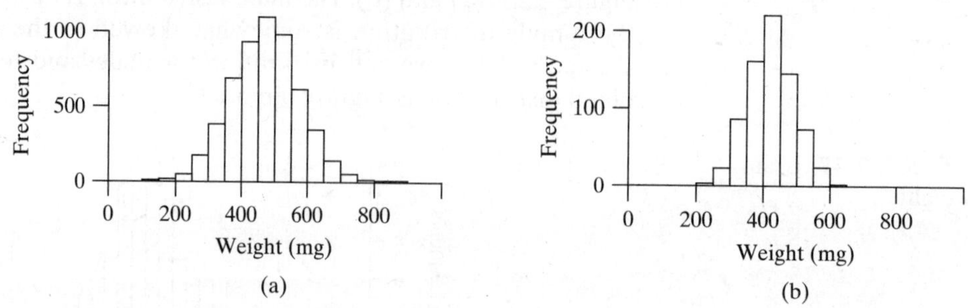

Figure 2.2.18 Weights of princess bean seeds: (a) from an open-bred population; (b) from an inbred line

Example 2.2.12 **Serum ALT** Alanine aminotransferase (ALT) is an enzyme found in most human tissues. Part (a) of Figure 2.2.19 shows the serum ALT concentrations for 129 adult volunteers. The following are potential sources of variability among the measurements:

1. Interindividual
   (a) Genetic
   (b) Environmental
2. Intraindividual
   (a) Biological: changes over time
   (b) Analytical: imprecision in assay

The effect of the last source—analytical variation—can be seen in part (b) of Figure 2.2.19, which shows the frequency distribution of 109 assays of the *same* specimen of serum; the figure shows that the ALT assay is fairly imprecise.[11] ■

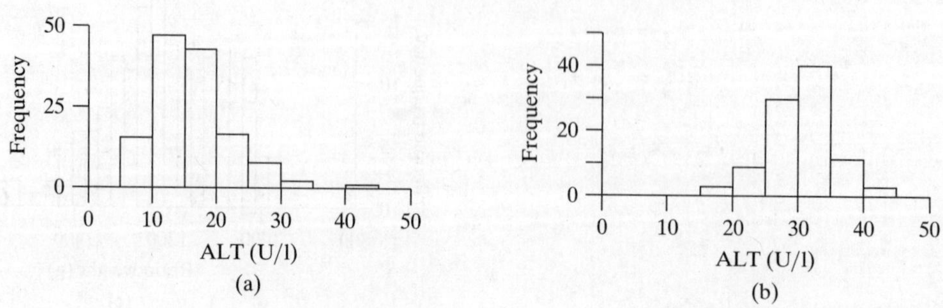

Figure 2.2.19 Distribution of serum ALT measurements (a) for 129 volunteers; (b) for 109 assays of the same specimen

Exercises 2.2.1–2.2.9

2.2.1 A paleontologist measured the width (in mm) of the last upper molar in 36 specimens of the extinct mammal *Acropithecus rigidus*. The results were as follows:[12]

6.1	5.7	6.0	6.5	6.0	5.7
6.1	5.8	5.9	6.1	6.2	6.0
6.3	6.2	6.1	6.2	6.0	5.7
6.2	5.8	5.7	6.3	6.2	5.7
6.2	6.1	5.9	6.5	5.4	6.7
5.9	6.1	5.9	5.9	6.1	6.1

(a) Construct a frequency distribution and display it as a table and as a histogram.

(b) Describe the shape of the distribution.

2.2.2 In a study of schizophrenia, researchers measured the activity of the enzyme monoamine oxidase (MAO) in the blood platelets of 18 patients. The results (expressed as nmoles benzylaldehyde product per 10^8 platelets) were as follows:[13]

6.8	8.4	8.7	11.9	14.2	18.8
9.9	4.1	9.7	12.7	5.2	7.8
7.8	7.4	7.3	10.6	14.5	10.7

Construct a dotplot of the data.

2.2.3 Consider the data presented in Exercise 2.2.2. Construct a frequency distribution and display it as a table and as a histogram.

2.2.4 A dendritic tree is a branched structure that emanates from the body of a nerve cell. As part of a study of brain development, 36 nerve cells were taken from the brains of newborn guinea pigs. The investigators counted the number of dendritic branch segments emanating from each nerve cell. The numbers were as follows:[14]

23	30	54	28	31	29	34	35	30
27	21	43	51	35	51	49	35	24
26	29	21	29	37	27	28	33	33
23	37	27	40	48	41	20	30	57

Construct a dotplot of the data.

2.2.5 Consider the data presented in Exercise 2.2.4. Construct a frequency distribution and display it as a table and as a histogram.

2.2.6 The total amount of protein produced by a dairy cow can be estimated from periodic testing of her milk. The following are the total annual protein production values (lb) for twenty-eight 2-year-old Holstein cows. Diet, milking procedures, and other conditions were the same for all the animals.[15]

425	481	477	434	410	397	438
545	528	496	502	529	500	465
539	408	513	496	477	445	546
471	495	445	565	499	508	426

Construct a frequency distribution and display it as a table and as a histogram.

2.2.7 For each of 31 healthy dogs, a veterinarian measured the glucose concentration in the anterior chamber of the right eye and also in the blood serum. The following data are the anterior chamber glucose measurements, expressed as a percentage of the blood glucose.[16]

81	85	93	93	99	76	75	84
78	84	81	82	89	81	96	82
74	70	84	86	80	70	131	75
88	102	115	89	82	79	106	

Construct a frequency distribution and display it as a table and as a histogram.

2.2.8 Agronomists measured the yield of a variety of hybrid corn in 16 locations in Illinois. The data, in bushels per acre, were[17]

241	230	207	219	266	167
204	144	178	158	153	
187	181	196	149	183	

(a) Construct a dotplot of the data.

(b) Describe the shape of the distribution.

2.2.9 (Computer problem) Trypanosomes are parasites that cause disease in humans and animals. In an early study of trypanosome morphology, researchers measured the lengths of 500 individual trypanosomes taken from the blood of a rat. The results are summarized in the accompanying frequency distribution.[18]

Length (μm)	Frequency (number of individuals)	Length (μm)	Frequency (number of individuals)
15	1	27	36
16	3	28	41
17	21	29	48
18	27	30	28
19	23	31	43
20	15	32	27
21	10	33	23
22	15	34	10
23	19	35	4
24	21	36	5
25	34	37	1
26	44	38	1

(a) Construct a histogram of the data using 24 classes (i.e., one class for each integer length, from 15 to 38).

(b) What feature of the histogram suggests the interpretation that the 500 individuals are a mixture of two distinct types?

(c) Construct a histogram of the data using only 6 classes. Discuss how this histogram gives a qualitatively different impression than the histogram from part (a).

2.3 Descriptive Statistics: Measures of Center

For categorical data, the frequency distribution provides a concise and complete summary of a sample. For numeric variables, the frequency distribution can usefully be supplemented by a few numerical measures. A numerical measure calculated from sample data is called a **statistic.*** **Descriptive statistics** are statistics that describe a set of data. Usually the descriptive statistics for a sample are calculated in order to provide information about a population of interest (see Section 2.8). In this section we discuss measures of the center of the data. There are several different ways to define the "center" or "typical value" of the observations in a sample. We will consider the two most widely used measures of center: the median and the mean.

THE MEDIAN

Perhaps the simplest measure of the center of a data set is the sample **median.** The sample median is the value that most nearly lies in the middle of the sample—it is the data value that splits the ordered data into two equal halves. To find the median, first arrange the observations in increasing order. In the array of ordered observations, the median is the middle value (if n is odd) or midway between the two middle values (if n is even). We denote the median of the sample by the symbol $\tilde{y}$ (read "y-tilde"). Example 2.3.1 illustrates these definitions.

Example 2.3.1 **Weight Gain of Lambs** The following are the 2-week weight gains (lb) of six young lambs of the same breed that had been raised on the same diet:[19]

11 13 19 2 10 1

The ordered observations are

1 2 10 11 13 19

The median weight gain is

$$\tilde{y} = \frac{10 + 11}{2} = 10.5 \text{ lb}$$

The median divides the sorted data into two equal pieces (the same number of observations fall above and below the median). Figure 2.3.1 shows a dotplot of the lamb weight-gain data, along with the location of $\tilde{y}$. ■

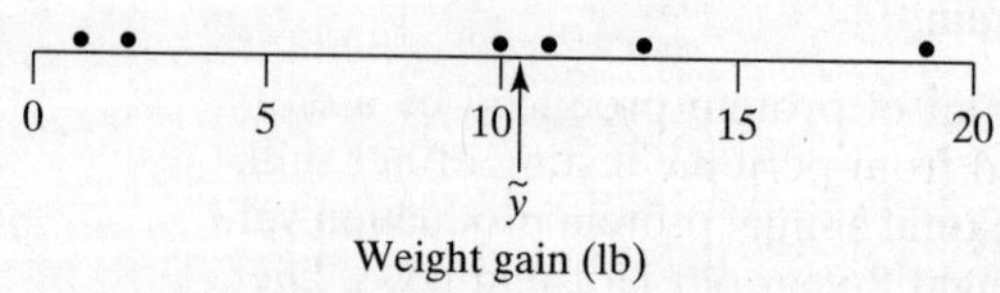

Figure 2.3.1 Plot of the lamb weight-gain data

*Numerical measures based on the entire population are called **parameters**, which are discussed in greater detail in Section 2.8.

Example 2.3.2

Weight Gain of Lambs Suppose the sample contained one more lamb, with the seven ranked observations as follows:

1 2 10 10 11 13 19

For this sample, the median weight gain is

$$\tilde{y} = 10 \text{ lb}$$

(Notice that in this example there are two lambs whose weight gain is equal to the median. The fourth observation—the second 10—is the median.) ■

A more formal way to define the median is in terms of rank position in the ordered array (counting the smallest observation as rank 1, the next as 2, and so on). The rank position of the median is equal to

$$(0.5)(n + 1)$$

Thus, if $n = 7$, we calculate $(0.5)(n + 1) = 4$, so that the median is the fourth largest observation; if $n = 6$, we have $(0.5)(n + 1) = 3.5$, so that the median is midway between the third and fourth largest observations. Note that the formula $(0.5)(n + 1)$ does not give the median, it gives the location of the median within the ordered list of the data.

THE MEAN

The most familiar measure of center is the ordinary average or **mean** (sometimes called the arithmetic mean). The mean of a sample (or "the sample mean") is the sum of the observations divided by the number of observations. If we denote a variable by Y, then we denote the observations in a sample by $y_1, y_2, \ldots, y_n$ and we denote the mean of the sample by the symbol $\bar{y}$ (read "y-bar"). Example 2.3.3 illustrates this notation.

Example 2.3.3

Weight Gain of Lambs The following are the data from Example 2.3.1:

11 13 19 2 10 1

Here $y_1 = 11, y_2 = 13$, and so on, and $y_6 = 1$. The sum of the observations is $11 + 13 + \cdots + 1 = 56$. We can write this using "summation notation" as $\sum_{i=1}^{n} y_i = 56$. The symbol $\sum_{i=1}^{n} y_i$ means to "add up the y_i's." Thus, when $n = 6$, $\sum_{i=1}^{n} y_i = y_1 + y_2 + y_3 + y_4 + y_5 + y_6$. In this case we get $\sum_{i=1}^{n} y_i = 11 + 13 + 19 + 2 + 10 + 1 = 56$.

The mean weight gain of the six lambs in this sample is

$$\begin{aligned}\bar{y} &= \frac{11 + 13 + 19 + 2 + 10 + 1}{6} \\ &= \frac{56}{6} \\ &= 9.33 \text{ lb}\end{aligned}$$

THE SAMPLE MEAN The general definition of the sample mean is

$$\bar{y} = \frac{\sum_{i=1}^{n} y_i}{n}$$

where the y_i's are the observations in the sample and n is the sample size (that is, the number of y_i's).

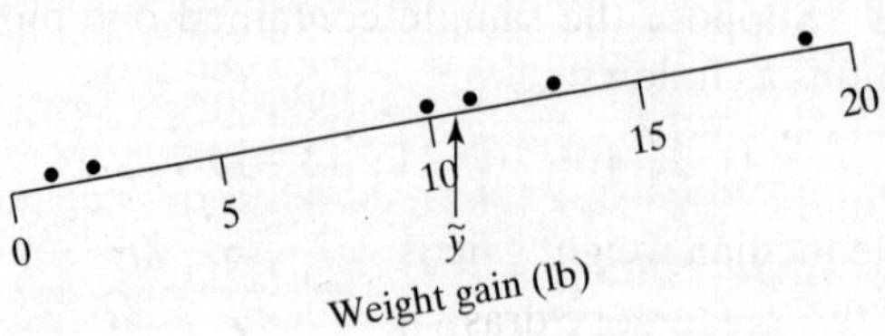

Figure 2.3.2 Plot of the lamb weight-gain data with the sample median as the fulcrum of a balance

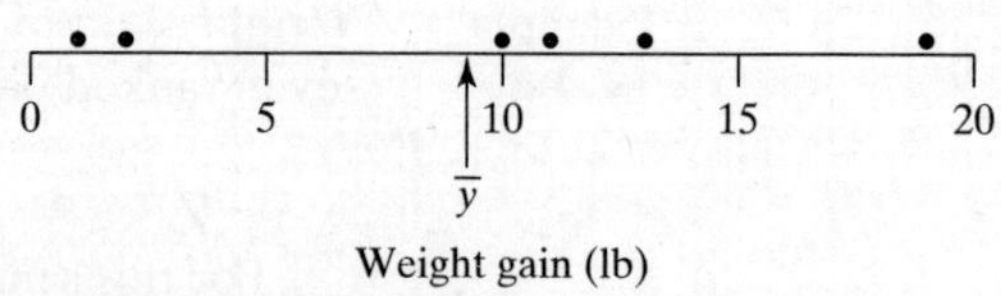

Figure 2.3.3 Plot of the lamb weight-gain data with the sample mean as the fulcrum of a balance

While the median divides the data into two equal pieces (i.e., the same number of observations above and below), the mean is the "point of balance" of the data. Figure 2.3.2 shows a dotplot of the lamb weight-gain data, along with the location of $\tilde{y}$. If the data points were children on a weightless seesaw, then the seesaw would tip if the fulcrum were placed at $\tilde{y}$ despite there being the same number of children on either side. The children on the left side (below $\tilde{y}$) tend to sit further from $\tilde{y}$ than the children on the right (above $\tilde{y}$) causing the seesaw to tip. However, if the fulcrum were placed at $\bar{y}$, the seesaw would exactly balance as in Figure 2.3.3. ■

The difference between a data point and the mean is called a **deviation**: $\text{deviation}_i = y_i - \bar{y}$. The mean has the property that the sum of the deviations from the mean is zero—that is, $\sum_{i=1}^{n}(y_i - \bar{y}) = 0$. In this sense, the mean is a center of the distribution—the positive deviations balance the negative deviations.

Example 2.3.4

Weight Gain of Lambs For the lamb weight-gain data, the deviations are as follows:

$$\text{deviation}_1 = y_1 - \bar{y} = 11 - 9.33 = 1.67$$

$$\text{deviation}_2 = y_2 - \bar{y} = 13 - 9.33 = 3.67$$

$$\text{deviation}_3 = y_3 - \bar{y} = 19 - 9.33 = 9.67$$

$$\text{deviation}_4 = y_4 - \bar{y} = 2 - 9.33 = -7.33$$

$$\text{deviation}_5 = y_5 - \bar{y} = 10 - 9.33 = 0.67$$

$$\text{deviation}_6 = y_6 - \bar{y} = 1 - 9.33 = -8.33$$

The sum of the deviations is $\sum_{i=1}^{n}(y_i - \bar{y}) = 1.67 + 3.67 + 9.67 - 7.33 + 0.67 - 8.33 = 0$. ■

Robustness A statistic is said to be **robust** if the value of the statistic is relatively unaffected by changes in a small portion of the data, even if the changes are dramatic ones. The median is a robust statistic, but the mean is not robust because it can be greatly shifted by changes in even one observation. Example 2.3.5 illustrates this behavior.

Example 2.3.5

Weight Gain of Lambs Recall that for the lamb weight-gain data

$$1 \quad 2 \quad 10 \quad 11 \quad 13 \quad 19$$

we found

$$\bar{y} = 9.33 \text{ and } \tilde{y} = 10.5$$

Suppose now that the observation 19 is changed. How would the mean and median be affected? You can visualize the effect by imagining moving the right-hand dot in Figure 2.3.3. Clearly the mean could change a great deal; the median would not be affected. For instance,

If the 19 is changed to 14, the mean becomes 8.5 and the median does not change.

If the 19 is changed to 29, the mean becomes 11 and the median does not change.

These changes are not wild ones; that is, the changed samples might well have arisen from the same feeding experiment. Of course, a huge change, such as changing the 19 to 100, would shift the mean very drastically. Note that it would not shift the median at all. ■

VISUALIZING THE MEAN AND MEDIAN

We can visualize the mean and the median in relation to the histogram of a distribution. The median divides the area under the histogram roughly in half because it divides the observations roughly in half ["roughly" because some observations may be tied at the median, as in Example 2.3.3(b), and because the observations within each class are not uniformly distributed across the class]. The mean can be visualized as the point of balance of the histogram: If the histogram were made out of plywood, it would balance if supported at the mean.

If the frequency distribution is symmetric, the mean and the median are equal and fall in the center of the distribution. If the frequency distribution is skewed, both measures are pulled toward the longer tail, but the mean is usually pulled farther than the median. The effect of skewness is illustrated by the following example.

Example 2.3.6 **Cricket Singing Times** Male Mormon crickets *(Anabrus simplex)* sing to attract mates. A field researcher measured the duration of 51 unsuccessful songs—that is, the time until the singing male gave up and left his perch.[20] Figure 2.3.4 shows the histogram of the 51 singing times. Table 2.3.1 gives the raw data. The median is 3.7 min and the mean is 4.3 min. The discrepancy between these measures is due largely to the long straggly tail of the distribution; the few unusually long singing times influence the mean, but not the median. ■

Table 2.3.1 Fifty-one cricket singing times (min)

4.3	3.9	17.4	2.3	0.8	1.5	0.7	3.7
24.1	9.4	5.6	3.7	5.2	3.9	4.2	3.5
6.6	6.2	2.0	0.8	2.0	3.7	4.7	
7.3	1.6	3.8	0.5	0.7	4.5	2.2	
4.0	6.5	1.2	4.5	1.7	1.8	1.4	
2.6	0.2	0.7	11.5	5.0	1.2	14.1	
4.0	2.7	1.6	3.5	2.8	0.7	8.6	

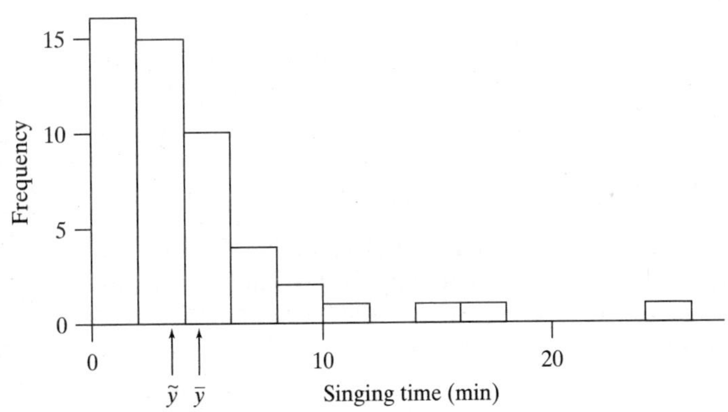

Figure 2.3.4 Histogram of cricket singing times

MEAN VERSUS MEDIAN

Both the mean and the median are usually reasonable measures of the center of a data set. The mean is related to the sum; for example, if the mean weight gain of 100 lambs is 9 lb, then the total weight gain is 900 lb, and this total may be of primary interest since it translates more or less directly into profit for the farmer. In some situations the mean makes very little sense. Suppose, for example, that the observations are survival times of cancer patients on a certain treatment protocol, and that most patients survive less than 1 year, while a few respond well and survive for 5 or even 10 years. In this case, the mean survival time might be greater than the survival time of most patients; the median would more nearly represent the experience of a "typical" patient. Note also that the mean survival time cannot be computed until the last patient has died; the median does not share this disadvantage. Situations in which the median can readily be computed, but the mean cannot, are not uncommon in bioassay, survival, and toxicity studies.

We have noted that the median is more robust than the mean. If a data set contains a few observations rather distant from the main body of the data—that is, a long, straggly tail—then the mean may be unduly influenced by these few unusual observations. Thus, the "tail" may "wag the dog"—an undesirable situation. In such cases, the robustness of the median may be advantageous.

An advantage of the mean is that in some circumstances it is more efficient than the median. Efficiency is a technical notion in statistical theory; roughly speaking, a method is efficient if it takes full advantage of all the information in the data. Partly because of its efficiency, the mean has played a major role in classical methods in statistics.

Exercises 2.3.1–2.3.14

2.3.1 Invent a sample of size 5 for which the sample mean is 20 and not all the observations are equal.

2.3.2 Invent a sample of size 5 for which the sample mean is 20 and the sample median is 15.

2.3.3 A researcher applied the carcinogenic (cancer-causing) compound benzo(a)pyrene to the skin of five mice, and measured the concentration in the liver tissue after 48 hours. The results (nmol/gm) were as follows:[21]

6.3 5.9 7.0 6.9 5.9

Determine the mean and the median.

2.3.4 Consider the data from Exercise 2.3.3. Do the calculated mean and median support the claim that, in general, liver tissue concentration after 48 hours differs from 6.3 nmol/gm?

2.3.5 Six men with high serum cholesterol participated in a study to evaluate the effects of diet on cholesterol level. At the beginning of the study their serum cholesterol levels (mg/dl) were as follows:[22]

366 327 274 292 274 230

Determine the mean and the median.

2.3.6 Consider the data from Exercise 2.3.5. Suppose an additional observation equal to 400 were added to the sample. What would be the mean and the median of the seven observations?

2.3.7 The weight gains of beef steers were measured over a 140-day test period. The average daily gains (lb/day) of 9 steers on the same diet were as follows:[23]

3.89 3.51 3.97 3.31 3.21
3.36 3.67 3.24 3.27

Determine the mean and median.

2.3.8 Consider the data from Exercise 2.3.7. Are the calculated mean and median consistent with the claim that, in general, steers gain 3.5 lb/day? Are they consistent with a claim of 4.0 lb/day?

2.3.9 Consider the data from Exercise 2.3.7. Suppose an additional observation equal to 2.46 were added to the sample. What would be the mean and the median of the 10 observations?

2.3.10 As part of a classic experiment on mutations, 10 aliquots of identical size were taken from the same culture of the bacterium *E. coli*. For each aliquot, the number of bacteria resistant to a certain virus was determined. The results were as follows:[24]

14 15 13 21 15
14 26 16 20 13

(a) Construct a frequency distribution of these data and display it as a histogram.

(b) Determine the mean and the median of the data and mark their locations on the histogram.

2.3.11 The accompanying table gives the litter size (number of piglets surviving to 21 days) for each of 36 sows (as in Example 2.2.4). Determine the median litter size. (*Hint*: Note that there is one 5, but there are two 7's, three 8's, etc.)

Number of piglets	Frequency (Number of sows)
5	1
6	0
7	2
8	3
9	3
10	9
11	8
12	5
13	3
14	2
Total	36

2.3.12 Consider the data from Exercise 2.3.11. Determine the mean of the 36 observations. (*Hint*: Note that there is one 5 but there are two 7's, three 8's, etc. Thus, $\Sigma y_i = 5 + 7 + 7 + 8 + 8 + 8 + \cdots = 5 + 2(7) + 3(8) + \cdots$)

2.3.13 Here is a histogram.

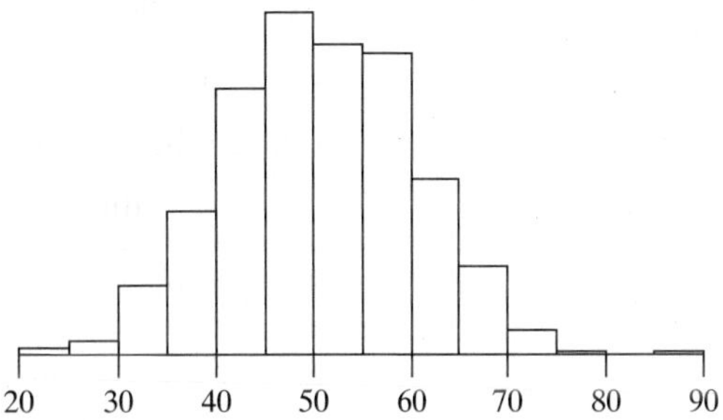

(a) Estimate the median of the distribution.

(b) Estimate the mean of the distribution.

2.3.14 Here is a histogram.

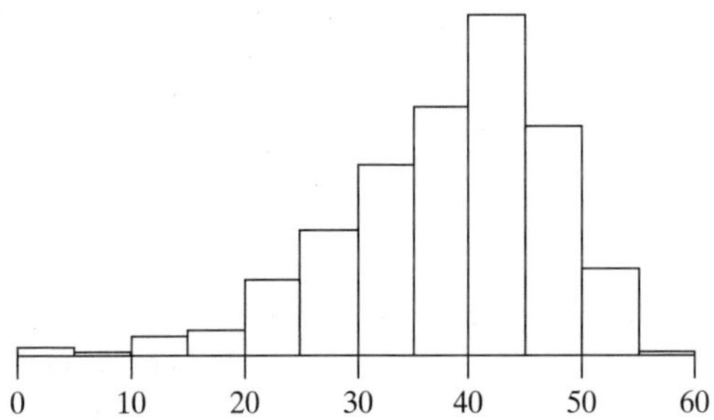

(a) Estimate the median of the distribution.

(b) Estimate the mean of the distribution.

2.4 Boxplots

One of the most efficient graphics, both for examining a single distribution and for making comparisons between distributions, is known as a boxplot, which is the topic of this section. Before discussing boxplots, however, we need to discuss quartiles.

QUARTILES AND THE INTERQUARTILE RANGE

The median of a distribution splits the distribution into two parts, a lower part and an upper part. The **quartiles** of a distribution divide each of these parts in half, thereby dividing the distribution into four quarters. The **first quartile**, denoted by Q_1, is the median of the data values in the lower half of the data set. The **third quartile**, denoted by Q_3, is the median of the data values in the upper half of the data set.* The following example illustrates these definitions.

*Some authors use other definitions of quartiles, as does some computer software. A common alternative definition is to say that the first quartile has rank position $(0.25)(n + 1)$ and that the third quartile has rank position $(0.75)(n + 1)$. Thus, if $n = 10$, the first quartile would have rank position $(0.25)(11) = 2.75$—that is, to find the first quartile we would have to interpolate between the second and third largest observations. If n is large, then there is little practical difference between the definitions that various authors use.

Example 2.4.1 **Blood Pressure** The systolic blood pressures (mm Hg) of seven middle-aged men were as follows:[25]

151 124 132 170 146 124 113

Putting these values in rank order, the sample is

113 124 124 132 146 151 170

The median is the fourth largest observation, which is 132. There are three data points in the lower part of the distribution: 113, 124, and 124. The median of these three values is 124. Thus, the first quartile, Q_1, is 124.

Likewise, there are three data points in the upper part of the distribution: 146, 151 and 170. The median of these three values is 151. Thus, the third quartile, Q_3, is 151.

113	124	124	132	146	151	170
	↑		⋮		↑	
	first quartile Q_1		median		third quartile Q_3	

■

Note that the median is not included in either the lower part or the upper part of the distribution. If the sample size, n, is even, then exactly one-half of the observations are in the lower part of the distribution and one-half are in the upper part.

The **interquartile range** is the difference between the first and third quartiles and is abbreviated as **IQR**: $\text{IQR} = Q_3 - Q_1$. For the blood pressure data in Example 2.4.1, the IQR is $151 - 124 = 27$. Note that the IQR is a *number,* not an interval; the IQR measures the spread of the middle 50% of the distribution.

Example 2.4.2 **Pulse** The pulses of 12 college students were measured.[26] Here are the data, arranged in order, with the position of the median indicated by a dashed line:

62 64 68 70 70 74 ¦ 74 76 76 78 78 80

The median is $\frac{74 + 74}{2} = 74$. There are six observations in the lower part of the distribution: 62, 64, 68, 70, 70, 74. Thus, the first quartile is the average of the third and fourth largest data values:

$$Q_1 = \frac{68 + 70}{2} = 69$$

There are six observations in the upper part of the distribution: 74, 76, 76, 78, 78, 80. Thus, the third quartile is the average of the ninth and tenth largest data values (the third and fourth values in the upper part of the distribution):

$$Q_3 = \frac{76 + 78}{2} = 77$$

Thus, the interquartile range is

$$\text{IQR} = 77 - 69 = 8$$

We have

$$\begin{array}{cccccccccccc} 62 & 64 & 68 & 70 & 70 & 74 \,\vdots & 74 & 76 & 76 & 78 & 78 & 80 \\ & & \uparrow & & & \text{median} & & & \uparrow & & & \\ & & \text{first quartile} & & & & & & \text{third quartile} & & & \\ & & Q_1 & & & & & & Q_3 & & & \end{array}$$

The minimum pulse value is 62 and the maximum is 80. ■

The minimum, the maximum, the median, and the quartiles, taken together, are referred to as the **five-number summary** of the data.

OUTLIERS

Sometimes a data point differs so much from the rest of the data that it doesn't seem to belong with the other data. Such a point is called an **outlier**. An outlier might occur because of a recording error or typographical error when the data are recorded, because of an equipment failure during an experiment, or for many other reasons. Outliers are the most interesting points in a data set. Sometimes outliers tell us about a problem with the experimental protocol (e.g., an equipment failure, a failure of a patient to take his or her medication consistently during a medical trial). At other times an outlier might alert us to the fact that a special circumstance has happened (e.g., an abnormally high or low value on a medical test could indicate the presence of a disease in a patient).

People often use the term "outlier" informally. There is, however, a common definition of "outlier" in statistical practice. To give a definition of outlier, we first discuss what are known as fences. The **lower fence** of a distribution is

$$\text{lower fence} = Q_1 - 1.5 \times \text{IQR}$$

The **upper fence** of a distribution is

$$\text{upper fence} = Q_3 + 1.5 \times \text{IQR}$$

Note that the fences need not be data values; indeed, there might be no data near the fences. The fences just locate limits within the sample distribution. These limits give us a way to define outliers. *An outlier is a data point that falls outside of the fences*. That is, if

$$\text{data point} < Q_1 - 1.5 \times \text{IQR}$$

or

$$\text{data point} > Q_3 + 1.5 \times \text{IQR}$$

then we call the point an outlier.

Example 2.4.3

Pulse In Example 2.4.2 we saw that $Q_1 = 69$, $Q_3 = 77$, and $\text{IQR} = 8$. Thus, the lower fence is $69 - 1.5 \times 8 = 69 - 12 = 57$. Any point less than 57 would be an outlier. The upper fence is $77 + 1.5 \times 8 = 77 + 12 = 89$. Any point greater than

89 would be an outlier. Since there are no points less than 57 or greater than 89, there are no outliers in this data set. ■

Example 2.4.4 **Radish Growth in Light** A common biology experiment involves growing radish seedlings under various conditions. In one experiment students grew 14 radish seedlings in constant light. The observations, in order, are

3 5 5 7 7 8 9 ⁞ 10 10 10 10 14 20 21

(arrow under the 4th value, 7: first quartile Q_1; dashed line between 9 and 10: median; arrow under the 11th value, 10: third quartile Q_3)

Thus, the median is $\frac{9 + 10}{2} = 9.5$, Q_1 is 7, and Q_3 is 10. The interquartile range is IQR $= 10 - 7 = 3$. The lower fence is $7 - 1.5 \times 3 = 7 - 4.5 = 2.5$, so any point less than 2.5 would be an outlier. The upper fence is $10 + 1.5 \times 3 = 10 + 4.5 = 14.5$, so any point greater than 14.5 is an outlier. Thus, the two largest observations in this data set are outliers: 20 and 21. ■

BOXPLOTS FOR DATA WITH NO OUTLIERS

A **boxplot** is a visual representation of the five-number summary. To make a boxplot for a data set with no outliers, we first make a number line; then we mark the positions minimum, Q_1, the median, Q_3, and the maximum:

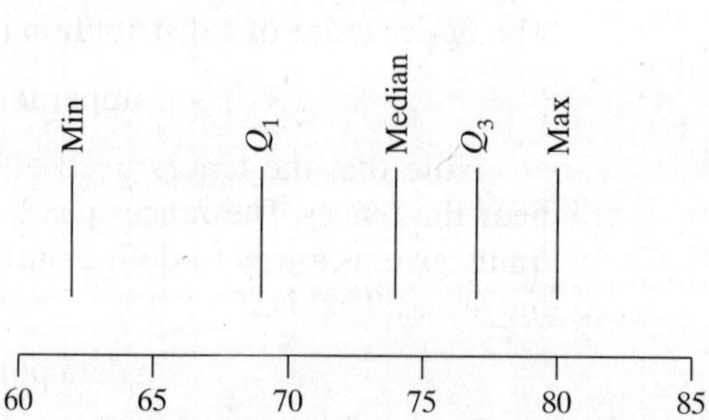

Next, we make a box connecting the quartiles:

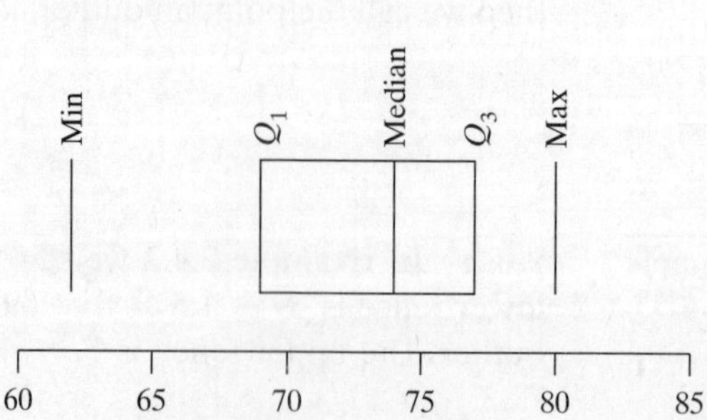

Note that the interquartile range is equal to the length of the box. Finally, provided there are no outliers* we extend "whiskers" from Q_1 down to the minimum and from Q_3 up to the maximum:

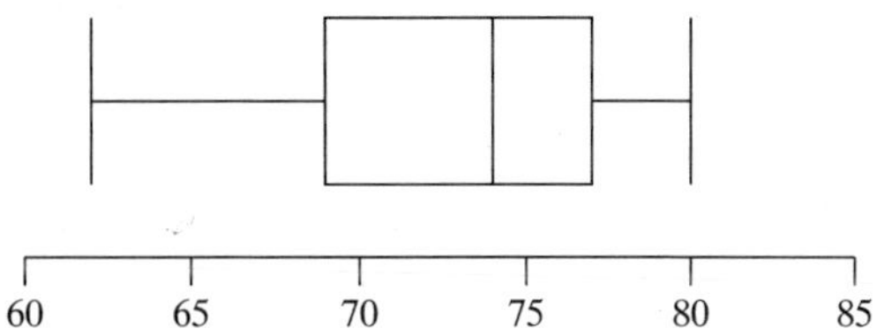

A boxplot gives a quick visual summary of the distribution. We can immediately see where the center of the data is from the line within the box that locates the median. We see the spread of the total distribution, from the minimum up to the maximum, as well as the spread of the middle half of the distribution—the interquartile range—from the length of the box. The boxplot also gives an indication of the shape of the distribution; the preceding boxplot has a long lower whisker, indicating that the distribution is skewed to the left. Example 2.4.5 shows a boxplot for data from a radish growth experiment that had no outliers.†

Example 2.4.5

Radish Growth In another version of the experiment in Example 2.4.4, a moist paper towel is put into a plastic bag. About one third of the way from the bottom of the bag a seam of staples was created; the radish seeds were placed along the seam. One group of students kept their radish seed bags in total darkness for 3 days and then measured the length, in mm, of each radish shoot at the end of the 3 days. They collected 14 observations; the data are shown in Table 2.4.1.[27]

Table 2.4.1 Radish growth, in mm, after three days in total darkness

15	20	11	30	33
20	29	35	8	10
22	37	15	25	

Here are the data in order from smallest to largest:

8 10 11 **15** 15 20 **20** ⋮ **22** 25 29 **30** 33 35 37

↑ first quartile Q_1 (at **15**); median (between **20** and **22**); ↑ third quartile Q_3 (at **30**)

The quartiles are $Q_1 = 15$ and $Q_3 = 30$. The median, $\tilde{y} = 21$, is the average of the two middle values of 20 and 22. Figure 2.4.1 shows a boxplot of the same data. ■

---

*We will consider situations with outliers after the next example.

†This and subsequent boxplots in our text are slightly stylized. Different computer packages present the plot somewhat differently, but all boxplots have the same basic five-number summary.

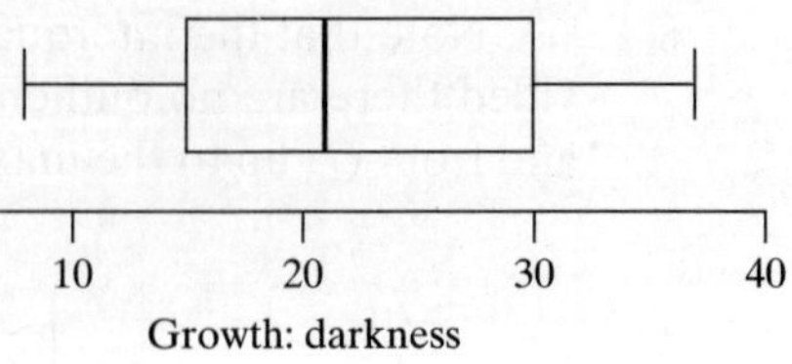

Figure 2.4.1 Boxplot of data on radish growth in darkness

BOXPLOTS FOR DATA WITH OUTLIERS

If there are outliers in the upper part of the distribution, then we can identify them with dots (or other plotting symbols) on the boxplot. We then extend a whisker from Q_3 up to the largest data point that is *not* an outlier. Likewise, if there are outliers in the lower part of the distribution, we identify them with dots and extend a whisker from Q_1 down to the smallest observation that is not an outlier. Figure 2.4.2 shows the distribution of radish seedlings grown under constant light. The area between the lower and upper fences is white, while the outlying region is blue.

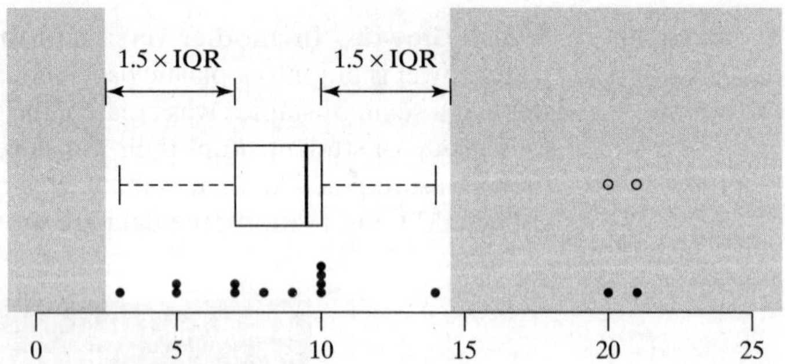

Figure 2.4.2 Dotplot and boxplot of data on radish growth in constant light. The points in the blue region are outliers.

Figure 2.4.3 shows a boxplot of the data on radish seedlings grown in constant light.*

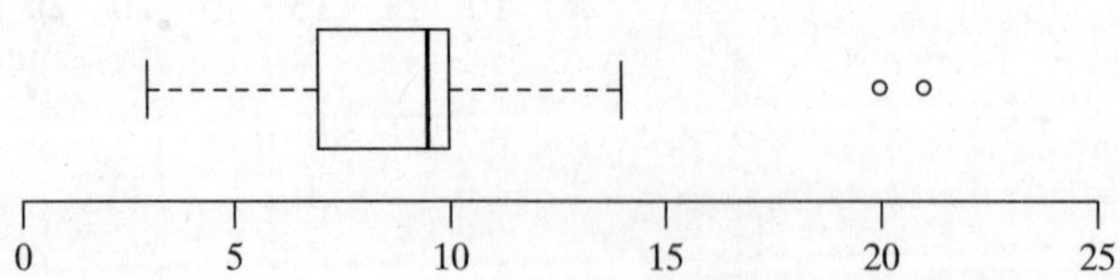

Figure 2.4.3 Boxplot of data on radish growth in constant light

The method we have defined for identifying outliers allows the bulk of the data to determine how extreme an observation must be before we consider it to be an

*Most computer software has options that can alter how outliers are determined and displayed.

outlier, since the quartiles and the IQR are determined from the data themselves. Thus, a point that is an outlier in one data set might not be an outlier in another data set. We label a point as an outlier if it is unusual relative to the inherent variability in the entire data set.

After an outlier has been identified, people are often tempted to remove the outlier from the data set. In general this is not a good idea. If we can identify that an outlier occurred due to an equipment error, for example, then we have good reason to remove the outlier before analyzing the rest of the data. However, quite often outliers appear in data sets without any identifiable, external reason for them. In such cases, we simply proceed with our analysis, aware that there is an outlier present. In some cases, we might want to calculate the mean, for example, with and without the outlier and then report both calculations to show the effect of the outlier in the overall analysis. This is preferable to removing the outlier, which obscures the fact that there was an unusual data point present.

Exercises 2.4.1–2.4.8

2.4.1 Here are the data from Exercise 2.3.10 on the number of virus-resistant bacteria in each of 10 aliquots:

14	15	13	21	15
14	26	16	20	13

(a) Determine the median and the quartiles.

(b) Determine the interquartile range.

(c) How large would an observation in this data set have to be in order to be an outlier?

2.4.2 Here are the 18 measurements of MAO activity reported in Exercise 2.2.2:

6.8	8.4	8.7	11.9	14.2	18.8
9.9	4.1	9.7	12.7	5.2	7.8
7.8	7.4	7.3	10.6	14.5	10.7

(a) Determine the median and the quartiles.

(b) Determine the interquartile range.

(c) How large would an observation in this data set have to be in order to be an outlier?

(d) Construct a boxplot of the data.

2.4.3 In a study of milk production in sheep (for use in making cheese), a researcher measured the 3-month milk yield for each of 11 ewes. The yields (liters) were as follows:[28]

56.5	89.8	110.1	65.6	63.7	82.6
75.1	91.5	102.9	44.4	108.1	

(a) Determine the median and the quartiles.

(b) Determine the interquartile range.

(c) Construct a boxplot of the data.

2.4.4 For each of the following histograms, use the histogram to estimate the median and the quartiles; then construct a boxplot for the distribution.

(a)

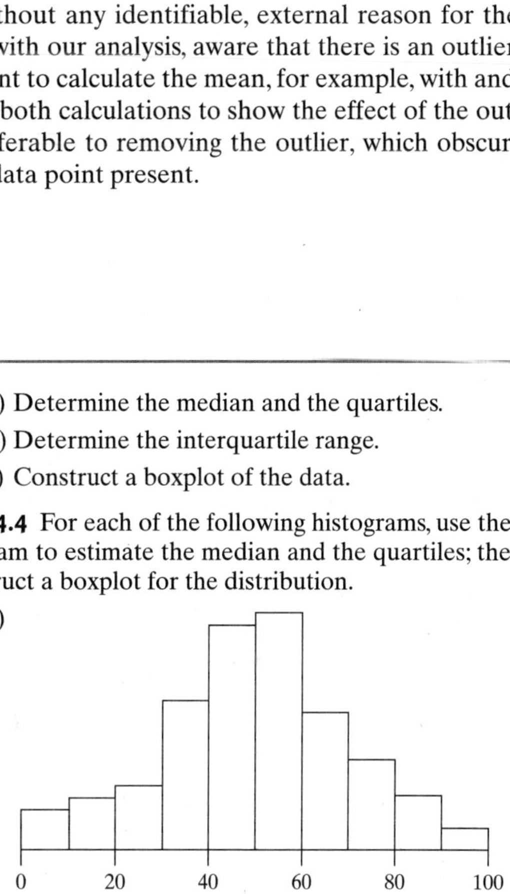

(b)

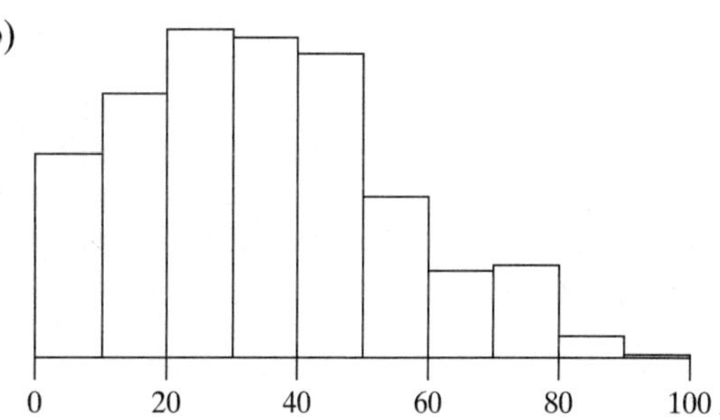

2.4.5 The following histogram shows the same data that are shown in one of the four boxplots. Which boxplot goes with the histogram? Explain your answer.

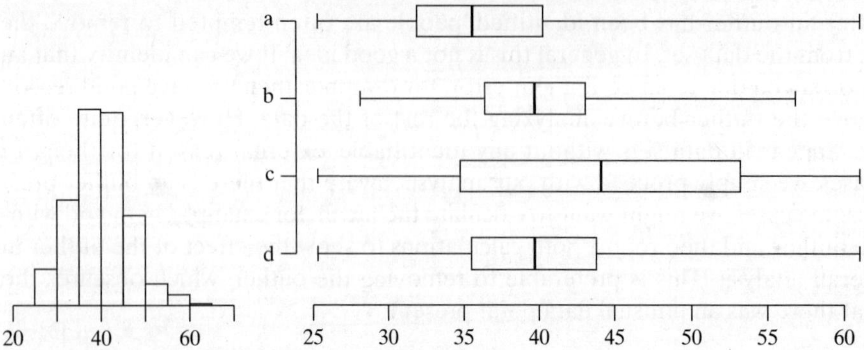

2.4.6 The following boxplot shows the five-number summary for a data set. For these data the minimum is 35, Q_1 is 42, the median is 49, Q_3 is 56, and the maximum is 65. Is it possible that no observation in the data set equals 42? Explain your answer.

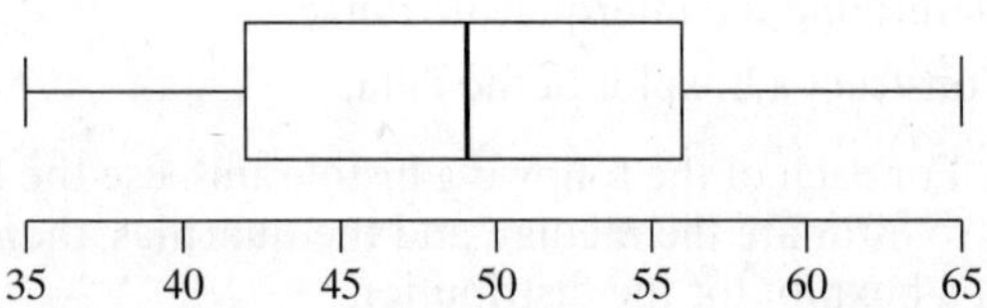

2.4.7 Statistics software can be used to find the five-number summary of a data set. Here is an example of MINITAB's descriptive statistics summary for a variable stored in column 1 (C1) of MINITAB's worksheet.

```
Variable  N   Mean    Median  TrMean  StDev  SEMean
C1        75  119.94  118.40  119.98  9.98   1.15

Variable  Min    Max     Q1      Q3
C1        95.16  145.11  113.59  127.42
```

(a) Use the MINITAB output to calculate the interquartile range.

(b) Are there any outliers in this set of data?

2.4.8 Consider the data from Exercise 2.4.7. Use the five-number summary that is given to create a boxplot of the data.

2.5 Relationships between Variables

In the previous sections we have studied **univariate** summaries of both numeric and categorical variables. A univariate summary is a graphical or numeric summary of a single variable.

The histogram, boxplot, sample mean, and median are all examples of univariate summaries for numeric data. The bar chart, frequency, and relative frequency tables are examples of univariate summaries for categorical data. In this section we present some common **bivariate** graphical summaries used to examine the *relationship* between pairs of variables.

CATEGORICAL–CATEGORICAL RELATIONSHIPS

To understand the relationship between two categorical variables, we first summarize the data in a **bivariate frequency table**. Unlike the frequency table presented in Section 2.2 (a univariate table), the bivariate frequency table has both rows and columns—one dimension for each variable. The choice of which variable to list with the rows and which to list with the columns is arbitrary. The following example considers the relationship between two categorical variables: *E. Coli* Source and Sampling Location.

Example 2.5.1

***E. Coli* Watershed Contamination** In an effort to determine if there are differences in the primary sources of fecal contamination at different locations in the Morro Bay watershed, $n = 623$ water specimens were collected at three primary locations that feed into Morro Bay: Chorro Creek ($n_1 = 241$), Los Osos Creek ($n_2 = 256$), and Baywood Seeps ($n_3 = 126$).[29] DNA fingerprinting techniques were used to determine the intestinal origin of the dominant *E. coli* strain in each water specimen. *E. coli* origins were classified into the following five categories: bird, domestic pet (e.g., cat or dog), farm animal (e.g., horse, cow, pig), human, or other terrestrial mammal (e.g., fox, mouse, coyote . . .). Thus, each water specimen had *two* categorical variables measured: location (Chorro, Los Osos, or Baywood) and *E. coli* source (bird, . . . , terrestrial mammal). Table 2.5.1 presents a frequency table of the data. ■

Table 2.5.1 Frequency table of *E. coli* source by location

	E. Coli Source					
Location	Bird	Domestic pet	Farm animal	Human	Terrestrial mammal	Total
Chorro Creek	46	29	106	38	22	**241**
Los Osos Creek	79	56	32	63	26	**256**
Baywood Seeps	35	23	0	60	8	**126**
Total	**160**	**108**	**138**	**161**	**56**	**623**

While Table 2.5.1 provides a concise summary of the data, it is difficult to discover any patterns in the data. Examining relative frequencies (row or column proportions) often helps us make meaningful comparisons as seen in the following example.

Example 2.5.2

***E. Coli* Watershed Contamination** Are domestic pets more of an *E. coli* problem (i.e., source) at Chorro Creek or Baywood Seeps? Table 2.5.1 shows that the domestic pet *E. coli* source count at Chorro (29) is higher than Baywood (23), so at first glance it seems that pets are more problematic at Chorro. However, as more water specimens were collected at Chorro ($n_1 = 241$) than Baywood ($n_2 = 126$), the relative frequency of domestic pet source *E. coli* is actually lower at Chorro ($29/241 = 0.120$) than Baywood ($23/126 = 0.183$). Table 2.5.2 displays row percentages and thus facilitates comparisons of *E. coli* sources among the locations. (Note that column percentages would not be meaningful in this context since the water was sampled by location and not by *E. coli* source.) ■

Table 2.5.2 Bivariate relative frequency table (row percentages) of *E. coli* source by location

	E. Coli Source					
Location	Bird	Domestic pet	Farm animal	Human	Terrestrial mammal	Total
Chorro Creek	19.1	12.0	44.0	15.8	9.1	**100**
Los Osos Creek	30.9	21.9	12.5	24.6	10.2	**100**
Baywood Seeps	27.8	18.3	0.0	47.6	6.3	**100**
All locations	**25.7**	**17.3**	**22.2**	**25.8**	**9.0**	**100**

To visualize the data in Tables 2.5.1 and 2.5.2, we can examine **stacked bar charts**. With a stacked frequency bar chart, the overall height of each bar reflects the sample size for a level of the X categorical variable (e.g., location), while the height or thickness of a slice that makes up a bar represents the count of the Y categorical variable (e.g., *E. coli* source) for that level of X. Figure 2.5.1 displays a stacked bar chart for the *E. coli* watershed count data in Table 2.5.1.

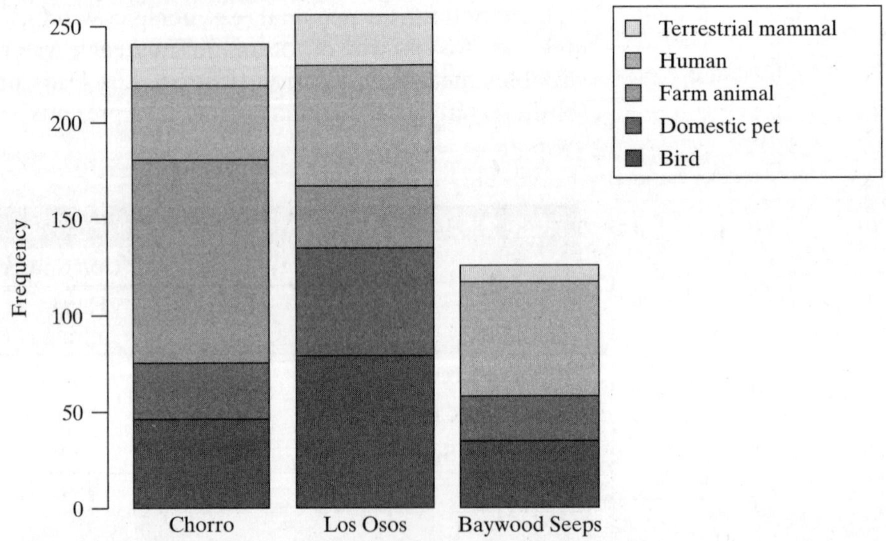

Figure 2.5.1 Stacked frequency chart of *E. coli* source by location

Like the frequency table, the stacked frequency bar chart is not conducive to making comparisons across the three locations as the sample sizes differ for these locations. (This graph does help highlight the difference in sample sizes; for example, it is very clear that many fewer water specimens were collected at Baywood Seeps.) A chart that better displays the distribution of one categorical variable across levels of another is a **stacked relative frequency** (or percentage) bar chart, which graphs the summaries from a bivariate relative frequency table such as Table 2.5.2. Figure 2.5.2 provides an example using the *E. coli* watershed contamination data. This plot normalizes the bars of Figure 2.5.1 to have the same height (100%) to facilitate comparisons across the three locations.

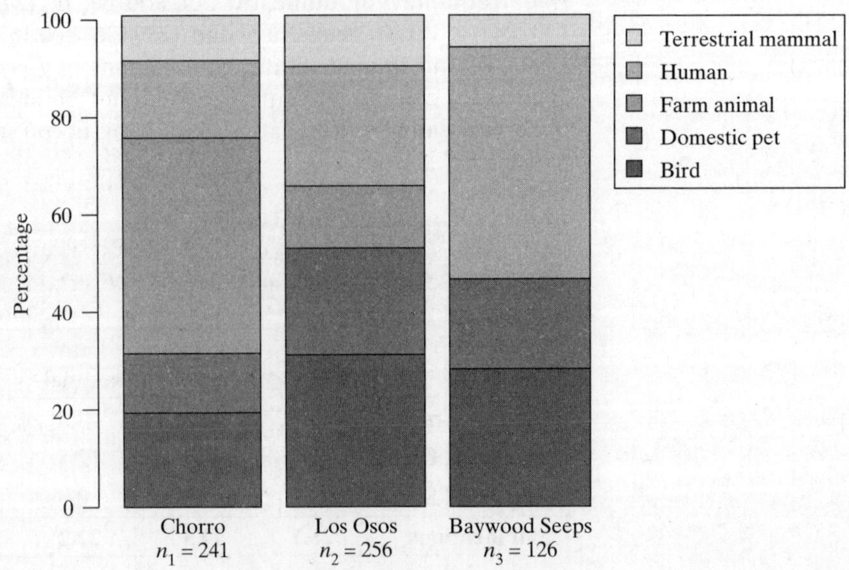

Figure 2.5.2 Stacked relative frequency (percentage) chart of *E. coli* source by location

Figure 2.5.2 makes it very easy to see that farm animals are the largest contributors of *E. coli* to Chorro Creek while humans are primarily responsible for the pollution at Baywood Seeps. The distribution of the slices in the three bars appears quite different, suggesting that the distribution of *E. coli* sources is not the same at the three locations. In Chapter 10 we will learn how to determine if these apparent differences are large enough to be compelling evidence for real differences in the distribution of *E. coli* source by location, or whether they are likely due to chance variation.

NUMERIC–CATEGORICAL RELATIONSHIPS

In Section 2.4 we learned that boxplots are graphs based on only five numbers: the minimum, first quartile, median, third quartile, and maximum. They are appealing plots because they are very simple and uncluttered, yet contain easy to read information about center, spread, skewness, and even outliers of a data set. By displaying **side-by-side boxplots** on the same graph, we are able to compare numeric data among several groups. We now consider an extension of the radish shoot growth problem in Example 2.4.3.

Example 2.5.3

Radish Growth Does light exposure alter initial radish shoot growth? The complete radish growth experiment of Examples 2.4.4 and 2.4.5 actually involved a total of 42 radish seeds randomly divided to receive one of three lighting conditions for germination (14 seeds in each lighting condition): 24-hour light, diurnal light (12 hours of light and 12 hours of darkness each day), and 24 hours of darkness. At the end of 3 days, shoot length was measured (mm). Thus, each shoot has two variables that are measured in this study: the categorical variable lighting condition (light, diurnal, dark) and the numeric variable sprout length (mm). Figure 2.5.3 displays side-by-side boxplots of the data. The boxplots make it very easy to compare the growth under the three conditions: It appears that light inhibits shoot growth. Are the observed differences in growth among the lighting conditions just due to chance variation, or is light really altering growth? We will learn how to numerically measure the strength of this evidence and answer this question in Chapters 7 and 11. ■

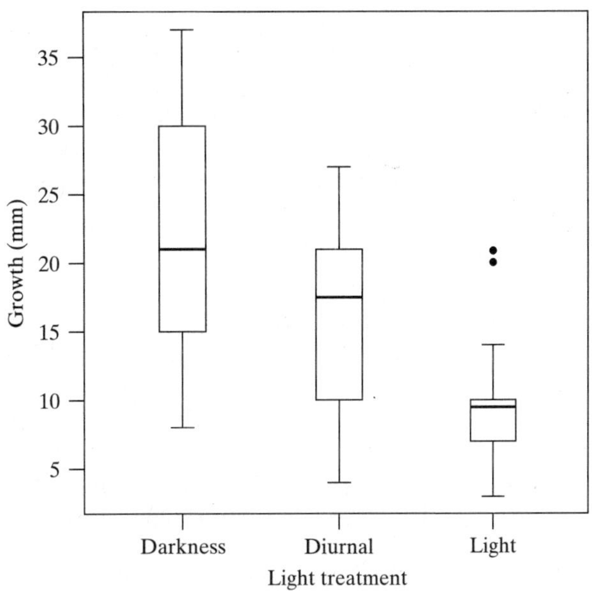

Figure 2.5.3 Side-by-side boxplots of radish growth under three conditions: constant darkness, half light–half darkness, and constant light

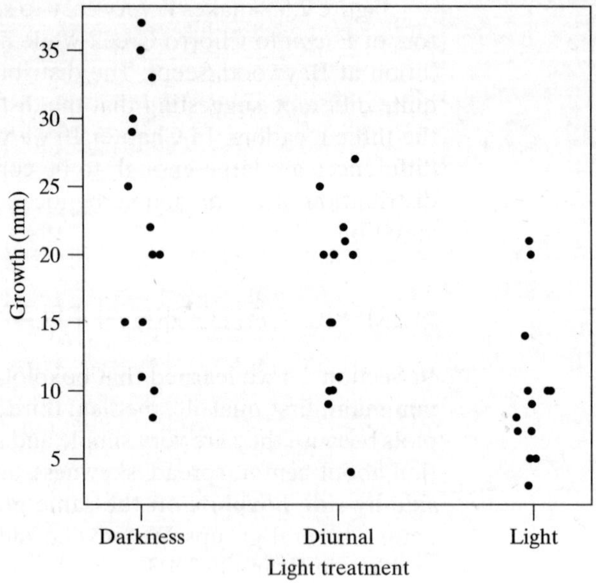

Figure 2.5.4 Side-by-side jittered dotplots of radish growth under three conditions: constant darkness, half light–half darkness, and constant light

For smaller data sets, we also may consider side-by-side dotplots of the data. Figure 2.5.4 displays a jittered side-by-side dotplot of the radish growth data of Example 2.5.3. The "jitter" is a common software option that adds horizontal scatter to the plot, helping to reduce the overlap of the dots. Choosing between side-by-side boxplots and dotplots is matter of personal preference. A good rule of thumb is to choose the plot that accurately reflects patterns in the data in the cleanest (least ink on the paper) way possible. For the radish growth example, the boxplot enables a very clean comparison of the growth under the three light treatments without hiding any information revealed by the dotplot.

NUMERIC–NUMERIC RELATIONSHIPS

Each of the previous examples considered comparing the distribution of one variable (either categorical or numeric) among several groups (i.e., across levels of a categorical variable). In the next example we illustrate the **scatterplot** as a tool to examine the relationship between two numeric variables, X and Y. A scatterplot plots each observed (x,y) pair as a dot on the x–y plane.

Example 2.5.4

Whale Selenium Can metal concentration in marine mammal teeth be used as a bioindicator for body burden? Selenium (Se) is an essential element that has been shown to play an important role in protecting marine mammals against the toxic effects of mercury (Hg) and other metals. Twenty beluga whales (*Delphinapterus leucas*) were harvested from the Mackenzie Delta, Northwest Territories, as part of an annual traditional Inuit hunt.[30] Each whale yielded two numeric measurements: Tooth Se (μg/g) and Liver Se (ng/g). Selenium concentrations for the whales are listed in Table 2.5.3. Liver Se concentration (Y) is graphed against Tooth Se concentration (X) in the scatterplot of Figure 2.5.5. ■

Table 2.5.3 Liver and tooth selenium concentrations of 20 belugas

Whale	Liver Se (μg/g)	Tooth Se (ng/g)	Whale	Liver Se (μg/g)	Tooth Se (ng/g)
1	6.23	140.16	11	15.28	112.63
2	6.79	133.32	12	18.68	245.07
3	7.92	135.34	13	22.08	140.48
4	8.02	127.82	14	27.55	177.93
5	9.34	108.67	15	32.83	160.73
6	10.00	146.22	16	36.04	227.60
7	10.57	131.18	17	37.74	177.69
8	11.04	145.51	18	40.00	174.23
9	12.36	163.24	19	41.23	206.30
10	14.53	136.55	20	45.47	141.31

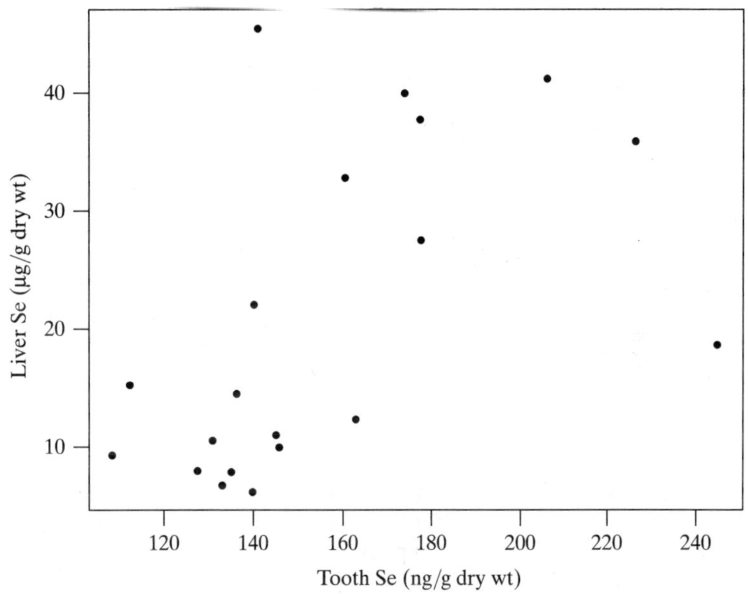

Figure 2.5.5 Scatterplot of liver selenium concentration against tooth selenium concentration for 20 belugas

Scatterplots are helpful in revealing relationships between numeric variables. In Figure 2.5.6 two lines have been added to the whale selenium scatterplot of Figure 2.5.5 to highlight the increasing trend in the data: Tooth Se concentration tends to increase with liver Se concentration. The dashed line is called a **lowess smooth**, whereas the straight solid line is called a **regression line**. Many software packages allow one to easily add these lines to a scatterplot. The lowess smooth is particularly helpful in visualizing curved or nonlinear relationships in data, while the regression line is used to highlight a linear trend. Generally speaking, we would choose only one of these to display on our graph. In this case, since the pattern is fairly linear (the lowess smooth is fairly straight), we would choose the solid regression line. In Chapter 12 we will learn how to identify the equation of the regression line that best summarizes the data and determine if the apparent trend in the data is likely to be just due to chance or if there is evidence for a real relationship between X and Y.

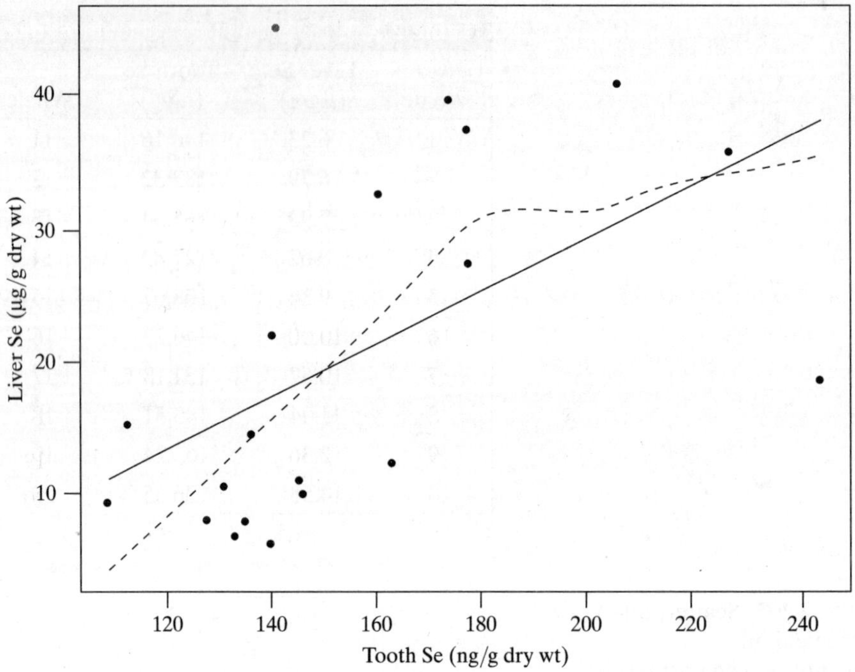

Figure 2.5.6 Scatterplot of liver selenium concentration against tooth selenium concentration for 20 belugas with regression (solid) and lowess (dashed) summary lines and outlier marked in blue

In addition to revealing relationships between two numeric variables, scatterplots also help reveal outliers that might otherwise be unnoticed in univariate plots (e.g., histograms, single boxplots). The colored point on Figure 2.5.6 falls far from the scatter of the other points. The X value of this point is not unusual in any way, and even the Y value, although large, doesn't appear extreme. The scatterplot, however, shows that the particular (x,y) pair for this whale is unusual.

Exercises 2.5.1–2.5.4

2.5.1 The two claws of the lobster (*Homarus americanus*) are identical in the juvenile stages. By adulthood, however, the two claws normally have differentiated into a stout claw called a "crusher" and a slender claw called a "cutter." In a study of the differentiation process, 26 juvenile animals were reared in smooth plastic trays and 18 were reared in trays containing oyster chips (which they could use to exercise their claws). Another 23 animals were reared in trays containing only one oyster chip. The claw configurations of all the animals as adults are summarized in the table.[31]

	Claw Configuration		
Treatment	Right crusher, left cutter	Right cutter, left crusher	Right and left cutter (no crusher)
Oyster chips	8	9	1
Smooth plastic	2	4	20
One oyster chip	7	9	7

(a) Create a stacked frequency bar chart to display these data.

(b) Create a stacked relative frequency bar chart to display these data.

(c) Of the two charts you created in parts (a) and (b), which is more useful for comparing the claw configurations across the three treatments? Why?

2.5.2 Does the length (mm) of the golden mantled ground squirrel (*Spermophilus lateralis*) differ by latitude in California? A graduate student captured squirrels at four locations across California. Listed from south to north the locations are Hemet, Big Bear, Susanville, and Loop Hill.[32]

Hemet	Big Bear	Susanville	Loop Hill
263	274	245	273
256	256	272	291
251	249	263	278
242	264	260	281
248		271	
281			

(a) Create side-by-side dotplots of the data. Consider the geography of these four locations when making your plot. Is alphabetic order of the locations the most appropriate, or is there a better way to order the location categories?

(b) Create side-by-side boxplots of the data. Again, consider the geography of these four locations when making your plot.

(c) Of the two plots created in parts (a) and (b), which do you prefer and why?

2.5.3 The rowan (*Sorbus aucuparia*) is a tree that grows in a wide range of altitudes. To study how the tree adapts to its varying habitats, researchers collected twigs with attached buds from 12 trees growing at various altitudes in North Angus, Scotland. The buds were brought back to the laboratory and measurements were made of the dark respiration rate. The accompanying table shows the altitude of origin (in meters) of each batch of buds and the dark respiration rate (expressed as μl of oxygen per hour per mg dry weight of tissue).[33]

Tree	Altitude of origin X (m)	Respiration rate Y (μl/hr · mg)
1	90	0.11
2	230	0.20
3	240	0.13
4	260	0.15
5	330	0.18
6	400	0.16
7	410	0.23
8	550	0.18
9	590	0.23
10	610	0.26
11	700	0.32
12	790	0.37

(a) Create a scatterplot of the data.

(b) If your software allows, add a regression line to summarize the trend.

(c) If your software allows, create a scatterplot with a lowess smooth to summarize the trend.

2.5.4 A group of college students were asked how many hours per week they exercise.[34]

The answers given by 12 men were as follows:

6 0 2 1 2 4.5 8 3 17 4.5 4 5

The answers given by 13 women were as follows:

5 13 3 2 6 14 3 1 1.5 1.5 3 8 4

(a) Construct parallel boxplots of the male and female distributions.

(b) Describe the two boxplots, including how they compare to each other.

2.6 Measures of Dispersion

We have considered the shapes and centers of distributions, but a good description of a distribution should also characterize how spread out the distribution is—are the observations in the sample all nearly equal, or do they differ substantially? In Section 2.4 we defined the interquartile range, which is one measure of dispersion. We will now consider other measures of dispersion: the range and the standard deviation.

THE RANGE

The sample **range** is the difference between the largest and smallest observations in a sample. Here is an example.

Example 2.6.1 **Blood Pressure** The systolic blood pressures (mm Hg) of seven middle-aged men were given in Example 2.4.1 as follows:

113 124 124 132 146 151 170

For these data, the sample range is

$$170 - 113 = 57 \text{ mm Hg}$$ ■

The range is easy to calculate, but it is very sensitive to extreme values; that is, it is not robust. If the maximum in the blood pressure sample had been 190 rather than 170, the range would have been changed from 57 to 77.

We defined the interquartile range (IQR) in Section 2.4 as the difference between the quartiles. Unlike the range, the IQR is robust. The IQR of the blood pressure data is $151 - 124 = 17$. If the maximum in the blood pressure sample had been 190 rather than 170, the IQR would not have changed; it would still be 17.

THE STANDARD DEVIATION

The standard deviation is the classical and most widely used measure of dispersion. Recall that a *deviation* is the difference between an observation and the sample mean:

$$\text{deviation} = \text{observation} - \bar{y}$$

The standard deviation of the sample, or sample **standard deviation**, is determined by combining the deviations in a special way, as described in the following box.

> **THE SAMPLE STANDARD DEVIATION** The sample standard deviation is denoted by s and is defined by the following formula:
>
> $$s = \sqrt{\frac{\sum_{i=1}^{n}(y_i - \bar{y})^2}{n - 1}}$$
>
> In this formula, the expression $\sum_{i=1}^{n}(y_i - \bar{y})^2$ denotes the sum of the squared deviations.

So, to find the standard deviation of a sample, first find the deviations. Then

1. square
2. add
3. divide by $n - 1$
4. take the square root

To illustrate the use of the formula, we have chosen a data set that is especially simple to handle because the mean happens to be an integer.

Example 2.6.2 **Chrysanthemum Growth** In an experiment on chrysanthemums, a botanist measured the stem elongation (mm in 7 days) of five plants grown on the same greenhouse bench. The results were as follows:[35]

76 72 65 70 82

The data are tabulated in the first column of Table 2.6.1. The sample mean is

$$\bar{y} = \frac{365}{5} = 73 \text{ mm}$$

The deviations $(y_i - \bar{y})$ are tabulated in the second column of Table 2.6.1; the first observation is 3 mm above the mean, the second is 1 mm below the mean, and so on.

The third column of Table 2.6.1 shows that the sum of the squared deviations is

$$= \sum_{i=1}^{n}(y_i - \bar{y})^2 = 164$$

Table 2.6.1 Illustration of the formula for the sample standard deviation

Observation (y_i)	Deviation ($y_i - \bar{y}$)	Squared deviation $(y_i - \bar{y})^2$
76	3	9
72	−1	1
65	−8	64
70	−3	9
82	9	81
Sum $365 = \sum_{i=1}^{n} y_i$	0	$164 = \sum_{i=1}^{n}(y_i - \bar{y})^2$

Since $n = 5$, the standard deviation is

$$\begin{aligned} s &= \sqrt{\frac{164}{4}} \\ &= \sqrt{41} \\ &= 6.4 \text{ mm} \end{aligned}$$

Note that the units of s (mm) are the same as the units of Y. This is because we have squared the deviations and then later taken the square root. ■

The sample **variance**, denoted by s^2, is simply the standard deviation squared: variance $= s^2$. Thus, $s = \sqrt{\text{variance}}$.

Example 2.6.3

Chrysanthemum Growth The variance of the chrysanthemum growth data is

$$s^2 = 41 \text{ mm}^2$$

Note that the units of the variance (mm^2) are not the same as the units of Y. ■

An abbreviation We will frequently abbreviate "standard deviation" as "SD"; the symbol "s" will be used in formulas.

INTERPRETATION OF THE DEFINITION OF s

The magnitude (disregarding sign) of each deviation $(y_i - \bar{y})$ can be interpreted as the *distance* of the corresponding observation from the sample mean $\bar{y}$. Figure 2.6.1 shows a plot of the chrysanthemum growth data (Example 2.6.2) with each distance marked.

Figure 2.6.1 Plot of chrysanthemum growth data with deviations indicated as distances

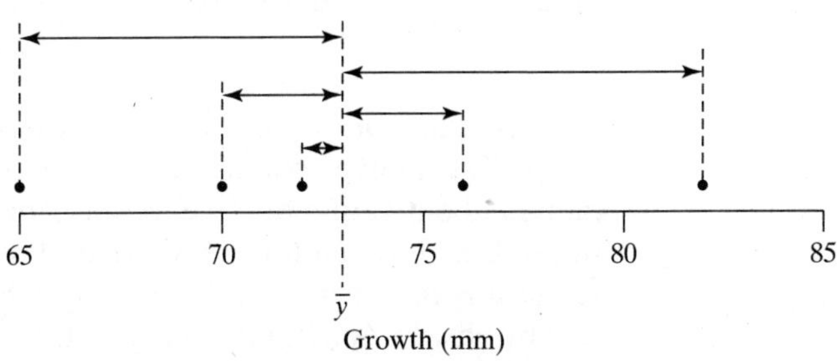

From the formula for s, you can see that each deviation contributes to the SD. Thus, a sample of the same size but with less dispersion will have a smaller SD, as illustrated in the following example.

Example 2.6.4

Chrysanthemum Growth If the chrysanthemum growth data of Example 2.6.2 are changed to

75 72 73 75 70

then the mean is the same ($\bar{y} = 73$ mm), but the SD is smaller ($s = 2.1$ mm), because the observations lie closer to the mean. The relative dispersion of the two samples can easily be seen from Figure 2.6.2. ■

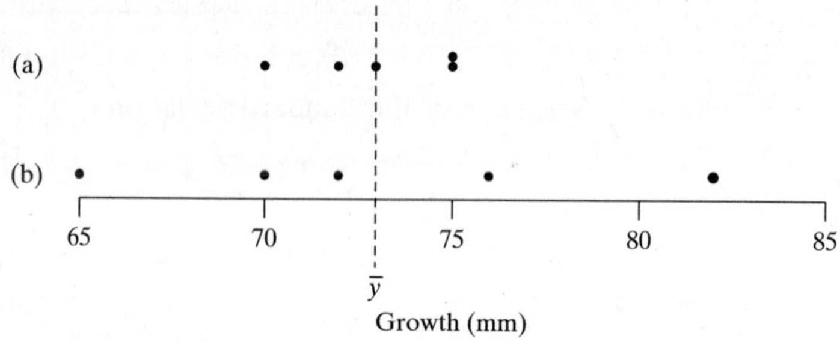

Figure 2.6.2 Two samples of chrysanthemum growth data with the same mean but different standard deviations: (a) $s = 2.1$ mm; (b) $s = 6.3$ mm

Let us look more closely at the way in which the deviations are combined to form the SD. The formula calls for dividing by $(n - 1)$. If the divisor were n instead of $(n - 1)$, then the quantity inside the square root sign would be the average (the mean) of the squared deviations. Unless n is very small, the inflation due to dividing by $(n - 1)$ instead of n is not very great, so that the SD can be interpreted approximately as

$$s \approx \sqrt{\text{sample average value of } (y_i - \bar{y})^2}$$

Thus, it is roughly appropriate to think of the SD as a "typical" distance of the observations from their mean.

Why $n - 1$? Since dividing by n seems more natural, you may wonder why the formula for the SD specifies dividing by $(n - 1)$. Note that the sum of the deviations $y_i - \bar{y}$ is always zero. Thus, once the first $n - 1$ deviations have been calculated, the last deviation is constrained. This means that in a sample with n observations, there are only $n - 1$ units of information concerning deviation from the average. The quantity $n - 1$ is called the **degrees of freedom** of the standard deviation or variance. We can also give an intuitive justification of why $n - 1$ is used by considering the extreme case when $n = 1$, as in the following example.

Example 2.6.5

Chrysanthemum Growth Suppose the chrysanthemum growth experiment of Example 2.6.2 had included only one plant, so that the sample consisted of the single observation

73

For this sample, $n = 1$ and $\bar{y} = 73$. However, the SD formula breaks down (giving $\frac{0}{0}$), so the SD cannot be computed. This is reasonable, because the sample gives no information about variability in chrysanthemum growth under the experimental conditions. If the formula for the SD said to divide by n, we would obtain an SD of zero, suggesting that there is little or no variability; such a conclusion hardly seems justified by observation of only one plant. ■

VISUALIZING MEASURES OF DISPERSION

The range and the interquartile range are easy to interpret. The range is the spread of all the observations, and the interquartile range is the spread of (roughly) the middle 50% of the observations. In terms of the histogram of a data set, the range can be visualized as (roughly) the width of the histogram. The quartiles are (roughly) the values that divide the area into four equal parts, and the interquartile range is the distance between the first and third quartiles. The following example illustrates these ideas.

Example 2.6.6

Daily Gain of Cattle The performance of beef cattle was evaluated by measuring their weight gain during a 140-day testing period on a standard diet. Table 2.6.2 gives the average daily gains (kg/day) for 39 bulls of the same breed (Charolais); the observations are listed in increasing order.[36] The values range from 1.18 kg/day to 1.92 kg/day. The quartiles are 1.29, 1.41, and 1.58 kg/day. Figure 2.6.3 shows a histogram of the data, the range, the quartiles, and the interquartile range (IQR). The shaded area represents the middle 50% (approximately) of the observations. ■

Table 2.6.2 Average daily gain (kg/day) of 39 Charolais bulls

1.18	1.24	1.29	1.37	1.41	1.51	1.58	1.72
1.20	1.26	1.33	1.37	1.41	1.53	1.59	1.76
1.23	1.27	1.34	1.38	1.44	1.55	1.64	1.83
1.23	1.29	1.36	1.40	1.48	1.57	1.64	1.92
1.23	1.29	1.36	1.41	1.50	1.58	1.65	

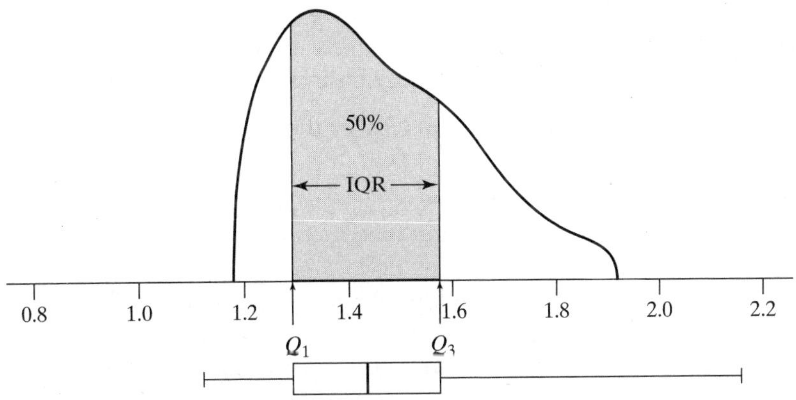

Figure 2.6.3 Smoothed histogram and boxplot of 39 daily gain measurements, showing the quartiles and the interquartile range (IQR). The shaded area represents about 50% of the observations.

VISUALIZING THE STANDARD DEVIATION

We have seen that the SD is a combined measure of the distances of the observations from their mean. It is natural to ask how many of the observations are within ± 1 SD of the mean, within ± 2 SDs of the mean, and so on. The following example explores this question.

Example 2.6.7

Daily Gain of Cattle For the daily-gain data of Example 2.6.6, the mean is $\bar{y} = 1.445$ kg/day and the SD is $s = 0.183$ kg/day. In Figure 2.6.4 the intervals $\bar{y} \pm s$, $\bar{y} \pm 2s$, and $\bar{y} \pm 3s$ have been marked on a histogram of the data. The interval $\bar{y} \pm s$ is

$$1.445 \pm 0.183 \text{ or } 1.262 \text{ to } 1.628$$

You can verify from Table 2.6.2 that this interval contains 25 of the 39 observations. Thus, $\frac{25}{39}$ or 64% of the observations are within ±1 SD of the mean; the corresponding area is shaded in Figure 2.6.4. The intervals $\bar{y} \pm 2s$ is

$$1.445 \pm 0.366 \text{ or } 1.079 \text{ to } 1.811$$

This interval contains $\frac{37}{39}$ or 95% of the observations. You may verify that the interval $y \pm 3s$ contains all the observations. ■

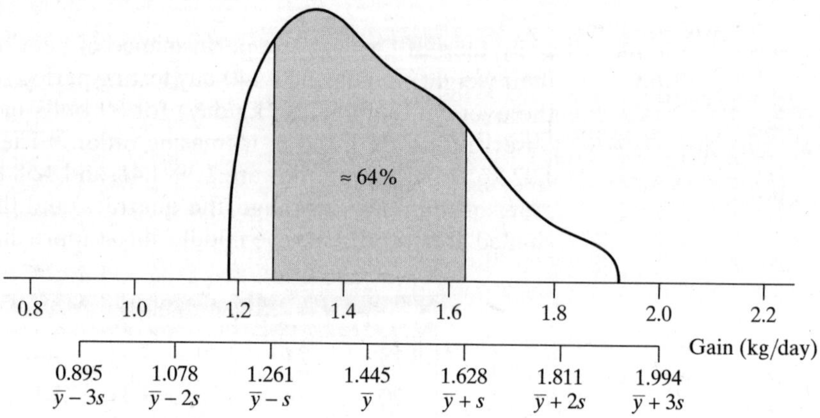

Figure 2.6.4 Histogram of daily-gain data showing intervals 1, 2, and 3 standard deviations from the mean. The shaded area represents about 64% of the observations.

It turns out that the percentages found in Example 2.6.7 are fairly typical of distributions that are observed in the life sciences.

Typical Percentages: The Empirical Rule

For "nicely shaped" distributions—that is, unimodal distributions that are not too skewed and whose tails are not overly long or short—we usually expect to find

about 68% of the observations within ±1 SD of the mean.

about 95% of the observations within ±2 SDs of the mean.

>99% of the observations within ±3 SDs of the mean.

The typical percentages enable us to construct a rough mental image of a frequency distribution if we know just the mean and SD. (The value 68% may seem to come from nowhere. Its origin will become clear in Chapter 4.)

ESTIMATING THE SD FROM A HISTOGRAM

The empirical rule gives us a way to construct a rough mental image of a frequency distribution if we know just the mean and SD: We can envision a histogram centered at the mean and extending out a bit more than 2 SDs in either direction. Of course, the actual distribution might not be symmetric, but our rough mental image will often be fairly accurate.

Thinking about this the other way around, we can look at a histogram and estimate the SD. To do this, we need to estimate the endpoints of an interval that is centered at the mean and that contains about 95% of the data. The empirical rule implies that this interval is roughly the same as $(\bar{y} - 2s, \bar{y} + 2s)$, so the length of the interval should be about 4 times the SD:

$$(\bar{y} - 2s, \bar{y} + 2s) \text{ has length of } 2s + 2s = 4s$$

This means

$$\text{length of interval} = 4s$$

so

$$\text{estimate of } s = \frac{\text{length of interval}}{4}$$

Of course, our visual estimate of the interval that covers the middle 95% of the data could be off. Moreover, the empirical rule works best for distributions that are symmetric. Thus, this method of estimating the SD will give only a general estimate. The method works best when the distribution is fairly symmetric, but it works reasonably well even if the distribution is somewhat skewed.

Example 2.6.8

Pulse after Exercise A group of 28 adults did some moderate exercise for 5 minutes and then measured their pulses. Figure 2.6.5 shows the distribution of the data.[37] We can see that about 95% of the observations are between about 75 and 125.* Thus, an interval of length 50(125 − 75) covers the middle 95% of the data. From this, we can estimate the SD to be $\frac{50}{4} = 12.5$. The actual SD is 13.4, which is not far off from our estimate.

Figure 2.6.5 Pulse after moderate exercise for a group of adults

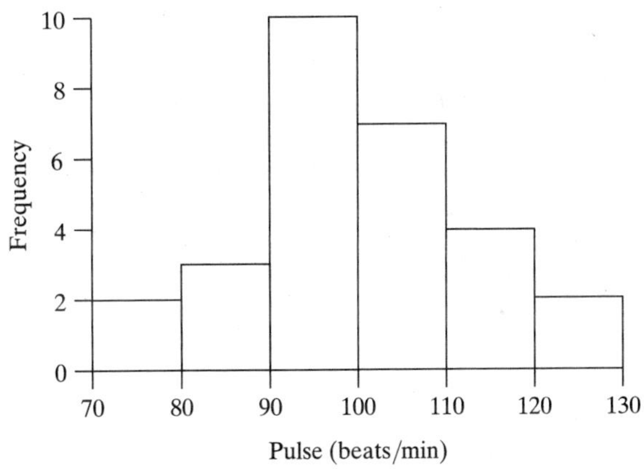

The typical percentages given by the empirical rule may be grossly wrong if the sample is small or if the shape of the frequency distribution is not "nice." For instance, the cricket singing time data (Table 2.3.1 and Figure 2.3.4) has $s = 4.4$ mm, and the interval $\bar{y} \pm s$ contains 90% of the observations. This is much higher than the "typical" 68% because the SD has been inflated by the long, straggly tail of the distribution.

COMPARISON OF MEASURES OF DISPERSION

The dispersion, or spread, of the data in a sample can be described by the standard deviation, the range, or the interquartile range.† The range is simple to understand, but it can be a poor descriptive measure because it depends only on the extreme tails of the distribution. The interquartile range, by contrast, describes the spread in the

---

*It is difficult to visually assess exactly where the middle 95% of the data lay using a histogram, but as this is only a visual estimate, we need not concern ourselves with producing an exact value. Our visual estimates of the SD might differ from one another, but they should all be relatively close.

†Another measure of dispersion is the **coefficient of variation**, which is the standard deviation expressed as a percentage of the mean: coefficient of variation $= \frac{s}{\bar{y}} * 100\%$. Because it is not affected by a change in scale (e.g., from inches to cm), the coefficient of variation is a useful measure for comparing the dispersions of two or more variables that are measured on different scales. See Exercises 2.6.13 and 2.6.14 for more information.

central "body" of the distribution. The standard deviation takes account of all the observations and can roughly be interpreted in terms of the spread of the observations around their mean. However, the SD can be inflated by observations in the extreme tails. The interquartile range is a robust measure, while the SD is not robust. Of course, the range is very highly nonrobust.

The descriptive interpretation of the SD is less straightforward than that of the range and the interquartile range. Nevertheless, the SD is the basis for most standard classical statistical methods. The SD enjoys this classic status for various technical reasons, including efficiency in certain situations.

The developments in later chapters will emphasize classical statistical methods, in which the mean and SD play a central role. Consequently, in this book we will rely primarily on the mean and SD rather than other descriptive measures.

Exercises 2.6.1–2.6.17

2.6.1 Calculate the SD of each of the following fictitious samples:

(a) 16, 13, 18, 13

(b) 38, 30, 34, 38, 35

(c) 1, −1, 5, −1

(d) 4, 6, −1, 4, 2

2.6.2 Calculate the SD of each of the following fictitious samples:

(a) 8, 6, 9, 4, 8

(b) 4, 7, 5, 4

(c) 9, 2, 6, 7, 6

2.6.3

(a) Invent a sample of size 5 for which the deviations $(y_i - \bar{y})$ are −3, −1, 0, 2, 2.

(b) Compute the SD of your sample.

(c) Should everyone get the same answer for part (b)? Why or why not?

2.6.4 Four plots of land, each 346 square feet, were planted with the same variety ("Beau") of wheat. The plot yields (lb) were as follows:[38]

35.1 30.6 36.9 29.8

Calculate the mean and the SD.

2.6.5 A plant physiologist grew birch seedlings in the greenhouse and measured the ATP content of their roots. (See Example 1.1.3.) The results (nmol ATP/mg tissue) were as follows for four seedlings that had been handled identically.[39]

1.45 1.19 1.05 1.07

Calculate the mean and the SD.

2.6.6 Ten patients with high blood pressure participated in a study to evaluate the effectiveness of the drug Timolol in reducing their blood pressure. The accompanying table shows systolic blood pressure measurements taken before and after 2 weeks of treatment with Timolol.[40] Calculate the mean and SD of the *change* in blood pressure (note that some values are negative).

	Blood pressure (mm HG)		
Patient	Before	After	Change
1	172	159	−13
2	186	157	−29
3	170	163	−7
4	205	207	2
5	174	164	−10
6	184	141	−43
7	178	182	4
8	156	171	15
9	190	177	−13
10	168	138	−30

2.6.7 Dopamine is a chemical that plays a role in the transmission of signals in the brain. A pharmacologist measured the amount of dopamine in the brain of each of seven rats. The dopamine levels (nmoles/g) were as follows:[41]

6.8 5.3 6.0 5.9 6.8 7.4 6.2

(a) Calculate the mean and SD.

(b) Determine the median and the interquartile range.

(c) Replace the observation 7.4 by 10.4 and repeat parts (a) and (b). Which of the descriptive measures display robustness and which do not?

2.6.8 In a study of the lizard *Sceloporus occidentalis*, biologists measured the distance (m) run in 2 minutes for each of 15 animals. The results (listed in increasing order) were as follows:[42]

18.4	22.2	24.5	26.4	27.5	28.7	30.6	32.9
32.9	34.0	34.8	37.5	42.1	45.5	45.5	

(a) Determine the quartiles and the interquartile range.

(b) Determine the range.

2.6.9 Refer to the running-distance data of Exercise 2.6.8. The sample mean is 32.23 m and the SD is 8.07 m. What percentage of the observations are within

(a) 1 SD of the mean?

(b) 2 SDs of the mean?

2.6.10 Compare the results of Exercise 2.6.9 with the predictions of the empirical rule.

2.6.11 Listed in increasing order are the serum creatine phosphokinase (CK) levels (U/l) of 36 healthy men (these are the data of Example 2.2.6):

25	62	82	95	110	139
42	64	83	95	113	145
48	67	84	100	118	151
57	68	92	101	119	163
58	70	93	104	121	201
60	78	94	110	123	203

The sample mean CK level is 98.3 U/l and the SD is 40.4 U/l. What percentage of the observations are within

(a) 1 SD of the mean?

(b) 2 SDs of the mean?

(c) 3 SDs of the mean?

2.6.12 Compare the results of Exercise 2.6.11 with the predictions of the empirical rule.

2.6.13 As part of the Berkeley Guidance Study[43] the heights (in cm) and weights (in kg) of 13 girls were measured at age 2 and again at age 9. Of course, the average height and weight were much greater at age 9 than at age 2. Likewise, the SDs of height and of weight were much greater at age 9 than they were at age 2. But what about the coefficient of variation, which gives the SD as a percentage of the mean? It turns out that the coefficient of variation for one of the variables (height or weight) went up only a moderate amount from age 2 to age 9, but for the other variable, the increase in the coefficient of variation was fairly large. For which variable, height or weight, would you expect the coefficient of variation to change more between age 2 and age 9? Why? [*Hint:* Think about how genetic and environmental factors each influence height and weight.]

2.6.14 Consider the 13 girls mentioned in Exercise 2.6.13. At age 18 their average height was 166.3 cm and the SD of their heights was 6.8 cm. Calculate the coefficient of variation.

2.6.15 Here is a histogram. Estimate the mean and the SD of the distribution.

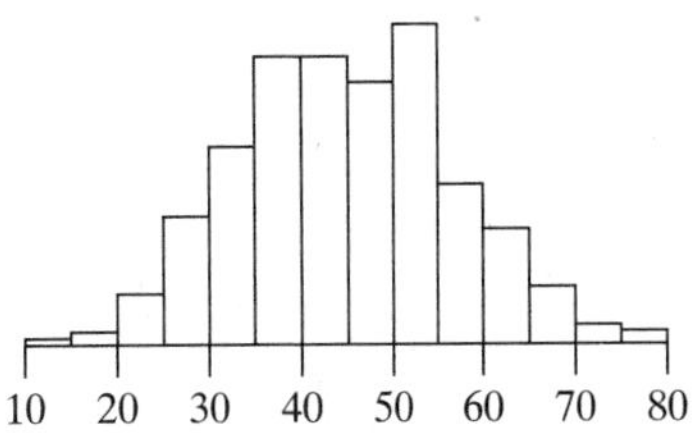

2.6.16 Here is a histogram. Estimate the mean and the SD of the distribution.

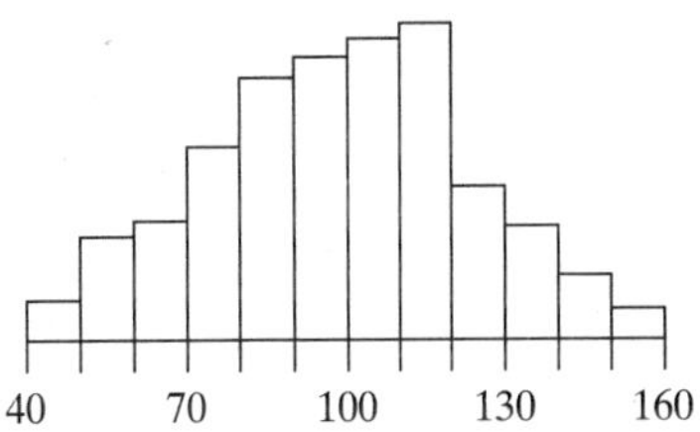

2.6.17 For which sample (i or ii) would you expect the SD of heights to be larger? Or, would they be about the same?

(a) (i) A sample of 10 women ages 18–24, or (ii) a sample of 100 women ages 18–24.

(b) (i) A sample of 20 male college basketball players, or (ii) a sample of 20 college-age men.

(c) (i) A sample of 15 professional male jockeys, or (ii) a sample of 15 professional male biologists.

2.7 Effect of Transformation of Variables (Optional)

Sometimes when we are working with a data set, we find it convenient to transform a variable. For example, we might convert from inches to centimeters or from °F to °C. Transformation, or reexpression, of a variable Y means replacing Y by a new variable, say Y'. To be more comfortable working with data, it is helpful to know how the features of a distribution are affected if the observed variable is transformed.

The simplest transformations are **linear** transformations, so called because a graph of Y against Y' would be a straight line. A familiar reason for linear transformation is a change in the scale of measurement, as illustrated in the following two examples.

Example 2.7.1 **Weight** Suppose Y represents the weight of an animal in kg, and we decide to reexpress the weight in lb. Then

$$Y = \text{Weight in kg}$$

$$Y' = \text{Weight in lb}$$

so

$$Y' = 2.2Y$$

This is a **multiplicative** transformation, because Y' is calculated from Y by multiplying by the constant value 2.2. ■

Example 2.7.2 **Body Temperature** Measurements of basal body temperature (temperature on waking) were made on 47 women.[44]

Typical observations Y, in °C, were

$$Y: \quad 36.23, \quad 36.41, \quad 36.77, \quad 36.15, \quad \ldots$$

Suppose we convert these data from °C to °F, and call the new variable Y':

$$Y': \quad 97.21, \quad 97.54, \quad 98.19, \quad 97.07, \quad \ldots$$

The relation between Y and Y' is

$$Y' = 1.8Y + 32$$

The combination of **additive** (+32) and multiplicative (×1.8) changes indicates a linear relationship. ■

As the foregoing examples illustrate, a linear transformation consists of (1) multiplying all the observations by a constant, or (2) adding a constant to all the observations, or (3) both.

HOW LINEAR TRANSFORMATIONS AFFECT THE FREQUENCY DISTRIBUTION

A linear transformation of the data does not change the essential shape of its frequency distribution; by suitably scaling the horizontal axis, you can make the transformed histogram identical to the original histogram. Example 2.7.3 illustrates this idea.

Example 2.7.3

Body Temperature Figure 2.7.1 shows the distribution of 47 temperature measurements that have been transformed by first subtracting 36 from each observation and then multiplying by 100 (as in Example 2.7.2). That is, $Y' = (Y - 36) \times 100$. The figure shows that the two distributions can be represented by the same histogram with different horizontal scales. ■

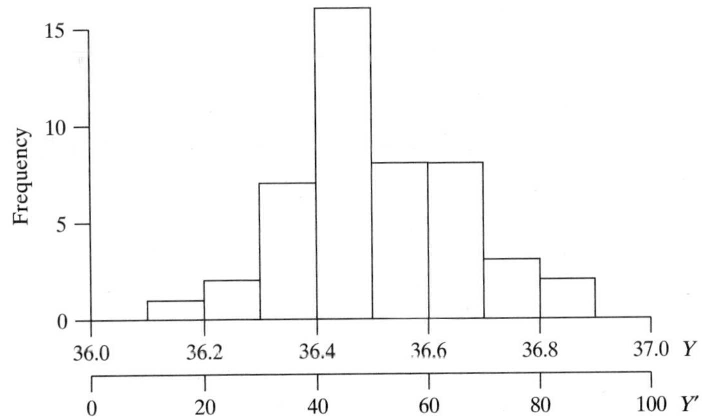

Figure 2.7.1 Distribution of 47 temperature measurements showing original and linearly transformed scales

HOW LINEAR TRANSFORMATIONS AFFECT $\bar{y}$ AND s

The effect of a linear transformation on $\bar{y}$ is "natural"; that is, **under a linear transformation, $\bar{y}$ changes like Y.** For instance, if temperatures are converted from °C to °F, then the mean is similarly converted:

$$Y' = 1.8Y + 32 \quad \text{so} \quad \bar{y}' = 1.8\bar{y}' + 32$$

The effect of multiplying Y by a positive constant on s is "natural"; if $Y' = c \times Y$, with $c > 0$, then $s' = c \times s$. For instance, if weights are converted from kg to lb, the SD is similarly converted: $s' = 2.2s$. If $Y' = c \times Y$ and $c < 0$, then $s' = -c \times s$. In general, if $Y' = c \times Y$ then $s' = |c| \times s$.

However, an additive transformation does not affect s. If we add or subtract a constant, we do not change how spread out the distribution is, so s does not change. Thus, for example, we would *not* convert the SD of temperature data from °C to °F in the same way as we convert each observation; we would multiply the SD by 1.8, but we would *not* add 32. The fact that the SD is unchanged by additive transformation will appear less surprising if you recall (from the definition) that s depends only on the deviations $(y_i - \bar{y})$, and these are not changed by an additive transformation. The following example illustrates this idea.

Example 2.7.4

Additive Transformation Consider the simple set of fictitious data shown in Table 2.7.1. The data were then transformed by subtracting 20 from each observation.

The SD for the original observations is

$$s = \sqrt{\frac{(-1)^2 + (0)^2 + (2)^2 + (-1)^2}{3}}$$

$$= 1.4$$

Table 2.7.1 Effect of additive transformation

	Original observations (y)	Deviations ($y_i - \bar{y}$)	Transformed observations (y')	Deviations ($y'_i - \bar{y}$)
	25	−1	5	−1
	26	0	6	0
	28	2	8	2
	25	−1	5	−1
Mean	26		6	

Because the deviations are unaffected by the transformation, the SD for the transformed observations is the same:

$$s' = 1.4$$

■

An additive transformation effectively picks up the histogram of a distribution and moves it to the left or to the right on the number line. The shape of the histogram does not change and the deviations do not change, so the SD does not change. A multiplicative transformation, on the other hand, stretches or shrinks the distribution, so the SD gets larger or smaller accordingly.

Other Statistics Under linear transformations, other measures of center (e.g., the median) change like $\bar{y}$, and other measures of dispersion (e.g., the interquartile range) change like s. The quartiles themselves change like $\bar{y}$.

NONLINEAR TRANSFORMATIONS

Data are sometimes reexpressed in a nonlinear way. Examples of nonlinear transformations are

$$Y' = \sqrt{Y}$$

$$Y' = \log(Y)$$

$$Y' = \frac{1}{Y}$$

$$Y' = Y^2$$

These transformations are termed "nonlinear" because a graph of Y' against Y would be a curve rather than a straight line. Computers make it easy to use nonlinear transformations. The logarithmic transformation is especially common in biology because many important relationships can be simply expressed in terms of logs. For instance, there is a phase in the growth of a bacterial colony when log(colony size) increases at a constant rate with time. [Note that logarithms are used in some familiar scales of measurement, such as pH measurement or earthquake magnitude (Richter scale).]

Nonlinear transformations can affect data in complex ways. For example, the mean does not change "naturally" under a log transformation; the log of the mean is *not* the same as the mean of the logs. Furthermore, nonlinear transformations (unlike linear ones) *do* change the essential shape of a frequency distribution.

In future chapters we will see that if a distribution is skewed to the right, such as the cricket singing-time distribution shown in Figure 2.7.2, then we may wish to apply a transformation that makes the distribution more symmetric, by pulling in the

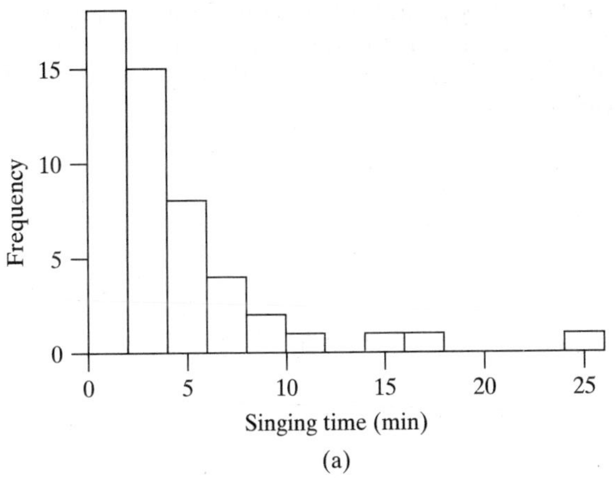

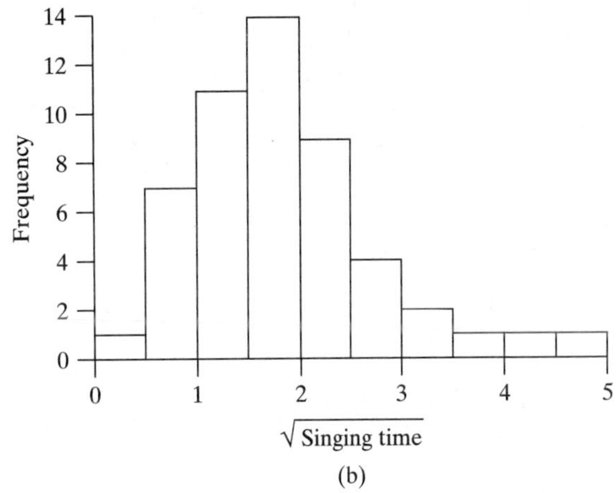

Figure 2.7.2 Distribution of Y, of $\sqrt{Y}$, and of $\log(Y)$ for 51 observations of Y = cricket singing time

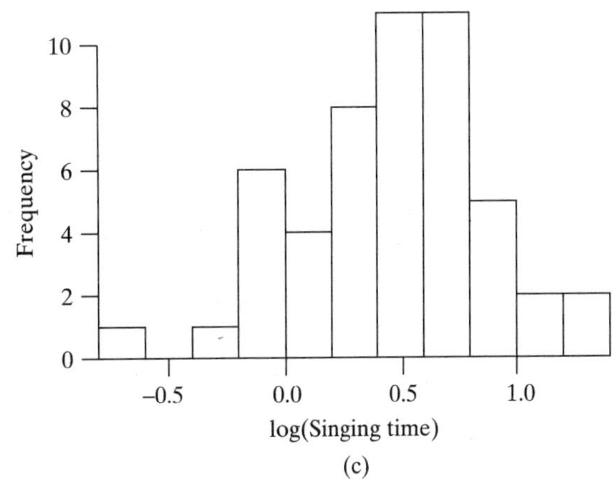

right-hand tail. Using $Y' = \sqrt{Y}$ will pull in the right-hand tail of a distribution and push out the left-hand tail. The transformation $Y' = \log(Y)$ is more severe than $\sqrt{Y}$ in this regard. The following example shows the effect of these transformations.

Example 2.7.5 **Cricket Singing Times** Figure 2.7.2(a) shows the distribution of the cricket singing-time data of Table 2.3.1. If we transform these data by taking square roots, the transformed data have the distribution shown in Figure 2.7.2(b). Taking logs (base 10) yields the distribution shown in Figure 2.7.2(c). Notice that the transformations have the effect of "pulling in" the straggly upper tail and "stretching out" the clumped values on the lower end of the original distribution.

Exercises 2.7.1–2.7.6

2.7.1 A biologist made a certain pH measurement in each of 24 frogs; typical values were[45]

7.43, 7.16, 7.51, ...

She calculated a mean of 7.373 and a SD of 0.129 for these original pH measurements. Next, she transformed the data by subtracting 7 from each observation and then multiplying by 100. For example, 7.43 was transformed to 43. The transformed data are

43, 16, 51, ...

What are the mean and SD of the transformed data?

2.7.2 The mean and SD of a set of 47 body temperature measurements were as follows:[46]

$$\bar{y} = 36.497\,°\text{C} \quad s = 0.172\,°\text{C}$$

If the 47 measurements were converted to °F,

(a) What would be the new mean and SD?

(b) What would be the new coefficient of variation?

2.7.3 A researcher measured the average daily gains (in kg/day) of 20 beef cattle; typical values were[47]

1.39, 1.57, 1.44, ...

The mean of the data was 1.461 and the SD was 0.178.

(a) Express the mean and SD in lb/day. (*Hint:* 1 kg = 2.20 lb.)

(b) Calculate the coefficient of variation when the data are expressed (i) in kg/day; (ii) in lb/day.

2.7.4 Consider the data from Exercise 2.7.3. The mean and SD were 1.461 and 0.178. Suppose we transformed the data from

1.39, 1.57, 1.44, ...

to

39, 57, 44, ...

What would be the mean and SD of the transformed data?

2.7.5 The following histogram shows the distribution for a sample of data:

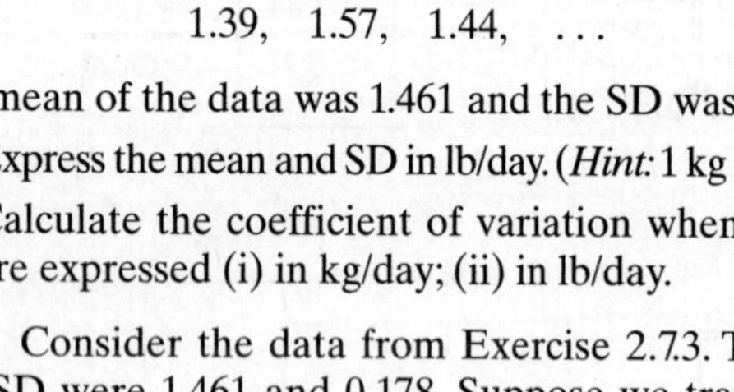

One of the following histograms is the result of applying a square root transformation, and the other is the result of applying a log transformation. Which is which? How do you know?

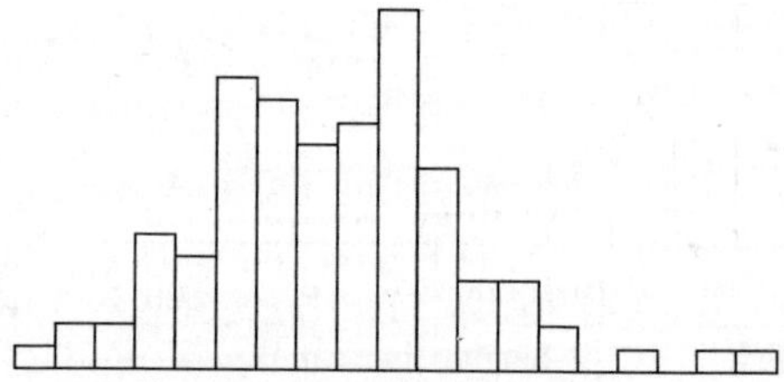

(a)

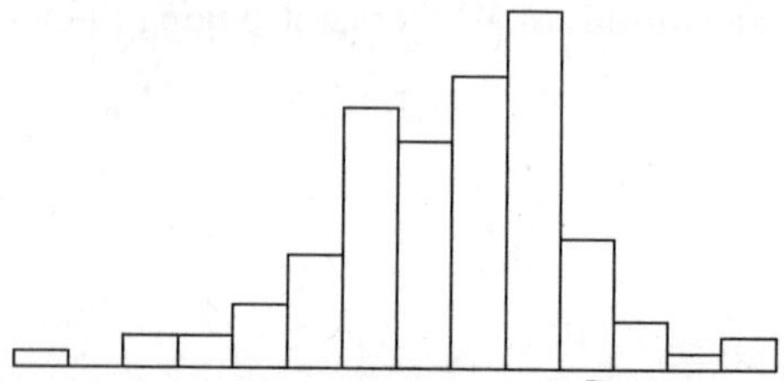

(b)

2.7.6 (Computer problem) The file "Exercise 2.7.6.csv" is included on the data disk packaged with this text. This file contains 36 observations on the number of dendritic branch segments emanating from nerve cells taken from the brains of newborn guinea pigs. (These data were used in Exercise 2.2.4.) Open the file and enter the data into a statistics package. Make a histogram of the data, which are skewed to the right. Now consider the following possible transformations: sqrt(Y), log(Y), and 1/sqrt(Y). Which of these transformations does the best job of meeting the goal of making the resulting distribution reasonably symmetric?

2.8 Statistical Inference

The description of a data set is sometimes of interest for its own sake. Usually, however, the researcher hopes to generalize, to extend the findings beyond the limited scope of the particular group of animals, plants, or other units that were actually observed. Statistical theory provides a rational basis for this process of generalization, building on the random sampling model from Section 1.3 and taking into account the variability of the data. The key idea of the statistical approach is to view the particular data in a study as a sample from a larger population; the population is the real focus of scientific and/or practical interest. The following example illustrates this idea.

Example 2.8.1

Blood Types In an early study of the ABO blood-typing system, researchers determined blood types of 3,696 persons in England. The results are given in Table 2.8.1.[48]

Table 2.8.1 Blood types of 3,696 persons

Blood type	Frequency
A	1,634
B	327
AB	119
O	1,616
Total	3,696

These data were not collected for the purpose of learning about the blood types of those particular 3,696 people. Rather, they were collected for their scientific value as a source of information about the distribution of blood types in a larger population. For instance, one might presume that the blood type distribution of all English people should resemble the distribution for these 3,696 people. In particular, the observed relative frequency of type A blood was

$$\frac{1634}{3696} \text{ or } 44\% \text{ type A}$$

One might conclude from this that approximately 44% of the people in England have type A blood. ■

The process of drawing conclusions about a population, based on observations in a sample from that population, is called **statistical inference**. For instance, in Example 2.8.1 the conclusion that approximately 44% of the people in England have type A blood would be a statistical inference. The inference is shown schematically in Figure 2.8.1. Of course, such an inference might be entirely wrong—perhaps the 3,696 people are not at all representative of English people in general. We might be worried about two possible sources of difficulty: (1) the 3,696 people might have been selected in a way that was systematically biased for (or against) type A people, and (2) the number of people examined might have been too small to permit generalization to a population of many millions. In general, it turns out that the population size being in the millions is *not* a problem, but bias in the way people are selected is a big concern.

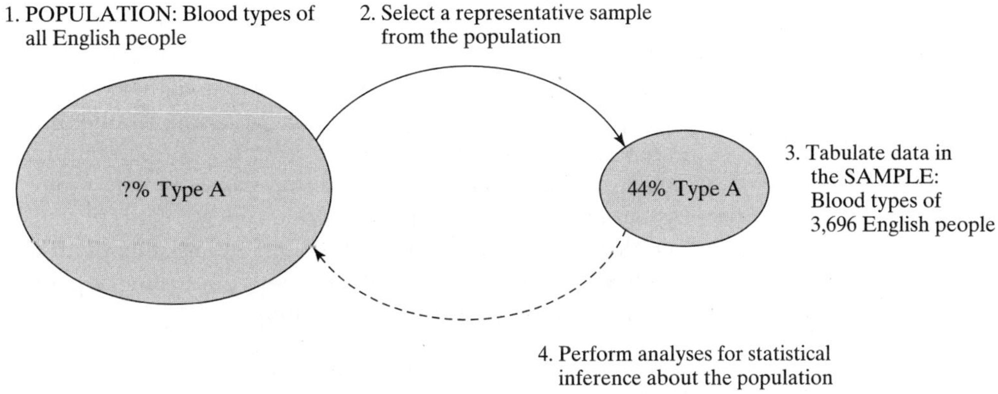

Figure 2.8.1 Schematic representation of inference from sample to population regarding prevalence of blood type A

In making a statistical inference, we hope that the sample resembles the population closely—that the sample is *representative* of the population. In Section 1.3 we saw how sampling bias can lead to nonrepresentative samples. However, even in the absence of bias, we must ask how likely it is that a particular sample will provide a good representation of the population. The important question is: *How representative (of the population) is a sample likely to be?* We will see in Chapter 5 how statistical theory can help to answer this question.

SPECIFYING THE POPULATION

In Section 1.3 we emphasized that the collection of individuals that comprise a sample should be representative of the population. In fact, this requirement is a bit stronger than what is actually necessary. Ultimately, what matters is that the measurements that we obtain on the variable of interest are representative of the values present in the population. The following provides an example of a case where the sample members might not be representative of the population, but one could argue that the measurements taken from this sample could be viewed as representative of the larger population.

Example 2.8.2 **Blood Types** How were the 3,696 English people of Example 2.8.1 actually chosen? It appears from the original paper that this was a "sample of convenience," that is, friends of the investigators, employees, and sundry unspecified sources. There is little basis for believing that the *people* themselves would be representative of the entire English population. Nevertheless, one might argue that their *blood types* might be (more or less) representative of the population. The argument would be that the biases that entered into the selection of those particular people were probably not related to blood type. [Nonetheless, an objection to this argument might be made on the basis of race. For example, the racial distribution of the sample could differ substantially from the racial distribution of England (the population), and there are known differences in blood type distributions among races.] The argument for representativeness would be much less plausible if the observed variable were blood pressure rather than blood type; we know that blood pressure tends to increase with age, and the selection procedure was undoubtedly biased against certain age groups (e.g., elderly people). ■

As Example 2.8.2 shows, whether the measurements obtained from a sample are likely to be representative of the measurements from a population depends not only on how the observational units (in this case people) were chosen, but also on the variable that was observed. Ideally we would always work with random samples, but we have noted that in some instances random samples are not possible or convenient. However, by turning our attention to the measurements themselves rather than the individuals from which they came, we can often make an argument for the generalizabiltity (or lack of generalizability) of our results to a larger population. We do this by thinking of the population as consisting of observations or a collection of values from a measurement process, rather than of people or other observational units. The following is another example.

Example 2.8.3 **Alcohol and MOPEG** The biochemical MOPEG plays a role in brain function. Seven healthy male volunteers participated in a study to determine whether drinking alcohol might elevate the concentration of MOPEG in the cerebrospinal fluid. The MOPEG concentration was measured twice for each man—once at the start of the experiment, and again after he drank 80 gm of ethanol. The results (in pmol/ml) are given in Table 2.8.2.[49]

Let us focus on the rightmost column, which shows the change in MOPEG concentration (i.e., the difference between the "after" and the "before" measurements). In thinking of these values as a sample from a population, we need to specify all the details of the experimental conditions—how the cerebrospinal specimens were obtained, the exact timing of the measurements and the alcohol consumption, and so

Table 2.8.2 Effect of alcohol on MOPEG

	MOPEG concentration		
Volunteer	Before	After	Change
1	46	56	10
2	47	52	5
3	41	47	6
4	45	48	3
5	37	37	0
6	48	51	3
7	58	62	4

on—as well as relevant characteristics of the volunteers themselves. Thus, the definition of the population might be something like this:

> **Population** Change in cerebrospinal MOPEG concentration in healthy young men when measured before and after drinking 80 gm of ethanol, both measurements being made at 8:00 A.M., . . . (other relevant experimental conditions are specified here).

There is no single "correct" definition of a population for an experiment like this. A scientist reading a report of the experiment might find this definition too narrow (e.g., perhaps it does not matter that the volunteers were measured at 8:00 A.M.) or too broad. She might use her knowledge of alcohol and brain chemistry to formulate her own definition, and she would then use that definition as a basis for interpreting these seven observations. ■

DESCRIBING A POPULATION

Because observations are made only on a sample, characteristics of biological populations are almost never known exactly. Typically, our knowledge of a population characteristic comes from a sample. In statistical language, we say that the sample characteristic is an estimate of the corresponding population characteristic. Thus, estimation is a type of statistical inference.

Just as each sample has a distribution, a mean, and an SD, so also we can envision a population distribution, a population mean, and a population SD. In order to discuss inference from a sample to a population, we will need a language for describing the population. This language parallels the language that describes the sample. A sample characteristic is called a **statistic**; a population characteristic is called a **parameter**.

PROPORTIONS

For a categorical variable, we can describe a population by simply stating the proportion, or relative frequency, of the population in each category. The following is a simple example.

Example 2.8.4 **Oat Plants** In a certain population of oat plants, resistance to crown rust disease is distributed as shown in Table 2.8.3.[50] ■

Table 2.8.3 Disease resistance in oats

Resistance	Proportion of plants
Resistant	0.47
Intermediate	0.43
Susceptible	0.10
Total	1.00

Remark The population described in Example 2.8.4 is realistic, but it is not a specific real population; the exact proportions for any real population are not known. For similar reasons, we will use fictitious but realistic populations in several other examples, here and in Chapters 3, 4, and 5

For categorical data, the sample proportion of a category is an estimate of the corresponding population proportion. Because these two proportions are not necessarily the same, it is essential to have a notation that distinguishes between them. We denote the population proportion of a category by p and the sample proportion by $\hat{p}$ (read "p-hat"):

$$p = \text{Population proportion}$$
$$\hat{p} = \text{Sample proportion}$$

The symbol "^" can be interpreted as "estimate of." Thus,

$$\hat{p} \text{ is an estimate of } p$$

We illustrate this notation with an example.

Example 2.8.5 **Lung Cancer** Eleven patients suffering from adenocarcinoma (a type of lung cancer) were treated with the chemotherapeutic agent Mitomycin. Three of the patients showed a positive response (defined as shrinkage of the tumor by at least 50%).[51] Suppose we define the population for this study as "responses of all adenocarcinoma patients." Then we can represent the sample and population proportions of the category "positive response" as follows:

$$p = \text{Proportion of positive responders among all adenocarcinoma patients}$$
$$\hat{p} = \text{Proportion of positive responders among the 11 patients in the study}$$
$$\hat{p} = \frac{3}{11} = 0.27$$

Note that p is unknown, and $\hat{p}$, which is known, is an estimate of p. ■

We should emphasize that an "estimate," as we are using the term, may or may not be a *good* estimate. For instance, the estimate $\hat{p}$ in Example 2.8.5 is based on very few patients; estimates based on a small number of observations are subject to considerable uncertainty. Of course, the question of whether an estimation procedure is good or poor is an important one, and we will show in later chapters how this question can be answered.

OTHER DESCRIPTIVE MEASURES

If the observed variable is quantitative, one can consider descriptive measures other than proportions—the mean, the quartiles, the SD, and so on. Each of these quantities can be computed for a sample of data, and each is an estimate of its corresponding population analog. For instance, the sample median is an estimate of the population median. In later chapters, we will focus especially on the mean and the SD, and so we will need a special notation for the population mean and SD. **The population mean is denoted by μ (mu), and the population SD is denoted by σ (sigma).** We may define these as follows for a quantitative variable Y:

$$\mu = \text{Population average value of } Y$$

$$\sigma = \sqrt{\text{Population average value of } (Y - \mu)^2}$$

The following example illustrates this notation.

Example 2.8.6

Tobacco Leaves An agronomist counted the number of leaves on each of 150 tobacco plants of the same strain (Havana). The results are shown in Table 2.8.4.[52]

The sample mean is

$$\bar{y} = 19.78 = \text{Mean number of leaves on the 150 plants}$$

Table 2.8.4 Number of leaves on tobacco plants

Number of leaves	Frequency (number of plants)
17	3
18	22
19	44
20	42
21	22
22	10
23	6
24	1
Total	150

The population mean is

$$\mu = \text{Mean number of leaves on Havana tobacco plants grown under these conditions}$$

We do not know μ, but we can regard $\bar{y} = 19.78$ as an estimate of μ. The sample SD is

$$s = 1.38 = \text{SD of number of leaves on the 150 plants}$$

The population SD is

$$\sigma = \text{SD of number of leaves on Havana tobacco plants grown under these conditions}$$

We do not know σ, but we can regard as an estimate of σ.* ■

*You may wonder why we use $\bar{y}$ and s instead of $\hat{\mu}$ and $\hat{\sigma}$. One answer is tradition. Another answer is that since "^" means estimate, you might have other estimates in mind.

2.9 Perspective

In this chapter we have considered various ways of describing a set of data. We have also introduced the notion of regarding features of a sample as estimates of corresponding features of a suitably defined population.

PARAMETERS AND STATISTICS

Some features of a distribution—for instance, the mean—can be represented by a single number, while some—for instance, the shape—cannot. We have noted that a numerical measure that describes a sample is called a statistic. Correspondingly, a numerical measure that describes a population is called a parameter. For the most important numerical measures, we have defined notations to distinguish between the statistic and the parameter. These notations are summarized in Table 2.9.1 for convenient reference.

Table 2.9.1 Notation for some important statistics and parameters

Measure	Sample value (statistic)	Population value (parameter)
Proportion	$\hat{p}$	p
Mean	$\bar{y}$	μ
Standard deviation	s	σ

A LOOK AHEAD

It is natural to view a sample characteristic (e.g., $\bar{y}$) as an estimate of the corresponding population characteristic (e.g., μ). But in taking such a view, one must guard against unjustified optimism. Of course, if the sample were perfectly representative of the population, then the estimate would be perfectly accurate. But this raises the central question: How representative (of the population) is a sample likely to be? Intuition suggests that, if the observational units are appropriately selected, then the sample should be more or less representative of the population. Intuition also suggests that larger samples should tend to be more representative than smaller samples. These intuitions are basically correct, but they are too vague to provide practical guidance for research in the life sciences. Practical questions that need to be answered are

1. How can an investigator judge whether a sample can be viewed as "more or less" representative of a population?
2. How can an investigator quantify "more or less" in a specific case?

In Section 1.3 we described a theoretical probability model based on random sampling that provides a framework for the judgment in question (1), and in Chapter 6 we will see how this model can provide a concrete answer to question (2). Specifically, in Chapter 6 we will see how to analyze a set of data so as to quantify how closely the sample mean ($\bar{y}$) estimates the population mean (μ). But before returning to data analysis in Chapter 6, we will need to lay some groundwork in Chapters 3, 4, and 5; the developments in these chapters are an essential prelude to understanding the techniques of statistical inference.

Supplementary Exercises 2.S.1–2.S.24

2.S.1 If 2, 7, 10, and 1 are the number of students who weigh (in kg) 35, 30, 42, and 40, respectively, what is the mean weight of these 20 students?

2.S.2 A botanist grew 15 pepper plants on the same greenhouse bench. After 21 days, she measured the total stem length (cm) of each plant, and obtained the following values:[53]

12.4	12.2	13.4
10.9	12.2	12.1
11.8	13.5	12.0
14.1	12.7	13.2
12.6	11.9	13.1

(a) Calculate all three quartiles.

(b) Compute the lower fence and the upper fence of the distribution.

(c) How large would an observation in this data set have to be in order to be an outlier?

2.S.3 In a behavioral study of the fruitfly *Drosophila melanogaster*, a biologist measured, for individual flies, the total time spent preening during a 6-minute observation period. The following are the preening times (sec) for 20 flies:[54]

34	24	10	16	52
76	33	31	46	24
18	26	57	32	25
48	22	48	29	19

(a) Determine the mode (s).

(b) Calculate the range.

(c) Construct a dotplot of the data.

2.S.4 To calibrate a standard curve for assaying protein concentrations, a plant pathologist used a spectrophotometer to measure the absorbance of light (wavelength 500 nm) by a protein solution. The results of 27 replicate assays of a standard solution containing 60 μg protein per ml water were as follows:[55]

0.111	0.115	0.115	0.110	0.099
0.121	0.107	0.107	0.100	0.110
0.106	0.116	0.098	0.116	0.108
0.098	0.120	0.123	0.124	0.122
0.116	0.130	0.114	0.100	0.123
0.119	0.107			

Construct a frequency distribution and display it as a table and as a histogram.

2.S.5 Refer to the absorbance data of Exercise 2.S.4.

(a) Determine the median, the quartiles, and the interquartile range.

(b) How large must an observation be to be an outlier?

2.S.6 The median splits data into two equal halves. Is the median a robust statistic? Why or why not?

2.S.7 Twenty patients with severe epilepsy were observed for 8 weeks. The following are the numbers of major seizures suffered by each patient during the observation period:[56]

5	0	9	6	0	0	5	0	6	1
5	0	0	0	0	7	0	0	4	7

(a) Is the distribution of seizures bimodal? Justify your answer.

(b) Calculate the SD of seizures.

(c) What percentage of the seizures is within 1 SD of the mean?

(d) What percentage of the seizures is within 2 SDs of the mean?

2.S.8 Consider the histogram from Exercise 2.3.13. By "reading" the histogram, estimate the percentage of observations that are less than 45. Is this percentage closest to 10%, 30%, 50%, 70%, 90%? (*Note*: The frequency scale is not given for this histogram, because there is no need to calculate the number of observations in each class. Rather, the percentage of observations that are less than 45 can be estimated by looking at area.)

2.S.9 Consider the histogram from Exercise 2.3.15. By "reading" the histogram, estimate the percentage of observations that are greater than 25. Is this percentage closest to 10%, 30%, 50%, 70%, 90%? (*Note*: The frequency scale is not given for this histogram, because there is no need to calculate the number of observations in each class. Rather, the percentage of observations that are greater than 25 can be estimated by looking at area.)

2.S.10 Calculate the variance of each of the following fictitious samples:

(a) 12, 6, 7, 3

(b) 9, 13, 8, 10

(c) 21, 15, −10, 6

2.S.11 To study the spatial distribution of Japanese beetle larvae in the soil, researchers divided a 12- × 12-foot section of a cornfield into 144 one-foot squares. They counted the number of larvae Y in each square, with the results shown in the following table.[57]

Number of larvae	Frequency (Number of squares)
0	13
1	34
2	50
3	18
4	16
5	10
6	2
7	1
Total	144

(a) The mean and SD of Y are $\bar{y} = 2.23$ and $s = 1.47$. What percentage of the observations are within

(i) 1 SD of the mean?

(ii) 2 SDs of the mean?

(b) Determine the total number of larvae in all 144 squares. How is this number related to $\bar{y}$?

(c) Determine the median value of the distribution.

2.S.12 One measure of physical fitness is maximal oxygen uptake, which is the maximum rate at which a person can consume oxygen. A treadmill test was used to determine the maximal oxygen uptake of nine college women before and after participation in a 10-week program of vigorous exercise. The accompanying table shows the before and after measurements and the change (after–before); all values are in ml O_2 per mm per kg body weight.[58]

	Maximal oxygen uptake		
Participant	Before	After	Change
1	48.6	38.8	−9.8
2	38.0	40.7	2.7
3	31.2	32.0	0.8
4	45.5	45.4	−0.1
5	41.7	43.2	1.5
6	41.8	45.3	3.5
7	37.9	38.9	1.0
8	39.2	43.5	4.3
9	47.2	45.0	−2.2

The following computations are to be done on the *change* in maximal oxygen uptake (the right-hand column).

(a) Calculate the mean and the SD.

(b) Determine the median.

(c) Eliminate participant 1 from the data and repeat parts (a) and (b). Which of the descriptive measures display robustness and which do not?

2.S.13 A veterinary anatomist investigated the spatial arrangement of the nerve cells in the intestine of a pony. He removed a block of tissue from the intestinal wall, cut the block into many equal sections, and counted the number of nerve cells in each of 23 randomly selected sections. The counts were as follows.[59]

35 19 33 34 17 26 16 40
28 30 23 12 27 33 22 31
28 28 35 23 23 19 29

(a) Determine the median, the quartiles, and the interquartile range.

(b) Construct a boxplot of the data.

2.S.14 Exercise 2.S.13 asks for a boxplot of the nerve-cell data. Does this graphic support the claim that the data came from a reasonably symmetric distribution?

2.S.15 A geneticist counted the number of bristles on a certain region of the abdomen of the fruitfly *Drosophila melanogaster*. The results for 119 individuals were as shown in the table.[60]

Number of bristles	Number of flies	Number of bristles	Number of flies
29	1	38	18
30	0	39	13
31	1	40	10
32	2	41	15
33	2	42	10
34	6	43	2
35	9	44	2
36	11	45	3
37	12	46	2

(a) Find the mean number of bristles.

(b) Find the SD of the sample.

(c) What percentage of the observations fall within 3 SDs of the mean?

(d) What is the coefficient of variation?

2.S.16 The carbon monoxide in cigarettes is thought to be hazardous to the fetus of a pregnant woman who smokes. In a study of this hypothesis, blood was drawn from pregnant women before and after smoking a cigarette. Measurements were made of the percent of blood hemoglobin bound to carbon monoxide as carboxyhemoglobin (COHb). The results for 10 women are shown in the table.[61]

Subject	Blood COHb (%) Before	After	Increase
1	1.2	7.6	6.4
2	1.4	4.0	2.6
3	1.5	5.0	3.5
4	2.4	6.3	3.9
5	3.6	5.8	2.2
6	0.5	6.0	5.5
7	2.0	6.4	4.4
8	1.5	5.0	3.5
9	1.0	4.2	3.2
10	1.7	5.2	3.5

(a) Calculate the mean and SD of the *increase* in COHb.

(b) Calculate the mean COHb before and the mean after. Is the mean increase equal to the increase in means?

(c) Determine the median increase in COHb.

(d) Repeat part (c) for the before measurements and for the after measurements. Is the median increase equal to the increase in medians?

2.S.17 (Computer problem) A medical researcher in India obtained blood specimens from 31 young children, all of whom were infected with malaria. The following data, listed in increasing order, are the numbers of malarial parasites found in 1 ml of blood from each child.[62]

100	140	140	271	400	435	455	770
826	1,400	1,540	1,640	1,920	2,280	2,340	3,672
4,914	6,160	6,560	6,741	7,609	8,547	9,560	10,516
14,960	16,855	18,600	22,995	29,800	83,200	134,232	

(a) Transform the data by taking the square root of each observation.

(b) Construct the frequency distribution of the data using a class width of 50.

(c) Determine the IQR of the original data and the IQR of the log-transformed data. Is the IQR of the logs equal to the log of the IQR?

(d) Determine the SD of the original data and the SD of the log-transformed data. Is the SD of the logs equal to the log of the SD?

2.S.18 Rainfall, measured in inches, for the month of June in Cleveland, Ohio, was recorded for each of 41 years.[63] The values had a minimum of 1.2, an average of 3.6, and an SD of 1.6. Which of the following is a rough histogram for the data? How do you know?

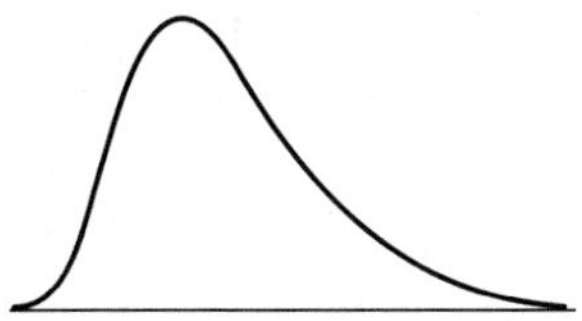

I

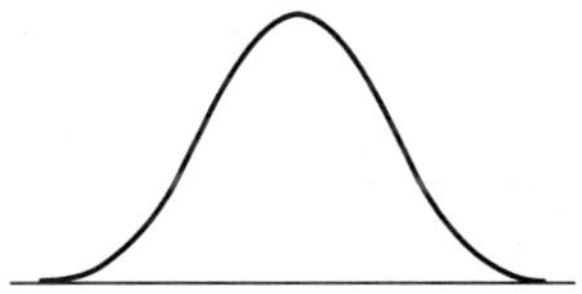

II

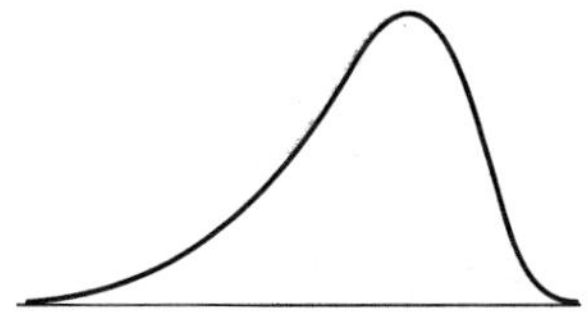

III

2.S.19 The following histograms (a), (b), and (c) show three distributions.

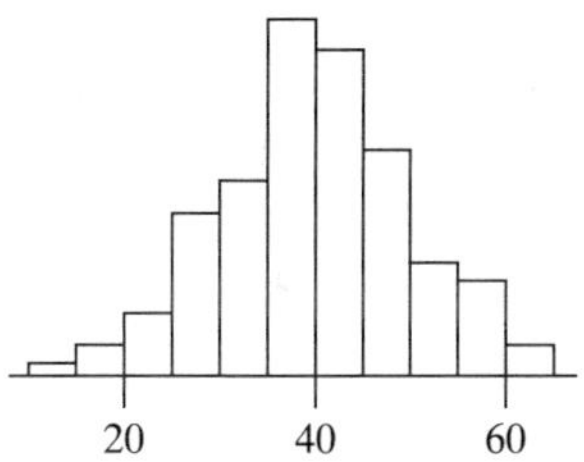

(a)

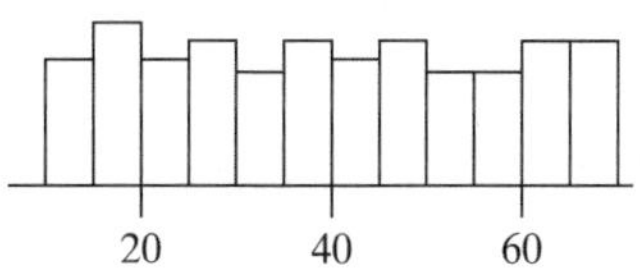

(b)

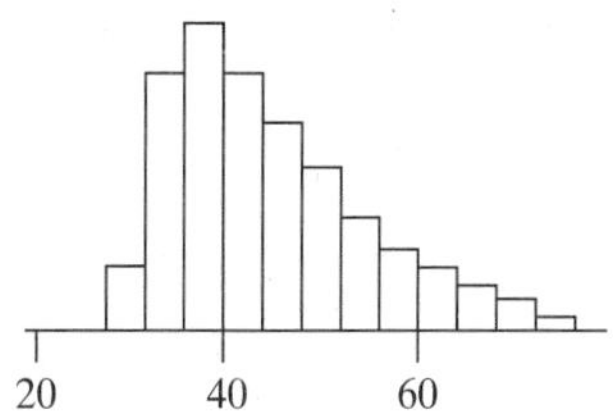

(c)

The accompanying computer output shows the mean, median, and SD of the three distributions, plus the mean, median, and SD for a fourth distribution. Match the histograms with the statistics. Explain your reasoning. (One set of statistics will not be used.)

1.			2.		
	Count	100		Count	100
	Mean	41.3522		Mean	39.6761
	Median	39.5585		Median	39.5377
	StdDev	13.0136		StdDev	10.0476

3.			4.		
	Count	100		Count	100
	Mean	37.7522		Mean	39.6493
	Median	39.5585		Median	39.5448
	StdDev	13.0136		StdDev	17.5126

2.S.20 The following boxplots show mortality rates (deaths within one year per 100 patients) for heart transplant patients at various hospitals. The low-volume hospitals are those that perform between 5 and 9 transplants per year. The high-volume hospitals perform 10 or more transplants per year.[64] Describe the distributions, paying special attention to how they compare to one another. Be sure to note the shape, center, and spread of each distribution.

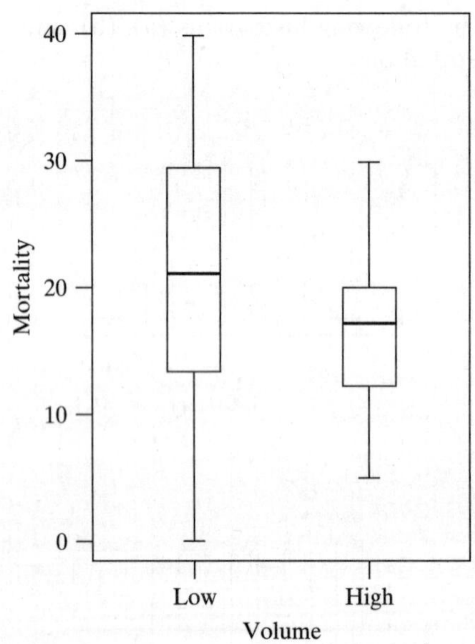

2.S.21 (Computer problem) Physicians measured the concentration of calcium (nM) in blood samples from 38 healthy persons. The data are listed as follows:[65]

95	110	135	120	88	125
112	100	130	107	86	130
122	122	127	107	107	107
88	126	125	112	78	115
78	102	103	93	88	110
104	122	112	80	121	126
90	96				

Calculate appropriate measures of the center and spread of the distribution. Describe the shape of the distribution and any unusual features in the data.

2.S.22 The following boxplot shows the same data that are shown in one of the three histograms. Which histogram goes with the boxplot? Explain your answer.

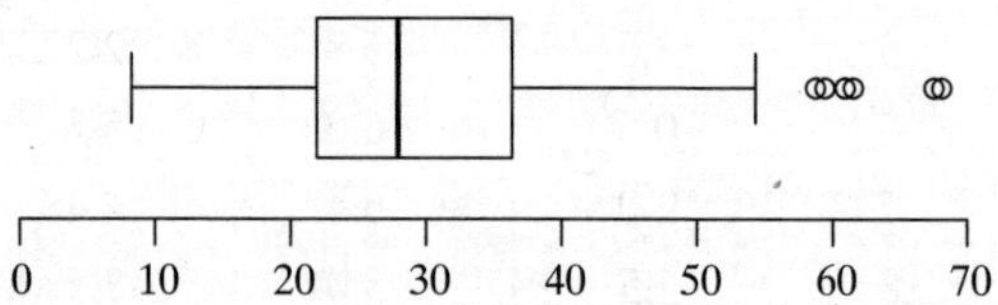

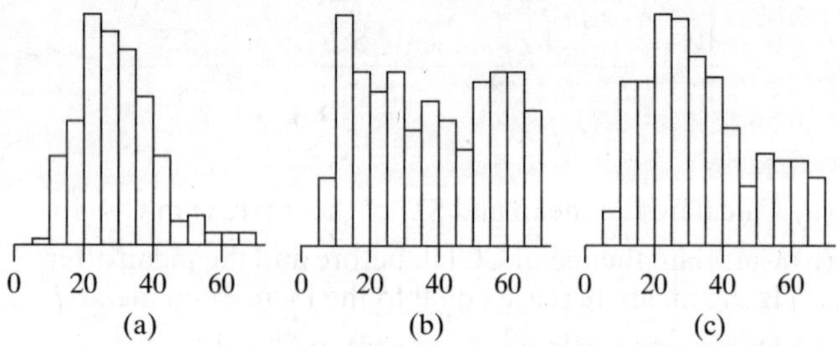

2.S.23 Here is a histogram.

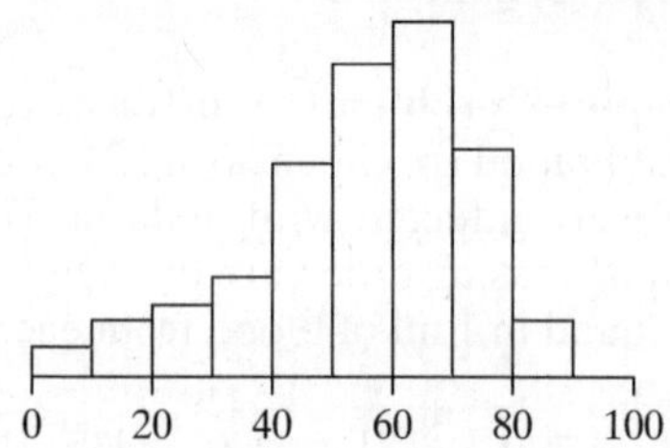

Explain why the mean is less than the median of the distribution.

2.S.24 Consider the histogram from Exercise 2.S.23. By "reading" the histogram, estimate the mode of the distribution. *Note:* The frequency scale is not given for this histogram, because there is no need to calculate the number of observations in each class.

Probability and the Binomial Distribution

Chapter 3

OBJECTIVES

In this chapter we will study the basic ideas of probability, including

- the "limiting frequency" definition of probability.
- the use of probability trees.
- the concept of a random variable.
- rules for finding means and standard deviations of random variables.
- the use of the binomial distribution.

3.1 Probability and the Life Sciences

Probability, or chance, plays an important role in scientific thinking about living systems. Some biological processes are affected directly by chance. A familiar example is the segregation of chromosomes in the formation of gametes; another example is the occurrence of mutations.

Even when the biological process itself does not involve chance, the results of an experiment are always somewhat affected by chance: chance fluctuations in environmental conditions, chance variation in the genetic makeup of experimental animals, and so on. Often, chance also enters directly through the design of an experiment; for instance, varieties of wheat may be randomly allocated to plots in a field. (Random allocation will be discussed in Chapter 11.)

The conclusions of a statistical data analysis are often stated in terms of probability. Probability enters statistical analysis not only because chance influences the results of an experiment, but also because probability models allow us to quantify how likely, or unlikely, an experimental result is, given certain modeling assumptions. In this chapter we will introduce the language of probability and develop some simple tools for calculating probabilities.

3.2 Introduction to Probability

In this section we introduce the language of probability and its interpretation.

BASIC CONCEPTS

A **probability** is a numerical quantity that expresses the likelihood of an event. The probability of an event E is written as

$$\Pr\{E\}$$

The probability $\Pr\{E\}$ is always a number between 0 and 1, inclusive.

We can speak meaningfully about a probability $\Pr\{E\}$ only in the context of a chance operation—that is, an operation whose outcome is not deterministic. The chance operation must be defined in such a way that *each time the chance operation is performed, the event E either occurs or does not occur.* The following two examples illustrate these ideas.

Example 3.2.1 **Coin Tossing** Consider the familiar chance operation of tossing a coin, and define the event

$$E\text{: Heads}$$

Each time the coin is tossed, either it falls heads or it does not. If the coin is equally likely to fall heads or tails, then

$$\Pr\{E\} = \frac{1}{2} = 0.5$$

Such an ideal coin is called a "fair" coin. If the coin is not fair (perhaps because it is slightly bent), then $\Pr\{E\}$ will be some value other than 0.5, for instance,

$$\Pr\{E\} = 0.6$$

■

Example 3.2.2 **Coin Tossing** Consider the event

$$E\text{: 3 heads in a row}$$

The chance operation "toss a coin" is *not* adequate for this event, because we cannot tell from one toss whether E has occurred. A chance operation that would be adequate is

Chance operation: Toss a coin 3 times. ■

The language of probability can be used to describe the results of random sampling from a population. The simplest application of this idea is a sample of size $n = 1$; that is, choosing one member at random from a population. The following is an illustration.

Example 3.2.3 **Sampling Fruitflies** A large population of the fruitfly *Drosophila melanogaster* is maintained in a lab. In the population, 30% of the individuals are black because of a mutation, while 70% of the individuals have the normal gray body color. Suppose one fly is chosen at random from the population. Then the probability that a black fly is chosen is 0.3. More formally, define

$$E\text{: Sampled fly is black}$$

Then

$$\Pr\{E\} = 0.3$$

■

The preceding example illustrates the basic relationship between probability and random sampling: *The probability that a randomly chosen individual has a certain characteristic is equal to the proportion of population members with the characteristic.*

FREQUENCY INTERPRETATION OF PROBABILITY

The **frequency interpretation** of probability provides a link between probability and the real world by relating the probability of an event to a measurable quantity, namely, the long-run relative frequency of occurrence of the event.*

*Some statisticians prefer a different view, namely that the probability of an event is a subjective quantity expressing a person's "degree of belief" that the event will happen. Statistical methods based on this "subjectivist" interpretation are rather different from those presented in this book.

According to the frequency interpretation, the probability of an event E is meaningful only in relation to a chance operation that can in principle be repeated indefinitely often. Each time the chance operation is repeated, the event E either occurs or does not occur. *The probability Pr{E} is interpreted as the relative frequency of occurrence of E in an indefinitely long series of repetitions of the chance operation.*

Specifically, suppose that the chance operation is repeated a large number of times, and that for each repetition the occurrence or nonoccurrence of E is noted. Then we may write

$$\Pr\{E\} \leftrightarrow \frac{\text{\# of times } E \text{ occurs}}{\text{\# of times chance operation is repeated}}$$

The arrow in the preceding expression indicates "equality in the long run"; that is, if the chance operation is repeated an unlimited number of times, the two sides of the expression will be approximately equal. Here is a simple example.

Example 3.2.4 **Coin Tossing** Consider again the chance operation of tossing a coin, and the event

$$E\text{: Heads}$$

If the coin is fair, then

$$\Pr\{E\} = 0.5 \leftrightarrow \frac{\text{\# of heads}}{\text{\# of tosses}}$$

The arrow in the preceding expression indicates that, in an infinitely long series of tosses of a fair coin, we expect to get heads about 50% of the time. ■

The following two examples illustrate the relative frequency interpretation for more complex events.

Example 3.2.5 **Coin Tossing** Suppose that a fair coin is tossed twice. For reasons that will be explained later in this section, the probability of getting heads both times is 0.25. This probability has the following relative frequency interpretation.

Chance operation: Toss a coin twice

$$E\text{: Both tosses are heads}$$

$$\Pr\{E\} = 0.25 \leftrightarrow \frac{\text{\# of times both tosses are heads}}{\text{\# of pairs of tosses}}$$ ■

Example 3.2.6 **Sampling Fruitflies** In the *Drosophila* population of Example 3.2.3, 30% of the flies are black and 70% are gray. Suppose that two flies are randomly chosen from the population. We will see later in this section that the probability that both flies are the same color is 0.58. This probability can be interpreted as follows:

Chance operation: Choose a random sample of size $n = 2$

$$E\text{: Both flies in the sample are the same color}$$

$$\Pr\{E\} = 0.58 \leftrightarrow \frac{\text{\# of times both flies are same color}}{\text{\# of times a sample of } n = 2 \text{ is chosen}}$$

We can relate this interpretation to a concrete sampling experiment. Suppose that the *Drosophila* population is in a very large container, and that we have some mechanism for choosing a fly at random from the container. We choose one fly at random, and then another; these two constitute the first sample of $n = 2$. After recording their colors, we put the two flies back into the container, and we are ready to repeat the sampling operation once again. Such a sampling experiment would be tedious to carry out physically, but it can readily be simulated using a computer. Table 3.2.1 shows a partial record of the results of choosing 10,000 random samples of size $n = 2$ from a simulated *Drosophila* population. After each repetition of the chance operation (i.e., after each sample of $n = 2$), the cumulative relative frequency of occurrence of the event E was updated, as shown in the rightmost column of the table.

Figure 3.2.1 shows the cumulative relative frequency plotted against the number of samples. Notice that, as the number of samples becomes large, the relative frequency of occurrence of E approaches 0.58 (which is $\Pr\{E\}$). In other words, the percentage of color-homogeneous samples among all the samples approaches 58% as the number of samples increases. It should be emphasized, however, that the

Table 3.2.1 Partial results of simulated sampling from a *Drosophila* population

Sample number	Color: 1st fly	Color: 2nd fly	Did E occur?	Relative frequency of E (cumulative)
1	G	B	No	0.000
2	B	B	Yes	0.500
3	B	G	No	0.333
4	G	B	No	0.250
5	G	G	Yes	0.400
6	G	B	No	0.333
7	B	B	Yes	0.429
8	G	G	Yes	0.500
9	G	B	No	0.444
10	B	B	Yes	0.500
⋮	⋮	⋮	⋮	⋮
20	G	B	No	0.450
⋮	⋮	⋮	⋮	⋮
100	G	B	No	0.540
⋮	⋮	⋮	⋮	⋮
1,000	G	G	Yes	0.596
⋮	⋮	⋮	⋮	⋮
10,000	B	B	Yes	0.577

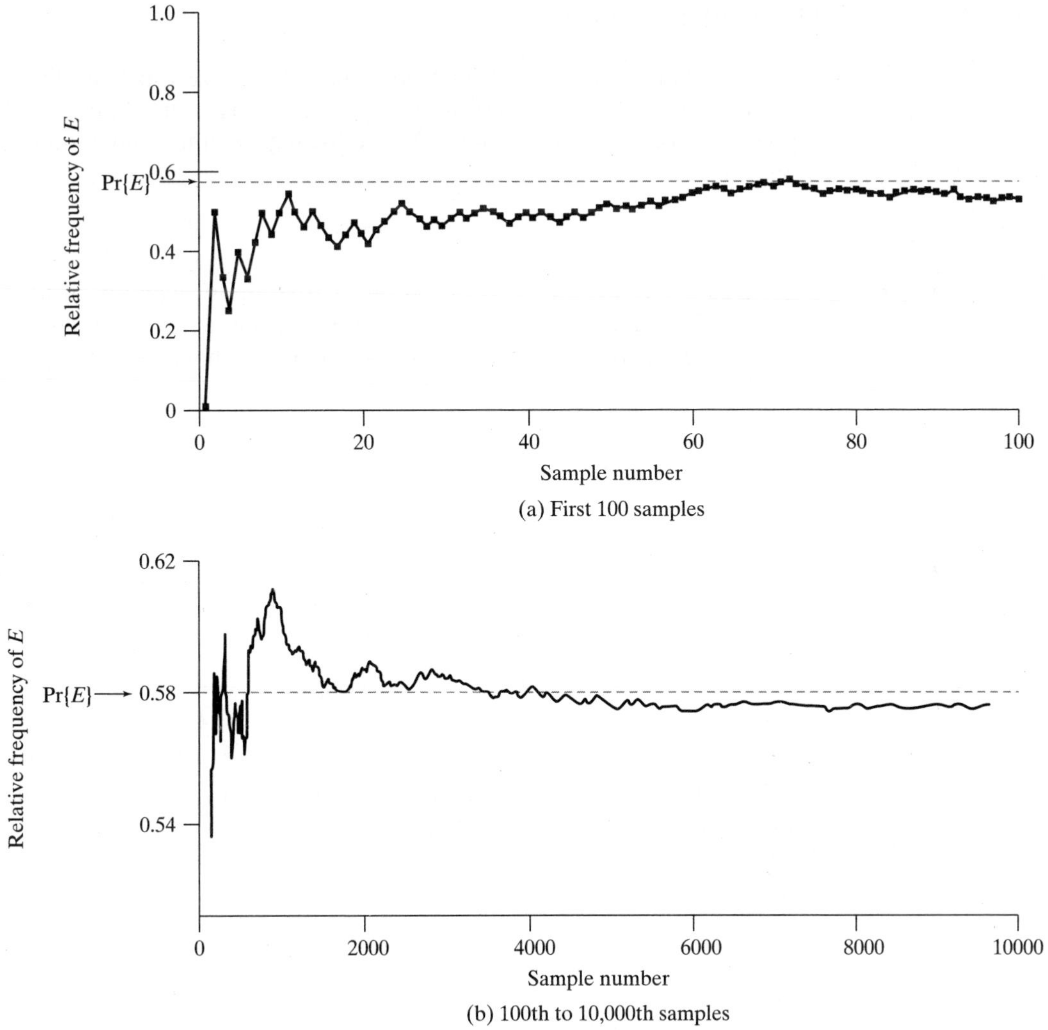

Figure 3.2.1 Results of sampling from fruitfly population. Note that the axes are scaled differently in (a) and (b).

absolute number of color-homogeneous samples generally does *not* tend to get closer to 58% of the total number. For instance, if we compare the results shown in Table 3.2.1 for the first 100 samples and the first 1,000 samples, we find the following:

	Color-Homogeneous			Deviation from 58% of Total		
First 100 samples:	54	or	54.0%	–4	or	–4.0%
First 1,000 samples:	596	or	59.6%	+16	or	+1.6%

Note that the deviation from 58% is larger in absolute terms, but smaller in relative terms (i.e., in percentage terms), for 1,000 samples than for 100 samples. Likewise, for 10,000 samples the deviation from 58% is rather larger (a deviation of –30), but the percentage deviation is quite small (30/10,000 is 0.3%). The deficit of 4 color-homogeneous samples among the first 100 samples is not *canceled* by a corresponding excess in later samples but rather is *swamped*, or overwhelmed, by a larger denominator. ■

PROBABILITY TREES

Often it is helpful to use a **probability tree** to analyze a probability problem. A probability tree provides a convenient way to break a problem into parts and to organize the information available. The following examples show some applications of this idea.

Example 3.2.7 **Coin Tossing** If a fair coin is tossed twice, then the probability of heads is 0.5 on each toss. The first part of a probability tree for this scenario shows that there are two possible outcomes for the first toss and that they have probability 0.5 each.

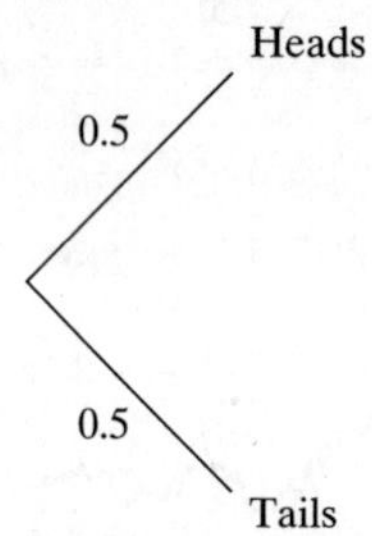

Then the tree shows that, for either outcome of the first toss, the second toss can be either heads or tails, again with probabilities 0.5 each.

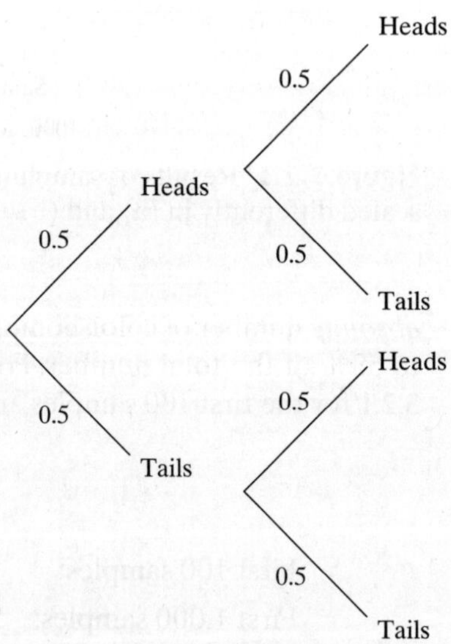

To find the probability of getting heads on both tosses, we consider the path through the tree that produces this event. We multiply together the probabilities that we encounter along the path. Figure 3.2.2 summarizes this example and shows that

$$\Pr\{\text{heads on both tosses}\} = 0.5 \times 0.5 = 0.25$$

■

Figure 3.2.2 Probability tree for two coin tosses

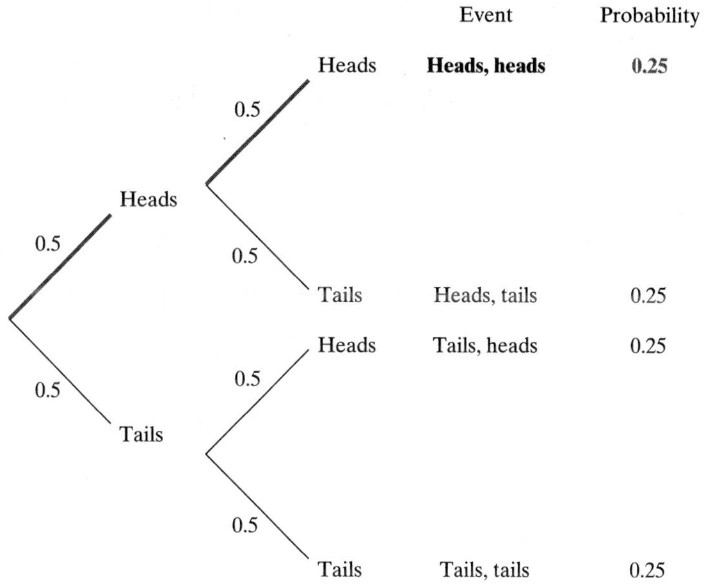

COMBINATION OF PROBABILITIES

If an event can happen in more than one way, the relative frequency interpretation of probability can be a guide to appropriate combinations of the probabilities of subevents. The following example illustrates this idea.

Example 3.2.8 **Sampling Fruitflies** In the *Drosophila* population of Examples 3.2.3 and 3.2.6, 30% of the flies are black and 70% are gray. Suppose that two flies are randomly chosen from the population. Suppose we wish to find the probability that both flies are the same color. The probability tree displayed in Figure 3.2.3 shows the four possible outcomes from sampling two flies. From the tree, we can see that the probability of getting two black flies is $0.3 \times 0.3 = 0.09$. Likewise, the probability of getting two gray flies is $0.7 \times 0.7 = 0.49$.

Figure 3.2.3 Probability tree for sampling two flies

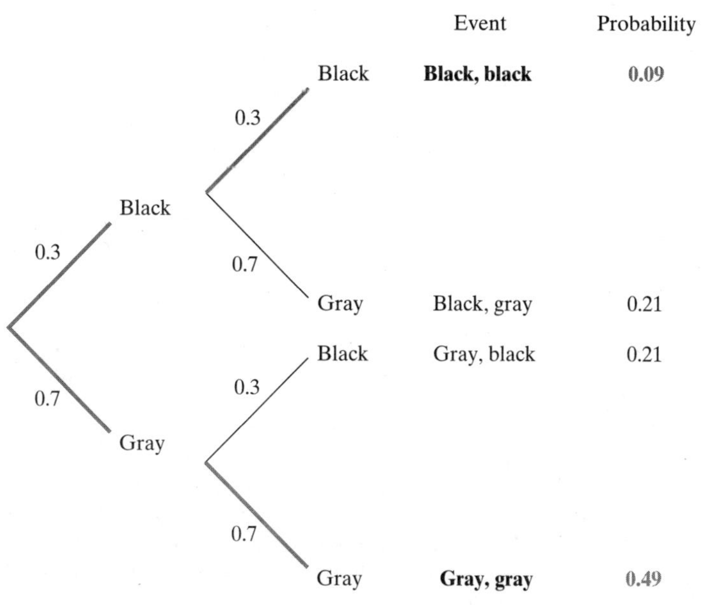

To find the probability of the event

E: Both flies in the sample are the same color

we add the probability of black, black to the probability of gray, gray to get 0.09 + 0.49 = 0.58. ■

In the coin tossing setting of Example 3.2.7, the second part of the probability tree had the same structure as the first part—namely, a 0.5 chance of heads and a 0.5 chance of tails—because the outcome of the first toss does not affect the probability of heads on the second toss. Likewise, in Example 3.2.8 the probability of the second fly being black was 0.3, regardless of the color of the first fly, because the population was assumed to be very large, so that removing one fly from the population would not affect the proportion of flies that are black. However, in some situations we need to treat the second part of the probability tree differently than the first part.

Example 3.2.9

Nitric Oxide Hypoxic respiratory failure is a serious condition that affects some newborns. If a newborn has this condition, it is often necessary to use extracorporeal membrane oxygenation (ECMO) to save the life of the child. However, ECMO is an invasive procedure that involves inserting a tube into a vein or artery near the heart, so physicians hope to avoid the need for it. One treatment for hypoxic respiratory failure is to have the newborn inhale nitric oxide. To test the effectiveness of this treatment, newborns suffering hypoxic respiratory failure were assigned at random to either be given nitric oxide or a control group.[1] In the treatment group 45.6% of the newborns had a negative outcome, meaning that either they needed ECMO or that they died. In the control group, 63.6% of the newborns had a negative outcome. Figure 3.2.4 shows a probability tree for this experiment.

Figure 3.2.4 Probability tree for nitric oxide example

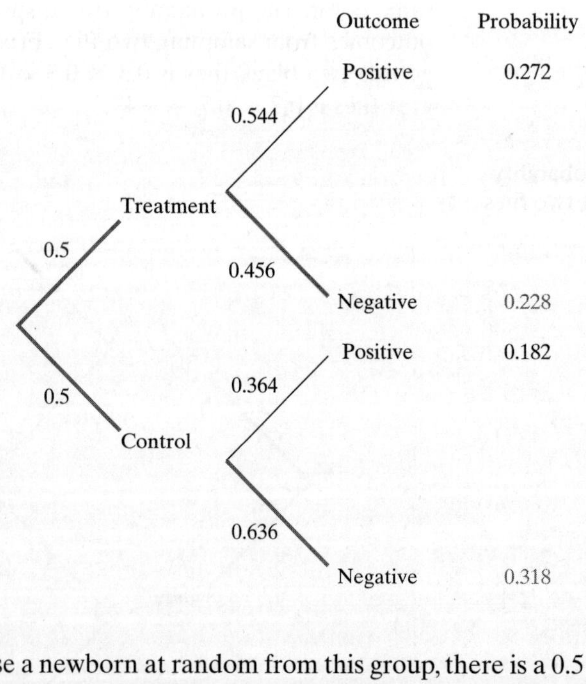

If we choose a newborn at random from this group, there is a 0.5 probability that the newborn will be in the treatment group and, if so, a probability of 0.456 of getting a negative outcome. Likewise, there is a 0.5 probability that the newborn will be in the control group and, if so, a probability of 0.636 of getting a negative outcome. Thus, the probability of a negative outcome is

$$0.5 \times 0.456 + 0.5 \times 0.636 = 0.228 + 0.318 = 0.546.$$

HYPOTHETICAL 1,000

It is often helpful to think about a probability question in terms of what we would expect to see in 1,000 repetitions of the situation. The following example illustrates this mode of thinking.

Example 3.2.10

Nitric Oxide Suppose that 1,000 infants experiencing hypoxic respiratory failure were enrolled in a study like the one described in Example 3.2.9. We would expect 500 of them to be given the treatment (nitric oxide) and 500 of them to be in the control group. Based on Figure 3.2.4, of those given the treatment we expect 54.4% to have a positive outcome. Thus, we expect $500 \times 0.544 = 272$ positive outcomes in the treatment group; likewise, we expect $500 \times 0.456 = 228$ negative outcomes in the treatment group. For the control group, the corresponding numbers are $500 \times 0.364 = 182$ positive outcomes and $500 \times 0.636 = 318$ negative outcomes.

We can put these numbers together to get Table 3.2.2. From the table we can see that there are 546 negative outcomes out of the 1,000 total cases. This agrees with the probability of 0.546 found in Example 3.2.9.

Table 3.2.2 Nitric oxide outcomes

	Outcome		
	Positive	Negative	Total
Treatment	272	228	500
Control	182	318	500
Total	454	546	1,000

Example 3.2.11

Medical Testing Suppose a medical test is conducted on someone to try to determine whether or not the person has a particular disease. If the test indicates that the disease is present, we say the person has "tested positive." If the test indicates that the disease is not present, we say the person has "tested negative." However, there are two types of mistakes that can be made. It is possible that the test indicates that the disease is present, but the person does not really have the disease; this is known as a false positive. It is also possible that the person has the disease, but the test does not detect it; this is known as a false negative.

Suppose that a particular test has a 95% chance of detecting the disease if the person has it (this is called the sensitivity of the test) and a 90% chance of correctly indicating that the disease is absent if the person really does not have the disease (this is called the specificity of the test). Suppose 8% of the population has the disease. What is the probability that a randomly chosen person will test positive?

Figure 3.2.5 shows a probability tree for this situation. The first split in the tree shows the division between those who have the disease and those who don't. If someone has the disease, then we use 0.95 as the chance of the person testing positive. If the person doesn't have the disease, then we use 0.10 as the chance of the

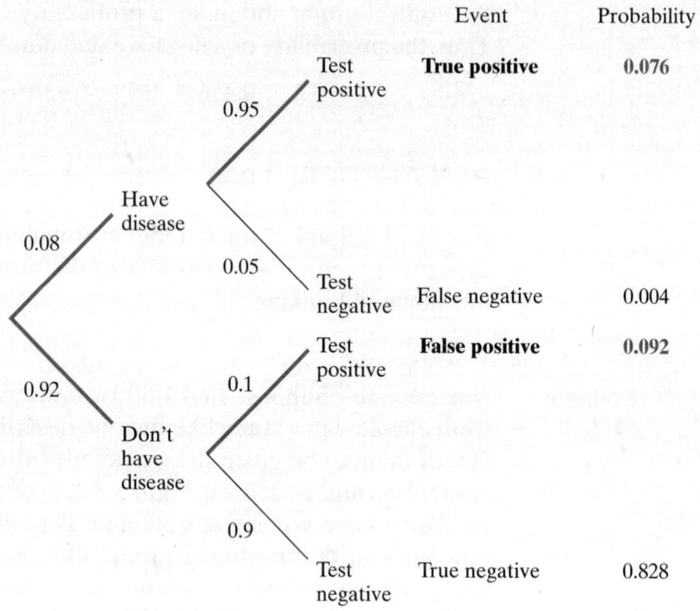

Figure 3.2.5 Probability tree for medical testing example

person testing positive. Thus, the probability of a randomly chosen person testing positive is

$$0.08 \times 0.95 + 0.92 \times 0.10 = 0.076 + 0.092 = 0.168.$$ ■

We can apply the Hypothetical 1,000 idea to Example 3.2.11. Table 3.2.3 shows that we expect 80 out of 1,000 people to have the disease ($0.08 \times 1{,}000$) and 76 of them to test positive (0.95×80). We expect 920 people to not have the disease and 92 of them (0.1×920) to test positive. From the table we see that 168 out of 1,000 people test positive; this agrees with the probability of 0.168 found in Example 3.2.11.

Table 3.2.3 Medical testing outcomes

	Test result		
	Positive	Negative	Total
Disease	76	4	80
No disease	92	828	920
Total	168	832	1,000

Example 3.2.12 **False Positives** Consider the medical testing scenario of Example 3.2.11. If someone tests positive, what is the chance the person really has the disease? Table 3.2.3 shows that we would expect 168 out of 1,000 to test positive. The probability of a true positive is 0.076, so we would expect 76 "true positives" out of 1,000 persons tested. Thus, we expect 76 true positives out of 168 total positives, which is to say that the probability that someone really has the disease, given that the person tests positive, is $\frac{76}{168} = \frac{0.076}{0.168} \approx 0.452$. This probability is quite a bit smaller than most people expect it to be, given that the sensitivity and specificity of the test are 0.95 and 0.90. We also see that out of the 168 positive test results only 76 are true positives, giving a rate of $76/168 = 0.452$. ■

Exercises 3.2.1–3.2.8

3.2.1 In a certain population of the freshwater sculpin, *Cottus rotheus,* the distribution of the number of tail vertebrae is as shown in the table.[2]

No. of vertebrae	Percent of fish
20	3
21	51
22	40
23	6
Total	100

Find the probability that the number of tail vertebrae in a fish randomly chosen from the population

(a) equals 21.

(b) is less than or equal to 22.

(c) is greater than 21.

(d) is no more than 21.

3.2.2 The following table shows the distribution of ages of Americans.[3]

Age distribution in reference population

Age	Proportion
0–19	0.27
20–29	0.14
30–39	0.13
40–49	0.14
50–64	0.19
65+	0.13

Find the probability that the age of a randomly chosen American

(a) is less than 20.

(b) is between 20 and 49.

(c) is greater than 49.

(d) is greater than 29.

3.2.3 In a certain college, 55% of the students are women. Suppose we take a sample of two students. Use a probability tree to find the probability

(a) that both chosen students are women.

(b) that at least one of the two students is a woman.

3.2.4 Suppose that a disease is inherited via a sex-linked mode of inheritance so that a male offspring has a 50% chance of inheriting the disease, but a female offspring has no chance of inheriting the disease. Further suppose that 51.3% of births are male. What is the probability that a randomly chosen child will be affected by the disease?

3.2.5 Suppose that a student who is about to take a multiple choice test has only learned 40% of the material covered by the exam. Thus, there is a 40% chance that she will know the answer to a question. However, even if she does not know the answer to a question, she still has a 20% chance of getting the right answer by guessing. If we choose a question at random from the exam, what is the probability that she will get it right?

3.2.6 If a woman takes an early pregnancy test, she will either test positive, meaning that the test says she is pregnant, or test negative, meaning that the test says she is not pregnant. Suppose that if a woman really is pregnant, there is a 98% chance that she will test positive. Also, suppose that if a woman really is *not* pregnant, there is a 99% chance that she will test negative.

(a) Suppose that 1,000 women take early pregnancy tests and that 100 of them really are pregnant. What is the probability that a randomly chosen woman from this group will test positive?

(b) Suppose that 1,000 women take early pregnancy tests and that 50 of them really are pregnant. What is the probability that a randomly chosen woman from this group will test positive?

3.2.7

(a) Consider the setting of Exercise 3.2.6, part (a). Suppose that a woman tests positive. What is the probability that she really is pregnant?

(b) Consider the setting of Exercise 3.2.6, part (b). Suppose that a woman tests positive. What is the probability that she really is pregnant?

3.2.8 Suppose that a medical test has a 92% chance of detecting a disease if the person has it (i.e., 92% sensitivity) and a 94% chance of correctly indicating that the disease is absent if the person really does not have the disease (i.e., 94% specificity). Suppose 10% of the population has the disease.

(a) What is the probability that a randomly chosen person will test positive?

(b) Suppose that a randomly chosen person does test positive. What is the probability that this person really has the disease?

3.3 Probability Rules (Optional)

We have defined the probability of an event, $\Pr\{E\}$, as the long-run relative frequency with which the event occurs. In this section we will briefly consider a few rules that help determine probabilities. We begin with three basic rules.

BASIC RULES

Rule (1) The probability of an event E is always between 0 and 1. That is, $0 \leq \Pr\{E\} \leq 1$.

Rule (2) The sum of the probabilities of all possible events equals 1. That is, if the set of all possible events is $E_1, E_2, \ldots, E_k$, then $\sum_{i=1}^{k} \Pr\{E_i\} = 1$.

Rule (3) The probability that an event E does not happen, denoted by E^C, is one minus the probability that the event happens. That is, $\Pr\{E^C\} = 1 - \Pr\{E\}$. (We refer to E^C as the complement of E.)

We illustrate these rules with an example.

Example 3.3.1 **Blood Type** In the United States, 44% of the population has type O blood, 42% has type A, 10% has type B, and 4% has type AB.[4] Consider choosing someone at random and determining the person's blood type. The probability of a given blood type will correspond to the population percentage.

(a) The probability that the person will have type O blood $= \Pr\{\text{O}\} = 0.44$.

(b) $\Pr\{\text{O}\} + \Pr\{\text{A}\} + \Pr\{\text{B}\} + \Pr\{\text{AB}\} = 0.44 + 0.42 + 0.10 + 0.04 = 1$.

(c) The probability that the person will *not* have type O blood $= \Pr\{\text{O}^C\} = 1 - 0.44 = 0.56$. This could also be found by adding the probabilities of the other blood types: $\Pr\{\text{O}^C\} = \Pr\{\text{A}\} + \Pr\{\text{B}\} + \Pr\{\text{AB}\} = 0.42 + 0.10 + 0.04 = 0.56$. ■

We often want to discuss two or more events at once; to do this we will find some terminology to be helpful. We say that two events are *disjoint** if they cannot occur simultaneously. Figure 3.3.1 is a *Venn diagram* that depicts a *sample space S* of all possible outcomes as a rectangle with two disjoint events depicted as nonoverlapping regions.

The *union* of two events is the event that one or the other occurs or both occur. The *intersection* of two events is the event that they both occur. Figure 3.3.2 is a Venn diagram that shows the union of two events as the total shaded area, with the intersection of the events being the overlapping region in the middle.

If two events are disjoint, then the probability of their union is the sum of their individual probabilities. If the events are not disjoint, then to find the probability of their union we take the sum of their individual probabilities and subtract the probability of their intersection (the part that was "counted twice").

ADDITION RULES

Rule (4) If two events E_1 and E_2 are disjoint, then $\Pr\{E_1 \text{ or } E_2\} = \Pr\{E_1\} + \Pr\{E_2\}$.

*Another term for disjoint events is "mutually exclusive" events.

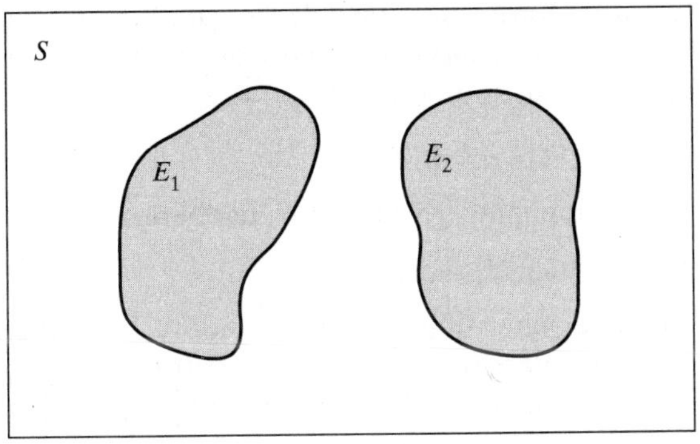

Figure 3.3.1 Venn diagram showing two disjoint events

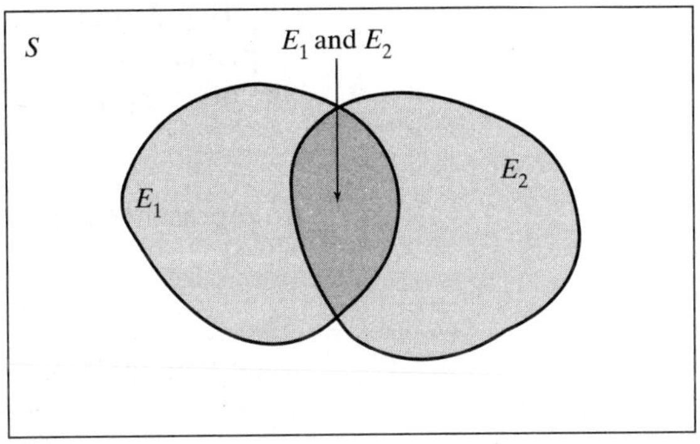

Figure 3.3.2 Venn diagram showing union (total shaded area) and intersection (middle area) of two events

Rule (5) For any two events E_1 and E_2,
$\Pr\{E_1 \text{ or } E_2\} = \Pr\{E_1\} + \Pr\{E_2\} - \Pr\{E_1 \text{ and } E_2\}$.

We illustrate these rules with an example.

Example 3.3.2

Hair Color and Eye Color Table 3.3.1 shows the relationship between hair color and eye color for a group of 1,770 German men.[5]

Table 3.3.1 Hair color and eye color

		Hair color			
		Brown	Black	Red	Total
Eye color	Brown	400	300	20	720
	Blue	800	200	50	1,050
	Total	1,200	500	70	1,770

(a) Because events "black hair" and "red hair" are disjoint, if we choose someone at random from this group then Pr{black hair or red hair} = Pr{black hair} + Pr{red hair} = 500/1,770 + 70/1,770 = 570/1,770.

(b) If we choose someone at random from this group, then Pr{black hair} = 500/1,770.

(c) If we choose someone at random from this group, then Pr{blue eyes} = 1,050/1,770.

(d) The events "black hair" and "blue eyes" are not disjoint, since there are 200 men with both black hair and blue eyes. Thus, Pr{black hair or blue eyes} = Pr{black hair} + Pr{blue eyes} − Pr{black hair and blue eyes} = 500/1,770 + 1,050/1,770 − 200/1,770 = 1,350/1,770. ■

Two events are said to be *independent* if knowing that one of them occurred does not change the probability of the other one occurring. For example, if a coin is tossed twice, the outcome of the second toss is independent of the outcome of the first toss, since knowing whether the first toss resulted in heads or in tails does not change the probability of getting heads on the second toss.

Events that are not independent are said to be *dependent*. When events are dependent, we need to consider the *conditional probability* of one event, given that the other event has happened. We use the notation

$$\Pr\{E_2 | E_1\}$$

to represent the probability of E_2 happening, given that E_1 happened.

Example 3.3.3

Hair Color and Eye Color Consider choosing a man at random from the group shown in Table 3.3.1. Overall, the probability of blue eyes is 1,050/1,770, or about 59.3%. However, if the man has black hair, then the conditional probability of blue eyes is only 200/500, or 40%; that is, Pr{blue eyes|black hair} = 0.40. Because the probability of blue eyes depends on hair color, the events "black hair" and "blue eyes" are dependent. ■

Refer again to Figure 3.3.2, which shows the intersection of two regions (for E_1 and E_2). If we know that the event E_1 has happened, then we can restrict our attention to the E_1 region in the Venn diagram. If we now want to find the chance that E_2 will happen, we need to consider the intersection of E_1 and E_2 relative to the entire E_1 region. In the case of Example 3.3.3, this corresponds to knowing that a randomly chosen man has black hair, so that we restrict our attention to the 500 men (out of 1,770 total in the group) with black hair. Of these men, 200 have blue eyes. The 200 are in the intersection of "black hair" and "blue eyes." The fraction 200/500 is the conditional probability of having blue eyes, given that the man has black hair.

This leads to the following formal definition of the conditional probability of E_2 given E_1:

> **DEFINITION** The conditional probability of E_2, given E_1, is
>
> $$\Pr\{E_2 | E_1\} = \frac{\Pr\{E_1 \text{ and } E_2\}}{\Pr\{E_1\}}$$
>
> provided that $\Pr\{E_1\} > 0$.

Example 3.3.4

Hair Color and Eye Color Consider choosing a man at random from the group shown in Table 3.3.1. The probability of the man having blue eyes given that he has black hair is

$$\Pr\{\text{blue eyes} | \text{black hair}\} = \Pr\{\text{black hair and blue eyes}\}/\Pr\{\text{black hair}\}$$

$$= \frac{200/1{,}770}{500/1{,}770} = \frac{200}{500} = 0.40.$$ ■

In Section 3.2 we used probability trees to study compound events. In doing so, we implicitly used multiplication rules that we now make explicit.

MULTIPLICATION RULES

Rule (6) If two events E_1 and E_2 are independent, then $\Pr\{E_1 \text{ and } E_2\} = \Pr\{E_1\} \times \Pr\{E_2\}$.

Rule (7) For any two events E_1 and E_2, $\Pr\{E_1 \text{ and } E_2\} = \Pr\{E_1\} \times \Pr\{E_2 | E_1\}$.

Example 3.3.5

Coin Tossing If a fair coin is tossed twice, the two tosses are independent of each other. Thus, the probability of getting heads on both tosses is

$$\begin{aligned}\Pr\{\text{heads twice}\} &= \Pr\{\text{heads on first toss}\} \times \Pr\{\text{heads on second toss}\}\\ &= 0.5 \times 0.5 = 0.25.\end{aligned}$$

■

Example 3.3.6

Blood Type In Example 3.3.1 we stated that 44% of the U.S. population has type O blood. It is also true that 15% of the population is Rh negative and that this is independent of blood group. Thus, if someone is chosen at random, the probability that the person has type O, Rh negative blood is

$$\begin{aligned}\Pr\{\text{group O and Rh negative}\} &= \Pr\{\text{group O}\} \times \Pr\{\text{Rh negative}\}\\ &= 0.44 \times 0.15 = 0.066.\end{aligned}$$

■

Example 3.3.7

Hair Color and Eye Color Consider choosing a man at random from the group shown in Table 3.3.1. What is the probability that the man will have red hair and brown eyes? Hair color and eye color are dependent, so finding this probability involves using a conditional probability. The probability that the man will have red hair is 70/1,770. Given that the man has red hair, the conditional probability of brown eyes is 20/70. Thus,

$$\begin{aligned}\Pr\{\text{red hair and brown eyes}\} &= \Pr\{\text{red hair}\} \times \Pr\{\text{brown eyes} \mid \text{red hair}\}\\ &= 70/1{,}770 \times 20/70 = 20/1{,}770.\end{aligned}$$

■

Sometimes a probability problem can be broken into two conditional "parts" that are solved separately and the answers combined.

Rule of Total Probability

Rule (8) For any two events E_1 and E_2,

$$\Pr\{E_1\} = \Pr\{E_2\} \times \Pr\{E_1 \mid E_2\} + \Pr\{E_2^C\} \times \Pr\{E_1 \mid E_2^C\}.$$

Example 3.3.8

Hand Size Consider choosing someone at random from a population that is 60% female and 40% male. Suppose that for a woman the probability of having a hand size smaller than 100 cm^2 is 0.31.[6] Suppose that for a man the probability of having a hand size smaller than 100 cm^2 is 0.08. What is the probability that the randomly chosen person will have a hand size smaller than 100 cm^2?

We are given that if the person is a woman, then the probability of a "small" hand size is 0.31 and that if the person is a man, then the probability of a "small" hand size is 0.08.

Thus,

$$\begin{aligned}\Pr\{\text{hand size} < 100\} &= \Pr\{\text{woman}\} \times \Pr\{\text{hand size} < 100 \mid \text{woman}\}\\ &\quad + \Pr\{\text{man}\} \times \Pr\{\text{hand size} < 100 \mid \text{man}\}\\ &= 0.6 \times 0.31 + 0.4 \times 0.08\\ &= 0.186 + 0.032\\ &= 0.218.\end{aligned}$$

■

We can apply the Hypothetical 1,000 idea here. Table 3.3.2 shows 600 women, of whom 31% have small hands ($0.31 \times 600 = 186$) and 69% don't. We also see

Table 3.3.2 Hand size

	Hand size		
	$< 100\text{ cm}^2$	$\geq 100\text{ cm}^2$	Total
Woman	186	414	600
Man	32	368	400
Total	218	782	1,000

400 men, of whom 8% have small hands ($0.08 \times 400 = 32$) and 92% don't. The column "$< 100\text{ cm}^2$" sums to 218; this agrees with the probability of 0.218.

Exercises 3.3.1–3.3.5

3.3.1 In a study of the relationship between health risk and income, a large group of people living in Massachusetts were asked a series of questions.[7] Some of the results are shown in the following table.

	Income			
	Low	Medium	High	Total
Smoke	634	332	247	1,213
Don't smoke	1,846	1,622	1,868	5,336
Total	2,480	1,954	2,115	6,549

(a) What is the probability that someone in this study smokes?

(b) What is the conditional probability that someone in this study smokes, given that the person has high income?

(c) Is being a smoker independent of having a high income? Why or why not?

3.3.2 Consider the data table reported in Exercise 3.3.1.

(a) What is the probability that someone in this study is from the low income group and smokes?

(b) What is the probability that someone in this study is not from the low income group?

(c) What is the probability that someone in this study is from the medium income group?

(d) What is the probability that someone in this study is from the low income group or from the medium income group?

3.3.3 The following data table is taken from the study reported in Exercise 3.3.1. Here "stressed" means that the person reported that most days are extremely stressful or quite stressful; "not stressed" means that the person reported that most days are a bit stressful, not very stressful, or not at all stressful.

	Income			
	Low	Medium	High	Total
Stressed	526	274	216	1,016
Not stressed	1,954	1,680	1,899	5,533
Total	2,480	1,954	2,115	6,549

(a) What is the probability that someone in this study is stressed?

(b) Given that someone in this study is from the high income group, what is the probability that the person is stressed?

(c) Compare your answers to parts (a) and (b). Is being stressed independent of having high income? Why or why not?

3.3.4 Consider the data table reported in Exercise 3.3.3.

(a) What is the probability that someone in this study has low income?

(b) What is the probability that someone in this study either is stressed or has low income (or both)?

(c) What is the probability that someone in this study is stressed and has low income?

3.3.5 Suppose that in a certain population of married couples, 30% of the husbands smoke, 20% of the wives smoke, and in 8% of the couples both the husband and the wife smoke. Is the smoking status (smoker or nonsmoker) of the husband independent of that of the wife? Why or why not?

3.4 Density Curves

The examples presented in Section 3.2 dealt with probabilities for discrete variables. In this section we will consider probability when the variable is continuous.

RELATIVE FREQUENCY HISTOGRAMS AND DENSITY CURVES

In Chapter 2 we discussed the use of a histogram to represent a frequency distribution for a variable. A *relative frequency histogram* is a histogram in which we indicate the proportion (i.e., the relative frequency) of observations in each category, rather than the count of observations in the category. We can think of the relative frequency histogram as an approximation of the underlying true population distribution from which the data came.

It is often desirable, especially when the observed variable is continuous, to describe a population frequency distribution by a smooth curve. We may visualize the curve as an idealization of a relative frequency histogram with very narrow classes. The following example illustrates this idea.

Example 3.4.1

Blood Glucose A glucose tolerance test can be useful in diagnosing diabetes. The blood level of glucose is measured one hour after the subject has drunk 50 mg of glucose dissolved in water. Figure 3.4.1 shows the distribution of responses to this test for a certain population of women.[8] The distribution is represented by histograms with class widths equal to (a) 10 and (b) 5, and by (c) a smooth curve. ■

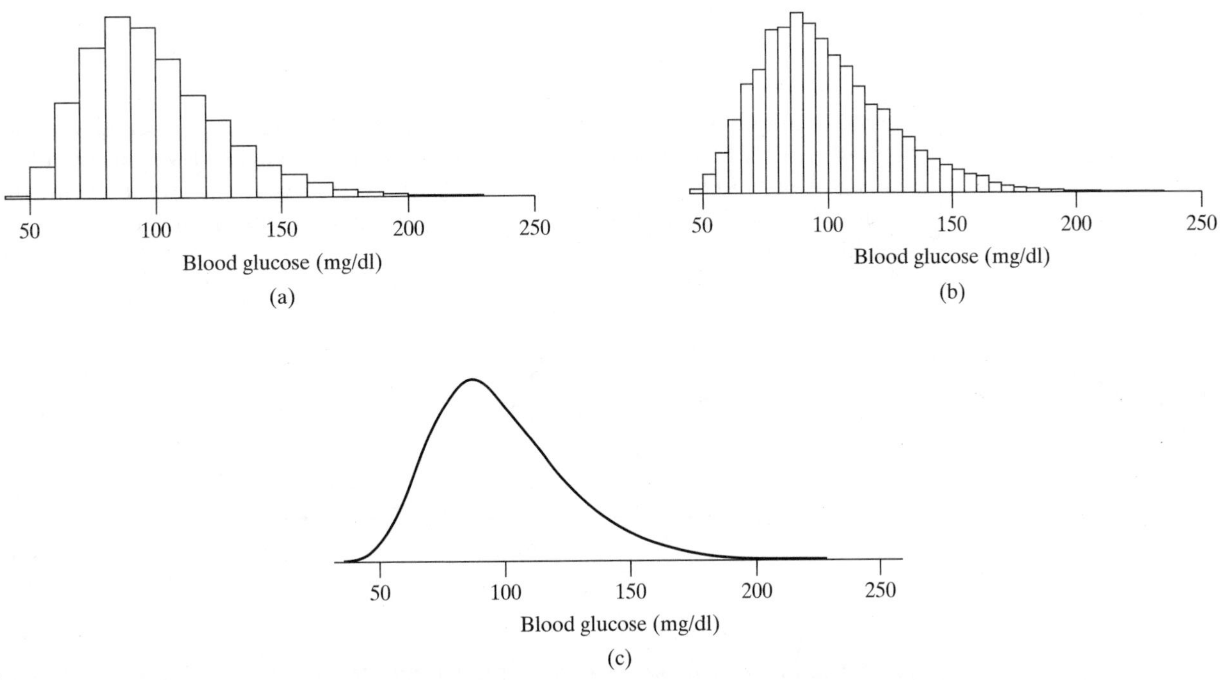

Figure 3.4.1 Different representations of the distribution of blood glucose levels in a population of women

A smooth curve representing a frequency distribution is called a **density curve**. The vertical coordinates of a density curve are plotted on a scale called a **density scale**. When the density scale is used, relative frequencies are represented as areas under the curve. Formally, the relation is as follows:

Interpretation of Density

For any two numbers a and b,

$$\frac{\text{Area under density curve}}{\text{between } a \text{ and } b} = \frac{\text{Proportion of } Y \text{ values}}{\text{between } a \text{ and } b}$$

This relation is indicated in Figure 3.4.2 for an arbitrary distribution

Because of the way the density curve is interpreted, the density curve is entirely above (or equal to) the x-axis and the area under the entire curve must be equal to 1, as shown in Figure 3.4.3.

The interpretation of density curves in terms of areas is illustrated concretely in the following example.

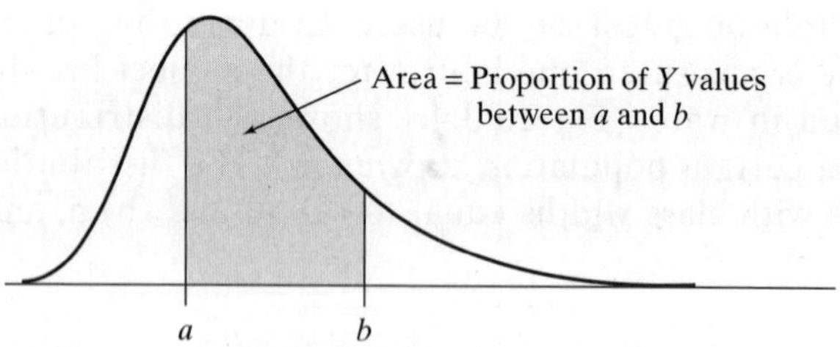

Figure 3.4.2 Interpretation of area under a density curve

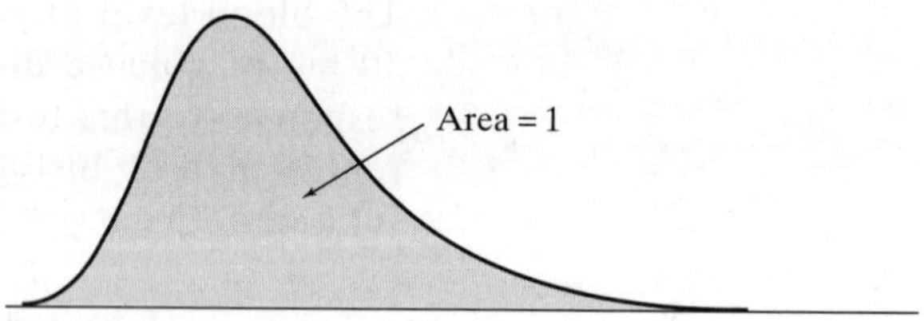

Figure 3.4.3 The area under an entire density curve must be 1

Example 3.4.2 **Blood Glucose** Figure 3.4.4 shows the density curve for the blood glucose distribution of Example 3.4.1, with the vertical scale explicitly shown. The shaded area is equal to 0.42, which indicates that about 42% of the glucose levels are between 100 mg/dl and 150 mg/dl. The area under the density curve to the left of 100 mg/dl is equal to 0.50; this indicates that the population median glucose level is 100 mg/dl. The area under the entire curve is 1. ■

Figure 3.4.4 Interpretation of an area under the blood glucose density curve

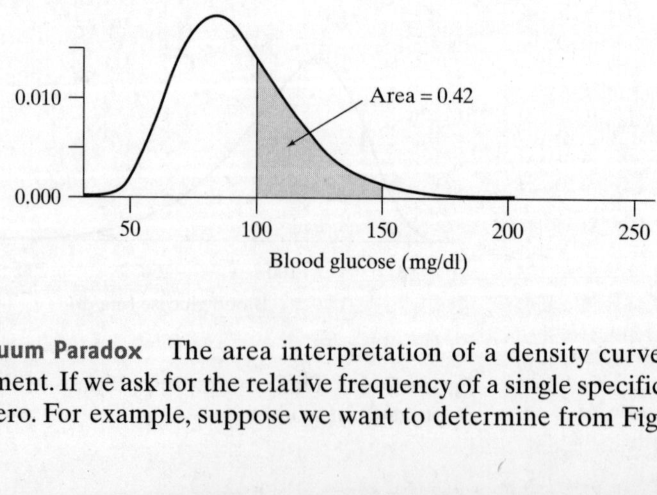

The Continuum Paradox The area interpretation of a density curve has a paradoxical element. If we ask for the relative frequency of a single specific Y value, the answer is zero. For example, suppose we want to determine from Figure 3.4.4 the

relative frequency of blood glucose levels *equal* to 150. The area interpretation gives an answer of zero. This seems to be nonsense—how can every value of Y have a relative frequency of zero? Let us look more closely at the question. If blood glucose is measured to the nearest mg/dl, then we are really asking for the relative frequency of glucose levels between 149.5 and 150.5 mg/dl, and the corresponding area is not zero. On the other hand, if we are thinking of blood glucose as an *idealized* continuous variable, then the relative frequency of any particular value (e.g., 150) *is* zero. This is admittedly a paradoxical situation. It is similar to the paradoxical fact that an idealized straight line can be 1 centimeter long, and yet each of the idealized points of which the line is composed has length equal to zero. In practice, the continuum paradox does not cause any trouble; we simply do not discuss the relative frequency of a single Y value (just as we do not discuss the length of a single point).

PROBABILITIES AND DENSITY CURVES

If a variable has a continuous distribution, then we find probabilities by using the density curve for the variable. A probability for a continuous variable equals the area under the density curve for the variable between two points.

Example 3.4.3

Blood Glucose Consider the blood glucose level, in mg/dl, of a randomly chosen subject from the population described in Example 3.4.2. We saw in Example 3.4.2 that 42% of the population glucose levels are between 100 mg/dl and 150 mg/dl. Thus, $\Pr\{100 \leq \text{glucose level} \leq 150\} = 0.42$.

We are modeling blood glucose level as being a continuous variable, which means that $\Pr\{\text{glucose level} = 100\} = 0$ and $\Pr\{\text{glucose level} = 150\} = 0$, as we noted above. Thus,

$$\Pr\{100 \leq \text{glucose level} \leq 150\} = \Pr\{100 < \text{glucose level} < 150\} = 0.42.$$ ■

Example 3.4.4

Tree Diameters The diameter of a tree trunk is an important variable in forestry. The density curve shown in Figure 3.4.5 represents the distribution of diameters (measured at breast height) in a population of 30-year-old Douglas fir trees; areas under the curve are shown in the figure.[9] Consider the diameter, in inches, of a randomly chosen tree. Then, for example, $\Pr\{4 < \text{diameter} < 6\} = 0.33$. If we want to find the probability that a randomly chosen tree has a diameter greater than 8 inches, we must add the last two areas under the curve in Figure 3.4.3: $\Pr\{\text{diameter} > 8\} = 0.12 + 0.07 = 0.19$. ■

Figure 3.4.5 Diameters of 30-year-old Douglas fir trees

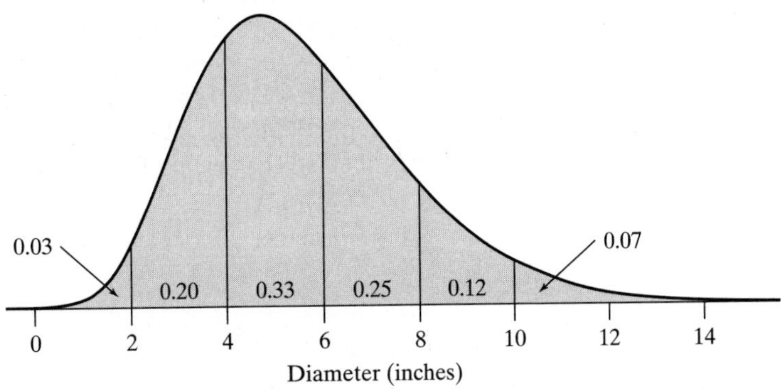

Exercises 3.4.1–3.4.5

3.4.1 Consider the density curve shown in Figure 3.4.5, which represents the distribution of diameters (measured 4.5 feet above the ground) in a population of 30-year-old Douglas fir trees. Areas under the curve are shown in the figure. What percentage of the trees have diameters

(a) between 4 inches and 10 inches?

(b) less than 4 inches?

(c) more than 6 inches?

3.4.2 Consider the diameter of a Douglas fir tree drawn at random from the population that is represented by the density curve shown in Figure 3.4.5. Find

(a) Pr{diameter < 10}

(b) Pr{diameter > 4}

(c) Pr{2 < diameter < 8}

3.4.3 In a certain population of the parasite *Trypanosoma*, the lengths of individuals are distributed as indicated by the density curve shown here. Areas under the curve are shown in the figure.[10]

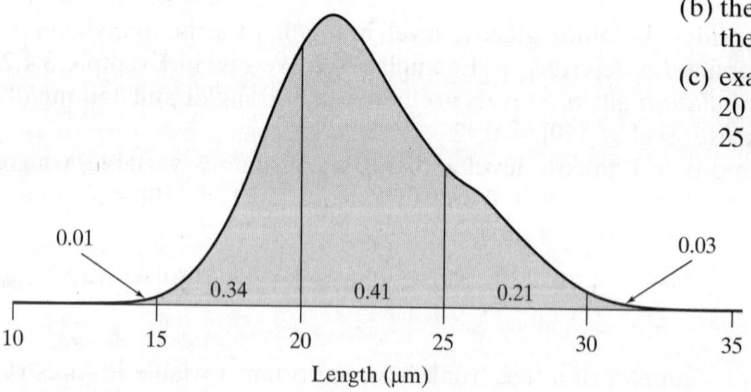

Consider the length of an individual trypanosome chosen at random from the population. Find

(a) Pr{20 < length < 30}

(b) Pr{length > 20}

(c) Pr{length < 20}

3.4.4 Consider the distribution of *Trypanosoma* lengths shown by the density curve in Exercise 3.4.3. Consider the length of an individual trypanosome chosen at random from the population. Find

(a) Pr{length < 25}

(b) Pr{length > 15}

(c) Pr{15 < length < 30}

3.4.5 Consider the distribution of *Trypanosoma* lengths shown by the density curve in Exercise 3.4.3. Suppose we take a sample of two trypanosomes. What is the probability that

(a) both trypanosomes will be shorter than 20 μm?

(b) the first trypanosome will be shorter than 20 μm and the second trypanosome will be longer than 25 μm?

(c) exactly one of the trypanosomes will be shorter than 20 μm and one trypanosome will be longer than 25 μm?

3.5 Random Variables

A **random variable** is simply a variable that takes on numerical values that depend on the outcome of a chance operation. The following examples illustrate this idea.

Example 3.5.1 **Dice** Consider the chance operation of tossing a die. Let the random variable Y represent the number of spots showing. The possible values of Y are $Y = 1, 2, 3, 4, 5$, or 6. We do not know the value of Y until we have tossed the die. If we know how the die is weighted, then we can specify the probability that Y has a particular value, say $\Pr\{Y = 4\}$, or a particular set of values, say $\Pr\{2 \le Y \le 4\}$. For instance, if the die is perfectly balanced so that each of the six faces is equally likely, then

$$\Pr\{Y = 4\} = \frac{1}{6} \approx 0.17$$

and

$$\Pr\{2 \le Y \le 4\} = \frac{3}{6} = 0.5$$ ■

Example 3.5.2 **Family Size** Suppose a family is chosen at random from a certain population, and let the random variable Y denote the number of children in the chosen family. The possible values of Y are $0, 1, 2, 3, \ldots$. The probability that Y has a particular value is equal to the percentage of families with that many children. For instance, if 23% of the families have 2 children, then

$$\Pr\{Y = 2\} = 0.23$$

■

Example 3.5.3 **Medications** After someone has heart surgery, the person is usually given several medications. Let the random variable Y denote the number of medications that a patient is given following cardiac surgery. If we know the distribution of the number of medications per patient for the entire population, then we can specify the probability that Y has a certain value or falls within a certain interval of values. For instance, if 52% of all patients are given 2, 3, 4, or 5 medications, then

$$\Pr\{2 \leq Y \leq 5\} = 0.52$$

■

Example 3.5.4 **Heights of Men** Let the random variable Y denote the height of a man chosen at random from a certain population. If we know the distribution of heights in the population, then we can specify the probability that Y falls in a certain range. For instance, if 46% of the men are between 65.2 and 70.4 inches tall, then

$$\Pr\{65.2 \leq Y \leq 70.4\} = 0.46$$

■

Each of the variables in Examples 3.5.1–3.5.3 is a *discrete random variable*, because in each case we can list the possible values that the variable can take on. In contrast, the variable in Example 3.5.4, height, is a *continuous random variable*: Height, at least in theory, can take on any of an infinite number of values in an interval. Of course, when we measure and record a person's height, we generally measure to the nearest inch or half inch. Nonetheless, we can think of true height as being a continuous variable. We use density curves to model the distributions of continuous random variables, such as blood glucose level or tree diameter, as discussed in Section 3.4.

MEAN AND VARIANCE OF A RANDOM VARIABLE

In Chapter 2 we briefly considered the concepts of population mean and population standard deviation. For the case of a discrete random variable, we can calculate the population mean and standard deviation if we know the probability distribution for the random variable. We begin with the mean.

The mean of a discrete random variable Y is defined as

$$\mu_Y = \Sigma y_i \Pr(Y = y_i)$$

where the y_i's are the values that the variable takes on and the sum is taken over all possible values.

The mean of a random variable is also known as the *expected value* and is often written as $E(Y)$; that is, $E(Y) = \mu_Y$.

Example 3.5.5 **Fish Vertebrae** In a certain population of the freshwater sculpin *Cottus rotheus*, the distribution of the number of tail vertebrae, Y, is as shown in Table 3.5.1.[2]

Table 3.5.1 Distribution of vertebrae

No. of vertebrae	Percent of fish
20	3
21	51
22	40
23	6
Total	100

The mean of Y is

$$\begin{aligned}\mu_Y &= 20 \times \Pr\{Y = 20\} + 21 \times \Pr\{Y = 21\} + 22 \times \Pr\{Y = 22\} + 23 \times \Pr\{Y = 23\}\\ &= 20 \times .03 + 21 \times .51 + 22 \times .40 + 23 \times .06\\ &= 0.6 + 10.71 + 8.8 + 1.38\\ &= 21.49.\end{aligned}$$

■

Example 3.5.6

Dice Consider rolling a die that is perfectly balanced so that each of the six faces is equally likely to come up and let the random variable Y represent the number of spots showing. The expected value, or mean, of Y is

$$E(Y) = \mu_Y = 1 \times \frac{1}{6} + 2 \times \frac{1}{6} + 3 \times \frac{1}{6} + 4 \times \frac{1}{6} + 5 \times \frac{1}{6} + 6 \times \frac{1}{6} = \frac{21}{6} = 3.5.$$ ■

To find the standard deviation of a random variable, we first find the variance, σ^2, of the random variable and then take the square root of the variance to get the the standard deviation, σ.

The variance of a discrete random variable Y is defined as

$$\sigma_Y^2 = \Sigma(y_i - \mu_Y)^2 \Pr(Y = y_i)$$

where the y_i's are the values that the variable takes on and the sum is taken over all possible values.

We often write VAR(Y) to denote the variance of Y.

Example 3.5.7

Fish Vertebrae Consider the distribution of vertebrae given in Table 3.5.1. In Example 3.5.5 we found that the mean of Y is $\mu_Y = 21.49$. The variance of Y is

$$\begin{aligned}\text{VAR}(Y) = \sigma_Y^2 &= (20 - 21.49)^2 \times \Pr\{Y = 20\}\\ &+ (21 - 21.49)^2 \times \Pr\{Y = 21\}\\ &+ (22 - 21.49)^2 \times \Pr\{Y = 22\}\\ &+ (23 - 21.49)^2 \times \Pr\{Y = 23\}\\ &= (-1.49)^2 \times 0.03 + (-.49)^2 \times 0.51\\ &+ (0.51)^2 \times 0.40 + (1.51)^2 \times 0.06\\ &= 2.2201 \times 0.03 + .2401 \times 0.51 + .2601 \times 0.40 + 2.2801 \times 0.06\\ &= 0.066603 + 0.122451 + 0.10404 + 0.136806\\ &= 0.4299.\end{aligned}$$

The standard deviation of Y is $\sigma_Y = \sqrt{0.4299} \approx 0.6557$.

■

Example 3.5.8

Dice In Example 3.5.6 we found that the mean number obtained from rolling a fair die is 3.5 (i.e., $\mu_Y = 3.5$). The variance of the number obtained from rolling a fair die is

$$\begin{aligned}\sigma_Y^2 &= (1 - 3.5)^2 \times \Pr\{Y = 1\} + (2 - 3.5)^2 \times \Pr\{Y = 2\} \\ &\quad + (3 - 3.5)^2 \times \Pr\{Y = 3\} + (4 - 3.5)^2 \times \Pr\{Y = 4\} \\ &\quad + (5 - 3.5)^2 \times \Pr\{Y = 5\} + (6 - 3.5)^2 \times \Pr\{Y = 6\} \\ &= (-2.5)^2 \times \frac{1}{6} + (-1.5)^2 \times \frac{1}{6} + (-0.5)^2 \times \frac{1}{6} + (0.5)^2 \times \frac{1}{6} \\ &\quad + (1.5)^2 \times \frac{1}{6} + (2.5)^2 \times \frac{1}{6} \\ &= (6.25) \times \frac{1}{6} + (2.25) \times \frac{1}{6} + (0.25) \times \frac{1}{6} + (0.25) \times \frac{1}{6} \\ &\quad + (2.25) \times \frac{1}{6} + (6.25) \times \frac{1}{6} \\ &= 17.5 \times \frac{1}{6} \\ &\approx 2.9167.\end{aligned}$$

The standard deviation of Y is $\sigma_Y = \sqrt{2.9167} \approx 1.708$. ■

The preceding definitions are appropriate for discrete random variables. There are analogous definitions for continuous random variables, but they involve integral calculus and won't be presented here.

ADDING AND SUBTRACTING RANDOM VARIABLES (OPTIONAL)

If we add two random variables, it makes sense that we add their means. Likewise, if we create a new random variable by subtracting two random variables, then we subtract the individual means to get the mean of the new random variable. If we multiply a random variable by a constant (e.g., if we are converting feet to inches so that we are multiplying by 12), then we multiply the mean of the random variable by the same constant. If we add a constant to a random variable, then we add that constant to the mean.

The following rules summarize the situation:

Rules for Means of Random Variables

Rule (1) If X and Y are two random variables, then $\mu_{X+Y} = \mu_X + \mu_Y$.

$$\mu_{X-Y} = \mu_X - \mu_Y$$

Rule (2) If Y is a random variable and a and b constants, then $\mu_{a+bY} = a + b\mu_Y$.

Example 3.5.9

Temperature The average summer temperature, μ_Y, in a city is 81°F. To convert °F to °C, we use the formula °C = (°F − 32) × (5/9) or °C = (5/9) × °F − (5/9) × 32. Thus, the mean in degrees Celsius is $(5/9) \times (81) - (5/9) \times 32 = 45 - 17.78 = 27.22$. ■

Dealing with standard deviations of functions of random variables is a bit more complicated. We work with the variance first and then take the square root, at the end, to get the standard deviation we want. If we *multiply* a random variable by a constant (e.g., if we are converting inches to centimeters by multiplying by 2.54), then we multiply the variance by the square of the constant. This has the effect of multiplying the standard deviation by the absolute value of the constant. If we *add* a constant to a random variable, then we are not changing the relative spread of the distribution, so the variance does not change.

Example 3.5.10 **Feet to Inches** Let Y denote the height, in feet, of a person in a given population; suppose the standard deviation of Y is $\sigma_Y = 0.35$ (feet). If we wish to convert from feet to inches, we can define a new variable X as $X = 12Y$. The variance of Y is 0.35^2 (the square of the standard deviation). The variance of X is $12^2 \times 0.35^2$, which means that the standard deviation of X is $\sigma_X = 12 \times 0.35 = 4.2$ (inches). ■

If we add two random variables *that are independent of one another*, then we add their variances.* Moreover, if we subtract two random variables *that are independent of one another*, then we *add* their variances. If we want to find the standard deviation of the sum (or difference) of two independent random variables, we first find the variance of the sum (or difference) and then take its square root.

Example 3.5.11 **Mass** Consider finding the mass of a 10-ml graduated cylinder. If several measurements are made, using an analytical balance, then in theory we would expect the measurements to all be the same. In reality, however, the readings will vary from one measurement to the next. Suppose that a given balance produces readings that have a standard deviation of 0.03g; let X denote the value of a reading made using this balance. Suppose that a second balance produces readings that have a standard deviation of 0.04g; let Y denote denote the value of a reading made using this second balance.[11]

If we use each balance to measure the mass of a graduated cylinder, we might be interested in the difference, $X - Y$, of the two measurements. The standard deviation of $X - Y$ is positive. To find the standard deviation of $X - Y$, we first find the variance of the difference. The variance of X is 0.03^2 and the variance of Y is 0.04^2. The variance of the difference is $0.03^2 + 0.04^2 = 0.0025$. The standard deviation of $X - Y$ is the square root of 0.0025, which is 0.05. ■

The following rules summarize the situation for variances:

Rules for Variances of Random Variables

Rule (3) If Y is a random variable and a and b constants, then $\sigma^2_{a+bY} = b^2\sigma^2_Y$.

Rule (4) If X and Y are two *independent* random variables, then

$$\sigma^2_{X+Y} = \sigma^2_X + \sigma^2_Y$$
$$\sigma^2_{X-Y} = \sigma^2_X + \sigma^2_Y$$

*If we add two random variables that are not independent of one another, then the variance of the sum depends on the degree of dependence between the variables. To take an extreme case, suppose that one of the random variables is the negative of the other. Then the sum of the two random variables will always be zero, so the variance of the sum will be zero. This is quite different from what we would get by adding the two variances together. As another example, suppose Y is the number of questions correct on a 20-question exam and X is the number of questions wrong. Then $Y + X$ is always equal to 20, so there is no variability at all. Hence, the variance of $Y + X$ is zero, even though the variance of Y is positive, as is the variance of X.

Exercises 3.5.1–3.5.10

3.5.1 In a certain population of the European starling, there are 5,000 nests with young. The distribution of brood size (number of young in a nest) is given in the accompanying table.[12]

Brood Size	Frequency (No. of Broods)
1	90
2	230
3	610
4	1,400
5	1,760
6	750
7	130
8	26
9	3
10	1
Total	5,000

Suppose one of the 5,000 broods is to be chosen at random, and let Y be the size of the chosen brood. Find

(a) $\Pr\{Y = 3\}$

(b) $\Pr\{Y \geq 7\}$

(c) $\Pr\{4 \leq Y \leq 6\}$

3.5.2 In the starling population of Exercise 3.5.1, there are 22,435 young in all the broods taken together. (There are 90 young from broods of size 1, there are 460 from broods of size 2, etc.) Suppose one of the young is to be chosen at random, and let Y' be the size of the chosen individual's brood.

(a) Find $\Pr\{Y' = 3\}$.

(b) Find $\Pr\{Y' \geq 7\}$.

(c) Explain why choosing a young at random and then observing its brood is not equivalent to choosing a brood at random. Your explanation should show why the answer to part (b) is greater than the answer to part (b) of Exercise 3.5.1.

3.5.3 Calculate the mean, μ_Y, of the random variable Y from Exercise 3.5.1.

3.5.4 Consider a population of the fruitfly *Drosophila melanogaster* in which 30% of the individuals are black because of a mutation, while 70% of the individuals have the normal gray body color. Suppose three flies are chosen at random from the population; let Y denote the number of black flies out of the three. Then the probability distribution for Y is given by the following table:

Y (No. Black)	Probability
0	0.343
1	0.441
2	0.189
3	0.027
Total	1.000

(a) Find $\Pr\{Y \geq 2\}$

(b) Find $\Pr\{Y \leq 2\}$

3.5.5 Calculate the mean, μ_Y, of the random variable Y from Exercise 3.5.4.

3.5.6 Calculate the standard deviation, σ_Y, of the random variable Y from Exercise 3.5.4.

3.5.7 The prevalence of mild myopia (nearsightedness) in adults over age 40 is 25% in the U.S.[13] Suppose four adults over age 40 are chosen at random from the population; let Y denote the number with myopia out of the four. Then the probability distribution for Y is given by the following table:

Y (No. Myopic)	Probability
0	0.316
1	0.422
2	0.211
3	0.047
4	0.004
Total	1.000

(a) Find $\Pr\{Y \geq 3\}$

(b) Find $\Pr\{Y \leq 1\}$

(c) Find $\Pr\{Y \geq 1\}$

3.5.8 Calculate the mean, μ_Y, of the random variable Y from Exercise 3.5.7.

3.5.9 A group of college students were surveyed to learn how many times they had visited a dentist in the previous year.[14] The probability distribution for Y, the number of visits, is given by the following table:

Y (No. Visits)	Probability
0	0.15
1	0.50
2	0.35
Total	1.00

Calculate the mean, μ_Y, of the number of visits.

3.5.10 Calculate the standard deviation, σ_Y, of the random variable Y from Exercise 3.5.9.

3.6 The Binomial Distribution

To add some depth to the notion of probability and random variables, we now consider a special type of random variable, the **binomial**. The distribution of a binomial random variable is a probability distribution associated with a special kind of chance operation. The chance operation is defined in terms of a set of conditions called the independent-trials model.

THE INDEPENDENT-TRIALS MODEL

The **independent-trials model** relates to a sequence of chance "trials." Each trial is assumed to have two possible outcomes, which are arbitrarily labeled "success" and "failure." The probability of success on each individual trial is denoted by the letter p and is assumed to be constant from one trial to the next. In addition, the trials are required to be independent, which means that the chance of success or failure on each trial does not depend on the outcome of any other trial. The total number of trials is denoted by n. These conditions are summarized in the following definition of the model.

Independent-Trials Model

A series of n independent trials is conducted. Each trial results in success or failure. The probability of success is equal to the same quantity, p, for each trial, regardless of the outcomes of the other trials.

The following examples illustrate situations that can be described by the independent-trials model.

Example 3.6.1 **Albinism** If two carriers of the gene for albinism marry, each of their children has probability $1/4$ of being albino. The chance that the second child is albino is the same $(1/4)$ whether or not the first child is albino; similarly, the outcome for the third child is independent of the first two, and so on. Using the labels "success" for albino and "failure" for nonalbino, the independent-trials model applies with $p = 1/4$ and $n =$ the number of children in the family. ■

Example 3.6.2 **Mutant Cats** A study of cats in Omaha, Nebraska, found that 37% of them have a certain mutant trait.[15] Suppose that 37% of all cats have this mutant trait and that a random sample of cats is chosen from the population. As each cat is chosen for the sample, the probability is 0.37 that it will be mutant. This probability is the same as each cat is chosen, regardless of the results of the other cats, because the percentage of mutants in the large population remains equal to 0.37 even when a few individual cats have been removed. Using the labels "success" for mutant and "failure" for nonmutant, the independent-trials model applies with $p = 0.37$ and $n =$ the sample size. ■

AN EXAMPLE OF THE BINOMIAL DISTRIBUTION

The binomial distribution specifies the probabilities of various numbers of successes and failures when the basic chance operation consists of n independent trials. Before giving the general formula for the binomial distribution, we consider a simple example.

Example 3.6.3 **Albinism** Suppose two carriers of the gene for albinism marry (see Example 3.6.1) and have two children. Then the probability that both of their children are albino is

$$\Pr\{\text{both children are albino}\} = \left(\frac{1}{4}\right)\left(\frac{1}{4}\right) = \frac{1}{16}$$

The reason for this probability can be seen by considering the relative frequency interpretation of probability. Of a great many such families with two children, $\frac{1}{4}$ would have the first child albino; furthermore, $\frac{1}{4}$ *of these* would have the second child albino; thus, $\frac{1}{4}$ of $\frac{1}{4}$, or $\frac{1}{16}$ of all the couples would have both albino children. A similar kind of reasoning shows that the probability that both children are not albino is

$$\Pr\{\text{both children are not albino}\} = \left(\frac{3}{4}\right)\left(\frac{3}{4}\right) = \frac{9}{16}$$

A new twist enters if we consider the probability that one child is albino and the other is not. There are two possible ways this can happen:

$$\Pr\{\text{first child is albino, second is not}\} = \left(\frac{1}{4}\right)\left(\frac{3}{4}\right) = \frac{3}{16}$$

$$\Pr\{\text{first child is not albino, second is}\} = \left(\frac{3}{4}\right)\left(\frac{1}{4}\right) = \frac{3}{16}$$

To see how to combine these possibilities, we again consider the relative frequency interpretation of probability. Of a great many such families with two children, the fraction of families with one albino and one nonalbino child would be the total of the two possibilities, or

$$\left(\frac{3}{16}\right) + \left(\frac{3}{16}\right) = \frac{6}{16}$$

Thus, the corresponding probability is

$$\Pr\{\text{one child is albino, the other is not}\} = \frac{6}{16}$$

Another way to see this is to consider a probability tree. The first split in the tree represents the birth of the first child; the second split represents the birth of the second child. The four possible outcomes and their associated probabilities are shown in Figure 3.6.1. These probabilities are collected in Table 3.6.1. ■

The probability distribution in Table 3.6.1 is called the binomial distribution with $p = \frac{1}{4}$ and $n = 2$. Note that the probabilities add to 1. This makes sense because all possibilities have been accounted for: We expect $\frac{9}{16}$ of the families to have no albino children, $\frac{6}{16}$ to have one albino child, and $\frac{1}{16}$ to have two albino children; there are no other possible compositions for a two-child family. The number of albino children, out of the two children, is an example of a binomial random variable. A **binomial random variable** is a random variable that satisfies the following four conditions, abbreviated as **BInS**:

Binary outcomes: There are two possible outcomes for each trial (success and failure).

Independent trials: The outcomes of the trials are independent of each other.

n is fixed: The number of trials, n, is fixed in advance.

Same value of p: The probability of a success on a single trial is the same for all trials.

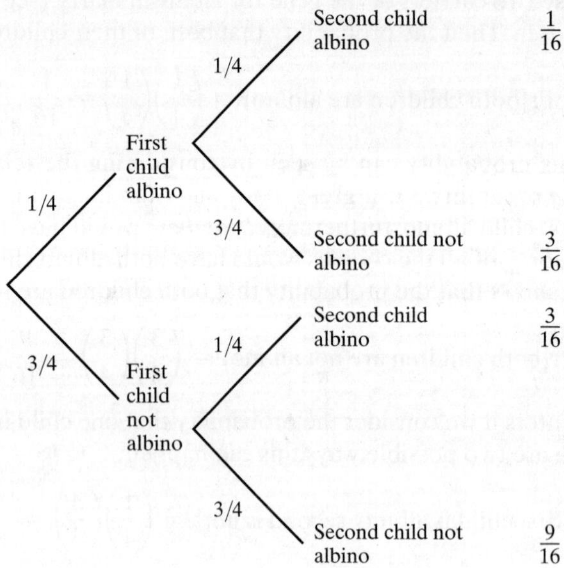

Figure 3.6.1 Probability tree for albinism among two children of carriers of the gene for albinism

Table 3.6.1 Probability distribution for number of albino children

Number of		
Albino	Nonalbino	Probability
0	2	$\frac{9}{16}$
1	1	$\frac{6}{16}$
2	0	$\frac{1}{16}$
	Total	1

THE BINOMIAL DISTRIBUTION FORMULA

A general formula is available that can be used to calculate probabilities associated with a binomial random variable for any values of n and p. This formula can be proved using logic similar to that in Example 3.6.3. (The formula is discussed further in Appendix 3.1.) The formula is given in the accompanying box.

The Binomial Distribution Formula

For a binomial random variable Y, the probability that the n trials result in j successes (and $n - j$ failures) is given by the following formula:

$$\Pr\{j \text{ successes}\} = \Pr\{Y = j\} = {}_nC_j p^j(1 - p)^{n-j}$$

The quantity ${}_nC_j$ appearing in the formula is called a **binomial coefficient**. Each binomial coefficient is an integer depending on n and on j. Values of binomial coefficients are given in Table 2 at the end of this book and can be found by the formula

$${}_nC_j = \frac{n!}{j!(n - j)!}$$

where $x!$ ("x-factorial") is defined for any positive integer x as

$$x! = x(x - 1)(x - 2) \ldots (2)(1)$$

and $0! = 1$. For more details, see Appendix 3.1.

For example, for $n = 5$ the binomial coefficients are as follows:

j:	0	1	2	3	4	5
${}_5C_j$:	1	5	10	10	5	1

Thus, for $n = 5$ the binomial probabilities are as indicated in Table 3.6.2. Notice the pattern in Table 3.6.2: The powers of p ascend (0, 1, 2, 3, 4, 5) and the powers of (1 −

Table 3.6.2 Binomial probabilities for $n = 5$

Number of		
Successes j	Failures $n - j$	Probability
0	5	$1p^0(1-p)^5$
1	4	$5p^1(1-p)^4$
2	3	$10p^2(1-p)^3$
3	2	$10p^3(1-p)^2$
4	1	$5p^4(1-p)^1$
5	0	$1p^5(1-p)^0$

p) descend (5, 4, 3, 2, 1, 0). (In using the binomial distribution formula, remember that $x^0 = 1$ for any nonzero x.)

Notes on Table 2 The following features in Table 2 are worth noting:

(a) The first and last entries in each row are equal to 1. This will be true for any row; that is, ${}_nC_0 = 1$ and ${}_nC_n = 1$ for any value of n.

(b) Each row of the table is symmetric; that is ${}_nC_j$ and ${}_nC_{n-j}$ are equal.

(c) The bottom rows of the table are left incomplete to save space, but you can easily complete them using the symmetry of the ${}_nC_j$'s; if you need to know ${}_nC_j$, you can look up ${}_nC_{n-j}$ in Table 2. For instance, consider $n = 18$; if you want to know ${}_{18}C_{15}$, you just look up ${}_{18}C_3$; both ${}_{18}C_3$ and ${}_{18}C_{15}$ are equal to 816.

The following example shows a specific application of the binomial distribution with $n = 5$.

Example 3.6.4

Mutant Cats Suppose we draw a random sample of five individuals from a large population in which 37% of the individuals are mutants (as in Example 3.6.2). The probabilities of the various possible samples are then given by the binomial distribution formula with $n = 5$ and $p = 0.37$; the results are displayed in Table 3.6.3. For instance, the probability of a sample containing two mutants and three nonmutants is

$$10(0.37)^2(0.63)^3 \approx 0.34$$

Table 3.6.3 Binomial distribution with $n = 5$ and $p = 0.37$

Number of		
Mutants	Nonmutants	Probability
0	5	0.10
1	4	0.29
2	3	0.34
3	2	0.20
4	1	0.06
5	0	0.01
	Total	1.00

Thus, $\Pr\{Y = 2\} \approx 0.34$. This means that about 34% of random samples of size 5 will contain two mutants and three nonmutants.

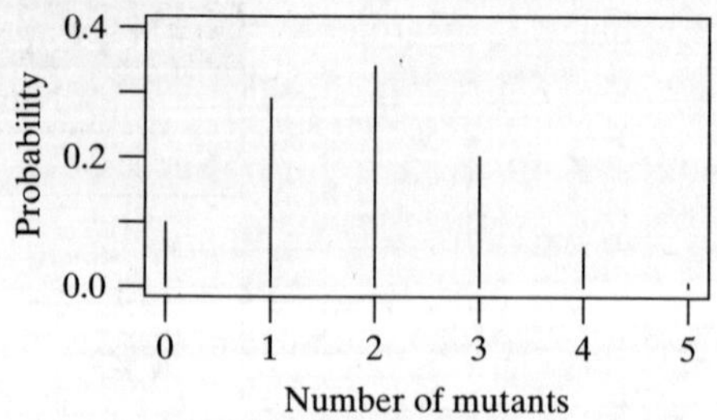

Figure 3.6.2 Binomial distribution with $n = 5$ and $p = 0.37$

Notice that the probabilities in Table 3.6.3 add to 1. The probabilities in a probability distribution must always add to 1, because they account for 100% of the possibilities. ■

The binomial distribution of Table 3.6.3 is pictured graphically in Figure 3.6.2. The spikes in the graph emphasize that the probability distribution is discrete.

Remark In applying the independent-trials model and the binomial distribution, we assign the labels "success" and "failure" arbitrarily. For instance, in Example 3.6.4, we could say "success" = "mutant" and $p = 0.37$; or, alternatively, we could say "success" = "nonmutant" and $p = 0.63$. Either assignment of labels is all right; it is only necessary to be consistent.

Computational Note Computer and calculator technology makes it fairly easy to handle the binomial distribution formula for small or moderate values of n. For large values of n, the use of the binomial formula gets to be tedious and even a computer will balk at being asked to calculate a binomial probability. However, the binomial formula can be approximated by other methods. One of these will be discussed in the optional Section 5.4.

Sometimes a binomial probability question involves combining two or more possible outcomes. The following example illustrates this idea.

Example 3.6.5 **Sampling Fruitflies** In a large *Drosophila* population, 30% of the flies are black (B) and 70% are gray (G). Suppose two flies are randomly chosen from the population (as in Example 3.2.3). The binomial distribution with $n = 2$ and $p = 0.3$ gives probabilities for the possible outcomes as shown in Table 3.6.4. (Using the binomial formula agrees with the results given by the probability tree shown in Figure 3.2.3.)

Table 3.6.4

Sample composition	Y	Probability
Both Gray	0	0.49
One Black, one Gray	1	0.42
Both Black	2	0.09
	Total	1.00

Let E be the event that both flies are the same color. Then E can happen in two ways: Both flies are gray or both are black. To find the probability of E, consider what would happen if we repeated the sampling procedure many times: Forty-nine percent of the samples would have both flies gray, and 9% would have both flies black. Consequently, the percentage of samples with both flies the same color would be 49% + 9% = 58%. Thus, we have shown that the probability of E is

$$\Pr\{E\} = 0.58$$

as we claimed in Example 3.2.3. ■

Whenever an event E can happen in two or more mutually exclusive ways, a rationale such as that of Example 3.6.5 can be used to find $\Pr\{E\}$.

Example 3.6.6

Blood Type In the United States, 85% of the population has Rh positive blood. Suppose we take a random sample of 6 persons and count the number with Rh positive blood. The binomial model can be applied here, since the BInS conditions are met: There is a binary outcome on each trial (Rh positive or Rh negative blood), the trials are independent (due to the random sampling), n is fixed at 6, and the same probability of Rh positive blood applies to each person ($p = 0.85$).

Let Y denote the number of persons, out of 6, with Rh positive blood. The probabilities of the possible values of Y are given by the binomial distribution formula with $n = 6$ and $p = 0.85$; the results are displayed in Table 3.6.5. For instance, the probability that $Y = 4$ is

$${}_6C_4(0.85)^4(0.15)^2 \approx 15(0.522)(0.0225) \approx 0.1762$$

If we want to find the probability that at least 4 persons (out of the 6 sampled) will have Rh positive blood, we need to find $\Pr\{Y \geq 4\} = \Pr\{Y = 4\} + \Pr\{Y = 5\} + \Pr\{Y = 6\} = 0.1762 + 0.3993 + 0.3771 = 0.9526$. This means that the probability of getting at least 4 persons with Rh positive blood in a sample of size 6 is 0.9526. ■

Table 3.6.5 Binomial distribution with $n = 6$ and $p = 0.85$

Number of successes	Probability
0	<0.0001
1	0.0004
2	0.0055
3	0.0415
4	0.1762
5	0.3993
6	0.3771
Total	1.0000

In some problems, it is easier to find the probability that an event *does not happen* rather than finding the probability of the event happening. To solve such problems we use the fact that the probability of an event happening is 1 minus the probability that the event does not happen: $\Pr\{E\} = 1 - \Pr\{E \text{ does not happen}\}$. The following is an example.

Example 3.6.7

Blood Type As in Example 3.6.6, let Y denote the number of persons, out of 6, with Rh positive blood. Suppose we want to find the probability that Y is less than 6 (i.e., the probability that there is *at least 1* person in the sample who has Rh *negative* blood). We could find this directly as $\Pr\{Y = 0\} + \Pr\{Y = 1\} + \cdots + \Pr\{Y = 5\}$. However, it is easier to find $\Pr\{Y = 6\}$ and subtract this from 1:

$$\Pr\{Y < 6\} = 1 - \Pr\{Y = 6\} = 1 - 0.3771 = 0.6229.$$ ■

MEAN AND STANDARD DEVIATION OF A BINOMIAL

If we toss a fair coin 10 times, then we expect to get 5 heads, on average. This is an example of a general rule: *For a binomial random variable, the mean (i.e., the average number of successes) is equal to np*. This is an intuitive fact: The probability of success on each trial is p, so if we conduct n trials, then np is the expected number of successes. In Appendix 3.2 we show that this result is consistent with the rule given in Section 3.5 for finding the mean of the sum of random variables. *The standard deviation for a binomial random variable is given by* $\sqrt{np(1-p)}$. This formula is not intuitively clear; a derivation of the result is given in Appendix 3.2. For the example of tossing a coin 10 times, the standard deviation of the number of heads is

$$\sqrt{10 \times 0.5 \times 0.5} = \sqrt{2.5} \approx 1.58.$$

Example 3.6.8 **Blood Type** As discussed in Example 3.6.6, if Y denotes the number of persons with Rh positive blood in a sample of size 6, then a binomial model can be used to find probabilities associated with Y. The single most likely value of Y is 5 (which has probability 0.3993). The average value of Y is $6 \times 0.85 = 5.1$, which means that if we take many samples, each of size 6, and count the number of Rh positive persons in each sample, and then average those counts, we expect to get 5.1. The standard deviation of those counts is $\sqrt{6 \times 0.85 \times .015} \approx 0.87$. ■

APPLICABILITY OF THE BINOMIAL DISTRIBUTION

A number of statistical procedures are based on the binomial distribution. We will study some of these procedures in later chapters. Of course, the binomial distribution is applicable only in experiments where the BInS conditions are satisfied in the real biological situation. We briefly discuss some aspects of these conditions.

Application to Sampling The most important application of the independent-trials model and the binomial distribution is to describe random sampling from a population when the observed variable is dichotomous—that is, a categorical variable with two categories (e.g., black and gray in Example 3.6.5). This application is valid if the sample size is a negligible fraction of the population size so that the population composition is not altered appreciably by the removal of the individuals in the sample (so that the S part of BInS is satisfied: The probability of a success remains the same from trial to trial). However, if the sample is *not* a negligibly small part of the population, then the population composition may be altered by the sampling process so that the "trials" involved in composing the sample are not independent and the probability of a success changes as the sampling progresses. In this case, the probabilities given by the binomial formula are not correct. In most biological studies, the population is so large that this kind of difficulty does not arise.

Contagion In some applications the phenomenon of contagion can invalidate the condition of independence between trials. The following is an example.

Example 3.6.9 **Chickenpox** Consider the occurrence of chickenpox in children. Each child in a family can be categorized according to whether he had chickenpox during a certain year. One can say that each child constitutes a "trial" and that "success" is having chickenpox during the year, but the trials are *not* independent because the chance of a particular child catching chickenpox depends on whether his sibling caught chickenpox. As a specific example, consider a family with five children, and suppose that the chance of an individual child catching chickenpox during the year is equal to 0.10. The binomial distribution gives the chance of all five children getting chickenpox as

$$\Pr\{5 \text{ children get chickenpox}\} = (0.10)^5 = 0.00001$$

However, this answer is not correct; because of contagion, the correct probability would be much larger. There would be many families in which one child caught chickenpox and then the other four children got chickenpox from the first child, so all five children would get chickenpox. ■

Exercises 3.6.1–3.6.12

3.6.1 The seeds of the garden pea *(Pisum sativum)* are either yellow or green. A certain cross between pea plants produces progeny in the ratio 3 yellow : 1 green.[16] If four randomly chosen progeny of such a cross are examined, what is the probability that

(a) three are yellow and one is green?

(b) all four are yellow?

(c) all four are the same color?

3.6.2 In Australia, 16% of the adult population is nearsighted.[17] If three Australians are chosen at random, what is the probability that

(a) two are nearsighted and one is not?

(b) exactly one is nearsighted?

(c) at most one is nearsighted?

(d) none of them are nearsighted?

3.6.3 In the United States, 44% of the population has type A blood. Consider taking a sample of size 4. Let Y denote the number of persons in the sample with type A blood. Find

(a) $\Pr\{Y = 0\}$.

(b) $\Pr\{Y = 1\}$.

(c) $\Pr\{Y = 2\}$.

(d) $\Pr\{0 \le Y \le 2\}$.

(e) $\Pr\{0 < Y \le 2\}$.

3.6.4 A certain drug treatment cures 90% of cases of hookworm in children.[18] Suppose that 20 children suffering from hookworm are to be treated, and that the children can be regarded as a random sample from the population. Find the probability that

(a) all 20 will be cured.

(b) all but 1 will be cured.

(c) exactly 18 will be cured.

(d) exactly 90% will be cured.

3.6.5 The shell of the land snail *Limocolaria martensiana* has two possible color forms: streaked and pallid. In a certain population of these snails, 60% of the individuals have streaked shells.[19] Suppose that a random sample of 10 snails is to be chosen from this population. Find the probability that the percentage of streaked-shelled snails in the *sample* will be

(a) 50%. (b) 60%. (c) 70%.

3.6.6 Consider taking a sample of size 10 from the snail population in Exercise 3.6.5.

(a) What is the mean number of streaked-shelled snails?

(b) What is the standard deviation of the number of streaked-shelled snails?

3.6.7 In Europe, 8% of men are colorblind.[20] Consider taking repeated samples of 20 European men.

(a) What is the mean number of colorblind men?

(b) What is the standard deviation of the number of colorblind men?

3.6.8 The sex ratio of newborn human infants is about 105 males : 100 females.[21] If four infants are chosen at random, what is the probability that

(a) two are male and two are female?

(b) all four are male?

(c) all four are the same sex?

3.6.9 Construct a binomial setting (different from any examples presented in this book) and a problem for which the following is the answer: ${}_7C_3(0.8)^3(0.2)^4$.

3.6.10 Neuroblastoma is a rare, serious, but treatable disease. A urine test, the VMA test, has been developed that gives a positive diagnosis in about 70% of cases of neuroblastoma.[22] It has been proposed that this test be used for large-scale screening of children. Assume that 300,000 children are to be tested, of whom 8 have the disease. We are interested in whether or not the test detects the disease in the 8 children who have the disease. Find the probability that

(a) all eight cases will be detected.

(b) only one case will be missed.

(c) two or more cases will be missed. [*Hint:* Use parts (a) and (b) to answer part (c).]

3.6.11 If two carriers of the gene for albinism marry, each of their children has probability $\frac{1}{4}$ of being albino (see

Example 3.6.1). If such a couple has six children, what is the probability that

(a) none will be albino?

(b) at least one will be albino? [*Hint:* Use part (a) to answer part (b); note that "at least one" means "one or more."]

3.6.12 Childhood lead poisoning is a public health concern in the United States. In a certain population, 1 child in 8 has a high blood lead level (defined as 30 μg/dl or more).[23] In a randomly chosen group of 16 children from the population, what is the probability that

(a) none has high blood lead?

(b) 1 has high blood lead?

(c) 2 have high blood lead?

(d) 3 or more have high blood lead? [*Hint:* Use parts (a)–(c) to answer part (d).]

3.7 Fitting a Binomial Distribution to Data (Optional)

Occasionally it is possible to obtain data that permit a direct check of the applicability of the binomial distribution. One such case is described in the next example.

Example 3.7.1 **Sexes of Children** In a classic study of the human sex ratio, families were categorized according to the sexes of the children. The data were collected in Germany in the nineteenth century, when large families were common. Table 3.7.1 shows the results for 6,115 families with 12 children.[24]

Table 3.7.1 Sex ratios in 6,115 families with 12 children

Number of Boys	Number of Girls	Observed frequency (number of families)
0	12	3
1	11	24
2	10	104
3	9	286
4	8	670
5	7	1,033
6	6	1,343
7	5	1,112
8	4	829
9	3	478
10	2	181
11	1	45
12	0	7
	Total	6,115

It is interesting to consider whether the observed variation among families can be explained by the independent-trials model. We will explore this question by fitting a binomial distribution to the data.

The first step in fitting the binomial distribution is to determine a value for $p = \Pr\{\text{boy}\}$. One possibility would be to assume that $p = 0.50$. However, since it is known that the human sex ratio at birth is not exactly 1 : 1 (in fact, it favors boys slightly), we will not make this assumption. Rather, we will "fit" p to the data; that is, we will determine a value for p that fits the data best. We observe that the total number of children in all the families is

$$(12)(6{,}115) = 73{,}380 \text{ children}$$

Among these children, the number of boys is

$$(3)(0) + (24)(1) + \cdots + (7)(12) = 38{,}100 \text{ boys}$$

Therefore, the value of p that fits the data best is

$$p = \frac{38{,}100}{73{,}380} = 0.519215$$

The next step is to compute probabilities from the binomial distribution formula with $n = 12$ and $p = 0.519215$. For instance, the probability of 3 boys and 9 girls is computed as

$$\begin{aligned} {}_{12}C_3(p)^3(1-p)^9 &= 220(0.519215)^3(0.480785)^9 \\ &\approx 0.042269 \end{aligned}$$

For comparison with the observed data, we convert each probability to a theoretical or "expected" frequency by multiplying by 6,115 (the total number of families). For instance, the expected number of families with 3 boys and 9 girls is

$$(6{,}115)(0.042269) \approx 258.5$$

The expected and observed frequencies are displayed together in Table 3.7.2. Table 3.7.2 shows reasonable agreement between the observed frequencies and the predictions of the binomial distribution. But a closer look reveals that the discrepancies, although not large, follow a definite pattern. The data contain more unisexual, or preponderantly unisexual, sibships than expected. In fact, the observed frequencies are higher than the expected frequencies for nine types of families in which one sex or the other predominates, while the observed frequencies are lower than the expected frequencies for four types of more "balanced" families. This pattern is

Table 3.7.2 Sex-ratio data and binomial expected frequencies

Number of				
Boys	Girls	Observed frequency	Expected frequency	Sign of (Obs. − Exp.)
0	12	3	0.9	+
1	11	24	12.1	+
2	10	104	71.8	+
3	9	286	258.5	+
4	8	670	628.1	+
5	7	1,033	1,085.2	−
6	6	1,343	1,367.3	−
7	5	1,112	1,265.6	−
8	4	829	854.2	−
9	3	478	410.0	+
10	2	181	132.8	+
11	1	45	26.1	+
12	0	7	2.3	+
		Total 6,115	6,115.0	

clearly revealed by the last column of Table 3.7.2, which shows the sign of the difference between the observed frequency and the expected frequency. Thus, the observed distribution of sex ratios has heavier "tails" and a lighter "middle" than the best-fitting binomial distribution.

The systematic pattern of deviations from the binomial distribution suggests that the observed variation among families cannot be entirely explained by the independent-trials model.* What factors might account for the discrepancy? This intriguing question has stimulated several researchers to undertake more detailed analysis of these data. We briefly discuss some of the issues.

One explanation for the excess of predominantly unisexual families is that the probability of producing a boy may vary among families. If p varies from one family to another, then sex will appear to "run" in families in the sense that the number of predominantly unisexual families will be inflated. In order to clearly visualize this effect, consider the fictitious data set shown in Table 3.7.3.

Table 3.7.3 Fictitious sex-ratio data and binomial expected frequencies

Number of		Observed	Expected	Sign of
Boys	Girls	frequency	frequency	(Obs. − Exp.)
0	12	2,940	0.9	+
1	11	0	12.1	−
2	10	0	71.8	−
3	9	0	258.5	−
4	8	0	628.1	−
5	7	0	1,085.2	−
6	6	0	1,367.3	−
7	5	0	1,265.6	−
8	4	0	854.3	−
9	3	0	410.0	−
10	2	0	132.8	−
11	1	0	26.1	−
12	0	3,175	2.3	+
	Total	6,115	6,115.0	

In the fictitious data set, there are (3,175)(12) = 38,100 males among 73,380 children, just as there are in the real data set. Consequently, the best-fitting p is the same ($p = 0.519215$) and the expected binomial frequencies are the same as in Table 3.7.2. The fictitious data set contains only unisexual sibships and so is an extreme example of sex "running" in families. The real data set exhibits the same phenomenon more weakly. One explanation of the fictitious data set would be that some families can have only boys ($p = 1$) and other families can have only girls ($p = 0$). In a parallel way, one explanation of the real data set would be that p varies slightly among families. Variation in p is biologically plausible, even though a mechanism causing the variation has not been discovered.

An alternative explanation for the inflated number of sexually homogeneous families would be that the sexes of the children in a family are literally dependent on

*A chi-square goodness-of-fit test of the binomial model shows that there is strong evidence that the differences between the observed and expected frequencies did not happen due to chance error in the sampling process. We will explore the topic of goodness-of-fit tests in Chapter 9.

one another, in the sense that the determination of an individual child's sex is somehow influenced by the sexes of the previous children. This explanation is implausible on biological grounds because it is difficult to imagine how the biological system could "remember" the sexes of previous offspring. ■

Example 3.7.1 shows that poorness of fit to the independent-trials model can be biologically interesting. We should emphasize, however, that most statistical applications of the binomial distribution proceed from the assumption that the independent-trials model is applicable*. In a typical application, the data are regarded as resulting from a *single* set of n trials. Data such as the family sex-ratio data, which refer to *many* sets of $n = 12$ trials, are not often encountered.

*In Example 3.6.1 we asserted that occurrences of albinism among siblings are independent, which is consistent with current understandings of human genetics.

Exercises 3.7.1–3.7.3

3.7.1 The accompanying data on families with 6 children are taken from the same study as the families with 12 children in Example 3.7.1. Fit a binomial distribution to the data. (Round the expected frequencies to one decimal place.) Compare with the results in Example 3.7.1. What features do the two data sets share?

Number of		
Boys	Girls	Number of families
0	6	1,096
1	5	6,233
2	4	15,700
3	3	22,221
4	2	17,332
5	1	7,908
6	0	1,579
	Total	72,069

3.7.2 An important method for studying mutation-causing substances involves killing female mice 17 days after mating and examining their uteri for living and dead embryos. The classical method of analysis of such data assumes that the survival or death of each embryo constitutes an independent binomial trial. The accompanying table, which is extracted from a larger study, gives data for 310 females, all of whose uteri contained 9 embryos; all of the animals were treated alike (as controls).[25]

(a) Fit a binomial distribution to the observed data. (Round the expected frequencies to one decimal place.)

(b) Interpret the relationship between the observed and expected frequencies. Do the data cast suspicion on the classical assumption?

Number of embryos		
Dead	Living	Number of female mice
0	9	136
1	8	103
2	7	50
3	6	13
4	5	6
5	4	1
6	3	1
7	2	0
8	1	0
9	0	0
	Total	310

3.7.3 Students in a large botany class conducted an experiment on the germination of seeds of the Saguaro cactus. As part of the experiment, each student planted five seeds in a small cup, kept the cup near a window, and checked every day for germination (sprouting). The class results on the seventh day after planting were as displayed in the table.[26]

Number of seeds		
Germinated	Not germinated	Number of students
0	5	17
1	4	53
2	3	94
3	2	79
4	1	33
5	0	4
	Total	280

(a) Fit a binomial distribution to the data. (Round the expected frequencies to one decimal place.)

(b) Two students, Fran and Bob, were talking before class. All of Fran's seeds had germinated by the seventh day, whereas none of Bob's had. Bob wondered whether he had done something wrong. With the perspective gained from seeing all 280 students' results, what would you say to Bob? (*Hint*: Can the variation among the students be explained by the hypothesis that some of the seeds were good and some were poor, with each student receiving a randomly chosen five seeds?)

(c) Invent a fictitious set of data for 280 students, with the same overall percentage germination as the observed data given in the table, but with all the students getting either Fran's results (perfect) or Bob's results (nothing). How would your answer to Bob differ if the actual data had looked like this fictitious data set?

Supplementary Exercises 3.S.1–3.S.12

3.S.1 In the United States, 10% of adolescent girls have iron deficiency.[27] Suppose two adolescent girls are chosen at random. Find the probability that

(a) all three girls have iron deficiency.

(b) one girl has iron deficiency and the other two do not.

3.S.2 In preparation for an ecological study of centipedes, the floor of a beech woods is divided into a large number of 1-foot squares.[28] At a certain moment, the distribution of centipedes in the squares is as shown in the table.

Number of centipedes	Percent frequency (% of squares)
0	45
1	36
2	14
3	4
4	1
Total	100

Suppose that a square is chosen at random, and let Y be the number of centipedes in the chosen square. Find

(a) $\Pr\{Y = 2\}$

(b) $\Pr\{Y \geq 3\}$

3.S.3 Refer to the distribution of centipedes given in Exercise 3.S.2. Suppose four squares are chosen at random. Find the probability that two of the squares contain centipedes and two do not.

3.S.4 Refer to the distribution of centipedes given in Exercise 3.S.2. Suppose four squares are chosen at random. Find the expected value (i.e., the mean) of the number of squares that contain at least two centipedes.

3.S.5 Wavy hair in mice is a recessive genetic trait. If mice with wavy hair are mated with straight-haired (heterozygous) mice, each offspring has probability $\frac{1}{2}$ of having wavy hair.[29] Consider a large number of such matings, each producing a litter of five offspring. What percentage of the litters will consist of

(a) three wavy-haired and three straight-haired offspring?

(b) four or more straight-haired offspring?

(c) all the same type (either all wavy- or all straight-haired) offspring?

3.S.6 A certain drug causes liver damage in 5% of patients. Suppose the drug is to be tested on 60 patients. Find the probability that

(a) none of the patients will experience liver damage.

(b) one or more of the patients will experience liver damage. [*Hint*: Use part (a) to answer part (b).]

3.S.7 Refer to Exercise 3.S.6. Suppose now that the drug is to be tested on n patients, and let E represent the event that liver damage occurs in one or more of the patients. The probability $\Pr\{E\}$ is useful in establishing criteria for drug safety.

(a) Find $\Pr\{E\}$ for $n = 100$.

(b) How large must n be in order for $\Pr\{E\}$ to exceed 0.94?

3.S.8 To study people's ability to deceive lie detectors, researchers sometimes use the "guilty knowledge" technique.[30] Certain subjects memorize six common words; other subjects memorize no words. Each subject is then tested on a polygraph machine (lie detector), as follows. The experimenter reads, in random order, 24 words: the six "critical" words (the memorized list) and, for each critical word, three "control" words with similar or related meanings. If the subject has memorized the six words, he or she tries to conceal that fact. The subject is scored a "failure" on a critical word if his or her electrodermal response is higher on the critical word than on any of the three control words. Thus, on each of the six critical words, even an innocent subject would have a 25% chance of failing. Suppose a subject is labeled "guilty" if the subject fails on five or more of the six critical words. If an innocent subject is tested, what is the probability that he or she will be labeled "guilty"?

3.S.9 The density curve shown here represents the distribution of systolic blood pressures in a population of middle-aged men.[31] Areas under the curve are shown in the figure. Suppose a man is selected at random from the population, and let Y be his blood pressure. Find

(a) $\Pr\{140 < Y < 180\}$.

(b) $\Pr\{Y < 140\}$.

(c) $\Pr\{Y > 160\}$.

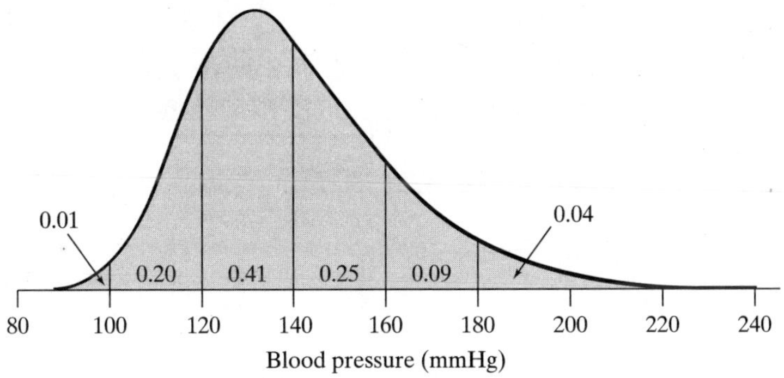

3.S.10 Refer to the blood pressure distribution of Exercise 3.S.9. Suppose four men are selected at random from the population. Find the probability that

(a) all four have blood pressures higher than 160 mm Hg.

(b) three have blood pressures higher than 160, and one has blood pressure 160 or less.

3.S.11 In the United States 9% of all people are left-handed.[32] If we take a random sample of six Americans what is the probability that

(a) none of them is left-handed?

(b) all six are left-handed?

(c) at least one is left-handed?

3.S.12 Refer to the information about left-handedness in Exercise 3.S.11. Consider taking repeated samples of 100 Americans.

(a) What is the mean number of left-handed persons?

(b) What is the standard deviation of the number of left-handed persons?

THE NORMAL DISTRIBUTION

Chapter 4

OBJECTIVES

In this chapter we will study the normal distribution, including

- the use of the normal curve in modeling distributions.
- finding probabilities using the normal curve.
- assessing normality of data sets with the use of normal quantile plots.

4.1 Introduction

In Chapter 2, we introduced the idea of regarding a set of data as a sample from a population. In Section 3.4 we saw that the population distribution of a quantitative variable Y can be described by its mean μ and its standard deviation σ and also by a density curve, which represents relative frequencies as areas under the curve. In this chapter we study the most important type of density curve: the **normal curve**. The normal curve is a symmetric "bell-shaped" curve whose exact form we will describe next. A distribution represented by a normal curve is called a **normal distribution.**

The family of normal distributions plays two roles in statistical applications. Its more straightforward use is as a convenient approximation to the distribution of an observed variable Y. The second, and more important, role of the normal distribution is more theoretical and will be explored in Chapter 5.

An example of a natural population distribution that can be approximated by a normal distribution follows.

Example 4.1.1

Serum Cholesterol The relationship between the concentration of cholesterol in the blood and the occurrence of heart disease has been the subject of much research. As part of a government health survey, researchers measured serum cholesterol levels for a large sample of Americans, including children. The distribution for children between 12 and 14 years of age can be fairly well approximated by a normal curve with mean $\mu = 155$ mg/dl and standard deviation $\sigma = 27$ mg/dl. Figure 4.1.1 shows a histogram based on a sample of 431 children between 12 and 14 years old, with the normal curve superimposed.[1] ■

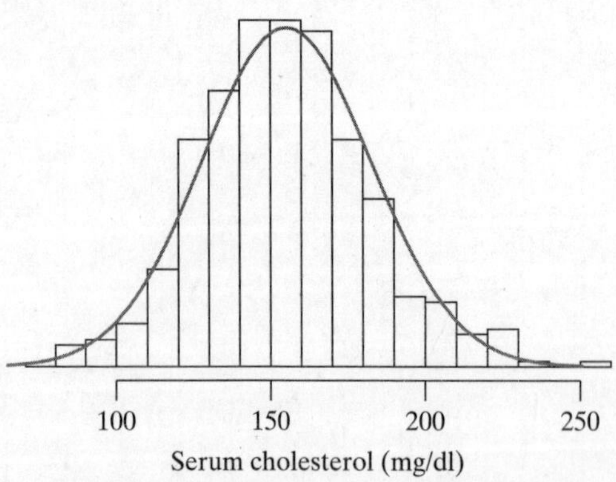

Figure 4.1.1 Distribution of serum cholesterol in 431 12- to 14-year-old children

To indicate how the mean μ and standard deviation σ relate to the normal curve, Figure 4.1.2 shows the normal curve for the serum cholesterol distribution of Example 4.1.1, with tick marks at 1, 2, and 3 standard deviations from the mean.

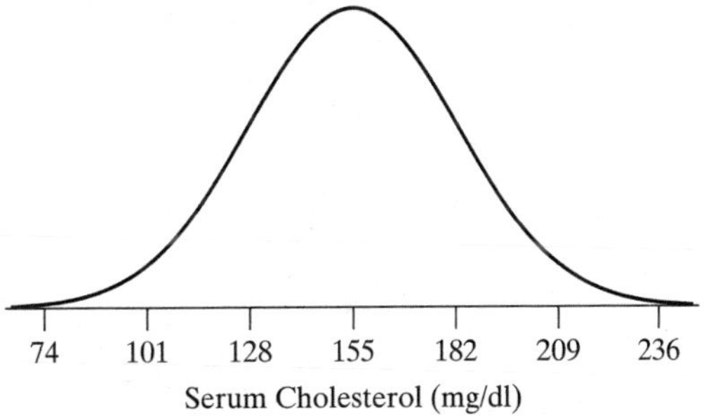

Figure 4.1.2 Normal distribution of serum cholesterol, with $\mu = 155$ mg/dl and $\sigma = 27$ mg/dl

The normal curve can be used to describe the distribution of an observed variable Y in two ways: (1) as a smooth approximation to a histogram based on a sample of Y values; and (2) as an idealized representation of the population distribution of Y. The normal curves in Figures 4.1.1 and 4.1.2 could be interpreted either way. For simplicity, in the remainder of this chapter we will consider the normal curve as representing a population distribution.

FURTHER EXAMPLES

We now give two more examples of normal curves that approximately describe real populations. In each figure, the horizontal axis is scaled with tick marks centered at the mean and one standard deviation apart.

Example 4.1.2

Blue Jay Bill Length Many biological measures of size follow normal curves. In one study of male Blue Jays in the Northeastern United States, the bill lengths were observed to follow approximately a normal distribution with mean $\mu = 25.4$ mm and standard deviation $\sigma = 0.8$ mm. This distribution is pictured in Figure 4.1.3.[2] ■

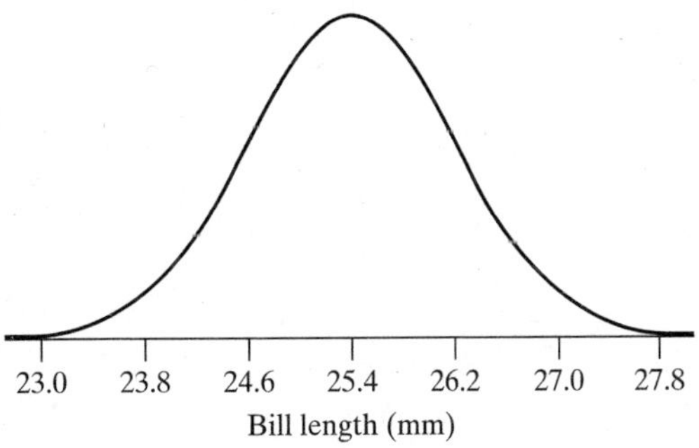

Figure 4.1.3 Normal distribution of male Blue Jay bill length, with $\mu = 25.4$ mm and $\sigma = 0.8$ mm

Example 4.1.3

Interspike Times in Nerve Cells In certain nerve cells, spontaneous electrical discharges are observed that are so rhythmically repetitive that they are called "clock-spikes." The timing of these spikes, even though remarkably regular, does exhibit variation. In one study, the interspike-time intervals (in milliseconds) for a single housefly *(Musca domestica)* were observed to follow approximately a normal distribution with mean $\mu = 15.6$ ms and standard deviation $\sigma = 0.4$ ms; this distribution is shown in Figure 4.1.4.[3] ■

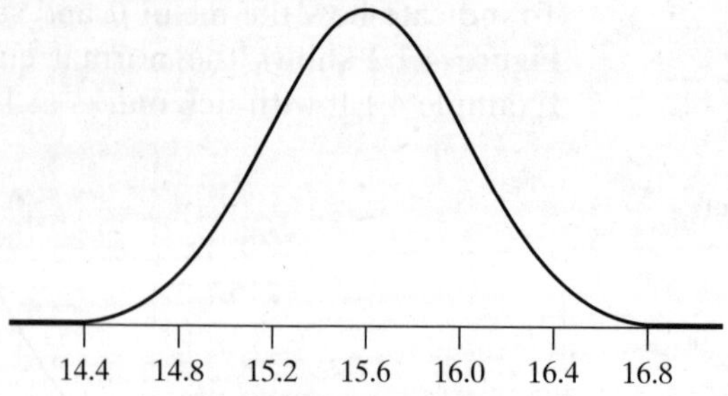

Figure 4.1.4 Normal distribution of interspike-time intervals, with $\mu = 15.6$ ms and $\sigma = 0.4$ ms

The preceding examples have illustrated very different kinds of populations. In Example 4.1.3, the entire population consists of measurements on only one fly. Still another type of population is a *measurement error* population, consisting of repeated measurements of exactly the same quantity. The deviation of an individual measurement from the "correct" value is called measurement error. Measurement error is not the result of a mistake but rather is due to lack of perfect precision in the measuring process or measuring instrument. Measurement error distributions are often approximately normal; in this case the mean of the distribution of repeated measurements of the same quantity is the true value of the quantity (assuming that the measuring instrument is correctly calibrated), and the standard deviation of the distribution indicates the precision of the instrument. One measurement error distribution was described in Example 2.2.12. The following is another example.

Example 4.1.4 **Measurement Error** When a certain electronic instrument is used for counting particles such as white blood cells, the measurement error distribution is approximately normal. For white blood cells, the standard deviation of repeated counts based on the same blood specimen is about 1.4% of the true count. Thus, if the true count of a certain blood specimen were 7,000 cells/mm^3, then the standard deviation would be about 100 cells/mm^3 and the distribution of repeated counts on that specimen would resemble Figure 4.1.5.[4] ■

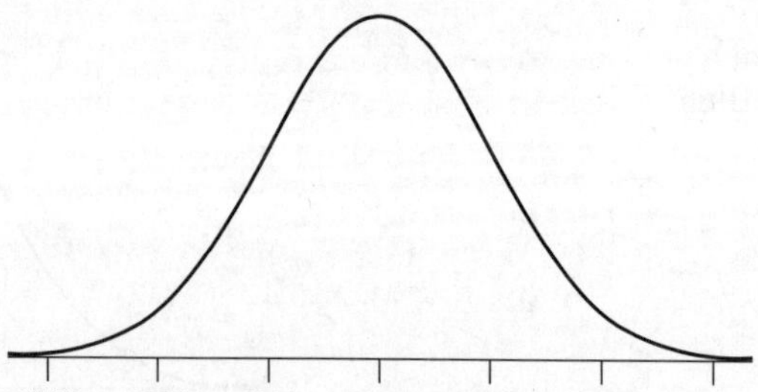

Figure 4.1.5 Normal distribution of repeated white blood cell counts of a blood specimen whose true value is $\mu = 7{,}000$ cells/mm^3. The standard deviation is $\sigma = 100$ cells/mm^3.

4.2 The Normal Curves

As the examples in Section 4.1 show, there are many normal curves; each particular normal curve is characterized by its mean and standard deviation. If a variable Y follows a normal distribution with mean μ and standard deviation σ, then it is common to write $Y \sim N(\mu, \sigma)$. All the normal curves can be described by a single formula. Even though we will not make any direct use of the formula in this book, we present it here, both as a matter of interest and also to emphasize that a normal curve is not just any symmetric curve, but rather a *specific* kind of symmetric curve.

If a variable Y follows a normal distribution with mean μ and standard deviation σ, then the density curve of the distribution of Y is given by the following formula:

$$f(y) = \frac{1}{\sigma\sqrt{2\pi}} e^{-\frac{1}{2}\left(\frac{y-\mu}{\sigma}\right)^2}$$

This function, $f(y)$, is called the *density function* of the distribution and expresses the height of the curve as a function of the position along the horizontal axis. The quantities e and π that appear in the formula are constants, with e approximately equal to 2.71 and π approximately equal to 3.14.

Figure 4.2.1 shows a graph of a normal curve. The shape of the curve is like a symmetric bell, centered at $y = \mu$. The direction of curvature is downward (like an inverted bowl) in the central portion of the curve, and upward in the tail portions. The points of inflection (i.e., where the curvature changes direction) are $y = \mu - \sigma$ and $y = \mu + \sigma$. In principle the curve extends to $+\infty$ and $-\infty$, never actually reaching the y-axis; however, the height of the curve is very small for y values more than three standard deviations from the mean. The area under the curve is exactly equal to 1. (*Note:* It may seem paradoxical that a curve can enclose a finite area, even though it never descends to touch the horizontal axis. This apparent paradox is clarified in Appendix 4.1.)

Figure 4.2.1 A normal curve with mean μ and standard deviation σ

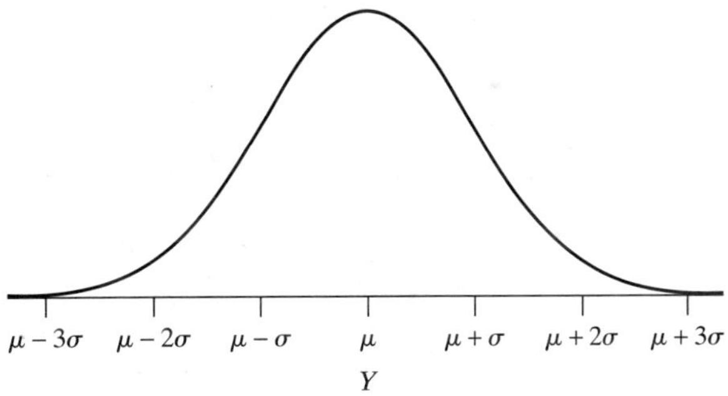

All normal curves have the same essential shape, in the sense that they can be made to look identical by suitable choice of the vertical and horizontal scales for each. (For instance, notice that the curves in Figures 4.1.2–4.1.5 look identical.) But normal curves with different values of μ and σ will not look identical if they are all plotted to the same scale, as illustrated by Figure 4.2.2. The location of the normal curve along the y-axis is governed by μ since the curve is centered at $y = \mu$; the width and the height of the curve (i.e., whether tall and thin or short and wide) are governed by σ. Since the area under each curve must be equal to 1, a curve with a smaller value of σ must be taller. This reflects the fact that the values of Y are more highly concentrated near the mean when the standard deviation is smaller.

Figure 4.2.2 Three normal curves with different means and standard deviations

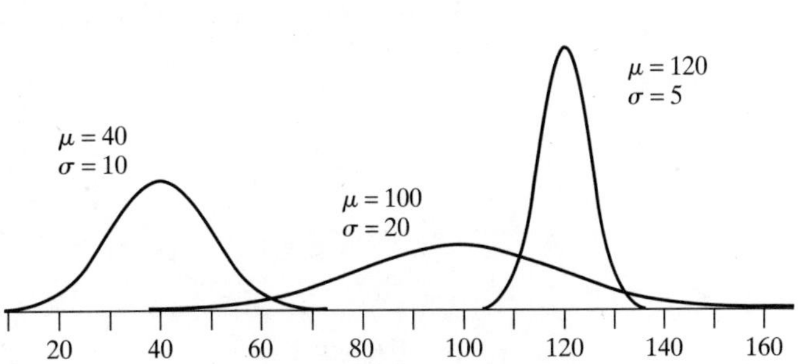

4.3 Areas under a Normal Curve

As explained in Section 3.4, a density curve can be quantitatively interpreted in terms of areas under the curve. While areas can be roughly estimated by eye, for some purposes it is desirable to have fairly precise information about areas.

THE STANDARDIZED SCALE

The areas under a normal curve have been computed mathematically and are tabulated in Table 3 at the end of this book for practical use. The use of this tabulated information is much simplified by the fact that all normal curves can be made equivalent with respect to areas under them by suitable rescaling of the horizontal axis. The rescaled variable is denoted by Z; the relationship between the two scales is shown in Figure 4.3.1.

Figure 4.3.1 A normal curve, showing the relationship between the natural scale (Y) and the standardized scale (Z)

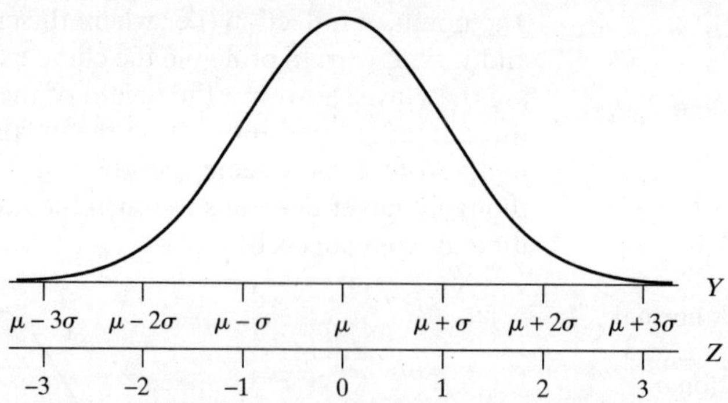

As Figure 4.3.1 indicates, the Z scale measures standard deviations from the mean: $z = 1.0$ corresponds to 1.0 standard deviation above the mean; $z = -2.5$ corresponds to 2.5 standard deviations below the mean, and so on. The Z scale is referred to as a **standardized scale**.

The correspondence between the Z scale and the Y scale can be expressed by the formula given in the following box.

Standardization Formula

$$Z = \frac{Y - \mu}{\sigma}$$

The variable Z is referred to as the **standard normal** and its distribution follows a normal curve with mean zero and standard deviation one. Table 3 gives areas under the standard normal curve, with distances along the horizontal axis measured in the Z scale. Each area tabled in the body of Table 3 is the area under the standard normal curve below a specified value of z, tabled in the margins. For example, for $z = 1.53$, the tabled area is 0.9370; this area is shaded in Figure 4.3.2.

Figure 4.3.2 Illustration of the use of Table 3

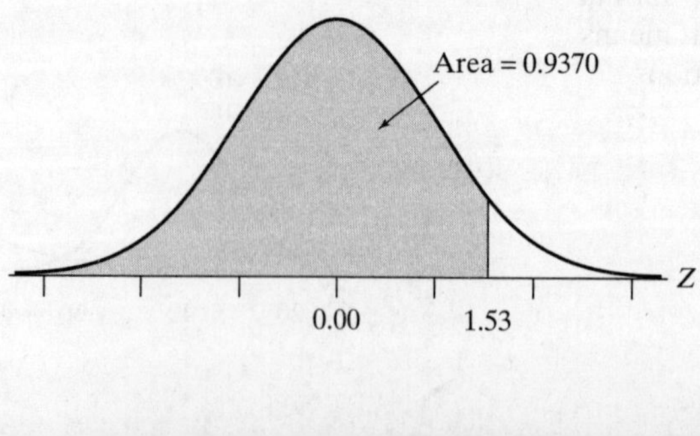

If we want to find the area above a given value of z, we subtract the tabulated area from 1. For example, the area above $z = 1.53$ is $1.0000 - 0.9370 = 0.0630$ (Figure 4.3.3).

To find the area between two z values (also commonly called z **scores**) we can subtract the areas given in Table 3. For example, to find the area under the Z curve between $z = -1.2$ and $z = 0.8$ (Figure 4.3.4), we take the area below 0.8, which is 0.7881, and subtract the area below -1.2, which is 0.1151, to get $0.7881 - 0.1151 = 0.6730$.

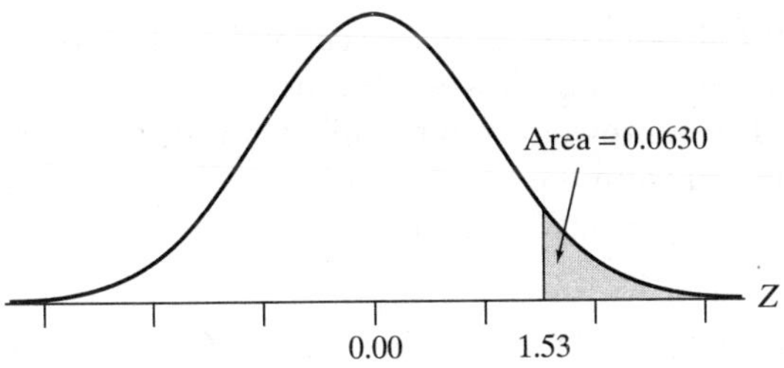

Figure 4.3.3 Area under a standard normal curve above 1.53

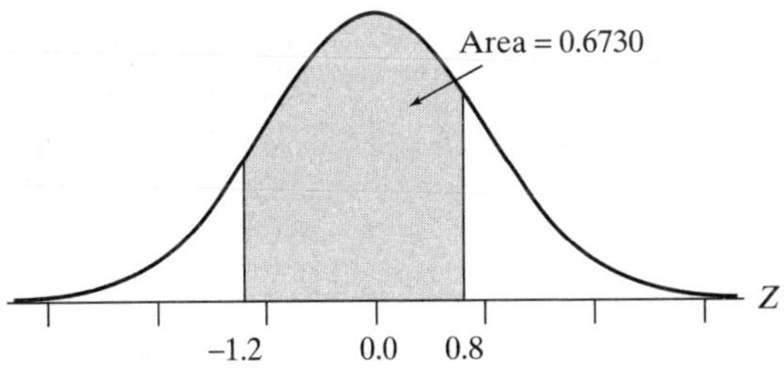

Figure 4.3.4 Area under a standard normal curve between -1.2 and 0.8

Using Table 3, we see that the area under the normal curve between $z = -1$ and $z = +1$ is $0.8413 - 0.1587 = 0.6826$. Thus, for any normal distribution, about 68% of the observations are within ± 1 standard deviation of the mean. Likewise, the area under the normal curve between $z = -2$ and $z = +2$ is $0.9772 - 0.0228 = 0.9544$ and the area under the normal curve between $z = -3$ and $z = +3$ is $0.9987 - 0.0013 = 0.9974$. This means that for any normal distribution about 95% of the observations are within ± 2 standard deviations of the mean and about 99.7% of the observations are within ± 3 standard deviations of the mean. (See Figure 4.3.5.) For example, about 68% of the serum cholesterol values in the idealized distribution of Figure 4.1.2 are between 128 mg/dl and 182 mg/dl, about 95% are between 101 mg/dl and 209 mg/dl, and virtually all are between 74 mg/dl and 236 mg/dl. Figure 4.3.6 shows these percentages.

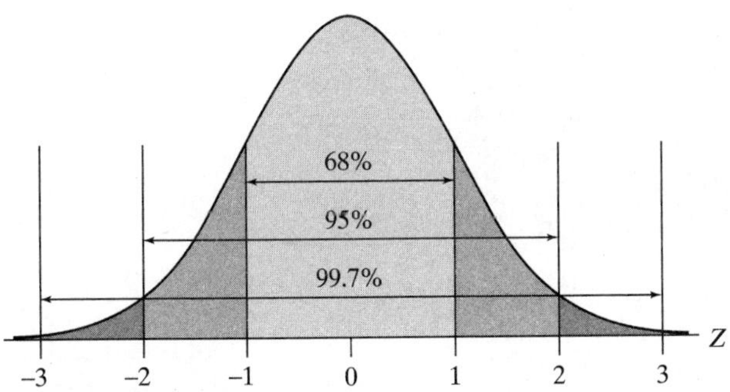

Figure 4.3.5 Areas under a standard normal curve between -1 and $+1$, between -2 and $+2$, and between -3 and $+3$

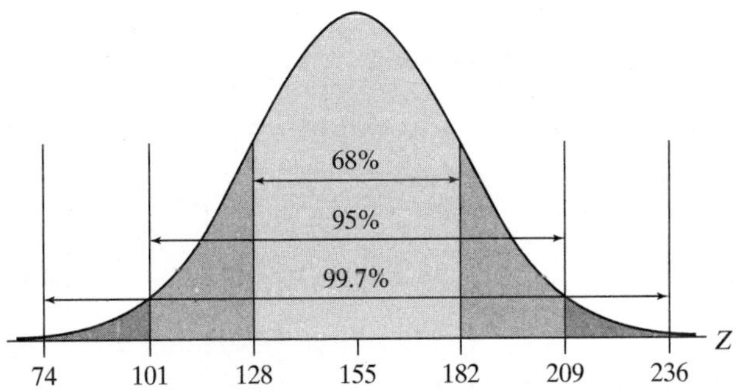

Figure 4.3.6 The 68/95/99.7 rule and the serum cholesterol distribution

If the variable Y follows a normal distribution, then

about 68% of the y's are within ± 1 SD of the mean.
about 95% of the y's are within ± 2 SDs of the mean.
about 99.7% of the y's are within ± 3 SDs of the mean.

These statements provide a very definite interpretation of the standard deviation in cases where a distribution is approximately normal. (In fact, the statements are often approximately true for moderately nonnormal distributions; that is why, in Section 2.6, these percentages—68%, 95%, and >99%—were described as "typical" for "nicely shaped" distributions.)

DETERMINING AREAS FOR A NORMAL CURVE

By taking advantage of the standardized scale, we can use Table 3 to answer detailed questions about any normal population when the population mean and standard deviation are specified. The following example illustrates the use of Table 3. (Of course, the population described in the example is an idealized one, since no actual population follows a normal distribution *exactly*.)

Example 4.3.1 **Lengths of Fish** In a certain population of the herring *Pomolobus aestivalis*, the lengths of the individual fish follow a normal distribution. The mean length of the fish is 54.0 mm, and the standard deviation is 4.5 mm.[5] We will use Table 3 to answer various questions about the population.

(a) What percentage of the fish are less than 60 mm long?

Figure 4.3.7 shows the population density curve, with the desired area indicated by shading. In order to use Table 3, we convert the limits of the area from the Y scale to the Z scale, as follows:

For $y = 60$, the z score is

$$z = \frac{y - \mu}{\sigma} = \frac{60 - 54}{4.5} = 1.33$$

Thus, the question "What percentage of the fish are less than 60 mm long?" is equivalent to the question "What is the area under the standard normal curve below the z value of 1.33?" Looking up $z = 1.33$ in Table 3, we find that the area is 0.9082; thus, 90.82% of the fish are less than 60 mm long.

Figure 4.3.7 Area under the normal curve in Example 4.3.1(a)

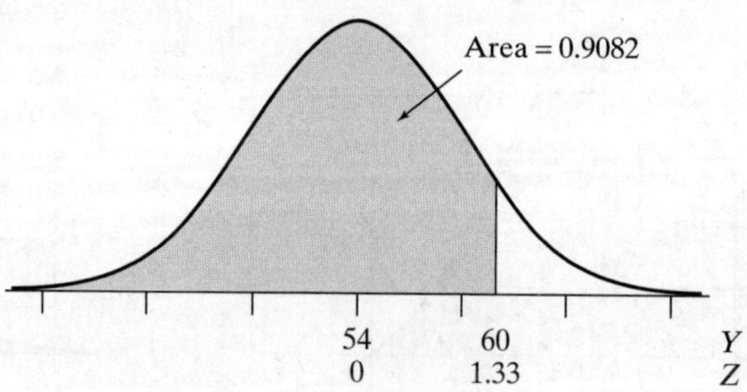

(b) What percentage of the fish are more than 51 mm long?

The standardized value for $y = 51$ is

$$z = \frac{y - \mu}{\sigma} = \frac{51 - 54}{4.5} = -0.67$$

Thus, the question "What percentage of the fish are more than 51 mm long?" is equivalent to the question "What is the area under the standard normal curve above the z value of -0.67?" Figure 4.3.8 shows this relationship. Looking up $z = -0.67$ in Table 3, we find that the area below $z = -0.67$ is 0.2514. This

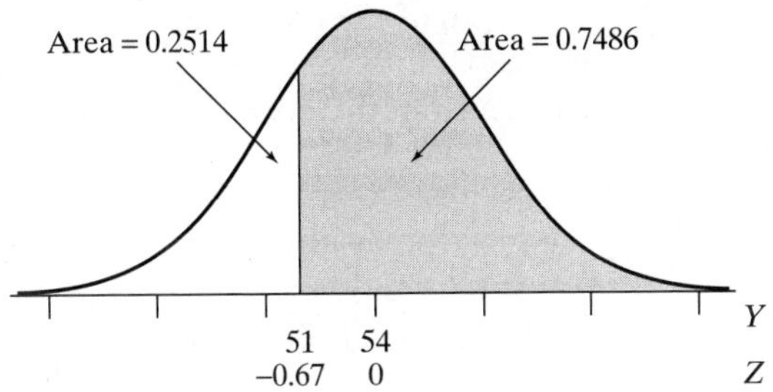

Figure 4.3.8 Area under the normal curve in Example 4.3.1(b)

means that the area above $z = -0.67$ is $1 - 0.2514 = 0.7486$. Thus, 74.86% of the fish are more than 51 mm long.

(c) What percentage of the fish are between 51 and 60 mm long?

Figure 4.3.9 shows the desired area. This area can be expressed as a difference of two areas found from Table 3. The area below $y = 60$ is 0.9082, as found in part (a), and the area below $y = 51$ is 0.2514, as found in part (b). Consequently, the desired area is computed as

$$0.9082 - 0.2514 = 0.6568$$

Thus, 65.68% of the fish are between 51 and 60 mm long.

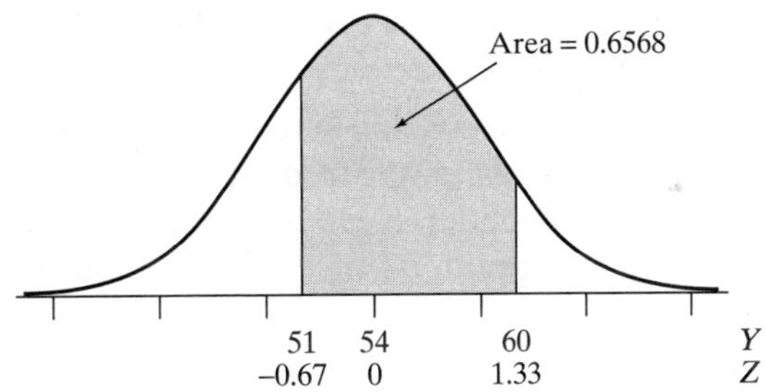

Figure 4.3.9 Area under the normal curve in Example 4.3.1(c)

(d) What percentage of the fish are between 58 and 60 mm long?

Figure 4.3.10 shows the desired area. This area can be expressed as a difference of two areas found from Table 3. The area below $y = 60$ is 0.9082, as was found in part (a). To find the area below $y = 58$, we first calculate the z value that corresponds to $y = 58$:

$$z = \frac{y - \mu}{\sigma} = \frac{58 - 54}{4.5} = 0.89$$

The area under the Z curve below $z = 0.89$ is 0.8133. Consequently, the desired area is computed as

$$0.9082 - 0.8133 = 0.0949$$

Thus, 9.49% of the fish are between 58 and 60 mm long. ■

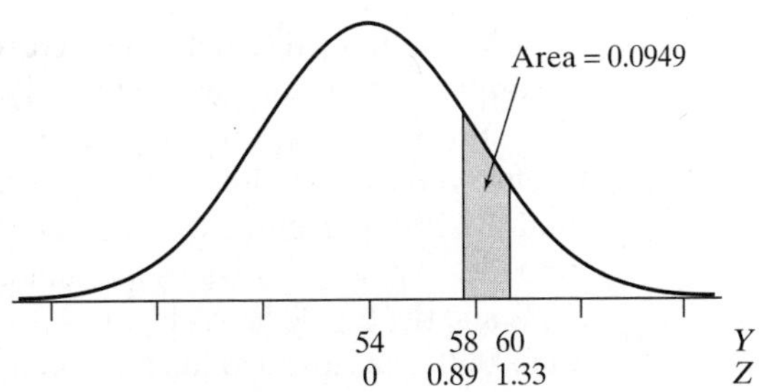

Figure 4.3.10 Area under the normal curve in Example 4.3.1(d)

Each of the percentages found in Example 4.3.1 can also be interpreted in terms of probability. Let the random variable Y represent the length of a fish randomly chosen from the population. Then the results in Example 4.3.1 imply that

$$\Pr\{Y < 60\} = 0.9082$$
$$\Pr\{Y > 51\} = 0.7486$$
$$\Pr\{51 < Y < 60\} = 0.6568$$

and

$$\Pr\{58 < Y < 60\} = 0.0949$$

In this way, the normal distribution can be interpreted as a continuous probability distribution.

Note that because the idealized normal distribution is perfectly continuous, probabilities such as

$$\Pr\{Y > 48\} \text{ and } \Pr\{Y \geq 48\}$$

are equal (see Section 3.4). That is,

$$\begin{aligned}\Pr\{Y \geq 48\} &= \Pr\{Y > 48\} + \Pr\{Y = 48\}\\ &= \Pr\{Y > 48\} + 0 \text{ (since } Y \text{ is taken to be continuous)}\\ &= \Pr\{Y > 48\}\end{aligned}$$

If, however, the length were measured only to the nearest mm, then the measured variable would actually be discrete, so that $\Pr\{Y > 48\}$ and $\Pr\{Y \geq 48\}$ would differ somewhat from each other. In cases where this discrepancy is important, the computation can be refined to take into account the discontinuity of the measured distribution (we will later see such an example in Section 5.4).

INVERSE READING OF TABLE 3

In determining facts about a normal distribution, it is sometimes necessary to read Table 3 in an "inverse" way—that is, to find the value of z corresponding to a given area rather than the other way around. For example, suppose we want to find the value on the Z scale that cuts off the top 2.5% of the distribution. This number is 1.96, as shown in Figure 4.3.11.

Figure 4.3.11 Area under the normal curve above 1.96

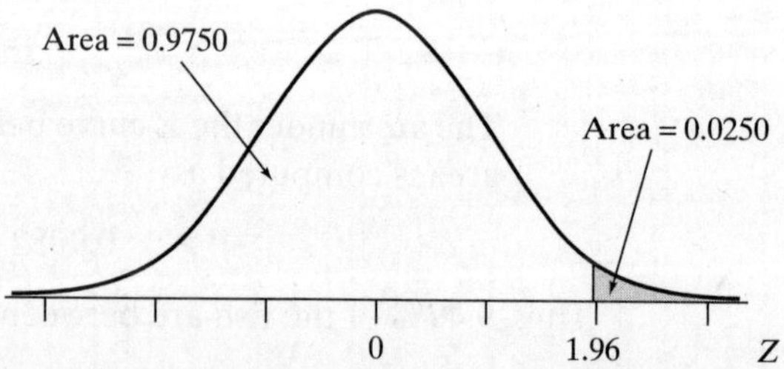

We often need to determine corresponding z-values when we want to determine a *percentile* of a normal distribution. The percentiles of a distribution divide the distribution into 100 equal parts, just as the quartiles divide it into 4 equal parts [from the Latin roots *centum* ("hundred") and *quartus* ("fourth")]. For example, suppose we want to find the 70th percentile of a standard normal distribution. That means that we want to find the z-value that divides the standard normal distribution into two parts: the bottom 70% and the top 30%. As Figure 4.3.12 illustrates, we need to look in Table 3 for an area of 0.7000. The closest value is an area of 0.6985, corresponding to a z value of 0.52.

Figure 4.3.12 Determining the 70th percentile of a normal distribution

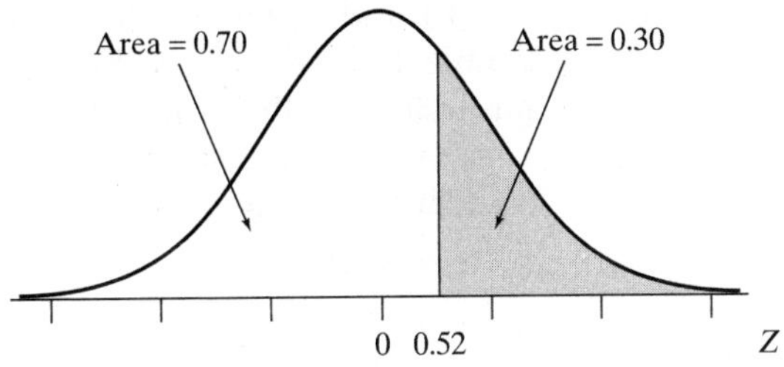

Example 4.3.2

Lengths of Fish

(a) Suppose we want to find the 70th percentile of the fish length distribution of Example 4.3.1. Let us denote the 70th percentile by y^*. By definition, y^* is the value such that 70% of the fish lengths are less than y^* and 30% are greater, as illustrated in Figure 4.3.13.

The z-value we just determined indicates that that y^* is 0.52 standard deviations above the mean. That is, if we were given the value of y^*, we could convert it to a standard normal (z scale) and the result would be 0.52. Thus, from the standardization formula we obtain the equation

$$0.52 = \frac{y^* - 54}{4.5}$$

which can be solved to give $y^* = 54 + 0.52 \times 4.5 = 56.3$. The 70th percentile of the fish length distribution is 56.3 mm.

Figure 4.3.13 Determining the 70th percentile of a normal distribution, Example 4.3.2(a)

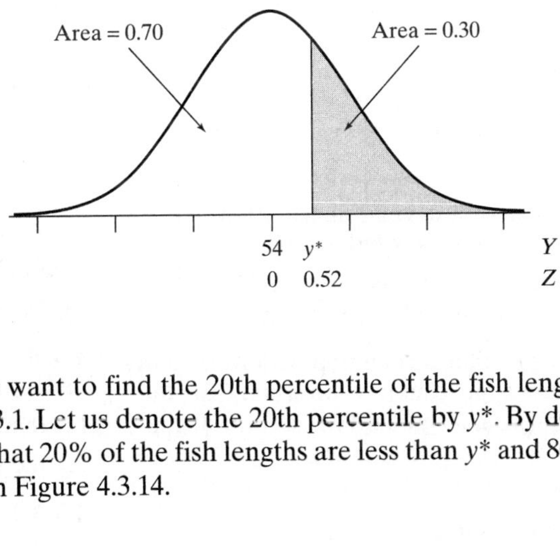

(b) Suppose we want to find the 20th percentile of the fish length distribution of Example 4.3.1. Let us denote the 20th percentile by y^*. By definition, y^* is the value such that 20% of the fish lengths are less than y^* and 80% are greater, as illustrated in Figure 4.3.14.

Figure 4.3.14 Determining the 20th percentile of a normal distribution, Example 4.3.2(b)

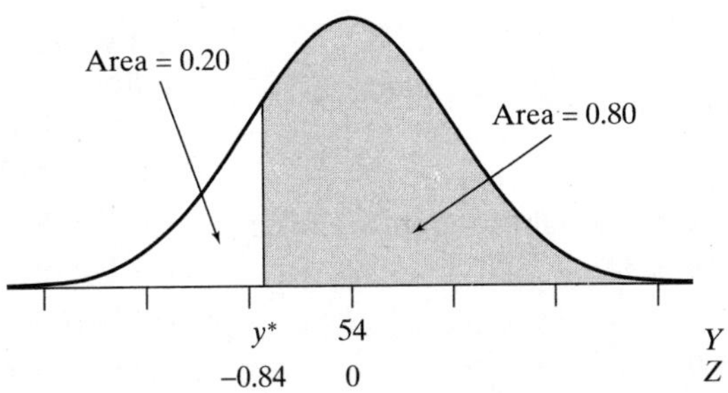

To find y^* we first determine the z-value corresponding to the 20th percentile in the Z scale. As Figure 4.3.14 illustrates, we need to look in Table 3 for an area of .2000. The closest value is an area of .2005, corresponding to $z = -0.84$. The next step is to convert this z value to the Y scale. From the standardization formula, we obtain the equation

$$-0.84 = \frac{y^* - 54}{4.5}$$

which can be solved to give $y^* = 54 - 0.84 \times 4.5 = 50.2$. The 20th percentile of the fish length distribution is 50.2 mm. ■

Problem-Solving Tip In solving problems that require the use of Table 3, a sketch of the distribution (as in Figures 4.3.7–4.3.10 and 4.3.13–4.3.14) is a very handy aid to straight thinking.

While Table 3 is handy for carrying out the sorts of computations discussed previously, computer software may also be used to find normal probabilities directly without the need for any standardization.

Exercises 4.3.1–4.3.17

4.3.1 Suppose a certain population of observations is normally distributed. What percentage of the observations in the population

(a) are within ± 1.5 standard deviations of the mean?

(b) are more than 2.5 standard deviations above the mean?

(c) are more than 3.5 standard deviations away from (above or below) the mean?

4.3.2

(a) The 90th percentile of a normal distribution is how many standard deviations above the mean?

(b) The 10th percentile of a normal distribution is how many standard deviations below the mean?

4.3.3 The brain weights of a certain population of adult Swedish males follow approximately a normal distribution with mean 1,400 gm and standard deviation 100 gm.[6] What percentage of the brain weights are

(a) 1,500 gm or less?

(b) between 1,325 and 1,500 gm?

(c) 1,325 gm or more?

(d) 1,475 gm or more?

(e) between 1,475 and 1,600 gm?

(f) between 1,200 and 1,325 gm?

4.3.4 Let Y represent a brain weight randomly chosen from the population of Exercise 4.3.3. Find

(a) $\Pr\{Y \le 1{,}325\}$

(b) $\Pr\{1{,}475 \le Y \le 1{,}600\}$

4.3.5 In an agricultural experiment, a large uniform field was planted with a single variety of wheat. The field was divided into many plots (each plot being 7 × 100 ft) and the yield (lb) of grain was measured for each plot. These plot yields followed approximately a normal distribution with mean 88 lb and standard deviation 7 lb.[7] What percentage of the plot yields were

(a) 80 lb or more?
(b) 90 lb or more?
(c) 75 lb or less?
(d) between 75 and 90 lb?
(e) between 90 and 100 lb?
(f) between 75 and 80 lb?

4.3.6 Refer to Exercise 4.3.5. Let Y represent the yield of a plot chosen at random from the field. Find

(a) $\Pr\{Y > 90\}$
(b) $\Pr\{75 < Y < 90\}$

4.3.7 Find the z-values corresponding to the following percentiles of the standard normal distribution.

(a) the 75th percentile
(b) the 90th percentile
(c) the 95th percentile
(d) the 99th percentile

4.3.8 For the wheat-yield distribution of Exercise 4.3.5, find

(a) the 65th percentile
(b) the 35th percentile

4.3.9 The serum cholesterol levels of 12- to 14-year-olds follow a normal distribution with mean 155 mg/dl and standard deviation 27 mg/dl. What percentage of 12 to 14-year-olds have serum cholesterol values

(a) 164 or more?
(b) 137 or less?
(c) 186 or less?
(d) 100 or more?
(e) between 159 and 186?
(f) between 100 and 132?
(g) between 132 and 159?

4.3.10 Refer to Exercise 4.3.9. Suppose a 13-year-old is chosen at random and let Y be the person's serum cholesterol value. Find

(a) $\Pr\{Y \geq 159\}$ (b) $\Pr\{159 < Y < 186\}$

4.3.11 For the serum cholesterol distribution of Exercise 4.3.9, find

(a) the 80th percentile (b) the 20th percentile

4.3.12 When red blood cells are counted using a certain electronic counter, the standard deviation of repeated counts of the same blood specimen is about 0.8% of the true value, and the distribution of repeated counts is approximately normal.[8] For example, this means that if the true value is 5,000,000 cells/mm^3, then the standard deviation is 40,000.

(a) If the true value of the red blood count for a certain specimen is 5,000,000 cells/mm^3, what is the probability that the counter would give a reading between 4,900,000 and 5,100,000?

(b) If the true value of the red blood count for a certain specimen is μ, what is the probability that the counter would give a reading between 0.98μ and 1.02μ?

(c) A hospital lab performs counts of many specimens every day. For what percentage of these specimens does the reported blood count differ from the correct value by 2% or more?

4.3.13 The amount of growth, in a 15-day period, for a population of sunflower plants was found to follow a normal distribution with mean 3.18 cm and standard deviation 0.53 cm.[9] What percentage of plants grow

(a) 4 cm or more?

(b) 3 cm or less?

(c) between 2.5 and 3.5 cm?

4.3.14 Refer to Exercise 4.3.13. In what range do the middle 90% of all growth values lie?

4.3.15 For the sunflower plant growth distribution of Exercise 4.3.13, what is the 25th percentile?

4.3.16 The ponderal index is a measure of overall size similar to a body mass index. The ponderal index of full-term newborns follows a normal distribution with a mean of 2.4 kg/m^3 and a standard deviation of 0.45 kg/m^3. Is a newborn with a ponderal index of 1.75 above or below the 10th percentile? Show your reasoning.

4.3.17 Many cities sponsor marathons each year. The following histogram shows the distribution of times that it took for 10,002 runners to complete the Rome marathon in 2008, with a normal curve superimposed. The fastest runner completed the 26.3-mile course in 2 hours and 9 minutes, or 129 minutes. The average time was 245 minutes and the standard deviation was 40 minutes. Use the normal curve to answer the following questions.[10]

(a) What percentage of times were greater than 200 minutes?

(b) What is the 60th percentile of the times?

(c) Notice that the normal curve approximation is fairly good except around the 240-minute mark. How can we explain this anomalous behavior of the distribution?

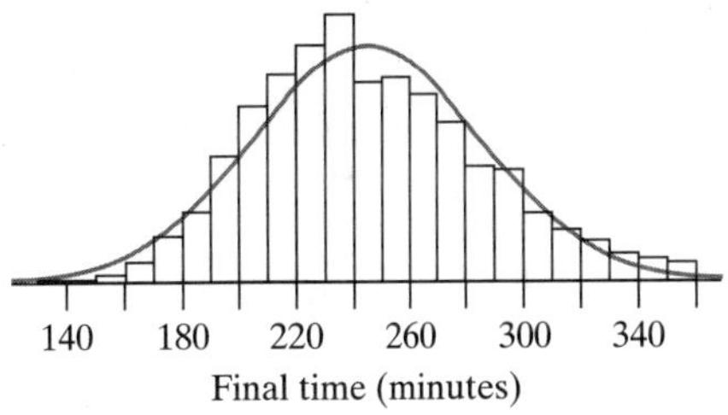

4.4 Assessing Normality

Many statistical procedures are based on having data from a normal population. In this section we consider ways to assess whether it is reasonable to use a normal curve model for a set of data and, if not, how we might proceed.

Recall from Section 4.3 that if the variable Y follows a normal distribution, then

about 68% of the y's are within ± 1 SD of the mean.
about 95% of the y's are within ± 2 SDs of the mean.
about 99.7% of the y's are within ± 3 SDs of the mean.

We can use these facts as a check of how closely a normal curve model fits a set of data.

Example 4.4.1

Serum Cholesterol For the serum cholesterol data of Example 4.1.1, the sample mean is 155 and the sample SD is 27. The interval "mean ± SD" is

$$(155 - 27, 155 + 27) \text{ or } (128, 182)$$

This interval contains 304 of the 431 observations, or 70.5% of the data. Likewise, the interval

$$(155 - 2 \times 27, 155 + 2 \times 27) \text{ is } (101, 209)$$

which contains 407, or 94.4%, of the 431 observations. Finally, the interval

$$(155 - 3 \times 27, 155 + 3 \times 27) \text{ is } (74, 236)$$

which contains 430, or 99.8%, of the 431 observations. The three observed percentages

70.5%, 94.4%, and 99.8%

agree quite well with the theoretical percentages of

68%, 95%, and 99.7%

This agreement supports the claim that serum cholesterol levels for 12- to 14-year-olds have a normal distribution. This reinforces the visual evidence of Figure 4.1.1. ■

Example 4.4.2

Moisture Content Moisture content was measured in each of 83 freshwater fruit.[11] Figure 4.4.1 shows that this distribution is strongly skewed to the left. The sample mean of these data is 80.7 and the sample SD is 12.7. The interval

$$(80.7 - 12.7, 80.7 + 12.7) \text{ is } (68.0, 93.4)$$

which contains 70, or 84.3%, of the 83 observations. The interval

$$(80.7 - 2 \times 12.7, 80.7 + 2 \times 12.7) \text{ is } (55.3, 106.1)$$

which contains 78, or 94.0%, of the 83 observations. Finally, the interval

$$(80.7 - 3 \times 12.7, 80.7 + 3 \times 12.7) \text{ is } (42.6, 118.8)$$

which contains 80, or 96.4%, of the 83 observations. The three percentages

84.3%, 94.0%, and 96.4%

differ from the theoretical percentages of

68%, 95%, and 99.7%

because the distribution is far from being bell-shaped. This reinforces the visual evidence of Figure 4.4.1. ■

Figure 4.4.1 Moisture content in freshwater fruit

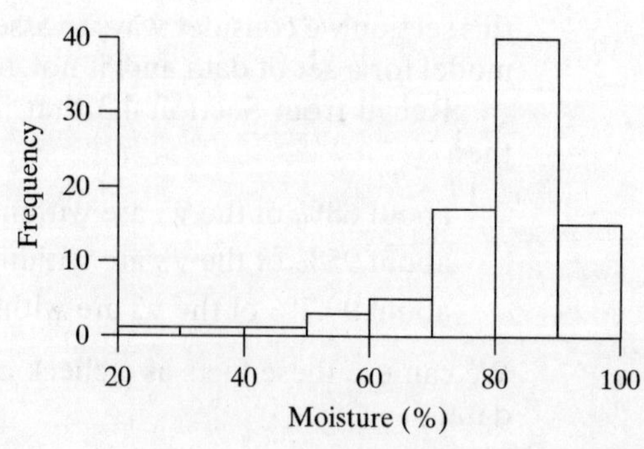

NORMAL QUANTILE PLOTS

A **normal quantile plot*** is a special statistical graph that is used to assess normality. We present this statistical tool with an example using the heights (in inches) of a sample of 11 women, sorted from smallest to largest:

$$61, 62.5, 63, 64, 64.5, 65, 66.5, 67, 68, 68.5, 70.5$$

Based on these data, does it make sense to use a normal curve to model the distribution of women's heights? Figure 4.4.2 is a histogram of the data with a normal curve superimposed, using the sample mean of 65.5 and the sample standard deviation of 2.9 as the parameters of the normal curve. This histogram is fairly symmetric, but when we have a small sample, it can be hard to tell the shape of the population distribution by looking at a histogram.

Figure 4.4.2 Histogram of the heights of 11 women

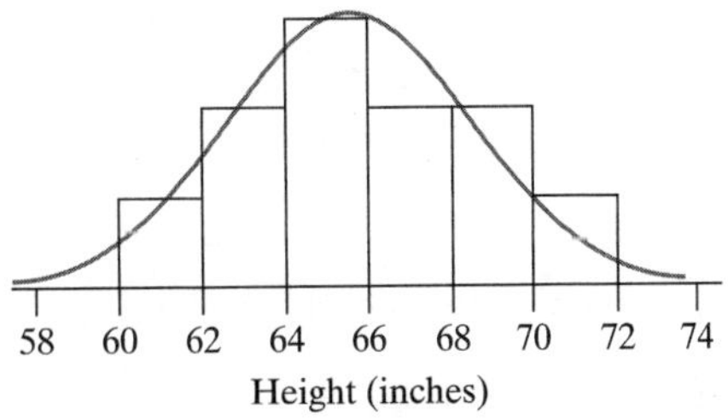

Because it is often difficult to visually examine a histogram and decide if it is bell-shaped or not, a visually simpler plot, the normal quantile plot, was developed.† A normal quantile plot is a scatterplot that compares our observed data values to values we would expect to see if the population were normal. If the data come from a normal population, the points in this plot should follow a straight line, which is much easier to visually recognize than a bell shape of a jagged histogram. As many statistical procedures are based on the condition that the data came from a normal population, it is important to be able to assess normality.

HOW NORMAL QUANTILE PLOTS WORK

In Examples 4.4.1 and 4.4.2 we compared the observed proportion of data that falls within 1, 2, and 3 SDs of the mean and then compared those values to the proportions we would expect to find if the data were from a normal population. Rather than focus on comparing proportions, we could instead focus on comparing actual observed women's heights to heights we would expect to see if the data were from a normal population. For example, the shortest woman in our sample is 61 inches tall; that is, 1/11th (or 0.0909) of the sample is 61 inches or shorter. If heights of women really follow a normal distribution, with mean 65.5 and standard deviation 2.9, then we would expect the 9.09th percentile to be $\mu + z_{(1-0.0909)}\sigma = 65.5 - 1.34 \times 2.9$ or 61.6 inches. This value is close to the observed value of 61 inches. We could repeat this sort of calculation for each of the 11 observed data values. A normal quantile plot provides a visual comparison of these values.

---

*Sometimes this is called a *normal probability plot.*

†Although visually simple, the construction of these plots is complex and typically performed using statistical software.

The first step in creating a normal quantile plot, therefore, is to compute the sample percentiles. Example 4.4.3 presents this computation, which is typically performed by statistical software.

Example 4.4.3 **Height of Eleven Women** Sorting the data from smallest to largest we observe that 1/11th (= 9.1%) of our sample is 61 inches or shorter, 2/11ths (= 18.2%) is 62.5 inches or shorter, . . . 10/11ths (90.9%) is 68.5 inches or shorter and 11/11ths (100%) is 70.5 inches or shorter. Unfortunately, computing percentages in this simplistic way (i.e., $100 \times i/n$, where i is the sorted observation number) creates some implausible population estimates. For example, it seems unreasonable to believe that 100% of the *population* is 70.5 inches or shorter when, after all, we are observing only a small sample; a larger sample would likely observe some taller women. To correct for this, an alternative and more reasonable percentage for each data value is computed as $100\left(i - \frac{1}{2}\right)/n$ where i is the index of the data value in the sorted list.* These adjusted percentiles are tabulated in Table 4.4.1. Note that these values actually do not depend on the data observed; they depend only on the number of data values in the sample. ■

Table 4.4.1 Computing indices and percentiles for the heights of 11 women

i	1	2	3	4	5	6	7	8	9	10	11
Observed height	61.0	62.5	63.0	64.0	64.5	65.0	66.5	67.0	68.0	68.5	70.5
Percentile $100(i/11)$	9.09	18.18	27.27	36.36	45.45	54.55	63.64	72.73	81.82	90.91	100.00
Adjusted percentile $100\left(i - \frac{1}{2}\right)/11$	4.55	13.64	22.73	31.82	40.91	50.00	59.09	68.18	77.27	86.36	95.45

Once we have the adjusted percentiles we find the corresponding z scores using Table 3 or a computer. Then, with these z scores we find the theoretical heights: $\mu + z \times \sigma$ as in Example 4.4.4.

Example 4.4.4 **Heights of Eleven Women** The shortest woman's adjusted percentile is 4.55%. The corresponding z score is $z_{(1-0.0455)} = z_{0.9545} = -1.69$. In this example, the sample mean and standard deviation are 65.5 and 2.9, respectively, so the expected height of the shortest woman in a sample of 11 women from a normal population is $65.5 - 1.69 \times 2.9 = 60.6$ inches. The z scores and theoretical heights for this woman and the remaining 10 women appear in Table 4.4.2.

Table 4.4.2 Computing theoretical z scores and heights for 11 women

i	1	2	3	4	5	6	7	8	9	10	11
Observed height	61.0	62.5	63.0	64.0	64.5	65.0	66.5	67.0	68.0	68.5	70.5
Adjusted percentile $100\left(i - \frac{1}{2}\right)/11$	4.55	13.64	22.73	31.82	40.91	50.00	59.09	68.18	77.27	86.36	95.45
z	−1.69	−1.10	−0.75	−0.47	−0.23	0.00	0.23	0.47	0.75	1.10	1.69
Theoretical height	60.6	62.3	63.4	64.1	64.8	65.5	66.2	66.9	67.6	68.7	70.4

*Different software packages may compute these proportions differently and may also modify the formula based on sample size. The preceding formula is used by the software package *R* when $n > 10$.

Next, by plotting the observed heights against the theoretical heights in a scatterplot as in Figure 4.4.3, we may visually compare the values. In this case our plot appears fairly linear, suggesting that the observed values generally agree with the theoretical values—that the normal model provides a reasonable approximation to the data. If the data do not agree with the normal model, then the plot will show strong nonlinear patterns such as curvature or S shapes.

Figure 4.4.3 Normal quantile plot of the heights of 11 women

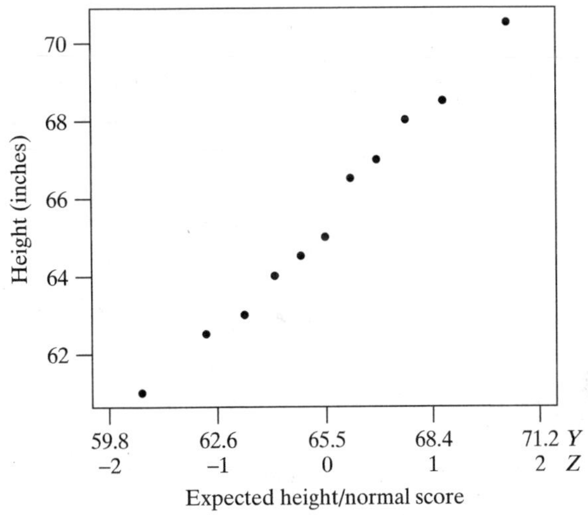

Because of the one-to-one correspondence between the z scores and theoretical values, it is not common to put both sets of labels on the x-axis as in Figure 4.4.3. Traditionally only the z scores are displayed.*

MAKING DECISIONS ABOUT NORMALITY

Of course, even when we sample from a perfectly normal distribution, we have to expect that there will be some variability between the sample we obtain and the theoretical normal scores. Figure 4.4.4 shows six normal quantile plots based on samples taken from a $N(0, 1)$ distribution. Notice that all six plots show a general linear pattern. It is true that there is a fair amount of "wiggle" in some of the plots, but the important feature of each of these plots is that we can draw a line that captures the trend in the bulk of the points, with little deviation away from this line, even at the extremes.

If the points in the normal quantile plot do not fall more or less along a straight line, then there is an indication that the data are not from a normal population. For example, if the top of the plot bends up, that means the y values at the upper end of the distribution are too large for the distribution to be bell-shaped; that is, the distribution is skewed to the right or has large outliers, as in Figure 4.4.5.

If the bottom of the plot bends down, that means the y values at the lower end of the distribution are too small for the distribution to be bell-shaped; that is, the distribution is skewed to the left or has small outliers. Figure 4.4.6 shows the distribution of moisture content in the freshwater fruit from Example 4.4.2, which is strongly skewed to the left.

*Some software programs create normal quantile plots with the normal scores on the vertical axis and the observed data on the horizontal axis.

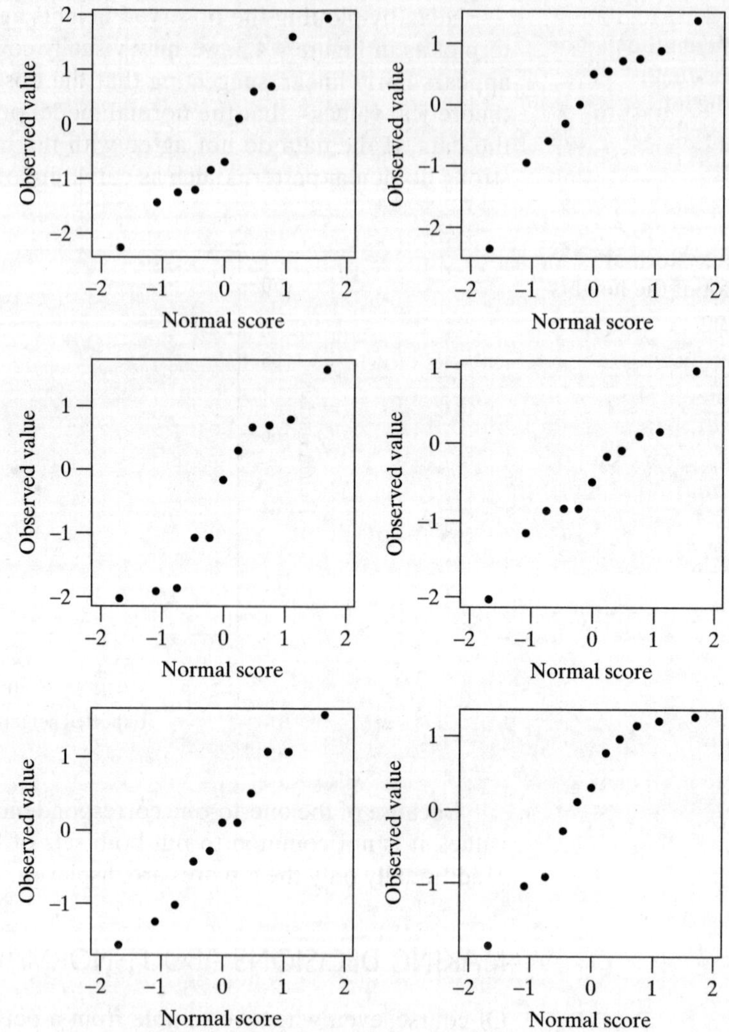

Figure 4.4.4 Normal quantile plots for normal data

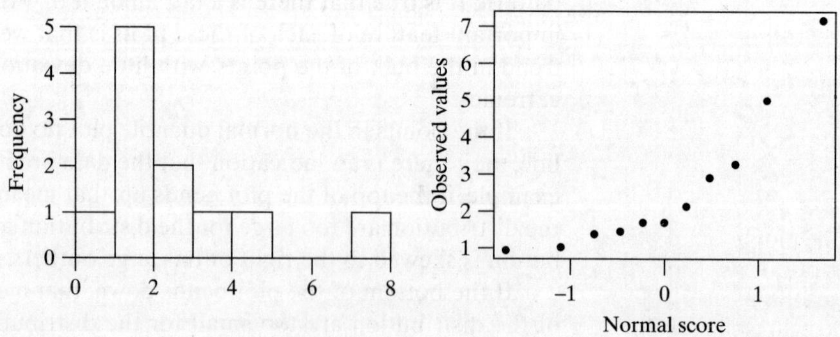

Figure 4.4.5 Histogram and normal quantile plot of a distribution that is skewed to the right

If a distribution has a very long left-hand tail and a long right-hand tail, when compared to a normal curve, then the normal quantile plot will have something of an S shape. Figure 4.4.7 shows such a distribution.

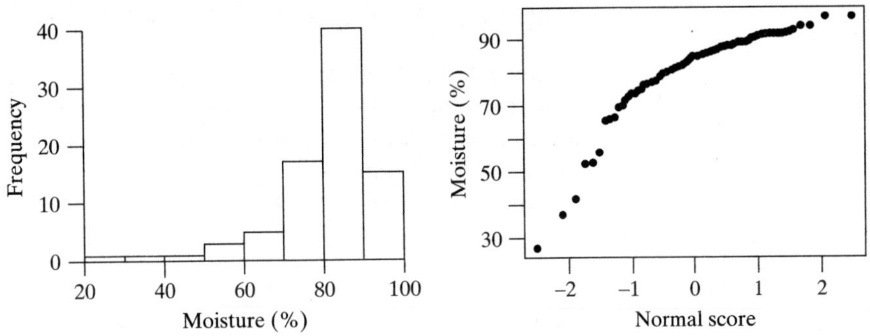

Figure 4.4.6 Histogram and normal quantile plot of a distribution that is skewed to the left

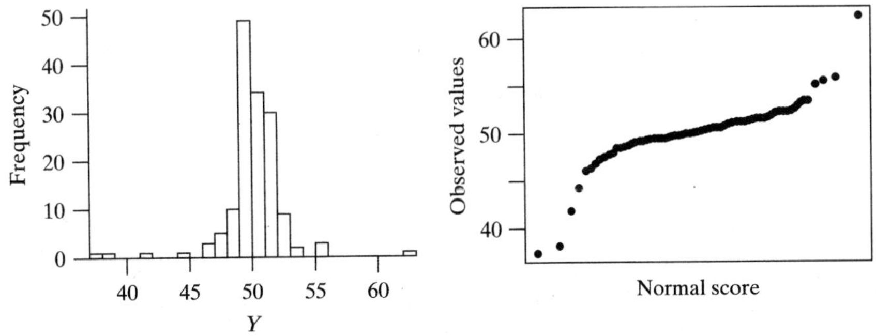

Figure 4.4.7 Histogram and normal quantile plot of a distribution that has long tails

Recall from Section 4.3 that if Y has a normal distribution with mean μ and standard deviation σ then $Z = \dfrac{Y - \mu}{\sigma}$ has a normal distribution with mean 0 and standard deviation 1. Solving for Y we have $Y = \mu + Z\sigma$, which is the equation of a straight line. That is, a plot of Y against Z will be straight when Y is normal. If Y is not normal then the transformation of $Z = \dfrac{Y - \mu}{\sigma}$ is not a simple rescaling of one normal (Y) into another normal (Z) and thus a plot of Y values versus corresponding Z values will not give a straight line.

Sometimes the same value shows up repeatedly in a sample, due to rounding in the measurement process. This leads to *granularity* in the normal quantile plot, as in Figure 4.4.8, but this does not stop us from inferring that the underlying distribution is normal.

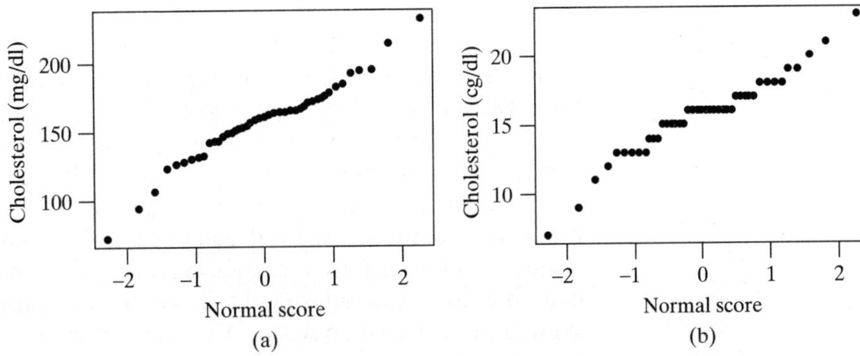

Figure 4.4.8 Normal quantile plots of cholesterol values of fifty 12- to 14-year-olds measured to (a) the nearest mg/dl and (b) the nearest cg/dl

TRANSFORMATIONS FOR NONNORMAL DATA

A normal quantile plot can help us assess whether or not the data came from a normal distribution. Sometimes a histogram or normal quantile plot shows that our data are nonnormal, but a transformation of the data gives us a symmetric, bell-shaped curve. In such a situation, we may wish to transform the data and continue our analysis in the new (transformed) scale.

Example 4.4.5 **Lentil Growth** The histogram and normal quantile plot in Figure 4.4.9 show the distribution of the growth rate, in cm per day, for a sample of 47 lentil plants.[12] This distribution is skewed to the right. If we take the logarithm of each observation, we get a distribution that is much more nearly symmetric. The plots in Figure 4.4.10 show that in log scale the growth rate distribution is approximately normal. (In Figure 4.4.10 the base 10 logarithm, $\log_{10}$, is used, but we could use any base, such as the natural log, $\log_e = \ln$, and the effect on the shape of the distribution would be the same.) ■

Figure 4.4.9 Histogram and normal quantile plot of growth rates of 47 lentil plants

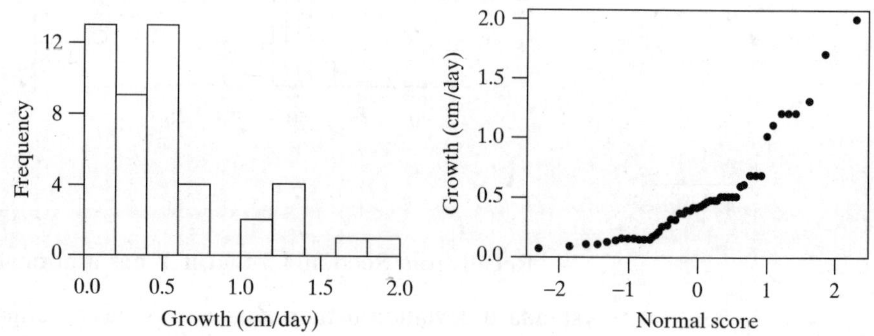

Figure 4.4.10 Histogram and normal quantile plot of the logarithms of the growth rates of 47 lentil plants

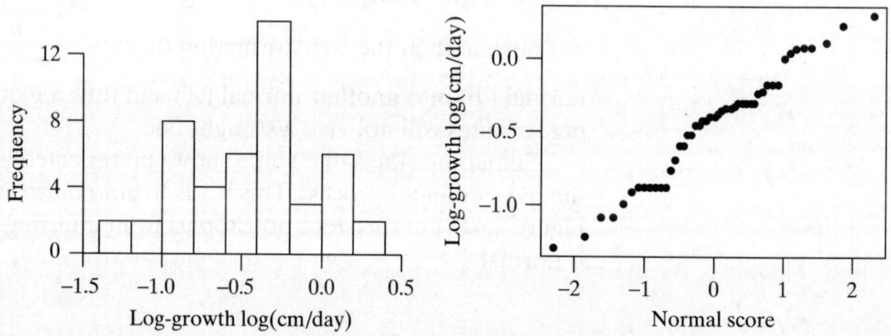

In general, if the distribution is skewed to the right then one of the following transformations should be considered: $\sqrt{Y}$, $\log Y$, $1/\sqrt{Y}$, $1/Y$. These transformations will pull in the long right-hand tail and push out the short left-hand tail, making the distribution more nearly symmetric. Each of these is more drastic than the one before. Thus, a square root transformation will change a mildly skewed distribution into a symmetric distribution, but a log transformation may be needed if the distribution is more heavily skewed, and so on. For example, we saw in Example 2.7.6 how a square root transformation pulls in a long right-hand tail and how a log transformation pulls in the right-hand tail even more. If the distribution of a variable Y is skewed to the left, then raising Y to a power greater than 1 can be helpful.

AN OBJECTIVE MEASURE OF ABNORMALITY: THE SHAPIRO–WILK TEST (OPTIONAL)

While normal quantile plots are better than histograms to visually assess departures of normality, our visual perception is still subjective. The data appearing in the quantile plots of Figure 4.4.4 come from a normal population, but to untrained eyes (and even to some trained ones) a few of the plots might be interpreted as being nonnormal. The **Shapiro–Wilk test** is a statistical procedure that numerically assesses evidence for certain types of nonnormality in data. As with the normal quantile plot, the mechanics of the procedure is complex, but fortunately many statistical software packages will perform this or similar tests of normality.*

The output of a Shapiro–Wilk test is a *P*-value† and is interpreted as follows:

P-value < 0.001	Very strong evidence for nonnormality
P-value < 0.01	Strong evidence for nonnormality
P-value < 0.05	Moderate evidence for nonnormality
P-value < 0.10	Mild or weak evidence for nonnormality
P-value ≥ 0.10	No compelling evidence for nonnormality

Example 4.4.6 illustrates the Shapiro–Wilk test for the lentil growth data of Example 4.4.5.

Example 4.4.6 **Lentil Growth** For the untransformed lentil data in Figure 4.4.9, the *P*-value (reported from the statistical software package *R*) for the Shapiro–Wilk test is 0.000006. Thus, there is very strong evidence that lentil growth does not follow a normal distribution. For the transformed data in Figure 4.4.10, however, the *P*-value for the Shapiro–Wilk test is 0.2090, indicating that there is no compelling evidence for nonnormality of the log-transformed growth data. ■

Caution The use of this test procedure and *P*-value is somewhat like the use of the "check engine light" on a car. When the *P*-value is small, there is an indication of nonnormality. This is like your engine light coming on: You pull over and assess the situation. Likewise, as we shall see in future chapters, when we have nonnormal data, we will carefully have to assess how to proceed with our analyses. On the other hand, when the *P*-value is not small (>0.10) we don't have evidence of nonnormality. This is similar to your engine light staying off: You continue to drive forward without worry, *but* this does not guarantee that your car is perfectly OK. Your car could break down at any time. Of course, if we were constantly worried about our car even when the check engine light were off, we would perpetually find ourselves paralyzed and pulled over at the side of the road. Analogously, when the *P*-value from the Shapiro–Wilk tests is not small (the light is off), this only means that there is no compelling evidence for nonnormality. It does not guarantee that the population is, in fact, normal.

---

*The Ryan–Joiner, Anderson–Darling, and Kolmogorov–Smirnoff tests are other tests of nonnormality commonly found in statistical software packages.

†As we shall see in much greater detail in Chapter 7, a *P*-value is not unique to testing for normality. In a test of all sorts of hypotheses, the weight of evidence for the hypothesis in question (in this case—the Shapiro–Wilk test—the hypothesis is that the data are nonnormal) can be reported using this term. Small *P*-values are interpreted as evidence for the hypothesis in question.

Exercises 4.4.1–4.4.8

4.4.1 In Example 4.1.2 it was stated that bill length in a population of Blue Jays follow a normal distribution with mean $\mu = 25.4$ mm and standard deviation $\sigma = 0.8$ mm. Use the 68%–95%–99.7% rule to determine intervals, centered at the mean, that include 68%, 95%, and 99.7% of the bill length in the distribution.

4.4.2 The following three normal quantile plots, (a), (b), and (c), were generated from the distributions shown by histograms I, II, and III. Which normal quantile plot goes with which histogram? How do you know?

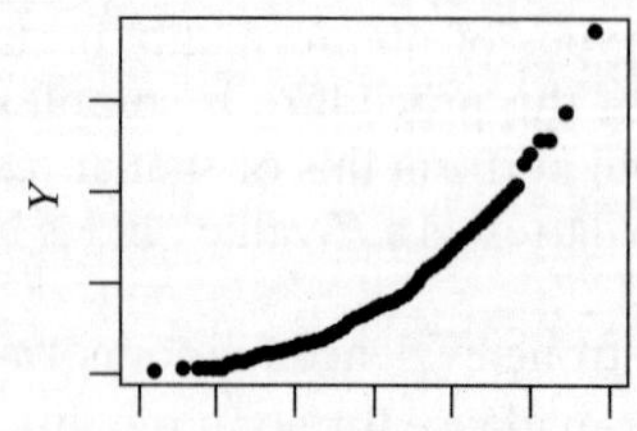

Normal scores
(a)

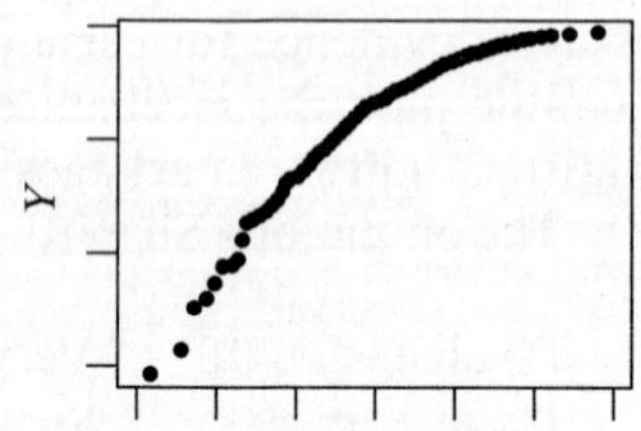

Normal scores
(b)

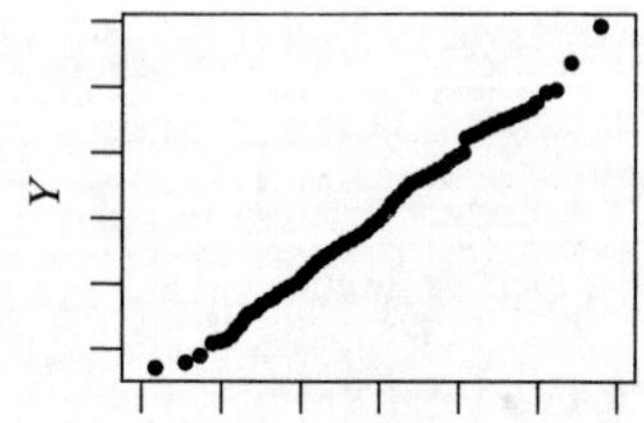

Normal scores
(c)

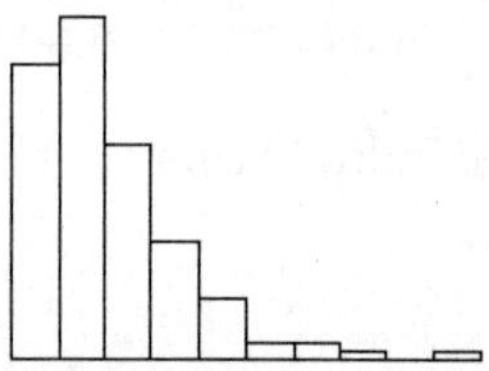

Y
I

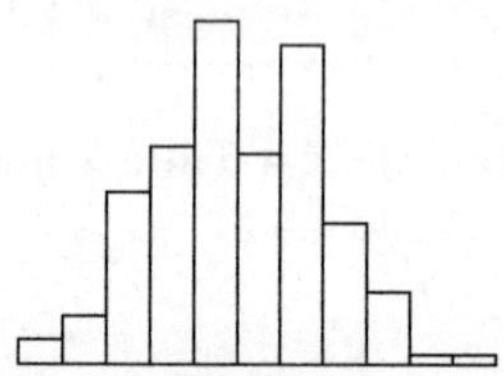

Y
II

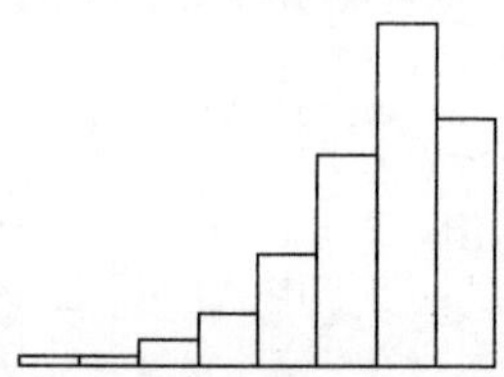

Y
III

4.4.3 For each of the following normal quantile plots, sketch the corresponding histogram of the data.

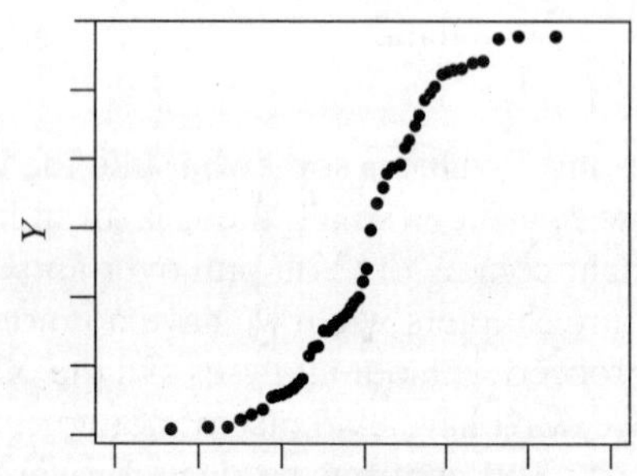

Normal scores
(a)

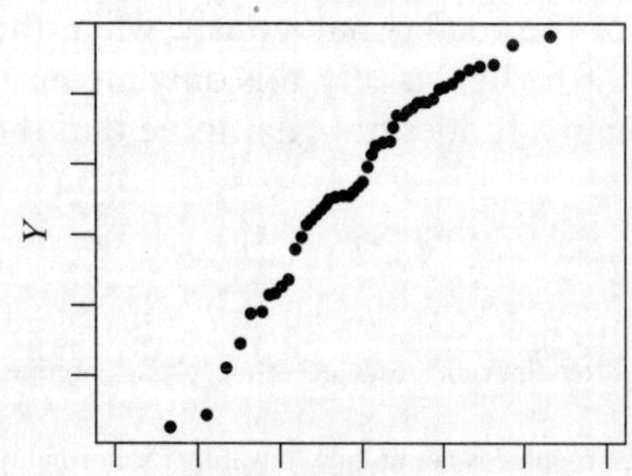

Normal scores
(b)

4.4.4 The mean daily rainfall between January 1, 2007, through January 1, 2009, at Pismo Beach, California, was 0.02 inches with a standard deviation of 0.11 inches. Based on this information, do you think it is reasonable to believe that daily rainfall at Pismo Beach follows a normal distribution? Explain. (*Hint*: Think about the possible values for daily rainfall.)[13]

4.4.5 The mean February 1 daily high temperature in Juneau, Alaska, between 1945 and 2005 was 1.1 °C with a standard deviation of 1.9 °C.[14]

(a) Based on this information, do you think it is reasonable to believe that the February 1 daily high temperatures in Juneau, Alaska, follow a normal distribution? Explain.

(b) Does this information provide compelling evidence that the February 1 daily high temperatures in Juneau, Alaska, follow a normal distribution? Explain.

4.4.6 The following normal quantile plot was created from the times that it took 166 bicycle riders to complete the stage 11 time trial, from Grenoble to Chamrousse, France, in the 2001 Tour de France cycling race.

(a) Consider the fastest riders. Are their times better than, worse than, or roughly equal to the times one would expect the fastest riders to have if the data came from a truly normal distribution?

(b) Consider the slowest riders. Are their times better than, worse than, or roughly equal to the times one would expect the slowest riders to have if the data came from a truly normal distribution?

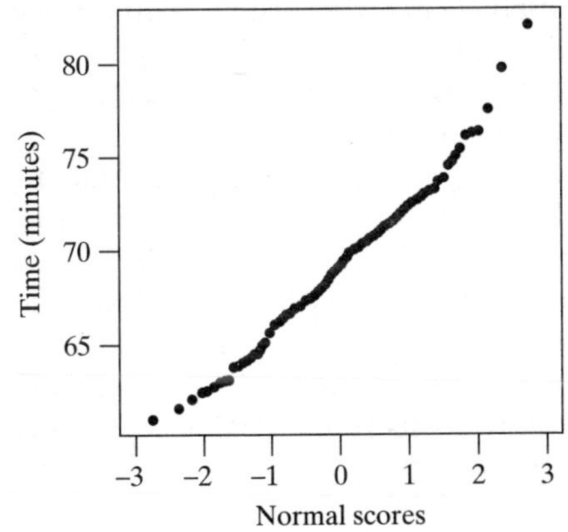

4.4.7 The *P*-values for the Shapiro–Wilk test for the data appearing in quantile plots (a) and (b) are 0.235 and 0.00015. Which *P*-value corresponds to which plot? What is the basis for your decision?

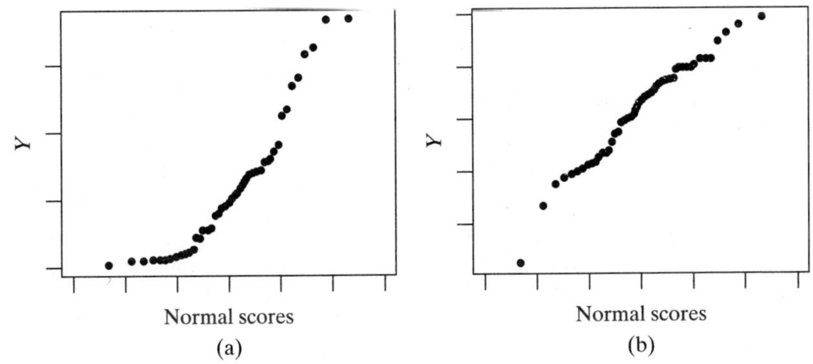

4.4.8

(a) The *P*-value for the Shapiro–Wilk test of normality for the data in Exercise 4.4.3(b) is 0.039. Using this value to justify your answer, does it seem reasonable to believe that these data came from a normal population?

(b) The *P*-value for the Shapiro–Wilk test of normality for the data in Exercise 4.4.2(c) is 0.770. Using this value to justify your answer, does it seem reasonable to believe that these data came from a normal population?

(c) Does the *P*-value in part (b) prove that the data come from a normal population?

4.5 Perspective

The normal distribution is also called the Gaussian distribution, after the German mathematician K. F. Gauss. The term *normal*, with its connotations of "typical" or "usual," can be seriously misleading. Consider, for instance, a medical context, where the primary meaning of "normal" is "not abnormal." Thus, confusingly, the phrase "the normal population of serum cholesterol levels" may refer to cholesterol levels in ideally "healthy" people, or it may refer to a Gaussian distribution such as the one in Example 4.1.1. In fact, for many variables the distribution in the normal (nondiseased) population is decidedly not normal (i.e., not Gaussian).

The examples of this chapter have illustrated one use of the normal distribution—as an approximation to naturally occurring biological distributions. If a natural distribution is well approximated by a normal distribution, then the mean and standard deviation provide a complete description of the distribution: The mean is the center of the distribution, about 68% of the values are within 1 standard deviation of the mean, about 95% are within 2 standard deviations of the mean, and so on.

As noted in Section 2.6, the 68% and 95% benchmarks can roughly be applicable even to distributions that are rather skewed. (But if the distribution is skewed, then the 68% is not symmetrically divided on both sides of the mean, and similarly for the 95%.) However, the benchmarks do not apply to a distribution (even a symmetric one) for which one or both tails are long and thin (see Figures 2.2.13 and 2.2.16).

We will see in later chapters that many classical statistical methods are specifically designed for, and function best with, data that have been sampled from normal populations. We will further see that in many practical situations these methods also work very well for samples from nonnormal populations.

The normal distribution is of central importance in spite of the fact that many, perhaps most, naturally occurring biological distributions could be described better by a skewed curve than by a normal curve. A major use of the normal distribution is not to describe natural distributions, but rather to describe certain theoretical distributions, called sampling distributions, that are used in the statistical analysis of data. We will see in Chapter 5 that many sampling distributions are approximately normal even when the underlying data are not; it is this property that makes the normal distribution so important in the study of statistics.

Supplementary Exercises 4.S.1–4.S.21

4.S.1 The activity of a certain enzyme is measured by counting emissions from a radioactively labeled molecule. For a given tissue specimen, the counts in consecutive 10-second time periods may be regarded (approximately) as repeated independent observations from a normal distribution.[15] Suppose the mean 10-second count for a certain tissue specimen is 1,200 and the standard deviation is 35. Let Y denote the count in a randomly chosen 10-second time period. Find

(a) $\Pr\{Y \geq 1{,}200\}$

(b) $\Pr\{Y \leq 1{,}270\}$

(c) $\Pr\{0 \leq Y \leq 1{,}305\}$

(d) $\Pr\{1{,}270 \leq Y \leq 1{,}400\}$

4.S.2 The bill lengths of a population of male Blue Jays follow approximately a normal distribution with mean equal to 25.4 mm and standard deviation equal to 0.8 mm (as in Example 4.1.2). Find the 95th percentile of the bill length distribution.

4.S.3 Refer to the bill length distribution of Exercise 4.S.2.

(a) What percentage of bill lengths is less than 23.9 mm?

(b) What percentage of bill lengths is greater than 27.1 mm?

4.S.4 The heights of a certain population of corn plants follow a normal distribution with mean 145 cm and standard deviation 22 cm.[16] What percentage of the plant heights are

(a) 100 cm or more?

(b) 120 cm or less?

(c) between 120 and 150 cm?

(d) between 100 and 120 cm?

(e) between 150 and 180 cm?

(f) 180 cm or more?

(g) 150 cm or less?

4.S.5 Suppose four plants are to be chosen at random from the corn plant population of Exercise 4.S.4. Find the probability that none of the four plants will be more than 80 cm tall.

4.S.6 Refer to the corn plant population of Exercise 4.S.4. Find the 90th percentile of the height distribution.

4.S.7 For the corn plant population described in Exercise 4.S.4, find the quartiles and the interquartile range.

4.S.8 Suppose a certain population of observations is normally distributed.

(a) Find the value of z^* such that 80% of the observations in the population are between $-z^*$ and $+z^*$ on the Z scale.

(b) Find the value of z^* such that 90% of the observations in the population are between $-z^*$ and $+z^*$ on the Z scale.

4.S.9 In the nerve-cell activity of a certain individual fly, the time intervals between "spike" discharges follow approximately a normal distribution with mean 15.6 ms and standard deviation 0.4 ms (as in Example 4.1.3). Let Y denote a randomly selected interspike interval. Find

(a) $\Pr\{Y > 15\}$ (b) $\Pr\{Y > 16.5\}$

(c) $\Pr\{15 < Y < 16.5\}$ (d) $\Pr\{15 < Y < 15.5\}$

4.S.10 For the distribution of interspike-time intervals described in Exercise 4.S.9, find the quartiles and the interquartile range.

4.S.11 Among American women aged 20 to 29 years, 10% are less than 60.8 inches tall, 80% are between 60.8 and 67.6 inches tall, and 10% are more than 67.6 inches tall.[17] Assuming that the height distribution can adequately be approximated by a normal curve, find the mean and standard deviation of the distribution.

4.S.12 The intelligence quotient (IQ) score, as measured by the Stanford-Binet IQ test, is normally distributed in a certain population of children. The mean IQ score is 100 points, and the standard deviation is 16 points.[18] What percentage of children in the population have IQ scores

(a) 116 or more? (b) 84 or less?

(c) between 84 and 116? (d) between 84 and 148?

(e) between 116 and 148?

4.S.13 Refer to the IQ distribution of Exercise 4.S.12. Let Y be the IQ score of a child chosen at random from the population. Find $\Pr\{80 \leq Y \leq 140\}$.

4.S.14 Refer to the IQ distribution of Exercise 4.S.12. Suppose five children are to be chosen at random from the population. Find the probability that exactly one of them will have an IQ score of 80 or less and four will have scores higher than 80. (*Hint:* First find the probability that a randomly chosen child will have an IQ score of 80 or less.)

4.S.15 A certain assay for serum alanine aminotransferase (ALT) is rather imprecise. The results of repeated assays of a single specimen follow a normal distribution with mean equal to the true ALT concentration for that specimen and standard deviation equal to 4 U/l (see Example 2.2.12). Suppose that a certain hospital lab measures many specimens every day, performing one assay for each specimen, and that specimens with ALT readings of 40 U/l or more are flagged as "unusually high." If a patient's true ALT concentration is 35 U/l, what is the probability that his specimen will be flagged as "unusually high"?

4.S.16 Resting heart rate was measured for a group of subjects; the subjects then drank 6 ounces of coffee. Ten minutes later their heart rates were measured again. The change in heart rate followed a normal distribution, with a mean increase of 7.3 beats per minute and a standard deviation of 11.1.[19] Let Y denote the change in heart rate for a randomly selected person. Find

(a) $\Pr\{Y > 10\}$ (b) $\Pr\{Y > 20\}$

(c) $\Pr\{5 < Y < 15\}$

4.S.17 Refer to the heart rate distribution of Exercise 4.S.16. The fact that the standard deviation is greater than the average and that the distribution is normal tells us that some of the data values are negative, meaning that the person's heart rate went down, rather than up. Find the probability that a randomly chosen person's heart rate will go down. That is, find $\Pr\{Y < 0\}$.

4.S.18 Refer to the heart rate distribution of Exercise 4.S.16. Suppose we take a random sample of size 400 from this distribution. How many observations do we expect to obtain that fall between 0 and 15?

4.S.19 Refer to the heart rate distribution of Exercise 4.S.16. If we use the $1.5 \times \text{IQR}$ rule, from Chapter 2, to identify outliers, how large would an observation need to be in order to be labeled an outlier on the upper end?

4.S.20 It is claimed that the heart rates of Exercise 4.S.16 follow a normal distribution. If this is true, which of the following Shapiro–Wilk's test P-values for a random sample of 15 subjects are consistent with this claim?

(a) P-value $= 0.0649$ (b) P-value $= 0.1545$

(c) P-value $= 0.2498$ (d) P-value $= 0.0005$

4.S.21 The following four normal quantile plots, (a), (b), (c), and (d), were generated from the distributions shown by histograms I, II, and III and another histogram that is not shown. Which normal quantile plot goes with which histogram? How do you know? (There will be one normal quantile plot that is not used.)

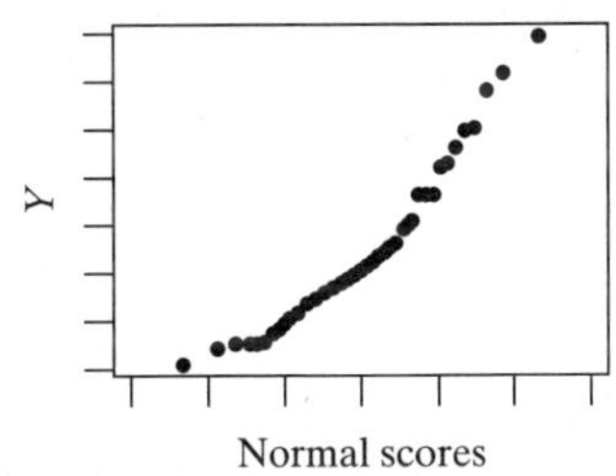

(a)

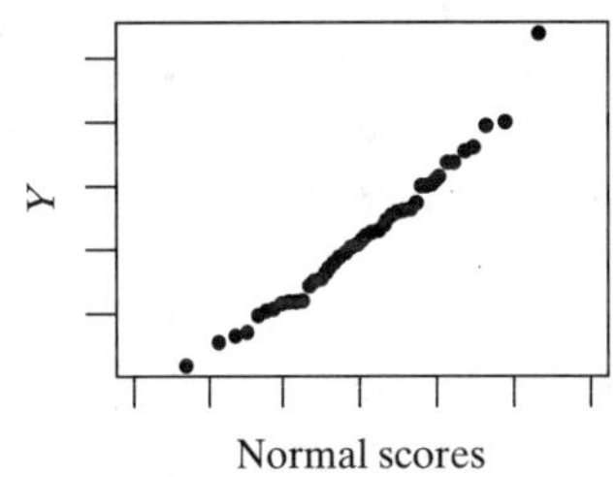

(b)

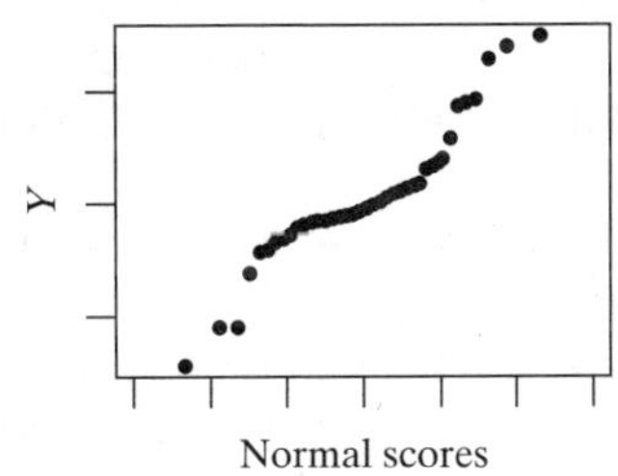

(c)

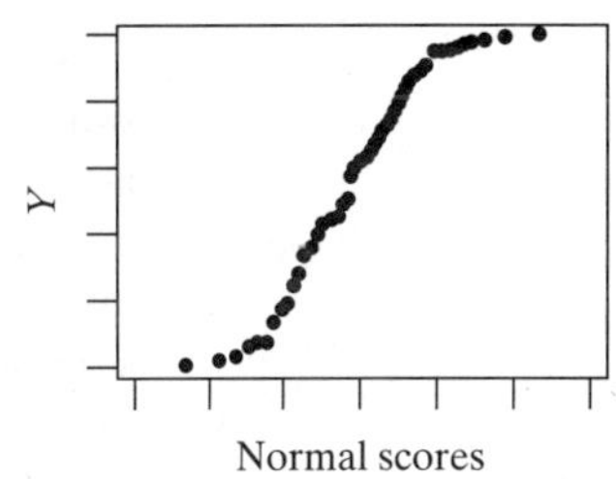

(d)

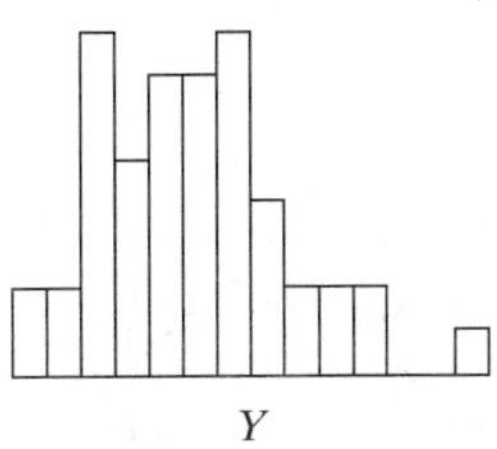

I

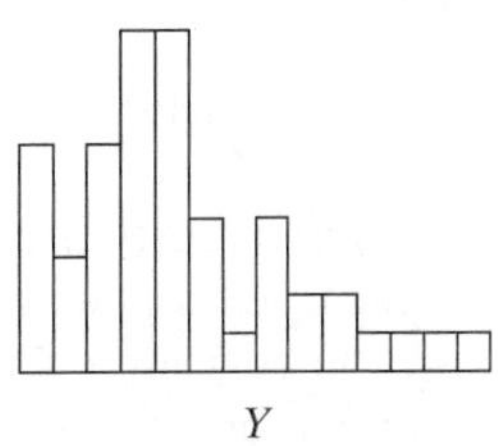

II

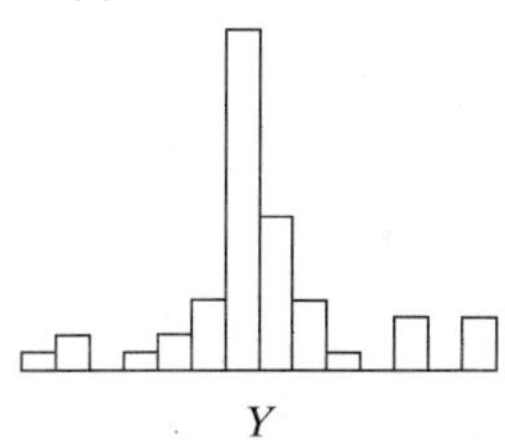

III

Chapter 5

SAMPLING DISTRIBUTIONS

OBJECTIVES

In this chapter we will develop the idea of a sampling distribution, which is central to classical statistical inference. In particular, we will

- describe sampling distributions.
- show how the sample size is related to the accuracy of the sample mean.
- explore the Central Limit Theorem.
- demonstrate how the normal distribution can be used to approximate the binomial distribution.

5.1 Basic Ideas

An important goal of data analysis is to distinguish between features of the data that reflect real biological facts and features that may reflect only chance effects. As explained in Sections 1.3 and 2.8, the random sampling model provides a framework for making this distinction. The underlying reality is visualized as a population, the data are viewed as a random sample from the population, and chance effects are regarded as sampling error—that is, discrepancy between the sample and the population.

In this chapter we develop the theoretical background that will enable us to place specific limits on the degree of sampling error to be expected in a study. (Although in Chapter 1 we distinguished between an experimental study and an observational study, for the present discussion we will call any scientific investigation a *study*.) As in earlier chapters, we continue to confine the discussion to the simple context of a study with only one group (one sample).

SAMPLING VARIABILITY

The variability among random samples from the same population is called **sampling variability**. A probability distribution that characterizes some aspect of sampling variability is termed a **sampling distribution**. Usually a random sample will resemble the population from which it came. Of course, we have to expect a certain amount of discrepancy between the sample and the population. A sampling distribution tells us how close the resemblance between the sample and the population is likely to be.

In this chapter we will discuss several aspects of sampling variability and study an important sampling distribution. From this point forward, we will assume that the sample size is a negligibly small fraction of the population size. This assumption simplifies the theory because it guarantees that the process of drawing the sample does not change the population composition in any appreciable way.

THE META-STUDY

According to the random sampling model, we regard the data in a study as a random sample from a population. Generally we obtain only a single random sample, which comes from a very large population. However, to visualize sampling variability we must broaden our frame of reference to include not merely one sample, but all the possible samples that might be drawn from the population. This wider frame of reference we will call the **meta-study**. A meta-study consists of indefinitely many repetitions, or replications, of the same study.* Thus, if the study consists of drawing a random sample of size n from some population, the corresponding meta-study involves drawing *repeated* random samples of size n from the same population. The process of repeated drawing is carried on indefinitely, with the members of each sample being replaced before the next sample is drawn. The study and the meta-study are schematically represented in Figure 5.1.1.

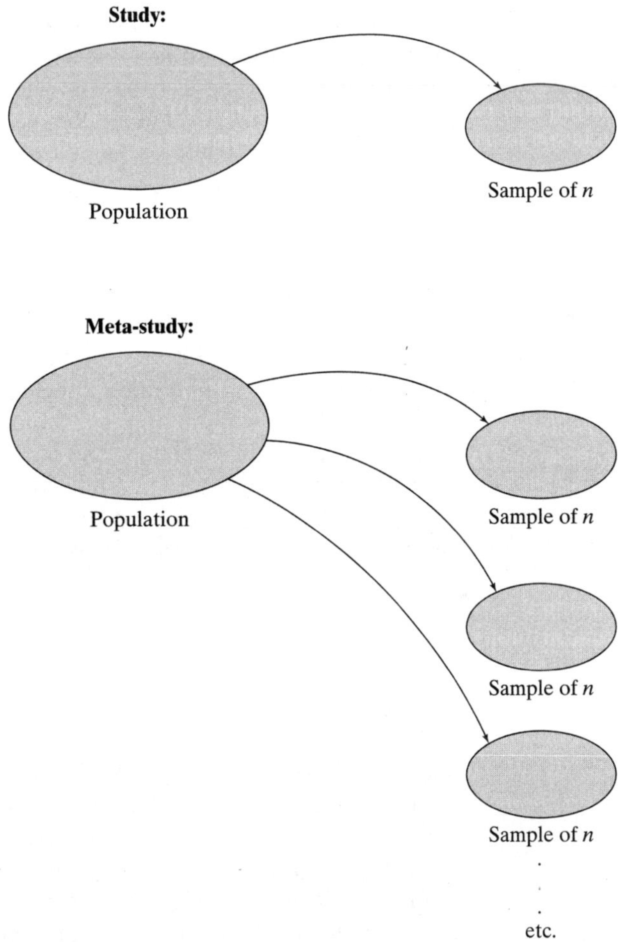

Figure 5.1.1 Schematic representation of study and meta-study

The following two examples illustrate the notion of a meta-study.

Example 5.1.1

Rat Blood Pressure A study consists of measuring the change in blood pressure in each of $n = 10$ rats after administering a certain drug. The corresponding meta-study would consist of repeatedly choosing groups of $n = 10$ rats from the same population and making blood pressure measurements under the same conditions. ■

*The term *meta-study* is not a standard term. It is unrelated to the term *meta-analysis*, which denotes a particular type of statistical analysis.

Example 5.1.2 **Bacterial Growth** A study consists of observing bacterial growth in $n = 5$ petri dishes that have been treated identically. The corresponding meta-study would consist of repeatedly preparing groups of five petri dishes and observing them in the same way. ■

Note that a meta-study is a theoretical construct rather than an operation that is actually performed by a researcher.

The meta-study concept provides a link between sampling variability and probability. Recall from Chapter 3 that the probability of an event can be interpreted as the long-run relative frequency of occurrence of the event. Choosing a random sample is a chance operation; the meta-study consists of many repetitions of this chance operation, and so *probabilities concerning a random sample can be interpreted as relative frequencies in a meta-study.* Thus, the meta-study is a device for explicitly visualizing a sampling distribution: The sampling distribution describes the variability, for a chosen statistic, among the many random samples in a meta-study.

We consider a small (and artificial) example to illustrate the idea of a sampling distribution.

Example 5.1.3 **Knee Replacement** Consider a population of women ages 65 to 75 who are experiencing pain in their knees and are candidates for knee replacement surgery. A woman might have replacement surgery done on one knee at a cost of $35,000, both knees at a cost of $60,000 (a "double replacement," which is less expensive than two single replacements), or neither knee. Consider the perspective of an insurance company regarding a sample of $n = 3$ women it insures: What is the total cost for treating these three? The smallest the total could be is zero—if all three women skip surgery—while the largest possible cost would be $180,000—if all three women have double replacements. To keep things relatively simple, suppose that one-fourth of women ages 65 to 75 elect a double knee replacement, one-half elect a single knee replacement, and one-fourth choose not to have surgery.

The complete list of possible samples is given in Table 5.1.1, along with the sample total (in thousands of dollars) in each case and the probability of each case arising. For example, the probability that all three women skip surgery ("None, None, None") is $(1/4) \times (1/4) \times (1/4) = 1/64$ while the probability that the first two women skip surgery and the third has a single knee operation ("None, None, Single") is $(1/4) \times (1/4) \times (2/4) = 2/64$. There are 10 possible values for the sample total: 0, 35, 60, 70, 95, 105, 120, 130, 155, and 180. The first and third columns of Table 5.1.2 give the sampling distribution of the sample total by combining the samples that yield the same total and summing their probabilities. For example, there are three ways for the total to be 70, each of which has probability $4/64$; these sum to $12/64$.

The second column of Table 5.1.2 shows the sample mean (rounded to one decimal place) so that the last two columns of the table give the sampling distribution of the sample mean. These two distributions, shown graphically in Figure 5.1.2, are scaled versions of each other. An insurance company might speak in terms of total cost, but this is equivalent to looking at average cost. ■

RELATIONSHIP TO STATISTICAL INFERENCE

Knowing a sampling distribution allows one to make probability statements about possible samples. For example, for the setting in Example 5.1.3 the insurance company might ask, What is the probability that the total knee replacement costs for a sample of three women will be less than $110,000? We can answer this question by adding the probabilities of the first six outcomes listed in Table 5.1.2; the sum is $42/64$. We will expand upon this idea as we formally develop ideas of statistical inference.

Table 5.1.1 Total knee replacement costs for all possible samples of size $n = 3$

Sample	Costs (in units of \$1,000)	Sample total	Probability
None, None, None	0, 0, 0	0	1/64
None, None, Single	0, 0, 35	35	2/64
None, None, Double	0, 0, 60	60	1/64
None, Single, None	0, 35, 0	35	2/64
None, Single, Single	0, 35, 35	70	4/64
None, Single, Double	0, 35, 60	95	2/64
None, Double, None	0, 60, 0	60	1/64
None, Double, Single	0, 60, 35	95	2/64
None, Double, Double	0, 60, 60	120	1/64
Single, None, None	35, 0, 0	35	2/64
Single, None, Single	35, 0, 35	70	4/64
Single, None, Double	35, 0, 60	95	2/64
Single, Single, None	35, 35, 0	70	4/64
Single, Single, Single	35, 35, 35	105	8/64
Single, Single, Double	35, 35, 60	130	4/64
Single, Double, None	35, 60, 0	95	2/64
Single, Double, Single	35, 60, 35	130	4/64
Single, Double, Double	35, 60, 60	155	2/64
Double, None, None	60, 0, 0	60	1/64
Double, None, Single	60, 0, 35	95	2/64
Double, None, Double	60, 0, 60	120	1/64
Double, Single, None	60, 35, 0	95	2/64
Double, Single, Single	60, 35, 35	130	4/64
Double, Single, Double	60, 35, 60	155	2/64
Double, Double, None	60, 60, 0	120	1/64
Double, Double, Single	60, 60, 35	155	2/64
Double, Double, Double	60, 60, 60	180	1/64

Table 5.1.2 Sampling distribution of total surgery costs for samples of size $n = 3$

Sample total	Sample mean	Probability
0	0.0	1/64
35	11.7	6/64
60	20.0	3/64
70	23.3	12/64
95	31.7	12/64
105	35.0	8/64
120	40.0	3/64
130	43.3	12/64
155	51.7	6/64
180	60.0	1/64

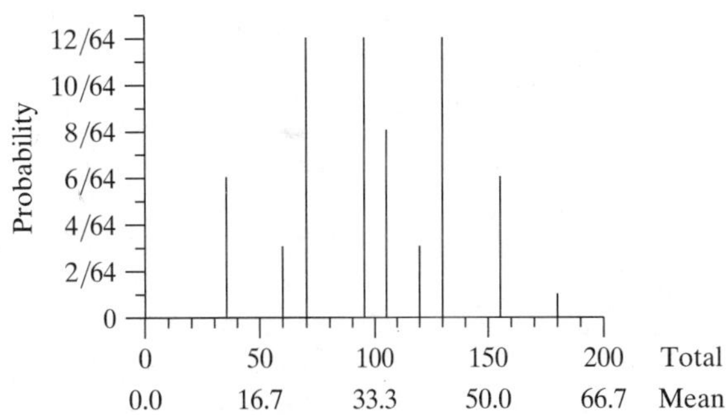

Figure 5.1.2 Graph of the sampling distribution of total surgery costs for samples of size $n = 3$

Exercises 5.1.1–5.1.5

5.1.1 Consider taking a random sample of size 3 from the knee replacement population of Example 5.1.3. What is the probability that the total cost for those in the sample will be greater than \$125,000?

5.1.2 Consider taking a random sample of size 3 from the knee replacement population of Example 5.1.3. What is the probability that the total cost for those in the sample will be between \$80,000 and \$125,000?

5.1.3 Consider taking a random sample of size 3 from the knee replacement population of Example 5.1.3. What is the probability that the mean cost for those in the sample will be between \$30,000 and \$45,000?

5.1.4 Consider taking a random sample of size 2 (rather than of size 3) from the knee replacement population of Example 5.1.3.

(a) How many possible samples of size $n = 2$ are there and what are those possible samples? (*Hint:* Consider the top third of Table 5.1.1 and ignore the "None" entry for the first of woman.)

(b) What is the sampling distribution of the total surgery costs for samples of size $n = 2$? That is, make a table similar to Table 5.1.2 but for $n = 2$ rather than $n = 3$.

(c) What is the probability that the total surgery costs will exceed \$75,000?

5.1.5 Consider a hypothetical population of dogs in which there are four possible weights, all of which are equally likely: 42, 48, 52, or 58 pounds. If a sample of size $n = 2$ is drawn from this population, what is the sampling distribution of the total weight of the two dogs selected? That is, what are the possible values for the total and what are the probabilities associated with each of those values?

5.2 The Sample Mean

For a quantitative variable, the sample and the population can be described in various ways—by the mean, the median, the standard deviation, and so on. The characteristics (e.g., shape, center, spread) of the sampling distributions for these descriptive measures are not all the same. In this section we will focus primarily on the sampling distribution of the sample mean.

THE SAMPLING DISTRIBUTION OF $\bar{Y}$

The sample mean $\bar{y}$ can be used, not only as a description of the data in the sample, but also as an estimate of the population mean μ. It is natural to ask, "How close to μ is $\bar{y}$?" We cannot answer this question for the mean $\bar{y}$ of a particular sample, but we can answer it if we think in terms of the random sampling model and regard the sample mean as a random variable $\bar{Y}$. The question then becomes: "How close to μ is $\bar{Y}$ *likely* to be?" and the answer is provided by the **sampling distribution of $\bar{Y}$**—that is, the probability distribution that describes sampling variability in $\bar{Y}$.

To visualize the sampling distribution of $\bar{Y}$, imagine the meta-study as follows: Random samples of size n are repeatedly drawn from a fixed population with mean μ and standard deviation σ; each sample has its own mean $\bar{y}$. The variation of the $\bar{y}$'s among the samples is specified by the sampling distribution of $\bar{Y}$. This relationship is indicated schematically in Figure 5.2.1.

When we think of $\bar{Y}$ as a random variable, we need to be aware of two basic facts. The first of these is intuitive: On average, the sample mean equals the population mean. That is, the average of the sampling distribution of $\bar{Y}$ is μ. The second fact is not obvious: The standard deviation of $\bar{Y}$ is equal to the standard deviation of Y divided by the square root of the sample size. That is, the standard deviation of $\bar{Y}$ is $\sigma/\sqrt{n}$. While the formula for the standard deviation of $\bar{Y}$ might not be intuitive, note that the formula indicates that as the sample size increases, the standard deviation of $\bar{Y}$ decreases. That is, for larger samples, $\bar{Y}$ will tend to vary less

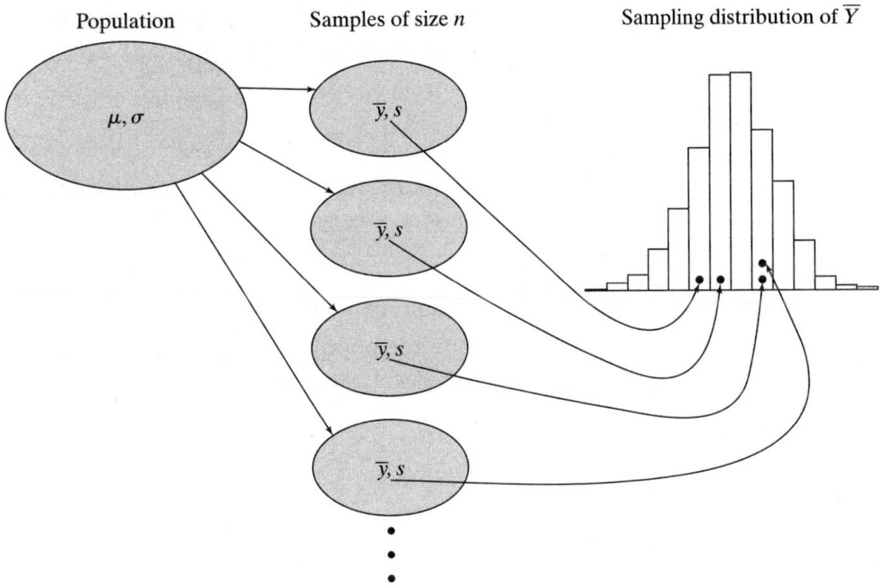

Figure 5.2.1 Schematic representation of the sampling distribution of $\overline{Y}$

from sample to sample; it will tend to be closer to the population mean than for smaller samples.

Example 5.2.1

Serum Cholesterol The serum cholesterol levels of 12- to 14-year-olds follow a normal distribution with mean $\mu = 155$ mg/dl and standard deviation $\sigma = 27$ mg/dl.[1] If we take a random sample, then we expect the sample mean to be near 155, with the means of some samples being larger than 155 and the means of some samples being smaller than 155. As the preceding formula indicates, the amount of variability in the sample mean depends on the variability of cholesterol levels of the population, σ. If the population is very homogeneous (everyone has nearly the same cholesterol value so that σ is small), then samples and hence sample means would all be very similar and thus exhibit low variability. If the population is very heterogeneous (σ is large), then samples (and hence sample mean values) would vary more. While researchers have little control over the value of σ, we can control the sample size, n, and n affects the amount of variability in the sample mean. If we take a sample of size $n = 9$, then the standard deviation of the sample mean is $\frac{27}{\sqrt{9}} = \frac{27}{3} = 9.0$. This means, loosely speaking, that the sample mean, $\overline{Y}$, will vary from one sample to the next by about 9.0 mg/dl.* If we took larger random samples of size $n = 25$, then the standard deviation of the sample mean would be smaller: $\frac{27}{\sqrt{25}} = \frac{27}{5} = 5.4$, which means that $\overline{Y}$ would vary from one sample to the next by about 5.4. As the sample size goes up, the variability in the sample mean $\overline{Y}$ goes down. ■

We now state as a theorem the basic facts about the sampling distribution of $\overline{Y}$. The theorem can be proved using the methods of mathematical statistics; we will

*Strictly speaking, the standard deviation measures deviation from the mean, not the difference between consecutive observations

state it without proof. The theorem describes the sampling distribution of $\overline{Y}$ in terms of its mean (denoted by $\mu_{\overline{Y}}$), its standard deviation (denoted by $\sigma_{\overline{Y}}$), and its shape.*

Theorem 5.2.1: The Sampling Distribution of $\overline{Y}$

1. **Mean** The mean of the sampling distribution of $\overline{Y}$ is equal to the population mean. In symbols,

$$\mu_{\overline{Y}} = \mu$$

2. **Standard deviation** The standard deviation of the sampling distribution of $\overline{Y}$ is equal to the population standard deviation divided by the square root of the sample size. In symbols,

$$\sigma_{\overline{Y}} = \frac{\sigma}{\sqrt{n}}$$

3. **Shape**
   (a) If the population distribution of Y is normal, then the sampling distribution of $\overline{Y}$ is normal, regardless of the sample size n.
   (b) *Central Limit Theorem* If n is large, then the sampling distribution of $\overline{Y}$ is approximately normal, even if the population distribution of Y is not normal.

Parts 1 and 2 of Theorem 5.2.1 specify the relationship between the mean and standard deviation of the population being sampled, and the mean and standard deviation of the sampling distribution of $\overline{Y}$. Part 3(a) of the theorem states that, if the observed variable Y follows a normal distribution in the population being sampled, then the sampling distribution of $\overline{Y}$ is also a normal distribution. These relationships are indicated in Figure 5.2.2.

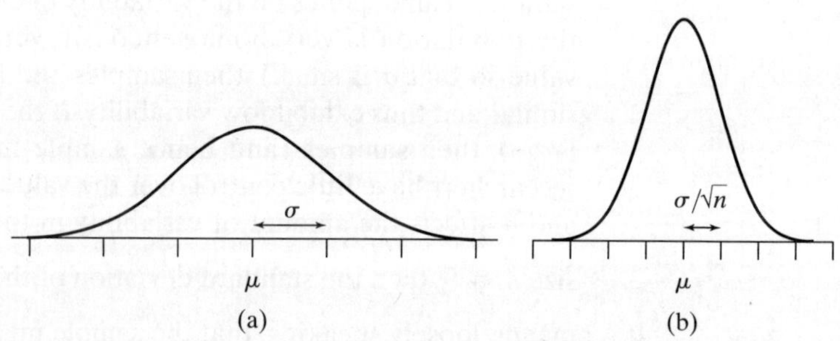

Figure 5.2.2 (a) The population distribution of a normally distributed variable Y; (b) the sampling distribution of $\overline{Y}$ in samples from the population of part (a)

The following example illustrates the meaning of parts 1, 2, and 3(a) of Theorem 5.2.1.

*We are assuming here that the population is infinitely large or, equivalently, that we are sampling with replacement so that we never exhaust the population. If we sample without replacement from a finite population then an adjustment is needed to get the right value for $\sigma_{\overline{Y}}$. Here $\sigma_{\overline{Y}}$ is given by $\frac{\sigma}{\sqrt{n}} \times \sqrt{\frac{N-n}{N-1}}$. The term $\sqrt{\frac{N-n}{N-1}}$ is called the **finite population correction factor**. Note that if the sample size n is 10% of the population size N, then the correction factor is $\sqrt{\frac{0.9N}{N-1}} \approx 0.95$, so the adjustment is small. Thus, if n is small, in comparison to N, then the finite population correction factor is close to 1 and can be ignored.

Example 5.2.2

Weights of Seeds A large population of seeds of the princess bean *Phaseotus vulgaris* is to be sampled. The weights of the seeds in the population follow a normal distribution with mean $\mu = 500$ mg and standard deviation $\sigma = 120$ mg.[2] Suppose now that a random sample of four seeds is to be weighed, and let $\overline{Y}$ represent the mean weight of the four seeds. Then, according to Theorem 5.2.1, the sampling distribution of $\overline{Y}$ will be a normal distribution with mean and standard deviation as follows:

$$\mu_{\overline{Y}} = \mu = 500 \text{ mg}$$

and

$$\sigma_{\overline{Y}} = \frac{\sigma}{\sqrt{n}} = \frac{120}{\sqrt{4}} = 60 \text{ mg}$$

Thus, on average the sample mean will equal 500 mg, but the variability from one sample of size 4 to the next sample of size 4 is such that about 68% of the time $\overline{Y}$ will be within 60 mg of 500 mg, that is, between $500 - 60 = 440$ mg and $500 + 60 = 560$ mg. Likewise, allowing for 2 SDs, we expect that $\overline{Y}$ will be within 120 mg of 500 mg or between $500 - 120 = 380$ mg and $500 + 120 = 620$ mg about 95% of the time. The sampling distribution of $\overline{Y}$ is shown in Figure 5.2.3; the ticks are 1 $\sigma_{\overline{Y}}$ apart. ■

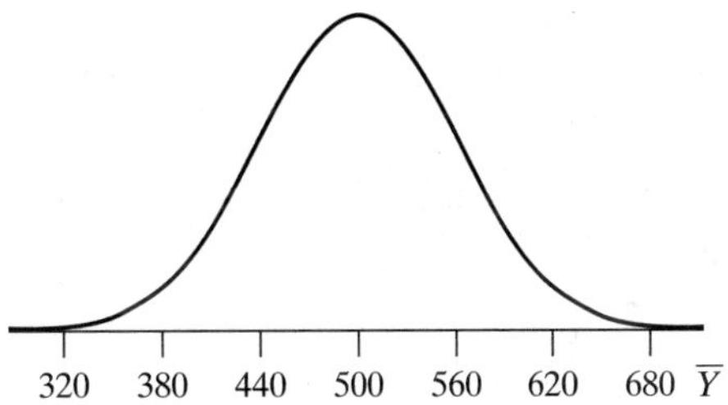

Figure 5.2.3 Sampling distribution of $\overline{Y}$ for Example 5.2.2

The sampling distribution of $\overline{Y}$ expresses the relative likelihood of the various possible values of $\overline{Y}$. For example, suppose we want to know the probability that the mean weight of the four seeds will be greater than 550 mg. This probability is shown as the shaded area in Figure 5.2.4. Notice that the value of $\bar{y} = 550$ must be converted to the Z scale using the standard deviation $\sigma_{\overline{Y}} = 60$, not $\sigma = 120$.

$$z = \frac{\bar{y} - \mu_{\overline{Y}}}{\sigma_{\overline{Y}}} = \frac{550 - 500}{60} = 0.83$$

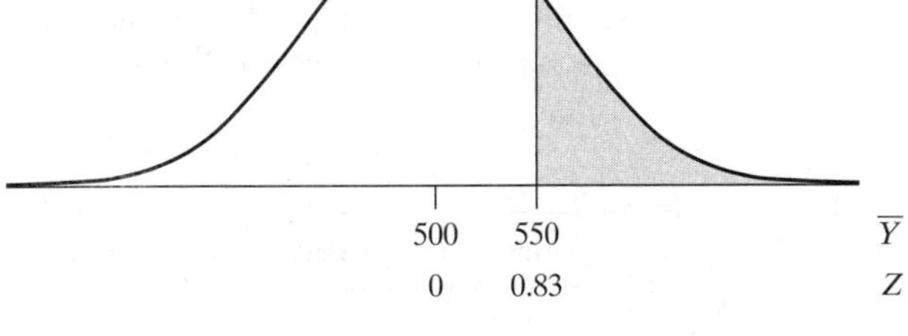

Figure 5.2.4 Calculation of $\Pr\{\overline{Y} > 550\}$ for Example 5.2.2

From Table 3, $z = 0.83$ corresponds to an area of 0.7967. Thus,

$$\Pr\{\bar{Y} > 550\} = \Pr\{Z > 0.83\} = 1 - 0.7967$$
$$= 0.2033 \approx 0.20$$

This probability can be interpreted in terms of a meta-study as follows: If we were to choose many random samples of four seeds each from the population, then about 20% of the samples would have a mean weight exceeding 550 mg.

Part 3(b) of Theorem 5.2.1 is known as the **Central Limit Theorem**. The Central Limit Theorem states that, *no matter what distribution Y may have in the population*,* if the sample size is large enough, then the sampling distribution of $\bar{Y}$ will be approximately a normal distribution.

The Central Limit Theorem is of fundamental importance because it can be applied when (as often happens in practice) the form of the population distribution is not known. It is because of the Central Limit Theorem (and other similar theorems) that the normal distribution plays such a central role in statistics.

It is natural to ask how "large" a sample size is required by the Central Limit Theorem: How large must n be in order that the sampling distribution of $\bar{Y}$ be well approximated by a normal curve? The answer is that the required n depends on the shape of the population distribution. If the shape is normal, any n will do. If the shape is moderately nonnormal, a moderate n is adequate. If the shape is highly nonnormal, then a rather large n will be required. (Some specific examples of this phenomenon are given in the optional Section 5.3.)

Remark We stated in Section 5.1 that the theory of this chapter is valid if the sample size is small compared to the population size. But the Central Limit Theorem is a statement about large samples. This may seem like a contradiction: How can a large sample be a small sample? In practice, there is no contradiction. In a typical biological application, the population size might be 10^6; a sample of size $n = 100$ would be a small fraction of the population but would nevertheless be large enough for the Central Limit Theorem to be applicable (in most situations).

DEPENDENCE ON SAMPLE SIZE

Consider the possibility of choosing random samples of various sizes from the same population. The sampling distribution of $\bar{Y}$ will depend on the sample size n in two ways. First, its standard deviation is

$$\sigma_{\bar{Y}} = \frac{\sigma}{\sqrt{n}}$$

and this is inversely proportional to $\sqrt{n}$. Second, if the population distribution is not normal, then the *shape* of the sampling distribution of $\bar{Y}$ depends on n, being more nearly normal for larger n. However, if the population distribution is normal, then the sampling distribution of $\bar{Y}$ is always normal, and only the standard deviation depends on n.

The more important of the two effects of sample size is the first: Larger n gives a smaller value of $\sigma_{\bar{Y}}$ and consequently a smaller expected sampling error if $\bar{y}$ is used as an estimate of μ. The following example illustrates this effect for sampling from a normal population.

*Technically, the Central Limit Theorem requires that the distribution of Y have a standard deviation. In practice this condition is always met.

Example 5.2.3

Weights of Seeds Figure 5.2.5 shows the sampling distribution of $\bar{Y}$ for samples of various sizes from the princess bean population of Example 5.2.2, which had a mean of 500 and standard deviation of 120 mg. Notice that for larger n the sampling distribution is more concentrated around the population mean. As a consequence, the probability that $\bar{Y}$ is close to it is larger for larger n. For instance, consider the probability that $\bar{Y}$ is within ± 50 mg of μ, that is, $\Pr\{450 \leq \bar{Y} \leq 550\}$. Table 5.2.1 shows how this probability depends on n. ■

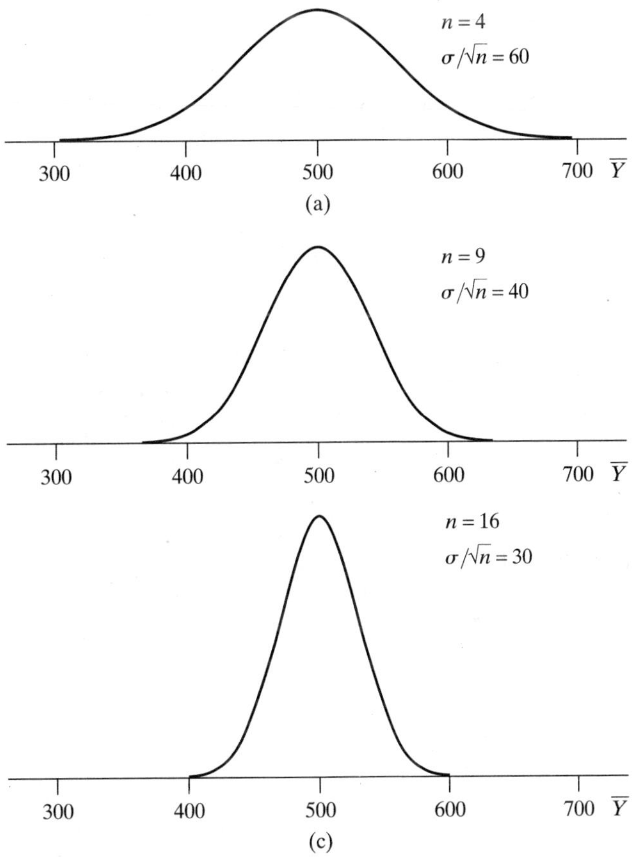

Figure 5.2.5 Sampling distribution of $\bar{Y}$ for various sample sizes n

Table 5.2.1

n	$\Pr\{450 \leq \bar{Y} \leq 550\}$
4	0.59
9	0.79
16	0.91
64	0.999

Example 5.2.3 illustrates how the closeness of $\bar{Y}$ to μ depends on sample size. The mean of a larger sample is not *necessarily* closer to it than the mean of a smaller sample, but it has a *greater probability* of being close. It is in this sense that a larger sample provides more information about the population mean than a smaller sample.

POPULATIONS, SAMPLES, AND SAMPLING DISTRIBUTIONS

In thinking about Theorem 5.2.1, it is important to distinguish clearly among three different distributions related to a quantitative variable Y: (1) the distribution of Y in the population; (2) the distribution of Y in a sample of data, and (3) the sampling distribution of $\bar{Y}$. The means and standard deviations of these distributions are summarized in Table 5.2.2.

Table 5.2.2

Distribution	Mean	Standard deviation
Y in population	μ	σ
Y in sample	$\bar{y}$	s
$\bar{Y}$ (in meta-study)	$\mu_{\bar{Y}} = \mu$	$\sigma_{\bar{Y}} = \dfrac{\sigma}{\sqrt{n}}$

The following example illustrates the distinction among the three distributions.

Example 5.2.4 **Weights of Seeds** For the princess bean population of Example 5.2.2, the population mean and standard deviation are $\mu = 500$ mg and $\sigma = 120$ mg; the population distribution of Y = weight is represented in Figure 5.2.6(a). Suppose we weigh a random sample of $n = 25$ seeds from the population and obtain the data in Table 5.2.3.

For the data in Table 5.2.3, the sample mean is $\bar{y} = 526.1$ mg and the sample standard deviation is $s = 113.7$ mg. Figure 5.2.6(b) shows a histogram of the data; this histogram represents the distribution of Y in the sample. The sampling distribution of $\bar{Y}$ is a theoretical distribution which relates, not to the particular sample shown in the histogram, but rather to the meta-study of repeated samples of size $n = 25$. The mean and standard deviation of the sampling distribution are

$$\mu_{\bar{Y}} = 500 \text{ mg and } \sigma_{\bar{Y}} = 120/\sqrt{25} = 24 \text{ mg}$$

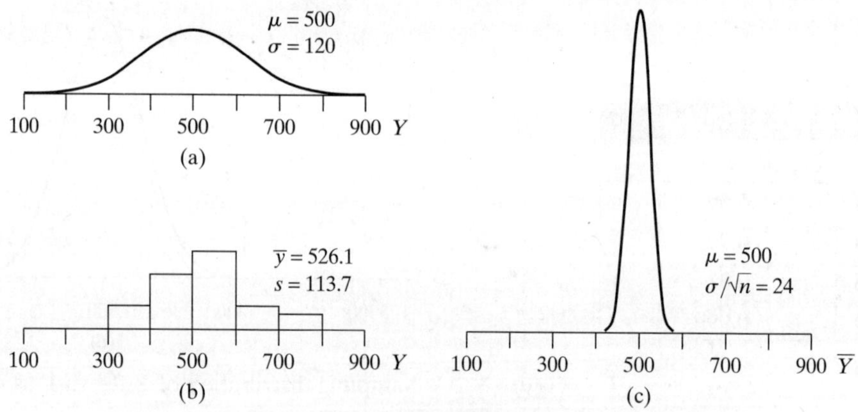

Figure 5.2.6 Three distributions related to Y = seed weight of princess beans: (a) population distribution of Y; (b) distribution of 25 observations of Y; (c) sampling distribution of $\bar{Y}$ for $n = 25$

Table 5.2.3 Weights of 25 princess bean seeds

Weight (mg)						
343	755	431	480	516	469	694
659	441	562	597	502	612	549
348	469	545	728	416	536	581
433	583	570	334			

The sampling distribution is represented in Figure 5.2.6(c). Notice that the distributions in Figures 5.2.6(a) and (b) are more or less similar; in fact, the distribution in (b) is an estimate (based on the data in Table 5.2.3) of the distribution in (a). By contrast, the distribution in (c) is much narrower, because it represents a distribution of *means* rather than of individual observations. ■

OTHER ASPECTS OF SAMPLING VARIABILITY

The preceding discussion has focused on sampling variability in the sample mean, $\bar{Y}$. Two other important aspects of sampling variability are (1) sampling variability in the sample standard deviation, s, and (2) sampling variability in the *shape* of the sample, as represented by the sample histogram. Rather than discuss these aspects formally, we illustrate them with the following example.

Example 5.2.5

Weights of Seeds In Figure 5.2.6(b) we displayed a random sample of 25 observations from the princess bean population of Example 5.2.2; now we display in Figure 5.2.7 eight additional random samples from the same population. (All nine samples were actually simulated using a computer.) Notice that, even though the samples were drawn from a normal population [pictured in Figure 5.2.6(a)], there is very substantial variation in the forms of the histograms. Notice also that there is considerable variation in the sample standard deviations. Of course, if the sample size were larger (say, $n = 100$ rather than $n = 25$), there would be less sampling variation; the histograms would tend to resemble a normal curve more closely, and the standard deviations would tend to be closer to the population value ($\sigma = 120$). ■

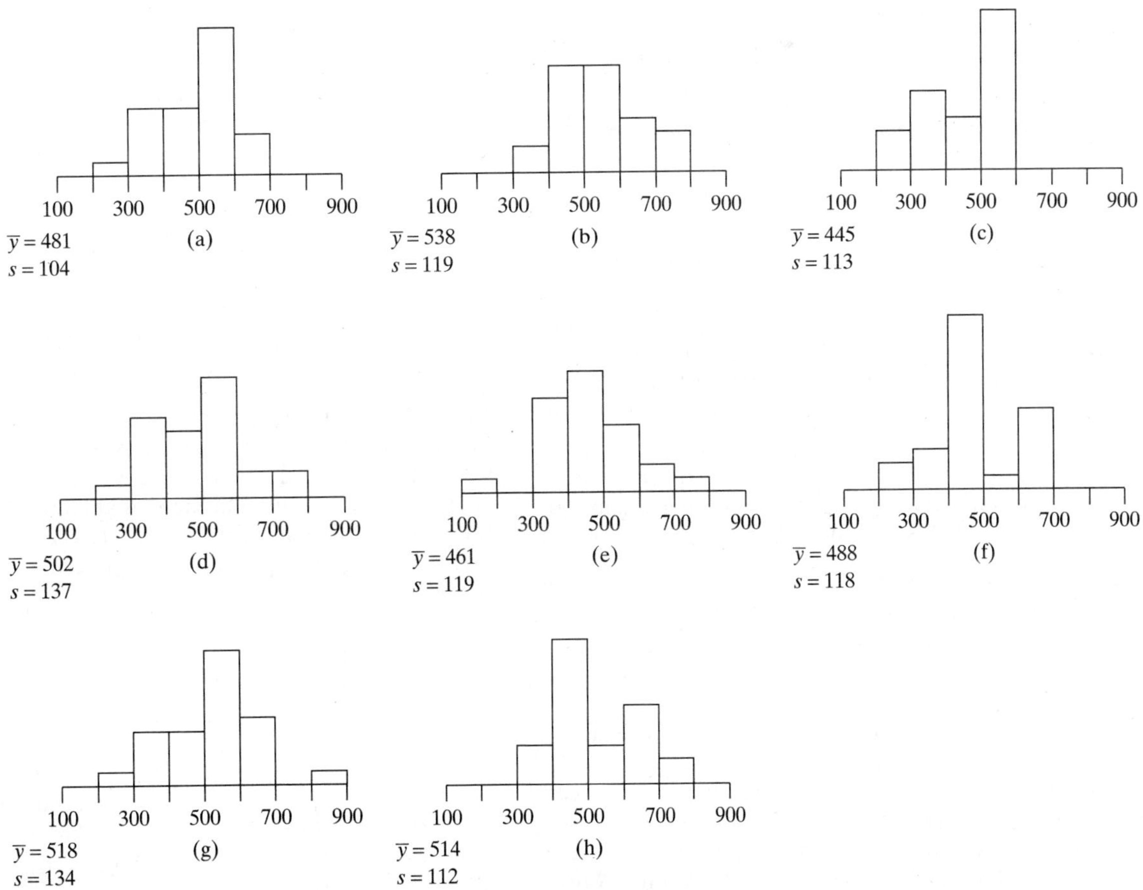

Figure 5.2.7 Eight random samples, each of size $n = 25$, from a normal population with $\mu = 500$ and $\sigma = 120$

Exercises 5.2.1–5.2.19

5.2.1 (Sampling exercise) Refer to Exercise 1.3.6. The collection of 100 ellipses shown there can be thought of as representing a natural population of the organism *C. ellipticus*. Use your judgment to choose a sample of 5 ellipses that you think should be reasonably representative of the population. (In order to best simulate the analogous judgment in a real-life setting, you should make your choice intuitively, without any detailed preliminary study of the population.) With a metric ruler, measure the length of each ellipse in your sample. Measure only the body, excluding any tail bristles; measurements to the nearest millimeter will be adequate. Compute the mean and standard deviation of the five lengths. To facilitate the pooling of results from the entire class, express the mean and standard deviation in millimeters, keeping two decimal places.

5.2.2 (Sampling exercise) Proceed as in Exercise 5.2.1, but use random sampling rather than "judgment" sampling. To do this, choose 10 random digits (from Table 1 or your calculator). Let the first 2 digits be the number of the first ellipse that goes into your sample, and so on. The 10 random digits will give you a random sample of five ellipses.

5.2.3 (Sampling exercise) Proceed as in Exercise 5.2.2, but choose a random sample of 20 ellipses.

5.2.4 The serum cholesterol levels of a population of 12- to 14-year-olds follow a normal distribution with mean 155 mg/dl and standard deviation 27 mg/dl (as in Example 4.1.1).

(a) What percentage of the 12- to 14-year-olds have serum cholesterol values between 145 and 165 mg/dl?

(b) Suppose we were to choose at random from the population a large number of groups of nine 12- to 14-year-olds each. In what percentage of the groups would the group mean cholesterol value be between 145 and 165 mg/dl?

(c) If $\bar{Y}$ represents the mean cholesterol value of a random sample of nine 12- to 14-year-olds from the population, what is $\Pr\{145 \le \bar{Y} \le 165\}$?

5.2.5 Refer to Exercise 5.2.4. Suppose we take a random sample of sixteen 12- to 14-year-olds from the population.

(a) What is the probability that the mean cholesterol value for the group will be between 145 and 165?

(b) What is the probability that the mean cholesterol value for the group will be between 140 and 170?

5.2.6 An important indicator of lung function is forced expiratory volume (FEV), which is the volume of air that a person can expire in one second. Dr. Hernandez plans to measure FEV in a random sample of n young women from a certain population, and to use the sample mean $\bar{y}$ as an estimate of the population mean. Let E be the event that Hernandez's sample mean will be within ± 100 ml of the population mean. Assume that the population distribution is normal with mean 3,000 ml and standard deviation 400 ml.[3] Find $\Pr\{E\}$ if

(a) $n = 15$

(b) $n = 60$

(c) How does $\Pr\{E\}$ depend on the sample size? That is, as n increases, does $\Pr\{E\}$ increase, decrease, or stay the same?

5.2.7 Refer to Exercise 5.2.6. Assume that the population distribution of FEV is normal with standard deviation 400 ml.

(a) Find $\Pr\{E\}$ if $n = 15$ and the population mean is 2,800 ml.

(b) Find $\Pr\{E\}$ if $n = 15$ and the population mean is 2,600 ml.

(c) How does $\Pr\{E\}$ depend on the population mean?

5.2.8 The heights of a certain population of corn plants follow a normal distribution with mean 145 cm and standard deviation 22 cm (as in Exercise 4.S.4).

(a) What percentage of the plants are between 135 and 155 cm tall?

(b) Suppose we were to choose at random from the population a large number of samples of 16 plants each. In what percentage of the samples would the sample mean height be between 135 and 155 cm?

(c) If $\bar{Y}$ represents the mean height of a random sample of 16 plants from the population, what is $\Pr\{135 \le \bar{Y} \le 155\}$?

(d) If $\bar{Y}$ represents the mean height of a random sample of 36 plants from the population, what is $\Pr\{135 \le \bar{Y} \le 155\}$?

5.2.9 The basal diameter of a sea anemone is an indicator of its age. The density curve shown here represents the distribution of diameters in a certain large population of anemones; the population mean diameter is 4.2 cm, and the standard deviation is 1.4 cm.[4] Let $\bar{Y}$ represent the mean diameter of 25 anemones randomly chosen from the population.

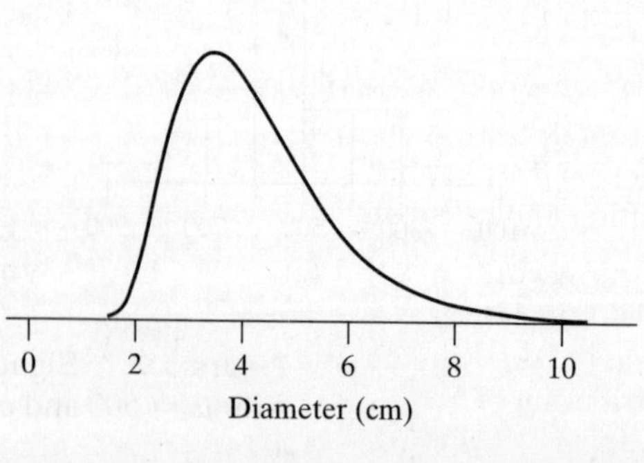

(a) Find the approximate value of $\Pr\{4 \le \bar{Y} \le 5\}$.

(b) Why is your answer to part (a) approximately correct even though the population distribution of diameters is clearly not normal? Would the same approach be equally valid for a sample of size 2 rather than 25? Why or why not?

5.2.10 In a certain population of fish, the lengths of the individual fish follow approximately a normal distribution with mean 54.0 mm and standard deviation 4.5 mm. We saw in Example 4.3.1 that in this situation, 65.68% of the fish are between 51 and 60 mm long. Suppose a random sample of four fish is chosen from the population. Find the probability that

(a) all four fish are between 51 and 60 mm long.

(b) the mean length of the four fish is between 51 and 60 mm.

5.2.11 In Exercise 5.2.10, the answer to part (b) was larger than the answer to part (a). Argue that this must necessarily be true, no matter what the population mean and standard deviation might be. [*Hint:* Can it happen that the event in part (a) occurs but the event in part (b) does not?]

5.2.12 Professor Smith conducted a class exercise in which students ran a computer program to generate random samples from a population that had a mean of 50 and a standard deviation of 9 mm. Each of Smith's students took a random sample of size n and calculated the sample mean. Smith found that about 68% of the students had sample means between 48.5 and 51.5 mm. What was n? (Assume that n is large enough that the Central Limit Theorem is applicable.)

5.2.13 A certain assay for serum alanine aminotransferase (ALT) is rather imprecise. The results of repeated assays of a single specimen follow a normal distribution with mean equal to the ALT concentration for that specimen and standard deviation equal to 4 U/l (as in Exercise 4.S.15). Suppose a hospital lab measures many specimens every day, and specimens with reported ALT values of 40 or more are flagged as "unusually high." If a patient's true ALT concentration is 35 U/l, find the probability that his specimen will be flagged as "unusually high"

(a) if the reported value is the result of a single assay.

(b) if the reported value is the mean of three independent assays of the same specimen.

5.2.14 The mean of the distribution shown in the following histogram is 162 and the standard deviation is 18. Consider taking random samples of size $n = 9$ from this distribution and calculating the sample mean, $\bar{y}$, for each sample.

(a) What is the mean of the sampling distribution of $\bar{Y}$?

(b) What is the standard deviation of the sampling distribution of $\bar{Y}$?

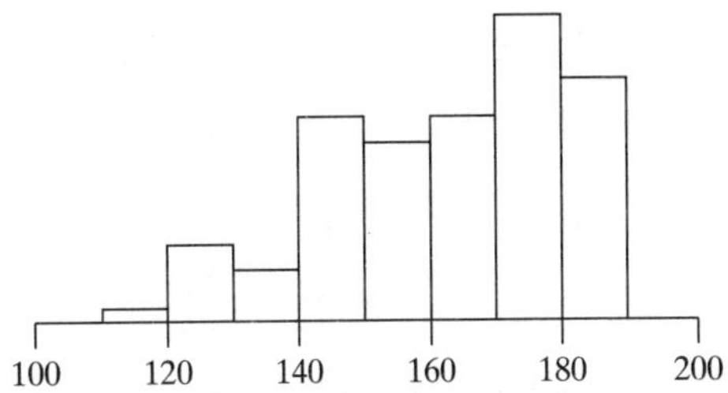

5.2.15 The mean of the distribution shown in the following histogram is 41.5 and the standard deviation is 4.7. Consider taking random samples of size $n = 4$ from this distribution and calculating the sample mean, $\bar{y}$, for each sample.

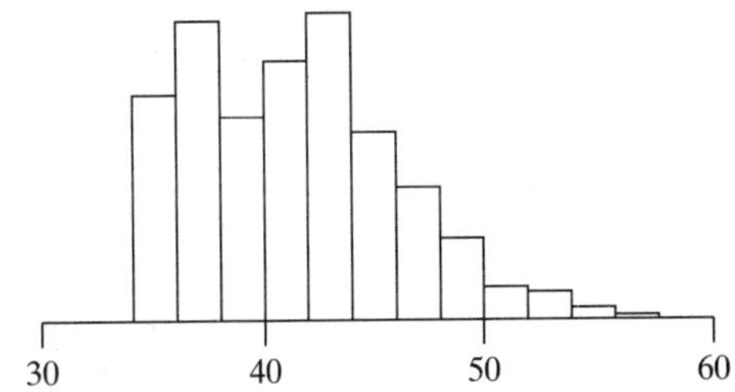

(a) What is the mean of the sampling distribution of $\bar{Y}$?

(b) What is the standard deviation of the sampling distribution of $\bar{Y}$?

5.2.16 Refer to the histogram in Exercise 5.2.15. Suppose that 100 random samples are taken from this population and the sample mean is calculated for each sample. If we were to make a histogram of the distribution of the sample means from 100 samples, what kind of shape would we expect the histogram to have

(a) if $n = 2$ for each random sample?

(b) if $n = 25$ for each random sample?

5.2.17 Refer to the histogram in Exercise 5.2.15. Suppose that 100 random samples are taken from this population and the sample mean is calculated for each sample. If we were to make a histogram of the distribution of the sample means from 100 samples, what kind of shape would we expect the histogram to have if $n = 1$ for each random sample? That is, what does the sampling distribution of the mean look like when the sample size is $n = 1$?

5.2.18 A medical researcher measured systolic blood pressure in 100 middle-aged men.[5] The results are displayed in the accompanying histogram; note that the distribution is rather skewed. According to the Central Limit Theorem, would we expect the distribution of blood pressure readings to be less skewed (and more bell shaped) if it were based on $n = 400$ rather than $n = 100$ men? Explain.

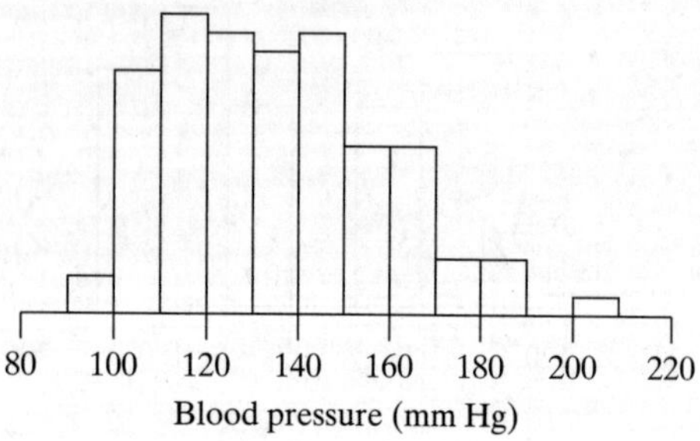

5.2.19 The partial pressure of oxygen, PaO_2, is a measure of the amount of oxygen in the blood. Assume that the distribution of PaO_2 levels among newborns has an average of 38 (mm Hg) and a standard deviation of 9.[6] If we take a sample of size $n = 25$,

(a) what is the probability that the sample average will be greater than 36?

(b) what is the probability that the sample average will be greater than 41?

5.3 Illustration of the Central Limit Theorem (Optional)

The importance of the normal distribution in statistics is due largely to the Central Limit Theorem and related theorems. In this section we take a closer look at the Central Limit Theorem. According to the Central Limit Theorem, the sampling distribution of $\overline{Y}$ is approximately normal if n is large. If we consider larger and larger samples from a fixed nonnormal population, then the sampling distribution of $\overline{Y}$ will be more nearly normal for larger n. The following examples show the Central Limit Theorem at work for two nonnormal distributions: a moderately skewed distribution (Example 5.3.1) and a highly skewed distribution (Example 5.3.2).

Example 5.3.1

Eye Facets The number of facets in the eye of the fruitfly *Drosophila melanogaster* is of interest in genetic studies. The distribution of this variable in a certain *Drosophila* population can be approximated by the density function shown in Figure 5.3.1. The distribution is moderately skewed; the population mean and standard deviation are $\mu = 64$ and $\sigma = 22$.[7]

Figure 5.3.1 Distribution of eye-facet number in a *Drosophila* population

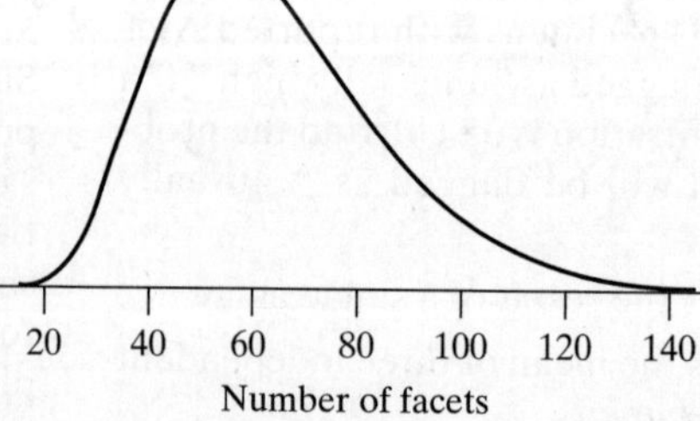

Figure 5.3.2 shows the sampling distribution of $\overline{Y}$ for samples of various sizes from the eye-facet population. In order to clearly show the shape of these distributions, we have plotted them to different scales; the horizontal scale is stretched more for larger n. Notice that the distributions are somewhat skewed to the right, but the skewness is diminished for larger n; for $n = 32$ the distribution looks very nearly normal. ■

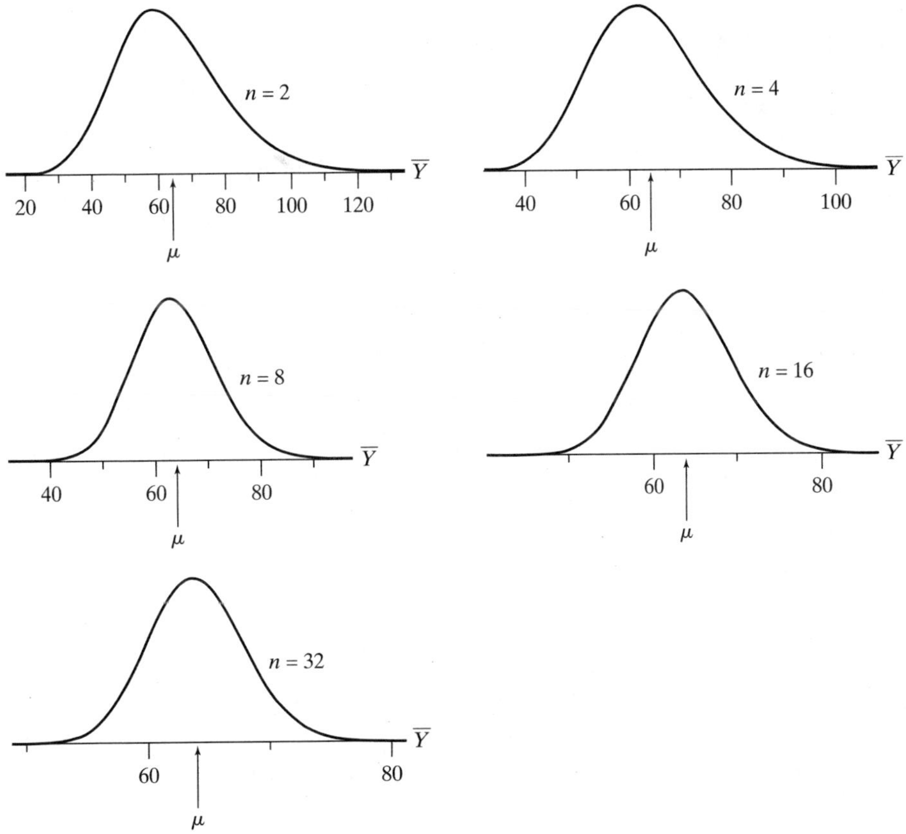

Figure 5.3.2 Sampling distributions of $\bar{Y}$ for samples from the *Drosophila* eye-facet population

Example 5.3.2

Reaction Time A psychologist measured the time required for a person to reach up from a fixed position and operate a pushbutton with his or her forefinger. The distribution of time scores (in milliseconds) for a single person is represented by the density shown in Figure 5.3.3. About 10% of the time, the subject fumbled, or missed the button on the first thrust; the resulting delayed times appear as the second peak of the distribution.[8] The first peak is centered at 115 ms and the second at 450 ms; because of the two peaks, the overall distribution is violently skewed. The population mean and standard deviation are $\mu = 148$ ms and $\sigma = 105$ ms, respectively.

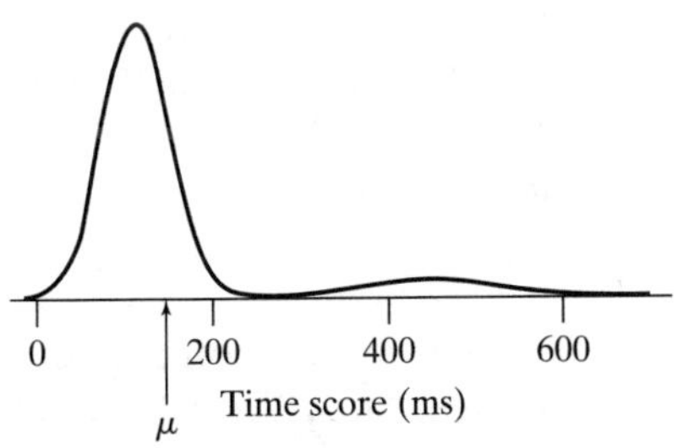

Figure 5.3.3 Distribution of time scores in a button-pushing task

Figure 5.3.4 shows the sampling distribution of $\bar{Y}$ for samples of various sizes from the time-score distribution. To show the shape clearly, the Y scale has been stretched more for larger n. Notice that for small n the distribution has several modes. As n increases, these modes are reduced to bumps and finally disappear, and the distribution becomes increasingly symmetric. ■

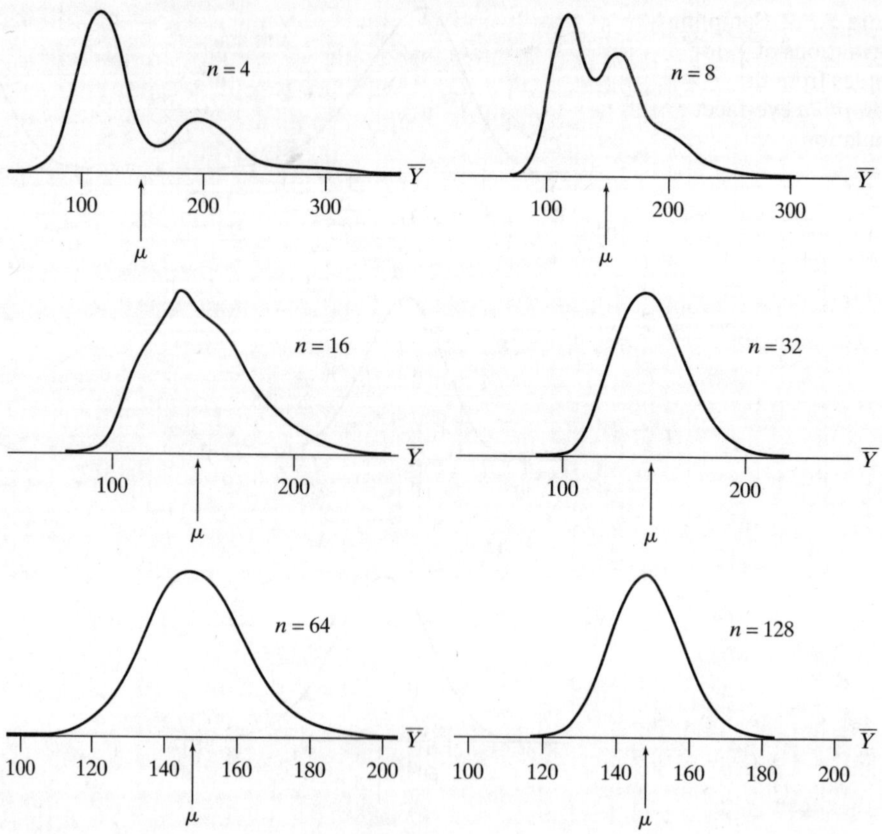

Figure 5.3.4 Sampling distributions of $\overline{Y}$ for samples from the time-score population

Examples 5.3.1 and 5.3.2 illustrate the fact, mentioned in Section 5.2, that the meaning of the requirement "n is large" in the Central Limit Theorem depends on the shape of the population distribution. Approximate normality of the sampling distribution of $\overline{Y}$ will be achieved for a moderate n if the population distribution is only moderately nonnormal (as in Example 5.3.1), while a highly nonnormal population (as in Example 5.3.2) will require a larger n.

Note, however, that Example 5.3.2 indicates the remarkable strength of the Central Limit Theorem. The skewness of the time-score distribution is so extreme that one might be reluctant to consider the mean as a summary measure. Even in this "worst case," you can see the effect of the Central Limit Theorem in the relative smoothness and symmetry of the sampling distribution for $n = 64$.

The Central Limit Theorem may seem rather like magic. To demystify it somewhat, we look at the time-score sampling distributions in more detail in the following example.

Example 5.3.3

Reaction Time Consider the sampling distributions of $\overline{Y}$ displayed in Figure 5.3.4. Consider first the distribution for $n = 4$, which is the distribution of the mean of four button-pressing times. The high peak at the left of the distribution represents cases in which the subject did not fumble any of the 4 thrusts, so all four times were about 115 ms; such an outcome would occur about 66% of the time [from the binomial distribution, because $(0.9)^4 = 0.66$]. The next lower peak represents cases in which 3 thrusts were not fumbled, so took around 115 ms each, while one was fumbled so took around 450 ms. (Notice that the average of three 115's and one 450 is about 200, which is the center of the second peak.) Similarly, the third peak (which is barely visible) represents cases in which the subject fumbled 2 of the 4 thrusts. The peaks representing 3 and 4 fumbles are too low to be visible in the plot.

Now consider the plot for $n = 8$. The first peak represents 8 good thrusts (no fumbles), the second represents 7 good thrusts and 1 fumble, the third represents 6 good thrusts and 2 fumbles, and so on. The fourth and later peaks are blended together. For $n = 16$ it is more likely to see 15 good thrusts and 1 fumble than 16 good thrusts (as you can verify from the binomial distribution) and thus there is a bump, corresponding to 16 good thrusts, below the overall peak, which corresponds to 15 good thrusts; the bump to the right of the peak corresponds to 14 good thrusts and 2 fumbles. For $n = 32$, the most likely outcome is 3 fumbles and 29 good thrusts; this outcome gives a mean time of about

$$\frac{(3)(450) + (29)(115)}{32} \approx 146 \text{ ms}$$

which is the location of the central peak. For similar reasons, the distribution for larger n is centered at about 148 ms, which is the population mean. ■

Exercises 5.3.1–5.3.3

5.3.1 Refer to Example 5.3.3. In the sampling distribution of $\overline{Y}$ for $n = 4$ (Figure 5.3.4), approximately what is the area under

(a) the first peak?

(b) the second peak?
(*Hint:* Use the binomial distribution.)

5.3.2 Refer to Example 5.3.3. Consider the sampling distribution of $\overline{Y}$ for $n = 2$ (which is not shown in Figure 5.3.4).

(a) Make a rough sketch of the sampling distribution. How many peaks does it have? Show the location (on the Y-axis) of each peak.

(b) Find the approximate area under each peak. (*Hint:* Use the binomial distribution.)

5.3.3 Refer to Example 5.3.3. Consider the sampling distribution of $\overline{Y}$ for $n = 1$ (which is not shown in Figure 5.3.4). Make a rough sketch of the sampling distribution. How many peaks does it have? Show the location (on the Y-axis) of each peak.

5.4 The Normal Approximation to the Binomial Distribution (Optional)

The Central Limit Theorem tells us that the sampling distribution of a mean becomes bell shaped as the sample size increases. Suppose we have a large dichotomous population for which we label the two types of outcomes as "1" (for "success") and "0" (for "failure"). If we take a sample and calculate the average outcome value, then this sample average is just the sample proportion of 1's—commonly labeled as $\hat{P}$—and is governed by the Central Limit Theorem. This means that if the sample size n is large, then the distribution of $\hat{P}$ will be approximately normal.

Note that if we know the number of 1's (i.e., the number of successes in n trials), then we know the proportion of 1's and vice versa. Thus, the normal approximation to the binomial distribution can be expressed in two equivalent ways: in terms of the number of successes, Y, or in terms of the proportion of successes, $\hat{P}$. We state both forms in the following theorem. In this theorem, n represents the sample size (or, more generally, the number of independent trials) and p represents the population proportion (or, more generally, the probability of success in each independent trial).

Theorem 5.4.1: Normal Approximation to Binomial Distribution

(a) If n is large, then the binomial distribution of the number of successes, Y, can be approximated by a normal distribution with

$$\text{Mean} = np$$

and

$$\text{Standard deviation} = \sqrt{np(1-p)}$$

(b) If n is large, then the sampling distribution of $\hat{P}$ can be approximated by a normal distribution with

$$\text{Mean} = p$$

and

$$\text{Standard deviation} = \sqrt{\frac{p(1-p)}{n}}$$

Remarks

1. Appendix 5.1 provides more detailed explanation of the relationship between the normal approximation to the binomial and the Central Limit Theorem.
2. As shown in Appendix 3.2, for a population of 0's and 1's, where the proportion of 1's is given by p, the standard deviation is $\sigma = \sqrt{p(1-p)}$. Theorem 5.2.1 stated that the standard deviation of a mean is given by $\frac{\sigma}{\sqrt{n}}$. We think of $\hat{P}$ in part (b) of Theorem 5.2.1 as a special kind of sample average, for the setting in which all of the data are 0's and 1's. Thus, Theorem 5.2.1 tells us that the standard deviation of $\hat{P}$ should be $\frac{\sqrt{p(1-p)}}{\sqrt{n}}$, or $\sqrt{\frac{p(1-p)}{n}}$, which agrees with the result stated in Theorem 5.4.1(b).

The following example illustrates the use of Theorem 5.4.1.

Example 5.4.1 **Normal Approximation to Binomial** We consider a binomial distribution with $n = 50$ and $p = 0.3$. Figure 5.4.1(a) shows this binomial distribution, using spikes to represent probabilities; superimposed is a normal curve with

$$\text{Mean} = np = (50)(0.3) = 15$$

and

$$\text{SD} = \sqrt{np(1-p)} = \sqrt{(50)(0.3)(0.7)} = 3.24$$

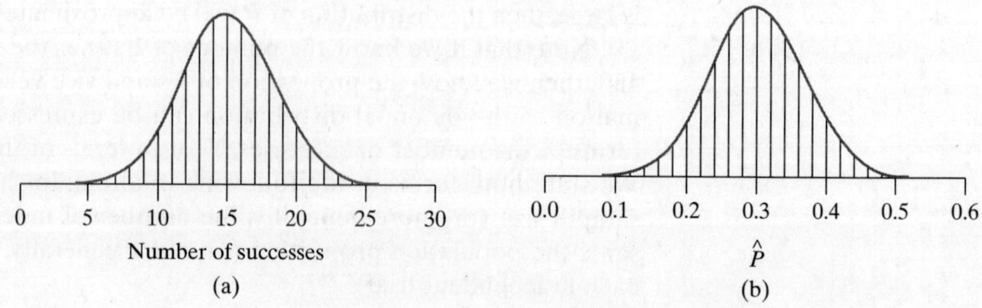

Figure 5.4.1 The normal approximation (blue curve) to the binomial distribution (black spikes) with $n = 50$ and $p = 0.3$

Note that the curve fits the distribution fairly well. Figure 5.4.1(b) shows the sampling distribution of $\hat{P}$ for $n = 50$ and $p = 0.3$; superimposed is a normal curve with

$$\text{Mean} = p = 0.3$$

and

$$\text{SD} = \sqrt{\frac{p(1-p)}{n}} = \sqrt{\frac{(0.3)(0.7)}{50}} = 0.0648$$

Note that Figure 5.4.1(b) is just a relabeled version of Figure 5.4.1(a).

To illustrate the use of the normal approximation, let us find the probability that 50 independent trials result in at least 18 successes. We could use the binomial formula to find the probability of exactly 18 successes in 50 trials and add this to the probability of exactly 19 successes, exactly 20 successes, and so on:

$$\begin{aligned}\Pr\{\text{at least 18 successes}\} &= {}_{50}C_{18}(0.3)^{18}(1-0.3)^{50-18} \\ &\quad + {}_{50}C_{19}(0.3)^{19}(1-0.3)^{50-19} + \ldots \\ &= 0.0772 + 0.0558 + \ldots = 0.2178\end{aligned}$$

This probability can be visualized as the area above and to the right of the "18" in Figure 5.4.2. The normal approximation to the probability is the corresponding area under the normal curve, which is shaded in Figure 5.4.2. The z value that corresponds to 18 is

$$z = \frac{18 - 15}{3.2404} = 0.93$$

Figure 5.4.2 Normal approximation to the probability of at least 18 successes

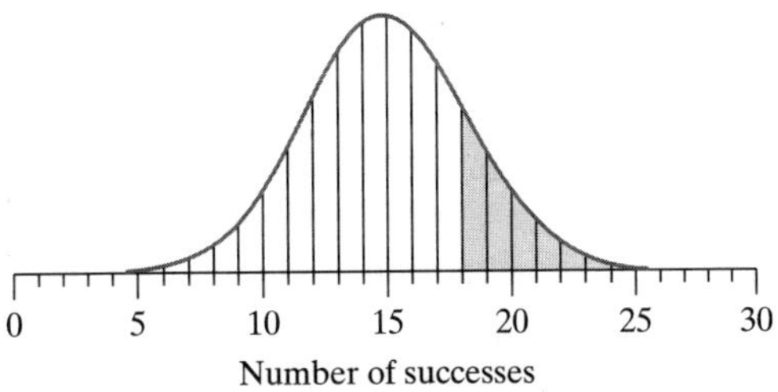

From Table 3, we find that the area is 1 0.8238 − 0.1762, which is reasonably close to the exact value of 0.2178. This approximation can be improved by accounting for the fact that the binomial distribution is discrete and the normal distribution is continuous, as we shall see below. ■

THE CONTINUITY CORRECTION

As we have seen in Chapter 4, because the normal distribution is continuous, probabilities are computed areas under the normal curve, rather than being the height of the normal curve at any particular value. Because of this, to compute $\Pr\{Y = 18\}$, the probability of 18 successes, we think of "18" as covering the space from 17.5 to 18.5 and thus we consider the area under the normal curve between 17.5 and 18.5; this is illustrated in Figure 5.4.3. Likewise, to get a more accurate approximation in Example 5.4.1, we can use 17.5 in place of 18 when finding the z value. Each of these is an example of a continuity correction.

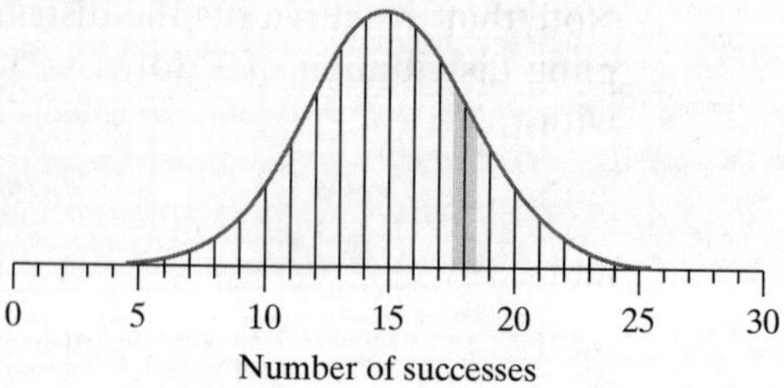

Figure 5.4.3 Normal approximation to the probability of exactly 18 successes

Example 5.4.2 Applying continuity correction within the normal approximation, the probability of at least 18 successes in 50 trials, when $p = 0.3$, is approximated by finding

$$z = \frac{17.5 - 15}{3.2404} = 0.77$$

From Table 3, we find that the area above 0.77 is $1 - 0.7794 = 0.2206$, which agrees quite well with the exact value of 0.2178. This area is displayed in Figure 5.4.4. ■

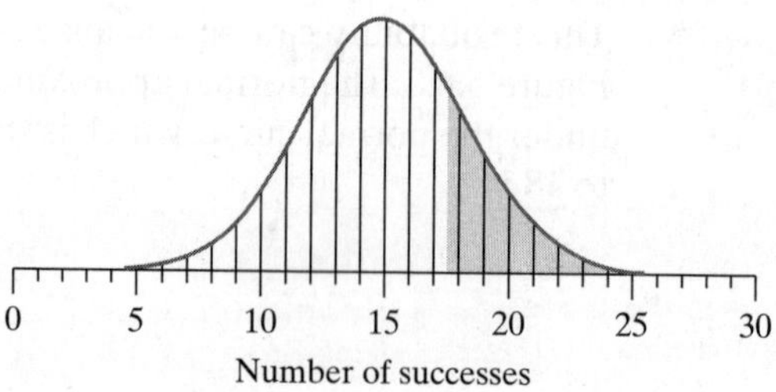

Figure 5.4.4 Improved normal approximation to the probability of at least 18 successes

Example 5.4.3 To illustrate part (b) of Theorem 5.4.1, we again assume that $n = 50$ and $p = 0.3$. Consider finding the probability that at least 40% of the 50 trials in a binomial experiment with $p = 0.3$ result in successes. That is, we wish to find $\Pr\{\hat{P} \geq 0.40\}$. The normal approximation to this probability is the shaded area in Figure 5.4.5. Using continuity correction, the boundary of the area is $\hat{P} = 19.5/50 = 0.39$, which corresponds on the Z scale to

$$z = \frac{0.39 - 0.30}{0.0648} = 1.39$$

The resulting approximation (from Table 3) is then

$$\Pr\{\hat{P} \geq 0.40\} \approx 1 - 0.9177 = 0.0823$$

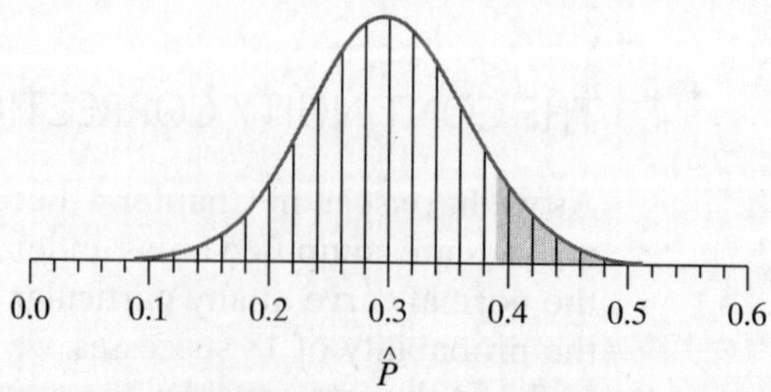

Figure 5.4.5 Normal approximation to $\Pr\{\hat{P} \geq 0.40\}$

which agrees very well with the exact value of 0.0848 (found by using the binomial formula). ■

Remark Any problem involving the normal approximation to the binomial can be solved in two ways: in terms of Y, using part (a) of Theorem 5.4.1, or in terms of $\hat{P}$, using part (b) of the theorem. Although it is natural to state questions in terms of proportions (e.g., "What is $\Pr\{\hat{P} > 0.70\}$?"), it is often easier to solve problems in terms of the binomial count Y (e.g., "What is $\Pr\{Y > 35\}$?"), particularly when using the continuity correction. The following example illustrates the approach of converting a question about a sample proportion into a question about the number of successes for a binomial random variable.

Example 5.4.4 Consider a binomial distribution with $n = 50$ and $p = 0.3$. The sample proportion of successes, out of the 50 trials, is $\hat{P}$. Figure 5.4.1(b) shows the sampling distribution of $\hat{P}$ with a normal curve superimposed.

Suppose we wish to find the probability that $0.24 \le \hat{P} \le 0.36$. Since $\hat{P} = Y/50$, this is the probability that $0.24 \le Y/50 \le 0.36$, which is the same as the probability that $12 \le Y \le 18$. That is, $\Pr\{0.24 \le \hat{P} \le 0.36\} = \Pr\{12 \le Y \le 18\}$.

We know that Y has a binomial distribution with mean $= np = (50)(0.3) = 15$ and SD $= \sqrt{np(1-p)} = \sqrt{(50)(0.3)(0.7)} = 3.24$. Using continuity correction, we would find the Z scale values of

$$z = \frac{11.5 - 15}{3.24} = -1.08$$

and

$$z = \frac{18.5 - 15}{3.24} = 1.08$$

Then, using Table 3, we have $\Pr\{0.24 \le \hat{P} \le 0.36\} = \Pr\{12 \le Y \le 18\} \approx 0.8599 - 0.1401 = 0.7198$. ■

HOW LARGE MUST n BE?

Theorem 5.4.1 states that the binomial distribution can be approximated by a normal distribution if n is "large." It is helpful to know how large n must be in order for the approximation to be adequate. The required n depends on the value of p. If $p = 0.5$, then the binomial distribution is symmetric and the normal approximation is quite good even for n as small as 10. However, if $p = 0.1$, the binomial distribution for $n = 10$ is quite skewed and is poorly fitted by a normal curve; for larger n the skewness is diminished and the normal approximation is better. A simple rule of thumb is the following:

> The normal approximation to the binomial distribution is fairly good if both np and $n(1 - p)$ are at least equal to 5.

For example, if $n = 50$ and $p = 0.3$, as in Example 5.4.4, then $np = 15$ and $n(1 - p) = 35$; since $15 \ge 5$ and $35 \ge 5$, the rule of thumb indicates that the normal approximation is fairly good.

Exercises 5.4.1–5.4.14

5.4.1 Consider interviewing a random sample of $n = 50$ adults. Let $\hat{P}$ denote the proportion of the 50 sampled adults who drink coffee. If the population proportion of coffee drinkers is 0.80, what is the appropriate approximate model for the distribution of $\hat{P}$ over many such samples of size 50? That is, what type of distribution is this, what is the mean, and what is the standard deviation?

5.4.2 A fair coin is to be tossed 20 times. Find the probability that 10 of the tosses will fall heads and 10 will fall tails,

(a) using the binomial distribution formula.

(b) using the normal approximation with the continuity correction.

5.4.3 In the United States, 44% of the population has type O blood. Suppose a random sample of 12 persons is taken. Find the probability that 6 of the persons will have type O blood (and 6 will not)

(a) using the binomial distribution formula.

(b) using the normal approximation.

5.4.4 Refer to Exercise 5.4.3. Find the probability that at most 6 of the persons will have type O blood by using the normal approximation

(a) without the continuity correction.

(b) with the continuity correction.

5.4.5 An epidemiologist is planning a study on the prevalence of oral contraceptive use in a certain population.[9] She plans to choose a random sample of n women and to use the sample proportion of oral contraceptive users ($\hat{P}$) as an estimate of the population proportion (p). Suppose that in fact $p = 0.12$. Use the normal approximation (with the continuity correction) to determine the probability that $\hat{P}$ will be within ± 0.03 of p if

(a) $n = 100$.

(b) $n = 200$.

[*Hint:* If you find using part (b) of Theorem 5.4.1 to be difficult here, try using part (a) of the theorem instead.]

5.4.6 In a study of how people make probability judgments, college students (with no background in probability or statistics) were asked the following question.[10] A certain town is served by two hospitals. In the larger hospital about 45 babies are born each day, and in the smaller hospital about 15 babies are born each day. As you know, about 50% of all babies are boys. The exact percentage of baby boys, however, varies from day to day. Sometimes it may be higher than 50%, sometimes lower.

For a period of 1 year, each hospital recorded the days on which at least 60% of the babies born were boys. Which hospital do you think recorded more such days?

- The larger hospital
- The smaller hospital
- About the same (i.e., within 5% of each other)

(a) Imagine that you are a participant in the study. Which answer would you choose, based on intuition alone?

(b) Determine the correct answer by using the normal approximation (without the continuity correction) to calculate the appropriate probabilities.

5.4.7 Consider random sampling from a dichotomous population with $p = 0.3$, and let E be the event that $\hat{P}$ is within ± 0.05 of p. Use the normal approximation (without the continuity correction) to calculate $\Pr\{E\}$ for a sample of size $n = 400$.

5.4.8 Refer to Exercise 5.4.7. Calculate $\Pr\{E\}$ for $n = 40$ (rather than 400) without the continuity correction.

5.4.9 Refer to Exercise 5.4.7. Calculate $\Pr\{E\}$ for $n = 40$ (rather than 400) *with* the continuity correction.

5.4.10 A certain cross between sweet-pea plants will produce progeny that are either purple flowered or white flowered;[11] the probability of a purple-flowered plant is $p = \frac{9}{16}$. Suppose n progeny are to be examined, and let $\hat{P}$ be the sample proportion of purple-flowered plants. It might happen, by chance, that $\hat{P}$ would be closer to $\frac{1}{2}$ than to $\frac{9}{16}$. Find the probability that this misleading event would occur if

(a) $n = 1$.

(b) $n = 64$.

(c) $n = 320$.

(Use the normal approximation without the continuity correction.)

5.4.11 Cytomegalovirus (CMV) is a (generally benign) virus that infects one-half of young adults.[12] If a random sample of 10 young adults is taken, find the probability that between 30% and 40% (inclusive) of those sampled will have CMV,

(a) using the binomial distribution formula.

(b) using the normal approximation with the continuity correction.

5.4.12 In a certain population of mussels (*Mytilus edulis*), 80% of the individuals are infected with an intestinal parasite.[13] A marine biologist plans to examine 100 randomly chosen mussels from the population. Find the probability that 85% or more of the sampled mussels will be infected, using the normal approximation without the continuity correction.

5.4.13 Refer to Exercise 5.4.12. Find the probability that 85% or more of the sampled mussels will be infected, using the normal approximation with the continuity correction.

5.4.14 Refer to Exercise 5.4.12. Suppose that the biologist takes a random sample of size 50. Find the probability that fewer than 35 of the sampled mussels will be infected, using the normal approximation

(a) without the continuity correction.

(b) with the continuity correction.

5.5 Perspective

In this chapter we have presented the concept of a sampling distribution and have focused on the sampling distribution of $\overline{Y}$. Of course, there are many other important sampling distributions, such as the sampling distribution of the sample standard deviation and the sampling distribution of the sample median.

Let us take another look at the random sampling model in the light of Chapter 5. As we have seen, a *random* sample is not necessarily a *representative* sample.* But using sampling distributions, one can specify the degree of representativeness to be expected in a random sample. For instance, it is intuitively plausible that a larger sample is likely to be more representative than a smaller sample from the same population. In Section 5.1 and Section 5.2 we saw how a sampling distribution can make this vague intuition precise by specifying the probability that a specified degree of representativeness will be achieved by a random sample. Thus, sampling distributions provide what has been called "certainty about uncertainty."[14]

In Chapter 6 we will see for the first time how the theory of sampling distributions can be put to practical use in the analysis of data. We will find that, although the calculations of Chapter 5 seem to require the knowledge of unknowable quantities (e.g., μ and σ), when analyzing data one can nevertheless estimate the probable magnitude of sampling error using only information contained in the sample itself.

In addition to their application to data analysis, sampling distributions provide a basis for comparing the relative merits of different methods of analysis. For example, consider sampling from a normal population with mean μ. Of course, the sample mean $\overline{Y}$ is an estimator of μ. But since a normal distribution is symmetric, it is also the population median, so the sample *median* is also an estimator of μ. How, then, can we decide which estimator is better? This question can be answered in terms of sampling distributions, as follows: Statisticians have determined that, if the population is normal, the sample median is inferior to the sample mean in the sense that its sampling distribution, while centered at μ, has a standard deviation larger than $\frac{\sigma}{\sqrt{n}}$.

Consequently, the sample median is less efficient (as an estimator of μ) than the sample mean; for a given sample size n, the sample median provides less information about μ than does the sample mean. (If the population is not normal, however, the sample median can be much more efficient than the mean.)

*It is true, however, that sometimes the investigator can force the sample to be representative with respect to some variable (not the one under study) whose population distribution is known; for example, a stratified random sample as discussed in Section 1.3. The methods of analysis given in this book, however, are only appropriate for *simple* random samples and cannot be applied without suitable modification.

Supplementary Exercises 5.S.1–5.S.13

(*Note:* Exercises preceded by an asterisk refer to optional sections.)

5.S.1 In an agricultural experiment, a large field of wheat was divided into many plots (each plot being 7 × 100 ft) and the yield of grain was measured for each plot. These plot yields followed approximately a normal distribution with mean 88 lb and standard deviation 7 lb (as in Exercise 4.3.5). Let $\overline{Y}$ represent the mean yield of five plots chosen at random from the field. Find $\Pr\{\overline{Y} > 80\}$.

5.S.2 Consider taking a random sample of size 20 from the population of students at a certain college and measuring the diastolic blood pressure each of the 20 students. In the context of this setting, explain what is meant by the sampling distribution of the sample mean.

5.S.3 Refer to the setting of Exercise 5.S.2. Suppose that the population mean is 100 mm Hg and the population standard deviation is 15 mm Hg. If the sample size is 20, what is the standard deviation of the sampling distribution of the sample mean?

5.S.4 The heights of men in a certain population follow a normal distribution with mean 69.7 inches and standard deviation 2.8 inches.[15]

(a) If a man is chosen at random from the population, find the probability that he will be more than 76 inches tall.

(b) If two men are chosen at random from the population, find the probability that (i) both of them will be more than 76 inches tall; (ii) their mean height will be more than 76 inches.

5.S.5 Suppose a botanist grows many individually potted eggplants, all treated identically and arranged in groups of four pots on the greenhouse bench. After 30 days of growth, she measures the total leaf area Y of each plant. Assume that the population distribution of Y is approximately normal with mean $= 800\ \text{cm}^2$ and SD $= 90\ \text{cm}^2$.[16]

(a) What percentage of the plants in the population will have leaf area between $790\ \text{cm}^2$ and $810\ \text{cm}^2$?

(b) Suppose each group of four plants can be regarded as a random sample from the population. What percentage of the groups will have a group mean leaf area between $790\ \text{cm}^2$ and $810\ \text{cm}^2$?

5.S.6 Refer to Exercise 5.S.5. In a real greenhouse, what factors might tend to invalidate the assumption that each group of plants can be regarded as a random sample from the same population?

***5.S.7** Consider taking a random sample of size 36 from a population in which 52% of the people have type A blood. What is the probability that the sample proportion with type A blood will be greater than 0.54? Use the normal approximation to the binomial with continuity correction.

5.S.8 The activity of a certain enzyme is measured by counting emissions from a radioactively labeled molecule. For a given tissue specimen, the counts in consecutive 10-second time periods may be regarded (approximately) as repeated independent observations from a normal distribution (as in Exercise 4.S.1). Suppose the mean 10-second count for a certain tissue specimen is 1,300 and the standard deviation is 40. For that specimen, let Y represent a 10-second count and let $\bar{Y}$ represent the mean of six 10-second counts. Both Y and $\bar{Y}$ are unbiased—they each have an average of 1,200—but that doesn't imply that they are equally good. Find $\Pr\{1{,}275 < Y < 1{,}325\}$ and $\Pr\{1{,}275 < \bar{Y} < 1{,}325\}$, and compare the two. Does the comparison indicate that counting for 1 minute and dividing by 6 would tend to give a more precise result than merely counting for a single 10-second time period? How?

5.S.9 In a certain lab population of mice, the weights at 20 days of age follow approximately a normal distribution with mean weight $= 8.3$ gm and standard deviation $= 1.7$ gm.[17] Suppose many litters of 10 mice each are to be weighed. If each litter can be regarded as a random sample from the population, what percentage of the litters will have a total weight of 80 gm or more? (*Hint:* How is the total weight of a litter related to the mean weight of its members?)

5.S.10 Refer to Exercise 5.S.9. In reality, what factors would tend to invalidate the assumption that each litter can be regarded as a random sample from the same population?

5.S.11 Consider taking a random sample of size 25 from a population of plants, measuring the weight of each plant, and adding the weights to get a sample total. In the context of this setting, explain what is meant by the sampling distribution of the sample total.

5.S.12 The skull breadths of a certain population of rodents follow a normal distribution with a standard deviation of 15 mm. Let $\bar{Y}$ be the mean skull breadth of a random sample of 81 individuals from this population, and let μ be the population mean skull breadth.

(a) Suppose $\mu = 50$ mm. Find $\Pr\{\bar{Y}$ is within ± 2 mm of $\mu\}$.

(b) Suppose $\mu = 100$ mm. Find $\Pr\{\bar{Y}$ is within ± 2 mm of $\mu\}$.

(c) Suppose μ is unknown. Can you find $\Pr\{\bar{Y}$ is within ± 2 mm of $\mu\}$? If so, do it. If not, explain why not.

5.S.13 Suppose that every day for 3 months Bill takes a random sample of 20 college students, records the number of calories they consume on that day, finds the average of the 20 observations, and adds the average to his histogram of the sampling distribution of the mean.

Suppose also that every day for 2 months Susan takes a random sample of 30 college students and records the number of calories they consume on that day (which is fairly symmetric), finds the average of the 30 observations, and adds the average to her histogram of the sampling distribution of the mean.

(a) Can we expect Bill's distribution and Susan's distribution to have the same shape? Why or why not? If not, how will the shapes differ?

(b) Can we expect Bill's distribution and Susan's distribution to have the same center? Why or why not? If not, how will the centers differ?

(c) Can we expect Bill's distribution and Susan's distribution to have the same spread? Why or why not? If not, how will the spreads differ?

Unit

I

HIGHLIGHTS AND STUDY

(REFLECTIONS ON CHAPTERS 1–5)

In Chapters 1–5 we discussed the building blocks of statistical practice:

- **Study design:** We learned that the way data are collected for a study determines the type and strength of evidence and may limit certain conclusions.
- **Data summarization:** We learned how to describe what kind of distribution we have as well as its shape, center, and spread.
- **Probability:** We learned how to mathematically describe uncertainty. Most important, we learned about the properties of the normal curve—a distribution that arises very frequently in statistical work.
- **Sampling distributions:** We learned how sample statistics (e.g., sample mean) vary from sample to sample, and we learned how to characterize this variability (shape, center, and spread) with the normal distribution (either exactly or approximately).

As we shall see in the upcoming chapters, the above topics (especially sampling distributions) serve as the foundation for statistical inference.

Reflections on Chapter 1

No matter how carefully one conducts a statistical analysis, if the wrong variable was measured, if the data were collected in a biased way, or if important factors were ignored, then the statistical calculations cannot be trusted to be informative.

Most research questions fundamentally involve a comparison of sorts. We often have to consider the question, "Compared to what?" As an example, if we see a medical treatment seeming to have a positive effect on patients, we need to ask, "What would have happened if those patients had not received the treatment (or if they had received a different treatment)?" This is a hypothetical question—at the conclusion of the study a patient either received the treatment or did not. We do not know what would have happened if the patient had not received the treatment (or received another treatment), yet this information is critically important in establishing the effect of the treatment. How do we know patients don't simply get better on their own? Even if a patient was given a placebo or sham treatment during the first week of an experiment and the experimental treatment during the second week, we need to wonder what would have happened if the order had been reversed: treatment given during the first week and the placebo or sham during the second.

Although we cannot know what would have happened, a careful study design can help us make comparisons that let us control for things as best we can. By making comparisons using systematically collected data rather than mere anecdotes, and, when possible, applying conditions (e.g., treatments) within an experiment rather than merely observing choices that people make (e.g., choosing treatment A vs. treatment B or choosing no treatment at all), we increase the quality of our evidence. Blinding can also strengthen our claims. We want measurements to be fair and to be accurate, so blinding of evaluators is important. We want to control for the placebo effect, so blinding of patients is important.

It might be that a new treatment works well but only on patients who are not severely ill and who are willing to try a new therapy. Thus, if we haven't taken a random sample from the population of interest and instead relied on a few volunteers, we cannot be sure that the results we saw in an experiment can be extended to the general population. It can be difficult to obtain a good sample, but sampling errors are only one type of problem. Nonsampling errors—things that can go wrong even if we have access to the entire population—are often more worrisome than errors associated with a sampling process.

In practice, one or more of the pitfalls listed above is usually present. This does not stop us from making scientific progress. Nonetheless, being aware of and honest with the limitations of our work is important.

Reflections on Chapters 2–4

When dealing with scientific data, we can think of a distribution as being the signature of the things we are studying. For example, we might give a drug to patients with high blood pressure and observe how much each person's blood pressure drops over 6 months. Ideally we would know the improvement in blood pressure for every possible patient; in reality, we see how the drug affects a sample of patients. The mean of a distribution is often a good summary number, but we need to know the shape of the distribution in order to know whether the mean is truly informative. To investigate shape, in Chapter 2 we examined several graphical tools, including histograms and boxplots. Simple graphs are often one of the best ways to reveal patterns in the data and communicate our findings, and their utility and importance cannot be overstated.

Graphical tools and numeric summaries can support research claims, but we shall see that understanding how graphics and summaries could vary from one sample to the next is extremely important. Statistical inference (discussed in the coming chapters) makes use of mathematical models of this variability and depends on an understanding of basic probability. In Chapter 3, we presented the fundamental ideas of probability together with the binomial distribution, which has many applications. The normal distribution (Chapter 4) can be used to model scientific phenomena directly, and some statistical procedures depend on having data from a normally distributed population.

Reflections on Chapter 5

A very important use of the normal curve is to describe how sample statistics, such as the sample mean, vary from one sample to the next. Examining how a statistic varies from one sample to the next is at the heart of statistical inference; it is the concept of a sampling distribution, presented in Chapter 5. When we ask, "Compared to what?" we often are wondering, "What would happen if this study were repeated? How do our data compare to typical results of an experiment like ours?" In order to understand whether the results of a study are consistent with a scientific model, we need to be able to think about the effect that inherent variability has on the data that we collect.

Typically, we settle on a particular statistic, such as the sample mean, as a useful representation of the variable we are studying. The sampling distribution of our statistic tells us what to expect under a given model. By knowing what kinds of results can be expected, we can make an inference about the model. In particular, we can determine if our data suggest that our model is wrong. This will become clearer when we put these ideas into practice in the chapters ahead. For now, our goal is to have a good working understanding of variability, of distributions in general, and of sampling distributions of statistics in particular.

Unit I Summary Exercises

I.1 Precipitation, measured in inches, for the month of March in Minneapolis, Minnesota, was recorded for 25 consecutive years. The values ranged from 0.3 up to 4.7, with a mean of 1.7 and an SD of 1.1.

(a) Which of the following is a rough histogram for the data? Explain your choice.

(I)

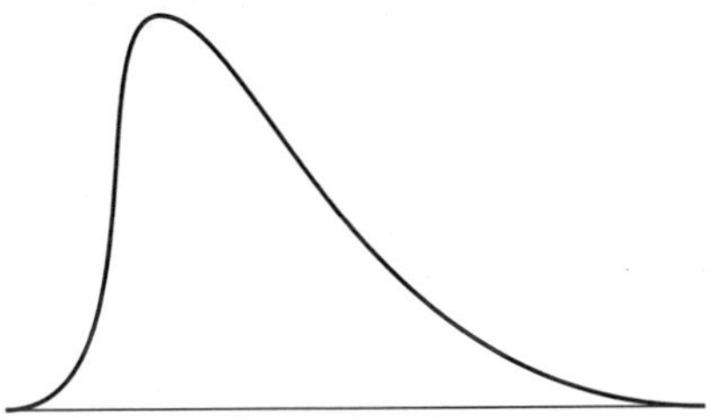

(II)

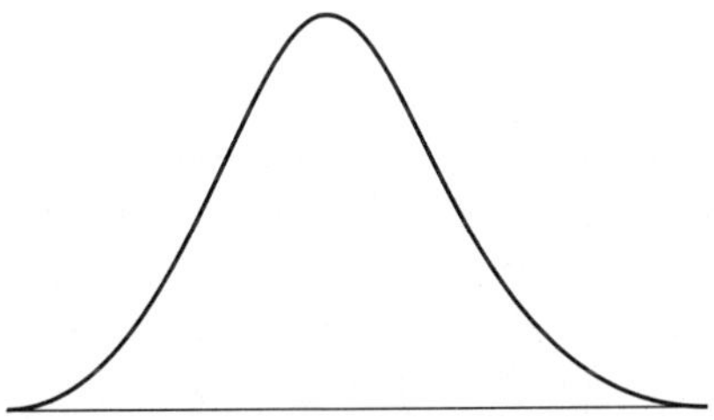

(III)

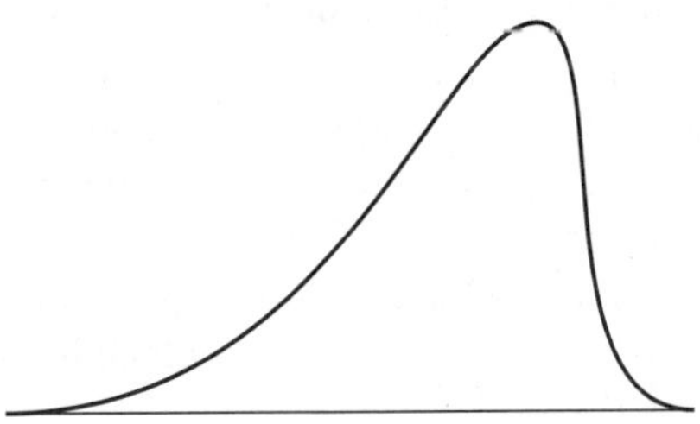

(b) The median is______________ the mean. Choose one and explain your choice.

(i) greater than

(ii) equal to

(iii) less than

I.2 Here is a list of life expectancies in 12 South American countries:

62, 64, 65, 66, 70, 71, 72, 73, 73, 74, 75, 75

The mean of these data is 70, and the SD is 4.6. (You do *not* need to verify this.) Without doing any calculations, which data point had the largest contribution to the SD? That is, if we could remove one of the data points, which data point should we remove if our goal is to make the SD of the remaining 11 points as small as possible? Why?

I.3 In a group of 18 patients, there were 8 men and 10 women. Suppose we were to choose two of them, at random and without replacement. What is the probability that they would be the same sex?

I.4 A researcher took a random sample of 20 mice and found that 5 of the 20 mice (25%) weighed more than 26 gm. *In the context of this setting*, explain what is meant by the sampling distribution of a percentage.

I.5 Tree diameters for a certain species of tree are normally distributed with a mean of 20 cm and a standard deviation of 5 cm.

(a) What is the probability that the diameter of a randomly chosen tree will be between 16 cm and 23 cm?

(b) Suppose we take a sample of $n = 5$ trees. What is the probability that the average of the 5 diameters will be between 16 cm and 23 cm?

I.6 Consider a hypothetical population of dogs in which there are four possible weights, all of which are equally likely: 40, 50, 65, or 70 pounds. A sample of size $n = 2$ is drawn from this population. We are interested in the sampling distribution of the total weight of the two dogs selected. How many possible values are there for the total?

I.7 In a study of distance runners, the mean weight was 63.1 kg. Weights followed a normal distribution. Also,

75% of the weights were between 58.1 and 68.1 kg. The standard deviation of weights is

(i) less than 5 kg

(ii) equal to 5 kg

(iii) more than 5 kg

Choose one of these and explain your choice.

I.8 Researchers wanted to compare two drugs, formoterol and salbutamol, in aerosol solution, for the treatment of patients who suffer from exercise-induced asthma. Patients were to take a drug, do some exercise, and then have their "forced expiratory volume" measured. There were 30 subjects available.[1]

(a) Should this be an experiment or an observational study? Why?

(b) Within the context of this setting, what is the placebo effect?

I.9 Heights of American women ages 18–24 follow a normal distribution with an average of 64.3 inches. (Assume that measurements are made to the nearest 0.1 inch.) Moreover, 50% of the heights are between 62.5 inches and 66.1 inches. What is the standard deviation of heights?

I.10 A certain cross between sweetpea plants will produce progeny that are either purple flowered or white flowered; the probability of a purple flowered plant is $p = 9/16$. Suppose 100 progeny are to be examined. Use the normal approximation to the binomial distribution to find the probability that at least 54 of them will be purple flowered.

I.11 For each of the following cases (a and b),

(i) state whether the study should be observational or experimental and why.

(ii) state whether blinding should be used. If the study should be run blind or double-blind, who should be blinded and why?

(a) An investigation of whether taking an aspirin every day decreases the chance of having a stroke.

(b) An investigation of whether attending religious services regularly reduces blood pressure.

I.12 For each of the following situations state whether or not a binomial would be an appropriate probability model for the variable Y and explain why.

(a) Seeds of the garden pea (*Pisum sativum*) are either yellow or green. A certain cross between pea plants produces progeny that are in the ratio 3 yellow:1 green. Suppose your goal is to get 3 yellow, but you don't care how many green you get. You sample, one at a time, until you have exactly 3 progeny that are yellow. Let Y be the number of progeny you have to observe in order to get 3 yellow. Is Y a binomial random variable? Why or why not?

(b) Some people exercise every day, some exercise occasionally, and some never exercise. Suppose you take a random sample of 45 people and ask each of them how often they exercise. Let Y be the number of people, out of 45, who exercise every day. Is Y a binomial random variable? Why or why not?

Problems **I.13–I.16** refer to the following Flaxseed and cyanide case study

The *Journal of Nutrition and Food Science* contained an article entitled "Flaxseed (Linum usitatissimum L.) consumption and blood thiocyanate concentration in rats".[2] The questions below are motivated from this study.

Flaxseed is a nutrient rich seed but contains cyanogenic glycosides, which can release hydrogen cyanide (HCN) into the body after consumption. This study aims to determine the cyanogenic content of raw and heated (170 °C, 15 min.) flaxseed as well as its effect on the blood thiocyanate (SCN2) concentration, a derivate of HCN, in rats.

I.13 One variable studied was the amount of thiocyanate (mg/L) in the blood after rats consumed raw or heated flaxseeds.

(a) Is this variable *numeric* or *categorical*?

(b) If numeric, is this variable inherently *discrete* or *continuous*? If categorical, is this variable *nominal* or *ordinal*?

I.14 This study aims to determine whether or not heating flaxseeds before consumption can alter the amount of thiocyanate in the blood. Researchers fed 14 rats a diet consisting of 30% flaxseeds for 30 days and then measured the amount of thiocyanate in the blood. Seven of the rats were randomly assigned to eat diets with raw flaxseeds, while the remaining seven were given diets with flaxseeds that were heated at 170°C for 15 minutes.

(a) Briefly explain how you can tell that the study described above is an experiment and not an observational study.

(b) Will this study design allow the researchers to investigate whether or not heating the seeds can actually affect blood thiocyanate levels in flaxseed rich diets? Very briefly explain.

I.15 Suppose that the thiocyanate blood concentration in rats fed a normal diet (free of flaxseeds) follows a normal distribution with population mean concentration of 53.3 mg/L and standard deviation of 14.6 mg/L.

(a) How high must a blood thiocyanate concentration be in order to be in the top 15% of the population? Show your work for full credit.

(b) The value you just computed in part (a) is called the ________ percentile. (Fill in the blank.)

(c) Would it be unusual to obtain a random sample of seven rats for which their sample mean thiocyanate concentration is more than 70.4 mg/L? Justify your answer, showing and briefly discussing a well-labeled illustration or appropriate computations.

I.16 The following graph is a normal probability plot of the blood thiocyanide concentration for the seven rats eating the raw 30% flaxseed diet. Statistical software reports the Shapiro-Wilk's normality test P-value is 0.2675.

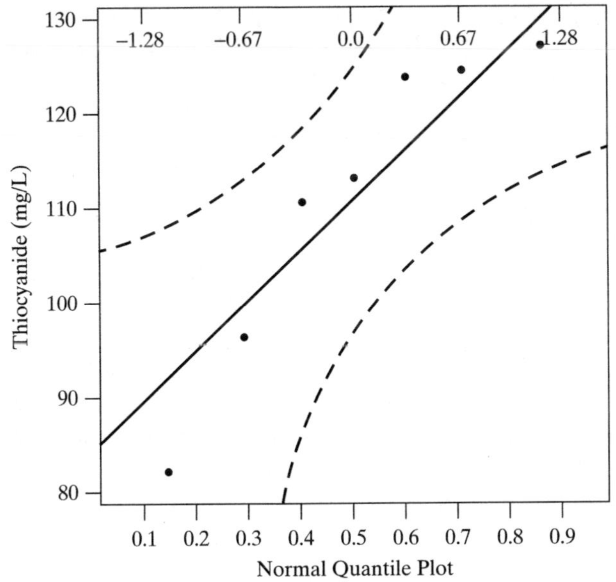

(a) Would it be reasonable to regard the population from which these data came as normal? Briefly explain using the graph and Shapiro-Wilk's test P-value to support your answer.

(b) Would it be reasonable to regard the sampling distribution of $\overline{Y}$ (for samples of size $n = 7$) to be normally distributed? Briefly explain.

(c) True or false? If the dots on the normal probability plot fell very closely to the line, we would have good evidence that blood thiocyanide concentrations are normally distributed.

I.17 Consider the following four boxplots (i, ii, iii, iv).

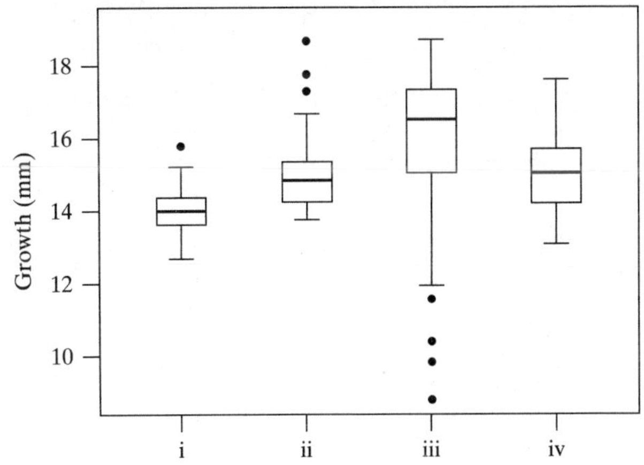

(a) Among the four sets of data, which is most strongly skewed to the left?

(b) Among the four sets of data, which has the smallest standard deviation?

(c) Among the four sets of data, which has the lowest third quartile?

(d) The interquartile range of boxplot iii is approximately _______ (2, 5, 7, 10, 17). Choose a value.

I.18 Bill lengths of a population of male blue jays have a normal distribution with mean 25.4 mm and standard deviation 0.8 mm. A bill is considered to be "short" if it is shorter than 24.0 mm. Suppose that a researcher has a large collection of these male blue jays and takes measurements each day on 10 of the birds. What is the percentage of days on which at least 1 blue jay out of the 10 has a short bill? Hint: Start by finding the probablity of a short bill for one blue jay.

CONFIDENCE INTERVALS

Chapter

6

OBJECTIVES

In this chapter we will begin a formal study of statistical inference. We will

- introduce the concept of the standard error to quantify the degree of uncertainty in an estimated quantity and compare it with the standard deviation.
- demonstrate the construction and interpretation of confidence intervals for means.
- provide a method to determine the sample size that is needed to achieve a desired level of accuracy.
- consider the conditions under which the use of a confidence interval is valid.
- introduce the standard error of a difference in sample means.
- demonstrate the construction and interpretation of confidence intervals for differences between means.

6.1 Statistical Estimation

In this chapter we undertake our first substantial adventure into statistical inference. Recall that statistical inference is based on the random sampling model: We view our data as a random sample from some population, and we use the information in the sample to infer facts about the population. Statistical estimation is a form of statistical inference in which we use the data to (1) determine an estimate of some feature of the population and (2) assess the precision of the estimate. Let us consider an example.

Example 6.1.1

Butterfly Wings As part of a larger study of body composition, researchers captured 14 male Monarch butterflies at Oceano Dunes State Park in California and measured wing area (in cm^2). The data are given in Table 6.1.1.[1]

Table 6.1.1 Wing areas of male monarch butterflies

Wing area (cm^2)				
33.9	33.0	30.6	36.6	36.5
34.0	36.1	32.0	28.0	32.0
32.2	32.2	32.3	30.0	

For these data, the mean and standard deviation are

$$\bar{y} = 32.8143 \approx 32.81 \text{ cm}^2 \quad \text{and} \quad s = 2.4757 \approx 2.48 \text{ cm}^2$$

Suppose we regard the 14 observations as a random sample from a population; the population could be described by (among other things) its mean, μ, and its standard deviation, σ. We might define μ and σ verbally as follows:

$\mu =$ the (population) mean wing area of male Monarch butterflies in the Oceano Dunes region

$\sigma =$ the (population) SD of wing area of male Monarch butterflies in the Oceano Dunes region

It is natural to estimate μ by the sample mean and σ by the sample standard deviation. Thus, from the data on the 14 butterflies,

32.81 is an estimate of μ.

2.48 is an estimate of σ.

We know that these estimates are subject to sampling error. Note that we are not speaking merely of measurement error; no matter how accurately each individual butterfly was measured, the sample information is imperfect due to the fact that only 14 butterflies were measured, rather than the entire population of butterflies. ■

In general, for a sample of observations on a quantitative variable Y, the sample mean and SD are estimates of the population mean and SD:

$\bar{y}$ is an estimate of μ.
s is an estimate of σ.

The notation for these means and SDs is summarized schematically in Figure 6.1.1.

Figure 6.1.1 Notation for means and SDs of sample and population

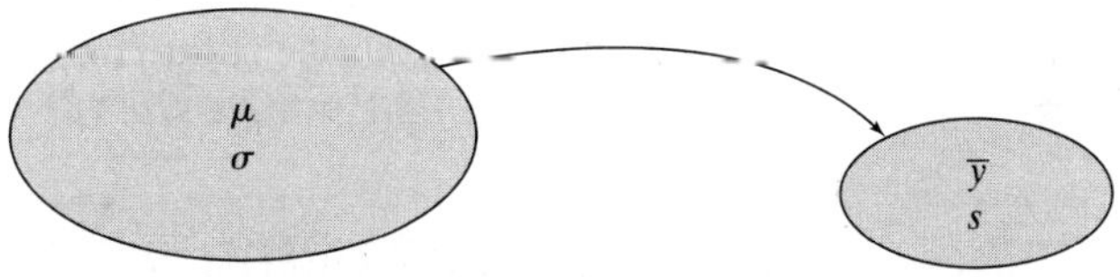

Our goal is to estimate μ. We will see how to assess the reliability or precision of this estimate, and how to plan a study large enough to attain a desired precision.

6.2 Standard Error of the Mean

It is intuitively reasonable that the sample mean $\bar{y}$ should be an estimate of μ. It is not so obvious how to determine the reliability of the estimate. As an estimate of μ, the sample mean $\bar{y}$ is imprecise to the extent that it is affected by sampling error. In Section 5.2 we saw that the magnitude of the sampling error—that is, the amount of discrepancy between $\bar{y}$ and μ—is described (in a probability sense) by the sampling distribution of $\bar{Y}$. The standard deviation of the sampling distribution of $\bar{Y}$ is

$$\sigma_{\bar{Y}} = \frac{\sigma}{\sqrt{n}}$$

Since s is an estimate of σ, a natural estimate of $\frac{\sigma}{\sqrt{n}}$ would be $\frac{s}{\sqrt{n}}$; this quantity is called the **standard error of the mean.** We will denote it as $\text{SE}_{\bar{Y}}$ or sometimes simply SE.*

> **DEFINITION** The **standard error of the mean** is defined as
>
> $$\text{SE}_{\bar{Y}} = \frac{s}{\sqrt{n}}$$

*Some statisticians prefer to reserve the term "standard error" for $\sigma/\sqrt{n}$ and to call $s/\sqrt{n}$ the "estimated standard error."

The following example illustrates the definition.

Example 6.2.1 **Butterfly Wings** For the Monarch butterfly data of Example 6.1.1, we have $n = 14$, $\bar{y} = 32.8143 \approx 32.81$ cm^2 and $s = 2.4757 \approx 2.48$ cm^2. The standard error of the mean is

$$\mathrm{SE}_{\bar{Y}} = \frac{s}{\sqrt{n}}$$

$$= \frac{2.4757}{\sqrt{14}} = 0.6617 \text{ cm}^2\text{, which we will round to } 0.66 \text{ cm}^2\text{*}$$

As we have seen, the SE is an estimate of $\sigma_{\bar{Y}}$. On a more practical level, the SE can be interpreted in terms of the expected sampling error: Roughly speaking, the difference between $\bar{y}$ and μ is rarely more than a few standard errors. Indeed, we expect $\bar{y}$ to be within about one standard error of μ quite often. Thus, the standard error is a measure of the reliability or precision of $\bar{y}$ as an estimate of μ; the smaller the SE, the more precise the estimate. Notice how the SE incorporates the two factors that affect reliability: (1) the inherent variability of the observations (expressed through s), and (2) the sample size (n).

STANDARD ERROR VERSUS STANDARD DEVIATION

The terms "standard error" and "standard deviation" are sometimes confused. It is extremely important to distinguish between standard error (SE) and standard deviation (s, or SD). These two quantities describe entirely different aspects of the data. The SD describes the dispersion of the data, while the SE describes the unreliability (due to sampling error) in the *mean* of the sample as an estimate of the *mean* of the population. Let us consider a concrete example.

Example 6.2.2 **Lamb Birthweights** A geneticist weighed 28 female lambs at birth. The lambs were all born in April, were all the same breed (Rambouillet), and were all single births (no twins). The diet and other environmental conditions were the same for all the parents. The birthweights are shown in Table 6.2.1.[2]

Table 6.2.1 Birthweights of 28 Rambouillet lambs

Birthweight (kg)						
4.3	5.2	6.2	6.7	5.3	4.9	4.7
5.5	5.3	4.0	4.9	5.2	4.9	5.3
5.4	5.5	3.6	5.8	5.6	5.0	5.2
5.8	6.1	4.9	4.5	4.8	5.4	4.7

*Rounding Summary Statistics

For reporting the mean, standard deviation, and standard error of the mean, the following procedure is recommended:

1. Round the SE to two significant digits.
2. Round $\bar{y}$ and s to match the SE with respect to the decimal position of the last significant digit. (The concept of significant digits is reviewed in Appendix 6.1.) For example, if the SE is rounded to the nearest hundredth, then $\bar{y}$ and s are also rounded to the nearest hundredth.

For these data, the mean is $\bar{y} = 5.17$ kg, the standard deviation is $s = 0.65$ kg, and the standard error is SE $= 0.12$ kg. The SD, s, describes the variability of birthweights among the lambs in the sample, while the SE indicates the variability associated with the sample mean (5.17 kg), viewed as an estimate of the population mean birthweight. This distinction is emphasized in Figure 6.2.1, which shows a histogram of the lamb birthweight data; the SD is indicated as a deviation from $\bar{y}$, while the SE is indicated as variability associated with $\bar{y}$ itself. ■

Figure 6.2.1 Birthweights of twenty-eight lambs

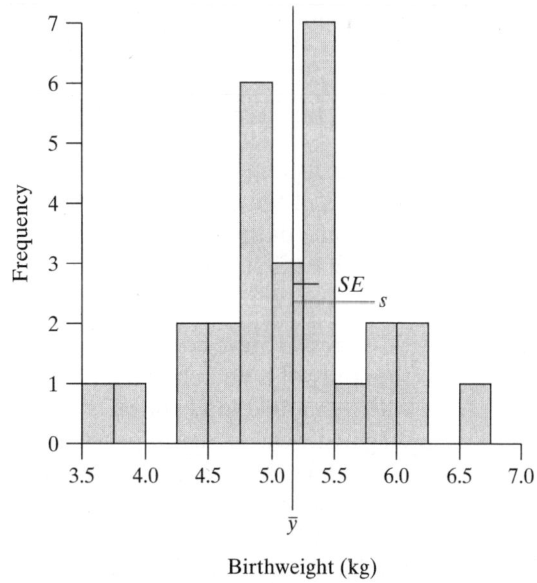

Another way to highlight the contrast between the SE and the SD is to consider samples of various sizes. As the sample size increases, the sample mean and SD tend to approach more closely the population mean and SD; indeed, the distribution of the data tends to approach the population distribution. The standard error, by contrast, tends to decrease as n increases; when n is very large, the SE is very small and so the sample mean is a very precise estimate of the population mean. The following example illustrates this effect.

Example 6.2.3

Lamb Birthweights Suppose we regard the birthweight data of Example 6.2.2 as a sample of size $n = 28$ from a population, and consider what would happen if we were to choose larger samples from the same population—that is, if we were to measure the birthweights of additional female Rambouillet lambs born under the specified conditions. Figure 6.2.2 shows the kind of results we might expect; the values given are fictitious but realistic. For very large n, $\bar{y}$ and s would be very close to μ and σ, where

$\mu =$ Mean birthweight of female Rambouillet lambs born under the conditions described

and

$\sigma =$ Standard deviation of birthweights of female Rambouillet lambs born under the conditions described. ■

Figure 6.2.2 Samples of various sizes from the lamb birthweight population

	$n = 28$	$n = 280$	$n = 2{,}800$	$n \to \infty$
$\bar{y}$	5.17	5.19	5.14	$\bar{y} \to \mu$
s	0.65	0.67	0.65	$s \to \sigma$
SE	0.12	0.040	0.012	$\text{SE} \to 0$

Sample distribution

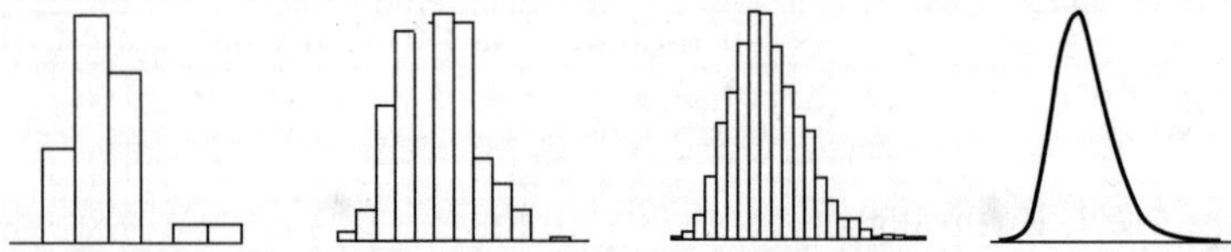

GRAPHICAL PRESENTATION OF THE SE AND THE SD

The clarity and impact of a scientific report can be greatly enhanced by well-designed displays of the data. Data can be displayed graphically or in a table. We briefly discuss some of the options.

Let us first consider graphical presentation of data. Here is an example.

Example 6.2.4 **MAO and Schizophrenia** The enzyme monoamine oxidase (MAO) is of interest in the study of human behavior. Figures 6.2.3 and 6.2.4 display measurements of MAO activity in the blood platelets in five groups of people: Groups I, II, and III are three diagnostic categories of patients with schizophrenia (see Example 1.1.4), and groups IV and V are healthy male and female controls.[3] The MAO activity values are expressed as nmol benzylaldehyde product per 10^8 platelets per hour. In both Figures 6.2.3 and 6.2.4, the dots (a) or bars (b) represent the group means; the vertical lines represent $\pm$ SE in Figure 6.2.3 and $\pm$ SD in Figure 6.2.4.

Figures 6.2.3 and 6.2.4 convey very different information. Figure 6.2.3 conveys (1) the mean MAO value in each group, and (2) the reliability of each group mean, viewed as an estimate of its respective population mean. Figure 6.2.4 conveys (1) the mean MAO value in each group, and (2) the variability of MAO within each group. For instance, group V shows greater variability of MAO than group I (Figure 6.2.4) but has a much smaller standard error (Figure 6.2.3). The standard error of group V is smaller than that of group I because the sample size of group V is greater. The standard deviation of group V is greater than group I simply because there exists more variability among individuals in group V, not because of differences in sample sizes.

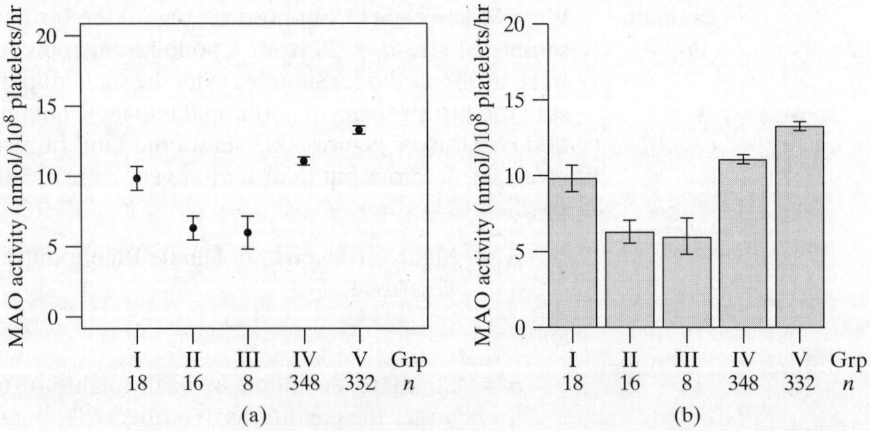

Figure 6.2.3 MAO data displayed as $\bar{y} \pm$ SE using (a) an interval plot and (b) a bargraph with standard error bars

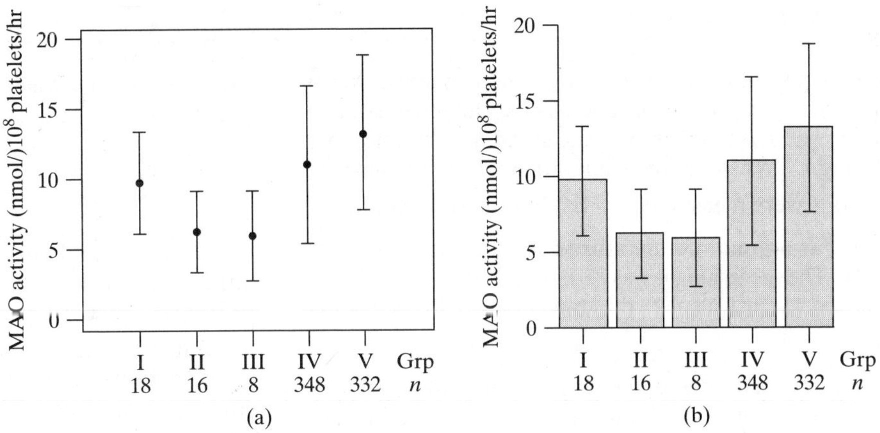

Figure 6.2.4 MAO data displayed as $\bar{y} \pm$ SD using (a) an interval plot and (b) a bargraph with standard deviation bars

Figure 6.2.3 invites the viewer to compare the means and gives some indication of the reliability of the comparisons. (But a full discussion of comparison of two or more means must wait until Chapter 7 and later chapters.) Figure 6.2.4 invites the viewer to compare the means and also to compare the standard deviations. Furthermore, Figure 6.2.4 gives the viewer some information about the extent of overlap of the MAO values in the various groups. For instance, consider groups IV and V; whereas they appear quite "separate" in Figure 6.2.3, we can easily see from Figure 6.2.4 that there is considerable overlap of individual MAO values in the two groups. ■

While we have displayed the MAO data using four individual plots in Figures 6.2.3 and 6.2.4, we typically would choose only one of these to publish in a report. Choosing between the interval plots and bargraphs is a matter of personal preference and style. And, as previously mentioned, choosing whether the interval bars represent the SD or SE will depend on whether we wish to emphasize a comparison of the means (SE), or more simply a summary of the variability in our observed data (SD).*

In some scientific reports, data are summarized in tables rather than graphically. Table 6.2.2 shows a tabular summary for the MAO data of Example 6.2.4. As with the preceding graphs, when formally presenting results, one typically displays either the SD or SE, but not both.

Table 6.2.2 MAO activity in five groups of people

MAO activity (nmol/10^8 platelets/hr)				
Group	n	Mean	SE	SD
I	18	9.81	0.85	3.62
II	16	6.28	0.72	2.88
III	8	5.97	1.13	3.19
IV	348	11.04	0.30	5.59
V	332	13.29	0.30	5.50

*To present a slightly simpler graphic, often only the "upper" error bars (SE or SD) on bargraphs are displayed.

Exercises 6.2.1–6.2.9

6.2.1 A pharmacologist measured the concentration of dopamine in the brains of several rats. The mean concentration was 1,269 ng/gm and the standard deviation was 145 ng/gm.[4] What was the standard error of the mean if

(a) 8 rats were measured? (b) 30 rats were measured?

6.2.2 An agronomist measured the heights of n corn plants.[5] The mean height was 220 cm and the standard deviation was 15 cm. Calculate the standard error of the mean if

(a) $n = 25$ (b) $n = 100$

6.2.3 In evaluating a forage crop, it is important to measure the concentration of various constituents in the plant tissue. In a study of the reliability of such measurements, a batch of alfalfa was dried, ground, and passed through a fine screen. Five small (0.3 gm) aliquots of the alfalfa were then analyzed for their content of insoluble ash.[6] The results (gm/kg) were as follows:

10.0 8.9 9.1 11.7 7.9

For these data, calculate the mean, the standard deviation, and the standard error of the mean.

6.2.4 A zoologist measured tail length in 86 individuals, all in the one-year age group, of the deermouse *Peromyscus*. The mean length was 60.43 mm and the standard deviation was 3.06 mm. The table presents a frequency distribution of the data.[7]

Tail length (mm)	Number of mice
[52, 54)	1
[54, 56)	3
[56, 58)	11
[58, 60)	18
[60, 62)	21
[62, 64)	20
[64, 66)	9
[66, 68)	2
[68, 70)	1
Total	86

(a) Calculate the standard error of the mean.

(b) Construct a histogram of the data and indicate the intervals $\bar{y} \pm$ SD and $\bar{y} \pm$ SE on your histogram. (See Figure 6.2.1.)

6.2.5 Refer to the mouse data of Exercise 6.2.4. Suppose the zoologist were to measure 500 additional animals from the same population. Based on the data in Exercise 6.2.4

(a) What would you predict would be the standard deviation of the 500 new measurements?

(b) What would you predict would be the standard error of the mean for the 500 new measurements?

6.2.6 In a report of a pharmacological study, the experimental animals were described as follows[8]: "Rats weighing 150 ± 10 gm were injected . . ." with a certain chemical, and then certain measurements were made on the rats. If the author intends to convey the degree of homogeneity of the group of experimental animals, then should the 10 gm be the SD or the SE? Explain.

6.2.7 For each of the following, decide whether the description fits the SD or the SE.

(a) This quantity is a measure of the accuracy of the sample mean as an estimate of the population mean.

(b) This quantity tends to stay the same as the sample size goes up.

(c) This quantity tends to go down as the sample size goes up.

6.2.8 Suppose that the population mean birthweight of human baby boys is 3.3 kg and that a sample of $n = 36$ baby boys resulted in an SE of 0.1 kg. Is it likely that the birthweight of a random baby boy will be between 3.2 and 3.4 kg? Why or why not?

6.2.9 Red blood cell counts (10^{-3} X cells per mm^3) of 15 lizards had an average of 843.4. The SD and the SE were, in random order, 64.9 and 251.2. Which is the SD, and which is the SE? How do you know?

6.3 Confidence Interval for μ

In Section 6.2 we said that the standard error of the mean (the SE) measures how far $\bar{y}$ is likely to be from the population mean μ. In this section we make that idea precise.

Figure 6.3.1 Invisible man walking his dog

CONFIDENCE INTERVAL FOR μ: BASIC IDEA

Figure 6.3.1 is a drawing of an invisible man walking his dog. The dog, which is visible, is on an invisible spring-loaded leash. The tension on the spring is such that the dog is within 1 SE of the man about two-thirds of the time. The dog is within 2 standard errors of the man 95% of the time. Only 5% of the time is the dog more than 2 SEs from the man—unless the leash breaks, in which case the dog could be anywhere. We can see the dog, but we would like to know where the man is. Since the man and the dog are usually within 2 SEs of each other, we can take the interval "dog $\pm$ 2 $\times$ SE" as an interval that typically would include the man. Indeed, we could say that we are 95% confident that the man is in this interval.

This is the basic idea of a confidence interval. We would like to know the value of the population mean μ—which corresponds to the man—but we cannot see it directly. What we *can* see is the sample mean $\bar{y}$—which corresponds to the dog. We use what we can see, $\bar{y}$, together with the standard error, which we can calculate from the data, as a way of constructing an interval that we hope will include what we cannot see, the population mean μ. We call the interval "position of the dog $\pm$ 2 $\times$ SE" a 95% confidence interval for the position of the man. [This all depends on having a model that is correct: We said that if the leash breaks, then knowing where the dog is doesn't tell us much about where the man is. Likewise, if our statistical model is wrong (e.g., if we have a biased sample), then knowing $\bar{y}$ doesn't tell us much about μ!]

CONFIDENCE INTERVAL FOR μ: MATHEMATICS

In the invisible man analogy,* we said that the dog is within 1 SE of the man about two-thirds of the time and within 2 SEs of the man 95% of the time. This is based on the idea of the sampling distribution of $\bar{Y}$ when we have a random sample from a normal distribution. If Z is a standard normal random variable, then the probability that Z is between ± 2 is about 95%. More precisely, $\Pr\{-1.96 < Z < 1.96\} = 0.95$. From Chapter 5 we know that if Y has a normal distribution, then $\dfrac{\bar{Y} - \mu}{\sigma/\sqrt{n}}$ has a standard normal (Z) distribution, so

$$\Pr\left\{-1.96 < \frac{\bar{Y} - \mu}{\sigma/\sqrt{n}} < 1.96\right\} = 0.95 \tag{6.3.1}$$

Thus,

$$\Pr\{-1.96 \times \sigma/\sqrt{n} < \bar{Y} - \mu < 1.96 \times \sigma/\sqrt{n}\} = 0.95$$

and

$$\Pr\{-\bar{Y} - 1.96 \times \sigma/\sqrt{n} < -\mu < -\bar{Y} + 1.96 \times \sigma/\sqrt{n}\} = 0.95$$

so

$$\Pr\{\bar{Y} - 1.96 \times \sigma/\sqrt{n} < \mu < \bar{Y} + 1.96 \times \sigma/\sqrt{n}\} = 0.95$$

That is, the interval

$$\bar{Y} \pm 1.96\frac{\sigma}{\sqrt{n}} \tag{6.3.2}$$

will contain μ for 95% of all samples.

The interval (6.3.2) cannot be used for data analysis because it contains a quantity—namely, σ—that cannot be determined from the data. If we replace σ by its estimate—namely, s—then we can calculate an interval from the data, but what happens to the 95% interpretation? Fortunately, it turns out that there is an escape from this dilemma. The escape was discovered by a British scientist named W. S. Gosset, who was employed by the Guinness Brewery. He published his findings in

*Credit for this analogy is due to Geoff Jowett.

1908 under the pseudonym "Student," and the method has borne his name ever since.[9] "Student" discovered that, *if the data come from a normal population* and if we replace σ in the interval (6.3.2) by the sample SD, s, then the 95% interpretation can be preserved if the multiplier of $\frac{\sigma}{\sqrt{n}}$ (i.e., 1.96) is replaced by a suitable quantity; the new quantity is denoted $t_{0.025}$ and is related to a distribution known as Student's t distribution.

STUDENT'S *t* DISTRIBUTION

The **Student's *t* distributions** are theoretical continuous distributions that are used for many purposes in statistics, including the construction of confidence intervals. The exact shape of a Student's t distribution depends on a quantity called "degrees of freedom," abbreviated "df." Figure 6.3.2 shows the density curves of two Student's t distributions with df $= 3$ and df $= 10$, and also a normal curve. A t curve is symmetric and bell shaped like the normal curve but has a larger standard deviation. As the df increase, the t curves approach the normal curve; thus, the normal curve can be regarded as a t curve with infinite df (df $= \infty$).

Figure 6.3.2 Two Student's t curves (dotted, df $= 3$ and dashed, df $= 10$) and a normal curve (df $= \infty$)

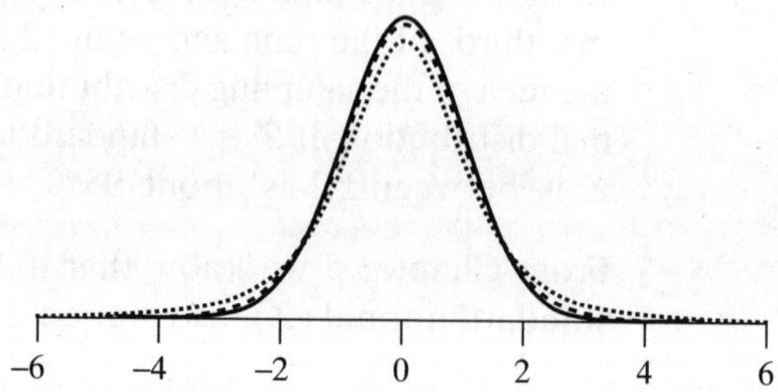

The quantity $t_{0.025}$ is called the "two-tailed 5% critical value" of Student's t distribution and is defined to be the value such that the interval between $-t_{0.025}$ and $+t_{0.025}$ contains 95% of the area under the curve, as shown in Figure 6.3.3.* That is, the combined area in the two tails—below $-t_{0.025}$ and above $+t_{0.025}$—is 5%. The total shaded area in Figure 6.3.3 is equal to 0.05; note that the shaded area consists of two "pieces" of area 0.025 each.

Figure 6.3.3 Definition of the critical value $t_{0.025}$

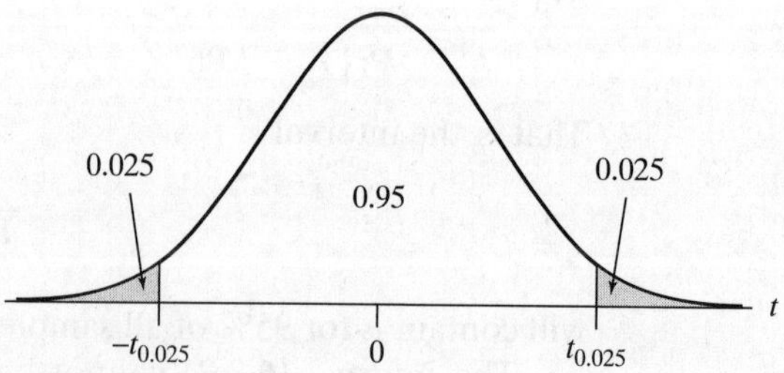

Critical values of Student's t distribution are tabulated in Table 4. The values of $t_{0.025}$ are shown in the column headed "Upper Tail Probability 0.025." If you glance down this column, you will see that the values of $t_{0.025}$ decrease as the df increase; for df $= \infty$ (i.e., for the standard normal distribution) the value is $t_{0.025} = 1.960$. You

*In some statistics textbooks, you may find other notations, such as $t_{0.05}$ or $t_{0.975}$, rather than $t_{0.025}$.

can confirm from Table 3 that the interval ± 1.96 (on the Z scale) contains 95% of the area under a normal curve.

Other columns of Table 4 show other critical values, which are defined analogously; for instance, the interval $\pm t_{0.05}$ contains 90% of the area under a Student's t curve.

CONFIDENCE INTERVAL FOR μ: METHOD

We describe Student's method for constructing a confidence interval for μ, based on a random sample from a normal population. First, suppose we have chosen a confidence level equal to 95% (i.e., we wish to be 95% confident). To construct a 95% confidence interval for μ, we compute the lower and upper limits of the interval as

$$\bar{y} - t_{0.025}\,\mathrm{SE}_{\bar{Y}} \quad \text{and} \quad \bar{y} + t_{0.025}\,\mathrm{SE}_{\bar{Y}}$$

that is,

$$\bar{y} \pm t_{0.025}\frac{s}{\sqrt{n}}$$

where the critical value $t_{0.025}$ is determined from Student's t distribution with

$$\mathrm{df} = n - 1$$

The following example illustrates the construction of a confidence interval.*

Example 6.3.1 **Butterfly Wings** For the Monarch butterfly data of Example 6.1.1, we have $n = 14$, $\bar{y} = 32.8143\ \mathrm{cm}^2$, and $s = 2.4757\ \mathrm{cm}^2$. Figure 6.3.4 shows a histogram and a normal probability plot of the data; these are consistent with the belief that the data came from a normal population. We have 14 observations, so the value of df is

$$\mathrm{df} = n - 1 = 14 - 1 = 13$$

From Table 4 we find

$$t_{0.025} = 2.160$$

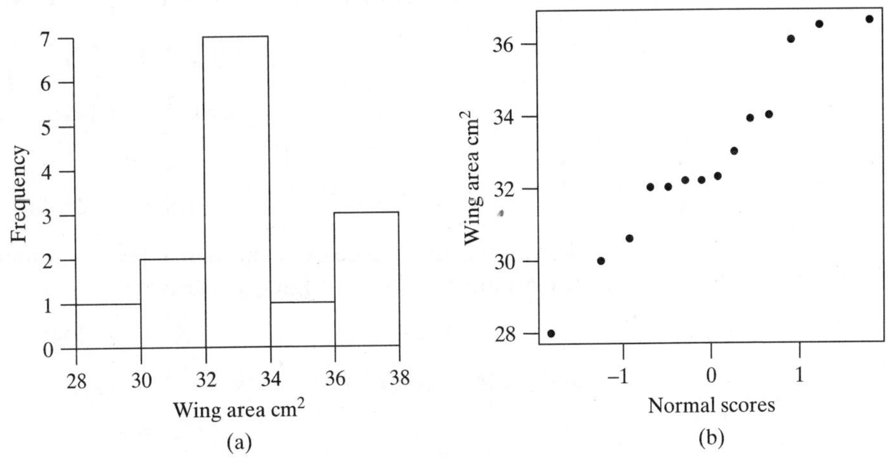

Figure 6.3.4 Histogram (a) and normal probability plot (b) of butterfly wings data

*In most cases, when computing 95% confidence intervals we will use the tabulated value of $t_{0.025}$ with the appropriate degrees of freedom; however, it is worth noting that $t_{0.025}$ is approximately 2 for all but the smallest sample sizes (< 15). Thus, a reasonably approximate 95% confidence interval for the population mean is given by $\bar{y} \pm 2 \times s/\sqrt{n}$.

The 95% confidence interval for μ is

$$32.8143 \pm 2.160 \frac{2.4757}{\sqrt{14}}$$

$$32.8143 \pm 2.160(0.6617)$$

$$32.8143 \pm 1.4293$$

or, approximately,

$$32.81 \pm 1.43$$

The confidence interval may be left in this form. Alternatively, the endpoints of the interval may be explicitly calculated as

$$32.81 - 1.43 = 31.38 \quad \text{and} \quad 32.81 + 1.43 = 34.24$$

and the interval may be written compactly as

$$(31.4, 34.2)$$

or in a more complete form as the following "confidence statement":

$$31.4 \text{ cm}^2 < \mu < 34.2 \text{ cm}^2$$

The confidence statement asserts that the population mean wing area of male Monarch butterflies in the Oceano Dunes region of California is between 31.4 cm^2 and 34.2 cm^2 with 95% confidence. ■

The interpretation of the "95% confidence" will be discussed after the next example.

Confidence coefficients other than 95% are used analogously. For instance, a 90% confidence interval for μ is constructed using $t_{0.05}$ instead of $t_{0.025}$ as follows:

$$\bar{y} \pm t_{0.05} \frac{s}{\sqrt{n}}$$

The following is an example.

Example 6.3.2 **Butterfly Wings** From Table 4, we find that $t_{0.05} = 1.771$ with df $= 13$. Thus, the 90% confidence interval for μ from the butterfly wings data is

$$32.8143 \pm 1.771 \frac{2.4757}{\sqrt{14}}$$

$$32.8143 \pm 1.1718$$

or

$$31.6 < \mu < 34.0$$

■

As you see, the choice of a confidence level is somewhat arbitrary. For the butterfly wings data, the 95% confidence interval is

$$32.81 \pm 1.43$$

and the 90% confidence interval is

$$32.81 \pm 1.17$$

Thus, the 90% confidence interval is narrower than the 95% confidence interval. If we want to be 95% confident that our interval contains μ, then we need a wider interval than we would need if we wanted to be only 90% confident: The higher the confidence level, the wider the confidence interval (for a fixed sample size; but note that as n increases the intervals tend to get smaller).

Remark The quantity $(n - 1)$ is referred to as "degrees of freedom" because the deviations $(y_i - \bar{y})$ must sum to zero, and so only $(n - 1)$ of them are "free" to vary. A sample of size n provides only $(n - 1)$ independent pieces of information about variability, that is, about σ. This is particularly clear if we consider the case $n = 1$; a sample of size 1 provides some information about μ, but no information about σ, and so no information about sampling error. It makes sense, then, that when $n = 1$, we cannot use Student's t method to calculate a confidence interval: the sample standard deviation does not exist (see Example 2.6.5) and there is no critical value with df $= 0$. A sample of size 1 is sometimes called an "anecdote"; for instance, an individual medical case history is an anecdote. Of course, a case history can contribute greatly to medical knowledge, but it does not (in itself) provide a basis for judging how closely the individual case resembles the population at large.

CONFIDENCE INTERVALS AND RANDOMNESS

In what sense can we be "confident" in a confidence interval? To answer this question, let us assume that we are dealing with a random sample from a normal population. Consider, for instance, a 95% confidence interval. One way to interpret the confidence level (95%) is to refer to the meta-study of repeated samples from the same population. If a 95% confidence interval for μ is constructed for each sample, then 95% of the confidence intervals will contain μ. Of course, the observed data in an experiment comprise only *one* of the possible samples; we can hope "confidently" that this sample is one of the lucky 95%, but we will never know.

The following example provides a more concrete visualization of the meta-study interpretation of a confidence level.

Example 6.3.3 **Blue Jay Bill Length** In a certain large population of Blue Jays (described in Example 4.1.2), the distribution of bill lengths is normal with mean $\mu = 25.4$ mm and standard deviation $\sigma = 0.08$ mm. Figure 6.3.5 shows some typical samples from this population; plotted on the right are the associated 95% confidence intervals. The sample sizes are $n = 5$ and $n = 20$. Notice that the second confidence interval with $n = 5$ does not contain μ. In the totality of potential confidence intervals, the percentage that would contain μ is 95% for either sample size; as Figure 6.3.5 shows, the larger samples tend to produce narrower confidence intervals. ■

A confidence level can be interpreted as a probability, but caution is required. If we consider 95% confidence intervals, for instance, then the following statement is correct:

$$\Pr\{\text{the next sample will give us a confidence interval that contains } \mu\} = 0.95$$

However, one should realize that it is *the confidence interval* that is the random item in this statement, and it is not correct to replace this item with its value from the data. Thus, for instance, we found in Example 6.3.1 that the 95% confidence interval for the mean butterfly wings is

$$31.4 \text{ cm}^2 < \mu < 34.2 \text{ cm}^2 \tag{6.3.3}$$

Nevertheless, it is *not* correct to say that

$$\Pr\{31.4 \text{ cm}^2 < \mu < 34.2 \text{ cm}^2\} = 0.95$$

because this statement has no chance element; either μ is between 20.6 and 22.1 or it is not. If $\mu = 32$, then $\Pr\{31.4 \text{ cm}^2 < \mu < 34.2 \text{ cm}^2\} = \Pr\{31.4 \text{ cm}^2 < 32 <$

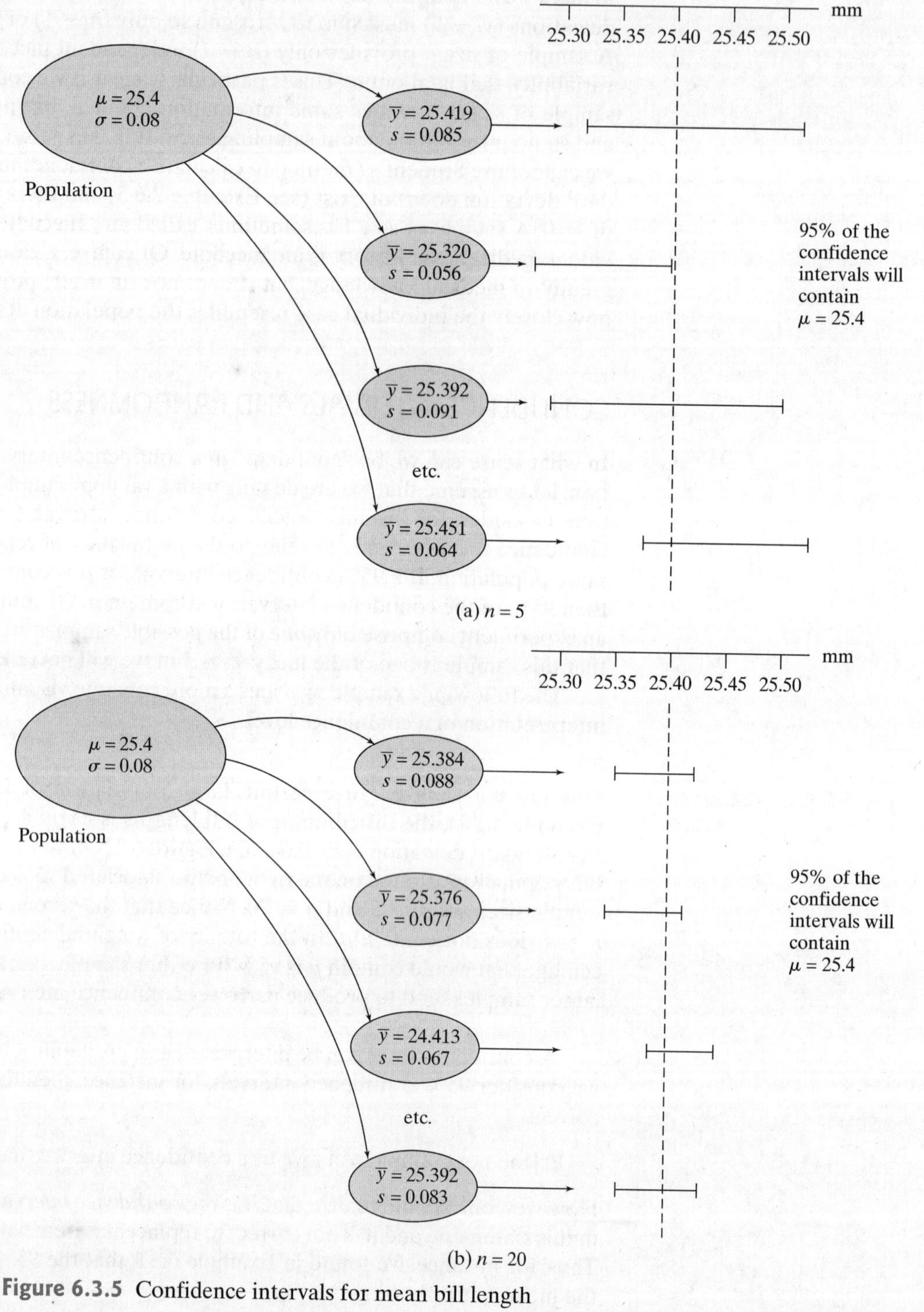

Figure 6.3.5 Confidence intervals for mean bill length

34.2 cm^2} = 1 (not 0.95). The following analogy may help to clarify this point. Suppose we let Y represent the number of spots showing when a balanced die is tossed; then

$$\Pr\{Y = 2\} = \frac{1}{6}$$

On the other hand, if we now toss the die and observe 5 spots, it is obviously *not* correct to substitute this "datum" in the probability statement to conclude that

$$\Pr\{5 = 2\} = \frac{1}{6}^{*}$$

As the preceding discussion indicates, the confidence level (e.g., 95%) is a property of the *method* rather than of a particular interval. An individual statement—such as (6.3.3)—is either true or false, but in the long run, if the researcher constructs 95% confidence intervals in various experiments, each time producing a statement such as (6.3.3), then 95% of the statements will be true.

INTERPRETATION OF A CONFIDENCE INTERVAL

Example 6.3.4

Bone Mineral Density Low bone mineral density often leads to hip fractures in the elderly. In an experiment to assess the effectiveness of hormone replacement therapy, researchers gave conjugated equine estrogen (CEE) to a sample of 94 women between the ages of 45 and 64.[10] After taking the medication for 36 months, the bone mineral density was measured for each of the 94 women. The average density was 0.878 g/cm^2, with a standard deviation of 0.126 g/cm^2.

The standard error of the mean is thus $\frac{0.126}{\sqrt{94}} = 0.013$. It is not clear that the distribution of bone mineral density is a normal distribution, but as we will see in Section 6.5, when the sample size is large, the condition of normality is not crucial. There were 94 observations, so there are 93 degrees of freedom. To find the t multiplier for a 95% confidence interval, we will use 100 degrees of freedom (since Table 4 doesn't list 93 degrees of freedom); the t multiplier is $t_{0.025} = 1.984$. A 95% confidence interval for μ is

$$0.878 \pm 1.984(0.013)$$

or, approximately,

$$0.878 \pm 0.026$$

or

$$(0.852, 0.904)^{\dagger}$$

Thus, *we are 95% confident that the mean hip bone mineral density of all women age 45 to 64 who take CEE for 36 months is between 0.852 g/cm^2 and 0.904 g/cm^2.* ■

Example 6.3.5

Seeds per Fruit The number of seeds per fruit for the freshwater plant *Vallisneria americana* varies considerably from one fruit to another. A researcher took a random sample of 12 fruit and found that the average number of seeds was 320, with a standard deviation of 125.[11] The researcher expected the number of seeds to follow, at least approximately, a normal distribution. A normal quantile plot of the data is shown in Figure 6.3.6. This supports the use of a normal distribution model for these data.

---

*Even if the die rolls under a chair and we can't immediately see that the top face of the die has 5 spots, it would be wrong (given our definition of probability) to say "The probability that the top of the die is showing 2 spots is 1/6."

†If we use a computer to calculate the confidence interval, we get (0.8522, 0.9038); there is very little difference between the t multipliers for 100 versus 93 degrees of freedom.

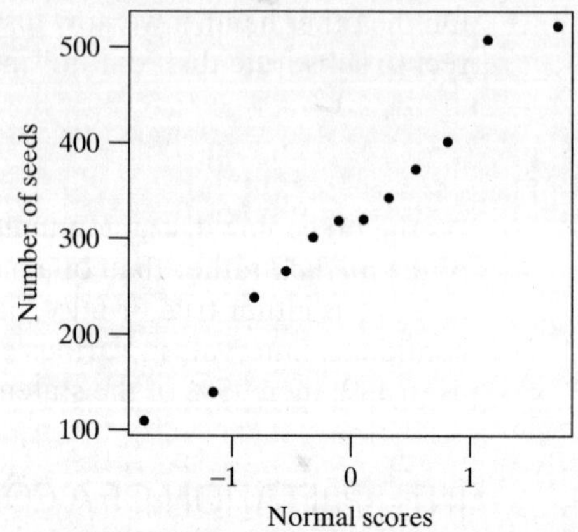

Figure 6.3.6 Normal quantile plot of seeds per fruit for *Vallisneria americana*

The standard error of the mean is $\frac{125}{\sqrt{12}} = 36$. There are 11 degrees of freedom. The t multiplier for a 90% confidence interval is $t_{0.05} = 1.796$. A 90% confidence interval for μ is

$$320 \pm 1.796(36)$$

or, approximately,

$$320 \pm 65$$

or

$$(255, 385)$$

Thus, *we are 90% confident that the (population) mean number of seeds per fruit for Vallisneria americana is between 255 and 385.* ■

RELATIONSHIP TO SAMPLING DISTRIBUTION OF $\overline{Y}$

At this point it may be helpful to look back and see how a confidence interval for μ is related to the sampling distribution of $\overline{Y}$. Recall from Section 5.2 that the mean of the sampling distribution is μ and its standard deviation is $\frac{\sigma}{\sqrt{n}}$. Figure 6.3.7 shows a particular sample mean ($\bar{y}$) and its associated 95% confidence interval for μ, superimposed on the sampling distribution of $\overline{Y}$. Notice that the particular confidence interval does contain μ; this will happen for 95% of samples.

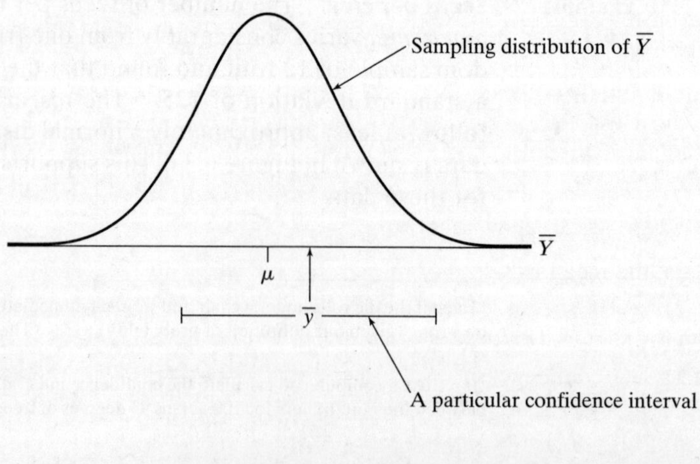

Figure 6.3.7 Relationship between a particular confidence interval for μ and the sampling distribution of $\overline{Y}$

ONE-SIDED CONFIDENCE INTERVALS

Most confidence intervals are of the form "estimate ± margin of error"; these are known as two-sided intervals. However, it is possible to construct a one-sided confidence interval, which is appropriate when only a lower bound, or only an upper bound, is of interest. The following two examples illustrate 90% and 95% one-sided confidence intervals.

Example 6.3.6 **Seeds per Fruit—One-Sided, 90%** Consider the seed data from Example 6.3.5, which are used to estimate the number of seeds per fruit for *Vallisneria americana*. It might be that we want a lower bound on μ, the population mean, but we are not concerned with how large μ might be. Whereas a two-sided 90% confidence interval is based on capturing the middle 90% of a t distribution and thus uses the t multipliers of $\pm t_{0.05}$, a one-sided 90% (lower) confidence interval uses the fact that $\Pr(-t_{0.10} < t < \infty) = 0.90$. Thus, the lower limit of the confidence interval is $\bar{y} - t_{0.10}\,\mathrm{SE}_{\bar{Y}}$ and the upper limit of the interval is infinity. In this case, with 11 degrees of freedom the t multiplier is $t_{11,\,0.10} = 1.363$ and we get

$$320 - 1.363(36) = 320 - 49 = 271$$

as the lower limit. The resulting interval is $(271, \infty)$. Thus, *we are 90% confident that the (population) mean number of seeds per fruit for Vallisneria americana is at least 271.* ■

Example 6.3.7 **Seeds per Fruit—One-Sided, 95%** A one-sided 95% confidence interval is constructed in the same manner as a one-sided 90% confidence interval, but with a different t multiplier. For the *Vallisneria americana* seeds data we have $t_{11,\,0.05} = 1.796$ and we get

$$320 - 1.796(36) = 320 - 65 = 255$$

as the lower limit. The resulting interval is $(255, \infty)$. Thus, *we are 95% confident that the (population) mean number of seeds per fruit for Vallisneria americana is at least 255.* ■

Exercises 6.3.1–6.3.22

6.3.1 (Sampling exercise) Refer to Exercise 5.2.1. Use your sample of five ellipse lengths to construct an 80% confidence interval for μ, using the formula $\bar{y} \pm (1.533)s/\sqrt{n}$.

6.3.2 (Sampling exercise) Refer to Exercise 5.2.3. Use your sample of 20 ellipse lengths to construct an 80% confidence interval for μ using the formula $\bar{y} \pm (1.328)s/\sqrt{n}$.

6.3.3 As part of a study of the development of the thymus gland, researchers weighed the glands of five chick embryos after 14 days of incubation. The thymus weights (mg) were as follows[12]:

29.6 21.5 28.0 34.6 44.9

For these data, the mean is 31.7 and the standard deviation is 8.7.

(a) Calculate the standard error of the mean.

(b) Construct a 90% confidence interval for the population mean.

6.3.4 Consider the data from Exercise 6.3.3.

(a) Construct a 95% confidence interval for the population mean.

(b) Interpret the confidence interval you found in part (a). That is, explain what the numbers in the interval mean. (See Examples 6.3.4 and 6.3.5.)

6.3.5 Six healthy 3-year-old female Suffolk sheep were injected with the antibiotic Gentamicin, at a dosage of 10 mg/kg body weight. Their blood serum concentrations (μg/ml) of Gentamicin 1.5 hours after injection were as follows[13]:

33 26 34 31 23 25

For these data, the mean is 28.7 and the standard deviation is 4.6.

(a) Construct a 95% confidence interval for the population mean.

(b) Define in words the population mean that you estimated in part (a). (See Example 6.1.1.)

(c) The interval constructed in part (a) nearly contains all of the observations; will this typically be true for a 95% confidence interval? Explain.

6.3.6 A zoologist measured tail length in 86 individuals, all in the 1-year age group, of the deermouse *Peromyscus.* The mean length was 60.43 mm and the standard deviation was 3.06 mm. A 95% confidence interval for the mean is (59.77, 61.09).

(a) True or false (and say why): We are 95% confident that the average tail length of the 86 individuals in the sample is between 59.77 mm and 61.09 mm.

(b) True or false (and say why): We are 95% confident that the average tail length of all the individuals in the population is between 59.77 mm and 61.09 mm.

6.3.7 Refer to Exercise 6.3.6.

(a) Without doing any computations, would an 80% confidence interval for the data in Exercise 6.3.6 be wider, narrower, or about the same? Explain.

(b) Without doing any computations, if 500 mice were sampled rather than 86, would the 95% confidence interval listed in Exercise 6.3.6 be wider, narrower, or about the same? Explain.

6.3.8 Researchers measured the bone mineral density of the spines of 94 women who had taken the drug CEE. (See Example 6.3.4, which dealt with hip bone mineral density.) The mean was 1.016 g/cm^2 and the standard deviation was 0.155 g/cm^2. A 95% confidence interval for the mean is (0.984, 1.048).

(a) True or false (and say why): 95% of the sampled bone mineral density measurements are between 0.984 and 1.048.

(b) True or false (and say why): 95% of the population bone mineral density measurements are between 0.984 and 1.048.

6.3.9 There was a control group in the study described in Example 6.3.4. The 124 women in the control group were given a placebo, rather than an active medication. At the end of the study they had an average bone mineral density of 0.840 g/cm^2. Shown are three confidence intervals: One is a 90% confidence interval, one is an 85% confidence interval, and the other is an 80% confidence interval. Without doing any calculations, match the intervals with the confidence levels and explain how you determined which interval goes with which level.

Confidence levels:

90% 85% 80%

Intervals (in scrambled order):

(0.826, 0.854) (0.824, 0.856) (0.822, 0.858)

6.3.10 Levels of insoluble ash (gm/kg) were measured in a sample of small (0.3 gm) aliquots of dried and ground alfalfa.[6] Below are three confidence intervals for the mean: One is a 95% confidence interval, one is a 90% confidence interval, and the other is an 85% confidence interval. Without doing any calculations, match the intervals with the confidence levels and explain how you determined which interval goes with which level.

Confidence levels:

95% 90% 85%

Intervals (in scrambled order):

(8.16, 10.88) (8.38, 10.66) (7.75, 11.29)

6.3.11 Human beta-endorphin (HBE) is a hormone secreted by the pituitary gland under conditions of stress. A researcher conducted a study to investigate whether a program of regular exercise might affect the resting (unstressed) concentration of HBE in the blood. He measured blood HBE levels, in January and again in May, from 10 participants in a physical fitness program. The results were as shown in the table.[14]

(a) Construct a 95% confidence interval for the population mean difference in HBE levels between January and May. (*Hint*: You need to use only the values in the right-hand column.)

	HBE Level (pg/ml)		
Participant	January	May	Difference
1	42	22	20
2	47	29	18
3	37	9	28
4	9	9	0
5	33	26	7
6	70	36	34
7	54	38	16
8	27	32	−5
9	41	33	8
10	18	14	4
Mean	37.8	24.8	13.0
SD	17.6	10.9	12.4

(b) Interpret the confidence interval from part (a). That is, explain what the interval tells you about HBE levels. (See Examples 6.3.4 and 6.3.5.)

(c) Using your interval to support your answer, is there evidence that HBE levels are lower in May than January? (*Hint*: Does your interval include the value zero?)

6.3.12 Consider the data from Exercise 6.3.11. If the sample size is small, as it is in this case, then in order for a confidence interval based on Student's *t* distribution to be valid, the data must come from a normally distributed population. Is it reasonable to think that difference in HBE level is normally distributed? How do you know?

6.3.13 Invertase is an enzyme that may aid in spore germination of the fungus *Colletotrichum graminicola.* A botanist incubated specimens of the fungal tissue in petri dishes and then assayed the tissue for invertase activity. The specific activity values for nine petri dishes incubated at 90% relative humidity for 24 hours are summarized as follows[15]:

Mean = 5,111 units SD = 818 units

(a) Assume that the data are a random sample from a normal population. Construct a 95% confidence interval for the mean invertase activity under these experimental conditions.

(b) Interpret the confidence interval you found in part (a). That is, explain what the numbers in the interval mean. (See Examples 6.3.4 and 6.3.5.)

(c) If you had the raw data, how could you check the condition that the data are from a normal population?

6.3.14 As part of a study of the treatment of anemia in cattle, researchers measured the concentration of selenium in the blood of 36 cows who had been given a dietary supplement of selenium (2 mg/day) for 1 year. The cows were all the same breed (Santa Gertrudis) and had borne their first calf during the year. The mean selenium concentration was 6.21 μg/dl and the standard deviation was 1.84 μg/dl.[16] Construct a 95% confidence interval for the population mean.

6.3.15 In a study of larval development in the tufted apple budmoth *(Platynota idaeusalis),* an entomologist measured the head widths of 50 larvae. All 50 larvae had been reared under identical conditions and had moulted six times. The mean head width was 1.20 mm and the standard deviation was 0.14 mm. Construct a 90% confidence interval for the population mean.[17]

6.3.16 In a study of the effect of aluminum intake on the mental development of infants, a group of 92 infants who had been born prematurely were given a special aluminum-depleted intravenous-feeding solution.[18] At age 18 months the neurologic development of the infants was measured using the Bayley Mental Development Index. (The Bayley Mental Development Index is similar to an IQ score, with 100 being the average in the general population.) A 95% confidence interval for the mean is (93.8, 102.1).

(a) Interpret this interval. That is, what does the interval tell us about neurologic development in the population of prematurely born infants who receive intravenous-feeding solutions?

(b) Does this interval indicate that the mean IQ of the sampled population is below the general population average of 100?

6.3.17 In Exercise 6.3.16 a 95% confidence interval of (93.8, 102.1) was given based on a sample of $n = 92$ infants. Suppose, for sake of argument, that we could change the sample size and we would still get a sample mean of 97.95 and a sample SD of 20.04. How large would the sample size, n, need to be in order for the 95% confidence interval to exclude 100?

6.3.18 A group of 101 patients with end-stage renal disease were given the drug epoetin.[19] The mean hemoglobin level of the patients was 10.3 (g/dl), with an SD of 0.9. Construct a 95% confidence interval for the population mean.

6.3.19 In Table 4 we find that $t_{0.025} = 1.960$ when df $= \infty$. Show how this value can be verified using Table 3.

6.3.20 Use Table 3 to find the value of $t_{0.0025}$ when df $= \infty$. (Do not attempt to interpolate in Table 4.)

6.3.21 Data are often summarized in this format: $\bar{y} \pm$ SE. Suppose this interval is interpreted as a confidence interval. If the sample size is large, what would be the confidence level of such an interval? That is, what is the chance that an interval computed as

$$\bar{y} \pm (1.00)\text{SE}$$

will actually contain the population mean? [*Hint*: Recall that the confidence level of the interval $\bar{y} \pm (1.96)$SE is 95%.]

6.3.22 (Continuation of Exercise 6.3.21)

(a) If the sample size is small but the population distribution is normal, is the confidence level of the interval $\bar{y} \pm$ SE larger or smaller than the answer to Exercise 6.3.21? Explain.

(b) How is the answer to Exercise 6.3.21 affected if the population distribution of Y is not approximately normal?

6.4 Planning a Study to Estimate μ

Before collecting data for a research study, it is wise to consider in advance whether the estimates generated from the data will be sufficiently precise. It can be painful indeed to discover after a long and expensive study that the standard errors are so large that the primary questions addressed by the study cannot be answered.

The precision with which a population mean can be estimated is determined by two factors: (1) the population variability of the observed variable Y, and (2) the sample size.

In some situations the variability of Y cannot, and perhaps should not, be reduced. For example, a wildlife ecologist may wish to conduct a field study of a natural population of fish; the heterogeneity of the population is not controllable and in fact is a proper subject of investigation. As another example, in a medical investigation, in addition to knowing the average response to a treatment, it may also be important to know how much the response varies from one patient to another, and so it may not be appropriate to use an overly homogeneous group of patients.

On the other hand, it is often appropriate, especially in comparative studies, to reduce the variability of Y by holding *extraneous* conditions as constant as possible. For example, physiological measurements may be taken at a fixed time of day; tissue may be held at a controlled temperature; all animals used in an experiment may be the same age.

Suppose, then, that plans have been made to reduce the variability of Y as much as possible, or desirable. What sample size will be sufficient to achieve a desired degree of precision in estimation of the population mean? If we use the standard error as our measure of precision, then this question can be approached in a straightforward manner. Recall that the SE is defined as

$$SE_{\bar{Y}} = \frac{s}{\sqrt{n}}$$

In order to decide on a value of n, one must (1) specify what value of the SE is considered desirable to achieve and (2) have available a preliminary guess of the SD, either from a pilot study or other previous experience, or from the scientific literature. The required sample size is then determined from the following equation:

$$\text{Desired SE} = \frac{\text{Guessed SD}}{\sqrt{n}}$$

The following example illustrates the use of this equation.

Example 6.4.1 **Butterfly Wings** The butterfly wing data of Example 6.1.1 yielded the following summary statistics:

$$\bar{y} = 32.81 \text{ cm}^2$$
$$s = 2.48 \text{ cm}^2$$
$$\text{SE} = 0.66 \text{ cm}^2$$

Suppose the researcher is now planning a new study of butterflies and has decided that it would be desirable that the SE be no more than 0.4 cm^2. As a preliminary guess of the SD, she will use the value from the old study, namely 2.48 cm^2. Thus, the desired n must satisfy the following relation:

$$\text{SE} = \frac{2.48}{\sqrt{n}} \leq 0.4$$

This equation is easily solved to give $n \geq 38.4$. Since one cannot have 38.4 butterflies, the new study should include at least 39 butterflies. ■

You may wonder how a researcher would arrive at a value such as 0.4 cm^2 for the desired SE. Such a value is determined by considering how much error one is willing to tolerate in the estimate of μ. For example, suppose the researcher in Example 6.4.1 has decided that she would like to be able to estimate the population mean, μ, to within ± 0.8 with 95% confidence. That is, she would like her 95% confidence interval for μ to be $\bar{y} \pm 0.8$. The "$\pm$ part" of the confidence interval, which is sometimes called the **margin of error for 95% confidence,** is $t_{0.025} \times$ SE. The precise value of $t_{0.025}$ depends on the degrees of freedom, but typically $t_{0.025}$ is approximately 2. Thus, the researcher wants 2 $\times$ SE to be no more than 0.8. This means that the SE should be no more than 0.4 cm^2.

In comparative studies, the primary consideration is usually the size of anticipated treatment effects. For instance, if one is planning to compare two experimental groups or distinct populations, the anticipated SE for each population or experimental group should be substantially smaller than (preferably less than one-fourth of) the anticipated difference between the two group means.* Thus, the butterfly researcher of Example 6.4.1 might arrive at the value 0.4 cm^2 if she were planning to compare male and female Monarch butterflies and she expected the wing areas for the sexes to differ (on the average) by about 1.6 cm^2. She would then plan to capture 39 male and 39 female butterflies.

To see how the required n depends on the specified precision, suppose the butterfly researcher specified the desired SE to be 0.2 cm^2 rather than 0.4 cm^2. Then the relation would be

$$\text{SE} = \frac{2.48}{\sqrt{n}} \leq 0.2$$

which yields $n = 153.76$, so she would plan to capture 154 butterflies of each sex. Thus, to double the precision (by cutting the SE in half) requires not twice as many but four times as many observations. This phenomenon of "diminishing returns" is due to the square root in the SE formula.

*This is a rough guideline for obtaining adequate sensitivity to discriminate between treatments. Such sensitivity, technically called *power,* is discussed in Chapter 7.

Exercises 6.4.1–6.4.6

6.4.1 An experiment is being planned to compare the effects of several diets on the weight gain of beef cattle, mea-sured over a 140-day test period.[20] In order to have enough precision to compare the diets, it is desired that the standard error of the mean for each diet should not exceed 5 kg.

(a) If the population standard deviation of weight gain is guessed to be about 20 kg on any of the diets, how many cattle should be put on each diet in order to achieve a sufficiently small standard error?

(b) If the guess of the standard deviation is doubled, to 40 kg, does the required number of cattle double? Explain.

6.4.2 A medical researcher proposes to estimate the mean serum cholesterol level of a certain population of middle-aged men, based on a random sample of the population. He asks a statistician for advice. The ensuing discussion reveals that the researcher wants to estimate the population mean to within ± 6 mg/dl or less, with 95% confidence. Thus, the standard error of the mean should be 3 mg/dl or less. Also, the researcher believes that the standard deviation of serum cholesterol in the population is probably about 40 mg/dl.[21] How large a sample does the researcher need to take?

6.4.3 A plant physiologist is planning to measure the stem lengths of soybean plants after 2 weeks of growth when using a new fertilizer. Previous experiments suggest that the standard deviation of stem length is around 1.2 cm.[22] Using this as a guess of σ, determine how many soybean plants the researcher should have if she wants the standard error of the group mean to be no more than 0.2 cm.

6.4.4 Suppose you are planning an experiment to test the effects of various diets on the weight gain of young turkeys. The observed variable will be Y = weight gain in 3 weeks (measured over a period starting 1 week after hatching and ending 3 weeks later). Previous experiments suggest that the standard deviation of Y under a standard diet is approximately 80 g.[23] Using this as a guess of σ, determine how many turkeys you should have in a treatment group, if you want the standard error of the group mean to be no more than

(a) 20 g

(b) 15 g

6.4.5 A study of 29 female Sumatran elephants provided a 95% confidence interval for the mean shoulder height as (197.2, 235.1) cm.[24] Consider a new study to estimate the mean shoulder height of male Sumatran elephants.

Assuming the standard deviations of shoulder heights are similar for males and females, how many male elephants should be sampled so that the 95% confidence interval for the mean shoulder height of males will have a margin of error of 10 cm?

6.4.6 A researcher is planning to compare the effects of two different types of lights on the growth of bean plants. She expects that the means of the two groups will differ by about 1 inch and that in each group the standard deviation of plant growth will be around 1.5 inches. Consider the guideline that the anticipated SE for each experimental group should be no more than one-fourth of the anticipated difference between the two group means. How large should the sample be (for each group) in order to meet this guideline?

6.5 Conditions for Validity of Estimation Methods

For any sample of quantitative data, one can use the methods of this chapter to compute the mean, its standard error, and various confidence intervals; indeed, computers can make this rather easy to carry out. However, the *interpretations* that we have given for these descriptions of the data are valid only under certain conditions.

CONDITIONS FOR VALIDITY OF THE SE FORMULA

First, the very notion of regarding the sample mean as an estimate of a population mean requires that the data be viewed "as if" they had been generated by random sampling from some population. To the extent that this is not possible, any inference beyond the actual data is questionable. The following example illustrates the difficulty.

Example 6.5.1 **Marijuana and Intelligence** Ten people who used marijuana heavily were found to be quite intelligent; their mean IQ was 128.4, whereas the mean IQ for the general population is known to be 100. The 10 people belonged to a religious group that uses marijuana for ritual purposes. Since their decision to join the group might very well be related to their intelligence, it is not clear that the 10 can be regarded (with respect to IQ) as a random sample from any particular population, and therefore there is no apparent basis for thinking of the sample mean (128.4) as an estimate of the mean IQ of a particular population (e.g., all heavy marijuana users). An inference about the *effect* of marijuana on IQ would be even more implausible, especially because data were not available on the IQs of the 10 people *before* they began marijuana use.[25] ■

Second, the use of the standard error formula $SE = s/\sqrt{n}$ requires two further conditions:

1. The population size must be large compared to the sample size. This requirement is rarely a problem in the life sciences; the sample can be as much as 5% of the population without seriously invalidating the SE formula.*
2. The observations must be independent of each other. This requirement means that the n observations actually give n independent pieces of information about the population.

Data often fail to meet the independence requirement if the experiment or sampling regime has a **hierarchical structure,** in which observational units are "nested" within sampling units, as illustrated by the following example.

*If the sample size, n, is a substantial fraction of the population size, N, then the "finite population correction factor" should be applied. This factor is $\sqrt{\frac{N-n}{N-1}}$. The standard error of the mean then becomes $\frac{s}{\sqrt{n}} \times \sqrt{\frac{N-n}{N-1}}$.

Example 6.5.2

Canine Anatomy The coccygeus muscle is a bilateral muscle in the pelvic region of the dog. As part of an anatomical study, the left side and the right side of the coccygeus muscle were weighed for each of 21 female dogs. There were thus $2 \times 21 = 42$ observations, but only 21 units chosen from the population of interest (female dogs). Because of the symmetry of the coccygeus, the information contained in the right and left sides is largely redundant, so the data contain not 42, but only 21, independent pieces of information about the coccygeus muscle of female dogs. It would therefore be incorrect to apply the SE formula as if the data comprised a sample of size $n = 42$. The hierarchical nature of the data set is indicated in Figure 6.5.1.[26] ■

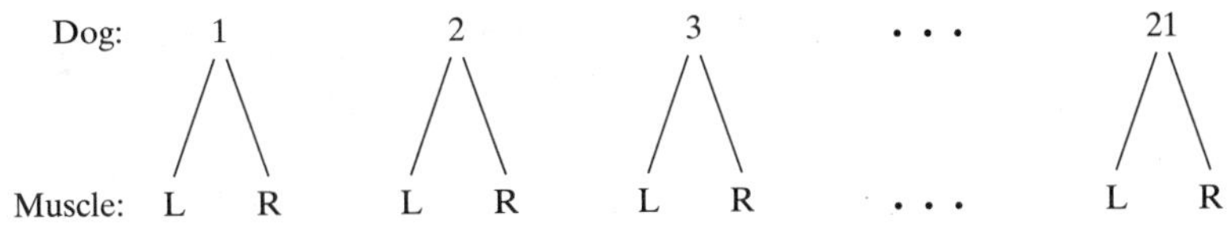

Figure 6.5.1 Hierarchical data structure of Example 6.5.2

Hierarchical data structures are rather common in the life sciences. For instance, observations may be made on 90 nerve cells that come from only three different cats; on 80 kernels of corn that come from only four ears; on 60 young mice who come from only 10 litters. A particularly clear example of nonindependent observations is replicated measurements on the same individual; for instance, if a physician makes triplicate blood pressure measurements on each of 10 patients, she clearly does not have 30 independent observations. In some situations a correct treatment of hierarchical data is obvious; for instance, the triplicate blood pressure measurements could be averaged to give a single value for each patient. In other situations, however, lack of independence can be more subtle. For instance, suppose 60 young mice from 10 litters are included in an experiment to compare two diets. Then the choice of a correct analysis depends on the *design* of the experiment—on such aspects as whether the diets are fed to the young mice themselves or to the mothers, and how the animals are allocated to the two diets.

Sometimes variation arises at several different hierarchical levels in an experiment, and it can be a challenge to sort these out, and particularly, to correctly identify the quantity n. Example 6.5.3 illustrates this issue.

Example 6.5.3

Germination of Spores In a study of the fungus that causes the anthracnose disease of corn, interest focused on the survival of the fungal spores.[27] Batches of spores, all prepared from a single culture of the fungus, were stored in chambers under various environmental conditions and then assayed for their ability to germinate, as follows. Each batch of spores was suspended in water and then plated on agar in a petri dish. Ten "plugs" of 3-mm diameter were cut from each petri dish and were incubated at 25 °C for 12 hours. Each plug was then examined with a microscope for germinated and ungerminated spores. The environmental conditions of storage (the "treatments") included the following:

T_1: Storage at 70% relative humidity for 1 week

T_2: Storage at 60% relative humidity for 1 week

T_3: Storage at 60% relative humidity for 2 weeks

and so on.

All together there were 43 treatments.

The design of the experiment is indicated schematically in Figure 6.5.2. There were 129 batches of spores, which were randomly allocated to the 43 treatments, three batches to each treatment. Each batch of spores resulted in one petri dish, and each petri dish resulted in 10 plugs.

Figure 6.5.2 Design of spore germination experiment

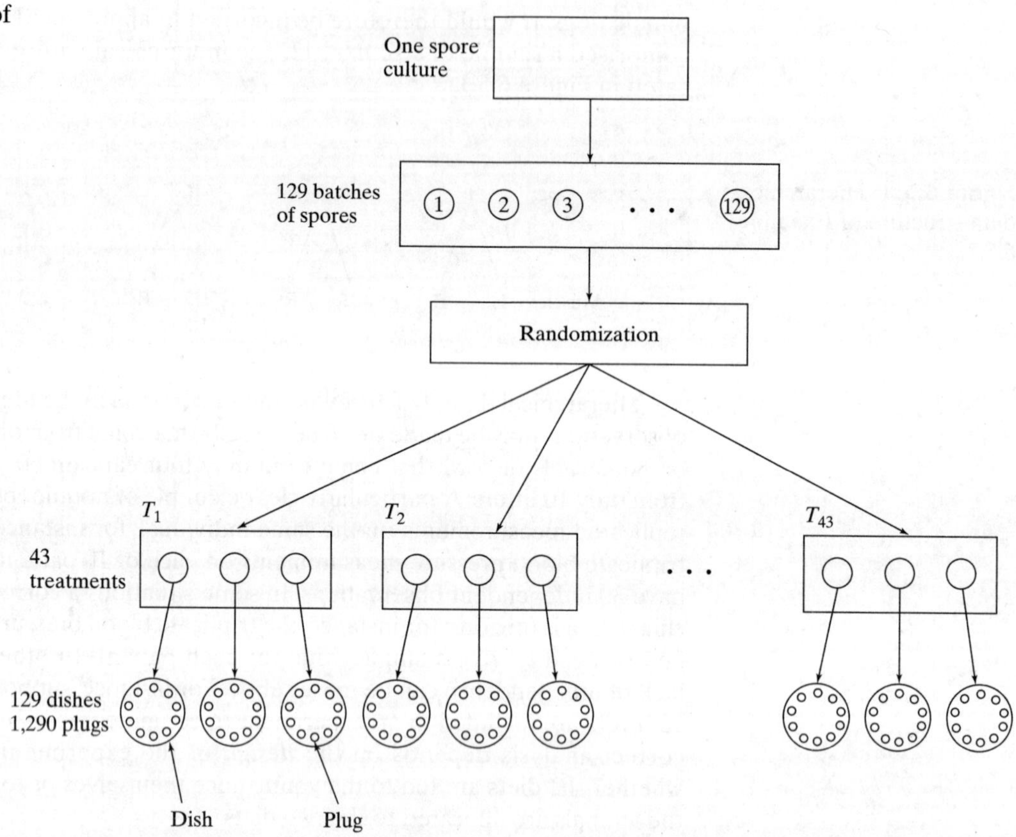

To get a feeling for the issues raised by this design, let us look at some of the raw data. Table 6.5.1 shows the percentage of the spores that had germinated for each plug assayed for treatment 1.

Table 6.5.1 shows that there is considerable variability both *within* each petri dish and *between* the dishes. The variability within the dishes reflects local variation in the percent germination, perhaps due largely to differences among the spores themselves (some of the spores were more mature than others). The variability between dishes is even larger, because it includes not only local variation, but also larger-scale variation such as the variability among the original batches of spores, and temperature and relative humidity variations within the storage chambers.

Now consider the problem of comparing treatment 1 to the other treatments. Would it be legitimate to take the point of view that we have 30 observations for each treatment? To focus this question, let us consider the matter of calculating the standard error for the mean of treatment 1. The mean and SD of all 30 observations are

$$\text{Mean} = 62.33$$

$$\text{SD} = 11.88$$

Table 6.5.1 Percentage germination under treatment 1

	Dish I	Dish II	Dish III
	49	66	49
	58	84	60
	48	83	54
	69	69	72
	45	72	57
	43	85	70
	60	59	65
	44	60	68
	44	75	66
	68	68	60
Mean	52.8	72.1	62.1
SD	10.1	9.5	7.4

Is it legitimate to calculate the SE of the mean as

$$\mathrm{SE}_{\bar{Y}} = \frac{s}{\sqrt{n}} = \frac{11.88}{\sqrt{30}} = 2.2?$$

As you may suspect, ***this is not legitimate.*** There is a hierarchical structure in the data, and so we cannot apply the SE formula so naively. An acceptable way to calculate the SE is to consider the mean for each dish as an observation; thus, we obtain the following*:

$$\text{Observations: } 52.8, 72.1, 62.1$$

$$n = 3$$

$$\text{Mean} = 62.33$$

$$\text{SD} = 9.65$$

$$\mathrm{SE}_{\bar{Y}} = \frac{s}{\sqrt{n}} = \frac{9.65}{\sqrt{3}} = 5.6$$

Notice that the incorrect analysis gave the same mean (62.33) as this analysis, but an inappropriately small SE (2.2 rather than 5.6). If we were comparing several treatments, the same pattern would tend to hold; the incorrect analysis would tend to produce SEs that were (individually and pooled) too small, which might cause us to "overinterpret" the data, in the sense of suggesting there is significant evidence of treatment differences where none exists.

We should emphasize that, even though the correct analysis requires combining the measurements on the 10 plugs in a dish into a single observation for that dish, the experimenter was not wasting effort by measuring 10 plugs per dish instead of, say, only one plug per dish. The mean of 10 plugs is a much better estimate of the average for the entire dish than is a measurement on one plug; the improved precision for measuring 10 plugs is reflected in a smaller between-dish SD. For instance, for treatment 1 the SD was 9.65; if fewer plugs per dish had been measured, this SD would probably have been larger. ■

*An alternative way to aggregate the data from the 10 plugs in a dish would be to combine the raw counts of germinated and ungerminated spores for the whole dish and express these as an overall percentage germination.

The pitfall illustrated by Example 6.5.3 has trapped many an unwary researcher. When hierarchical structures result from repeated measurements on the same individual organism (as in Example 6.5.2), they are relatively easy to recognize. But the hierarchical structure in Example 6.5.3 has a different origin; it is due to the fact that the unit of observation is an individual plug, but individual plugs are not randomly allocated to the treatment groups. Rather, the unit that is randomly allocated to treatment is a batch of spores, which later is plated in a petri dish, which then gives rise to 10 plugs. In the language of experimental design, plugs are **nested** within petri dishes. *Whenever observational units are nested within the units that are randomly allocated to treatments, a hierarchical structure may potentially exist in the data.* Note that the difficulty is only "potential"; in some cases a nonhierarchical analysis may be acceptable. For instance, if experience had shown that the differences between petri dishes were negligible, then we might ignore the hierarchical structure in analyzing the data. The decision can be a difficult one and may require expert statistical advice.

The issue of hierarchical data structures has important implications for the design of an experiment as well as its analysis. The sample size (n) must be appropriately identified in order to determine whether the experiment includes enough replication. As a simple example, suppose it is proposed to do a spore germination experiment such as that of Example 6.5.3, but with only *one* dish per treatment, rather than three. To see the flaw in this proposal, suppose that the proposed experiment is to include three treatments, with one dish per treatment. With this design, would we then be able to distinguish treatment differences from inherent differences between the dishes? No. The intertreatment differences and the interdish differences would be mutually entangled, or confounded. You can easily visualize this situation if you look at the data in Table 6.5.1 and pretend that those data came from the proposed experiment; that is, pretend that dishes I, II, and III had received different treatments, and that we had no other data. It would be difficult to extract meaningful information about intertreatment differences unless we knew for *certain* that interdish variation was negligible.

We saw in Section 6.4 how to use a preliminary estimate of the SD to determine the sample size (n) required to attain a desired degree of precision, as expressed by the SE. These ideas carry over to experiments involving hierarchical data structures. For example, suppose a botanist is planning a spore germination experiment such as that of Example 6.5.3. If she has already decided to use 10 plugs per dish, the remaining problem would be to decide on the number of dishes per treatment. This question could be approached as in Section 6.4, considering the dish as the experimental unit, and using a preliminary estimate of the SD between dishes (which was 9.65 in Example 6.5.3). If, however, she wants to choose optimal values for *both* the number of plugs per dish *and* the number of dishes per treatment, she may wish to consult a statistician.

CONDITIONS FOR VALIDITY OF A CONFIDENCE INTERVAL FOR μ

A confidence interval for μ provides a definite quantitative interpretation for $SE_{\bar{Y}}$. Note that the data must be a random sample from the population of interest. If there is bias in the sampling process, then the sampling distribution concepts on which the confidence interval method is based do not hold: Knowing the mean of a biased sample does not provide information about the population mean μ. The validity of Student's t method for constructing confidence intervals also depends on the form of the population distribution of the observed variable Y. If Y follows a normal distribution in the population, then Student's t method is exactly **valid**—that is to say, the probability that the confidence interval will contain μ is actually equal to the confidence level (e.g., 95%). By the same token, this interpretation is approximately valid if the population distribution is approximately normal. Even if the population distribution is not normal, the Student's t confidence interval is approximately valid *if* the sample size is large. This fact can often be used to justify the use of the confidence interval even in situations where the population distribution cannot be assumed to be approximately normal.

From a practical point of view, the important question is: How large must the sample be in order for the confidence interval to be approximately valid? Not surprisingly, the answer to this question depends on the *degree* of nonnormality of the population distribution: If the population is only moderately nonnormal, then n need not be very large. Table 6.5.2 shows the actual probability that a Student's t confidence interval will contain μ for samples from three different populations.[28] The forms of the population distributions are shown in Figure 6.5.3. Population 1 is a normal population, population 2 is moderately skewed, and population 3 is an extremely skewed, "L-shaped" distribution. (Populations 2 and 3 were discussed in optional Section 5.3.)

Table 6.5.2 Actual probability that confidence intervals will contain the population mean

(a) 95% confidence interval							
	Sample size						
	2	4	8	16	32	64	Very large
Population 1	0.95	0.95	0.95	0.95	0.95	0.95	0.95
Population 2	0.94	0.93	0.94	0.94	0.95	0.95	0.95
Population 3	0.87	0.53	0.57	0.80	0.88	0.92	0.95
(b) 99% confidence interval							
	Sample size						
	2	4	8	16	32	64	Very large
Population 1	0.99	0.99	0.99	0.99	0.99	0.99	0.99
Population 2	0.99	0.98	0.98	0.98	0.99	0.99	0.99
Population 3	0.97	0.82	0.60	0.81	0.93	0.96	0.99

For population 1, Table 6.5.2 shows that the confidence interval method is exactly valid for all sample sizes, even $n = 2$. For population 2, the method is

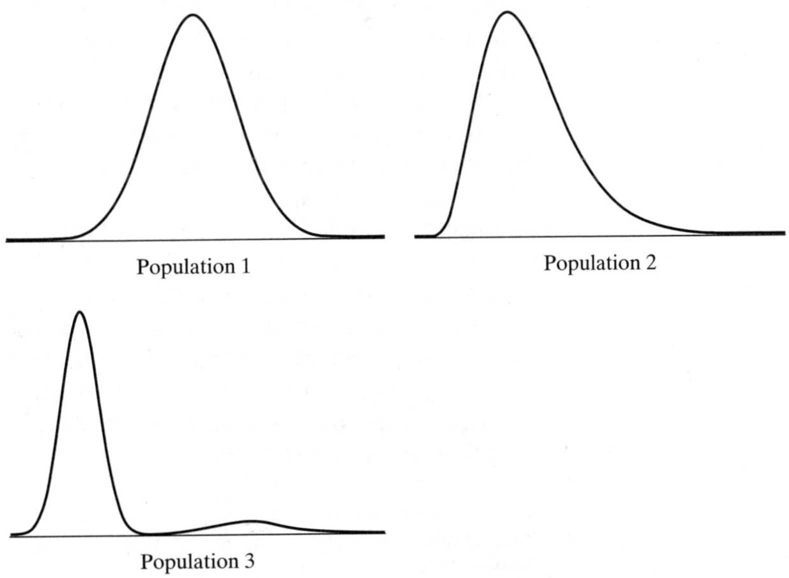

Figure 6.5.3 Three population distributions: (1) normal, (2) slightly skewed right, (3) heavily skewed right

approximately valid even for fairly small samples. For population 3 the approximation is very poor for small samples and is only fair for samples as large as $n = 64$. In a sense, population 3 is a "worst case"; it could be argued that the mean is not a meaningful measure for population 3, because of its bizarre shape.

SUMMARY OF CONDITIONS

In summary, Student's t method of constructing a confidence interval for μ is appropriate if the following conditions hold.

1. **Conditions on the design of the study**
   (a) It must be reasonable to regard the data as a random sample from a large population.
   (b) The observations in the sample must be independent of each other.
2. **Conditions on the form of the population distribution**
   (a) If n is small, the population distribution must be approximately normal.
   (b) If n is large, the population distribution need not be approximately normal.

The requirement that the data are a random sample is the most important condition.

The required "largeness" in condition 2(b) depends (as shown in Example 6.5.3) on the degree of nonnormality of the population. In many practical situations, moderate sample sizes (say, $n = 20$ to 30) are large enough.

VERIFICATION OF CONDITIONS

In practice, the preceding "conditions" are often "assumptions" rather than known facts. However, it is always important to check whether the conditions are reasonable in a given case.

To determine whether the random sampling model is applicable to a particular study, the design of the study should be scrutinized, with particular attention to possible biases in the choice of experimental material and to possible nonindependence of the observations due to hierarchical data structures.

As to whether the population distribution is approximately normal, information on this point may be available from previous experience with similar data. If the only source of information is the data at hand, then normality can be roughly checked by making a histogram and normal quantile plot of the data. Unfortunately, for a small or moderate sample size, this check is fairly crude; for instance, if you look back at Figure 5.2.7, you will see that even samples of size 25 from a normal population often do not appear particularly normal.* Of course, if the sample is large, then the sample histogram gives us good information about the population shape; however, if n is large, the requirement of normality is less important anyway.

In any case, a crude check is better than none, and *every* data analysis should begin with inspection of a graph of the data, with special attention to any observations that lie very far from the center of the distribution.

---

*We could aid our graphical assessment of normality by using a more objective method such as the Shapiro–Wilk test of Section 4.4.

Sometimes a histogram or normal quantile plot of the data indicates that the data did not come from a normal population. If the sample size is small, then Student's t method will not give valid results. However, it may be possible to transform the data to achieve approximate normality and then analyze the data in the transformed scale.

Example 6.5.4

Sediment Yield Sediment yield, which is a measure of the amount of suspended sediment in water, is a measure of water quality for a river. The distribution of sediment yield often has a skewed distribution. However, taking the logarithm of each observation can produce a distribution that follows a normal curve quite well. Figure 6.5.4 shows normal quantile plots of sediment yields of water samples from the Black River in northern Ohio for $n = 9$ days (a) in mg/l and (b) in log scale [i.e., ln(mg/l)].[29]

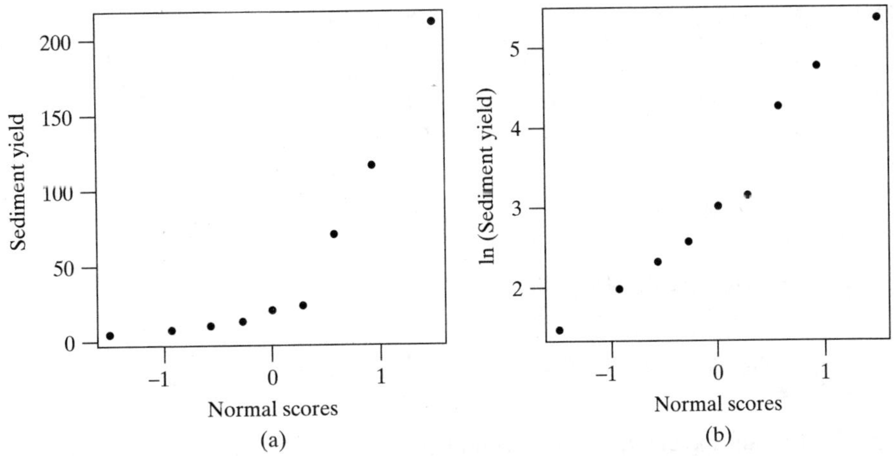

Figure 6.5.4 Normal probability plots of sediment yields of water samples from the Black River for 9 days (a) in mg/l and (b) after taking the natural logarithm of each observation*

The natural logarithms of the sediment yields have an average of $\bar{y} = 3.21$ and a standard deviation of $s = 1.33$. Thus, the standard error of the mean is $\frac{1.33}{\sqrt{9}} = 0.44$. The t multiplier for a 95% confidence interval is $t_{8,\,0.025} = 2.306$. A 95% confidence interval for μ is

$$3.21 \pm 2.306(0.44)$$

or, approximately,

$$3.21 \pm 1.01$$

or

$$(2.20, 4.22)$$

Thus, *we are 95% confident that the mean natural logarithm of sediment yield for the Black River is between 2.20 and 4.22.*[†] ■

*The Shapiro–Wilk test of normality (from Section 4.4) for the raw data yields a P-value of 0.0039 providing strong evidence of abnormality for the untransformed data. In contrast, for the natural-log transformed data, the Shapiro–Wilk P-value is 0.6551, showing no significant evidence for non-normality. Note that we could also have taken the base 10 log to normalize the data.

[†]Note that we have constructed a confidence interval for the population average logarithm of sediment yield. Because the logarithm transformation is not linear, the mean of the logarithms is not the logarithm of the mean, so applying the inverse transformation to the endpoints of the confidence interval will not convert it properly into a confidence interval for the population mean in the original scale of mg/l. However, we can get an approximate confidence interval by taking $\exp(2.2 + 1.33^2/2)$ and $\exp(4.22 + 1.33^2/2)$. [This is based on the fact that the mean of a log normal distribution (which is bell shaped after taking logarithms) is $\exp(\mu + \sigma^2/2)$.]

Exercises 6.5.1–6.5.8

6.5.1 SGOT is an enzyme that shows elevated activity when the heart muscle is damaged. In a study of 31 patients who underwent heart surgery, serum levels of SGOT were measured 18 hours after surgery.[30] The mean was 49.3 U/l and the standard deviation was 68.3 U/l. If we regard the 31 observations as a sample from a population, what feature of the data would cause one to doubt that the population distribution is normal?

6.5.2 A dendritic tree is a branched structure that emanates from the body of a nerve cell. In a study of brain development, researchers examined brain tissue from seven adult guinea pigs. The investigators randomly selected nerve cells from a certain region of the brain and counted the number of dendritic branch segments emanating from each selected cell. A total of 36 cells was selected, and the resulting counts were as follows[31]:

38	42	25	35	35	33	48	53	17
24	26	26	47	28	24	35	38	26
38	29	49	26	41	26	35	38	44
25	45	28	31	46	32	39	59	53

The mean of these counts is 35.67 and the standard deviation is 9.99.

Suppose we want to construct a 95% confidence interval for the population mean. We could calculate the standard error as

$$SE_{\bar{Y}} = \frac{9.99}{\sqrt{36}} = 1.67$$

and obtain the confidence interval as

$$35.67 \pm (2.042)(1.67)$$

or

$$32.3 < \mu < 39.1$$

(a) On what grounds might the above analysis be criticized? (*Hint*: Are the observations independent?)

(b) Using the classes [15,20), [20,25), and so on, construct a histogram of the data. Does the shape of the distribution support the criticism you made in part (a)? If so, explain how.

6.5.3 In an experiment to study the regulation of insulin secretion, blood samples were obtained from seven dogs before and after electrical stimulation of the vagus nerve. The following values show, for each animal, the increase (after minus before) in the immunoreactive insulin concentration (μU/ml) in pancreatic venous plasma.[32]

30	100	60	30	130	1,060	30

For these data, Student's t method yields the following 95% confidence interval for the population mean:

$$-145 < \mu < 556$$

Is Student's t method appropriate in this case? Why or why not?

6.5.4 In a study of parasite–host relationships, 242 larvae of the moth *Ephestia* were exposed to parasitization by the Ichneumon fly. The following table shows the number of Ichneumon eggs found in each of the *Ephestia* larva.[33]

Number of eggs (Y)	Number of larvae
0	21
1	77
2	52
3	41
4	23
5	13
6	9
7	1
8	2
9	0
10	2
11	0
12	0
13	0
14	0
15	1
Total	242

For these data, $\bar{y} = 2.368$ and $s = 1.950$. Student's t method yields the following 95% confidence interval for μ, the population mean number of eggs per larva:

$$2.12 < \mu < 2.61$$

(a) Does it appear reasonable to assume that the population distribution of Y is approximately normal? Explain.

(b) In view of your answer to part (a), on what grounds can you defend the application of Student's t method to these data?

6.5.5 The following normal quantile plot shows the distribution of the diameters, in cm, of each of nine American Sycamore trees.[34]

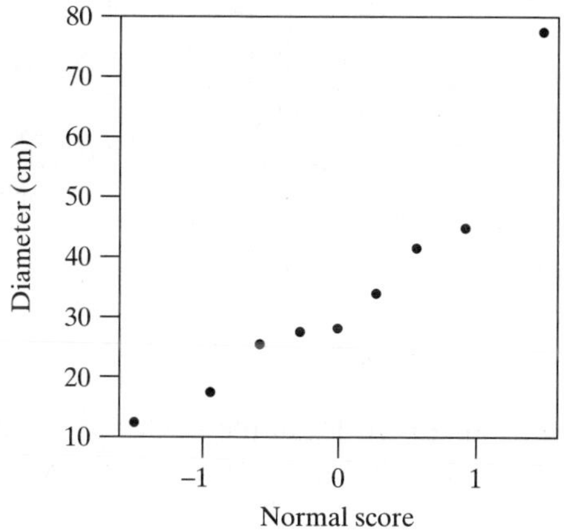

The normal quantile plot is not linear, which suggests that a transformation of the data is needed before a confidence interval can be constructed using Student's t method. The raw data are

12.4 44.8 28.2 77.6 34 17.5 41.5 25.5 27.5

(a) Take the square root of each observation and construct a 90% confidence interval for the mean.

(b) Interpret the confidence interval from part (a). That is, explain what the interval tells you about the square root of the diameters of these trees.

6.5.6 Four treatments were compared for their effect on the growth of spinach cells in cell culture flasks. The experimenter randomly allocated two flasks to each treatment. After a certain time on treatment, he randomly drew three aliquots (1 cc each) from each flask and measured the cell density in each aliquot; thus, he had six cell density measurements for each treatment. In calculating the standard error of a treatment mean, the experimenter calculated the standard deviation of the six measurements and divided by $\sqrt{6}$. On what grounds might an objection be raised to this method of calculating the SE?

6.5.7 In an experiment on soybean varieties, individually potted soybean plants were grown in a greenhouse, with 10 plants of each variety used in the experiment. From the harvest of each plant, five seeds were chosen at random and individually analyzed for their percentage of oil. This gave a total of 50 measurements for each variety. To calculate the standard error of the mean for a variety, the experimenter calculated the standard deviation of the 50 observations and divided by $\sqrt{50}$. Why would this calculation be of doubtful validity?

6.5.8 In a plant mitigation project, an entire local (endangered) population of 255 Congdon's tarplants was transplanted to a new location.[35] One year after transplant, 30 of the 255 plants were randomly selected and the diameter at the root caudix junction (the top of the root just beneath the surface of the soil) was measured. If the population of plants under consideration consists of only the local 255 plants, explain why it would be inappropriate to use Student's t method of constructing a confidence interval for μ, the population mean root caudix junction diameter.

6.6 Comparing Two Means

In previous sections we have considered the analysis of a single sample of quantitative data. In practice, however, much scientific research involves the comparison of two or more samples from different populations. When the observed variable is quantitative, the comparison of two samples can include several aspects, notably (1) comparison of means, (2) comparison of standard deviations, and (3) comparison of shapes. In this section, and indeed throughout this book, the primary emphasis will be on comparison of means and on other comparisons related to shift. We will begin by discussing the confidence interval approach to comparing means, which is a natural extension of the material in Section 6.3; in Chapter 7 we will consider an approach known as *hypothesis testing*.

NOTATION

Figure 6.6.1 presents our notation for comparison of two samples. The notation is exactly parallel to our earlier notation, but now a subscript (1 or 2) is used to differentiate between the two samples. The two "populations" can be naturally occurring populations (as in Example 6.1.1) or they can be conceptual populations

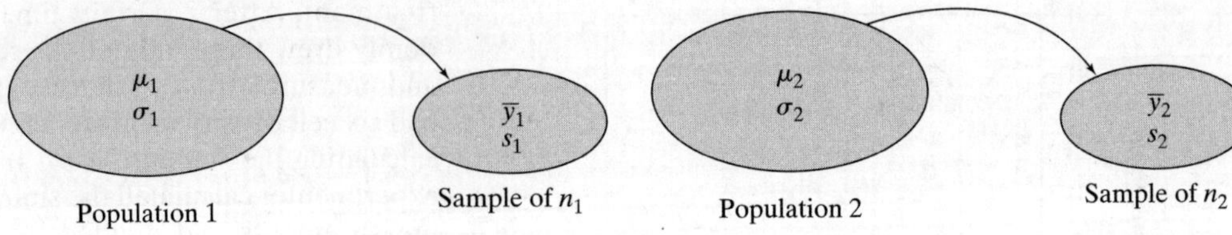

Figure 6.6.1 Notation for comparison of two samples

defined by certain experimental conditions (as in Example 6.3.4). In either case, the data in each sample are viewed as a random sample from the corresponding population.

We begin by describing, in the next section, some simple computations that are used for both confidence intervals and hypothesis testing.

STANDARD ERROR OF $(\bar{Y}_1 - \bar{Y}_2)$

In this section we introduce a fundamental quantity for comparing two samples: the standard error of the difference between two sample means.

BASIC IDEAS

We saw in Section 6.2 that the precision of a sample mean $\bar{Y}$ can be expressed by its standard error, which is equal to

$$\mathrm{SE}_{\bar{Y}} = \frac{s}{\sqrt{n}}$$

To compare two sample means, it is natural to consider the difference between them:

$$\bar{Y}_1 - \bar{Y}_2$$

which is an estimate of the quantity $(\mu_1 - \mu_2)$. To characterize the sampling error of estimation, we need to be concerned with the standard error of the difference $(\bar{Y}_1 - \bar{Y}_2)$. We illustrate this idea with an example.

Example 6.6.1 **Vital Capacity** Vital capacity is a measure of the amount of air that someone can exhale after taking a deep breath. One might expect that musicians who play brass instruments would have greater vital capacities, on average, than would other persons of the same age, sex, and height. In one study the vital capacities of eight brass players were compared to the vital capacities of seven control subjects; Table 6.6.1 shows the data.[36]

The difference between the sample means is

$$\bar{y}_1 - \bar{y}_2 = 4.83 - 4.74 = 0.09$$

We know that both $\bar{y}_1$ and $\bar{y}_2$ are subject to sampling error, and consequently the difference (0.09) is subject to sampling error. The standard error of $\bar{Y}_1 - \bar{Y}_2$ tells us how much precision to attach to this difference between $\bar{Y}_1$ and $\bar{Y}_2$. ■

Table 6.6.1 Vital capacity (liters)

	Brass player	Control
	4.7	4.2
	4.6	4.7
	4.3	5.1
	4.5	4.7
	5.5	5.0
	4.9	
	5.3	
n	7	5
$\bar{y}$	4.83	4.74
s	0.435	0.351

DEFINITION The **standard error of $\bar{Y}_1 - \bar{Y}_2$** is defined as

$$\mathrm{SE}_{(\bar{Y}_1-\bar{Y}_2)} = \sqrt{\frac{s_1^2}{n_1} + \frac{s_2^2}{n_2}}$$

The following alternative form of the formula shows how the SE of the difference is related to the individual SEs of the means:

$$\mathrm{SE}_{(\bar{Y}_1-\bar{Y}_2)} = \sqrt{\mathrm{SE}_1^2 + \mathrm{SE}_2^2}$$

where

$$\mathrm{SE}_1 = \mathrm{SE}_{\bar{Y}_1} = \frac{s_1}{\sqrt{n_1}}$$

$$\mathrm{SE}_2 = \mathrm{SE}_{\bar{Y}_2} = \frac{s_2}{\sqrt{n_2}}$$

Notice that this version of the formula shows that "SEs add like Pythagoras." When we have two independent samples, we take the SE of each mean, square them, add them, and then take the square root of the sum. Figure 6.6.2 illustrates this idea.

Figure 6.6.2 SE for a difference

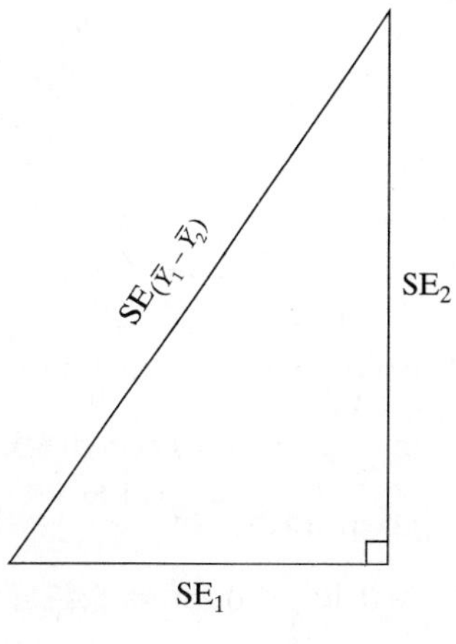

It may seem odd that in calculating the SE of a difference we *add* rather than subtract within the formula $SE_{(\bar{Y}_1-\bar{Y}_2)} = \sqrt{SE_1^2 + SE_2^2}$. However, as was discussed in Section 3.5, the variability of the difference depends on the variability of each part. Whether we add $\bar{Y}_2$ to $\bar{Y}_1$ or subtract $\bar{Y}_2$ from $\bar{Y}_1$, the "noise" associated with $\bar{Y}_2$ (i.e., SE_2) adds to the overall uncertainty. The greater the variability in $\bar{Y}_2$, the greater the variability in $\bar{Y}_1 - \bar{Y}_2$. The formula $SE_{(\bar{Y}_1-\bar{Y}_2)} = \sqrt{SE_1^2 + SE_2^2}$ accounts for this variability.

We illustrate the formulas in the following example.

Example 6.6.2

Vital Capacity For the vital capacity data, preliminary computations yield the results in Table 6.6.2.

The SE of $(\bar{Y}_1 - \bar{Y}_2)$ is

$$SE_{(\bar{Y}_1-\bar{Y}_2)} = \sqrt{\frac{0.1892}{7} + \frac{0.1232}{5}} = 0.227 \approx 0.23$$

Note that

$$0.227 = \sqrt{(0.164)^2 + (0.157)^2}$$

Notice that the SE of the difference is greater than either of the individual SEs but less than their sum. ■

Table 6.6.2

	Brass player	Control
s^2	0.1892	0.1232
n	7	5
SE	0.164	0.157

Example 6.6.3

Tonsillectomy An experiment was conducted to compare conventional surgery to a newer procedure called Coblation-assisted intracapsular tonsillectomy for children who needed to have their tonsils removed. A key measurement taken during the study was the pain score that each child reported, on a scale of 0–10, four days after surgery. Table 6.6.3 gives the means and standard deviations of pain scores for the two groups.[37]

Table 6.6.3 Pain score

	Type of surgery	
	Conventional	Coblation
Mean	4.3	1.9
SD	2.8	1.8
n	49	52

The data in Table 6.6.3 show that the standard deviation of pain scores in 49 children given conventional surgery was 2.8. Thus, the SE for the conventional mean is $\frac{2.8}{\sqrt{49}} = 0.40$. For the 52 children in the coblation group, the SD was 1.8, which gives an SE of $\frac{1.8}{\sqrt{52}} = 0.2496$. The SE for the difference in the two means is $\sqrt{0.40^2 + 0.25^2} = 0.4717 \approx 0.47$. ■

THE POOLED STANDARD ERROR (OPTIONAL)

The preceding standard error is known as the "unpooled" standard error. Many statistics software packages allow the user to specify use of what is known as the "pooled" standard error, which we will discuss briefly.

Recall that the square of the standard deviation, s, is the sample variance, s^2, defined as

$$s^2 = \frac{\Sigma(y_i - \bar{y})^2}{n - 1}$$

The pooled variance is a weighted average of s_1^2, the variance of the first sample, and s_2^2, the variance of the second sample, with weights equal to the degrees of freedom from each sample, $n_i - 1$:

$$s^2_{\text{pooled}} = \frac{(n_1 - 1)s_1^2 + (n_2 - 1)s_2^2}{(n_1 - 1) + (n_2 - 1)} = \frac{(n_1 - 1)s_1^2 + (n_2 - 1)s_2^2}{(n_1 + n_2 - 2)}.$$

The pooled standard error is defined as

$$\text{SE}_{\text{pooled}} = \sqrt{s^2_{\text{pooled}}\left(\frac{1}{n_1} + \frac{1}{n_2}\right)}.$$

We illustrate with an example.

Example 6.6.4 **Vital Capacity** For the vital capacity data we found that $s_1^2 = 0.1892$ and $s_2^2 = 0.1232$. The pooled variance is

$$s^2_{\text{pooled}} = \frac{(7 - 1)0.1892 + (5 - 1)0.1232}{(7 + 5 - 2)} = 0.1628$$

and the pooled SE is

$$\text{SE}_{\text{pooled}} = \sqrt{0.1628\left(\frac{1}{7} + \frac{1}{5}\right)} = 0.236.$$

Recall from Example 6.6.2 that the unpooled SE for the same data was 0.227. ■

If the sample sizes are equal ($n_1 = n_2$) or if the sample standard deviations are equal ($s_1 = s_2$), then the unpooled and the pooled method will give the same answer for $\text{SE}_{(\bar{Y}_1 - \bar{Y}_2)}$. The two answers will not differ substantially unless both the sample sizes and the sample SDs are quite discrepant.

To show the analogy between the two SE formulas, we can write them as follows:

$$\text{SE}_{(\bar{Y}_1 - \bar{Y}_2)} = \sqrt{\frac{s_1^2}{n_1} + \frac{s_2^2}{n_2}}$$

and

$$\text{SE}_{\text{pooled}} = \sqrt{\frac{s^2_{\text{pooled}}}{n_1} + \frac{s^2_{\text{pooled}}}{n_2}}$$

In the pooled method, the separate variances—s_1^2 and s_2^2—are replaced by the single variance s^2_{pooled}, which is calculated from both samples.

Both the unpooled and the pooled SE have the same purpose—to estimate the standard deviation of the sampling distribution of $(\bar{Y}_1 - \bar{Y}_2)$. In fact, it can be shown that the standard deviation is

$$\sigma_{(\bar{Y}-\bar{Y}_2)} = \sqrt{\frac{\sigma_1^2}{n_1} + \frac{\sigma_2^2}{n_2}}$$

Note the resemblance between this formula and the formula for $SE_{(\bar{Y}_1-\bar{Y}_2)}$.

In analyzing data when the sample sizes are unequal ($n_1 \neq n_2$), one needs to decide whether to use the pooled or unpooled method for calculating the standard error. The choice depends on whether one is willing to assume that the population SDs (σ_1 and σ_2) are equal. It can be shown that if $\sigma_1 = \sigma_2$, then the pooled method should be used, because in this case s_{pooled} is the best estimate of the population SD. However, in this case the unpooled method will typically give an SE that is quite similar to that given by the pooled method. If $\sigma_1 \neq \sigma_2$, then the unpooled method should be used, because in this case s_{pooled} is not an estimate of either σ_1 or σ_2, so pooling would accomplish nothing. Because the two methods substantially agree when $\sigma_1 = \sigma_2$ and the pooled method is not valid when $\sigma_1 \neq \sigma_2$, most statisticians prefer the unpooled method. There is little to be gained by pooling when pooling is appropriate and there is much to be lost when pooling is not appropriate. Many software packages use the unpooled method by default; the user must specify use of the pooled method if she or he wishes to pool the variances.

Exercises 6.6.1–6.6.9

6.6.1 Data from two samples gave the following results:

	Sample 1	Sample 2
n	6	12
$\bar{y}$	40	50
s	4.3	5.7

Compute the standard error of $(\bar{Y}_1 - \bar{Y}_2)$.

6.6.2 Compute the standard error of $(\bar{Y}_1 - \bar{Y}_2)$ for the following data:

	Sample 1	Sample 2
n	10	10
$\bar{y}$	125	217
s	44.2	28.7

6.6.3 Compute the standard error of $(\bar{Y}_1 - \bar{Y}_2)$ for the following data:

	Sample 1	Sample 2
n	5	7
$\bar{y}$	44	47
s	6.5	8.4

6.6.4 Consider the data from Exercise 6.6.3. Suppose the sample sizes were doubled, but the means and SDs stayed the same, as follows. Compute the standard error of $(\bar{Y}_1 - \bar{Y}_2)$.

	Sample 1	Sample 2
n	10	14
$\bar{y}$	44	47
s	6.5	8.4

6.6.5 Data from two samples gave the following results:

	Sample 1	Sample 2
$\bar{y}$	96.2	87.3
SE	3.7	4.6

Compute the standard error of $(\bar{Y}_1 - \bar{Y}_2)$.

6.6.6 Data from two samples gave the following results:

	Sample 1	Sample 2
n	22	21
$\bar{y}$	1.7	2.4
SE	0.5	0.7

Compute the standard error of $(\bar{Y}_1 - \bar{Y}_2)$.

6.6.7 Example 6.6.3 reports measurements of pain for children who have had their tonsils removed. Another variable measured in that experiment was the number of doses of Tylenol taken by the children in the two groups. Those data are

	Type of surgery	
	Conventional	Coblation
n	49	52
$\bar{y}$	3.0	2.3
SD	2.4	2.0

Compute the standard error of $(\bar{Y}_1 - \bar{Y}_2)$.

6.6.8 Two varieties of lettuce were grown for 16 days in a controlled environment. The following table shows the total dry weight (in grams) of the leaves of nine plants of the variety "Salad Bowl" and six plants of the variety "Bibb."[38]

	Salad bowl	Bibb
	3.06	1.31
	2.78	1.17
	2.87	1.72
	3.52	1.20
	3.81	1.55
	3.60	1.53
	3.30	
	2.77	
	3.62	
$\bar{y}$	3.259	1.413
s	.400	.220

Compute the standard error of $(\bar{Y}_1 - \bar{Y}_2)$ for these data.

6.6.9 Some soap manufacturers sell special "antibacterial" soaps. However, one might expect ordinary soap to also kill bacteria. To investigate this, a researcher prepared a solution from ordinary, nonantibiotic soap and a control solution of sterile water. The two solutions were placed onto petri dishes and *E. coli* bacteria were added. The dishes were incubated for 24 hours and the number of bacteria colonies on each dish were counted.[39] The data are given in the following table.

	Control (Group I)	Soap (Group 2)
	30	76
	36	27
	66	16
	21	30
	63	26
	38	46
	35	6
	45	
n	8	7
$\bar{y}$	41.8	32.4
s	15.6	22.8
SE	5.5	8.6

Compute the standard error of $(\bar{Y}_1 - \bar{Y}_2)$ for these data.

6.7 Confidence Interval for $(\mu_1 - \mu_2)$

One way to compare two sample means is to construct a confidence interval for the difference in the population means—that is, a confidence interval for the quantity $(\mu_1 - \mu_2)$. Recall from Section 6.3 that a 95% confidence interval for the mean μ of a single population that is normally distributed is constructed as

$$\bar{y} \pm t_{0.025}\mathrm{SE}_{\bar{Y}}$$

Analogously, a 95% confidence interval for $(\mu_1 - \mu_2)$ is constructed as

$$(\bar{y}_1 - \bar{y}_2) \pm t_{0.025}\mathrm{SE}_{(\bar{Y}_1 - \bar{Y}_2)}$$

The critical value $t_{0.025}$ is determined from Student's t distribution using degrees of freedom* given as

$$\mathrm{df} = \frac{(\mathrm{SE}_1^2 + \mathrm{SE}_2^2)^2}{\mathrm{SE}_1^4/(n_1 - 1) + \mathrm{SE}_2^4/(n_2 - 1)} \qquad (6.7.1)$$

where $\mathrm{SE}_1 = s_1/\sqrt{n_1}$ and $\mathrm{SE}_2 = s_2/\sqrt{n_2}$.

*Strictly speaking, the distribution needed to construct a confidence interval here depends on the unknown population standard deviations σ_1 and σ_2 and is not a Student's t distribution. However, Student's t distribution with degrees of freedom given by formula (6.7.1) is a very good approximation. This is sometimes known as Welch's method or Satterthwaite's method.

Of course, calculating the degrees of freedom from formula (6.7.1) is complicated and time consuming. Most computer software uses formula (6.7.1), as do some graphing calculators. A simpler method to obtain the approximate degrees of freedom is to use the smaller of $(n_1 - 1)$ and $(n_2 - 1)$. This option gives a confidence interval that is somewhat conservative, in the sense that the true confidence level is a bit larger than 95% when $t_{0.025}$ is used. A third approach is to approximate the degrees of freedom as $n_1 + n_2 - 2$. This approach is somewhat liberal, in the sense that the true confidence level is a bit smaller than 95% when $t_{0.025}$ is used.

Intervals with other confidence coefficients are constructed analogously; for example, for a 90% confidence interval one would use $t_{0.05}$ instead of $t_{0.025}$.

The following example illustrates the construction of a confidence interval for $(\mu_1 - \mu_2)$.

Example 6.7.1

Fast Plants The Wisconsin Fast Plant, *Brassica campestris*, has a very rapid growth cycle that makes it particularly well suited for the study of factors that affect plant growth. In one such study, seven plants were treated with the substance Ancymidol (ancy) and were compared to eight control plants that were given ordinary water. Heights of all of the plants were measured, in cm, after 14 days of growth.[40] The data are given in Table 6.7.1.

Parallel dotplots and normal quantile plots (Figure 6.7.1) show that both sample distributions are reasonably symmetric and bell shaped. Moreover, we would expect that a distribution of plant heights might well be normally distributed, since height distributions often follow a normal curve. The dotplots show that the ancy distribution is shifted down a bit from the control distribution; the difference in sample means is 15.9 − 11.0 = 4.9. The SE for the difference in sample means is

$$SE_{(\bar{Y}_1 - \bar{Y}_2)} = \sqrt{\frac{4.8^2}{8} + \frac{4.7^2}{7}} = 2.46$$

Table 6.7.1 Fourteen-day height of control and of ancy plants (cm)

	Control (Group 1)	Ancy (Group 2)
	10.0	13.2
	13.2	19.5
	19.8	11.0
	19.3	5.8
	21.2	12.8
	13.9	7.1
	20.3	7.7
	9.6	
n	8	7
$\bar{y}$	15.9	11.0
s	4.8	4.7
SE	1.7	1.8

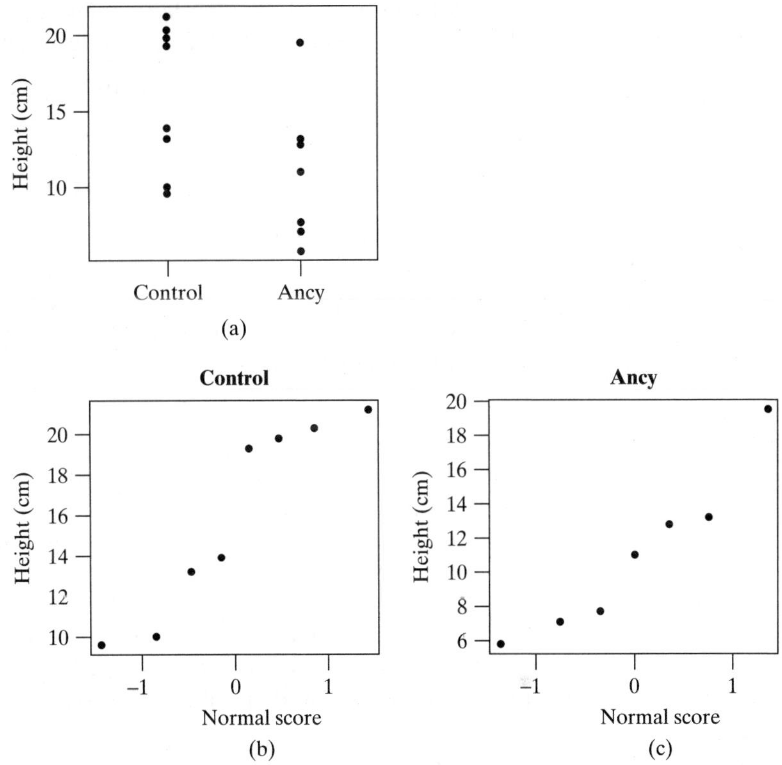

Figure 6.7.1 Parallel dotplots (a) and normal quantile plots of heights of fast plants receiving Control (b) and Ancy (c)

Using Formula (6.7.1), we find the degrees of freedom to be 12.8:

$$\text{df} = \frac{(1.7^2 + 1.8^2)^2}{1.7^4/7 + 1.8^4/6} = 12.8$$

Using a computer, we can find that for a 95% confidence interval the t multiplier for 12.8 degrees of freedom is $t_{12.8,\,0.025} = 2.164$. (Without a computer, we could round down the degrees of freedom to 12, in which case the t multiplier is 2.179. This change from 12.8 to 12 degrees of freedom has little effect on the final answer. Without a computer or a table, we could approximate $t_{0.025} \approx 2$.) The confidence interval formula gives

$$(15.9 - 11.0) \pm (2.164)(2.46)$$

or

$$4.9 \pm 5.32$$

The 95% confidence interval for $(\mu_1 - \mu_2)$ is

$$(-0.42, 10.22)$$

Rounding off, we have

$$(-0.4, 10.2)$$

Thus, we are 95% confident that the population average 14-day height of fast plants when water is used (μ_1) is between 0.4 cm lower and 10.2 cm higher than the average 14-day height of fast plants when ancy is used (μ_2). ■

Example 6.7.2

Fast Plants We said that a conservative method of constructing a confidence interval for a difference in means is to use the smaller of $n_1 - 1$ and $n_2 - 1$. For the data given in Example 6.7.1, this method would use 6 degrees of freedom and a t multiplier of 2.447. In this case, the 95% confidence interval for $(\mu_1 - \mu_2)$ is

$$(15.9 - 11.0) \pm (2.447)(2.46)$$

or

$$4.9 \pm 6.02$$

The 95% confidence interval for $(\mu_1 - \mu_2)$ is

$$(-1.1, 10.9)$$

This interval is a bit conservative in the sense that the interval is wider than the interval found in Example 6.7.1. ■

Example 6.7.3

Thorax Weight Biologists have theorized that male Monarch butterflies have, on average, a larger thorax than do females. A sample of seven male and eight female Monarchs yielded the data in Table 6.7.2, which are displayed in Figure 6.7.2. (These data come from another part of the study described in Example 6.1.1.)

The dotplots in Figure 6.7.2 show reasonably symmetry; thus, we can be comfortable constructing a standard confidence interval using the method of this section. Entering the data of Table 6.7.2 into a computer and asking for a 95% confidence interval gives the confidence interval for $(\mu_1 - \mu_2)$ as

$$(3.3, 21.4).$$

According to the confidence interval, we can be 95% confident that the population mean thorax weight for male Monarch butterflies (μ_1) is larger than that for females (μ_2) by an amount that might be as small as 3.3 mg or as large as 21.4 mg.

Likewise, the 90% confidence interval for $(\mu_1 - \mu_2)$ is

$$(5.0, 19.7).$$

According to the confidence interval, we can be 90% confident that the population mean thorax weight for male Monarch butterflies (μ_1) is larger than that for females (μ_2) by an amount that might be as small as 5.0 mg or as large as 19.7 mg. ■

Table 6.7.2 Thorax weight (mg)

	Male	Female
	67	73
	73	54
	85	61
	84	63
	78	66
	63	57
	80	75
		58
n	7	8
$\bar{y}$	75.7	63.4
s	8.4	7.5
SE	3.2	2.7

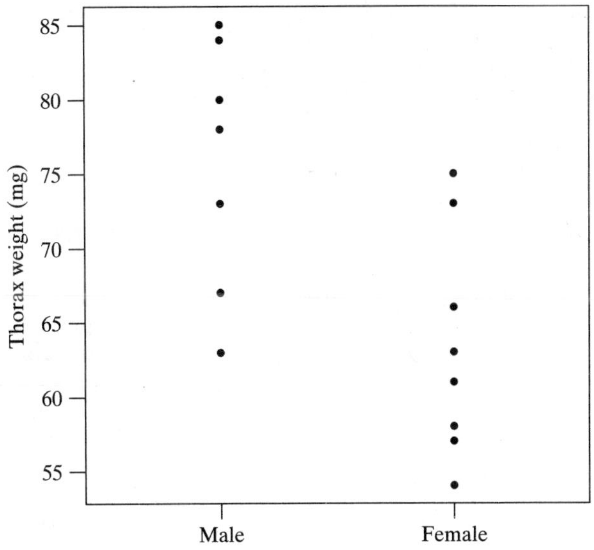

Figure 6.7.2 Parallel dotplots of thorax weights

Conditions for Validity In Section 6.5 we stated the conditions that make a confidence interval for a mean valid: We require that the data can be thought of as (1) a random sample from (2) a normal population. Likewise, when comparing two means, we require two independent, random samples from normal populations. If the sample sizes are large, then the condition of normality is not crucial (due to the Central Limit Theorem).

Exercises 6.7.1–6.7.15

6.7.1 In Table 6.6.3, data were presented from an experiment that compared two types of surgery. The average pain score of the 49 children given conventional tonsillectomies was 4.3, with an SD of 2.8. For the 52 children in the Coblation group the average was 1.9 with an SD of 1.8. Use these data to construct a 95% confidence interval for the difference in population average pain scores. [*Note*: Formula (6.7.1) yields 81.1 degrees of freedom for these data.]

6.7.2 Ferulic acid is a compound that may play a role in disease resistance in corn. A botanist measured the concentration of soluble ferulic acid in corn seedlings grown in the dark or in a light/dark photoperiod. The results (nmol acid per gm tissue) were as shown in the table.[41]

	Dark	Photoperiod
n	4	4
$\bar{y}$	92	115
s	13	13

(a) Construct a 95% confidence interval for the difference in ferulic acid concentration under the two lighting conditions. (Assume that the two populations from which the data came are normally distributed.) [*Note*: Formula (6.7.1) yields 6 degrees of freedom for these data.]

(b) Repeat part (a) for a 90% level of confidence.

6.7.3 (Continuation of 6.7.2) Using your work from Exercise 6.7.2(a), fill in the blank: "We are 95% confident that the difference in population means is at least __________ nmol/g."

6.7.4 A study was conducted to determine whether relaxation training, aided by biofeedback and meditation, could help in reducing high blood pressure. Subjects were randomly allocated to a biofeedback group or a control group. The biofeedback group received training for 8 weeks. The table reports the reduction in systolic blood pressure (mm Hg) after eight weeks.[42] [*Note*: Formula (6.7.1) yields 190 degrees of freedom for these data.]

(a) Construct a 95% confidence interval for the difference in mean response.

(b) Interpret the confidence interval from part (a) in the context of this setting.

	Biofeedback	Control
n	99	93
$\bar{y}$	13.8	4.0
SE	1.34	1.30

6.7.5 Consider the data in Exercise 6.7.4. Suppose we are worried that the blood pressure data do not come from normal distributions. Does this mean that the confidence interval found in Exercise 6.7.3 is not valid? Why or why not?

6.7.6 Prothrombin time is a measure of the clotting ability of blood. For 10 rats treated with an antibiotic and 10 control rats, the prothrombin times (in seconds) were reported as follows[43]:

	Antibiotic	Control
n	10	10
$\bar{y}$	25	23
s	10	8

(a) Construct a 90% confidence interval for the difference in population means. (Assume that the two populations from which the data came are normally distributed.) [*Note*: Formula (6.7.1) yields 17.2 degrees of freedom for these data.]

(b) Why is it important that we assume that the two populations are normally distributed in part (a)?

(c) Interpret the confidence interval from part (a) in the context of this setting.

6.7.7 The accompanying table summarizes the sucrose consumption (mg in 30 minutes) of black blowflies injected with Pargyline or saline (control).[44]

	Control	Pargyline
n	900	905
$\bar{y}$	14.9	46.5
s	5.4	11.7

(a) Construct a 95% confidence interval for the difference in population means. [*Note*: Formula (6.7.1) yields 1,274 degrees of freedom for these data.]

(b) Repeat part (a) using a 99% level of confidence.

6.7.8 In a field study of mating behavior in the Mormon cricket *(Anabrus simplex)*, a biologist noted that some females mated successfully while others were rejected by the males before coupling was complete. The question arose whether some aspect of body size might play a role in mating success. The accompanying table summarizes measurements of head width (mm) in the two groups of females.[45]

(a) Construct a 95% confidence interval for the difference in population means. [*Note*: Formula (6.7.1) yields 35.7 degrees of freedom for these data.]

(b) Interpret the confidence interval from part (a) in the context of this setting.

(c) Using your interval computed in (a) to support your answer, is there strong evidence that the population mean head width is indeed larger for successful maters than unsuccessful maters?

	Successful	Unsuccessful
n	22	17
$\bar{y}$	8.498	8.440
s	0.283	0.262

6.7.9 In an experiment to assess the effect of diet on blood pressure, 154 adults were placed on a diet rich in fruits and vegetables. A second group of 154 adults was placed on a standard diet. The blood pressures of the 308 subjects were recorded at the start of the study. Eight weeks later, the blood pressures of the subjects were measured again and the change in blood pressure was recorded for each person. Subjects on the fruits-and-vegetables diet had an average drop in systolic blood pressure of 2.8 mm Hg more than did subjects on the standard diet. A 97.5% confidence interval for the difference between the two population means is (0.9, 4.7).[46] Interpret this confidence interval. That is, explain what the numbers in the interval mean. (See Examples 6.7.1 and 6.7.3.)

6.7.10 Consider the experiment described in Exercise 6.7.9. For the same subjects, the change in diastolic blood pressure was 1.1 mm Hg greater, on average, for the subjects on the fruits-and-vegetables diet than for subjects on the standard diet. A 97.5% confidence interval for the difference between the two population means is (−0.3, 2.4). Interpret this confidence interval. That is, explain what the numbers in the interval mean. (See Examples 6.7.1 and 6.7.3.)

6.7.11 Researchers were interested in the short-term effect that caffeine has on heart rate. They enlisted a group of volunteers and measured each person's resting heart rate. Then they had each subject drink 6 ounces of coffee. Nine of the subjects were given coffee containing caffeine, and 11 were given decaffeinated coffee. After 10 minutes each person's heart rate was measured again. The data in the table show the change in heart rate; a positive number means that heart rate went up, and a negative number means that heart rate went down.[47]

(a) Use these data to construct a 90% confidence interval for the difference in mean effect that caffeinated coffee has on heart rate, in comparison to decaffeinated coffee. [*Note*: Formula (6.7.1) yields 17.3 degrees of freedom for these data.]

(b) Using the interval computed in part (a) to justify your answer, is it reasonable to believe that caffeine may not affect heart rates?

(c) Using the interval computed in part (a) to justify your answer, is it reasonable to believe that caffeine may affect heart rates? If so, by how much?

(d) Are your answers to (b) and (c) contradictory? Explain.

	Caffeine	Decaf
	28	26
	11	1
	−3	0
	14	−4
	−2	−4
	−4	14
	18	16
	2	8
	2	0
		18
		−10
n	9	11
$\bar{y}$	7.3	5.9
s	11.1	11.2
SE	3.7	3.4

6.7.12 Consider the data from Exercise 6.7.11. Given that there are only a small number of observations in each group, the confidence interval calculated in Exercise 6.7.11 is only valid if the underlying populations are normally distributed. Is the normality condition reasonable here? Support your answer with appropriate graphs.

6.7.13 A researcher investigated the effect of green light, in comparison to red light, on the growth rate of bean plants. The following table shows data on the heights of plants (in inches) from the soil to the first branching stem, 2 weeks after germination.[48] Use these data to construct a 95% confidence interval for the difference in mean effect that red light has on bean plant growth, in comparison to green light and interpret the interval in the context of the study. [*Note*: Formula (6.7.1) yields 38 degrees of freedom for these data.]

6.7.14 The distributions of the data from Exercise 6.7.13 are somewhat skewed, particularly the red group. Does this mean that the confidence interval calculated in Exercise 6.7.13 is not valid? Why or why not?

6.7.15 Researchers reported the mean ± SE foot circumferences of many male and female Sumatran elephants as 125.6 ± 4.4 cm for the males and 151.7 ± 3.3 for the females.[24] You may assume that at least 20 of each sex were sampled.

(a) Compute an approximate 95% confidence interval for the difference between the mean foot circumference for males and females.

(b) Interpret the confidence interval from part (a) in the context of the problem.

(c) Is there compelling evidence that the population mean foot circumference for male and females is not the same? Briefly justify your answer.

	Red	Green
	8.4	8.6
	8.4	5.9
	10.0	4.6
	8.8	9.1
	7.1	9.8
	9.4	10.1
	8.8	6.0
	4.3	10.4
	9.0	10.8
	8.4	9.6
	7.1	10.5
	9.6	9.0
	9.3	8.6
	8.6	10.5
	6.1	9.9
	8.4	11.1
	10.4	5.5
		8.2
		8.3
		10.0
		8.7
		9.8
		9.5
		11.0
		8.0
n	17	25
$\bar{y}$	8.36	8.94
s	1.50	1.78
SE	0.36	0.36

6.8 Perspective and Summary

In this section we place Chapter 6 in perspective by relating it to other chapters and also to other methods for analyzing a single sample of data. We also present a condensed summary of the methods of Chapter 6.

SAMPLING DISTRIBUTIONS AND DATA ANALYSIS

The theory of the sampling distribution of $\overline{Y}$(Section 5.2) seemed to require knowledge of quantities—μ and σ—that in practice are unknown. In Chapter 6, however, we have seen how to make an inference about μ and $(\mu_1 - \mu_2)$, including an assessment of the precision of that inference, using only information provided by the sample. Thus, the theory of sampling distributions has led to a practical method of analyzing data.

In later chapters we will study more complex methods of data analysis. Each method is derived from an appropriate sampling distribution; in most cases, however, we will not study the sampling distribution in detail.

CHOICE OF CONFIDENCE LEVEL

In illustrating the confidence interval methods, we have often chosen a confidence level equal to 95%. However, you should remember that the confidence level is arbitrary. It is true that in practice the 95% level is the confidence level that is most widely used; however, there is nothing wrong with an 80% confidence interval, for example.

CHARACTERISTICS OF OTHER MEASURES

This chapter has primarily discussed estimation of a population mean, μ, and the difference of two population means $(\mu_1 - \mu_2)$. In some situations, one may wish to estimate other parameters of a population such as a population proportion (which we shall address in Chapter 9). The methods in this chapter can be extended to even more complex situations; for example, in evaluating a measurement technique, interest may focus on the repeatability of the technique, as indicated by the standard deviation of repeated determinations. As another example, in defining the limits of health, a medical researcher might want to estimate the 95th percentile of serum cholesterol levels in a certain population. Just as the precision of the mean can be indicated by a standard error or a confidence interval, statistical techniques are also available to specify the precision of estimation of parameters such as the population standard deviation or 95th percentile.

SUMMARY OF ESTIMATION METHODS

For convenient reference, we summarize in the box the confidence interval methods presented in this chapter.

STANDARD ERROR OF THE MEAN

$$\mathrm{SE}_{\overline{Y}} = \frac{s}{\sqrt{n}}$$

CONFIDENCE INTERVAL FOR μ

$$95\% \text{ confidence interval: } \bar{y} \pm t_{0.025}\mathrm{SE}_{\overline{Y}}$$

Critical value $t_{0.025}$ from Student's t distribution with df $= n - 1$.

Intervals with other confidence levels (e.g., 90%, 99%) are constructed analogously (using $t_{0.05}$, $t_{0.005}$, etc.).

The confidence interval formula is valid if (1) the data can be regarded as a random sample from a large population, (2) the observations are independent, and (3) the population is normal. If n is large then condition (3) is less important.

STANDARD ERROR OF $\bar{y}_1 - \bar{y}_2$

$$SE_{(\bar{Y}_1 - \bar{Y}_2)} = \sqrt{\frac{s_1^2}{n_1} + \frac{s_2^2}{n_2}} = \sqrt{SE_1^2 + SE_2^2}$$

CONFIDENCE INTERVAL FOR $\mu_1 - \mu_2$

95% confidence interval:

$$(\bar{y}_1 - \bar{y}_2) \pm t_{0.025} SE_{(\bar{Y}_1 - \bar{Y}_2)}$$

Critical value $t_{0.025}$ from Student's t distribution with

$$df = \frac{(SE_1^2 + SE_2^2)^2}{SE_1^4/(n_1 - 1) + SE_2^4/(n_2 - 1)}$$

where $SE_1 = s_1/\sqrt{n_1}$ and $SE_2 = s_2/\sqrt{n_2}$.

Confidence intervals with other confidence levels (90%, 99%, etc.) are constructed analogously (using $t_{0.05}$, $t_{0.005}$, etc.).

The confidence interval formula is valid if (1) the data can be regarded as coming from two independently chosen random samples, (2) the observations are independent within each sample, and (3) each of the populations is normally distributed. If n_1 and n_2 are large, condition (3) is less important.

Supplementary Exercises 6.S.1–6.S.23

6.S.1 To study the conversion of nitrite to nitrate in the blood, researchers injected four rabbits with a solution of radioactively labeled nitrite molecules. Ten minutes after injection, they measured for each rabbit the percentage of the nitrite that had been converted to nitrate. The results were as follows[49]:

51.1 55.4 48.0 49.5

(a) For these data, calculate the mean, median, and the standard error of the mean.

(b) Assuming the percentage of nitrite converted to nitrate follows a normal distribution, construct a 90% confidence interval for the population mean percentage.

(c) Without doing any calculations, would a 95% confidence interval be wider, narrower, or the same width as the confidence interval you found in part (b)? Why?

6.S.2 The diameter of the stem of a wheat plant is an important trait because of its relationship to breakage of the stem, which interferes with harvesting the crop. An agronomist measured stem diameter in eight plants of the Tetrastichon cultivar of soft red winter wheat. All observations were made 3 weeks after flowering of the plant. The stem diameters (mm) were as follows[50]:

2.3 2.6 2.4 2.2 2.3 2.5 1.9 2.0

The mean of these data is 2.275 and the standard deviation is 0.238.

(a) Calculate the standard error of the mean.

(b) Construct a 95% confidence interval for the population mean.

(c) Define in words the population mean that you estimated in part (b). (See Example 6.1.1.)

6.S.3 Refer to Exercise 6.S.2.

(a) What conditions are needed for the confidence interval to be valid?

(b) Are these conditions met? How do you know?

(c) Which of these conditions is most important?

6.S.4 Refer to Exercise 6.S.2. Suppose that the data on the eight plants are regarded as a pilot study, and the agronomist now wishes to design a new study for which he wants the standard error of the mean to be only 0.05 mm. How many plants should be measured in the new study?

6.S.5 A sample of 20 fruitfly (*Drosophila melanogaster*) larva were incubated at 37 °C for 30 minutes. It is theorized that such exposure to heat causes polytene chromosomes located in the salivary glands of the fly to unwind, creating puffs on the chromosome arm that are visible under a microscope. The following normal quantile plot supports the use of a normal curve to model the distribution of puffs.[51]

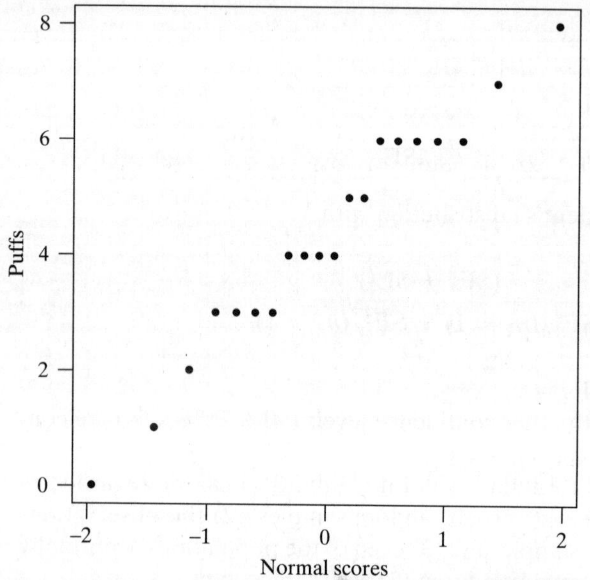

The average number of puffs for the 20 observations was 4.30, with a standard deviation of 2.03.

(a) Construct a 95% confidence interval for μ.

(b) In the context of this problem, describe what μ represents. That is, the confidence interval from part (a) is a confidence interval for what quantity?

(c) The normal probability plot shows the dots lining up on horizontal bands. Is this sort of behavior surprising for this type of data? Explain.

6.S.6 Over a period of about 9 months, 1,353 women reported the timing of each of their menstrual cycles. For the first cycle reported by each woman, the mean cycle time was 28.86 days, and the standard deviation of the 1,353 times was 4.24 days.[52]

(a) Construct a 99% confidence interval for the population mean cycle time.

(b) Because environmental rhythms can influence biological rhythms, one might hypothesize that the population mean menstrual cycle time is 29.5 days, the length of the lunar month. Is the confidence interval of part (a) consistent with this hypothesis?

6.S.7 Refer to the menstrual cycle data of Exercise 6.S.6.

(a) Over the entire time period of the study, the women reported a total of 12,247 cycles. When all of these cycles are included, the mean cycle time is 28.22 days. Explain why one would expect that this mean would be smaller than the value 28.86 given in Exercise 6.S.6. (*Hint*: If each woman reported for a fixed time period, which women contributed more cycles to the total of 12,247 observations?)

(b) Instead of using only the first reported cycle as in Exercise 6.S.6, one could use the first four cycles for each woman, thus obtaining $1{,}353 \times 4 = 5{,}412$ observations. One could then calculate the mean and standard deviation of the 5,412 observations and divide the SD by $\sqrt{5412}$ to obtain the SE; this would yield a much smaller value than the SE found in Exercise 6.S.6. Why would this approach not be valid?

6.S.8 For the 28 lamb birthweights of Example 6.2.2, the mean is 5.1679 kg, the SD is 0.6544 kg, and the SE is 0.1237 kg.

(a) Construct a 95% confidence interval for the population mean.

(b) Construct a 99% confidence interval for the population mean.

(c) Interpret the confidence interval you found in part (a). That is, explain what the numbers in the interval mean. (*Hint*: See Examples 6.3.4 and 6.3.5.)

(d) Often researchers will summarize their data in reports and articles by writing $\bar{y} \pm$ SD (5.17 ± 0.65) or $\bar{y} \pm$ SE (5.17 ± 0.12). If the researcher of this study is planning to compare the mean birthweight of these Rambouillet lambs to another breed, Booroolas, which style of presentation should she use?

6.S.9 Refer to Exercise 6.S.8.

(a) What conditions are required for the validity of the confidence intervals?

(b) Which of the conditions of part (a) can be checked (roughly) from the histogram of Figure 6.2.1?

(c) Twin births were excluded from the lamb birthweight data. If twin births had been included, would the confidence intervals be valid? Why or why not?

6.S.10 Researchers measured the number of tree species in each of 69 vegetational plots in the Lama Forest of Benin, West Africa.[53] The number of species ranged from a low of 1 to a high of 12. The sample mean was 6.8 and the sample SD was 2.4, which results in a 95% confidence interval of (6.2, 7.4). However, the number of tree species in a plot takes on only integer values. Does this mean that the confidence interval should be (7, 7)? Or does it mean that we should round off the endpoints of the confidence interval and report it as (6, 7)? Or should the confidence interval really be (6.2, 7.4)? Explain.

6.S.11 As part of a study of natural variation in blood chemistry, serum potassium concentrations were measured in 84 healthy women. The mean concentration was 4.36 mEq/l, and the standard deviation was 0.42 mEq/l. The table presents a frequency distribution of the data.[54]

Serum potassium (mEq/I)	Number of women
[3.1, 3.4)	1
[3.4, 3.7)	2
[3.7, 4.0)	7
[4.0, 4.3)	22
[4.3, 4.6)	28
[4.6, 4.9)	16
[4.9, 5.2)	4
[5.2, 5.5)	3
[5.5, 5.8)	1
Total	84

(a) Calculate the standard error of the mean.

(b) Construct a histogram of the data and indicate the intervals $\bar{y} \pm$ SD and $\bar{y} \pm$ SE on the histogram. (See Figure 6.2.1.)

(c) Construct a 95% confidence interval for the population mean.

(d) Interpret the confidence interval you found in part (c). That is, explain what the numbers in the interval mean. (*Hint*: See Examples 6.3.4 and 6.3.5.)

6.S.12 Refer to Exercise 6.S.11. In medical diagnosis, physicians often use "reference limits" for judging blood chemistry values; these are the limits within which we would expect to find 95% of healthy people. Would a 95% confidence interval for the mean be a reasonable choice of "reference limits" for serum potassium in women? Why or why not?

6.S.13 Refer to Exercise 6.S.11. Suppose a similar study is to be conducted next year, to include serum potassium measurements on 200 healthy women. Based on the data in Exercise 6.S.11, what would you predict would be

(a) the 95% confidence interval for the population mean?

(b) the 99% confidence interval for the population mean?

6.S.14 An agronomist selected six wheat plants at random from a plot, and then, for each plant, selected 12 seeds from the main portion of the wheat head; by weighing, drying, and reweighing, she determined the percentage moisture in each batch of seeds. The results were as follows[55]:

62.7 63.6 60.9 63.0 62.7 63.7

(a) Calculate the mean, the standard deviation, and the standard error of the mean.

(b) Assuming the percent moisture in each batch follows a normal distribution, construct a 90% confidence interval for the population mean.

6.S.15 As part of the National Health and Nutrition Examination Survey (NHANES), hemoglobin levels were checked for a sample of 1139 men age 70 and over.[56] The sample mean was 145.3 g/l and the standard deviation was 12.87 g/l.

(a) Calculate the standard error of the mean.

(b) Assuming that the hemoglobin level follows a normal distribution, construct a 99% confidence interval for the population mean of hemoglobin levels.

6.S.16 The following data are 16 weeks of weekly fecal coliform counts (MPN/100 ml) at Dairy Creek in San Luis Obispo County, California.[57]

203	215	240	236	217	296	301	190
197	203	210	215	270	290	310	287

(a) Counts above 225 MPN/100ml are considered unsafe. What type of one-sided interval (upper- or lower-bound) would be appropriate to assess the safety of this creek? Explain your reasoning.

(b) Using 95% confidence, construct the interval chosen in part (a).

(c) Based on your interval in part (b), what conclusions can you make regarding the safety of the water?

6.S.17 The blood pressure (average of systolic and diastolic measurements) of each of 38 persons were measured.[58] The average was 94.5 (mm Hg). A histogram of the data is shown.

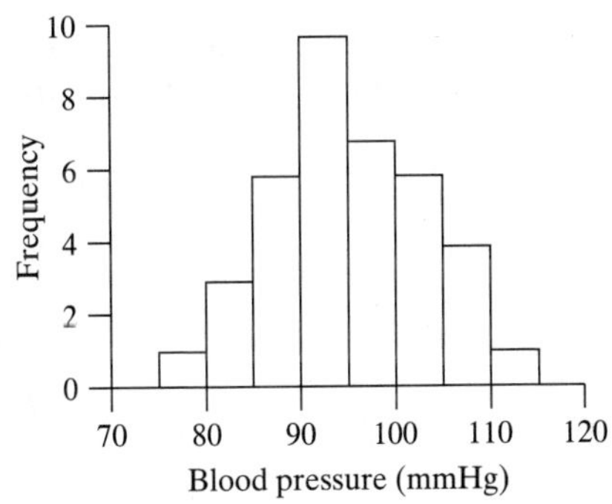

Which of the following is an approximate 95% confidence interval for the population mean blood pressure? Explain.

(i) 94.5 ± 16

(ii) 94.5 ± 8

(iii) 94.5 ± 2.6

(iv) 94.5 ± 1.3

6.S.18 Suppose you wished to estimate the mean blood pressure of students at your school to within 3 mmHg with 99% confidence.

(a) Using the data displayed in Exercise 6.S.17 as pilot data for your study, determine the (approximate) sample size necessary to achieve your goals. (*Hint*: You will need to use the graph to make some visual estimates.)

(b) Suppose your school is a small private college that only has 100 students. Would the interval based on your sample size be valid? Explain. Do you think it would be too wide or too narrow?

6.S.19 It is known that alcohol consumption during pregnancy can harm the fetus. To study this phenomenon, 10 pregnant mice will receive a low dose of alcohol. When each mouse gives birth, the birthweight of each pup will be measured. Suppose the mice give birth to a total of 85 pups, so the experimenter has 85 observations of Y = birthweight. To calculate the standard error of the mean of these 85 observations, the experimenter could calculate the standard deviation of the 85 observations and divide by $\sqrt{85}$. On what grounds might an objection be raised to this method of calculating the SE?

6.S.20 Is the nutrition information on commercially produced food accurate? In one study, researchers sampled 13 packages of a certain frozen reduced-calorie chicken entrée with a reported calorie content of 252 calories per package. The mean calorie count of the sampled entrées was 306 with a sample standard deviation of 51 calories.[59]

(a) Compute a 90% confidence interval for the population mean calorie content of the frozen entrée.

(b) Compute a 99% confidence interval for the population mean calorie content of the frozen entrée.

(c) Based on the two intervals computed in parts (a) and (b), what do you think about the reported calorie content for this entrée?

6.S.21 Nitric oxide is sometimes given to newborns who experience respiratory failure. In one experiment, nitric oxide was given to 114 infants with respiratory failure. This group was compared to a control group of 121 infants with respiratory failure. The length of hospitalization (in days) was recorded for each of the 235 infants. The mean in the nitric oxide sample was $\overline{Y}_1 = 36.4$; the mean in the control sample was $\overline{Y}_2 = 29.5$. A 95% confidence interval for $\mu_1 - \mu_2$ is $(-2.3, 16.1)$, where μ_1 is the population mean length of hospitalization for infants who get nitric oxide and μ_2 is the mean length of hospitalization for infants in the control population.[60]

(a) True or false (and say why): We are 95% confident that μ_1 is greater than μ_2, since most of the confidence interval is greater than zero.

(b) True or false (and say why): We are 95% confident that the difference between μ_1 and μ_2 is between -2.3 days and 16.1 days.

(c) True or false (and say why): 95% of the nitric oxide infants were hospitalized longer than the mean of the control infants.

6.S.22 Consider the data from Exercise 6.S.20(a) which led to a 95% confidence interval for $\mu_1 - \mu_2$ of $(-2.3, 16.1)$.

(a) True or false (and say why): We are 95% confident that the difference between $\overline{Y}_1$ and $\overline{Y}_2$ is between -2.3 days and 16.1 days.

(b) True or false (and say why): If this experiment were repeated many times (with the same sample sizes), then 95% of the time $\bar{y}_1 - \bar{y}_2$ would be between -2.3 and 16.1.

(c) True or false (and say why): We are 95% confident that nitric oxide has no effect on the mean length of hospitalization because zero is in the confidence interval.

6.S.23 Suppose you are discussing confidence intervals with someone who has a limited statistics background. Briefly explain, without doing any calculations, why a 100% confidence interval for a mean is infinitely wide.

COMPARISON OF TWO INDEPENDENT SAMPLES

Chapter 7

OBJECTIVES

In this chapter we continue our study of comparisons of two independent samples by introducing hypothesis testing. We will

- explore how randomization can be used to form the basis of a statistical inference.
- demonstrate how to a conduct a two-sample t test to compare sample means and explain how this test relates to the confidence interval for the difference of two means.
- discuss the interpretation of P-values.
- take a closer look at how confounding and spurious association can limit the utility of a study.
- compare causal versus associative inferences and their relationships to experiments and observational studies.
- discuss the concepts of significance level, effect size, Type I and II errors, and power.
- distinguish between directional and nondirectional tests and examine how the P-values of these tests compare.

7.1 Hypothesis Testing: The Randomization Test

Consider taking a sample from a population and then randomly dividing the sample into two parts. We would expect the two parts of the sample to look similar, but not exactly alike. Now suppose that we have samples from two populations. If the two samples look quite similar to each other, we might infer that the two populations are identical; if the samples look quite different, we would infer that the populations differ. The question is, "How different do two samples have to be in order for us to infer that the populations that generated them are actually different?"

One way to approach this question is to compare the two sample means and to see how much they differ in comparison to the amount of difference we would expect to see due to chance.* The randomization test gives us a way to measure the variability in the difference of two sample means.

Example 7.1.1

Flexibility A researcher studied the flexibility of each of seven women, four of whom were in an aerobics class and three of whom were dancers. One measure she recorded was the "trunk flexion"—how far forward each of the women could stretch while seated on the floor.† The measures (in centimeters) are shown in Table 7.1.1.[1]

Table 7.1.1 Trunk flexion measurements

Aerobics	Dance
38	48
45	59
58	61
64	
Mean 51.25	56.00

*One could compare the two sample medians rather than the means. We compare means so that we have a process similar to the t test, which is introduced in the next section and is based on means.

†These data are part of a larger study—we are working with a subset of the full study in order to simplify matters.

OBJECTIVES (*Continued*)

- consider the conditions under which the use of a *t* test is valid.
- show how to compare distributions using the Wilcoxon-Mann-Whitney test.

Do the data provide evidence that the flexibility is associated with being a dancer?

If being a dancer has no effect on flexibility, then one could argue that the seven data points in the study came from a common population: Some women have greater trunk flexion than others, but this has nothing to do with being a dancer.

Another way of saying this is

Claim: The seven trunk flexion measures came from a single population; the labels "aerobics" and "dance" are arbitrary and have nothing to do with flexibility (as measured by trunk flexion). ■

If the claim stated in Example 7.1.1 is true, then any rearrangement of the seven observations into two groups, with four "aerobics" and three "dance" women, is as likely as any other rearrangement. Indeed, we could imagine writing the seven observations onto seven cards, shuffling the cards, and then drawing four of them to be the observations for the "aerobics" group, with the other three being the observations for the "dance" group.

Example 7.1.2

Flexibility There are 35 possible ways to divide the trunk flexion measures of the seven observations into two groups, of sizes 4 and 3. Table 7.1.2 lists each of the 35 possibilities, along with the difference in sample means for each. (We report the means to two decimal places, since we will be using these values in future calculations.) The two samples obtained in the study are listed first, followed by the other 34 ways that the samples might have turned out.

Table 7.1.2 All 35 possible arrangements of the seven sample trunk flexion observations into two groups of sizes 4 and 3

Sample 1 ("aerobics")	Sample 2 ("dance")	Mean of sample 1	Mean of sample 2	Difference in means
38 45 58 64	48 59 61	51.25	56.00	**−4.75**
38 45 58 48	64 59 61	47.25	61.33	**−14.08**
38 45 58 59	64 48 61	50.00	57.67	**−7.67**
38 45 58 61	64 48 59	50.50	57.00	**−6.50**
38 45 64 48	58 59 61	48.75	59.33	**−10.58**
38 45 64 59	58 48 61	51.50	55.67	−4.17
38 45 64 61	58 48 59	52.00	55.00	−3.00
38 45 48 59	58 64 61	47.50	61.00	**−13.50**
38 45 48 61	58 64 59	48.00	60.33	**−12.33**
38 45 59 61	58 64 48	50.75	56.67	**−5.92**
38 58 64 48	45 59 61	52.00	55.00	−3.00
38 58 64 59	45 48 61	54.75	51.33	3.42
38 58 64 61	45 48 59	55.25	50.67	4.58
38 58 48 59	45 64 61	50.75	56.67	**−5.92**
38 58 48 61	45 64 59	51.25	56.00	**−4.75**
38 58 59 61	45 64 48	54.00	52.33	1.67
38 64 48 59	45 58 61	52.25	54.67	−2.42

Sample 1 ("aerobics")	Sample 2 ("dance")	Mean of sample 1	Mean of sample 2	Difference in means
38 64 48 61	45 58 59	52.75	54.00	−1.25
38 64 59 61	45 58 48	55.50	50.33	**5.17**
38 48 59 61	45 58 64	51.50	55.67	−4.17
45 58 64 48	38 59 61	53.75	52.67	1.08
45 58 64 59	38 48 61	56.50	49.00	**7.50**
45 58 64 61	38 48 59	57.00	48.33	**8.67**
45 58 48 59	38 64 61	52.50	54.33	−1.83
45 58 48 61	38 64 59	53.00	53.67	−0.67
45 58 59 61	38 64 48	55.75	50.00	**5.75**
45 64 48 59	38 58 61	54.00	52.33	1.67
45 64 48 61	38 58 59	54.50	51.67	2.83
45 64 59 61	38 58 48	57.25	48.00	**9.25**
45 48 59 61	38 58 64	53.25	53.33	−0.08
58 64 48 59	38 45 61	57.25	48.00	**9.25**
58 64 48 61	38 45 59	57.75	47.33	**10.42**
58 64 59 61	38 45 48	60.50	43.67	**16.83**
58 48 59 61	38 45 64	56.50	49.00	**7.50**
64 48 59 61	38 45 58	58.00	47.00	**11.00**

Figure 7.1.1 gives a visual display of these 35 possible values. The observed result of −4.75, which is highlighted, falls not far from the middle of the distribution.

Suppose that the labels "aerobics" and "dance" are, in fact, arbitrary and have nothing to do with trunk flexion. Then each of the 35 outcomes listed in Table 7.1.2, and shown in Figure 7.1.1, is equally likely. This means that the differences, shown in the last column of the table, are equally likely. Of the 35 differences, 20 of them are at least as large in magnitude as the −4.75 obtained in the study; these are shown in bold type in the table and filled in gray in the figure. Thus, if the claim is true (that the labels "aerobics" and "dance" are arbitrary), there is a 20/35 chance of obtaining a difference in sample means at least as large, in magnitude, as the difference that was observed.

The fraction 20/35 is approximately equal to 0.57, which is rather large. Thus, the observed data are consistent with the claim that the labels "aerobics" and "dance" are arbitrary and have nothing to do with flexibility. If the claim were true, we would expect to see a difference in sample means of 4.75 or more over half of the time, just due to chance alone. Therefore, these data provide little evidence that flexibility is associated with dancing. ■

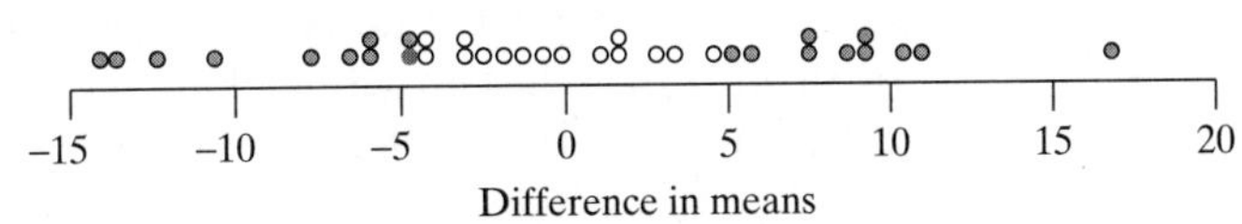

Figure 7.1.1 Distribution of "Difference in means" values, with the observed result of −4.75 colored blue, and values with observed results as or more extreme (in magnitude) than 4.75 colored gray

The process shown in Example 7.1.2 is called the **randomization test.*** In a randomization test one randomly divides the observed data into groups in order to see how likely it is that the observed difference will arise due to chance alone.

Note: In Section 7.2 we will introduce a procedure known as the *t* test, which often provides a good approximation to the randomization test. The value of 20/35 (0.57) computed in Example 7.1.2 is called a *P*-value. (We have seen this term used earlier for the decision making in the context of the Shapiro–Wilk test for normality in Section 4.4. The general use of this term, and others, will be explained more fully in Section 7.2.) For the data in Example 7.1.1 the *t* test yields a *P*-value of 0.54. We can think of the 0.54 *P*-value from the *t* test as an approximation to the 0.57 *P*-value found with the randomization test.

LARGER SAMPLES

When we are dealing with small samples, such as in Example 7.1.1, we can list all of the possible outcomes from randomly assigning observations to groups. The following example shows how to handle large samples, where no such listing is possible.

Example 7.1.3 **Leaf Area** A plant physiologist investigated the effect of mechanical stress on the growth of soybean plants. Individually potted seedlings were divided into two groups. Those in the first group were stressed by shaking for 20 minutes twice daily, while those in the second group (the control group) were not shaken. After 16 days of growth the plants were harvested and total leaf area (cm^2) was measured for each plant. The data are given in Table 7.1.3 and are graphed in Figure 7.1.2.[2]

Table 7.1.3 Leaf areas of soybean plants

Control	Stressed
314	283
320	312
310	291
340	259
299	216
268	201
345	267
271	326
285	241
Mean 305.8	266.2

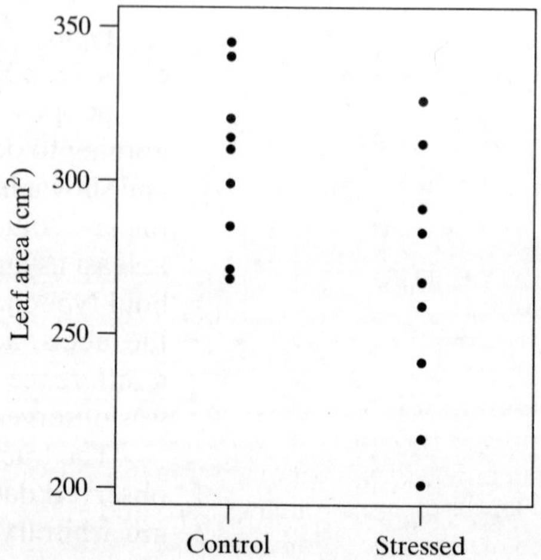

Figure 7.1.2 Parallel dotplots of leaf areas

The mean for the stressed plants is lower than for the control plants, and Figure 7.1.2 provides some visual evidence of a difference between the two groups. On the other hand, the dotplots overlap quite a bit. Perhaps stressing the seedlings by shaking them has no actual effect on leaf area and the difference observed in this experiment (305.8 − 266.2 = 39.6) was simply due to chance. That is, it might be

*Some authors use the label "permutation test" when there are random samples from the appropriate population(s) and "randomization test" when the experimental design justifies the inference. This distinction seems not to be a widely observed practice.

that the "control" and "stressed" conditions have nothing to do with leaf area. If this is the case, then we can think of the 18 seedlings as having come from one population, with the division into "control" and "stressed" groups being arbitrary.

In Example 7.1.2 we could list all of the possible ways that the two groups could have been formed. However, in the current example there are 48,620 possible ways to select 9 of the 18 seedlings as the control group (and the other 9 as the stressed group). Thus, it is not feasible to create a table similar to Table 7.1.2 and list all the possibilities. What we can do, however, is to randomly sample from the 48,620 possibilities. One way to do this would be to (1) write the 18 observations on each of 18 cards; (2) shuffle the cards; (3) randomly deal out 9 of them as the control group, with the other 9 being the stress group; (4) calculate the difference in sample means; (5) record whether the magnitude of the difference in sample means is at least 39.6; (6) repeat steps (1)—(5) many times.

Consider the fraction of times that the magnitude of the difference in sample means is at least as large as the value of 39.6 obtained in the experiment. This is a measure of the evidence against the claim that "Stressing the seedlings by shaking them has no actual effect on leaf area."

Rather than use 18 cards, we could use a computer simulation to accomplish the same thing. In one simulation with 1,000 trials there were only 36 trials that gave a difference in sample means as large in magnitude as 39.6.* This indicates that the observed difference of 39.6 is unlikely to arise by chance—the chance is only 36/1,000 or 3.6%—so we have evidence that stressing the plants has an effect. Indeed, it appears that shaking the seedlings led to a reduction in average leaf area. ■

Note: The t test procedure (to be introduced in Section 7.2) yields a P-value of 0.033, which is a good approximation to the 0.036 P-value given by the randomization test.

---

*In this instance, we could also use a computer to consider the difference in means for each of the 48,620 possibilities and note how many of these yield differences larger than 39.6 in magnitude. However, as samples grow larger, listing all possibilities can be computationally expensive (even with fast computers) and only marginally more accurate than conducting simulations as we have described.

Exercises 7.1.1–7.1.5

7.1.1 Suppose we have samples of five men and of five women and have conducted a randomization test to compare the sexes on the variable Y = pulse. Further, suppose we have found that in 120 out of the 252 possible outcomes under randomization the difference in means is at least as large as the difference in the two observed sample means. Does the randomization test provide evidence that the sexes differ with regard to pulse? Justify your answer using the randomization results.

7.1.2 In an investigation of the possible influence of dietary chromium on diabetic symptoms, some rats were fed a low-chromium diet, and others were fed a normal diet. One response variable was activity of the liver enzyme GITH, which was measured using a radioactively labeled molecule. The accompanying table shows the results, expressed as thousands of counts per minute per gram of liver.[3] The sample means are 49.17 for the low-chromium diet and 51.90 for the normal diet; thus the difference in sample means is −2.73. There are 10 possible randomizations of the five observations into two groups, of sizes 3 and 2.

(a) Create a list of these 10 randomizations (one of which is the original assignment of observations to the two groups) and for each case calculate the low-chromium diet mean minus the normal diet mean.

(b) How many of the 10 randomizations yield a difference in sample means as far from zero as −2.73, the difference in sample means for our observed samples?

(c) Is there evidence that dietary chromium affects GITH liver enzyme activity? Justify your answer using the randomization results.

Low-chromium diet	Normal diet
42.3	53.1
51.5	50.7
53.7	

7.1.3 The following table shows the number of bacteria colonies present in each of several petri dishes, after *E. coli* bacteria were added to the dishes and they were incubated for 24 hours. The "soap" dishes contained a solution prepared from ordinary soap; the "control" dishes contained a solution of sterile water. (These data are a subset of the larger data set seen in Exercise 6.6.9.) The sample means are 44 for the control group and 39.7 for the soap group; thus the difference in sample means is 4.3, with the control mean being larger, as would be expected if the soap were effective. There are 20 possible randomizations of the six observations into two groups, each of size 3.

(a) Create a list of these 20 randomizations (one of which is the original assignment of observations to the two groups) and for each case calculate the control mean minus the soap mean.

(b) How many of the 20 randomizations produce a difference in means at least as large as 4.3?

(c) Is there evidence that the soap inhibits *E. coli* growth? Justify your answer using the randomization results.

Control	Soap
30	76
36	27
66	16

7.1.4 Consider the randomization test presented in Example 7.1.3. It is reported that in one simulation of 1,000 trials there were only 36 trials that gave a sample difference as extreme as 39.6 (the difference observed in the original data). Suppose a simulation were done with 100,000 trials. How many of the 100,000 would we expect to be as extreme as 39.6?

7.1.5 Suppose that Sample 1 contains the numbers {32, 34, 36, 37, 38}, for a sample average of 35.4, and Sample 2 contains the numbers {21, 23, 25, 28, 29}, for a sample average of 25.2. With samples of this size, there are 252 possible ways to randomize the 10 total observations into two groups of 5 each. How many of those 252 ways will result in a difference of group means as large as 10.2 or as small as −10.2?

PREVIEW OF THE *t* TEST (SECTION 7.2)

In the discussion above, we have talked about the difference in two means obtained after randomly allocating 9 out of 18 seedlings to a group called "Control" and the other 9 to a group called "Stressed." That is, we have calculated

$$\bar{y}_{Control} - \bar{y}_{Stressed}$$

for each of many randomizations and have compared this difference to 39.6, the actual observed difference in average leaf areas.

Comparing means is a simple and natural thing to do. However, as we will see in Section 7.2, it is common to divide the difference in means by the standard error of the difference. This ratio describes the magnitude of the difference in means relative to the magnitude of the inherent variability of this difference due to random sampling. If we do this with the original data (and report values to four decimal places) we get

$$ratio = \frac{305.7778 - 266.2222}{16.7254} = 2.365$$

Let's consider taking 18 cards that have leaf areas written on them, shuffling the cards into two piles that we will call "Control" and "Stressed," calculating the mean for each pile and the SE of the difference between the two means (as in Section 6.6), and then calculating the ratio:

$$ratio = \frac{\bar{y}_{Control} - \bar{y}_{Stressed}}{SE_{(\bar{Y}_{Control} - \bar{Y}_{Stressed})}}$$

For example, if we do this once we might get the following: $\bar{y}_{Control} = 295.5556$, $\bar{y}_{Stressed} = 276.4444$, SE $= 18.8335$, and ratio $= \frac{295.5556 - 276.4444}{18.8335} = 1.0147$.

If we do this 1,000 times and make a histogram of the 1,000 ratios, we get Figure 7.1.2. In the figure, the area above 2.365 is shaded in, as is the area below −2.365, because this corresponds to times in which the shuffling of cards into two piles resulted in a ratio at least as far from zero as was produced originally.

The shaded areas add up to 0.036 because 36 out of the 1,000 shufflings resulted in ratios as far from zero as 2.365. In Section 7.2 we present the t test procedure, which yields a P-value of 0.033, a very good approximation to the randomization P-value of 0.036. In fact, we can think of the t test of Section 7.2 as a kind of short-cut way to get a P-value. Notice that the histogram in Figure 7.1.3 is symmetric and looks like the t distribution curves we saw in Chapter 6.

Figure 7.1.3 Histogram of ratios

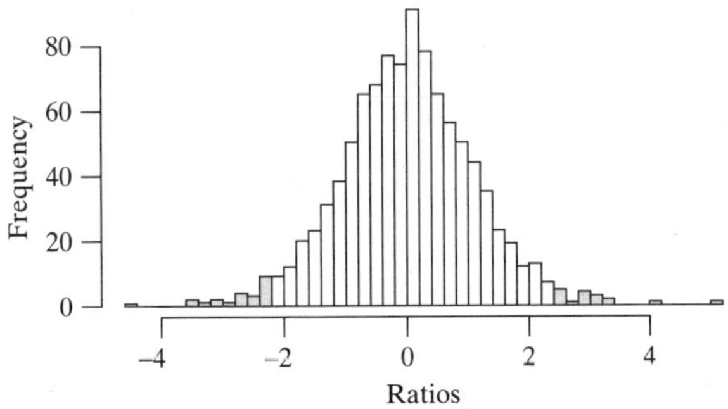

7.2 Hypothesis Testing: The *t* Test

In Chapter 6 we saw that two means can be compared by using a confidence interval for the difference $(\mu_1 - \mu_2)$. Now we will explore another approach to the comparison of means: the procedure known as *hypothesis testing*. The general idea is to formulate as a hypothesis the statement that μ_1 and μ_2 differ and then to see whether the data provide sufficient evidence in support of that hypothesis.

THE NULL AND ALTERNATIVE HYPOTHESES

The hypothesis that μ_1 and μ_2 are *not* equal is called an **alternative hypothesis** (or a research hypothesis) and is abbreviated H_A. It can be written as

$$H_A: \mu_1 \neq \mu_2$$

Its antithesis is the **null hypothesis,**

$$H_0: \mu_1 = \mu_2$$

which asserts that μ_1 and μ_2 are equal. A researcher would usually express these hypotheses more informally, as in the following example.

Example 7.2.1

Toluene and the Brain Abuse of substances containing toluene (e.g., glue) can produce various neurological symptoms. In an investigation of the mechanism of these toxic effects, researchers measured the concentrations of various chemicals in the brains of rats that had been exposed to a toluene-laden atmosphere, and also in unexposed control rats. The concentrations of the brain chemical norepinephrine (NE) in the medulla region of the brain, for six toluene-exposed rats and five control rats, are given in Table 7.2.1 and displayed in Figure 7.2.1.[4]

The observed mean NE in the toluene group ($\bar{y}_1 = 540.8$ ng/gm) is substantially higher than the mean in the control group ($\bar{y}_2 = 444.2$ ng/gm). One might ask

Table 7.2.1 NE concentration (ng/gm)

	Toluene (Group 1)	Control (Group 2)
	543	535
	523	385
	431	502
	635	412
	564	387
	549	
n	6	5
$\bar{y}$	540.8	444.2
s	66.1	69.6
SE	27	31

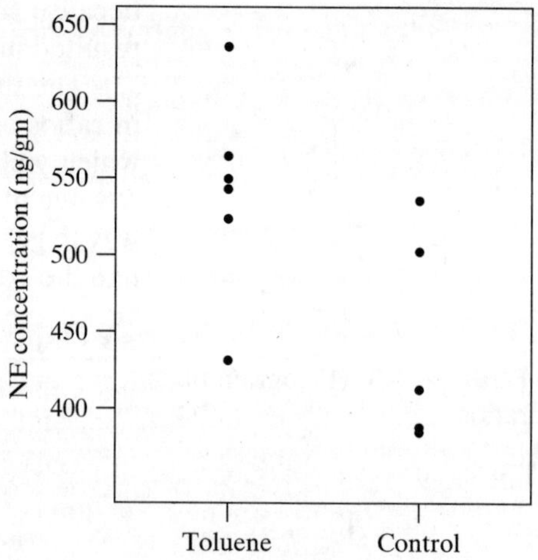

Figure 7.2.1 Parallel dotplots of NE concentration

whether this observed difference indicates a real biological phenomenon—the effect of toluene—or whether the truth might be that toluene has no effect and that the observed difference between $\bar{y}_1$ and $\bar{y}_2$ reflects only chance variation. Corresponding hypotheses, informally stated, would be

H_0^*: Toluene has no effect on NE concentration in rat medulla.

H_A^*: Toluene has some effect on NE concentration in rat medulla. ■

We denote the informal statements by different symbols (H_0^* and H_A^* rather than H_0 and H_A) because they make different assertions. In Example 7.2.1 the informal alternative hypothesis makes a very strong claim—not only that there is a difference, but that the difference is *caused* by toluene.*

A statistical **test of hypothesis** is a procedure for assessing the strength of evidence present in the data in support of H_A. The data are considered to demonstrate evidence for H_A if any discrepancies from H_0 (the opposite of H_A) could not be readily attributed to chance (i.e., to sampling error).

THE t STATISTIC

We consider the problem of testing the null hypothesis

$$H_0: \mu_1 = \mu_2$$

against the alternative hypothesis

$$H_A: \mu_1 \neq \mu_2$$

Note that the null hypothesis says that the two population means are equal, which is the same as saying that the difference between them is zero:

$$H_0: \mu_1 = \mu_2 \leftrightarrow H_0: \mu_1 - \mu_2 = 0$$

*Of course, our statements of H_0^* and H_A^* are abbreviated. Complete statements would include all relevant conditions of the experiment—adult male rats, toluene 1,000 ppm atmosphere for 8 hours, and so on. Our use of abbreviated statements should not cause any confusion.

The alternative hypothesis asserts that the difference is not zero:

$$H_A: \mu_1 \neq \mu_2 \leftrightarrow H_A: \mu_1 - \mu_2 \neq 0$$

The ***t* test** is a standard method of choosing between these two hypotheses. To carry out the t test, the first step is to compute the **test statistic,** which for a t test is defined as

$$t_s = \frac{(\bar{y}_1 - \bar{y}_2) - 0}{\mathrm{SE}_{(\bar{Y}_1 - \bar{Y}_2)}}$$

Notice the structure of t_s: It is a measure of how far the difference between the sample means ($\bar{y}$'s) is from the difference we would expect to see if H_0 were true (zero difference), expressed in relation to the SE of the difference—the amount of variation we expect to see in differences of means from random samples. The subscript "s" on t_s serves as a reminder that this value is calculated from the data ("s" for "sample"). The quantity t_s is the test statistic for the t test; that is, t_s provides the data summary that is the basis for the test procedure. We illustrate with an example.

Example 7.2.2

Toluene and the Brain For the brain NE data of Example 7.2.1, the SE for $(\bar{Y}_1 - \bar{Y}_2)$ is

$$\mathrm{SE}_{(\bar{Y}_1 - \bar{Y}_2)} = \sqrt{\frac{66.1^2}{6} + \frac{69.6^2}{5}} = 41.195$$

and the value of t_s is

$$t_s = \frac{(540.8 - 444.2) - 0}{41.195} = 2.34$$

The t statistic shows that the difference between $\bar{y}_1$ and $\bar{y}_2$ is about 2.3 SEs from zero, the difference we'd expect to see if toluene had no effect on NE. ■

How shall we judge whether our data are sufficient evidence for H_A? A complete lack of evidence for H_A (*perfect* agreement with H_0) would be expressed by sample means that were identical and a resulting t statistic equal to zero ($t_s = 0$). But, even if the null hypothesis H_0 were true, we would not expect t_s to be exactly zero; we expect the sample means to differ from one another because of sampling variability (measured via $\mathrm{SE}_{(\bar{Y}_1 - \bar{Y}_2)}$). Fortunately, we know what to expect regarding this sampling variability; in fact, the chance difference in the $\bar{Y}$'s is not likely to exceed a couple of standard errors when the null hypothesis is true. To put this more precisely, it can be shown mathematically that

> If H_0 is true, then the sampling distribution of t_s is well approximated by a Student's t distribution with degrees of freedom given by formula (6.7.1).*

The preceding statement is true if certain conditions are met. Briefly: We require independent random samples from normally distributed populations. These conditions will be considered in detail in Section 7.9.

The essence of the t test procedure is to identify where the observed value t_s falls in the Student's t distribution, as indicated in Figure 7.2.2. If t_s is near the center, as in Figure 7.2.2(a), then the data are regarded as compatible with H_0 because the observed difference between $(\bar{Y}_1 - \bar{Y}_2)$ and the null difference of zero can readily be attributed

*As we stated in Section 6.7, a conservative approximation to formula (6.7.1) is to use degrees of freedom given by the smaller of $n_1 - 1$ and $n_2 - 1$.

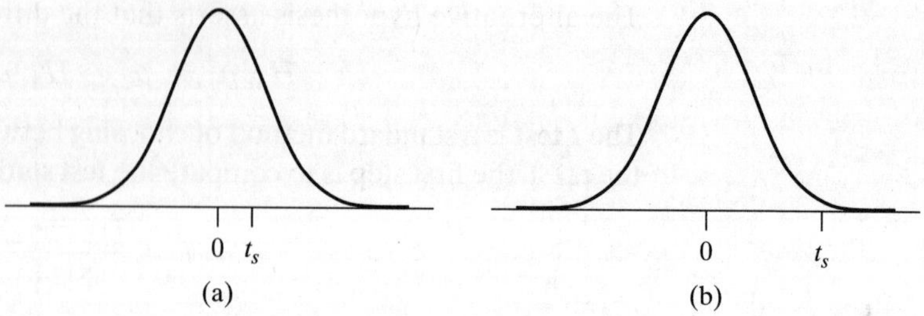

Figure 7.2.2 Essence of the t test. (a) Data compatible with H_0 (and thus a lack of significant evidence for H_A); (b) data incompatible with H_0 (and thus significant evidence for H_A).

to chance variation caused by sampling error. (H_0 predicts that the sample means will be equal, since H_0 says that the population means are equal.) If, on the other hand, t_s falls in the far tail of the t distribution, as in Figure 7.2.2(b), then the data are regarded as evidence for H_A, because the observed deviation cannot be readily explained as being due to chance variation. To put this another way, if H_0 is true, then it is unlikely that t_s would fall in the far tails of the t distribution.

THE P-VALUE

To judge whether an observed value t_s is "far" in the tail of the t distribution, we need a quantitative yardstick for locating t_s within the distribution. This yardstick is provided by the P-value, which can be defined (in the present context) as follows:

> The ***P*-value** of the t test is the area under Student's t curve in the double tails beyond $-t_s$ and $+t_s$.

Thus, the P-value, which is sometimes abbreviated as simply "P," is the shaded area in Figure 7.2.3. Note that we have defined the P-value as the total area in *both* tails; this is sometimes called the "two-tailed" P-value.

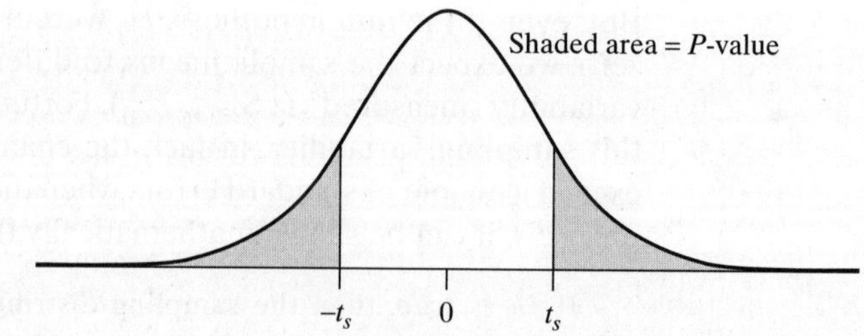

Figure 7.2.3 The two-tailed P-value for the t test

Example 7.2.3 **Toluene and the Brain** For the brain NE data of Example 7.2.1, the value of t_s is 2.34. We can ask, "If H_0 were true so that one would expect $\bar{Y}_1 - \bar{Y}_2 = 0$, on average, what is the probability that $\bar{Y}_1 - \bar{Y}_2$ would differ from zero by as many as 2.34 SEs?" The P-value answers this question. Formula (6.7.1) yields 8.47 degrees of freedom for these data. Thus, the P-value is the area under the t curve (with 8.47 degrees of freedom) beyond ± 2.34. This area, which was found using a computer,* is shown in Figure 7.2.4 to be 0.0454. ■

*For example, in the program R the command is t.test(Toluene, Control).

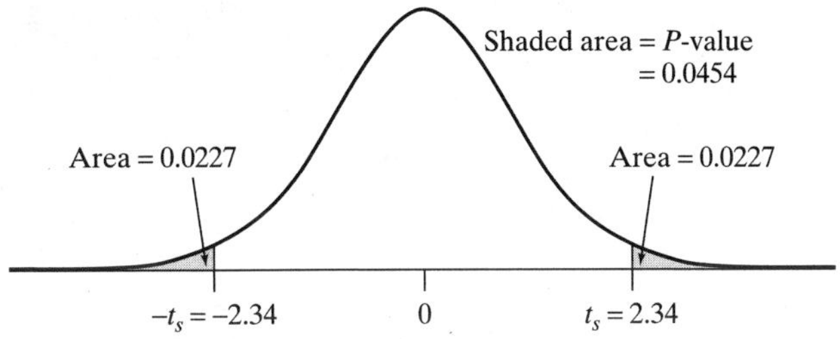

Figure 7.2.4 The two-tailed *P*-value for the toluene data

> **DEFINITION** The ***P*-value** for a hypothesis test is the probability, computed under the condition that the null hypothesis is true, of the test statistic being at least as extreme as the value of the test statistic that was actually obtained.

From the definition of *P*-value, it follows that the ***P*-value is a measure of compatibility between the data and H_0** and thus measures the ***evidence for*** H_A: A large *P*-value (close to 1) indicates a value of t_s near the center of the t distribution (lack of evidence for H_A), whereas a small *P*-value (close to 0) indicates a value of t_s in the far tails of the t distribution (evidence for H_A).

DRAWING CONCLUSIONS FROM A t TEST

The *P*-value is a measure of the evidence in the data for H_A, but where does one draw the line in determining how much evidence is sufficient? Most people would agree that *P*-value $= 0.0001$ indicates very strong evidence, and that *P*-value $= 0.80$ indicates a lack of evidence, but what about intermediate values? For example, should *P*-value $= 0.10$ be regarded as sufficient evidence for H_A? The answer is not intuitively obvious.

In much scientific research, it is not necessary to draw a sharp line. However, in many situations a *decision* must be reached. For example, the Food and Drug Administration (FDA) must decide whether the data submitted by a pharmaceutical manufacturer are sufficient to justify approval of a medication. As another example, a fertilizer manufacturer must decide whether the evidence favoring a new fertilizer is sufficient to justify the expense of further research.

Making a decision requires drawing a definite line between sufficient and insufficient evidence. The threshold value, on the *P*-value scale, is called the **significance level** of the test and is denoted by the Greek letter α (alpha). The value of α is chosen by whoever is making the decision. Common choices are $\alpha = 0.10, 0.05$, and 0.01. *If the P-value of the data is less than or equal to* α*, the data are judged to provide statistically significant evidence in favor of* H_A*; we also may say that* H_0 *is* **rejected.** If the *P*-value of the data is greater than α, we say that the data provide insufficient evidence to claim that H_A is true, and thus H_0 is **not rejected.** We can think of α as a preset threshold of statistical significance. If $\alpha = 0.05$, for example, then we are saying in advance that we will have sufficient evidence for H_A if we get a result that would happen no more than 5% of the time by chance variation under H_0. This is the same as saying that we will reject H_0 if the *P*-value is 0.05 or less.

The following example illustrates the use of the t test to make a decision.

Example 7.2.4 **Toluene and the Brain** For the brain NE experiment of Example 7.2.1, the data are summarized in Table 7.2.2. Suppose we choose to make a decision at the 5% significance level, $\alpha = 0.05$. In Example 7.2.3 we found that the *P*-value of these data is

Table 7.2.2 NE concentration (ng/gm)

	Toluene	Control
n	6	5
$\bar{y}$	540.8	444.2
s	66.1	69.6

0.0454. This means that one of two things happened: Either (1) H_0 is true and we got a strange set of data just by chance or (2) H_0 is false. If H_0 is true, the kind of discrepancy we observed between $\bar{y}_1$ and $\bar{y}_2$ would happen only about 4.5% of the time. Because the P-value, 0.0454, is less than 0.05, we reject H_0 and conclude that the data provide statistically significant evidence in favor of H_A. The strength of the evidence is expressed by the statement that the P-value is 0.0454.

Conclusion: The data provide sufficient evidence at the 0.05 level of significance (P-value = 0.0454) that toluene increases NE concentration.* ■

The next example illustrates a t test in which there is a lack of sufficient evidence at the 0.05 level of significance for H_A.

Example 7.2.5

Fast Plants In Example 6.7.1 we saw that the mean height of fast plants was smaller when ancy was used than when water (the control) was used. Table 7.2.3 summarizes the data. The difference between the sample means is 15.9 − 11.0 = 4.9. The SE for the difference is

$$SE_{(\bar{Y}_1 - \bar{Y}_2)} = \sqrt{\frac{4.8^2}{8} + \frac{4.7^2}{7}} = 2.46$$

Table 7.2.3 Fourteen-day height of control and of ancy plants

	Control	Ancy
n	8	7
$\bar{y}$	15.9	11.0
s	4.8	4.7

Suppose we choose to use $\alpha = 0.05$ in testing

$$H_0: \mu_1 = \mu_2 \text{ (i.e., } \mu_1 - \mu_2 = 0)$$

against the alternative hypothesis

$$H_A: \mu_1 \neq \mu_2 \text{ (i.e., } \mu_1 - \mu_2 \neq 0)$$

The value of the test statistic is

$$t_s = \frac{(15.9 - 11.0) - 0}{2.46} = 1.99$$

Formula (6.7.1) gives 12.8 degrees of freedom for the t distribution. The P-value for the test is the probability of getting a t statistic that is at least as far away from zero as 1.99. Figure 7.2.5 shows that this probability is 0.0678. (This 4-digit P-value was

*Because the alternative hypothesis was $H_A: \mu_1 \neq \mu_2$, some authors would say, "We conclude that toluene affects NE concentration," rather than saying that toluene increases NE concentration.

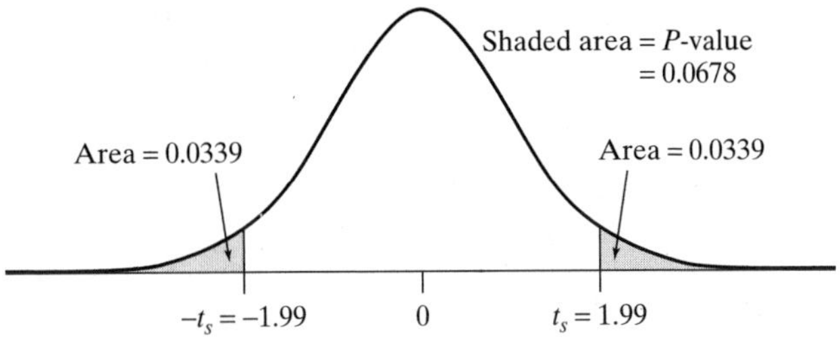

Figure 7.2.5 The two-sided *P*-value for the ancy data

found using a computer.) Because the *P*-value is greater than α, we have insufficient evidence for H_A; thus, we do not reject H_0. That is, these data do not provide sufficient evidence to conclude that μ_1 and μ_2 differ; the difference we observed between $\bar{y}_1$ and $\bar{y}_2$ could easily have happened by chance.

Conclusion: The data do *not* provide sufficient evidence (*P*-value $= 0.0678$) at the 0.05 level of significance to conclude that ancy and water differ in their effects on fast plant growth (under the conditions of the experiment that was conducted). ■

Note carefully the phrasing of the conclusion in Example 7.2.5. We do *not* say that there is evidence *for* the null hypothesis, but only that there is insufficient evidence *against* it. When we do not reject H_0, this indicates a lack of evidence that H_0 is false, which is *not* the same thing as evidence that H_0 is true. The astronomer Carl Sagan (in another context) summed up this principle of evidence in this succinct statement[5]:

> Absence of evidence is not evidence of absence.

In other words, nonrejection of H_0 is *not* the same as *acceptance* of H_0. (To avoid confusion, it may be best not to use the phrase "accept H_0" at all.)

Nonrejection of H_0 indicates that the data are compatible with H_0, but the data may *also* be quite compatible with H_A. For instance, in Example 7.2.5 we found that the observed difference between the sample means could be due to sampling variation, but this finding does not rule out the possibility that the observed difference is actually due to a real effect caused by ancy. (Methods for such ruling out of possible alternatives will be discussed in Section 7.7 and optional Section 7.8.)

In testing a hypothesis, the researcher starts out with the assumption that H_0 is true and then asks whether the data contradict that assumption. This logic can make sense even if the researcher regards the null hypothesis as implausible. For instance, in Example 7.2.5 it could be argued that there is almost certainly *some* difference (perhaps very small) between using ancy and not using ancy. The fact that we did not reject H_0 does not mean that we accept H_0.

USING TABLES VERSUS USING TECHNOLOGY

In analyzing data, how do we determine the *P*-value of a test? Statistical computer software, and some calculators, will provide exact *P*-values. If such technology is not available, then we can use formula (6.7.1) to find the degrees of freedom but round down to make the value an integer. A conservative alternative to using formula (6.7.1) is to use the smaller of $n_1 - 1$ and $n_2 - 1$ as the degrees of freedom for the test. A liberal approach is to use $n_1 + n_2 - 2$ as the degrees of freedom. (Formula (6.7.1) will always give degrees of freedom between the conservative value of the smaller of $n_1 - 1$ and $n_2 - 1$ and the liberal value of $n_1 + n_2 - 2$.) We can rely on the limited information in Table 4 to *bracket* the *P*-value, rather than to determine it exactly. The *P*-value found using the conservative approach will be somewhat larger than the exact *P*-value; the *P*-value found using the liberal approach will be somewhat smaller than the exact *P*-value. The following example illustrates the bracketing process.

Example 7.2.6 **Fast Plants** For the fast plant growth data, the value of the *t* statistic (as determined in Example 7.2.5) is $t_s = 1.99$. The smaller of $n_1 - 1$ and $n_2 - 1$ is $7 - 1 = 6$, so the conservative degrees of freedom are 6. The liberal degrees of freedom are $8 + 7 - 2 = 13$. Here is a copy of part of Table 4, with key numbers highlighted.

Upper Tail Probability			
df	.05	.04	.03
6	**1.943**	**2.104**	2.313
7	1.895	2.046	2.241
8	1.860	2.004	2.189
9	1.833	1.973	2.150
10	1.812	1.948	2.120
11	1.796	1.928	2.096
12	1.782	1.912	2.076
13	1.771	**1.899**	**2.060**

We begin with the conservative degrees of freedom, 6. From the preceding table (or from Table 4), we find $t_{6,\,0.05} = 1.943$ and $t_{6,\,0.04} = 2.104$. The corresponding conservative *P*-value, based on a *t* distribution with 6 degrees of freedom, is shaded in Figure 7.2.6. Because t_s is between the 0.04 and 0.05 critical values, the upper tail area must be between 0.04 and 0.05; thus, the conservative *P*-value must be between 0.08 and 0.10.

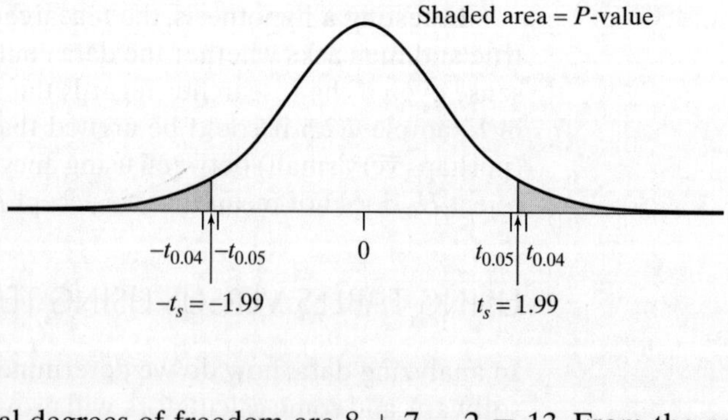

Figure 7.2.6 Conservative *P*-value for Example 7.2.6

The liberal degrees of freedom are $8 + 7 - 2 = 13$. From the preceding table (or from Table 4) we find $t_{13,\,0.04} = 1.899$ and $t_{13,\,0.03} = 2.060$. Because t_s is between these 0.03 and 0.04 critical values, the upper tail area must be between 0.06 and 0.08; thus, the liberal *P*-value must be between 0.06 and 0.08.

Putting these two together, we have

$$0.06 < P\text{-value} < 0.10$$

■

If the observed t_s is not within the boundaries of Table 4, then the P-value is bracketed on only one side. For example, if t_s is greater than $t_{0.0005}$, then the two-sided P-value is bracketed as

$$P\text{-value} < 0.001$$

REPORTING THE RESULTS OF A t TEST

In reporting the results of a t test, a researcher may choose to make a definite decision (to claim there is significant evidence for H_A or not significant evidence to support H_A) at a specified significance level α, or the researcher may choose simply to describe the results in phrases such as "There is very strong evidence that . . ." or "The evidence suggests that . . ." or "There is virtually no evidence that . . .". In writing a report for publication, it is very desirable to state the P-value so that the reader can make a decision on his or her own.

The term *significant* is often used in reporting results. For instance, an observed difference is said to be "statistically significant at the 5% level" if it is large enough to justify significant evidence for H_A at $\alpha = 0.05$. In Example 7.2.4 we saw that the observed difference between the two sample means in the toluene data is statistically significant at the 5% level, since the P-value is 0.0454, which is less than 0.05. In contrast, the fast plant data of Example 7.2.5 do not show a statistically significant difference at the 5% level, since the P-value for the fast plant data is 0.0678. However, the difference in sample means in the fast plant data *is* statistically significant at the $\alpha = 0.10$ level, since the P-value is less than 0.10. When α is not specified, it is usually understood to be 0.05; we should emphasize, however, that *α is an arbitrarily chosen value and there is nothing "official" about 0.05*. Unfortunately, the term "significant" is easily misunderstood and should be used with care; we will return to this point in Section 7.6.

Note: In this section we have considered tests of the form $H_0: \mu_1 = \mu_2$ (i.e., $\mu_1 - \mu_2 = 0$) versus $H_A: \mu_1 \neq \mu_2$ (i.e., $\mu_1 - \mu_2 \neq 0$); this is the most common pair of hypotheses. However, it may be that we wish to test that μ_1 differs from μ_2 by some specific, nonzero amount, say c. To test $H_0: |\mu_1 - \mu_2| = c$ versus $H_A: \mu_1 - \mu_2 \neq c$ we use the t test with test statistic given by

$$t_s = \frac{|\bar{y}_1 - \bar{y}_2| - c}{\mathrm{SE}_{(\bar{Y}_1 - \bar{Y}_2)}}$$

From this point on, the test proceeds as before (i.e., as for the case when $c = 0$).

Exercises 7.2.1–7.2.18

[*Note*: Answers to hypothesis testing questions should include a statement of the conclusion in the context of the setting. (See Examples 7.2.4 and 7.2.5.)]

7.2.1 For each of the following data sets, use Table 4 to bracket the two-tailed P-value of the data as analyzed by the t test.

(a)

	Sample 1	Sample 2
n	4	3
$\bar{y}$	735	854
$\mathrm{SE}_{(\bar{Y}_1 - \bar{Y}_2)} = 38$ with df $= 4$		

(b)

	Sample 1	Sample 2
n	7	7
$\bar{y}$	5.3	5.0
$\mathrm{SE}_{(\bar{Y}_1 - \bar{Y}_2)} = 0.24$ with df $= 12$		

(c)

	Sample 1	Sample 2
n	15	20
$\bar{y}$	36	30
$\mathrm{SE}_{(\bar{Y}_1 - \bar{Y}_2)} = 1.3$ with df $= 30$		

7.2.2 For each of the following data sets, use Table 4 to bracket the two-tailed *P*-value of the data as analyzed by the *t* test.

(a)

	Sample 1	Sample 2
n	8	5
$\bar{y}$	100.2	106.8
$SE_{(\bar{Y}_1-\bar{Y}_2)} = 5.7$ with df $= 10$		

(b)

	Sample 1	Sample 2
n	8	8
$\bar{y}$	49.8	44.3
$SE_{(\bar{Y}_1-\bar{Y}_2)} = 1.9$ with df $= 13$		

(c)

	Sample 1	Sample 2
n	10	15
$\bar{y}$	3.58	3.00
$SE_{(\bar{Y}_1-\bar{Y}_2)} = 0.12$ with df $= 19$		

7.2.3 For each of the following situations, suppose $H_0: \mu_1 = \mu_2$ is being tested against $H_A: \mu_1 \neq \mu_2$. State whether or not there is significant evidence for H_A.

(a) *P*-value $= 0.085$, $\alpha = 0.10$.
(b) *P*-value $= 0.065$, $\alpha = 0.05$.
(c) $t_s = 3.75$ with 19 degrees of freedom, $\alpha = 0.01$.
(d) $t_s = 1.85$ with 12 degrees of freedom, $\alpha = 0.05$.

7.2.4 For each of the following situations, suppose $H_0: \mu_1 = \mu_2$ is being tested against $H_A: \mu_1 \neq \mu_2$. State whether or not there is significant evidence for H_A.

(a) *P*-value $= 0.046$, $\alpha = 0.02$.
(b) *P*-value $= 0.033$, $\alpha = 0.05$.
(c) $t_s = 2.26$ with 5 degrees of freedom, $\alpha = 0.10$.
(d) $t_s = 1.94$ with 16 degrees of freedom, $\alpha = 0.05$.

7.2.5 In a study of the nutritional requirements of cattle, researchers measured the weight gains of cows during a 78-day period. For two breeds of cows, Hereford (HH) and Brown Swiss/Hereford (SH), the results are summarized in the following table.[6]

	HH	SH
n	33	51
$\bar{y}$	18.3	13.9
s	17.8	19.1

(a) What is the value of the *t* test statistic for comparing the means?

(b) In the context of this study, state the null and alternative hypotheses.

(c) The *P*-value for the *t* test is 0.29. If $\alpha = 0.10$, what is your conclusion regarding the hypotheses in (b)?

7.2.6 Backfat thickness is a variable used in evaluating the meat quality of pigs. An animal scientist measured backfat thickness (cm) in pigs raised on two different diets, with the results given in the table.[7]

	Diet 1	Diet 2
$\bar{y}$	3.49	3.05
s	0.40	0.40

Consider using the *t* test to compare the diets. Bracket the *P*-value, assuming that the number of pigs on each diet was

(a) 5
(b) 10
(c) 15

Use $n_1 + n_2 - 2$ as the approximate degrees of freedom.

7.2.7 Heart disease patients often experience spasms of the coronary arteries. Because biological amines may play a role in these spasms, a research team measured amine levels in coronary arteries that were obtained postmortem from patients who had died of heart disease and also from a control group of patients who had died from other causes. The accompanying table summarizes the concentration of the amine serotonin.[8]

	Serotonin (NG/GM)	
	Heart disease	Controls
n	8	12
$\bar{y}$	3,840	5,310
SE	850	640

(a) For these data, the SE of $(\bar{Y}_1 - \bar{Y}_2)$ is 1,064 and df $= 14.3$ (which can be rounded to 14). Use a *t* test to compare the means at the 5% significance level.

(b) Verify the value of $SE_{(\bar{Y}_1-\bar{Y}_2)}$ given in part (a).

7.2.8 In a study of the periodical cicada (*Magicicada septendecim*), researchers measured the hind tibia lengths of the shed skins of 110 individuals. Results for males and females are shown in the accompanying table.[9]

	Tibia length (μm)		
Group	n	Mean	SD
Males	60	78.42	2.87
Females	50	80.44	3.52

(a) Use a *t* test to investigate the association of tibia length on gender in this species. Use the 5% significance level. [*Note*: Formula (6.7.1) yields 94.3 df.]

(b) Given the preceding data, if you were told the tibia length of an individual of this species, could you make a fairly confident prediction of its sex? Why or why not?

(c) Repeat the *t* test of part (a), assuming that the means and standard deviations were as given in the table, but that they were based on only one-tenth as many individuals (6 males and 5 females). [*Note*: Formula (6.7.1) yields 7.8 df.]

7.2.9 Myocardial blood flow (MBF) was measured for two groups of subjects after 5 minutes of bicycle exercise. The normoxia ("normal oxygen") group was provided normal air to breath whereas the hypoxia group was provided with a gas mixture with reduced oxygen, to simulate high altitude. The results (ml/min/g) are as follows and are plotted below[10]:

	Normoxia	Hypoxia
	3.45	6.37
	3.09	5.69
	3.09	5.58
	2.65	5.27
	2.49	5.11
	2.33	4.88
	2.28	4.68
	2.24	3.50
	2.17	
	1.34	
n	10	8
$\bar{y}$	2.51	5.14
s	0.60	0.84

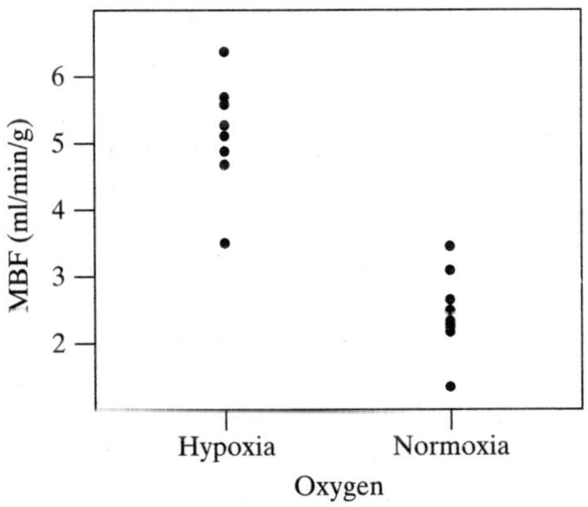

(a) What is the null hypothesis for a *t* test here?

(b) What do the dotplots suggest about H_0?

(c) Here is computer output for a *t* test. Explain what the *P*-value means in the context of this study. [*Note*: 7.307e-06 means 7.307*10^{-6} which is 0.000007307.]

```
t = 7.417, df = 12.212, p-value = 7.307e-06

alternative hypothesis: true difference in
means is not equal to 0
```

(d) If $\alpha = 0.05$, what is your conclusion regarding H_0? State your conclusion in the context of the research.

7.2.10 In a study of the development of the thymus gland, researchers weighed the glands of 10 chick embryos. Five of the embryos had been incubated 14 days, and five had been incubated 15 days. The thymus weights were as shown in the table.[11] [*Note*: Formula (6.7.1) yields 7.7 df.]

	Thymus weight (MG)	
	14 Days	15 Days
	29.6	32.7
	21.5	40.3
	28.0	23.7
	34.6	25.2
	44.9	24.2
n	5	5
$\bar{y}$	31.72	29.22
s	8.73	7.19

(a) Use a *t* test to compare the means at $\alpha = 0.10$.

(b) Note that the chicks that were incubated longer had a smaller mean thymus weight. Is this "backward" result surprising, or could it easily be attributed to chance? Explain.

7.2.11 As part of an experiment on root metabolism, a plant physiologist grew birch tree seedlings in the greenhouse. He flooded four seedlings with water for one day and kept four others as controls. He then harvested the seedlings and analyzed the roots for ATP content. The results (nmol ATP per mg tissue) are shown in the table.[12] [*Note*: Formula (6.7.1) yields 5.6 df.]

	Flooded	Control
	1.45	1.70
	1.19	2.04
	1.05	1.49
	1.07	1.91
n	4	4
$\bar{y}$	1.190	1.785
s	0.184	0.241

Use a *t* test to investigate the effect of flooding. Use $\alpha = 0.05$.

7.2.12 After surgery a patient's blood volume is often depleted. In one study, the total circulating volume of blood plasma was measured for each patient immediately after surgery. After infusion of a "plasma expander" into the bloodstream, the plasma volume was measured again and the increase in plasma volume (ml) was calculated. Two of the plasma expanders used were albumin (25 patients) and polygelatin (14 patients). The accompanying table reports the increase in plasma volume.[13]

	Albumin	Polygelatin
n	25	14
mean increase	490	240
SE	60	30

(a) What is the value of the t test statistic for comparing the means?

(b) The P-value for the t test is 0.0007. If $\alpha = 0.01$, what is your conclusion regarding H_0?

7.2.13 Nutritional researchers conducted an investigation of two high-fiber diets intended to reduce serum cholesterol level. Twenty men with high serum cholesterol were randomly allocated to receive an "oat" diet or a "bean" diet for 21 days. The table summarizes the fall (before minus after) in serum cholesterol levels.[14] Use a t test to compare the diets at the 5% significance level. [*Note*: Formula (6.7.1) yields 17.9 df.]

Fall in cholesterol (MG/DL)			
Diet	n	Mean	SD
Oat	10	53.6	31.1
Bean	10	55.5	29.4

7.2.14 Suppose we have conducted a t test, with $\alpha = 0.05$, and the P-value is 0.03. For each of the following statements, say whether the statement is true or false and explain why.

(a) We reject H_0 with $\alpha = 0.05$.

(b) We have significant evidence for H_A with $\alpha = 0.05$.

(c) We would reject H_0 if α were 0.10.

(d) We do not have significant evidence for H_A with $\alpha = 0.10$.

(e) If H_0 is true, the probability of getting a test statistic at least as extreme as the value of the t_s that was actually obtained is 3%.

(f) There is a 3% probability that H_0 is true.

7.2.15 Suppose we have conducted a t test, with $\alpha = 0.10$, and the P-value is 0.07. For each of the following statements, say whether the statement is true or false and explain why.

(a) We reject H_0 with $\alpha = 0.10$.

(b) We have significant evidence for H_A with $\alpha = 0.10$.

(c) We would reject H_0 if α were 0.05.

(d) We do not have significant evidence for H_A with $\alpha = 0.05$.

(e) The probability that $\bar{Y}_1$ is greater than $\bar{Y}_2$ is 0.07.

7.2.16 The following table shows the number of bacteria colonies present in each of several petri dishes after *E. coli* bacteria were added to the dishes and they were incubated for 24 hours. The "soap" dishes contained a solution prepared from ordinary soap; the "control" dishes contained a solution of sterile water. (These data were seen in Exercise 6.6.9.)

	Control	Soap
	30	76
	36	27
	66	16
	21	30
	63	26
	38	46
	35	6
	45	
n	8	7
$\bar{y}$	41.8	32.4
s	15.6	22.8
SE	5.5	8.6

Here is computer output for a t test.

```
t = 0.9095, df = 10.43, p-value = 0.3836
alternative hypothesis: true difference in
means is not equal to 0
```

(a) State the null and alternative hypotheses in context.

(b) If $\alpha = 0.10$, what is your conclusion regarding H_0?

7.2.17 Researchers studied the effect of a houseplant fertilizer on radish sprout growth. They randomly selected some radish seeds to serve as controls, while others were planted in aluminum planters to which fertilizer sticks were added. Other conditions were held constant between the two groups. The following table shows data on the heights of plants (in cm) 2 weeks after germination.[15]

Use a t test to investigate whether the fertilizer has an effect on average radish sprout growth. Use $\alpha = 0.05$. [*Note*: Formula (6.7.1) yields 53.5 degrees of freedom for these data.]

Control		Fertilized	
3.4	1.6	2.8	1.9
4.4	2.9	1.9	2.7
3.5	2.3	3.6	2.3
2.9	2.8	1.2	1.8
2.7	2.5	2.4	2.7
2.6	2.3	2.2	2.6
3.7	1.6	3.6	1.3
2.7	1.6	1.2	3.0
2.3	3.0	0.9	1.4
2.0	2.3	1.5	1.2
1.8	3.2	2.4	2.6
2.3	2.0	1.7	1.8
2.4	2.6	1.4	1.7
2.5	2.4	1.8	1.5
n	28		28
$\bar{y}$	2.58		2.04
s	0.65		0.72

7.2.18 A scientist recorded the tail lengths (in cm) of two varieties of woodland salamanders: redbacked and leadbacked. The table below shows summary values.[16]

	Redbacked	Leadbacked
Mean	3.2	4.0
SD	0.9	0.7
n	60	27

(a) What is the value of the t test statistic for comparing the means?

(b) In the context of this study, state the null and alternative hypotheses.

(c) The P-value for the t test is 0.00003. If $\alpha = 0.05$, what is your conclusion regarding the hypotheses in (b)?

7.3 Further Discussion of the t Test

In this section we discuss more fully the method and interpretation of the t test.

RELATIONSHIP BETWEEN TEST AND CONFIDENCE INTERVAL

There is a close connection between the confidence interval approach and the hypothesis testing approach to the comparison of μ_1 and μ_2. Consider, for example, a 95% confidence interval for $(\mu_1 - \mu_2)$ and its relationship to the t test at the 5% significance level. The t test and the confidence interval use the same three quantities—$(\bar{Y}_1 - \bar{Y}_2)$, $SE_{(\bar{Y}_1 - \bar{Y}_2)}$, and $t_{0.025}$—but manipulate them in different ways.

In the t test, when $\alpha = 0.05$, we have significant evidence for H_A (and so we reject H_0) if the P-value is less than or equal to 0.05. This happens if and only if the test statistic, t_s, is in the tail of the t distribution, at or beyond $\pm t_{0.025}$. If the magnitude of t_s (symbolized as $|t_s|$) is greater than or equal to $t_{0.025}$, then the P-value is less than or equal to 0.05 and we have significant evidence for H_A; if $|t_s|$ is less than $t_{0.025}$, then the P-value is greater than 0.05 and we do *not* have significant evidence for H_A. Figure 7.3.1 shows this relationship.

Thus, we lack significant evidence for H_A: $\mu_1 - \mu_2 \neq 0$ if and only if $|t_s| < t_{0.025}$. That is, we lack significant evidence for H_A when

$$\frac{|\bar{y}_1 - \bar{y}_2|}{SE_{(\bar{Y}_1 - \bar{Y}_2)}} < t_{0.025}$$

This is equivalent to

$$|\bar{y}_1 - \bar{y}_2| < t_{0.025}\, SE_{(\bar{Y}_1 - \bar{Y}_2)}$$

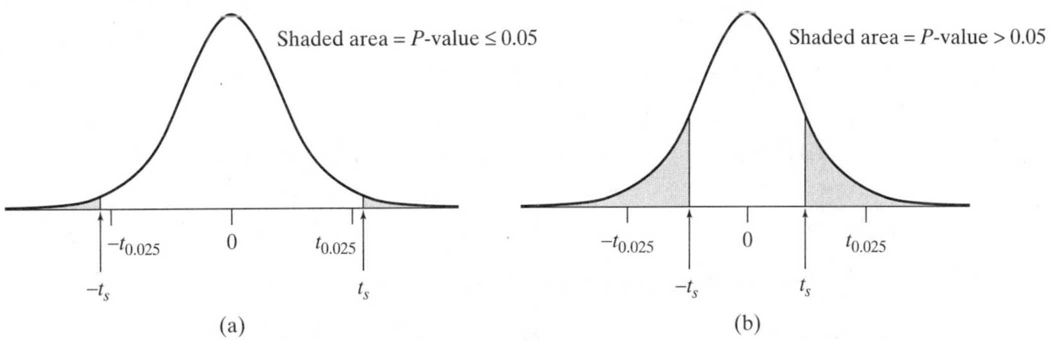

Figure 7.3.1 Possible outcomes of the t test at $\alpha = 0.05$. (a) If $|t_s| \geq t_{0.025}$ then P-value ≤ 0.05 and there is significant evidence for H_A (so H_0 is rejected). (b) If $|t_s| < t_{0.025}$, then P-value > 0.05 and there is a lack of significant evidence for H_A.

or

$$-t_{0.025}\,\mathrm{SE}_{(\bar{Y}_1-\bar{Y}_2)} < (\bar{y}_1 - \bar{y}_2) < t_{0.025}\,\mathrm{SE}_{(\bar{Y}_1-\bar{Y}_2)}$$

which is equivalent to

$$-(\bar{y}_1 - \bar{y}_2) - t_{0.025}\mathrm{SE}_{(\bar{Y}_1-\bar{Y}_2)} < 0 < -(\bar{y}_1 - \bar{y}_2) + t_{0.025}\mathrm{SE}_{(\bar{Y}_1-\bar{Y}_2)}$$

or

$$(\bar{y}_1 - \bar{y}_2) + t_{0.025}\mathrm{SE}_{(\bar{Y}_1-\bar{Y}_2)} > 0 > (\bar{y}_1 - \bar{y}_2) - t_{0.025}\mathrm{SE}_{(\bar{Y}_1-\bar{Y}_2)}$$

or

$$(\bar{y}_1 - \bar{y}_2) - t_{0.025}\mathrm{SE}_{(\bar{Y}_1-\bar{Y}_2)} < 0 < (\bar{y}_1 - \bar{y}_2) + t_{0.025}\mathrm{SE}_{(\bar{Y}_1-\bar{Y}_2)}$$

Thus, we have shown that we lack significant evidence for $H_A\colon \mu_1 - \mu_2 \neq 0$ if and only if the confidence interval for $(\mu_1 - \mu_2)$ includes zero. Conversely, if the 95% confidence interval for $(\mu_1 - \mu_2)$ does not cover zero, then we have significant evidence for $H_A\colon \mu_1 - \mu_2 \neq 0$ when $\alpha = 0.05$. (The same relationship holds between the 90% confidence interval and the test at $\alpha = 0.10$, and so on.) We illustrate with an example.

Example 7.3.1 **Crawfish Lengths** Biologists took samples of the crawfish species *Orconectes sanborii* from two rivers in central Ohio, the Upper Cuyahoga River (CUY) and East Fork of Pine Creek (EFP), and measured the length (mm) of each crawfish captured.[17] Table 7.3.1 shows the summary statistics; Figure 7.3.2 shows parallel boxplots of the data. The EFP sample distribution is shifted down from the CUY distribution; both distributions are reasonably symmetric.

Table 7.3.1 Crawfish data: length (mm)

	CUY	EFP
n	30	30
$\bar{y}$	22.91	21.97
s	3.78	2.90

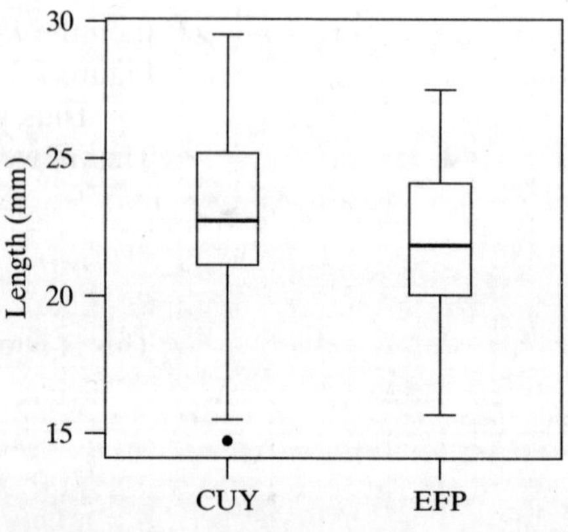

Figure 7.3.2 Boxplots of the crawfish data

For these data the two SEs are $3.78/\sqrt{30} = 0.69$ and $2.90/\sqrt{30} = 0.53$ for CUY and EFP, respectively. The degrees of freedom are

$$\mathrm{df} = \frac{(0.69^2 + 0.53^2)^2}{0.69^4/29 + 0.53^4/29} = 54.4$$

The quantities needed for a t test with $\alpha = 0.05$ are

$$\bar{y}_1 - \bar{y}_2 = 22.91 - 21.97 = 0.94$$

and

$$\mathrm{SE}_{(\bar{Y}_1-\bar{Y}_2)} = \sqrt{0.69^2 + 0.532} = 0.87$$

The test statistic is

$$t_s = \frac{(22.91 - 21.97) - 0}{0.87} = \frac{0.94}{0.87} = 1.08$$

The P-value for this test (found using a computer) is 0.2850, which is greater than 0.05, so we do not reject H_0. (A quick look at Table 4, using df $= 50$, shows that the P-value is between 0.20 and 0.40.)

If we construct a 95% confidence interval for $(\mu_1 - \mu_2)$ we get

$$0.94 \pm 2.004 \times 0.87$$

or $(-2.68, 0.80)$.*

The confidence interval includes zero, which is consistent with not having significant evidence for H_A: $\mu_1 - \mu_2 \neq 0$ in the t test. Note that this equivalence between the test and the confidence interval makes common sense; according to the confidence interval, μ_1 may be as much as 2.68 less, or as much as 0.81 more, than μ_2; it is natural, then, to say that we are uncertain as to whether μ_1 is greater than (or less than, or equal to) μ_2. ■

In the context of the Student's t method, the confidence interval approach and hypothesis testing approach are different ways of using the same basic information. The confidence interval has the advantage that it indicates the magnitude of the difference between μ_1 and μ_2. The testing approach has the advantage that the P-value describes on a continuous scale the strength of the evidence that μ_1 and μ_2 are really different. In Section 7.7 we will explore further the use of a confidence interval to supplement the interpretation of a t test. In later chapters we will encounter other hypothesis tests that cannot so readily be supplemented by a confidence interval.

INTERPRETATION OF α

In analyzing data or making a decision based on data, you will often need to choose a significance level α. How do you know whether to choose $\alpha = 0.05$ or $\alpha = 0.01$ or some other value? To make this judgment, it is helpful to have an *operational* interpretation of α. We now give such an interpretation.

Recall from Section 7.2 that the sampling distribution of t_s, if H_0 is true, is a Student's t distribution. Let us assume for definiteness that df $= 60$ and that α is chosen equal to 0.05. The critical value (from Table 4) is $t_{0.025} = 2.000$. Figure 7.3.3 shows the Student's t distribution and the values ± 2.000. The total shaded area in the figure is 0.05; it is split into two equal parts of area 0.025 each. We can think of Figure 7.3.3 as a formal guide for deciding whether the evidence is strong enough to significantly support H_A: If the observed value of t_s falls in the hatched regions of the t_s axis, then

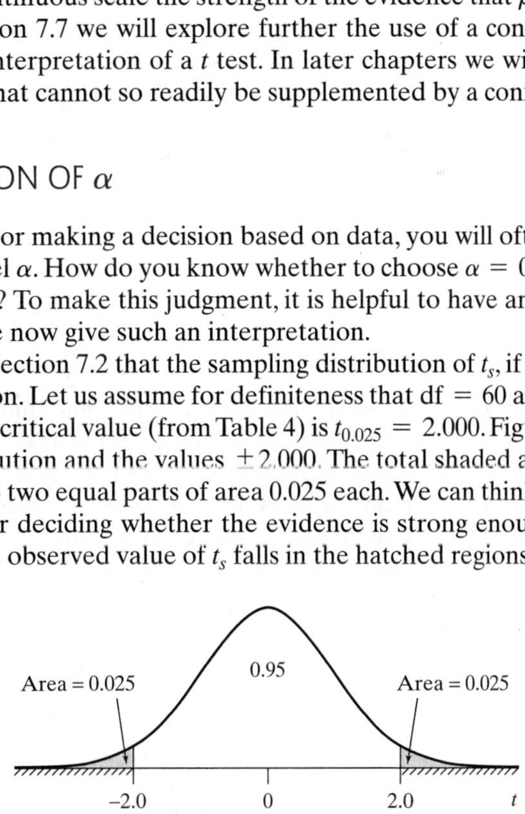

Figure 7.3.3 A t test at $\alpha = 0.05$. There is significant evidence for H_A if t_s falls in the hatched region

*The value of $t_{0.025} = 2.004$ is based on 54.4 degrees of freedom. If we were to use 50 degrees of freedom (i.e., if we had to rely on Table 4, rather than a computer) the t multiplier would be 2.009. This makes almost no difference in the resulting confidence interval.

there is significant evidence for H_A. But the chance of this happening is 5%, if H_0 is true. Thus, we can say that

$$\Pr\{\text{data provide significant evidence for } H_A\} = 0.05 \text{ if } H_0 \text{ is true}$$

This probability has meaning in the context of a meta-study (depicted in Figure 7.3.4) in which we repeatedly sample from two populations and calculate a value of t_s. It is important to realize that the probability refers to a situation in which H_0 is true. In order to concretely picture such a situation, you are invited to suspend disbelief for a moment and come on an imaginary trip in Example 7.3.2.

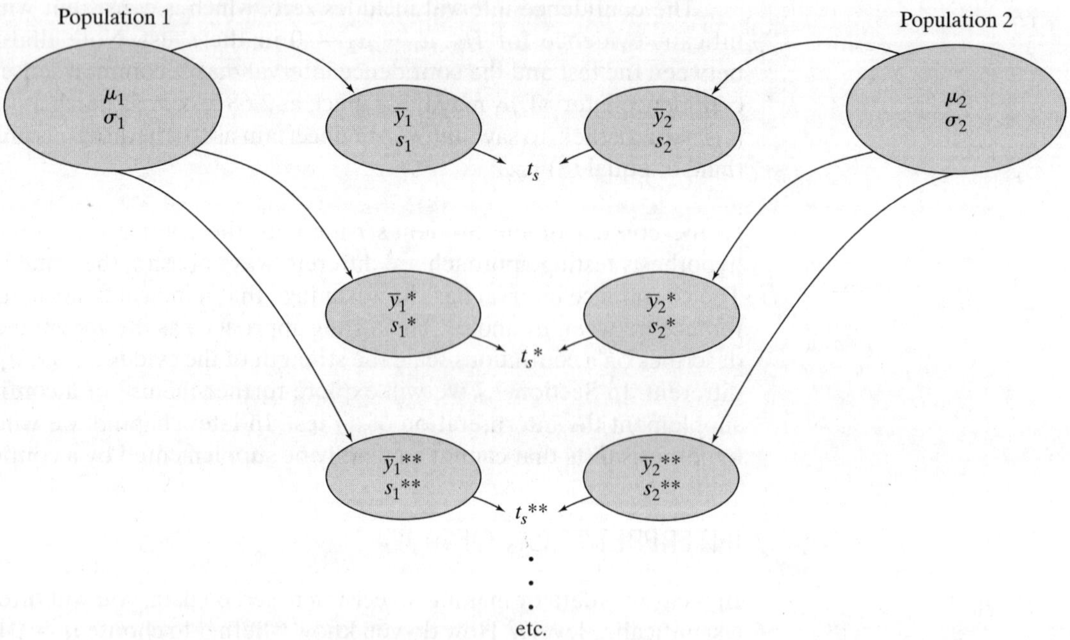

Figure 7.3.4 Meta-study for the *t* test

Example 7.3.2

Music and Marigolds* Imagine that the scientific community has developed great interest in the influence of music on the growth of marigolds. One school of investigation centers on whether music written by Bach or Mozart produces taller plants. Plants are randomly allocated to listen to Bach (treatment 1) or Mozart (treatment 2) and, after a suitable period of listening, data are collected on plant height. The null hypothesis is

$$H_0\text{: Marigolds respond equally well to Bach or Mozart.}$$

or

$$H_0: \mu_1 = \mu_2$$

where

$$\mu_1 = \text{Mean height of marigolds if exposed to Bach}$$

$$\mu_2 = \text{Mean height of marigolds if exposed to Mozart}$$

*This example is intentionally fanciful.

Assume for the sake of argument that H_0 is in fact true. Imagine now that many investigators perform the Bach versus Mozart experiment, and that each experiment results in data with 60 degrees of freedom. Suppose each investigator analyzes his or her data with a t test at $\alpha = 0.05$. What conclusions will the investigators reach? In the meta-study of Figure 7.3.4, suppose each pair of samples represents a different investigator. Since we are assuming that μ_1 and μ_2 are actually equal, the values of t_s will deviate from 0 only because of chance sampling error. If all the investigators were to get together and make a frequency distribution of their t_s values, that distribution would follow a Student's t curve with 60 degrees of freedom. The investigators would make their decisions as indicated by Figure 7.3.3, so we would expect them to have the following experiences:

95% of them would (correctly) not find significant evidence for H_A;

2.5% of them would find significant evidence for H_A and conclude (incorrectly) that the plants prefer Bach.

2.5% of them would find significant evidence for H_A and conclude (incorrectly) that the plants prefer Mozart.

Thus, a total of 5% of the investigators would find significant evidence for the alternative hypothesis. ■

Example 7.3.2 provides an image for interpreting α. Of course, in analyzing data, we are not dealing with a meta-study but rather with a single experiment. When we perform a t test at the 5% significance level, we are playing the role of one of the investigators in Example 7.3.2, and the others are imaginary. If we find significant evidence for H_A, there are two possibilities:

1. H_A is in fact true; or
2. H_0 is in fact true, but we are one of the unlucky 5% who obtained data that provided significant evidence for H_A anyway. In this case, we can think of the significant evidence for H_A as "setting off a false alarm."

We feel "confident" in claiming our evidence for H_A is significant because the second possibility is unlikely (assuming that we regard 5% as a small percentage). Of course, we never know (unless someone replicates the experiment) whether or not we are one of the unlucky 5%.

Significance Level versus P-Value Students sometimes find it hard to distinguish between significance level (α) and P-value.* For the t test, both α and the P-value are tail areas under Student's t curve. But α is an arbitrary prespecified value; it can be (and should be) chosen before looking at the data. By contrast, the P-value is determined from the data; indeed, giving the P-value is a way of describing the data. You may find it helpful at this point to compare Figure 7.2.3 with Figure 7.3.3. The shaded area represents P-value in the former and α in the latter figure.

TYPE I AND TYPE II ERRORS

We have seen that α can be interpreted as a probability:

$$\alpha = \Pr\{\text{finding significant evidence for } H_A\} \text{ if } H_0 \text{ is true}$$

*Unfortunately, the term "significance level" is not used consistently by all people who write about statistics. A few authors use the terms "significance level" or "significance probability" where we have used "P-value."

Claiming that data provide evidence that significantly supports H_A when H_0 is true is called a **Type I error.** In choosing α, we are choosing our level of protection against Type I error. Many researchers regard 5% as an acceptably small risk. If we do not regard 5% as small enough, we might choose to use a more conservative value of α such as $\alpha = 0.01$; in this case the percentage of true null hypotheses that we reject would be not 5% but 1%.

In practice, the choice of α may depend on the context of the particular experiment. For example, a regulatory agency might demand more exacting proof of efficacy for a toxic drug than for a relatively innocuous one. Also, a person's choice of α may be influenced by his or her prior opinion about the phenomenon under study. For instance, suppose an agronomist is skeptical of claims for a certain soil treatment; in evaluating a new study of the treatment, he might express his skepticism by choosing a very conservative significance level (say, $\alpha = 0.001$), thus indicating that it would take a lot of evidence to convince him that the treatment is effective. For this reason, written reports of an investigation should include a P-value so that each reader is free to choose his or her own value of α in evaluating the reported results.

If H_A is true, but we do not observe sufficient evidence to support H_A, then we have made a **Type II error.** Table 7.3.2 displays the situations in which Type I and Type II errors can occur. For example, if we find significant evidence for H_A, then we eliminate the possibility of a Type II error, but by rejecting H_0 we may have made a Type I error.

Table 7.3.2 Possible outcomes of testing H_0

		True situation	
		H_0 true	H_A true
OUR DECISION	Lack of significant evidence for H_A	Correct	Type II error
	Significant evidence for H_A	Type I error	Correct

The consequences of Type I and Type II errors can be very different. The following two examples show some of the variety of these consequences.

Example 7.3.3 **Marijuana and the Pituitary** Cannabinoids, which are substances contained in marijuana, can be transmitted from mother to young through the placenta and through the milk. Suppose we conduct the following experiment on pregnant mice: We give one group of mice a dose of cannabinoids and keep another group as controls. We then evaluate the function of the pituitary gland in the offspring. The hypotheses would be

H_0: Cannabinoids do not affect pituitary of offspring.

H_A: Cannabinoids do affect pituitary of offspring.

If in fact cannabinoids do not affect the pituitary of the offspring, but we conclude that our data provide significant evidence for H_A, we would be making a Type I error; the consequence might be unnecessary alarm if the conclusion were made public. On the other hand, if cannabinoids do affect the pituitary of the offspring, but our t test results in a lack of significant evidence for H_A, this would be a Type II error; one consequence might be unjustifiable complacency on the part of marijuana-smoking mothers. ■

Example 7.3.4

Immunotherapy Chemotherapy is standard treatment for a certain cancer. Suppose we conduct a clinical trial to study the efficacy of supplementing the chemotherapy with immunotherapy (stimulation of the immune system). Patients are given either chemotherapy or chemotherapy plus immunotherapy. The hypotheses would be

H_0: Immunotherapy has no effect on survival.

H_A: Immunotherapy does affect survival.

If immunotherapy is actually not effective, but we conclude that our data provide significant evidence for H_A and thus conclude that immunotherapy is effective, then we have made a Type I error. The consequence, if this conclusion is acted on by the medical community, might be the widespread use of unpleasant, dangerous, and worthless immunotherapy. If, on the other hand, immunotherapy is actually effective, but our data do not enable us to detect that fact (perhaps because our sample sizes are too small), then we have made a Type II error, with consequences quite different from those of a Type I error: The standard treatment will continue to be used until someone provides convincing evidence that supplementary immunotherapy is effective. If we still "believe" in immunotherapy, we can conduct another trial (perhaps with larger samples) to try again to establish its effectiveness. ■

As the foregoing examples illustrate, the consequences of a Type I error are usually quite different from those of a Type II error. The likelihoods of the two types of error may be very different, also. The significance level α is the probability of obtaining significant evidence for H_A if H_0 is true. Because α is chosen at will, the hypothesis testing procedure "protects" you against Type I error by giving you control over the risk of such an error. This control is independent of the sample size and other factors. The chance of a Type II error, by contrast, depends on many factors, and may be large or small. In particular, an experiment with small sample sizes often has a high risk of Type II error.

We are now in a position to reexamine Carl Sagan's aphorism that "Absence of evidence is not evidence of absence." Because the risk of Type I error is controlled and that of Type II error is not, our state of knowledge is much stronger after rejection of a null hypothesis than after nonrejection. For example, suppose we are testing whether a certain soil additive is effective in increasing the yield of field corn. If we find significant evidence for H_A and claim the additive is effective, then either (1) we are right; or (2) we have made a Type I error. Since the risk of a Type I error is controlled, we can be relatively confident of our conclusion that the additive is effective (although not necessarily very effective). Suppose, on the other hand, that the data are such that there is a lack of evidence for the additive's effectiveness—we do not have evidence for H_A. Then either (1) we are right (that is, H_0 is true), or (2) we have made a Type II error. Since the risk of a Type II error may be quite high, we cannot say confidently that the additive is ineffective. In order to justify a claim that the additive is ineffective, we would need to supplement our test of hypothesis with further analysis, such as a confidence interval or an analysis of the chance of Type II error. We will consider this in more detail in Sections 7.6 and 7.7.

POWER

As we have seen, Type II error is an important concept. The probability of making a Type II error is denoted by β:

$$\beta = \Pr\{\text{lack of significant evidence for } H_A\} \text{ if } H_A \text{ is true}$$

The chance of not making a Type II error when H_A is true—that is, the chance of having significant evidence for H_A when H_A is true—is called the **power** of a statistical test:

$$\text{Power} = 1 - \beta = \Pr\{\text{significant evidence for } H_A\} \text{ if } H_A \text{ is true}$$

Thus, the power of a t test is a measure of the sensitivity of the test, or the ability of the test procedure to detect a difference between μ_1 and μ_2 when such a difference really *does* exist. In this way the power is analogous to the resolving power of a microscope.

The power of a statistical test depends on many factors in an investigation, including the sample sizes, the inherent variability of the observations, and the magnitude of the difference between μ_1 and μ_2. All other things being equal, using larger samples gives more information and thereby increases power. In addition, we will see that some statistical tests can be more powerful than others, and that some study designs can be more powerful than others.

The planning of a scientific investigation should always take power into consideration. No one wants to emerge from lengthy and perhaps expensive labor in the lab or the field, only to discover upon analyzing the data that the sample sizes were insufficient or the experimental material too variable, so experimental effects that were considered important were not detected. Two techniques are available to aid the researcher in planning for adequate sample sizes. One technique is to decide how small each standard error ought to be and choose n using an analysis such as that of Section 6.4. A second technique is a quantitative analysis of the power of the statistical test. Such an analysis for the t test is discussed in Section 7.7.

Exercises 7.3.1–7.3.11

7.3.1 (Sampling exercise) Refer to the collection of 100 ellipses shown with Exercise 1.3.6, which can be thought of as representing a natural population of the organism *C. ellipticus.* Use random digits (from Table 1 or your calculator) to choose two random samples of five ellipses each. Use a metric ruler to measure the body length of each ellipse; measurements to the nearest millimeter will be adequate.

(a) Compare the means of your two samples, using a t test at $\alpha = 0.05$.

(b) Did the analysis of part (a) lead you to a Type I error, a Type II error, or no error?

7.3.2 (Sampling exercise) Simulate choosing random samples from two different populations, as follows. First, proceed as in Exercise 7.3.1 to choose two random samples of five ellipses each and measure their lengths. Then add 6 mm to *each* measurement in one of the samples.

(a) Compare the means of your two samples, using a t test at $\alpha = 0.05$.

(b) Did the analysis of part (a) lead you to a Type I error, a Type II error, or no error?

7.3.3 (Sampling exercise) Prepare simulated data as follows. First, proceed as in Exercise 7.3.1 to choose two random samples of five ellipses each and measure their lengths. Then, toss a coin. If the coin falls heads, add 6 mm to *each* measurement in one of the samples. If the coin falls tails, do not modify either sample.

(a) Prepare two copies of the simulated data. On the Student Copy, show the data only; on the Instructor Copy, indicate also which sample (if any) was modified.

(b) Give your Instructor Copy to the instructor and trade your Student Copy with another student when you are told to do so.

(c) After you have received another student's paper, compare the means of his or her two samples using a two-tailed t test at $\alpha = 0.05$. If you reject H_0, decide which sample was modified.

7.3.4 Suppose a new drug is being considered for approval by the Food and Drug Administration. The null hypothesis is that the drug is not effective. If the FDA approves the drug, what type of error, Type I or Type II, could not possibly have been made?

7.3.5 In Example 7.3.1, the null hypothesis was not rejected. What type of error, Type I or Type II, might have been made in that t test?

7.3.6 Suppose that a 95% confidence interval for $(\mu_1 - \mu_2)$ is calculated to be (1.4, 6.7). If we test H_0: $\mu_1 - \mu_2 = 0$ versus H_A: $\mu_1 - \mu_2 \neq 0$ using $\alpha = 0.05$, will we reject H_0? Why or why not?

7.3.7 Suppose that a 95% confidence interval for $(\mu_1 - \mu_2)$ is calculated to be $(-7.4, -2.3)$. If we test H_0: $\mu_1 = \mu_2$ versus H_A: $\mu_1 \neq \mu_2$ using $\alpha = 0.10$, will we reject H_0? Why or why not?

7.3.8 A dairy researcher has developed a new technique for culturing cheese that is purported to age cheese in substantially less time than traditional methods without affecting other properties of the cheese. Retrofitting cheese manufacturing plants with this new technology will initially cost millions of dollars, but if it indeed reduces aging time—even marginally—it will lead to higher company profits in the long run. If, on the other hand, the new method is no better than the old, the retrofit would be a financial mistake. Before making the decision to retrofit, an experiment will be performed to compare culture times of the new and old methods.

(a) In plain English, what are the null and alternative hypotheses for this experiment?

(b) In the context of this scenario, what would be the consequence of a Type I error?

(c) In the context of this scenario, what would be the consequence of a Type II error?

(d) In your opinion, which type of error would be more serious? Justify your answer. (It is possible to argue both sides.)

7.3.9 A scientist conducted a t test using $\alpha = 0.05$ and arrived at a P-value of 0.03, which led him to reject H_0: $\mu_1 = \mu_2$. However, he made a mistake in his calculations; the correct P-value is 0.23. True or false (and say why): When the scientist rejected H_0 he clearly made a Type I error.

7.3.10 Patients with heart disease were randomly assigned to receive either treatment A or treatment B. At the end of the experiment it was found that systolic volume (a measure of health) "did not differ between groups (P-value = 0.77)."[18] Choose one of the following and explain your choice:

(i) a Type I error might have been made.

(ii) a Type II error might have been made.

7.3.11 Consider the study in Exercise 7.3.10. Since the P-value was so large, do the data provide compelling evidence that the treatments are equally effective?

7.4 Association and Causation

When we are comparing two populations we often focus on the nature of the relationship between a **response variable,** Y—a variable that measures an outcome of interest—and an **explanatory variable** X—a variable used to explain or predict an outcome. As we will explore next, with data collected from an **experiment** we can assess whether or not there is evidence that X *affects* the mean value of Y. That is, we can ask, Do changes in X *cause* changes in Y? (For example, does toluene affect the mean amount of norepinephrine in the brain?) With **observational studies** our conclusions are more limited—we are not able to make causal claims, but rather only conclusions regarding association between X and Y. For example, we can ask, Are changes in X associated with changes in the mean value of Y? Or, Is there evidence that the mean values of Y differ for two populations? (For example, do crawfish captured from two different locations have different mean lengths?)

Thus, our ability to investigate such questions depends on how the data were collected: experimentally or with an observational study. Below are examples of each type of study as they pertain to comparing the means of two samples, followed by a more formal discussion of these study types.

Example 7.4.1 **Mosquito Weight** A researcher measured the body weights of two samples of a certain type of mosquito. Table 7.4.1 gives the sample means and standard deviations for 20 males and 20 females.[19]

Table 7.4.1 Weight (mg)

	Males	Females
Mean	0.94	1.53
SD	0.30	0.28

Example 7.4.2

Exercise During Pregnancy A group of 74 pregnant women were randomly divided into two groups. Half of the women participated in three 60-minute exercise classes per week over a period of 3 months. The other 37 women were the control group and did not participate in an exercise program.

The researchers assessed each woman for symptoms of depression using a 20-item scale both before and after the 3-month period. The response variable is change in depression score, with a positive change being good.[20] The results are shown in Table 7.4.2. ■

Table 7.4.2 Change in depression score

	Exercise	Control
Mean	5.2	1.4
SD	5.0	7.8

Examples 7.4.1 and 7.4.2 both involve two-sample comparisons, but notice that the two studies differ in a fundamental way. In Example 7.4.1 the samples come from populations that occur naturally; the investigator is merely an observer:

Population 1: Weights of male mosquitoes

Population 2: Weights of female mosquitoes

By contrast, the two populations in Example 7.4.2 do not actually exist but rather are defined in terms of specific experimental conditions; in a sense, the populations are created by experimental intervention:

Population 1: Change in depression score for pregnant women who participate in an exercise program like the one used here

Population 2: Change in depression score for women who do not participate in an exercise program

These two types of two-sample comparisons—the observational and the experimental—are both widely used in research. The formal methods of analysis are often the same for the two types, but the interpretation of the results is often somewhat different. For instance, in Example 7.4.2 it might be reasonable to say that the particular exercise program *causes* the improvement in depression score, while no such notion applies in Example 7.4.1.

OBSERVATIONAL VERSUS EXPERIMENTAL STUDIES

A major consideration in interpreting the results of a biological study is whether the study was observational or experimental. In an *experiment*, the researcher intervenes in or manipulates the assignment of experimental conditions to the subjects or experimental units.* In an *observational study*, the researcher merely observes an existing situation, as in the following example.

Example 7.4.3

Cigarette Smoking In studies of the effects of smoking cigarettes, both experimental and observational approaches have been used. Effects in animals can be studied experimentally, because animals (e.g., dogs) can be allocated to treatment groups and the groups can be given various doses of cigarette smoke. Effects in humans are usually

*The conditions being manipulated must be those defining the populations being compared. For example, if five men and five women are given the same drug and then the sexes are compared, the comparison of men to women is observational, not experimental.

studied observationally. In one study, for example, pregnant women were questioned about their smoking habits, dietary habits, and so on.[21] When the babies were born, their physical and mental development was followed. One striking finding related to the babies' birthweights: The smokers tended to have smaller babies than the nonsmokers. The difference was not attributable to chance (the P-value was less than 10^{-5}). Nevertheless, it was far from clear that the difference was *caused* by smoking, because the women who smoked differed from the nonsmokers in many other aspects of their lifestyle besides smoking—for instance, they had very different dietary habits. ■

As Example 7.4.3 illustrates, it can be difficult to determine the exact nature of a cause–effect relationship in an observational study. In an experiment, on the other hand, a cause–effect relationship may be easy to see, based on the way in which the researcher assigned the experimental conditions. To help fix the ideas, consider studying cholesterol level. Suppose a group of patients with high cholesterol levels enrolls in a clinical trial—that is, in a medical experiment—in which some of the patients are randomly chosen to receive a new drug and others are given a standard drug that has shown only modest effects in the past. If a two-sample t test shows that the mean cholesterol level decreased more for those on the new drug than for those on the standard drug, then the researcher can conclude that the new drug *caused* the superior outcome and is better than the standard drug.

Now consider a two-sample t test to compare average cholesterol level in a random sample of 50-year-olds to average cholesterol level in a random sample of 25-year-olds. Suppose a two-sample t test gives a small P-value, with the 50-year-olds having higher cholesterol than the 25-year-olds. We could be fairly confident that cholesterol level tends to increase with age. However, it would be *possible* that some other explanation were at work. For example, maybe diets have changed over time and the 25-year-olds are eating foods that the 50-year-olds don't eat, causing the 25-year-olds to have low cholesterol; perhaps if the 25-year-olds keep the same diet until they are 50, they will still have low cholesterol at age 50.

As a third example, consider comparing a random sample of home owners to a random sample of renters. Suppose a two-sample t test shows a significantly higher mean cholesterol level among the home owners than among the renters. We should not conclude that buying a home causes one's cholesterol level to rise. Rather, we should consider that people who own homes tend to be older than are renters. It might very well be the case that age is the causal factor, which explains why the home owners have higher cholesterol than do the renters.

All three of these cases might involve a two-sample t test and the rejection of H_0. Indeed, we might get the same P-value in each test. However, the conclusions we can draw from the three situations are quite different. The scope of the inference we can draw depends on the way in which the data are collected. Experiments allow us to infer cause–effect relationships that can only be guessed at in observational studies. Sometimes an observational study will leave us feeling reasonably confident that we understand the causal mechanism at work; however, we will see that drawing such conclusions is fraught with danger. For this reason, researchers interested in drawing causal conclusions should make great efforts to conduct controlled experiments rather than observational studies.

MORE ON OBSERVATIONAL STUDIES

The difficulties in interpreting observational studies arise from two primary sources:

Nonrandom selection from populations

Uncontrolled extraneous variables

The following example illustrates both of these.

Example 7.4.4 **Race and Brain Size** In the nineteenth century, much effort was expended in the attempt to show "scientifically" that certain human races were inferior to others. A leading researcher on this subject was the American physician S. G. Morton, who won widespread admiration for his studies of human brain size. Throughout his life, Morton collected human skulls from various sources, and he carefully measured the cranial capacities of hundreds of these skulls. His data appeared to suggest that (as he suspected) the "inferior" races had smaller cranial capacities. Table 7.4.3 gives a summary of Morton's data comparing Caucasian skulls to those of Native Americans.[22] According to a t test, the difference between these two samples is "statistically significant" (P-value < 0.001). But is it *meaningful*?

Table 7.4.3 Cranial capacity (in^3)

	Caucasian	Native American
Mean	87	82
SD	8	10
n	52	144

In the first place, the notion that cranial capacity is a measure of intelligence is no longer taken seriously. Leaving that question aside, one can still ask whether it is true that the mean cranial capacity of Native Americans is less than that of Caucasians. Such an inference beyond the actual data requires that the data be viewed as random samples from their respective populations. Of course, in actuality, Morton's data are not random samples but "samples of convenience," because Morton measured those skulls that he happened to obtain. But might the data be viewed "as if" they were generated by random sampling? One way to approach this question is to look for sources of bias. In 1977, the noted biologist Stephen Jay Gould reexamined Morton's data with this goal in mind, and indeed Gould found several sources of bias. For instance, the 144 Native American skulls represent many different groups of Native Americans; as it happens, 25% of the skulls (i.e., 36 of them) were from Inca Peruvians, who were a small-boned people with small skulls, while relatively few were from large-skulled tribes such as the Iroquois. Clearly, a comparison between Native Americans and Caucasians is meaningless unless somehow adjusted for such imbalances. When Gould made such an adjustment, he found that the difference between Native Americans and Caucasians vanished. ■

Even though the story of Morton's skulls is more than 100 years old, it can still serve to alert us to the pitfalls of inference. Morton was a conscientious researcher and took great care to make accurate measurements; Gould's reexamination did not reveal any suggestion of conscious fraud on Morton's part. Morton may have overlooked the biases in his data because they were *invisible* biases; that is, they related to aspects of the selection process rather than aspects of the measurements themselves.

When we look at a set of observational data, we can sometimes become so hypnotized by its apparent *solidity* and *objectivity* that we forget to ask how the observational units—the persons or things that were observed—were selected. The question should always be asked. If the selection was haphazard rather than truly random, the results can be severely distorted.

CONFOUNDING

Many observational studies are aimed at discovering some kind of causal relationship. Such discovery can be very difficult because of extraneous variables that enter in an uncontrolled (and perhaps unknown) way. The investigator must be guided by the maxim:

> Association does not imply causation.

For instance, it is known that some populations whose diets are high in fiber enjoy a reduced incidence of colon cancer. But this observation does not in itself show that it is the high-fiber diet, rather than some other factor, that provides the protection against colon cancer.

The following example shows how uncontrolled extraneous variables can cloud an observational study, and what kinds of steps can be taken to clarify the picture.

Example 7.4.5 **Smoking and Birthweight** In a large observational study of pregnant women, it was found that the women who smoked cigarettes tended to have smaller babies than the nonsmokers.[21] (This study was mentioned in Example 7.4.3.) It is plausible that smoking could cause a reduction in birthweight, for instance, by interfering with the flow of oxygen and nutrients across the placenta. But, of course, plausibility is not proof. In fact, the investigators found that the smokers differed from the nonsmokers with respect to many other variables. For instance, the smokers drank more whiskey than the nonsmokers. Alcohol consumption might plausibly be linked to a deficit in growth. ■

In Example 7.4.5 three variables are presented; let us refer to these as X = smoking, Y = birthweight, and Z = alcohol consumption. There is an association between X and Y, but is there a *causal* link between them? Or is there a causal link between Z and Y? Figure 7.4.1 gives a schematic representation of the situation. Changes in X are associated with changes in Y. However, changes in Z are also associated with changes in Y. We say that the effect that X has on Y is **confounded** with the effect that Z has on Y. In the context of Example 7.4.5, we say that the effect that smoking has on birthweight is confounded with the effect that alcohol consumption has on birthweight. In observational studies, confounding of effects is a common problem.

(a) (b) The effect of X on Y is confounded with the effect of Z on Y

Figure 7.4.1 Schematic representation of causation (a) and of confounding (b)

Example 7.4.6 **Smoking and Birthweight** The study presented in Example 7.4.5 uncovered many confounding variables. For example, the smokers drank more coffee than the nonsmokers. In addition—and this is especially puzzling—it was found that the smokers began to menstruate at younger ages than the nonsmokers. This phenomenon (early onset of menstruation) could not possibly have been *caused* by smoking, because it occurred (in almost all instances) *before* the woman began to smoke. One interpretation that has been proposed is that the two populations—women who choose to smoke and those who do not—are different in some biological way; thus, it has been suggested that the reduced birthweight is due "to the *smoker,* not the *smoking.*"[23]

A number of more recent studies have attempted to shed some light on the relationship between maternal smoking and infant development. Researchers in one study observed, in addition to smoking habits, about 50 extraneous variables, including the mother's age, weight, height, blood type, upper arm circumference, religion, education, income, and so on.[24] After applying complex statistical methods of adjustment, they concluded that birthweight varies with smoking even when these extraneous factors are held constant. This says that there quite likely is a link between X = smoking and Y = birthweight as shown in Figure 7.4.1, although several other variables also affect birthweight. The point is that the presence of confounding doesn't mean that a link does not exist between X and Y, only that it is tangled up with other effects, so we have to be cautious when interpreting the findings of an observational study.

In another study of pregnant women, researchers measured various quantities related to the functioning of the placenta.[25] They found that, compared to nonsmokers, women who smoked had more abnormalities of the placenta, and that their infants had very much higher blood levels of cotinine, a substance derived from nicotine. They also found evidence that, in the women who smoked, the circulation of blood in the placenta was notably improved by abstaining from smoking for 3 hours.

A third study used a matched design to try to isolate the effect of smoking behavior. The investigators identified 159 women who had smoked during one pregnancy but quit smoking before the next pregnancy.[26] These women were individually matched with 159 women who smoked during two consecutive pregnancies; pairs were matched with respect to the birthweight of the first child, amount of smoking during the first pregnancy, and several other factors. Thus, the members of a pair were believed to have identical "reproductive potential." The researchers then considered the birthweight of the second child; they found that the women who had quit smoking gave birth to infants who weighed more than the infants of their matched controls who continued to smoke. Of course, we cannot rule out the possibility that the women who quit smoking also quit other harmful habits, such as drinking too much alcohol, and that the increased birthweight was not really caused by giving up smoking. ■

Example 7.4.6 shows that observational studies can provide information about causality but must be interpreted cautiously. Researchers generally agree that a causal interpretation of an observed association requires extra support—for instance, that the association be observed consistently in observational studies conducted under various conditions and taking various extraneous factors into account, and also, ideally, that the causal link be supported by experimental evidence. We do not mean to say that an observed association *cannot* be causally interpreted, but only that such interpretation requires particular caution.

SPURIOUS ASSOCIATION

Example 7.4.7 **Ultrasound** It is quite common for a physician to use ultrasound examination of the fetus of a pregnant woman. However, when ultrasound technology was first used, there were concerns that the procedure might be harmful to the baby. An early study seemed to bear this out: On average, babies exposed to ultrasound in the womb were lighter at birth than were babies not exposed to ultrasound.[27] Later, a study was done in which some women were randomly chosen to have ultrasounds and others were not given ultrasounds. This study found no difference in birthweight between the two groups.[28] It seems that the reason a difference appeared in the first study was that ultrasound was being used mostly for women who were experiencing problem pregnancies. The complications with the pregnancy were leading to low birthweight, not the use of ultrasound. ■

Figure 7.4.2 gives a schematic representation of the situation in Example 7.4.7. Changes in X (having an ultrasound examination) are associated with changes in Y (lower birthweight). However, X and Y are both dependent on a third variable Z (whether or not there are problems with the pregnancy), which is the variable that is driving the relationship. Changes in X and changes in Y are a common response to the third variable Z. We say that the association between X and Y is **spurious**: When we control for the "lurking variable" Z, the link between X and Y disappears. In the case of Example 7.4.7, it was not having an ultrasound that influenced birthweight; what mattered was whether or not there were problems with the pregnancy.

Figure 7.4.2 Schematic representation of spurious association

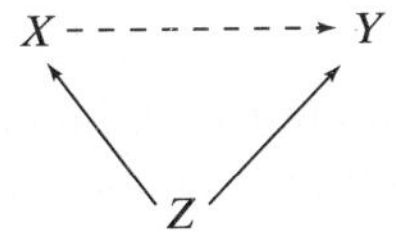

The association between X and Y is spurious; controlling for the lurking variable Z eliminates the X–Y link.

MORE ON EXPERIMENTS

An experiment is a study in which the researcher intervenes and imposes treatment conditions. The following is a simple example.

Example 7.4.8

Headache Pain Suppose a researcher gives ibuprofen to some people who have headaches and aspirin to others and then measures how long it takes for each person's headache to disappear. In this case, there are two treatments: ibuprofen and aspirin. By assigning people to treatment groups—ibuprofen and aspirin—the researcher is conducting an experiment. ■

When we are discussing an experiment, we refer to the units to which the treatments are assigned as **experimental units.** In an agricultural experiment, an experimental unit might be a plot of land. In general, an experimental unit is the smallest unit to which a treatment is applied in an experiment. Thus, in Example 7.4.8 the experimental units are individual people, since treatment is assigned on a person-by-person basis.

If treatments are assigned at random, for example, by tossing a coin and letting heads mean the person gets ibuprofen, while tails means the person gets aspirin, then the experiment is a *randomized* experiment. Sometimes an experiment is conducted in which one group is given a treatment and a second—the control group—is given nothing. For example, one could investigate the effectiveness of ibuprofen in treating headache pain by giving it to some people, while giving no painkiller to others. In contrast, the experiment in which some people are given ibuprofen and others are given aspirin is said to have an "active" control—the aspirin group.

RANDOMIZATION DISTRIBUTIONS

In Section 5.2 we developed the concept of a sampling distribution for the sample mean, $\overline{Y}$, by considering how $\overline{Y}$ varies from one random sample to another. Strictly speaking, this provides the foundation for inference when analyzing an observational study, but not when the data arise from an experiment—in which treatments are assigned to experimental units, rather than a random sample being taken from a population. However, the concepts of Section 5.2 can be extended in a natural way to develop the **randomization distribution** of $\overline{Y}$, which is the distribution that $\overline{Y}$ takes on under all possible random assignments within an experiment. Randomization distributions then form the foundation for inference for experiments.

ONLY STATISTICAL?

The term "statistical" is sometimes used—or, rather, misused—as an epithet. For instance, some people say that the evidence linking dietary cholesterol and heart disease is "only statistical." What they really mean is "only observational." Statistical evidence can be very strong indeed, if it flows from a randomized experiment rather than an observational study. As we have seen in the preceding examples, statistical evidence from an observational study must be interpreted with great care, because of potential distortions caused by extraneous variables.

Exercises 7.4.1–7.4.10

7.4.1 In 2005, 5.3% of the deaths in the United States were caused by chronic lower respiratory diseases (e.g., asthma and emphysema). In Arizona, 6.2% of deaths were due to chronic lower respiratory diseases.[29] Does this mean that living in Arizona exacerbates respiratory problems? If not, how can we explain the Arizona rate being above the national rate?

7.4.2 It has been hypothesized that silicone breast implants cause illness. In one study it was found that women with implants were more likely to smoke, to be heavy drinkers, to use hair dye, and to have had an abortion than were women in a comparison group who did not have implants.[30] Use the language of statistics to explain why this study casts doubt on the claim that implants cause illness.

7.4.3 Consider the setting of Exercise 7.4.2.

(a) What is the explanatory variable?

(b) What is the response variable?

(c) What are the observational units?

7.4.4 In a study of 1,040 subjects, researchers found that the prevalence of coronary heart disease increased as the number of cups of coffee consumed per day increased.[31]

(a) What is the explanatory variable?

(b) What is the response variable?

(c) What are the observational units?

7.4.5 For an early study of the relationship between diet and heart disease, the investigator obtained data on heart disease mortality in various countries and on national average dietary compositions in the same countries. The accompanying graph shows, for six countries, the 1948–1949 death rate from degenerative heart disease (among men ages 55–59 years) plotted against the amount of fat in the diet.[32]

In what ways might this graph be misleading? Which extraneous variables might be relevant here? Discuss.

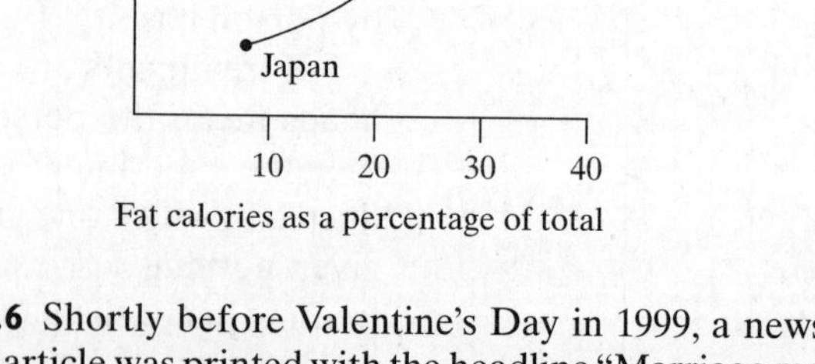

7.4.6 Shortly before Valentine's Day in 1999, a newspaper article was printed with the headline "Marriage makes for healthier, longer life, studies show." The headline was based on studies that showed that married persons live longer and have lower rates of cancer, heart disease, and stroke than do those who never marry.[33] Use the language of statistics to discuss the headline. Use a schematic diagram similar to Figure 7.4.1 or Figure 7.4.2 to support your explanation of the situation.

7.4.7 In June 2009, the *New York Times* published an article entitled "Alcohol's Good for You? Some Scientists Doubt It." The author wrote, "Study after study suggests that alcohol in moderation may promote heart health and even ward off diabetes and dementia. The evidence is so plentiful that some experts consider moderate drinking—about one drink a day for women, about two for men—a central component of a healthy lifestyle." Later in the article, the author wrote, "For some scientists, the question will not go away. No study, these critics say, has ever proved a causal relationship between moderate drinking and lower risk of death." Explain using the language of statistics and a schematic diagram similar to Figure 7.4.1 or Figure 7.4.2 why the critics say no study has ever proved a causal relationship.

7.4.8 In a study of the relationship between birthweight and race, birth records of babies born in Illinois were examined. The researchers found that the percentage of low birthweight babies among babies born to U.S.-born white women was much lower than the percentage of low birthweight babies among babies born to U.S.-born black women. This suggests that race plays an important role in determining the chance that a baby will have a low birthweight. However, the percentage of low birthweight babies among babies born to African-born black women was roughly equal to the percentage among babies born to U.S.-born white women.[34] Use the language of statistics to discuss what these data say about the relationships between low birthweight, race, and mother's birthplace. Use a schematic diagram similar to Figure 7.4.1 or Figure 7.4.2 to support your explanation.

7.4.9 Does the release of a Harry Potter book lead children to spend more time reading and thus reduce the number of accidents they have? Doctors in England compared the number of emergency room visits due to musculoskeletal injuries to children ages 7 to 15 during two types of weekends: (1) following the release dates of two books in the Harry Potter series and (2) during 24 "control" weekends, for one hospital. The following table shows the data, with the "Harry Potter weekends" in italics.[35]

Weekend	Injuries	Weekend	Injuries
6/7/03	63	7/10/04	57
6/14/03	77	7/17/04	66
6/21/03	*36**	7/24/04	62
6/28/03	63	6/4/05	51
7/5/03	75	6/11/05	83
7/12/03	71	6/18/05	60
7/19/03	60	6/25/05	66
7/26/03	52	7/2/05	74
6/5/04	78	7/9/05	75
6/12/04	84	*7/16/05*	*37**
6/19/04	70	7/23/05	46
6/26/04	75	7/30/05	68
7/3/04	81	8/6/05	60

(a) Given the nature of the data, can we make an inference about the release of Harry Potter books *causing* a change in accidents? Why or why not?

(b) The average for the Harry Potter weekends is 36.5, with a standard deviation of 0.7. The corresponding numbers for the other (control) weekends are 67.4 and 10.4. Use a *t* test to investigate the claim that the small number of injuries during Harry Potter weekends is consistent with chance variation. Use $\alpha = 0.01$. [*Note*: Formula (6.7.1) yields 23.9 degrees of freedom for these data.]

7.4.10 Suppose an observational study that compares two groups (group 1 and group 2) has sample means $\bar{y}_1 = 115$ and $\bar{y}_2 = 108$. Further suppose that we reject H_0 in favor of H_A when doing a hypothesis test. Which of the following statements are true?

(i) We have evidence that the difference in sample means is unlikely to arise by chance.

(ii) The difference between the two sample means is important.

(iii) The data show that being in group 1 *causes* the mean to go up.

7.5 One-Tailed *t* Tests

The *t* test described in the preceding sections is called a **two-tailed *t* test** or a **two-sided *t* test** because the null hypothesis is rejected if t_s falls in either tail of the Student's *t* distribution and the *P*-value of the data is a two-tailed area under Student's *t* curve. A two-tailed *t* test is used to test the null hypothesis

$$H_0\colon \mu_1 = \mu_2$$

against the alternative hypothesis

$$H_A\colon \mu_1 \neq \mu_2$$

This alternative H_A is called a **nondirectional alternative.**

DIRECTIONAL ALTERNATIVE HYPOTHESES

In some studies it is apparent from the beginning—*before* the data are collected—that there is only one reasonable direction of deviation from H_0. In such situations it is appropriate to formulate a directional alternative hypothesis. The following is a directional alternative:

$$H_A\text{: } \mu_1 < \mu_2$$

Another directional alternative is

$$H_A\text{: } \mu_1 > \mu_2$$

The following example illustrates a situation in which a directional alternative is appropriate.

Example 7.5.1 **Niacin Supplementation** Consider a feeding experiment with lambs. The observation Y will be weight gain in a 2-week trial. Ten animals will receive diet 1, and 10 animals will receive diet 2, where

$$\text{Diet 1} = \text{Standard ration} + \text{Niacin}$$

$$\text{Diet 2} = \text{Standard ration}$$

On biological grounds it is expected that niacin may increase weight gain; there is no reason to suspect that it could possibly decrease weight gain. An appropriate formulation would be

H_0: Niacin is not effective in increasing weight gain ($\mu_1 = \mu_2$).

H_A: Niacin is effective in increasing weight gain ($\mu_1 > \mu_2$). ■

Note: If H_A is directional, then some people would rewrite H_0 to include the "opposite direction." For example, if H_A is H_A: $\mu_1 > \mu_2$, then we could write H_0 as H_0: $\mu_1 \leq \mu_2$. Thus, the null hypothesis is stating that the mean of population 1 *is* not greater than the mean of population 2, whereas the alternative hypothesis asserts that the mean of population 1 *is* greater than the mean of population 2. Between these two hypotheses, all possibilities are covered.

THE ONE-TAILED TEST PROCEDURE

When the alternative hypothesis is directional, the t test procedure must be modified. The modified procedure is called a **one-tailed t test** and is carried out in two steps as follows:

Step 1. Check directionality—see if the data deviate from H_0 in the direction specified by H_A:

(a) If not, the P-value is greater than 0.50.

(b) If so, proceed to step 2.

Step 2. The P-value of the data is the *one-tailed* area beyond t_s.

To conclude the test, one can make a decision at a prespecified significance level α: H_0 is rejected if P-value $\leq \alpha$.

The rationale of the two-step procedure is that the P-value measures deviation from H_0 in the direction specified by H_A. The one-tailed P-value is illustrated in

Figure 7.5.1 for two cases in which the data deviate from H_0 in the direction specified by H_A. Figure 7.5.2 illustrates the P-value for (a) a case in which the data are consistent with H_A: $\mu_1 > \mu_2$ and (b) a case in which the data are inconsistent with H_A: $\mu_1 > \mu_2$. The two-step testing procedure is demonstrated in Example 7.5.3.

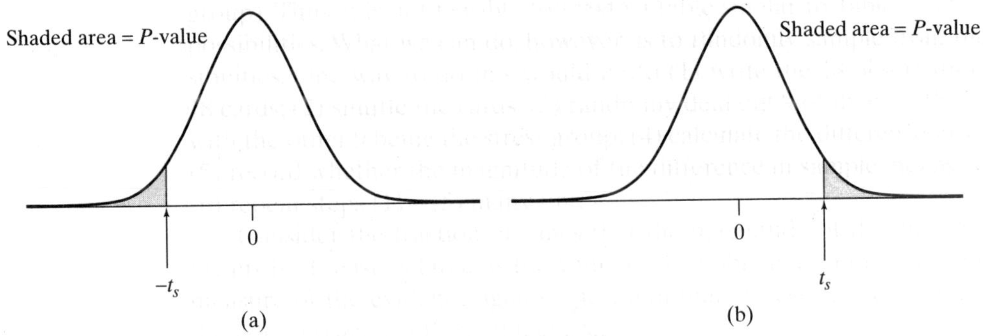

Figure 7.5.1 One-tailed P-value for a t test, (a) if the alternative is H_A: $\mu_1 < \mu_2$ and t_s is negative; (b) if the alternative is H_A: $\mu_1 > \mu_2$ and t_s is positive

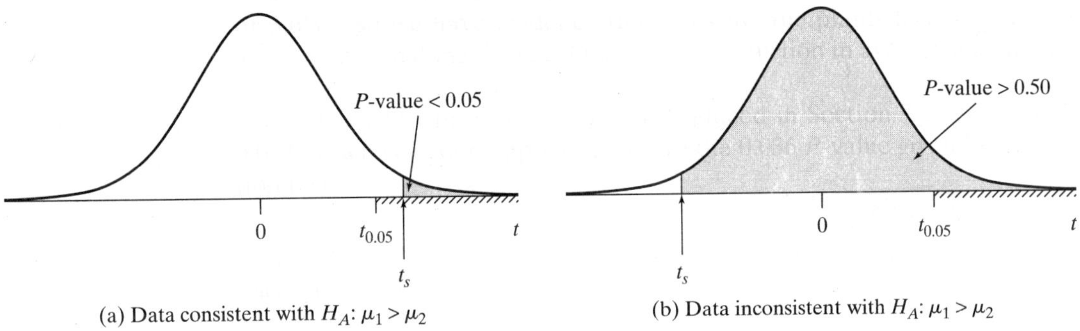

Figure 7.5.2 One-tailed P-value for a t test, (a) in which the data are consistent with H_A: $\mu_1 > \mu_2$; (b) in which the data are inconsistent with H_A: $\mu_1 > \mu_2$

Example 7.5.2 **Niacin Supplementation** Consider the lamb feeding experiment of Example 7.5.1. The alternative hypothesis is

$$H_A: \mu_1 > \mu_2$$

We will claim significant evidence for H_A if $\bar{Y}_1$ is sufficiently greater than $\bar{Y}_2$. Suppose formula (6.7.1) yields df = 18. The critical values from Table 4 are reproduced in Table 7.5.1.

Table 7.5.1 Critical values with df = 18

Tail area	0.20	0.10	0.05	0.04	0.03	0.025	0.02	0.01	0.005	0.0005
Critical value	0.862	1.330	1.734	1.855	2.007	2.101	2.214	2.552	2.878	3.922

To illustrate the one-tailed test procedure, suppose that we have[36]

$$SE_{(\bar{Y}_1 - \bar{Y}_2)} = 2.2 \text{ lb}$$

and that we choose $\alpha = 0.05$. Let us consider various possibilities for the two sample means.

(a) Suppose the data give $\bar{y}_1 = 10$ lb and $\bar{y}_2 = 13$ lb. This deviation from H_0 is opposite to the assertion of H_A: We have $\bar{y}_1 < \bar{y}_2$, but H_A asserts that $\mu_1 > \mu_2$. Consequently, P-value > 0.50, so we would not find significant evidence for H_A at any significance level. (We would never use an α greater than 0.50.) We conclude that the data provide no evidence that niacin is effective in increasing weight gain.

(b) Suppose the data give $\bar{y}_1 = 14$ lb and $\bar{y}_2 = 10$ lb. This deviation from H_0 is in the direction of H_A (because $\bar{y}_1 > \bar{y}_2$), so we proceed to step 2. The value of t_s is

$$t_s = \frac{(14 - 10) - 0}{2.2} = 1.82$$

The (one-tailed) P-value for the test is the probability of getting a t statistic, with 18 degrees of freedom, that is as large or larger than 1.82. This upper tail probability (found with a computer*) is 0.043, as shown in Figure 7.5.3.

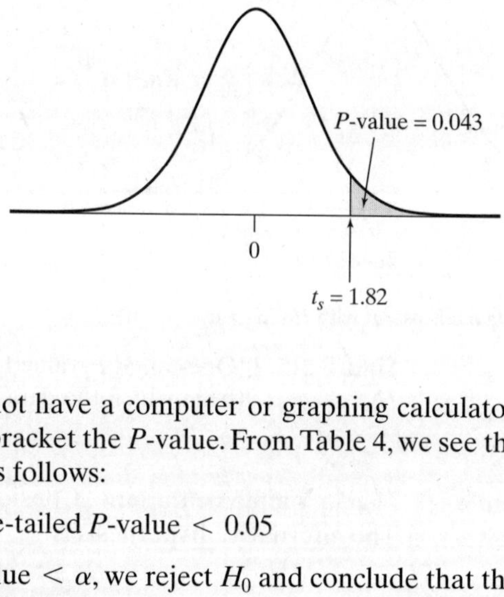

Figure 7.5.3 One-tailed P-value for the t test in Example 7.5.2

If we did not have a computer or graphing calculator available, we could use Table 4 to bracket the P-value. From Table 4, we see that the P-value would be bracketed as follows:

$$0.04 < \text{one-tailed } P\text{-value} < 0.05$$

Since P-value $< \alpha$, we reject H_0 and conclude that there is some evidence that niacin is effective.

(c) Suppose the data give $\bar{y}_1 = 11$ lb and $\bar{y}_2 = 10$ lb. Then, proceeding as in part (b), we compute the test statistic as $t_s = 0.45$. The P-value is 0.329.

If we did not have a computer or graphing calculator available, we could use Table 4 to bracket the P-value as

$$P\text{-value} > 0.20$$

Since P-value $> \alpha$, we do not find significant evidence for H_A; we conclude that there is insufficient evidence to claim that niacin is effective. Thus, although these data deviate from H_0 in the direction of H_A, the amount of deviation is not great enough to justify significant evidence for H_A. ■

*For example, in the program R one would enter the command t.test(Diet1, Diet2, alternative = "greater").

Notice that what distinguishes a one-tailed from a two-tailed t test is the way in which the P-value is determined, but not the directionality or nondirectionality of the conclusion. If we find significant evidence for H_A, our conclusion may be considered directional even if our H_A is nondirectional.* (For instance, in Example 7.2.4 we concluded that toluene increases NE concentration.)

DIRECTIONAL VERSUS NONDIRECTIONAL ALTERNATIVES

The same data will give a different P-value depending on whether the alternative hypothesis is directional or nondirectional. Indeed, if the data deviate from H_0 in the direction specified by H_A, the P-value for a directional alternative hypothesis will be 1/2 of the P-value for the test that uses a nondirectional alternative. It can happen that the same data will provide significant evidence for H_A using the one-tailed procedure but not using the two-tailed procedure, as Example 7.5.3 shows.

Example 7.5.3

Niacin Supplementation Consider part (b) of Example 7.5.2. In that example we chose $\alpha = 0.05$ and tested

$$H_0: \mu_1 = \mu_2$$

against the directional alternative hypothesis

$$H_A: \mu_1 > \mu_2$$

With $\bar{y}_1 = 14$ lb and $\bar{y}_2 = 10$ lb, the test statistic was $t_s = 1.82$ and the P-value was 0.043, as indicated in Figure 7.5.3. Our conclusion was to claim there is significant evidence for H_A.

However, suppose we had wished to test

$$H_0: \mu_1 = \mu_2$$

against the nondirectional alternative hypothesis

$$H_A: \mu_1 \neq \mu_2$$

With the same data of $\bar{y}_1 = 14$ lb and $\bar{y}_2 = 10$ lb, the test statistic is still $t_s = 1.82$. The P-value, however, is 0.086, as shown in Figure 7.5.4. Thus, P-value $> \alpha$ and we do not reject H_0.

Hence, the one-tailed procedure finds significant evidence for H_A, but the two-tailed procedure does not. In this sense, it is "easier" to find evidence that significantly supports H_A with the one-tailed procedure than with the two-tailed procedure. ■

Figure 7.5.4 Two-tailed P-value for the t test in Example 7.5.3

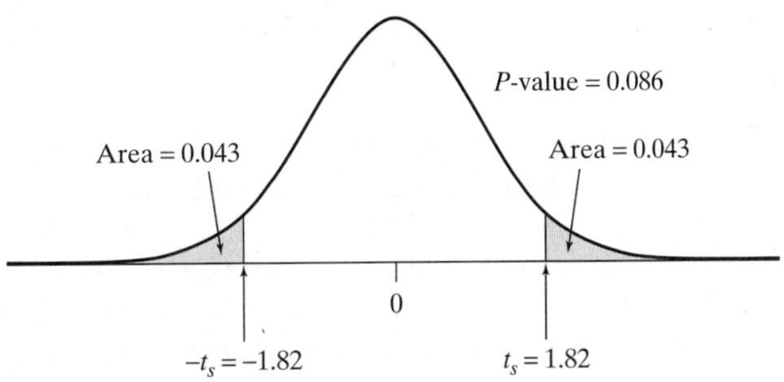

*Some authors prefer not to draw a directional conclusion if H_A is nondirectional.

Why is the two-tailed P-value cut in half when the alternative hypothesis is directional? In Example 7.5.3, the researcher would conclude by saying, "The data suggest that niacin increases weight gain. But if niacin has no effect, then the kind of data I got in my experiment—having two sample means that differ by 1.82 SEs or more—would happen fairly often (P-value $= 0.086$). Sometimes the niacin diet would come out on top; sometimes the standard diet would come out on top. I cannot find significant evidence for H_A on the basis of what I have seen in these data." In Example 7.5.2(b), the researcher would conclude by saying, "*Before the experiment was run*, I suspected that niacin increases weight gain. The data provide evidence in support of this theory. If niacin has no effect, then the kind of data I got in my experiment—having the niacin diet sample mean exceed the standard diet that differ by 1.82 SEs or more—would rarely happen (P-value 0.043). (*Before the experiment was run I dismissed the possibility that the niacin diet mean could be less than the standard diet mean.*) Thus, I can claim my evidence significantly supports H_A." The researcher in Example 7.5.2(b) is using *two* sources of information to claim the significance of evidence for H_A: (1) what the data have to say (as measured by the tail area) and (2) previous expectations (which allow the researcher to ignore the lower tail area—the 0.043 area under the curve below -1.82 in Figure 7.5.4).

Note that the modification in procedure, when going from a two-tailed to a one-tailed test, preserves the interpretation of significance level α as given in Section 7.3, that is,

$$\alpha = \Pr\{\text{reject } H_0\} \text{ if } H_0 \text{ is true}$$

For instance, consider the case $\alpha = 0.05$. Figure 7.5.5 shows that the total shaded area—the probability of rejecting H_0—is equal to 0.05 in both a two-tailed test and a one-tailed test. This means that, if a great many investigators were to test a true H_0, then 5% of them would find significant evidence for H_A and commit a Type I error; this statement is true whether the alternative H_A is directional or nondirectional.

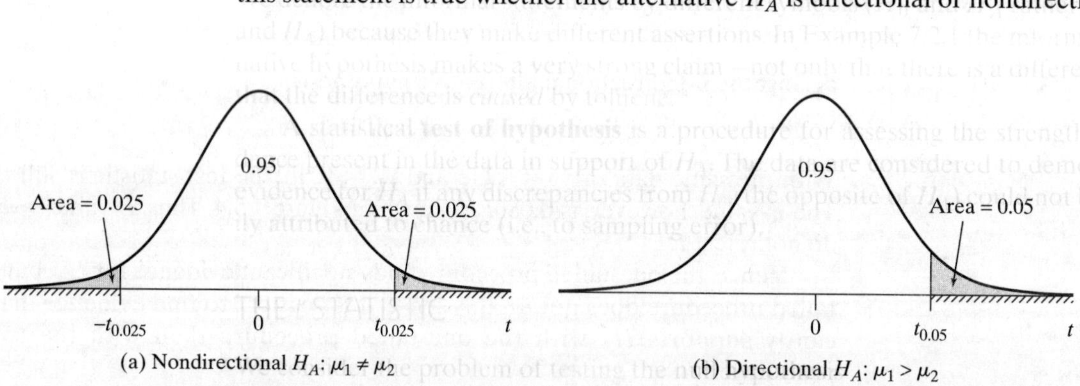

Figure 7.5.5 Two-tailed and one-tailed t test with $\alpha = 0.05$. The data provide significant evidence for H_A if t_s falls in the hatched region of the t-axis

The crucial point in justification of the modified procedure for testing against a directional H_A is that *if* the direction of deviation of the data from H_0 is *not* as specified by H_A, then we will not claim that the evidence significantly supports H_A. For example, in the niacin experiment of Example 7.5.1, (1) if the lambs given niacin gained *less* weight than those given only the standard ration, we might simply conclude that the data do not indicate a niacin effect, or (2) if the niacin group had *substantially* smaller weight gain so that the test statistic t_s was very far in the wrong tail of the t distribution, we might look for methodological errors in the experiment—for example, mistakes in lab technique or in recording the data, nonrandom allocation of the lambs to the two groups, and so on. We would not claim significant evidence for H_A.

A one-tailed t test is especially natural when only one direction of deviation from H_0 is believed to be plausible. However, one-tailed tests are also used in situations where deviation in both directions is possible, but only one direction is of interest. For instance, in the niacin experiment of Example 7.5.2, it is not necessary that the experimenter believe that it is *impossible* for niacin to reduce weight gain rather than increase it. Deviations in the wrong direction (less weight gain on niacin) would not lead to claiming there is significant evidence for H_A, and thus we would not make claims about the effect of niacin; this is the essential feature that distinguishes a directional from a nondirectional formulation.

CHOOSING THE FORM OF H_A

When is it legitimate to use a directional H_A, and therefore perform a one-tailed test? The answer to this question is linked to the directionality check—step 1 of the two-step test procedure given previously. Clearly such a check makes sense only if H_A was formulated before the data were inspected. (If we were to formulate a directional H_A that was "inspired" by the data, then of course the data would always deviate from H_0 in the "right" direction and the test procedure would always proceed to step 2.) This is the rationale for the following rule.

Rule for Directional Alternatives

It is legitimate to use a directional alternative H_A *only if* H_A is formulated before seeing the data and there is no scientific interest in results that deviate in a manner opposite to that specified by H_A.

In research, investigators often get more pleasure from finding significant evidence for an alternative hypothesis than not finding evidence. In fact, research reports often contain phrases such as "we are unable to find significant evidence for the alternative hypothesis" or "the results failed to reach statistical significance." Under these circumstances, one might wonder what the consequences would be if researchers succumbed to the natural temptation to ignore the preceding rule for using directional alternatives. After all, very often one can think of a rationale for an effect *ex post facto*—that is, after the effect has been observed. A return to the imaginary experiment on plants' musical tastes will illustrate this situation.

Example 7.5.4 **Music and Marigolds** Recall the imaginary experiment of Example 7.3.2, in which investigators measure the heights of marigolds exposed to Bach or Mozart. Suppose, as before, that the null hypothesis is true, that df $= 60$, and that the investigators all perform t tests at $\alpha = 0.05$. Now suppose in addition that all of the investigators violate the rule for use of directional alternatives, and that they formulate H_A after seeing the data. Half of the investigators would obtain data for which $\bar{y}_1 > \bar{y}_2$, and they would formulate the alternative

$$H_A\colon \mu_1 > \mu_2 \text{ (plants prefer Bach)}$$

The other half would obtain data for which $\bar{y}_1 < \bar{y}_2$, and they would formulate the alternative

$$H_A\colon \mu_1 < \mu_2 \text{ (plants prefer Mozart)}$$

Now envision what would happen. Since the investigators are using directional alternatives, they will all compute *P*-values using only one tail of the distribution. We would expect them to have the following experiences:

90% of them would get a t_s in the middle 90% of the distribution and would not find significant evidence for H_A.

5% of them would get a t_s in the top 5% of the distribution and would conclude that the plants prefer Bach.

5% of them would get a t_s in the bottom 5% of the distribution and would conclude that the plants prefer Mozart.

Thus, a total of 10% of the investigators would claim there is significant evidence for H_A. Of course each investigator individually never realizes that the overall percentage of Type I errors is 10% rather than 5%. And the conclusions that plants prefer Bach or Mozart could be supported by *ex post facto* rationales that would be limited only by the imagination of the investigators. ■

As Example 7.5.4 illustrates, a researcher who uses a directional alternative when it is not justified pays the price of a doubled risk of Type I error. Moreover, those who read the researcher's report will not be aware of this doubling of risk, which is why some scientists advocate never using a directional alternative. If a nondirectional test suggests an effect but the data are not statistically significant, a follow-up study could be conducted in which new data are collected and a directional alternative is tested.

Exercises 7.5.1–7.5.16

7.5.1 For each of the following data sets, use Table 4 to bracket the one-tailed *P*-value of the data as analyzed by the *t* test, assuming that the alternative hypothesis is $H_A: \mu_1 > \mu_2$.

(a)

	Sample 1	Sample 2
n	10	10
$\bar{y}$	10.8	10.5
$SE_{(\bar{Y}_1-\bar{Y}_2)} = 0.23$ with df $= 18$		

(b)

	Sample 1	Sample 2
n	100	100
$\bar{y}$	750	730
$SE_{(\bar{Y}_1-\bar{Y}_2)} = 11$ with df $= 180$		

7.5.2 For each of the following data sets, use Table 4 to bracket the one-tailed *P*-value of the data as analyzed by the *t* test, assuming that the alternative hypothesis is $H_A: \mu_1 > \mu_2$.

(a)

	Sample 1	Sample 2
n	10	10
$\bar{y}$	3.24	3.00
$SE_{(\bar{Y}_1-\bar{Y}_2)} = 0.61$ with df $= 17$		

(b)

	Sample 1	Sample 2
n	6	5
$\bar{y}$	560	500
$SE_{(\bar{Y}_1-\bar{Y}_2)} = 45$ with df $= 8$		

(c)

	Sample 1	Sample 2
n	20	20
$\bar{y}$	73	79
$SE_{(\bar{Y}_1-\bar{Y}_2)} = 2.8$ with df $= 35$		

7.5.3 For each of the following situations, suppose $H_0: \mu_1 = \mu_2$ is being tested against $H_A: \mu_1 > \mu_2$. State whether or not there is significant evidence for H_A.

(a) $t_s = 3.75$ with 19 degrees of freedom, $\alpha = 0.01$.

(b) $t_s = 2.6$ with 5 degrees of freedom, $\alpha = 0.10$.

(c) $t_s = 2.1$ with 7 degrees of freedom, $\alpha = 0.05$.

(d) $t_s = 1.8$ with 7 degrees of freedom, $\alpha = 0.05$.

7.5.4 For each of the following situations, suppose $H_0: \mu_1 = \mu_2$ is being tested against $H_A: \mu_1 < \mu_2$. State whether or not there is significant evidence for H_A.

(a) $t_s = -1.6$ with 23 degrees of freedom, $\alpha = 0.05$.

(b) $t_s = -2.3$ with 5 degrees of freedom, $\alpha = 0.10$.

(c) $t_s = 0.4$ with 16 degrees of freedom, $\alpha = 0.10$.

(d) $t_s = -2.8$ with 27 degrees of freedom, $\alpha = 0.01$.

7.5.5 Ecological researchers measured the concentration of red cells in the blood of 27 field-caught lizards (*Sceloporus occidetitalis*). In addition, they examined each lizard for infection by the malarial parasite *Plasmodium*. The red cell counts ($10^{-3} \times$ cells per mm^3) were as reported in the table.[37]

	Infected animals	Noninfected animals
n	12	15
$\bar{y}$	972.1	843.4
s	245.1	251.2

One might expect that malaria would reduce the red cell count, and in fact previous research with another lizard species had shown such an effect. Do the data support this expectation? Assume that the data are normally distributed. Test the null hypothesis of no difference against the alternative that the infected population has a lower red cell count. Use a t test at

(a) $\alpha = 0.05$ (b) $\alpha = 0.10$

[*Note*: Formula (6.7.1) yields 24 df.]

7.5.6 A study was undertaken to compare the respiratory responses of hypnotized and nonhypnotized subjects to certain instructions. The 16 male volunteers were allocated at random to an experimental group to be hypnotized or to a control group. Baseline measurements were taken at the start of the experiment. In analyzing the data, the researchers noticed that the baseline breathing patterns of the two groups were different; this was surprising, since all the subjects had been treated the same up to that time. One explanation proposed for this unexpected difference was that the experimental group were more excited in anticipation of the experience of being hypnotized. The accompanying table presents a summary of the baseline measurements of total ventilation (liters of air per minute per square meter of body area). Parallel dotplots of the data are given in the following graph.[38] [*Note*: Formula (6.7.1) yields 14 df.]

	Experimental	Control
	5.32	4.50
	5.60	4.78
	5.74	4.79
	6.06	4.86
	6.32	5.41
	6.34	5.70
	6.79	6.08
	7.18	6.21
n	8	8
$\bar{y}$	6.169	5.291
s	0.621	0.652

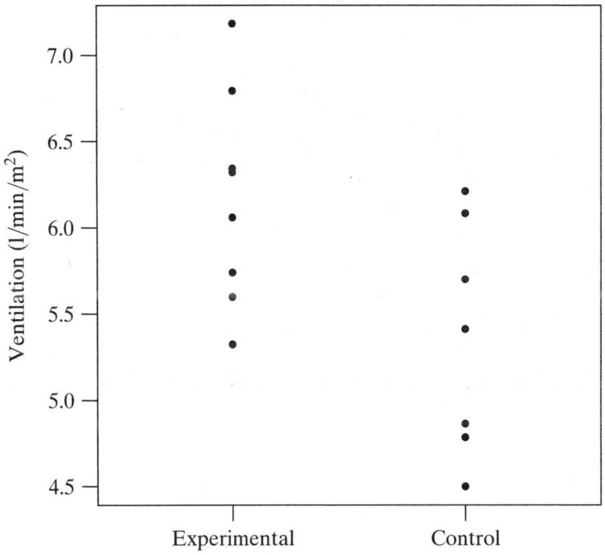

(a) Use a t test to test the hypothesis of no difference against a nondirectional alternative. Let $\alpha = 0.05$.

(b) Use a t test to test the hypothesis of no difference against the alternative that the experimental conditions produce a larger mean than the control conditions. Let $\alpha = 0.05$.

(c) Which of the two tests, that of part (a) or part (b), is more appropriate? Explain.

7.5.7 In a study of lettuce growth, 10 seedlings were randomly allocated to be grown in either standard nutrient solution or in a solution containing extra nitrogen. After 22 days of growth, the plants were harvested and weighed, with the results given in the table.[39] Are the data sufficient to conclude that the extra nitrogen enhances plant growth under these conditions? Use a t test at $\alpha = 0.10$ against a directional alternative. (Assume that the data are normally distributed.)

Nutrient solution	Leaf dry weight (gm)		
	n	Mean	SD
Standard	5	3.62	0.54
Extra nitrogen	5	4.17	0.67

(a) What is the value of the t test statistic for comparing the means?

(b) The P-value for the t test is 0.096. If $\alpha = 0.10$, what is your conclusion regarding H_0?

7.5.8 Research has shown that for mammals giving birth to a son versus a daughter places a greater strain on mothers. Does this affect the health of their next child? A study compared the birthweights of humans born after a male versus after a female. Summary statistics for the sample of size 76 are given in the following table; the data were normally distributed.[40] Consider a t test, with a directional alternative, to investigate the research hypothesis that birthweight is lower when the elder sibling is male.

Sex of elder sibling	Birthweight (kg)		
	n	Mean	SD
Male	33	3.32	0.62
Female	43	3.63	0.63

(a) Write the null hypothesis and the alternative hypothesis in symbols.

(b) Here is computer output for the t test. Explain what the P-value means in the context of this study.

```
t = -2.145, df = 69.54, p-value = 0.0177
alternative hypothesis: true difference in
means is less than 0
```

(c) If $\alpha = 0.05$, what is your conclusion regarding H_0?

7.5.9 An entomologist conducted an experiment to see if wounding a tomato plant would induce changes that improve its defense against insect attack. She grew larvae of the tobacco hornworm (*Manduca sexta*) on wounded plants or control plants. The accompanying table shows the weights (mg) of the larvae after 7 days of growth.[41] (Assume that the data are normally distributed.) How strongly do the data support the researcher's expectation? Use a t test at the 5% significance level. Let H_A be that wounding the plant tends to diminish larval growth. [*Note:* Formula (6.7.1) yields 31.8 df.]

	Wounded	Control
n	16	18
$\bar{y}$	28.66	37.96
s	9.02	11.14

7.5.10 A pain-killing drug was tested for efficacy in 50 women who were experiencing uterine cramping pain following childbirth. Twenty-five of the women were randomly allocated to receive the drug, and the remaining 25 received a placebo (inert substance). Capsules of drug or placebo were given before breakfast and again at noon. A pain relief score, based on hourly questioning throughout the day, was computed for each woman. The possible pain relief scores ranged from 0 (no relief) to 56 (complete relief for 8 hours). Summary results are shown in the table.[42] [*Note:* Formula (6.7.1) yields 47.2 df.]

Treatment	n	Pain relief score	
		Mean	SD
Drug	25	31.96	12.05
Placebo	25	25.32	13.78

(a) Test for evidence of efficacy using a t test. Use a directional alternative and $\alpha = 0.05$.

(b) If the alternative hypothesis were nondirectional, how would the answer to part (a) change?

7.5.11 Postoperative ileus (POI) is a form of gastrointestinal dysfunction that commonly occurs after abdominal surgery and results in absent or delayed gastrointestinal motility. Does rocking in a chair after abdominal surgery reduce postoperative ileus (POI) duration? Sixty-six postoperative abdominal surgery patients were randomly divided into two groups. The experimental group ($n = 34$) received standard care plus the use of a rocking chair while the control group ($n = 32$) received only standard care. For each patient, the postoperative time until first flatus (days) (an indication that the POI has ended) was measured. The results are tabulated below.[43]

	Time until first flatus (days)		
	n	Mean (days)	SD
Rocking	34	3.16	0.86
Control	32	3.88	0.80

Here is computer output for a t test.

```
t = -3.524, df = 63.99, p-value = 0.000396
alternative hypothesis: true difference in
means is less than 0
```

(a) State the null and alternative hypotheses in context.

(b) Is there evidence at the level of $\alpha = 0.05$ that use of the rocking chair reduces POI duration (i.e., the time until first flatus)?

(c) Although the researchers hypothesized that the use of a rocking chair could reduce POI duration, it is not unreasonable to hypothesize that the use of a rocking chair could increase POI duration. Based on this possibility, discuss the appropriateness of using a directional versus nondirectional test. (*Hint:* Consider what medical recommendations might be made based on this research.)

7.5.12 In Example 7.2.6 we considered testing $H_0: \mu_1 = \mu_2$ against the nondirectional alternative hypothesis $H_A: \mu_1 \neq \mu_2$ and found that the P-value could be bracketed as $0.06 < P\text{-value} < 0.10$. Recall that the sample mean for the group 1 (the control group) was 15.9, which was less than the sample mean of 11.0 for group 2 (the group treated with Ancymidol). However, Ancymidol is considered to be a growth inhibitor, which means that one would expect the control group to have a larger mean than the treatment group if ancy has any effect on the type of plant being studied (in this case, the Wisconsin Fast Plant). Suppose the researcher had expected ancy to retard growth—before conducting the experiment—and had conducted a test of H_0: $\mu_1 = \mu_2$ against the nondirectional alternative hypothesis H_A: $\mu_1 > \mu_2$, using $\alpha = 0.05$. What would be the bounds on the *P*-value? Would H_0 be rejected? Why or why not? What would be the conclusion of the experiment? (*Note:* This problem requires almost no calculation.)

7.5.13 (Computer exercise) An ecologist studied the habitat of a marine reef fish, the six bar wrasse (*Thalassoma hardwicke*), near an island in French Polynesia that is surrounded by a barrier reef. He examined 48 patch reef settlements at each of two distances from the reef crest: 250 meters from the crest and 800 meters from the crest. For each patch reef, he calculated the "settler density," which is the number of settlers (juvenile fish) per unit of settlement habitat. Before collecting the data, he hypothesized that the settler density might decrease as distance from the reef crest increased, since the way that waves break over the reef crest causes resources (i.e., food) to tend to decrease as distance from the reef crest increases. Here are the data[44]:

250 Meters			800 Meters		
0.318	0.758	0.318	0.941	0.289	0.399
0.637	0.372	0.524	0.279	0.392	0.955
0.196	0.637	1.404	1.021	0.725	0.531
0.624	1.560	0.000	0.108	1.318	0.252
0.909	0.207	1.061	0.738	0.612	1.179
0.295	0.685	0.590	0.907	0.637	0.442
0.594	0.000	0.363	0.503	0.181	0.291
0.442	1.303	1.567	0.637	0.941	0.579
1.220	0.898	1.577	1.498	0.265	0.252
1.303	1.157	0.312	0.866	0.979	0.373
0.187	0.970	0.758	0.588	0.909	0.000
1.560	0.624	0.505	0.606	0.283	0.463
0.849	1.592	0.909	0.490	0.337	1.248
2.411	1.019	0.362	0.163	0.813	2.010
1.705	0.829	0.329	0.277	0.000	1.213
1.019	0.884	0.909	0.293	0.544	0.808

For 250 meters, the sample mean is 0.818 and the sample SD is 0.514. For 800 meters, the sample mean is 0.628 and the sample SD is 0.413. Do these data provide statistically significant evidence, at the 0.10 level, to support the ecologist's theory? Investigate with an appropriate graph and test.

7.5.14 Improvement in quality of life was measured for a group of heart disease patients after 8 weeks in an exercise program. This experimental group was compared to a control group who were not in the exercise program. Quality of life was measured using a 21-item questionnaire, with scores of 1–5 on each item. The improvement data are as follows and are plotted below[45]:

	Experimental	Control
	4	6
	6	−1
	14	1
	2	1
	3	0
	7	5
	1	0
	19	1
	1	−3
	12	4
	−4	6
	14	−42
	4	1
	1	−17
	1	−4
	6	2
	0	13
	−2	−6
	0	1
	0	−4
	10	5
	−1	−6
n	22	22
$\bar{y}$	4.9	−1.7
s	6.1	10.7

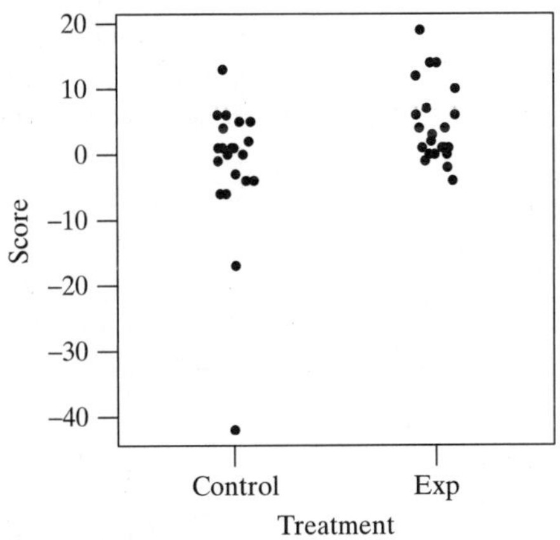

(a) What is the null hypothesis for a *t* test here?

(b) What do the dotplots suggest about H_0?

(c) Here is computer output for a t test. Explain what the P-value means in the context of this study.

```
t = 2.505, df = 33.23, p-value = 0.00866
alternative hypothesis: true difference in
means is greater than 0
```

(d) If $\alpha = 0.01$, what is your conclusion regarding H_0? State your conclusion in the context of the research.

(e) The computer output in part (c) is for the directional test. What is the P-value for the nondirectional test?

(f) If the test were nondirectional, and $\alpha = 0.01$, what conclusions would we make?

(g) Since the nondirectional P-value is not less than 0.01, we do not reject H_0. There is no significant evidence that the mean quality of life score differs for the two groups. There is no significant evidence that exercise affects quality of life.

7.5.15 Refer to the data in Exercise 7.5.14. There is one extremely low point in the control group. If that point is deleted, then the summary statistics change, as shown in the following table.

	Experimental	Control
n	22	21
$\bar{y}$	4.9	0.2
s	6.1	6.0

(a) What is the value of the t test statistic now?

(b) Does your conclusion about H_0 change when the outlier is deleted? (Note: The degrees of freedom for the test do not change very much.)

7.5.16 A researcher performed a t test of the null hypothesis that two means are equal. He stated that he chose an alternative hypothesis of $H_A: \mu_1 > \mu_2$ because he observed $\bar{y}_1 > \bar{y}_2$.

(a) Explain what is wrong with this procedure and why it is wrong.

(b) Suppose he reported $t = 1.97$ on 25 degrees of freedom and a P-value of 0.03. What is the proper P-value? Why?

7.6 More on Interpretation of Statistical Significance

Ideally, statistical analysis should aid the researcher by helping to clarify whatever message is contained in the data. For this purpose, it is not enough that the statistical calculations be correct; the results must also be correctly interpreted. In this section we explore some principles of interpretation that apply not only to the t test, but also to other statistical tests to be discussed later.

SIGNIFICANT DIFFERENCE VERSUS IMPORTANT DIFFERENCE

The term *significant* is often used in describing the results of a statistical analysis. For example, if an experiment to compare a drug against a placebo gave data with a very small P-value, then the conclusion might be stated as "The effect of the drug was highly significant." As another example, if two fertilizers for wheat gave a yield comparison with a large P-value, then the conclusion might be stated as "The wheat yields did not differ significantly between the two fertilizers" or "The difference between the fertilizers was not significant." As a third example, suppose a substance is tested for toxic effects by comparing exposed animals and control animals, and that the null hypothesis of no difference is not rejected. Then the conclusion might be stated as "No significant toxicity was found."

Clearly such phraseology using the term *significant* can be seriously misleading. After all, in ordinary English usage, the word significant connotes "substantial" or "important." In statistical jargon, however, the statement

> "The difference was significant"

means nothing more or less than

> "The null hypothesis of no difference was rejected."

This is to say, "We found sufficient evidence that the difference in sample means was not caused by chance error alone."

By the same token, the statement

> "The difference was not significant"

means

> "There was not sufficient evidence that the observed difference in means was due to anything other than chance variation."

It would perhaps be preferable if a different word were used in place of "significant," such as "discernible" (meaning that the test discerned a difference). Alas, the specialized usage of the word *significant* has become quite common in scientific writing and understandably is the source of much confusion.

It is essential to recognize that a statistical test provides information about only one question: Is the difference observed in the data large enough to infer that a difference in the same direction exists in the population? The question of whether a difference is *important,* as opposed to (statistically) significant, cannot be decided on the basis of the *P*-values alone but must also include an examination of the magnitude of the estimated population difference as well as specific expertise in the research area or practical situation. The following two examples illustrate this fact.

Example 7.6.1 **Serum LD** Lactate dehydrogenase (LD) is an enzyme that may show elevated activity following damage to the heart muscle or other tissues. A large study of serum LD levels in healthy young people yielded the results shown in Table 7.6.1.[46]

Table 7.6.1 Serum LD (U/l)

	Males	Females
n	270	264
$\bar{y}$	60	57
s	11	10

The difference between males and females is quite (statistically) significant; in fact, $t_s = 3.3$, which gives a *P*-value ≈ 0.001. However, this does not imply that the difference $(60 - 57 = 3$ U/*l*) is large or important in any practical sense. ■

Example 7.6.2 **Body Weight** Imagine that we are studying the body weight of men and women, and we obtain the fictitious but realistic data shown in Table 7.6.2.[47]

Table 7.6.2 Body weight (lb)

	Males	Females
n	2	2
$\bar{y}$	175	143
s	35	34

For these data the t test gives $t_s = 0.93$ and a *P*-value ≈ 0.45. The observed difference between males and females is not small (it is $175 - 143 = 32$ lb), yet it is not statistically significant for any reasonable choice of α. The lack of statistical significance does not imply that the sex difference in body weight is small or unimportant. It means only that the data are inadequate to characterize the difference in the population means. A sample difference of 32 lb could easily happen by chance if the two populations are identical, especially with such small sample sizes. ■

EFFECT SIZE

The preceding examples show that the statistical significance or nonsignificance of a difference does not indicate whether the difference is important. Nevertheless, the question of "importance" can and should be addressed in most data analyses. To assess importance, one needs to consider the *magnitude* of the difference. In Example 7.6.1 the male versus female difference is "statistically significant," but this is largely due to the sample sizes being quite large. A t test uses the test statistic

$$t_s = \frac{(\bar{y}_1 - \bar{y}_2)}{SE_{(\bar{Y}_1 - \bar{Y}_2)}}$$

If n_1 and n_2 are large, then $SE_{(\bar{Y}_1 - \bar{Y}_2)}$ will be small and the test statistic will tend to be large even when the difference in observed means $(\bar{Y}_1 - \bar{Y}_2)$ is very small. Thus, one might find significant evidence for H_A due to the sample size being large, even if μ_1 and μ_2 are nearly equal. The sample size acts like a magnifying glass: *The larger the sample size, the smaller the difference that can be detected in a hypothesis test.*

The **effect size** in a study is the difference between μ_1 and μ_2, expressed relative to the standard deviation of one of the populations. If the two populations have the same standard deviation, σ, then the effect size is*

$$\text{Effect size} = \frac{|\mu_1 - \mu_2|}{\sigma}$$

Of course, when working with sample data we can only calculate an *estimated* effect size by using sample values in place of the unknown population values.

Example 7.6.3 **Serum LD** For the data given in Example 7.6.1 the difference in sample means, $60 - 57 = 3$, is less than one-third of a standard deviation. Using the larger sample SD we can calculate a sample effect size of

$$\text{Effect size} = \frac{|\bar{y}_1 - \bar{y}_2|}{s} = \frac{60 - 57}{11} = 0.27$$

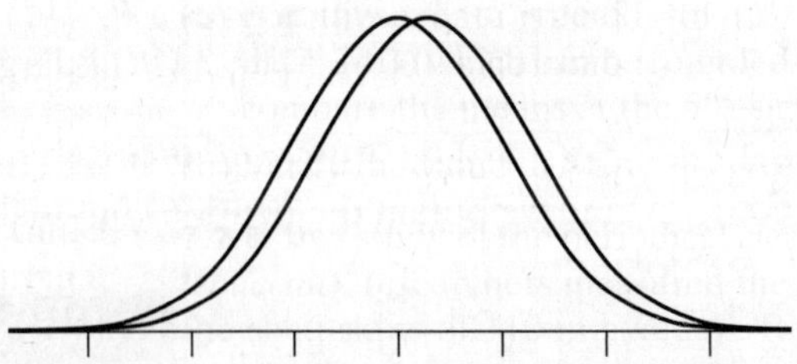

Figure 7.6.1 Overlap between two normally distributed populations when the effect size is 0.27

This indicates that there is a lot of overlap between the two groups. Figure 7.6.1 shows the extent of the overlap that occurs if two normally distributed populations differ on average by 0.27 SDs. ■

Example 7.6.4 **Body Weight** For the data given in Example 7.6.2 the difference in sample means, $175 - 143 = 32$, is roughly one standard deviation. The sample effect size is

$$\text{Effect size} = \frac{|\bar{y}_1 - \bar{y}_2|}{s} = \frac{175 - 143}{35} = 0.91$$

*If the standard deviations are not equal, we can use the larger SD in defining the effect size.

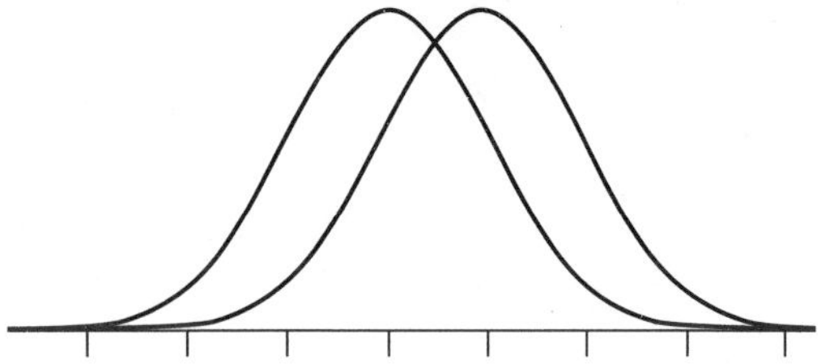

Figure 7.6.2 Overlap between two normally distributed populations when the effect size is 0.91

Figure 7.6.2 shows the extent of the overlap that occurs if two normally distributed populations differ on average by 0.91 SD. ■

The definition of effect size that we are using is probably unfamiliar to the biologically oriented reader. It is more common in biology to "standardize" a difference of two quantities by expressing it as a percentage of one of them. For example, the weight difference given in Table 7.6.2 between males and females, expressed as a percentage of mean female weight, is

$$\frac{\bar{y}_1 - \bar{y}_2}{\bar{y}_2} = \frac{175 - 143}{143} = 0.22 \text{ or } 22\%$$

Thus, the males are about 22% heavier than the females. However, from a statistical viewpoint it is often more relevant that the average weights for males and females are 0.91 SD apart.

CONFIDENCE INTERVALS TO ASSESS IMPORTANCE

Calculating the effect size is one way to quantify how far apart two sample means are. Another reasonable approach is to use the observed difference $(\bar{Y}_1 - \bar{Y}_2)$ to construct a confidence interval for the population difference $(\mu_1 - \mu_2)$. In interpreting the confidence interval, the judgment of what is "important" is made on the basis of experience with the particular practical situation. The following three examples illustrate this use of confidence intervals.

Example 7.6.5 **Serum LD** For the LD data of Example 7.6.1, a 95% confidence interval for $(\mu_1 - \mu_2)$ is

$$3 \pm 1.8$$

or

$$(1.2, 4.8)$$

This interval implies (with 95% confidence) that the population difference in means between the sexes does not exceed 4.8 U/l. As an expert, a physician evaluating this information would know that typical day-to-day fluctuation in a person's LD level is around 6.5 U/l, which is higher than 4.8 U/l, the highest we estimate the mean sex difference to be, and therefore this difference is negligible from the medical standpoint. Consequently, the physician might conclude that it is unnecessary to differentiate between the sexes in establishing clinical thresholds for diagnosis of illness. In this case, the sex difference in LD may be said to be statistically significant but medically unimportant. To put this another way, the data suggest that men do in fact tend to have higher levels than women, but not higher in any clinically useful way. ■

Example 7.6.6 **Body Weight** For the body-weight data of Example 7.6.2, a 95% confidence interval for $(\mu_1 - \mu_2)$ is

$$32 \pm 149$$

or

$$(-117, 181)$$

From this confidence interval we cannot tell whether the true difference (between the population means) is large favoring females, is small, or is large favoring males. Because the confidence interval contains numbers of both small and large magnitude, it does not tell us whether the difference between the sexes is important or unimportant. With such a wide confidence interval a researcher would likely wish to conduct a larger study to better assess the importance of the difference. Suppose, for example, that the means and standard deviations were as given in Table 7.6.2, but that they were based on 2,000 rather than 2 people of each sex. Then the 95% confidence interval would be

$$32 \pm 2$$

or

$$(30, 34)$$

This interval would imply (with 95% confidence) that the difference is at least 30 lb, an amount that might reasonably be regarded as important, at least for some purposes. ■

Example 7.6.7 **Yield of Tomatoes** Suppose a horticulturist is comparing the yields of two varieties of tomatoes; yield is measured as pounds of tomatoes per plant. On the basis of practical considerations, the horticulturist has decided that a difference between the varieties is "important" only if it exceeds 1 pound per plant, on the average. That is, the difference is important if

$$|\mu_1 - \mu_2| > 1.0 \text{ lb}$$

Suppose the horticulturist's data give the following 95% confidence interval:

$$(0.2, 0.3)$$

Because the largest estimate for the population difference is only 0.3 lb (all values in the interval are less than 1.0 lb), the data support (with 95% confidence) the assertion that the difference is *not* important, using the horticulturist's criterion. ■

In many investigations, statistical significance and practical importance are both of interest. The following example shows how the relationship between these two concepts can be visualized using confidence intervals.

Example 7.6.8 **Yield of Tomatoes** Let us return to the tomato experiment of Example 7.6.7. The confidence interval was

$$(0.2, 0.3)$$

Recall from Section 7.3 that the confidence interval can be interpreted in terms of a t test. Because all values within the confidence interval are positive, a t test (two-tailed) at $\alpha = 0.05$ finds significant evidence for H_A. Thus, the difference between the two varieties is statistically significant, although it is not horticulturally important: The data indicate that variety 1 is better than variety 2, but also that it is not much better. The distinction between significance and importance for this example can be seen in Figure 7.6.3, which shows the confidence interval plotted on the $(\mu_1 - \mu_2)$-axis. Note that the confidence interval lies entirely to one side of zero and also entirely to one side of the "importance" threshold of 1.0.

Figure 7.6.3 Confidence interval for Example 7.6.8

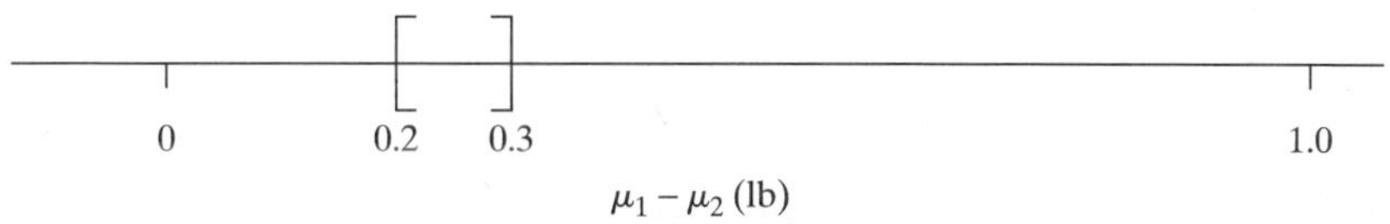

To further explore the relationship between significance and importance, let us consider other possible outcomes of the tomato experiment. Table 7.6.3 shows how the horticulturist would interpret various possible confidence intervals, still using the criterion that a difference must exceed 1.0 lb in order to be considered important.

Table 7.6.3 Interpretation of confidence intervals

95% confidence interval	Is the difference significant?	Is the difference important?
(0.2, 0.3)	Yes	No
(1.2, 1.3)	Yes	Yes
(0.2, 1.3)	Yes	Cannot tell
(−0.2, 0.3)	No	No
(−1.2, 1.3)	No	Cannot tell

Table 7.6.3 shows that a significant difference may or may not be important, and an important difference may or may not be significant. In practice, the assessment of importance using confidence intervals is a simple and extremely useful supplement to a test of hypothesis.* ■

*In some situations, the objective of the researcher is to show that the difference between two population means is negligible. Perhaps a new variety of tomatoes is much easier to grow than an older variety. The researcher does not need to determine if the new variety has a greater (or even different) yield than the old variety. If the mean yields are shown to be similar, that is still very useful information. For example, suppose the 95% confidence interval for the difference in mean yields is (−0.2, 0.1) lb. This interval indicates that if there is a difference in population means, the largest we estimate the magnitude of the difference to be is 0.2 lb. This difference is negligibly small, and we would conclude that the two varieties have mean yields that are practically similar.

Exercises 7.6.1–7.6.11

7.6.1 A field trial was conducted to evaluate a new seed treatment that was supposed to increase soybean yield. When a statistician analyzed the data, the statistician found that the mean yield from the treated seeds was 40 lb/acre greater than that from control plots planted with untreated seeds. However, the statistician declared the difference to be "not (statistically) significant." Proponents of the treatment objected strenuously to the statistician's statement, pointing out that, at current market prices, 40 lb/acre would bring a tidy sum, which would be highly significant to the farmer. How would you answer this objection?[48]

7.6.2 In a clinical study of treatments for rheumatoid arthritis, patients were randomly allocated to receive either a standard medication or a newly designed medication. After a suitable period of observation, statistical analysis showed that there was no significant difference in the therapeutic response of the two groups, but that the incidence of undesirable side effects was significantly lower in the group receiving the new medication. The researchers concluded that the new medication should be regarded as clearly preferable to the standard medication, because it had been shown to be equally effective therapeutically and to produce fewer side effects. In what respect is the researchers' reasoning faulty? (Assume that the term "significant" refers to rejection of H_0 at $\alpha = 0.05$.)

7.6.3 There is an old folk belief that the sex of a baby can be guessed before birth on the basis of its heart rate. In an investigation to test this theory, fetal heart rates were observed for mothers admitted to a maternity ward. The results (in beats per minute) are summarized in the table.[49]

	Heart rate (bpm)		
	n	Mean	SE
Males	250	137.21	0.62
Females	250	137.18	0.53

Construct a 95% confidence interval for the difference in population means. Does the confidence interval support the claim that the population mean sex difference (if any) in fetal heart rates is small and unimportant? (Use your own "expert" knowledge of heart rate to make a judgment of what is "unimportant.")

7.6.4 Coumaric acid is a compound that may play a role in disease resistance in corn. A botanist measured the concentration of coumaric acid in corn seedlings grown in the dark or in a light/dark photoperiod. The results (nmol acid per gm tissue) are given in the accompanying table.[50] [*Note:* Formula (6.7.1) yields 5.7 df.]

	Dark	Photoperiod
n	4	4
$\bar{y}$	106	102
s	21	27

Suppose the botanist considers the effect of lighting conditions to be "important" if the difference in means is 20%, that is, about 20 nmol/g. Based on a 95% confidence interval, do the preceding data indicate whether the true difference is "important"?

7.6.5 Repeat Exercise 7.6.4, assuming that the means and standard deviations are as given in the table, but that the sample sizes are 10 times as large (that is, $n = 40$ for "dark" and $n = 40$ for "photoperiod"). [*Note:* Formula (6.7.1) yields 73.5 df.]

7.6.6 Researchers measured the breadths, in mm, of the ankles of 460 youth (ages 11–16); the results are shown in the table.[51]

	Males	Females
n	244	216
$\bar{y}$	55.3	53.3
s	6.1	5.4

Calculate the sample effect size from these data.

7.6.7 As part of a large study of serum chemistry in healthy people, the following data were obtained for the serum concentration of uric acid in men and women ages 18–55 years.[52]

	Serum uric acid (mmol/l)	
	Men	Women
n	530	420
$\bar{y}$	0.354	0.263
s	0.058	0.051

Construct a 95% confidence interval for the true difference in population means. Suppose the investigators feel that the difference in population means is "clinically important" if it exceeds 0.08 mmol/l. Does the confidence interval indicate whether the difference is "clinically important"? [Note: Formula (6.7.1) yields 934 df.]

7.6.8 Repeat Exercise 7.6.7, assuming that the means and standard deviations are as given in the table, but that the sample sizes are only one-tenth as large (i.e., 53 men and 42 women). [Note: Formula (6.7.1) yields 92 df.]

7.6.9 Scientists recorded the dry weight (in mg) of a species of flower growing in two habitats: unfragmented (natural) and fragmented [on land disturbed by human development (e.g., roadside or near a parking lot)]. The table below shows summary values.[53]

	Fragmented	Unfragmented
Mean	86.2	92.6
SD	17.2	18.3
n	86	116

Calculate the sample effect size from these data.

7.6.10 Consider the data from Exercise 7.6.9. Here is computer output from an analysis of the data:

```
t = 2.52, df = 188, p-value = 0.01273

alternative hypothesis: true difference in
means is not equal to 0
95 percent confidence interval: 1.36 11.28
```

Suppose the scientists believe that a difference in population means is "important" if it is at least 5 mg. Is there evidence here that the difference is "important"?

7.6.11 Patients with nonchronic low back pain were randomly assigned to either a treatment group that received health coaching via telephone plus physiotherapy care or a control group that received only the physiotherapy care. One variable measured was improvement, over 4 weeks, in "modified Oswestry Disability Index," which is scored as a percentage, ranging from 0 to 100, with high scores representing high levels of disability. Summary statistics are shown in the table.[54]

	Experimental	Control
Mean	16.8	11.9
SD	18.7	11.6
n	12	14

(a) Calculate the sample effect size from these data.

(b) A 95% confidence interval for the difference in population mean Oswestry Disability Index for the two groups ($\mu_{Experimental} - \mu_{Control}$) is $(-8.2, 18.0)$ percentage points. The researchers were hoping to find a difference of at least 5 percentage points, a difference that they thought would be clinically important. Based on the confidence interval, would they be ill-advised to conduct a larger study in the hope of finding convincing evidence of a 5 percentage point difference?

(c) How would your answer to (b) change if the confidence interval was $(-2.4, 3.2)$ percentage points.

7.7 Planning for Adequate Power (Optional)

We have defined the power of a statistical test as

$$\text{Power} = \Pr\{\text{significant evidence for } H_A\} \text{ if } H_A \text{ is true}$$

To put this another way, the power of a test is the probability of obtaining data that provide statistically significant evidence for H_A when H_A is true.

Since the power is the probability of *not* making an error (of Type II), high power is desirable: If H_A is true, a researcher would like to find that out when conducting a study. But power comes at a price. All other things being equal, more observations (larger samples) bring more power, but observations cost time and money. In this section we explain how a researcher can rationally plan an experiment to have adequate power for the purposes of the research project while holding down costs.

Specifically, we will consider the power of the two-sample t test, conducted at significance level α. We will assume that the populations are normal with equal SDs, and we denote the common value of the SD by σ (i.e., $\sigma_1 = \sigma_2 = \sigma$). It can be shown that in this case, for a given total sample size of $2n$, the power is maximized if the sample sizes are equal; thus we will assume that n_1 and n_2 are equal and denote the common value by n (i.e., $n_1 = n_2 = n$).

Under the above conditions, the power of the t test depends on the following factors: (a) α; (b) σ; (c) n; and (d) $(\mu_1 - \mu_2)$. After briefly discussing each of these factors, we will address the all-important question of choosing the value of n.

DEPENDENCE OF POWER ON α

In choosing α, one chooses a level of protection against Type I error. However, this protection is traded for vulnerability to Type II error. If, for example, one chooses $\alpha = 0.01$ rather than $\alpha = 0.05$, then one is requiring stronger evidence before claiming that there is significant support for H_A, and so is (perhaps unwittingly) also choosing to increase the risk of Type II error and reduce the power. Thus, there is an unavoidable trade-off between the risk of Type I error and the risk of Type II error.

DEPENDENCE ON σ

The larger the σ is, the smaller the power (all other things being equal). Recall from Chapter 5 that the reliability of a sample mean is determined by the quantity

$$\sigma_{\bar{Y}} = \frac{\sigma}{\sqrt{n}}$$

The larger σ is, the more variability there is in the sample mean. Thus, having a larger σ implies having samples that produce less reliable information about each population mean, and so less power to discern a difference between them. In order to increase power, then, a researcher usually tries to design the investigation so as to have σ as small as possible. For example, a botanist will try to hold light conditions constant throughout a greenhouse area, a pharmacologist will use genetically identical experimental animals, and so on. Usually, however, σ cannot be reduced to zero; there will still be variation in the observations.

DEPENDENCE ON n

The larger the n, the higher the power (all other things being equal). If we increase n, we decrease $\sigma/\sqrt{n}$; this improves the precision of the sample means ($\bar{Y}_1$ and $\bar{Y}_2$). In addition, larger n gives more information about σ; this is reflected in a reduced critical value for the test (reduced because of more df). Thus, increasing n increases the power of the test in two ways.

DEPENDENCE ON $(\mu_1 - \mu_2)$

In addition to the factors we have discussed, the power of the t test also depends on the actual difference between the population means, that is, on $(\mu_1 - \mu_2)$. This dependence is very natural, as illustrated by the following example.

Example 7.7.1 **Heights of People** In order to clearly illustrate the concepts, we consider a familiar variable, body height of people. Imagine what would happen if an investigator were to measure the heights of two random samples of 11 people each ($n = 11$), and then conduct a two-tailed t test at $\alpha = 0.05$.

(a) First, suppose that sample 1 consisted of 17-year-old males and sample 2 consisted of 17-year-old females. The two population means differ substantially; in fact, $(\mu_1 - \mu_2)$ is about 5 inches ($\mu_1 \approx 69.1$ and $\mu_2 \approx 64.1$ inches).[55] It can be shown (as we will see) that in this case the investigator has about a 99% chance of obtaining significant evidence for a difference (i.e., H_A) and correctly concluding that the males in the population of 17-year-olds are taller (on average) than the females.

(b) By contrast, suppose that sample 1 consisted of 17-year-old females and sample 2 consisted of 14-year-old females. The two population means differ, but by a modest amount; the difference is $(\mu_1 - \mu_2) = 0.6$ inches ($\mu_1 \approx 64.1$ and $\mu_2 \approx 63.5$ inches). It can be shown that in this case the investigator has less than a 10% chance of obtaining significant evidence of a difference (i.e., H_A); in other words, there is more than a 90% chance that the investigator will fail to detect the fact that 17-year-old girls are taller than 14-year-old girls. (In fact, it can be shown that there is a 29% chance that $\bar{Y}_1$ will be less than $\bar{Y}_2$—that is, there is a 29% chance that eleven 17-year-old girls chosen at random will be shorter on the average than eleven 14-year-old girls chosen at random!)

The contrast between cases (a) and (b) is not due to any change in the SDs; in fact, for each of the three populations the value of σ is about 2.5 inches. Rather, the contrast is due to the simple fact that, with a fixed n and σ, it is easier to detect a large difference than a small difference. ■

PLANNING A STUDY

Suppose an investigator is planning a study for which the t test will be appropriate. How shall she take into account all the factors that influence the power of the test? First consider the choice of significance level α. A simple approach is to begin by determining the cost of an adequately powerful study using a somewhat liberal choice (say, $\alpha = 0.05$ or 0.10). If that cost is not high, the investigator can consider reducing α (say, to 0.01) and see if an adequately powerful study is still affordable.

Suppose, then, that the investigator has chosen a working value of α. Suppose also that the experiment has been designed to reduce σ as far as practicable, and that the investigator has available an estimate or guess of the value of σ.

At this point, the investigator needs to ask herself about the magnitude of the difference she wants to detect. As we saw in Example 7.7.1, a given sample size may be adequate to detect a large difference in population means, but entirely inadequate to detect a small difference. As a more realistic example, an experiment using 5 rats in a treatment group and 5 rats in a control group might be large enough to detect a substantial treatment effect, while detection of a subtle treatment effect would require more rats (perhaps 30) in each group.

The preceding discussion suggests that choosing a sample size for adequate power is somewhat analogous to choosing a microscope: We need high resolving power if we want to see a very tiny structure; for large structures a hand lens will do. In order to proceed with planning the experiment, the investigator needs to decide how large an effect she is looking for.

Recall that in Section 7.6, we defined the effect size in a study as the difference between μ_1 and μ_2, expressed relative to the standard deviation of one of the populations. If, as we are assuming here, the two populations have the same standard deviation, σ, then the effect size is

$$\text{Effect size} = \frac{|\mu_1 - \mu_2|}{\sigma}$$

That is, the effect size is the difference in population means expressed relative to the common population SD. The effect size is a kind of "signal to noise ratio," where $(\mu_1 - \mu_2)$ represents the signal we want to detect and σ represents the background noise that tends to obscure the signal. Figure 7.7.1(a) shows two normal curves for which the effect size is 0.5; Figure 7.7.1(b) shows two normal curves for which the effect size is 4. Clearly, at a fixed sample size it is easier to detect the difference between the curves in graph (b) than it is in graph (a).

If α and the effect size have been specified, then the power of the t test depends only on the sample sizes (n). Table 5 at the end of the book shows the value of n required in order to achieve a specified power against a specified effect size. Let us see how Table 5 applies to our familiar example of body height.

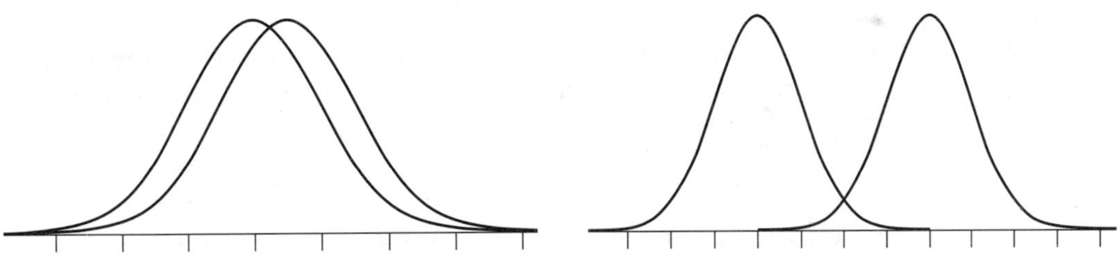

(a) (b)

Figure 7.7.1 Normal distributions with an effect size (a) of 0.5 and (b) of 4

Example 7.7.2

Heights of People In Example 7.7.1, case (a), we considered samples of 17-year-old males and 17-year-old females. The effect size is

$$\frac{|\mu_1 - \mu_2|}{\sigma} = \frac{|69.1 - 64.1|}{2.5} = \frac{5}{2.5} = 2.0$$

For a two-tailed t test at $\alpha = 0.05$, Table 5 shows that the sample size required for a power of 0.99 is $n = 11$; this is the basis for the claim in Example 7.7.1 that the investigator has a 99% chance of detecting the difference between males and females. Figure 7.7.2 shows the two distributions being considered in Example 7.7.2. Suppose 100 researchers each conduct the following study. Take a random sample of eleven 17-year-old males and a random sample of eleven 17-year-old females, find the sample average heights of the two groups, and then conduct a two-tailed t test of H_0: $\mu_1 = \mu_2$ using $\alpha = 0.05$. We would expect 99 of the 100 researchers to find statistically significant evidence that the average heights of 17-year-old males and females differ (i.e., significant evidence for H_A). We would expect one of the 100 researchers not to find sufficient evidence for a difference, at the 0.05 level of significance. (So one researcher would make a Type II error*.) ■

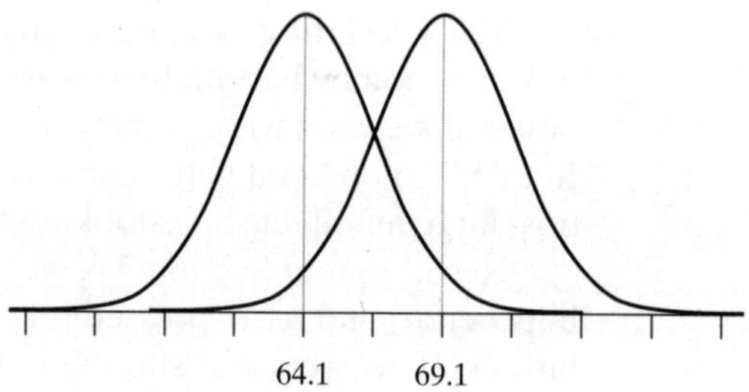

Figure 7.7.2 Height distributions for Example 7.7.2

As we have seen, in order to choose a sample size the researcher needs to specify not only the size of the effect she wishes to detect, but also how certain she wants to be of detecting it; that is, it is necessary to specify how much power is wanted. Since the power measures the protection against Type II error, the choice of a desired power level depends on the consequences that would result from a Type II error. If the consequences of a Type II error would be very unfortunate (e.g., if a promising but risky cancer treatment is being tested on humans and a negative result would discredit the treatment so that it would never be tested again), then the researcher might specify a high power, say 0.95 or 0.99. But, of course, high power is expensive in terms of n. For much research, a Type II error is not a disaster, and a lower power such as 0.80 is considered adequate.

The following example illustrates a typical use of Table 5 in planning an experiment.

Example 7.7.3

Postpartum Weight Loss A group of scientists wished to investigate whether or not an Internet-based intervention program would help women lose weight after giving birth. One group of postpartum women was to be enrolled in an Internet-based program that provides weekly exercise and dietary guidance appropriate to their time

*Here is another way to think about effect sizes. Among 17-year-olds the probability that a female is taller than the average male is $\Pr\{Y_1 > 69.1\}$, which is the area under the first curve to the right of 69.1. This is equal to $\Pr\{Z > 2\}$, that is, the chance that a female is at least 2 SDs taller than average. In general, if the effect size is D then $\Pr\{Y_1 > \mu_2\} = \Pr\{Z > D\}$.

since giving birth, track their weight-loss progress, and establish an online forum for nutrition and exercise discussion with other recent mothers. Another group of postpartum women (the "control group") was to be given traditional written dietary and exercise guidelines by their doctors. The response variable for the study was to be the amount of weight lost at 12 months postpartum in kg. Previous studies have shown that at 12 months postpartum, the mean weight loss is about 3.6 kg with a standard deviation of 4.0 kg. (*Note*: A negative weight loss is a weight gain.) The research team wanted to show at least a 50% improvement in weight loss for the Internet-intervention group; that is, they would like to show that the Internet-based program women lose at least 1.8 kg (50% of 3.6 kg) more weight than the controls. They planned to conduct a one-tailed t-test at the 5% significance level. The team had to decide how many women (n) to put in each group.

The effect size that the team wanted to consider is

$$\frac{|\mu_1 - \mu_2|}{\sigma} = \frac{1.8}{4.0} = 0.45$$

For this effect size, and for a power of 0.80 with a one-tailed test at the 5% significance level, Table 5 yields $n = 62$, which means 62 women were needed in each group.

At this point, the research team had to consider questions, such as (1) Is it feasible to enroll 124 postpartum women (62 in each group) in the study? If not, then (2) Would they perhaps be willing to redefine the size of the difference between the groups that they considered to be important, in order to reduce the required n? With questions such as these, and repeated use of Table 5, they could finally decide on a firm value for n, or possibly decide to abandon the project because an adequate study would be too costly.

Normally the story ends here, but there was an extra wrinkle in the planning of this study: The research team knew from experience that about 20% of the women enrolled in these types of studies would drop out, for one reason or another, before the study ended. (There is no formula or table that tells one how many subjects will drop out of a study such as this. Here the only guide is experience.) In this case, the research team planned to enroll 150 women (a little more than 20% extra, 13 women in each group), in order to allow for some attrition and still end up with enough data so that they would have the power they wanted.[56]

Exercises 7.7.1–7.7.12

7.7.1 One measure of the meat quality of pigs is backfat thickness. Suppose two researchers, Jones and Smith, are planning to measure backfat thickness in two groups of pigs raised on different diets. They have decided to use the same number (n) of pigs in each group, and to compare the mean backfat thickness using a two-tailed t test at the 5% significance level. Preliminary data indicate that the SD of backfat thickness is about 0.3 cm.

When the researchers approach a statistician for help in choosing n, she naturally asks how much difference they want to detect. Jones replies, "If the true difference is 1/4 cm or more, I want to be reasonably sure of rejecting H_0." Smith replies, "If the true difference is 1/2 cm or more, I want to be very sure of rejecting H_0."

If the statistician interprets "reasonably sure" as 80% power, and "very sure" as 95% power, what value of n will she recommend

(a) to satisfy Jones's requirement?

(b) to satisfy Smith's requirement?

7.7.2 Refer to the brain NE data of Example 7.2.1. Suppose you are planning a similar experiment; you will study the effect of LSD (rather than toluene) on brain NE. You anticipate using a two-tailed t test at $\alpha = 0.05$. Suppose you have decided that a 10% effect (increase or decrease in mean NE) of LSD would be important, and so you want to have good power (80%) to detect a difference of this magnitude.

(a) Using the data of Example 7.2.1 as a "pilot study," determine how many rats you should have in each group. (The mean NE in the control group in Example 7.2.1 is 444.2 ng/g and the SD is = 69.6 ng/g.)

(b) If you were planning to use a one-tailed t test, what would be the required number of rats?

7.7.3 Suppose you are planning a greenhouse experiment on growth of pepper plants. You will grow n individually potted seedlings in standard soil and another n seedlings in specially treated soil. After 21 days, you will measure Y = total stem length (cm) for each plant. If the effect of the soil treatment is to increase the population mean stem length by 2 cm, you would like to have a 90% chance of rejecting H_0 with a one-tailed t test. Data from a pilot study (e.g., the data in Exercise 2.S.2) on 15 plants grown in standard soil give $\bar{y} = 12.5$ cm and $s = 0.8$ cm.

(a) Suppose you plan to test at $\alpha = 0.05$. Use the pilot information to determine what value of n you should use.

(b) What conditions are necessary for the validity of the calculation in part (a)? Which of these can be checked (roughly) from the data of the pilot study?

(c) Suppose you decide to adopt a more conservative posture and test at $\alpha = 0.01$. What value of n should you use?

7.7.4 Diastolic blood pressure measurements on American men ages 18–44 years follow approximately a normal curve with $\mu = 81$ mm Hg and $\sigma = 11$ mm Hg. The distribution for women ages 18–44 is also approximately normal with the same SD but with a lower mean: $\mu = 75$ mm Hg.[57] Suppose we are going to measure the diastolic blood pressure of n randomly selected men and n randomly selected women in the age group 18–44 years. Let E be the event that the difference between men and women will be found statistically significant by a t test. How large must n be in order to have $\Pr\{E\} = 0.9$

(a) if we use a two-tailed test at $\alpha = 0.05$?

(b) if we use a two-tailed test at $\alpha = 0.01$?

(c) if we use a one-tailed test (in the correct direction) at $\alpha = 0.05$?

7.7.5 Suppose you are planning an experiment to test the effect of a certain drug treatment on drinking behavior in the rat. You will use a two-tailed t test to compare a treated group of rats against a control group; the observed variable will be Y = 1-hour water consumption after 23-hour deprivation. You have decided that, if the effect of the drug is to shift the population mean consumption by 2 ml or more, then you want to have at least an 80% chance of finding significant evidence for H_A at the 5% significance level.

(a) Preliminary data indicate that the SD of Y under control conditions is approximately 2.5 ml. Using this as a guess of σ, determine how many rats you should have in each group.

(b) Suppose that, because the calculation of part (a) indicates a rather large number of rats, you consider modifying the experiment so as to reduce σ. You find that, by switching to a better supplier of rats and by improving lab procedures, you could cut the SD in half; however, the cost of each observation would be doubled. Would these measures be cost-effective; that is, would the modified experiment be less costly?

7.7.6 Data from a large study indicate that the serum concentration of lactate dehydrogenase (LD) is higher in men than in women. (The data are summarized in Example 7.6.1.) Suppose Dr. Sanchez proposes to conduct his own study to replicate this finding; however, because of limited resources Sanchez can enlist only 35 men and 35 women for his study. Supposing that the true difference in population means is 4 U/l and each population SD is 10 U/l, what is the probability that Sanchez will be successful? Specifically, find the probability that Sanchez will reject H_0 with a one-tailed t test at the 5% significance level.

7.7.7 Refer to the painkiller study of Exercise 7.5.10. That study included 25 observations in each treatment group and showed an effect size of about 0.5. If this is the true population effect size, what is the (approximate) chance of finding a significant difference between the mean effectiveness of the two drugs in an experiment of this size (i.e., with samples of 25 each)?

7.7.8 Refer to the painkiller study of Exercise 7.5.10. In that study, the evidence favoring the drug was marginally significant ($0.025 < P < 0.05$). Suppose Dr. Williams is planning a new study on the same drug in order to try to replicate the original findings, that is, to show the drug to be effective. She will consider this study successful if she rejects H_0 with a one-tailed test at $\alpha = 0.05$. In the original study, the difference between the treatment means was about half a standard deviation [$(32 - 25)/13 \approx 0.5$]. Taking this as a provisional value for the effect size, determine how many patients Williams should have in each group in order for her chance of success to be

(a) 80% (b) 90%

(*Note:* This problem illustrates that surprisingly large sample sizes may be required to make a replication study worthwhile, especially if the original findings were only marginally significant.)

7.7.9 Consider comparing two normally distributed distributions for which the effect size of the difference is

(a) 3 (b) 1

In each case, draw a sketch that shows how the distributions overlap. (See Figure 7.2.1.)

7.7.10 An animal scientist is planning an experiment to evaluate a new dietary supplement for beef cattle. One group of cattle will receive a standard diet and a second group will receive the standard diet plus the supplement. The researcher wants to have 90% power to detect an increase in mean weight gain of 20 kg, using a one-tailed t test at $\alpha = 0.05$. Based on previous experience, he expects the SD to be 17 kg. How many cattle does he need for each group?

7.7.11 A researcher is planning to conduct a study that will be analyzed with a two-tailed t test at the 5% significance level. She can afford to collect 20 observations in each of the two groups in her study. What is the smallest effect size for which she has at least 95% power?

7.7.12 Researchers in Norway examined the records of 85,176 children to look for a possible association between use of folic acid during pregnancy and development of autism. The null hypothesis of interest is that use of folic acid is not related to risk of autism. The researchers set α at 0.05 and stated that they had 93% power to detect a certain difference in autism rates. Explain what "93% power" means in this context. In particular, what is the probability of a Type II error?[58]

7.8 Student's *t*: Conditions and Summary

In the preceding sections we have discussed the comparison of two means using classical methods based on Student's t distribution. In this section we describe the conditions on which these methods are based. In addition, we summarize the methods for convenient reference.

CONDITIONS

The t test and confidence interval procedures we have described are appropriate if the following conditions* hold:

1. *Conditions on the design of the study*
   (a) It must be reasonable to regard the data as random samples from their respective populations. The populations must be large relative to their sample sizes. The observations within each sample must be independent.
   (b) The two samples must be independent of each other.
2. *Condition on the form of the population distributions*
   The sampling distributions of $\bar{Y}_1$ and $\bar{Y}_2$ must be (approximately) normal. This can be achieved via normality of the populations or by appealing to the Central Limit Theorem (recall Section 6.5) if the populations are nonnormal but the sample sizes are large, where "largeness" depends on the degree of nonnormality of the populations. In many practical situations, moderate sample sizes (say, $n_1 = 20$, $n_2 = 20$) are quite "large" enough. However, we always need to be aware that one or two extreme outliers can have a great effect on the results of any statistical procedure, including the t test.

VERIFICATION OF CONDITIONS

A check of the preceding conditions should be a part of every data analysis.

A check of condition 1(a) would proceed as for a confidence interval (Section 6.5), with the researcher looking for biases in the experimental design and verifying that there is no hierarchical structure within each sample.

Condition 1(b) means that there must be no pairing or dependency between the two samples. The full meaning of this condition will become clear in Chapter 8.

Sometimes it is known from previous studies whether the populations can be considered to be approximately normal. In the absence of such information, the normality

*Many authors use the word "assumptions" where we are using the word "conditions."

requirement can be checked by making histograms, normal probability plots, or Shapiro–Wilk normality tests for each sample separately. Fortunately, the *t* test is fairly robust against departures from normality.[59] Usually, only a rather conspicuous departure from normality (outliers, or long straggly tails) should be cause for concern. Moderate skewness has very little effect on the *t* test, even for small samples.

CONSEQUENCES OF INAPPROPRIATE USE OF STUDENT'S *t*

Our discussion of the *t* test and confidence interval (in Sections 7.3–7.8) was based on the conditions (1) and (2). Violation of the conditions may render the methods inappropriate.

If the conditions are not satisfied, then the *t* test may be inappropriate in two possible ways:

1. It may be invalid in the sense that the actual risk of Type I error is larger than the nominal significance level α. (To put this another way, the *P*-value yielded by the *t* test procedure may be inappropriately small.)
2. The *t* test may be valid, but less powerful than a more appropriate test.

If the design includes hierarchical structures that are ignored in the analysis, the *t* test may be seriously invalid. If the samples are not independent of each other, the usual consequence is a loss of power.

One fairly common type of departure from the condition of normality is for one or both populations to have long straggly tails. The effect of this form of nonnormality is to inflate the SE, and thus to rob the *t* test of power.

Inappropriate use of confidence intervals is analogous to that for *t* tests. If the conditions are violated, then the confidence interval may not be valid (i.e., too narrow for the prescribed level of confidence), or it may be valid but wider than necessary.

OTHER APPROACHES

When methods based on Student's *t* distribution are not appropriate (e.g., the data have long straggly tails), a randomization test as discussed in Section 7.1 may be used. Alternatively, statisticians have devised other methods that do not require simulations to be run on a computer. One of these is the Wilcoxon–Mann–Whitney test, which we will describe in Section 7.10. Another approach to the difficulty is to transform the data, for instance, to analyze log (Y) or ln (Y) instead of Y itself.

Example 7.8.1 **Tissue Inflammation** Researchers took skin samples from 10 patients who had breast implants and from a control group of 6 patients. They recorded the level of interleukin-6 (in pg/ml/10 g of tissue), a measure of tissue inflammation, after each tissue sample was cultured for 24 hours. Table 7.8.1 shows the data.[60] Parallel dotplots of these data shown in Figure 7.8.1(a) and normal probability plots shown in Figure 7.8.2(a) indicate that the distributions are severely skewed, so a transformation is needed before Student's *t* procedure can be used. Taking the base 10 logarithm of each observation produces the values shown in the right-hand columns of Table 7.8.1 and in Figure 7.8.1(b). The normal probability plots in Figure 7.8.2(b) show that the condition of normality is met after the data have been transformed to log scale. Thus, we will conduct an analysis of the data in log scale. That is, we will test

$$H_0: \mu_1 = \mu_2$$

Table 7.8.1 Interleukin-6 levels of breast implant patients and control patients

	Original data		Log scale	
	Breast implant patients	Control patients	Breast implant patients	Control patients
	231	35,324	2.364	4.548
	308,287	12,457	5.489	4.095
	33,291	8,276	4.522	3.918
	124,550	44	5.095	1.643
	17,075	278	4.232	2.444
	22,955	840	4.361	2.924
	95,102		4.978	
	5,649		3.752	
	840,585		5.925	
	58,924		4.770	
$\bar{y}$	150,665	9,537	4.549	3.262
s	259,189	13,613	0.992	1.111

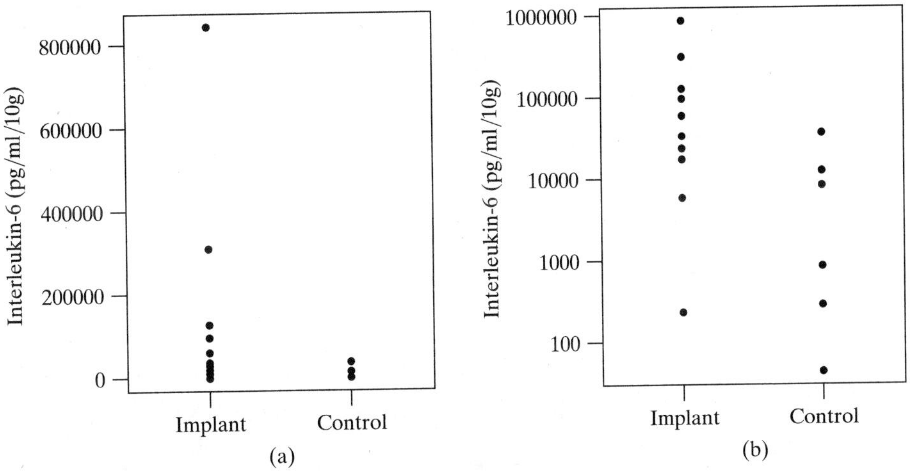

Figure 7.8.1 Dotplots of tissue inflammation data from Example 7.8.1 (a) in the original scale; (b) in log scale

against

$$H_A: \mu_1 \neq \mu_2$$

where μ_1 is the population mean of the log of interleukin-6 level for breast implant patients and μ_2 is the population mean of the log of interleukin-6 level for control patients. Suppose we choose $\alpha = 0.10$. The test statistic is

$$t_s = \frac{(4.549 - 3.262)}{0.553} = 2.33$$

Formula (6.7.1) yields df = 9.7. The *P*-value for the test is 0.045. Thus, we have evidence, at the 0.10 level of significance (and at the 0.05 level, as well), that the mean log interleukin-6 level is higher in the breast implant population than in the control population. ■

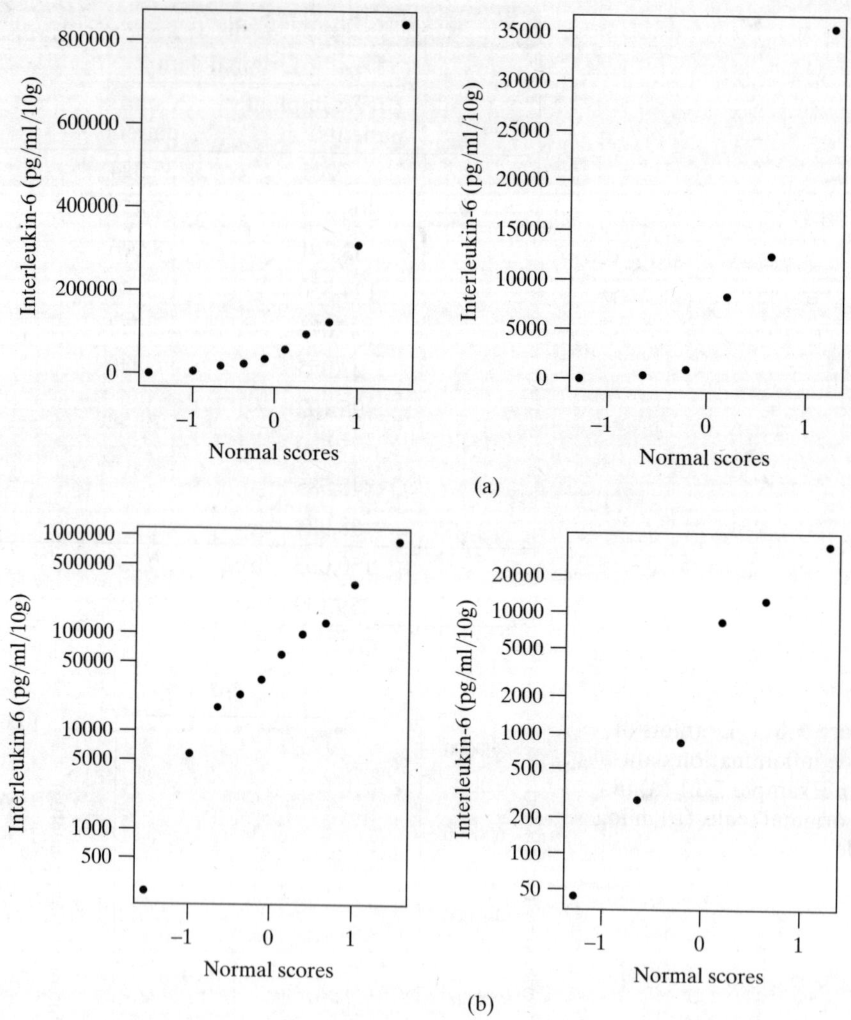

Figure 7.8.2 Normal probability plots of tissue inflammation data from Example 7.8.1 (a) in the original scale and (b) in log scale

SUMMARY OF t TEST MECHANICS

For convenient reference, we summarize the mechanics for Student's t test of equality of the means of independent samples.

t Test

$$H_0: \mu_1 = \mu_2$$
$$H_A: \mu_1 \neq \mu_2 \text{ (nondirectional)}$$
$$H_A: \mu_1 < \mu_2 \text{ (directional)}$$
$$H_A: \mu_1 > \mu_2 \text{ (directional)}$$

$$\text{Test statistic: } t_s = \frac{(\bar{y}_1 - \bar{y}_2) - 0}{SE_{(\bar{Y}_1 - \bar{Y}_2)}}$$

P-value = tail area under Student's t curve with

$$df = \frac{(SE_1^2 + SE_2^2)^2}{SE_1^4/(n_1 - 1) + SE_2^4/(n_2 - 1)}$$

Nondirectional H_A: P-value $=$ two-tailed area beyond t_s and $-t_s$

Directional H_A: Step 1. Check directionality.

Step 2. P-value $=$ single-tail area beyond t_s

Decision: Significant evidence for H_A if P-value $\leq \alpha$

Exercises 7.8.1–7.8.3

7.8.1 To determine if the environment can affect sperm quality and production in cattle, a researcher randomly assigned 13 bulls to one of two environments. Six were raised in an open range environment while 7 were reared in a smaller penned environment. The following plot displays the sperm concentrations (millions of sperm/ml) of semen samples from the 13 bulls.[61]

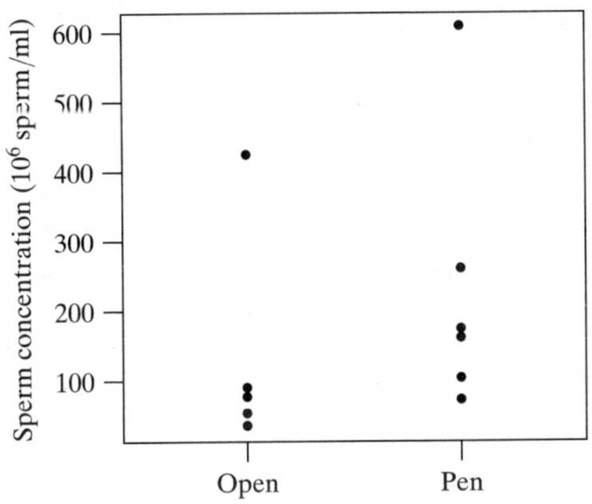

(a) Using the preceding graph to justify your answer, would the use of Student's t method be appropriate to compare the mean sperm concentrations under these two experimental conditions?

(b) How would your answer to (a) change if the data consisted of 60 and 70 specimens rather than 6 and 7?

(c) The Shapiro–Wilk test of normality yields P-values of 0.0012 and 0.0139 for the Open and Pen data, respectively. How do these results support or refute your response to part (a)?

(d) How might a transformation help you analyze these data?

7.8.2 Refer to the serotonin data of Exercise 7.2.7. On what grounds might an objection be raised to the use of the t test on these data? (*Hint:* For each sample, calculate the SD and compare it to the sample mean.)

7.8.3 Adults who broke their wrists were tested to see how much their grip strength (kg) changed over 6 weeks.[62]

Subject	Baseline	6 Weeks later
1	8.3	19.3
2	5.7	10.7
3	3.3	8.3
4	4.6	9
5	5.6	13.6
6	2.3	9.3
7	11.7	16.6
8	33.7	47.3
9	3.3	9
10	1.3	18
11	5.3	12
12	32.3	43
13	2	10.3
14	0.8	7
15	2.7	9
16	5.7	10
17	3	7.7
18	0	23.7
19	3.7	10.3
20	4.7	15

Explain why the methods of this chapter cannot be used to analyze these data.

7.9 More on Principles of Testing Hypotheses

Our study of the t test has illustrated some of the general principles of statistical tests of hypotheses. In the remainder of this book we will introduce several other types of tests besides the t test.*

*We will not address testing the hypothesis H_0: $\sigma_1 = \sigma_2$. Most of the techniques for testing that hypothesis are very sensitive to the condition that the underlying distributions are normal, which limits their use in practice.

A GENERAL VIEW OF HYPOTHESIS TESTS

A typical statistical test involves a null hypothesis H_0, an alternative hypothesis, or research hypothesis, H_A, and a test statistic that measures deviation or discrepancy of the data from H_0. The sampling distribution of the test statistic, under the assumption that H_0 is true, is called the **null distribution** of the test statistic. (If we are conducting a randomization test as in Section 7.1, then the null distribution is the distribution of all possible differences in sample means due to random assignment of observations to groups, such as that shown in Table 7.1.2; as another example, if we are conducting a t test, then the null distribution of the t statistic t_s is—under certain conditions—a Student's t distribution.) The null distribution indicates how much the test statistic can be expected to deviate from H_0 because of chance alone.

In testing a hypothesis, we assess the evidence against H_0 (and in favor of H_A) by locating the test statistic within the null distribution; the P-value is a measure of this location, which indicates the degree of compatibility between the data and H_0. The dividing line between compatibility and incompatibility is specified by an arbitrarily chosen significance level α. The decision whether to claim there is significant evidence for H_A is made according to the following rule:

Reject H_0 if P-value $\leq \alpha$.

When a computer is not available, we will not be able to calculate the P-value exactly but will bracket it using a table of critical values. If H_A is directional, the bracketing of P-value is a two-step procedure.

Every test of a null hypothesis H_0 has its associated risks of Type I error (finding significant evidence for H_A when H_0 is true) and Type II error (not finding significant evidence for H_A when H_A is true). The risk of Type I error is always limited by the chosen significance level, α:

$\Pr\{\text{reject } H_0\} \leq \alpha$ if H_0 is true

Thus, the hypothesis testing procedure treats the Type I error as the one to be most stringently guarded against. By contrast, the power of a test can be quite low, and equivalently the risk of Type II error can be quite large, if the samples are small.

HOW ARE H_0 AND H_A CHOSEN?

A common difficulty when first studying hypothesis testing is figuring out what the null and alternative hypotheses should be. In general, the null hypothesis represents the status quo—what one would believe, by default, unless the data showed otherwise.* Typically the alternative hypothesis is a statement that the researcher is trying to establish; thus H_A is also referred to as the research hypothesis. For example, if we are testing a new drug against a standard drug, the research hypothesis is that the new drug is better than the standard drug, while the null hypothesis is that the new drug is no different than the standard—in the absence of evidence, we would expect the two drugs to be equally effective. The typical null hypothesis, H_0: $\mu_1 = \mu_2$, states that the two population means are equal and that any difference between the sample means is simply due to chance error in the sampling process. The alternative hypothesis is that there *is* a difference between the drugs, so any observed difference in sample means is due to a real effect, rather than being due to chance error alone. We conclude that we have statistically significant evidence for the research hypothesis if the data show a difference in sample means beyond what can reasonably be attributed to chance.

*This general rule is not always true; it is provided only as a guideline.

Here are other examples: If we are comparing men and women on some attribute, the usual null hypothesis is that there is no difference, on average, between men and women; if we are studying a measure of biodiversity in two environments, the usual null hypothesis is that the biodiversities of the two environments are equal, on average; if we are studying two diets, the usual null hypothesis is that the diets produce the same average response.

ANOTHER LOOK AT *P*-VALUE

In order to place *P*-value in a general setting, let us consider some verbal interpretations of *P*-value.

First we revisit the randomization test. For a nondirectional H_A the *P*-value is the proportion of all randomizations that results in a difference of sample means that is as large, or larger than, the difference that was observed in the actual study. Thus we can define the *P*-value as follows:

> The *P*-value of the data is the probability (assuming H_0 is true) of getting a result as extreme as, or more extreme than, the result that was actually observed.

To put this another way,

> The *P*-value is the probability that, if H_0 were true, a result would be obtained that would deviate from H_0 as much as (or more than) the actual data do.

Now consider the *t* test. For a nondirectional H_A, we have defined the *P*-value to be the two-tailed area under the Student's *t* curve beyond the observed value of t_s.

Actually, these descriptions of *P*-value are a bit too limited. The *P*-value actually depends on the nature of the alternative hypothesis. When we are performing a *t* test against a *directional* alternative, the *P*-value of the data is (if the observed deviation is in the direction of H_A) only a *single-tailed* area beyond the observed value of t_s. The more general definition of *P*-value is the following:

> The *P*-value of the data is the probability (assuming H_0 is true) of getting a result as deviant as, or more deviant than, the result actually observed—where deviance is measured as discrepancy from H_0 in the direction of H_A.

The *P*-value measures how easily the observed deviation could be explained as chance variation rather than by the alternative explanation provided by H_A. For example, if the *t* test yields a *P*-value of $P = 0.036$ for our data, then we may say that *if* H_0 *were true* we would expect data to deviate from H_0 as much as our data did only 3.6% of the time (in the meta-study).

Another definition of *P*-value that is worth thinking about is the following:

> The *P*-value of the data is the smallest value of α for which H_0 would be rejected, using those data.

To interpret this definition, imagine that a research report that includes a *P*-value is read by a number of interested scientists. The scientists who are quite skeptical of H_A might require very strong evidence before being convinced and thus would use a very conservative decision threshold, such as $\alpha = 0.001$; the scientists who are more favorably disposed toward H_A might require only weak evidence and thus use a liberal value such as $\alpha = 0.10$. The *P*-value of the data determines the point, within this spectrum of opinion, that separates those who find the data to be convincing in favor of H_A and those who do not. Of course, if the *P*-value is large, for instance $P = 0.40$, then presumably no reasonable person would reject H_0 and be convinced of H_A.

As the preceding discussion shows, the P-value does not describe all facets of the data, but relates only to a test of a particular null hypothesis against a particular alternative. In fact, we will see that the P-value of the data also depends on which statistical test is used to test a given null hypothesis. For this reason, when describing in a scientific report the results of a statistical test, it is best to report the P-value (exactly, if possible), the name of the statistical test, degrees of freedom, and whether the alternative hypothesis was directional or nondirectional.

We repeat here, because it applies to any statistical test, the principle expounded in Section 7.6: The P-value is a measure of the strength of the evidence against H_0, but the P-value does *not* reflect the *magnitude* of the discrepancy between the data and H_0. The data may deviate from H_0 only slightly, yet if the samples are large, the P-value may be quite small. By the same token, data that deviate substantially from H_0 can nevertheless yield a large P-value. The P-value alone does *not* indicate whether a scientific finding is important.

INTERPRETATION OF ERROR PROBABILITIES

A common mistake is to interpret the P-value as the probability that the null hypothesis is true. A related misconception is the belief that, if we find significant evidence for H_A (for example) at the 5% significance level, then the probability that H_0 is true is 5%. These interpretations are not correct.* This point can be illustrated by an analogy with medical diagnosis.

In applying a diagnostic test for an illness, the null hypothesis is that the person is healthy—this is what we will believe unless the medical test indicates otherwise. Two types of error are possible: A healthy individual may be diagnosed as ill (false positive) or an ill individual may be diagnosed as healthy (false negative). Trying out a diagnostic test on individuals *known* to be healthy or ill will enable us to estimate the proportions of these groups who will be misdiagnosed; yet this information alone will not tell us what proportion of all positive diagnoses are false diagnoses. These ideas are illustrated numerically in the next example.

Example 7.9.1 **Medical Testing** Suppose a medical test is conducted to detect an illness. Further, suppose that 1% of the population has the illness in question. If the test indicates that the disease is present, we reject the null hypothesis that the person is healthy. If H_0 is true, then this is a Type I error—a false positive. If the test indicates that the disease is not present, we have a lack of significant evidence for H_A (illness). Suppose that the test has an 80% chance of detecting the disease if the person has it (this is analogous to the power of a hypothesis test being 80%) and a 95% chance of correctly indicating that the disease is absent if the person really does not have the disease (this is analogous to a 5% Type I error rate). Figure 7.9.1 shows a probability tree for this situation, with bold lines indicating the two ways in which the test result can be positive (i.e., the two ways that H_0 can be rejected).

Now suppose that 100,000 persons are tested and that 1,000 of them (1%) actually have the illness. Then we would expect results like those given in Table 7.9.1, with 5,750 persons testing positive (which is like finding significant evidence for H_A 5,750 times). Of these, 4,950 are false positives. Put another way, the proportion of the time that H_0 is true, given that we found significant evidence for H_A, is $\frac{4{,}950}{5{,}750} \approx 0.86$,

*In fact, the probability that H_0 is true cannot be calculated at all within the standard, "frequentist" approach to hypothesis testing. $\Pr\{H_0 \text{ is true}\}$ *can* be calculated if one uses what are known as Bayesian methods, which are beyond the scope of this book.

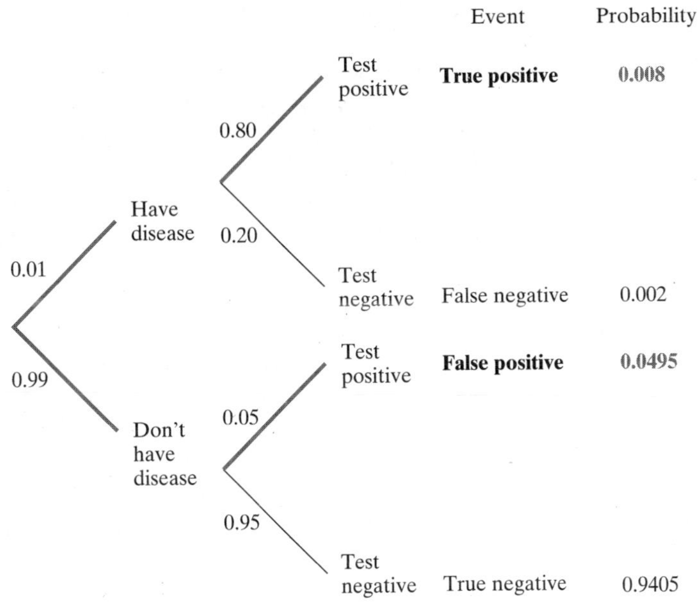

Figure 7.9.1 Probability tree for medical testing example

Table 7.9.1 Hypothetical results of medical test of 100,000 persons

		True situation		
		Healthy (H_0 true)	Ill (H_A true)	Total
TEST RESULT	Negative (lack of significant evidence for H_A)	94,050	200	94,250
	Positive (significant evidence for H_A)	4,950	800	5,750
	Total	99,000	1,000	100,000

which is quite different from 0.05; this startlingly high proportion of false positives is due to the rarity of the disease. (The proportion of times that there is significant evidence for H_A, given that H_0 is true, is $\frac{4{,}950}{99{,}000} = 0.05$, as expected, but that is a different conditional probability. Pr$\{A \text{ given } B\} \neq$ Pr$\{B \text{ given } A\}$: The probability of rainfall, given that there are thunder and lightning, is not the same as the probability of thunder and lightning, given that it is raining.) ■

The risk of Type I error is a probability computed *under the assumption that H_0 is true*; similarly, the risk of a Type II error is computed assuming that H_A is true. If we have a well-designed study with adequate sample sizes, both of these probabilities will be small. We then have a good test procedure in the same sense that the medical test is a good diagnostic procedure. But this does not in itself guarantee that most of the null hypotheses we reject are in fact false, or that most of those we do not reject are in fact true. The validity or nonvalidity of such guarantees would depend on an unknown and unknowable quantity—namely, the proportion of true null hypotheses among all null hypotheses that are tested (which is analogous to the incidence of the illness in the medical test scenario).

AN IMPLICIT ASSUMPTION

In discussing the tests of this chapter, we have made an unspoken assumption that we now bring to light. When interpreting the comparison of two distributions, we have assumed that the relationship between the two distributions is relatively simple—that if the distributions differ, then one of the two variables has a consistent tendency to be larger than the other. For instance, suppose we are comparing the effects of two diets on the weight gain of mice, with

Y_1 = Weight gain of mice on diet 1
Y_2 = Weight gain of mice on diet 2

Our implicit assumption has been that, if the two diets differ at all, then that difference is in a consistent direction for all individual mice. To appreciate the meaning of this assumption, suppose the two distributions are as pictured in Figure 7.9.2. In this case, even though the mean weight gain is higher on diet 1, it would be an oversimplification to say that mice tend to gain more weight on diet 1 than on diet 2; apparently *some* mice gain *less* on diet 1. Paradoxical situations of this kind do occasionally occur, and then the simple analysis typified by the *t* test (or the randomization test, or the Wilcoxon-Mann-Whitney test of Section 7.10) may be inadequate.

Figure 7.9.2 Weight gain distributions on two diets

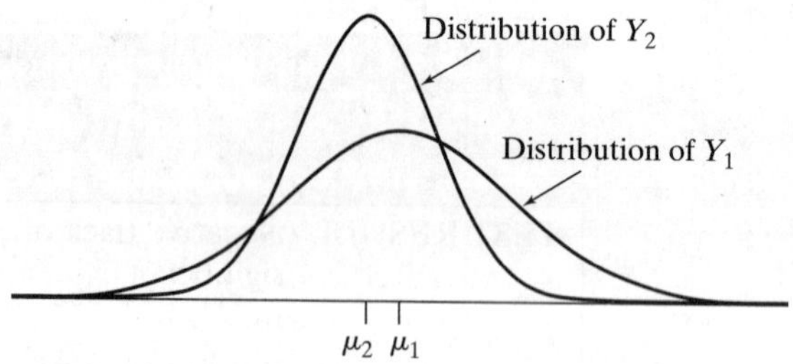

It is relatively easy to compare two distributions that have the same general shape and similar standard deviations. However, if either the shapes or the SDs of two distributions are very different from one another, then making a meaningful comparison of the distributions is difficult. In particular, a comparison of the two means might not be appropriate.*

PERSPECTIVE

We should mention that the philosophy of statistical hypothesis testing that we have explained in this chapter is not shared by all statisticians. The view presented here, which is called the **frequentist view,** is widely used in scientific research. An alternative view, the **Bayesian view,** incorporates not only the data observed in the study at hand, but also the information that the researcher has from previous, related studies. In the past, many Bayesian techniques were not practical due to the complexity of the mathematics that they require. However, greater computing power and improved software have made Bayesian methods more popular in recent years.

*Randomization tests are more general than *t* tests. However, we note that although randomization tests do not require a parametric family, they do require a condition known as "exchangeability," which implies that a randomization test of a difference in means requires equal variances.

Exercise 7.9.1

7.9.1 Suppose we have conducted a t test, with $\alpha = 0.05$, and the P-value is 0.04. For each of the following statements, say whether the statement is true or false and explain why.

(a) There is a 4% chance that H_0 is true.

(b) We reject H_0 with $\alpha = 0.05$.

(c) We should reject H_0, and if we repeated the experiment, there is a 4% chance that we would reject H_0 again.

(d) If H_0 is true, the probability of getting a test statistic at least as extreme as the value of the t_s that was actually obtained is 4%.

7.10 The Wilcoxon-Mann-Whitney Test

The **Wilcoxon-Mann-Whitney test** is used to compare two independent samples.* It is a competitor to the t test, but unlike the t test, the Wilcoxon-Mann-Whitney test is valid even if the population distributions are not normal. The Wilcoxon-Mann-Whitney test is therefore called a **distribution-free** type of test. In addition, the Wilcoxon-Mann-Whitney test does not focus on any particular parameter such as a mean or a median; for this reason it is called a **nonparametric** type of test.

STATEMENT OF H_0 AND H_A

Let us denote the observations in the two samples by Y_1 and Y_2. A general statement of the null and alternative hypotheses of a Wilcoxon-Mann-Whitney test are

H_0: The population distributions of Y_1 and Y_2 are the same.

H_A: The population distribution of Y_1 is shifted from the population distribution of Y_2 (i.e., Y_1 tends to be either greater or less than Y_2).

In practice, it is more natural to state H_0 and H_A in words suitable to the particular application, as illustrated in Example 7.10.1.

Example 7.10.1

Soil Respiration Soil respiration is a measure of microbial activity in soil, which affects plant growth. In one study, soil cores were taken from two locations in a forest: (1) under an opening in the forest canopy (the "gap" location) and (2) at a nearby area under heavy tree growth (the "growth" location). The amount of carbon dioxide given off by each soil core was measured (in mol CO_2/g soil/hr). Table 7.10.1 contains the data.[63]

Table 7.10.1 Soil respiration data (mol CO_2/g soil/hr) from example 7.10.1

Growth	Gap
17 20 170 315	22 29 13 16
22 190 64	15 18 14 6

*The test presented here was developed by Wilcoxon in a 1945 article. Mann and Whitney, in a 1947 article, elaborated on the test, which can be conducted in two mathematically equivalent ways. Thus, some books and some computer programs implement the test in a different fashion than the way it is presented here. Also note that some books refer to this as the Wilcoxon test, some as the Mann–Whitney test, and some (including this text) as the Wilcoxon-Mann-Whitney test.

An appropriate null hypothesis could be stated as

H_0: The populations from which the two samples were drawn have the same distribution of soil respiration.

or, more informally, as

H_0: The gap and growth areas do not differ with respect to soil respiration.

A nondirectional alternative could be stated as

H_A: The distribution of soil respiration rates tends to be higher in one of the two populations.

or the alternative hypothesis might be directional, for example,

H_A: Soil respiration rates tend to be greater in the growth area than they are in the gap area. ■

APPLICABILITY OF THE WILCOXON-MANN-WHITNEY TEST

Figure 7.10.1 shows dotplots of the soil respiration data from Example 7.10.1; Figure 7.10.2 shows normal probability plots of these data. The growth distribution is skewed to the right, whereas the gap distribution is slightly skewed to the left. If both distributions were skewed to the right, we could apply a transformation to the data. However, any attempt to transform the growth distribution, such

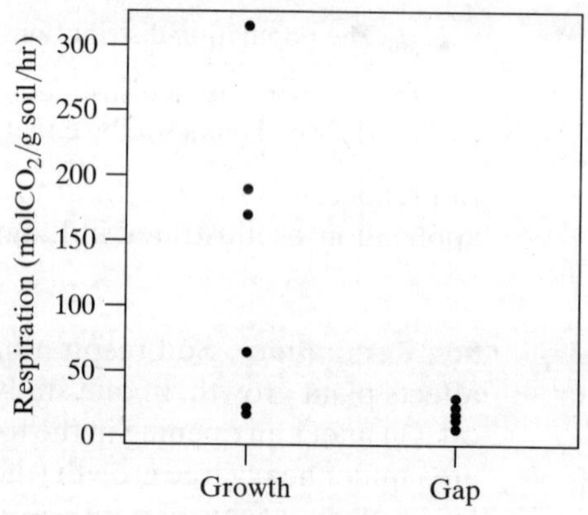

Figure 7.10.1 Dotplots of the soil respiration data from Example 7.10.1

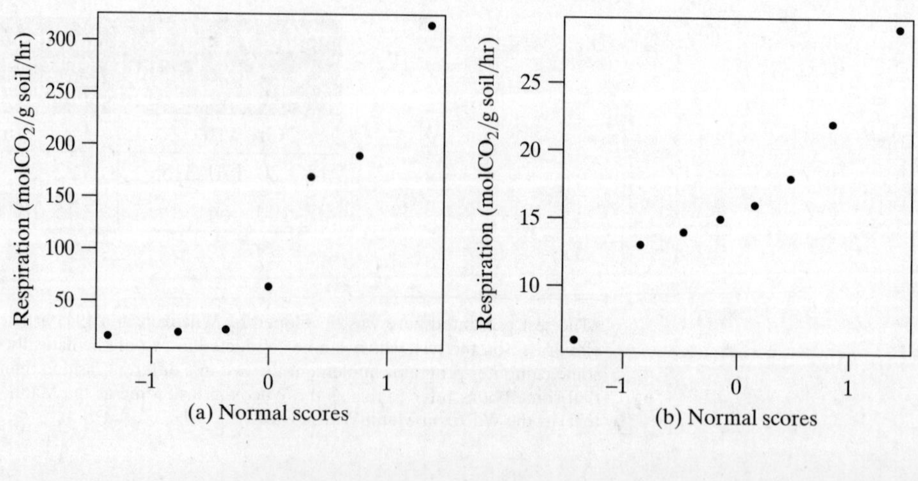

Figure 7.10.2 Normal probability plots of (a) the growth data and (b) the gap data from Example 7.10.1

as taking logarithms of the data, will make the skewness of the gap distribution worse. Hence, the t test is not applicable here. The Wilcoxon-Mann-Whitney test does not require normality of the distributions.

METHOD

The Wilcoxon-Mann-Whitney test statistic, which is denoted U_s, measures the degree of separation or shift between two samples. A large value of U_s indicates that the two samples are well separated, with relatively little overlap between them. Critical values for the Wilcoxon-Mann-Whitney test are given in Table 6 at the end of this book. The following example illustrates the Wilcoxon-Mann-Whitney test.

Example 7.10.2 **Soil Respiration** Let us carry out a Wilcoxon-Mann-Whitney test on the biodiversity data of Example 7.10.1.

1. The value of U_s depends on the relative positions of the Y_1's and the Y_2's. The first step in determining U_s is to arrange the observations in increasing order, as is shown in Table 7.10.2.
2. We next determine two counts, K_1 and K_2, as follows:
   (a) *The K_1 count* For each observation in sample 1, we count the number of observations in sample 2 that are smaller in value (i.e., to the left). We count 1/2 for each tied observation. In the above data, there are five Y_2's less than the first Y_1, there are six Y_2's less than the second Y_1, there are six Y_2's less than the third Y_1 and one equal to it, so we count 6 1/2. So far we have counts of 5, 6, and 6.5. Continuing in a similar way, we get further counts of 8, 8, 8, and 8. All together there are seven counts, one for each Y_1. The sum of all seven counts is $K_1 = 49.5$.
   (b) *The K_2 count* For each observation in sample 2, we count the number of observations in sample 1 that are smaller in value, counting 1/2 for ties. This gives counts of 0, 0, 0, 0, 0, 1, 2.5, and 3. The sum of these counts is $K_2 = 6.5$.
   (c) *Check* If the work is correct, the sum of K_1 and K_2 should be equal to the product of the sample sizes:

$$K_1 + K_2 = n_1 n_2$$

$$49.5 + 6.5 = 7 \times 8$$

Table 7.10.2 Wilcoxon-Mann-Whitney calculations for example 7.10.2

Number of gap observations that are smaller	Y_1 Growth data	Y_2 Gap data	Number of growth observations that are smaller
5	17	6	0
6	20	13	0
6.5	22	14	0
8	64	15	0
8	170	16	0
8	190	18	1
8	315	22	2.5
		29	3
$K_1 = 49.5$			$K_2 = 6.5$

3. The test statistic U_s is the larger of K_1 and K_2. In this example, $U_s = 49.5$.
4. To determine the P-value, we consult Table 6 with $n =$ the larger sample size, and $n' =$ the smaller sample size. In the present case, $n = 8$ and $n' = 7$. Values from Table 6 are reproduced in Table 7.10.3.

Table 7.10.3 Values from table 6 for $n = 8, n' = 7$

40 *0.189*	**44** *0.093*	**46** *0.054*	**47** *0.040*	**48** *0.021*	**49** *0.014*	**50** *0.009*

Let us test H_0 against a nondirectional alternative at significance level $\alpha = 0.05$. From Table 7.10.3, we note that when $U_s = 49$, the P-value is 0.014 and when $U_s = 50$, the P-value is 0.009; since $49 < U_s < 50$, the P-value is between 0.009 and 0.014 and thus there is significant evidence for H_A. There is sufficient evidence to conclude that soil respiration rates are different in the gap and growth areas. ■

As Example 7.10.2 illustrates, Table 6 can be used to bracket the P-value for the Wilcoxon-Mann-Whitney test just as Table 4 is used for the t test. If the observed U_s value is not given, then one simply locates the values that bracket the observed U_s. One then brackets the P-value by the corresponding column headings.

Directionality For the t test, one determines the directionality of the data by seeing whether $\bar{Y}_1 > \bar{Y}_2$ or $\bar{Y}_1 < \bar{Y}_2$. Similarly, one can check directionality for the Wilcoxon-Mann-Whitney test by comparing K_1 and K_2: $K_1 > K_2$ indicates a trend for the Y_1's to be larger than the Y_2's, while $K_1 < K_2$ indicates the opposite trend. Often, however, this formal comparison is unnecessary; a glance at a graph of the data is enough.

Directional Alternative If the alternative hypothesis H_A is directional rather than nondirectional, the Wilcoxon-Mann-Whitney procedure must be modified. As with the t test, the modified procedure has two steps and the second step involves halving the nondirectional P-value to obtain the directional P-value.

Step 1. Check directionality—see if the data deviate from H_0 in the direction specified by H_A.
(a) If not, the P-value is greater than 0.50.
(b) If so, proceed to step 2.

Step 2. The P-value of the data is half as much as it would be if H_A were nondirectional.

To make a decision at a prespecified significance level α, one claims significant evidence for H_A if P-value $\leq \alpha$.

The following example illustrates the two-step procedure.

Example 7.10.3 **Directional H_A** Suppose $n = 8, n' = 7$, and H_A is directional. Suppose further that the data do deviate from H_0 in the direction specified by H_A. The values shown in Table 7.10.3 can be used to find the P-value as follows:

If $U_s = 40$, then P-value $= 0.189/2 = 0.0945$.

If $U_s = 46$, then P-value $= 0.054/2 = 0.027$.

If $U_s = 49.5$, then $0.009/2 < P\text{-value} < 0.014/2$ so $0.0045 < P\text{-value} < 0.007$.

If $U_s = 50$ (or larger), then P-value $< 0.009/2 = 0.0045$. ■

RATIONALE

Let us see why the Wilcoxon-Mann-Whitney test procedure makes sense. To take a specific case, suppose the sample sizes are $n_1 = 5$ and $n_2 = 4$, so there are $5 \times 4 = 20$ comparisons that can be made between a data point in the first sample and a data point in the second sample. Thus, regardless of what the data look like, we must have

$$K_1 + K_2 = 5 \times 4 = 20$$

The relative magnitudes of K_1 and K_2 indicate the amount of overlap of the Y_1's and the Y_2's. Figure 7.10.3 shows how this works. For the data of Figure 7.10.3(a), the two samples do not overlap at all; the data are *least* compatible with H_0 and show the *strongest* evidence for H_A and thus U_s has its maximum value, $U_s = 20$. Similarly, $U_s = 20$ for Figure 7.10.3(b). On the other hand, the arrangement *most* compatible with H_0 and shows a lack of evidence for H_A is the one with maximal overlap, shown in Figure 7.10.3(c); for this arrangement $K_1 = 10$, $K_2 = 10$, and $U_s = 10$.

Figure 7.10.3 Three data arrays for a Wilcoxon-Mann-Whitney Test

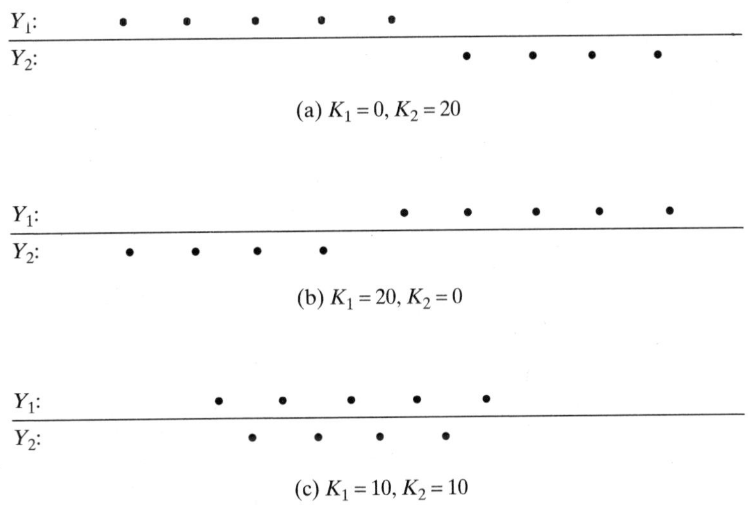

All other possible arrangements of the data lie somewhere between the three arrangements shown in Figure 7.10.3; those with much overlap have U_s close to 10, and those with little overlap have U_s closer to 20. Thus, large values of U_s indicate evidence for the research hypothesis, H_A, or equivalently the incompatibility of the data with H_0.

We now briefly consider the null distribution of U_s and indicate how the critical values of Table 6 were determined. (Recall from Section 7.9 that, for any statistical test, the reference distribution for critical values is always the null distribution of the test statistic—that is, its sampling distribution under the condition that H_0 is true.) To determine the null distribution of U_s, it is necessary to calculate the probabilities associated with various arrangements of the data, assuming that all the Y's were actually drawn from the same population.* (The method for calculating the probabilities is briefly described in Appendix 7.2.)

---

*In calculating the probabilities used in this section, it has been assumed that the chance of tied observations is negligible. This will be true for a continuous variable that is measured with high precision. If the number of ties is large, a correction can be made; see Noether (1967).[64]

Figure 7.10.4(a) shows the null distribution of K_1 and K_2 for the case $n = 5, n' = 4$. For example, it can be shown that, if H_0 is true, then

$$\Pr\{K_1 = 0, K_2 = 20\} = 0.008$$

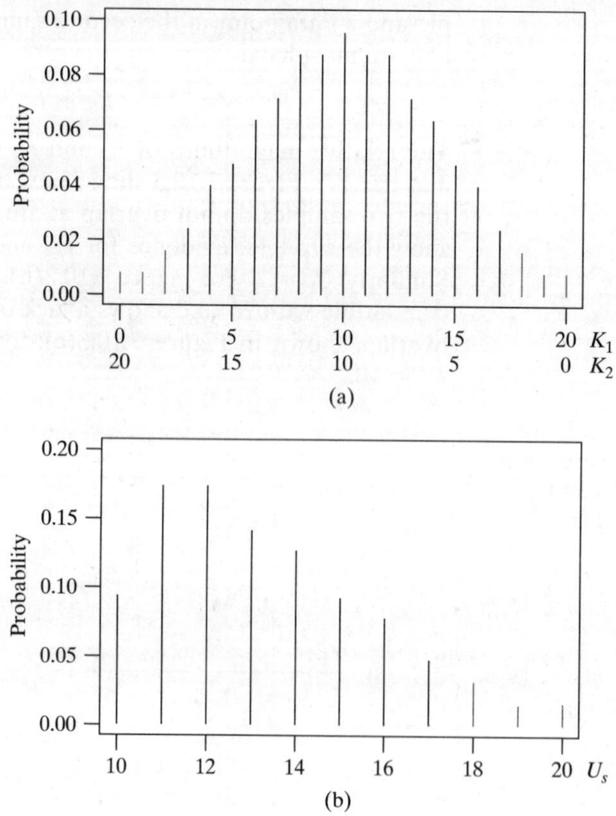

Figure 7.10.4 Null distributions for the Wilcoxon-Mann-Whitney test when $n = 5, n' = 4$. (a) Null distribution of K_1 and K_2; (b) Null distribution of U_s. Shading corresponds to the P-value when $U_s = 18$.

This is the first probability plotted in Figure 7.10.4(a). Note that Figure 7.10.4(a) is roughly analogous to a t distribution; large values of K_1 (right tail) represent evidence that the Y_1's tend to be larger than the Y_2's and large values of K_2 (left tail) represent evidence that the Y_2's tend to be larger than the Y_1's.

Figure 7.10.4(b) shows the null distribution of U_s, which is derived directly from the distribution in Figure 7.10.4(a). For instance, if H_0 is true, then

$$\Pr\{K_1 = 0, K_2 = 20\} = 0.008$$

and

$$\Pr\{K_1 = 20, K_2 = 0\} = 0.008$$

so that

$$\Pr\{U_s = 20\} = 0.008 + 0.008 = 0.016$$

which is the rightmost probability plotted in Figure 7.10.4(b). Thus, both tails of the K distribution have been "folded" into the upper tail of the U distribution; for instance, the one-tailed shaded area in Figure 7.10.4(b) is equal to the two-tailed shaded area in Figure 7.10.4(a).

P-values for the Wilcoxon-Mann-Whitney test are upper-tail areas in the U_s distribution. For instance, it can be shown that the blue shaded area in Figure 7.10.4(b) is equal to 0.064; this means that if H_0 is true, then

$$\Pr\{U_s \geq 18\} = 0.064$$

Thus, a data set that yielded $U_s = 18$ would have an associated P-value 0.064 (assuming a nondirectional H_A).

The values in Table 6 have been determined from the null distribution of U_s. Because the U_s distribution is discrete, only a few P-values are possible for any given sample sizes n_1 and n_2. Table 6 shows selected values of U_s in bold type, with the P-value given in italics. For example, if the sample sizes are 5 and 4, then a U_s value of 17 gives a P-value of 0.111, a U_s value of 18 gives a P-value of 0.064, and a U_s value of 19 gives a P-value of 0.032. Thus, to achieve statistical significance at the $\alpha = 0.05$ level requires a test statistic (U_s) value of at least 19. The smallest possible P-value when the sample sizes are 5 and 4 is 0.016, when $U_s = 20$, which means that statistical significance at the $\alpha = 0.01$ level cannot be obtained with a nondirectional test.

CONDITIONS FOR USE OF THE WILCOXON-MANN-WHITNEY TEST

In order for the Wilcoxon-Mann-Whitney test to be applicable, it must be reasonable to regard the data as random samples from their respective populations, with the observations within each sample being independent, and the two samples being independent of each other. Under these conditions, the Wilcoxon-Mann-Whitney test is valid no matter what the form of the population distributions, provided that the observed variable Y is continuous.[65]

The critical values given in Table 6 have been calculated assuming that ties do not occur. If the data contain only a few ties, then the P-values are approximately correct.*

THE WILCOXON-MANN-WHITNEY TEST VERSUS THE *t* TEST AND THE RANDOMIZATION TEST

While the Wilcoxon-Mann-Whitney test and the t test are aimed at answering the same basic question—Are the locations of the two population distributions different or does one population tend to have larger (or smaller) values than the other?—they treat the data in very different ways. Unlike the t test, the Wilcoxon-Mann-Whitney test does not use the actual values of the Y's but only their relative positions in a rank ordering. This is both a strength and a weakness of the Wilcoxon-Mann-Whitney test. On the one hand, the test is distribution-free because the null distribution of U_s relates only to the various rankings of the Y's, and therefore does not depend on the form of the population distribution. On the other hand, the Wilcoxon-Mann-Whitney test can be inefficient: It can lack power because it does not use all the information in the data. This inefficiency is especially evident for small samples.

The randomization test is similar in spirit to the Wilcoxon-Mann-Whitney test in that it does not depend on normality, yet the power of the randomization test is often similar to that of the t test. Conducting a randomization test can be difficult, which is a primary reason that randomization tests were not more widely used until computing power became more prevalent.

None of the competitors—the randomization test, the t test, or the Wilcoxon-Mann-Whitney test—is clearly superior to the others. If the population distributions are not approximately normal, the t test may not even be valid. In addition, the Wilcoxon-Mann-Whitney test can be much more powerful than the t test, especially

*Actually, the Wilcoxon-Mann-Whitney test need not be restricted to continuous variables; it can be applied to any ordinal variable. However, if Y is discrete or categorical, then the data may contain many ties, and the test should not be used without appropriate modification of the critical values.

if the population distributions are highly skewed. If the population distributions are approximately normal with equal standard deviations, then the t test is best, but its properties are similar to those of the randomization test. For moderate sample sizes, the Wilcoxon-Mann-Whitney test can be nearly as powerful as the t test.[66]

There is a confidence interval procedure for population medians that is associated with the Wilcoxon-Mann-Whitney test in the same way that the confidence interval for $(\mu_1 - \mu_2)$ is associated with the t test. The procedure is beyond the scope of this book.

Exercises 7.10.1–7.10.10

7.10.1 Consider two samples of sizes $n_1 = 5, n_2 = 7$. Use Table 6 to find the P-value, assuming that H_A is nondirectional and that

(a) $U_s = 26$
(b) $U_s = 30$
(c) $U_s = 35$

7.10.2 Consider two samples of sizes $n_1 = 4, n_2 = 8$. Use Table 6 to find the P-value, assuming that H_A is nondirectional and that

(a) $U_s = 25$
(b) $U_s = 31$
(c) $U_s = 32$

7.10.3 In a pharmacological study, researchers measured the concentration of the brain chemical dopamine in six rats exposed to toluene and six control rats. (This is the same study described in Example 7.2.1.) The concentrations in the striatum region of the brain were as shown in the table.[4]

Dopamine (ng/gm)	
Toluene	Control
3,420	1,820
2,314	1,843
1,911	1,397
2,464	1,803
2,781	2,539
2,803	1,990

(a) Use a Wilcoxon-Mann-Whitney test to compare the treatments at $\alpha = 0.05$. Use a nondirectional alternative.
(b) Proceed as in part (a), but let the alternative hypothesis be that toluene increases dopamine concentration.

7.10.4 In a study of hypnosis, breathing patterns were observed in an experimental group of subjects and in a control group. The measurements of total ventilation (liters of air per minute per square meter of body area) are shown.[67] (These are the same data that were summarized in Exercise 7.5.6.) Use a Wilcoxon-Mann-Whitney test to compare the two groups at $\alpha = 0.10$. Use a nondirectional alternative.

Experimental	Control
5.32	4.50
5.60	4.78
5.74	4.79
6.06	4.86
6.32	5.41
6.34	5.70
6.79	6.08
7.18	6.21

7.10.5 In an experiment to compare the effects of two different growing conditions on the heights of greenhouse chrysanthemums, all plants grown under condition 1 were found to be taller than any of those grown under condition 2 (i.e., the two height distributions did not overlap). Calculate the value of U_s and find the P-value if the number of plants in each group was

(a) 3
(b) 4
(c) 5

(Assume that H_A is nondirectional.)

7.10.6 In a study of preening behavior in the fruitfly *Drosophila melanogaster,* a single experimental fly was observed for 3 minutes while in a chamber with 10 other flies of the same sex. The observer recorded the timing of each episode ("bout") of preening by the experimental fly. This experiment was replicated 15 times with male flies and 15 times with female flies (different flies each time). One question of interest was whether there is a sex difference in preening behavior. The observed preening times (average time per bout, in seconds) were as follows[68]:

Male: 1.2, 1.2, 1.3, 1.9, 1.9, 2.0, 2.1, 2.2, 2.2, 2.3, 2.3, 2.4, 2.7, 2.9, 3.3

$\bar{y} = 2.127 \quad s = 0.5936$

Female: 2.0, 2.2, 2.4, 2.4, 2.4, 2.8, 2.8, 2.8, 2.9, 3.2, 3.7, 4.0, 5.4, 10.7, 11.7

$\bar{y} = 4.093 \quad s = 3.014$

(a) For these data, the value of the Wilcoxon-Mann-Whitney statistic is $U_s = 189.5$. Use a Wilcoxon-Mann-Whitney test to investigate the sex difference in preening behavior. Let H_A be nondirectional and let $\alpha = 0.01$.

(b) For these data, the standard error of $(\bar{Y}_1 - \bar{Y}_2)$ is SE = 0.7933 sec. Use a t test to investigate the sex difference in preening behavior. Let H_A be nondirectional and let $\alpha = 0.01$.

(c) What condition is required for the validity of the t test but not for the Wilcoxon-Mann-Whitney test? What feature or features of the data suggest that this condition may not hold in this case?

(d) Verify the value of U_s given in part (a).

7.10.7 Substances to be tested for cancer-causing potential are often painted on the skin of mice. The question arose whether mice might get an additional dose of the substance by licking or biting their cagemates. To answer this question, the compound benzo(a)pyrene was applied to the backs of 10 mice: 5 were individually housed, and 5 were group-housed in a single cage. After 48 hours, the concentration of the compound in the stomach tissue of each mouse was determined. The results (nmol/gm) were as follows[69]:

Singly housed	Group-housed
3.3	3.9
2.4	4.1
2.5	4.8
3.3	3.9
2.4	3.4

(a) What is the value of the Wilcoxon-Mann-Whitney test statistic for comparing the distributions?

(b) Let the alternative hypothesis be that benzo(a)pyrene concentrations tend to be high in group-housed mice than in singly housed mice. The P-value for the directional Wilcoxon-Mann-Whitney test is 0.004. If $\alpha = 0.01$, what is your conclusion regarding H_0?

(c) Why is a directional alternative valid in this case?

7.10.8 Human beta-endorphin (HBE) is a hormone secreted by the pituitary gland under conditions of stress. An exercise physiologist measured the resting (unstressed) blood concentration of HBE in two groups of men: Group 1 consisted of 11 men who had been jogging regularly for some time, and group 2 consisted of 15 men who had just entered a physical fitness program. The results are given in the following table.[70]

Joggers	Fitness program entrants
39 40 32 60	70 47 54 27 31
19 52 41 32	42 37 41 9 18
13 37 28	33 23 49 41 59

Use a Wilcoxon-Mann-Whitney test to compare the two distributions at $\alpha = 0.10$. Use a nondirectional alternative.

7.10.9 (Continuation of 7.10.8) Below are normal probability plots of the HBE data from Exercise 7.10.8.

(a) Using the plots to support your answer, is there evidence of abnormality in either of the samples?

(b) Considering your answer to (a) and the preceding plots, should we conclude that the data are indeed normally distributed? Explain.

(c) If the data are indeed normally distributed, explain in the context of this problem what the drawback would be with using the Wilcoxon-Mann-Whitney test over the two-sample t test to analyze these data.

(d) If the data are not normally distributed, explain in the context of this problem what the drawback would be with using the two-sample t test over the Wilcoxon-Mann-Whitney test to analyze this data.

(e) Considering your answers to the above, argue which test should be used with these data. Note there is more than one correct answer.

Joggers

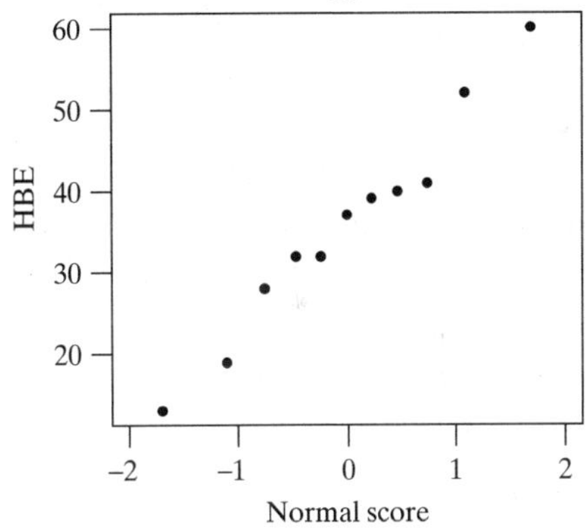

Fitness program

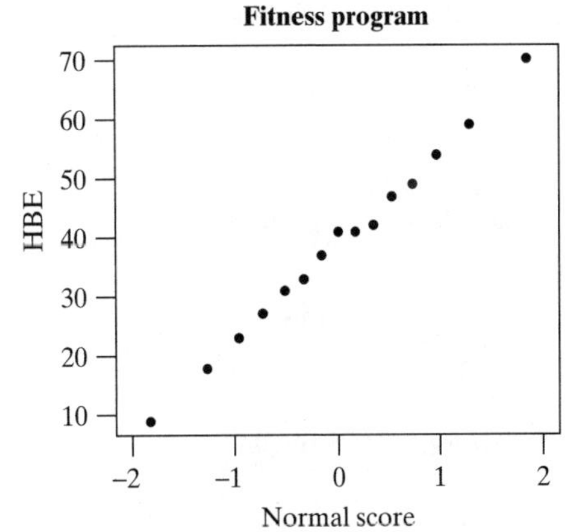

7.10.10 Recall from Exercise 7.6.11 that patients with non-chronic low back pain were randomly assigned to either a treatment group that received health coaching via telephone, in addition to physiotherapy care, or a control group that received only the physiotherapy care. One variable measured was improvement, over 12 weeks, in "modified

Oswestry Disability Index," which is scored as a percentage, ranging from 0 to 100, with high scores representing high levels of disability. Data are shown in the table.

Experimental	Control
58	18
−4	−6
10	−22
−6	8
52	20
26	12
10	38
38	18
26	42
−2	−44
12	56
44	6
28	4

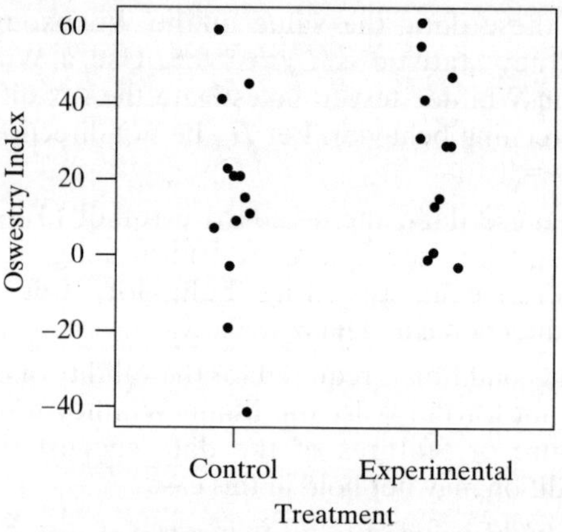

The Wilcoxon-Mann-Whitney test statistic for these data is 104.5. If the alternative hypothesis is nondirectional, do the data provide evidence of a difference in the two distributions?

Supplementary Exercises 7.S.1–7.S.35

(*Note:* Exercises preceded by an asterisk refer to optional sections.)

Answers to hypothesis testing questions should include a statement of the conclusion in the context of the setting. (See Examples 7.2.4 and 7.2.5.)

7.S.1 For each of the following pairs of samples, compute the standard error of $(\bar{Y}_1 - \bar{Y}_2)$.

(a)

	Sample 1	Sample 2
n	14	16
$\bar{y}$	47	49
s	9.8	10.4

(b)

	Sample 1	Sample 2
n	26	23
$\bar{y}$	114	130
s	3.1	2.3

(c)

	Sample 1	Sample 2
n	8	10
$\bar{y}$	17	19
SE	1.5	1.7

7.S.2 To investigate the relationship between intracellular calcium and blood pressure, researchers measured the free calcium concentration in the blood platelets of 38 people with normal blood pressure and 45 people with high blood pressure. The results are given in the table and the distributions are shown in the boxplots.[71] Use the t test to compare the means. Let $\alpha = 0.05$ and let H_A be nondirectional. [*Note:* Formula (6.7.1) yields 67.5 df.]

		Platelet calcium (nM)	
Blood pressure	n	Mean	SD
Normal	38	107.9	16.1
High	45	168.2	31.7

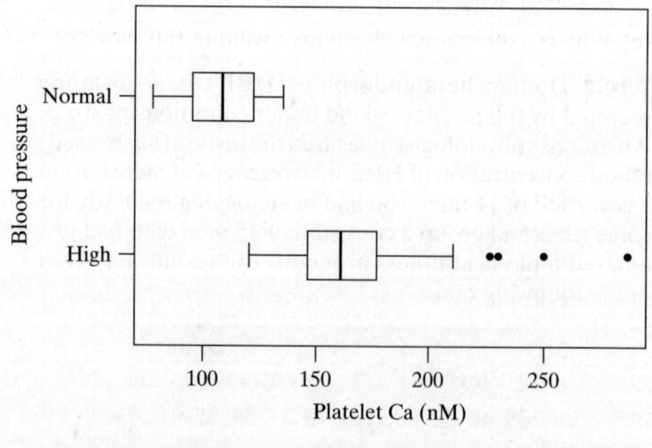

7.S.3 Refer to Exercise 7.S.2. Construct a 98% confidence interval for the difference between the population means.

7.S.4 Refer to Exercise 7.S.2. The boxplot for the high blood pressure group is skewed to the right and includes outliers. Does this mean that the t test is not valid for these data? Why or why not?

7.S.5 In a study of methods of producing sheep's milk for use in cheese manufacture, ewes were randomly allocated to either a mechanical or a manual milking method. The investigator suspected that the mechanical method might irritate the udder and thus produce a higher concentration of somatic cells in the milk. The accompanying data show the average somatic cell count for each animal.[72]

	Somatic count ($10^{-3} \times$ cells/ml)	
	Mechanical milking	Manual milking
	2,966	186
	269	107
	59	65
	1,887	126
	3,452	123
	189	164
	93	408
	618	324
	130	548
	2,493	139
n	10	10
Mean	1,215.6	219.0
SD	1,342.9	156.2

(a) Do the data support the investigator's suspicion? Use a t test against a directional alternative at $\alpha = 0.05$. The standard error of $(\bar{Y}_1 - \bar{Y}_2)$ is SE $= 427.54$ and formula (6.7.1) yields 9.2 df.

(b) Do the data support the investigator's suspicion? Use a Wilcoxon-Mann-Whitney test against a directional alternative at $\alpha = 0.05$. (The value of the Wilcoxon-Mann-Whitney statistic is $U_s = 69$.) Compare with the result of part (a).

(c) What condition is required for the validity of the t test but not for the Wilcoxon-Mann-Whitney test? What features of the data cast doubt on this condition?

(d) Verify the value of U_s given in part (b).

7.S.6 A plant physiologist conducted an experiment to determine whether mechanical stress can retard the growth of soybean plants. Young plants were randomly allocated to two groups of 13 plants each. Plants in one group were mechanically agitated by shaking for 20 minutes twice daily, while plants in the other group were not agitated. After 16 days of growth, the total stem length (cm) of each plant was measured, with the results given in the accompanying table.[73]

Consider a t test to compare the treatments with the alternative hypothesis that stress tends to retard growth.

	Control	Stress
n	13	13
$\bar{y}$	30.59	27.78
s	2.13	1.73

(a) What is the value of the t test statistic for comparing the means?

(b) The P-value for the t test is 0.0006. If $\alpha = 0.05$, what is your conclusion regarding H_0?

7.S.7 Refer to Exercise 7.S.6. Construct a 95% confidence interval for the population mean reduction in stem length. Does the confidence interval indicate whether the effect of stress is "horticulturally important," if "horticulturally important" is defined as a reduction in population mean stem length of at least

(a) 1 cm

(b) 2 cm

(c) 5 cm

7.S.8 Refer to Exercise 7.S.6. The observations (cm), in increasing order, are shown. Compare the treatments using a Wilcoxon-Mann-Whitney test at $\alpha = 0.01$. Let the alternative hypothesis be that stress tends to retard growth.

Control	Stress
25.2	24.7
29.5	25.7
30.1	26.5
30.1	27.0
30.2	27.1
30.2	27.2
30.3	27.3
30.6	27.7
31.1	28.7
31.2	28.9
31.4	29.7
33.5	30.0
34.3	30.6

7.S.9 One measure of the impact of pollution along a river is the diversity of species in the river floodplain. In one study, two rivers, the Black River and the Vermilion River, were compared. Random 50 m $\times$ 20 m plots were sampled along each river and the number of species of trees in each plot was recorded. The following table contains the data.[74]

Vermilion River	Black River
9 9 16 13 12	13 10 6 9
13 13 13 8 11	10 7 6 18
9 9 10	6

The Black River was considered to have been polluted quite a bit more than the Vermilion River, and this was expected to lead to lower biodiversity along the Black River. Consider a Wilcoxon-Mann-Whitney test, with $\alpha = 0.10$, of the null hypothesis that the populations from which the two samples were drawn have the same biodiversity (distribution of tree species per plot) versus an appropriate directional alternative. The *P*-value for the test is 0.078. State your conclusion regarding H_0 in the context of this setting.

7.S.10 A developmental biologist removed the oocytes (developing egg cells) from the ovaries of 24 frogs (*Xenopus laevis*). For each frog the oocyte pH was determined. In addition, each frog was classified according to its response to a certain stimulus with the hormone progesterone. The pH values were as follows[75]:

Positive response:

7.06, 7.18, 7.30, 7.30, 7.31, 7.32, 7.33, 7.34, 7.36, 7.36, 7.40, 7.41, 7.43, 7.48, 7.49, 7.53, 7.55, 7.57

No response:

7.55, 7.70, 7.73, 7.75, 7.75, 7.77

Investigate the relationship of oocyte pH to progesterone response using a Wilcoxon-Mann-Whitney test at $\alpha = 0.05$. Use a nondirectional alternative.

7.S.11 Refer to Exercise 7.S.10. Summary statistics for the pH measurements are given in the following table. We wish to investigate the relationship of oocyte pH to progesterone response using a *t* test at $\alpha = 0.01$ and a nondirectional alternative.

	Positive response	No response
n	18	6
$\bar{y}$	7.373	7.708
s	0.129	0.081

(a) Write the null hypothesis and the alternative hypothesis in symbols.

(b) Here is computer output for the *t* test. Explain what the *P*-value means in the context of this study. [*Note:* 3.042e-06 means 3.042×10^{-6} which is 0.000003042.]

```
t = -7.46, df = 14.003, p-value = 3.042e-06
alternative hypothesis: true difference in
means is not equal to 0
```

(c) If $\alpha = 0.01$, what is your conclusion regarding H_0?

7.S.12 A proposed new diet for beef cattle is less expensive than the standard diet. The proponents of the new diet have conducted a comparative study in which one group of cattle was fed the new diet and another group was fed the standard. They found that the mean weight gains in the two groups were not statistically significantly different at the 5% significance level, and they stated that this finding supported the claim that the new cheaper diet was as good (for weight gain) as the standard diet. Criticize this statement.

***7.S.13** Refer to Exercise 7.S.12. Suppose you discover that the study used 25 animals on each of the two diets, and that the coefficient of variation of weight gain under the conditions of the study was about 20%. Using this additional information, write an expanded criticism of the proponents' claim, indicating how likely such a study would be to detect a 10% deficiency in weight gain on the cheaper diet (using a two-tailed test at the 5% significance level).

7.S.14 In a study of hearing loss, endolymphatic sac tumors (ELSTs) were discovered in 13 patients. These 13 patients had a total of 15 tumors (i.e., most patients had a single tumor, but two of the patients had 2 tumors each). Ten of the tumors were associated with the loss of functional hearing in an ear, but for 5 of the ears with tumors the patient had no hearing loss.[76] A natural question is whether hearing loss is more likely with large tumors than with small tumors. Thus, the sizes of the tumors were measured. Suppose that the sample means and standard deviations were given and that a comparison of average tumor size (hearing loss versus no hearing loss) was being considered.

(a) Explain why a *t* test to compare average tumor size is not appropriate here.

(b) If the raw data were given, could a Wilcoxon-Mann-Whitney test be used?

7.S.15 (Computer exercise) In an investigation of the possible influence of dietary chromium on diabetic symptoms, 14 rats were fed a low-chromium diet and 10 were fed a normal diet. One response variable was activity of the liver enzyme GITH, which was measured using a radioactively labeled molecule. The accompanying table shows the results, expressed as thousands of counts per minute per gram of liver.[77] Use a *t* test to compare the diets at $\alpha = 0.01$. Use a nondirectional alternative. [*Note:* Formula (6.7.1) yields 21.9 df.]

Low-chromium diet		Normal diet	
42.3	52.8	53.1	53.6
51.5	51.3	50.7	47.8
53.7	58.5	55.8	61.8
48.0	55.4	55.1	52.6
56.0	38.3	47.5	53.7
55.7	54.1		
54.8	52.1		

7.S.16 (Computer exercise) Refer to Exercise 7.S.15. Use a Wilcoxon-Mann-Whitney test to compare the diets at $\alpha = 0.05$. Use a nondirectional alternative.

7.S.17 (Computer exercise) Refer to Exercise 7.S.15.

(a) Construct a 95% confidence interval for the difference in population means.

(b) Suppose the investigators believe that the effect of the low-chromium diet is "unimportant" if it shifts mean GITH activity by less than 15%—that is, if the population mean difference is less than about 8000 cpm/gm. According to the confidence interval of part (a), do the data support the conclusion that the difference is "unimportant"?

(c) How would you answer the question in part (b) if the criterion were 4000 rather than 8000 cpm/gm?

7.S.18 (Computer exercise) In a study of the lizard *Sceloporus occidentalis*, researchers examined field-caught lizards for infection by the malarial parasite *Plasmodium*. To help assess the ecological impact of malarial infection, the researchers tested 15 infected and 15 noninfected lizards for stamina, as indicated by the distance each animal could run in 2 minutes. The distances (meters) are shown in the table.[78]

Infected animals		Uninfected animals	
16.4	36.7	22.2	18.4
29.4	28.7	34.8	27.5
37.1	30.2	42.1	45.5
23.0	21.8	32.9	34.0
24.1	37.1	26.4	45.5
24.5	20.3	30.6	24.5
16.4	28.3	32.9	28.7
29.1		37.5	

Do the data provide evidence that the infection is associated with decreased stamina? Investigate this question using

(a) a t test.

(b) a Wilcoxon-Mann-Whitney test.

Let H_A be directional and $\alpha = 0.05$.

7.S.19 In a study of the effect of amphetamine on water consumption, a pharmacologist injected four rats with amphetamine and four with saline as controls. She measured the amount of water each rat consumed in 24 hours. The following are the results, expressed as ml water per kg body weight[79]:

Amphetamine	Control
118.4	122.9
124.4	162.1
169.4	184.1
105.3	154.9

(a) Use a t test to compare the treatments at $\alpha = 0.10$. Let the alternative hypothesis be that amphetamine tends to suppress water consumption.

(b) Use a Wilcoxon-Mann-Whitney test to compare the treatments at $\alpha = 0.10$, with the directional alternative that amphetamine tends to suppress water consumption.

(c) Why is it important that some of the rats received saline injections as a control? That is, why didn't the researchers simply compare rats receiving amphetamine injections to rats receiving no injection?

7.S.20 Researchers studied subjects who had pneumonia and classified them as being in one of two groups: those who were given medical therapy that is consistent with American Thoracic Society (ATS) guidelines and those who were given medical therapy that is inconsistent with ATS guidelines. Subjects in the "consistent" group were generally able to return to work sooner than were subjects in the "inconsistent" group. A Wilcoxon-Mann-Whitney test was applied to the data; the P-value for the test was 0.04.[80] For each of the following, say whether the statement is true or false and say why.

(a) There is a 4% chance that the "consistent" and "inconsistent" population distributions really are the same.

(b) If the "consistent" and "inconsistent" population distributions really are the same, then a difference between the two samples at least as large as the difference that these researchers observed would only happen 4% of the time.

(c) If a new study were done that compared the "consistent" and "inconsistent" populations, there is a 4% probability that H_0 would be rejected again.

7.S.21 A student recorded the number of calories in each of 56 entrees—28 vegetarian and 28 nonvegetarian—served at a college dining hall.[81] The following table summarizes the data. Graphs of the data (not given here) show that both distributions are reasonably symmetric and bell shaped. A 95% confidence interval for $\mu_1 - \mu_2$ is $(-27, 85)$. For each of the following, say whether the statement is true or false and say why.

	n	Mean	SD
Vegetarian	28	351	119
Nonvegetarian	28	322	87

(a) 95% of the data are between -27 and 85 calories.

(b) We are 95% confident that $\mu_1 - \mu_2$ is between -27 and 85 calories.

(c) 95% of the time $\bar{Y}_1 - \bar{Y}_2$ will be between -27 and 85 calories.

(d) 95% of the vegetarian entrees have between 27 fewer calories and 85 more calories than the average nonvegetarian entree.

7.S.22 Refer to Exercise 7.S.21. True or false (and say why): 95% of the time, when conducting a study of this size, the difference in sample means ($\bar{Y}_1 - \bar{Y}_2$) will be within approximately $\frac{(85-(-27))}{2} = 56$ calories of the difference in population means ($\mu_1 - \mu_2$).

7.S.23 (Computer exercise) Lianas are woody vines that grow in tropical forests. Researchers measured liana abundance (stems/ha) in several plots in the central Amazon region of Brazil. The plots were classified into two types: plots that were near the edge of the forest (less than 100 meters from the edge) or plots far from the edge of the forest. The raw data are given and summarized in the table.[82]

	n	Mean	SD
Near	34	438	125
Far	34	368	114

Near			Far		
639	601	600	470	339	384
605	581	555	309	395	393
535	531	466	236	252	407
437	423	380	241	215	427
376	362	350	320	228	445
349	346	337	325	267	451
320	317	310	352	294	493
285	271	265	275	356	502
250	450	441	181	418	540
436	432	420	250	425	590
419	407		266	495	
702	676		338	648	

(a) Make normal probability plots of the data to confirm that the distributions are mildly skewed.

(b) Conduct a *t* test to compare the two types of plots at $\alpha = 0.05$. Use a nondirectional alternative.

(c) Apply a logarithm transformation to the data and repeat parts (a) and (b).

(d) Compare the *t* tests from parts (b) and (c). What do these results indicate about the effect on a *t* test of mild skewness when the sample sizes are fairly large?

7.S.24 Androstenedione (andro) is a steroid that is thought by some athletes to increase strength. Researchers investigated this claim by giving andro to one group of men and a placebo to a control group of men. One of the variables measured in the experiment was the increase in "lat pulldown" strength (in pounds) of each subject after 4 weeks. (A lat pulldown is a type of weightlifting exercise.) The raw data are given below and are summarized in the table.[83]

	n	Mean	SD
Andro	9	14.4	13.3
Control	10	20.0	12.5

Control				Andro			
30	10	10	30	0	10	0	10
40	20	30	20	10	40	20	10
10	0			30			

(a) Conduct a *t* test to compare the two groups at $\alpha = 0.10$. Use a nondirectional alternative. [*Note:* Formula (6.7.1) yields 16.5 df.]

(b) Prior to the study it was expected that andro would increase strength, which means that a directional alternative might have been used. Redo the analysis in part (a) using the appropriate directional alternative.

7.S.25 The following is a sample of computer output from a study.[84] Describe the problem and the conclusion, based on the computer output.

```
Y = number of drinks in the previous 7 days

Two-sample T for treatment vs. control:

              n     Mean     SD
Treatment    244   13.62   12.39
Control      238   16.86   13.49

95% CI for mu1 - mu2:(-5.56, - 0.92)

T-test mu1 = mu2 (vs  < ):
T = -2.74 P = .0031 DF = 474.3
```

7.S.26 In a controversial study to determine the effectiveness of AZT, a group of HIV-positive pregnant women were randomly assigned to get either AZT or a placebo. Some of the babies born to these women were HIV-positive, while others were not.[85]

(a) What is the explanatory variable?

(b) What is the response variable?

(c) What are the experimental units?

7.S.27 Patients suffering from acute respiratory failure were randomly assigned to either be placed in a prone (face down) position or a supine (face up) position. In the prone group, 21 out of 152 patients died. In the supine group, 25 out of 152 patients died.[86]

(a) What is the explanatory variable?

(b) What is the response variable?

(c) What are the experimental units?

7.S.28 A study of postmenopausal women on hormone replacement therapy (HRT) reported that they had a reduced heart attack rate, but had even greater reductions in death from homicide and accidents—two causes of death that cannot be linked to HRT. It seems that the women on HRT differ from others in many other aspects of their lives—for instance, they exercise more; they also tend to be wealthier and to be better educated.[87] Use the language of statistics to discuss what these data say about the relationships among HRT, heart attack risk, and variables such as exercise, wealth, and education. Use a schematic diagram similar to Figure 7.4.1 or Figure 7.4.2 to support your explanation.

7.S.29 Alice did a two-sample t test of the hypotheses H_0: $\mu_1 = \mu_2$ versus H_A: $\mu_1 \neq \mu_2$, using samples sizes of $n_1 = n_2 = 15$. The P-value for the test was 0.13, and α was 0.10. It happened that $\bar{y}_1$ was less than $\bar{y}_2$. Unbeknownst to Alice, Linda was interested in the same data. However, Linda had reason to believe, based on an earlier study of which Alice was not aware, that either $\mu_1 = \mu_2$ or else $\mu_1 < \mu_2$. Thus, Linda did a test of the hypotheses H_0: $\mu_1 = \mu_2$ versus H_A: $\mu_1 < \mu_2$. So, for Linda's test

(i) the P-value would still be 0.13 and H_0 would not be rejected if $\alpha = 0.10$.
(ii) the P-value would still be 0.13 and H_0 would be rejected if $\alpha = 0.10$.
(iii) the P-value would be less than 0.13 and H_0 would not be rejected if $\alpha = 0.10$.
(iv) the P-value would be less than 0.13 and H_0 would be rejected if $\alpha = 0.10$.
(v) the P-value would be greater than 0.13 and H_0 would not be rejected if $\alpha = 0.10$.
(vi) the P-value would be greater than 0.13 and H_0 would be rejected if $\alpha = 0.10$.

Choose one of these and briefly explain why your choice is true.

7.S.30 Jayden did a two-sample t test of the hypotheses H_0: $\mu_1 = \mu_2$ versus H_A: $\mu_1 \neq \mu_2$, using samples sizes of $n_1 = n_2 = 12$. The P-value for the test was 0.12, and α was 0.10. Suppose that everything except the sample sizes were the same (i.e., $\bar{y}_1$ was the same, $\bar{y}_2$ was the same, etc.), but the sample sizes had been $n_1 = n_2 = 7$. Then

(i) the P-value would still be 0.12 and α would be less than 0.10.
(ii) the P-value would still be 0.12 and α would be greater than 0.10.
(iii) the P-value would be less than 0.12 and α would still be 0.10.
(iv) the P-value would be less than 0.12 and α would be less than 0.10.
(v) the P-value would be less than 0.12 and α would be greater than 0.10
(vi) the P-value would be greater than 0.12 and α would still be 0.10.
(vii) the P-value would be greater than 0.12 and α would be less than 0.10
(viii) the P-value would be greater than 0.12 and α would be greater than 0.10

Choose one of these and explain why your choice is true.

7.S.31 Omar collects data from two populations. Sahar uses those data to construct a (two-sided) 90% confidence interval for $\mu_1 - \mu_2$ and gets $(-12.5, -3.7)$. At the same time, Miguel uses the data to test H_0: $\mu_1 = \mu_2$ versus H_A: $\mu_1 < \mu_2$ with $\alpha = 0.05$. Can we tell from the information if Miguel rejects H_0? Explain your reasoning.

7.S.32 A group of 477 healthy children, aged 5–11 years, were randomly divided into two groups. Children in the "sugar-free" group were given one can (250 mL) of an artificially sweetened beverage each day for 18 months. Children in the "sugar-containing" group were given one can each day for 18 months of a beverage sweetened with sugar. The researchers developed the two drinks specifically for this experiment and made them taste and look as similar as possible.

The researchers recorded body mass index (BMI, which is weight [kg] divided by the square of height [m].) In order to control for age and sex effects, the response variable was change in BMI z score; that is, the number of SDs that a BMI differed from the national average for the child's age and sex. The table below shows the means and standard deviations of change in BMI z score for the two groups.[88] Consider a t test with a directional alternative to investigate the research hypothesis that change in BMI is lower for the sugar-free condition.

	Sugar-free	Sugar-containing
Mean	0.02	0.15
SD	0.41	0.42
n	225	252

(a) Write the null hypothesis and the alternative hypothesis in symbols.
(b) The P-value for the t test is 0.0003. Using $\alpha = 0.01$, state your conclusion regarding the hypotheses in the context of this problem.

7.S.33 Adults who broke their wrists were randomly divided into two groups. The "intervention" group was given exercises to do under the direction of a physiotherapist. The "control" group did not receive such intervention. The primary variable of interest was improvement in wrist extension, measured in degrees with a device called a goniometer, after 6 weeks.[62] Consider a t test, with a directional alternative, to investigate the research hypothesis that the intervention improves wrist extension.

	Intervention	Control
Mean	25.6	20.5
SD	17.5	9.0
n	27	20

(a) Write the null hypothesis and the alternative hypothesis in symbols.

(b) Here is computer output for the t test. Explain what the P-value means in the context of this study.

```
t = 1.29, df = 40.8, p-value = 0.1022

alternative hypothesis: true difference in
means is greater than 0
```

(c) If $\alpha = 0.10$, what is your conclusion regarding H_0?

(d) If the research hypothesis were nondirectional, how would your answer to part (c) change, if at all?

7.S.34 Parents of newborn babies were divided into three groups as part of an experiment to see if the age at which a baby first walks can be affected by special exercises. For simplicity we will call the groups Grp1, Grp2, and Grp3.[89] This computer output compares Grp1 to Grp2:

```
t = 1.28, df = 9.35, p-value = 0.2301
```

This computer output compares Grp1 to Grp3:

```
t = 3.04, df = 8.66, p-value = 0.01453
```

It seems that we are being told that $\mu_1 = \mu_2$ and that $\mu_1 \neq \mu_3$. This might lead us to think that $\mu_2 \neq \mu_3$. However, when we conduct a test to compare Grp2 to Grp3 we get a large P-value:

```
t = 1.1, df = 7.65, p-value = 0.3042
```

All three t tests used a nondirectional alternative. Discuss how it is possible that the three tests, taken together, seem to say that $\mu_1 = \mu_2$ and $\mu_2 = \mu_3$ but $\mu_1 \neq \mu_3$.

7.S.35 Patients with pleural infection were randomly assigned to be given either an active medical treatment ($n = 52$) or a placebo ($n = 55$).[90] One outcome that was measured on each patient was duration of hospital stay. Consider conducting a t test of the null hypothesis that average hospital stay is the same for treatment and for control. Assuming that there are 57 degrees of freedom for the test, what is the distribution of the test statistic if the null hypothesis is true?

Chapter 8

COMPARISON OF PAIRED SAMPLES

OBJECTIVES

In this chapter we study comparisons of paired samples. We will

- demonstrate how to conduct a paired *t* test.
- demonstrate how to construct and interpret a confidence interval for the mean of a paired difference.
- discuss ways in which paired data arise and how pairing can be advantageous.
- consider the conditions under which a paired *t* test is valid.
- show how paired data may be analyzed using the sign test and the Wilcoxon signed-rank test.

8.1 Introduction

In Chapter 7 we considered the comparison of two independent samples when the response variable Y is a quantitative variable. In the present chapter we consider the comparison of two samples that are not independent but are paired. In a **paired design**, the observations (Y_1, Y_2) occur in pairs; the observational units in a pair are linked in some way, so they have more in common with each other than with members of another pair. The following is an example of a paired design.

Example 8.1.1

Hunger Rating During a weight loss study, each of nine subjects was given (1) the active drug *m*-chlorophenylpiperazine (mCPP) for 2 weeks and then a placebo for another 2 weeks, or (2) the placebo for the first 2 weeks and then mCPP for the second 2 weeks. As part of the study, the subjects were asked to rate how hungry there were at the end of each 2-week period. The hunger rating data are shown in Table 8.1.1, which also shows the difference in hunger ratings (mCPP – placebo),[1] and Figure 8.1.1. ■

A suitable analysis of the data should take advantage of this pairing. That is, we could imagine an experiment in which some subjects rate hunger after taking mCPP and others after taking a placebo; such an experiment would provide two independent samples of data and could be analyzed using the methods of Chapter 7. But the current experiment uses a paired design. We expect to see hunger, and hunger rating, vary from person to person. Knowing a subject's hunger rating after mCPP tells us something about that subject's hunger rating after placebo, and vice versa. We want to use this information when we analyze the data.

Example 8.1.2

Hunger Rating Randomization Test Suppose the drug mCPP has no actual effect on how hungry a person feels. Taking mCPP would be equivalent to taking a placebo, which means that we would expect the mCPP and placebo columns of Table 8.1.1 to be the same, on average. The observed mean difference (mCPP – placebo) was −29.6. We would expect this to be zero if mCPP only acted like a placebo. We want to investigate the following null hypothesis:

H_0: Hunger when taking mCPP is no different from hunger when taking a placebo

Table 8.1.1 Hunger rating for nine women

	Hunger rating		Difference
Subject	Drug (mCPP)	Placebo	mCPP – Placebo
1	79	78	1
2	48	54	−6
3	52	142	−90
4	15	25	−10
5	61	101	−40
6	107	99	8
7	77	94	−17
8	54	107	−53
9	5	64	−59
Mean	55.3	84.9	−29.6
SD	31.5	34.1	32.8

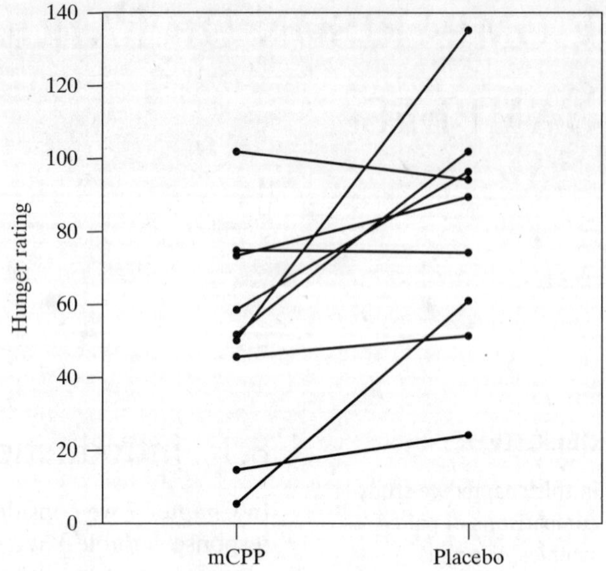

Figure 8.1.1 Dotplots of hunger ratings after mCPP and placebo, with line segments connecting readings on each subject

If H_0 were true, then the observed sample mean difference of −29.6 would have happened by chance. Further, if H_0 were true and the experiment were repeated many times, sometimes the mean difference would be negative, sometimes it would be positive, and on average it would be zero. If H_0 is true, then each of the nine differences in Table 8.1.1 is equally likely to be positive or negative. For example, the first subject had hunger ratings of 79 (on cMPP) and 78 (on placebo) for a difference of $79 - 78 = 1$; but if H_0 is true then the ratings could just as well have been 78 (on cMPP) and 79 (on placebo) for a difference of $78 - 79 = -1$. The same reasoning applies to each of the other eight differences.

If H_0 is true, then the experiment might have generated the results in Table 8.1.2, which was created by tossing a coin nine times to determine which hunger rating to

Table 8.1.2 Randomized hunger ratings for nine women

	Hunger rating		Difference
Subject	Drug (mCPP)	Placebo	mCPP – Placebo
1	**78**	**79**	**−1**
2	**48**	**54**	**−6**
3	**52**	**142**	**−90**
4	25	15	10
5	**61**	**101**	**−40**
6	**99**	**107**	**−8**
7	94	77	17
8	107	54	53
9	64	5	59
Mean			−0.7

put in the mCPP column and which to put in the Placebo column for each of the nine subjects. Subjects for which the coin toss came up tails, which meant that the smaller hunger rating went in the mCPP column, are shown in bold type.

Notice that the only thing that changed between the original differences in Table 8.1.1 and the differences in Table 8.1.2 is the sign, positive or negative, attached to each difference.

We could toss a coin nine times, assign hunger ratings to the two columns, and then calculate the mean difference over and over again. If we were to carry out this process repeatedly, with the aid of a computer, then we would get a mean difference as extreme as −29.6 only 2.3% of the time.* Thus, we can say that the randomization P-value is 0.023 for testing

H_0: Hunger when taking mCPP is no different from hunger when taking a placebo

versus

H_A: Hunger when taking mCPP is different from hunger when taking a placebo ■

In Section 8.2 we show how to analyze paired data using methods based on Student's t distribution. In Sections 8.4 and 8.5 we describe two nonparametric tests for paired data. Sections 8.3, 8.6, and 8.7 contain more examples and discussion of the paired design.

*In this example, there are only $2^9 = 512$ possible ways to assign plus and minus signs to the nine differences of which values at least as extreme as −29.6 occur 12 times. We talk about computer simulation of many repetitions because this approach can be used with arbitrary sample sizes.

Exercise 8.1.1

8.1.1 Cyclic adenosine monophosphate (cAMP) is a substance that can mediate cellular response to hormones. In a study of maturation of egg cells in the frog *Xenopus laevis,* oocytes from each of four females were divided into two batches; one batch was exposed to progesterone and the other was not. After 2 minutes, each batch was assayed for its cAMP content, with the results given in the table.[2] There are $2^4 = 16$ possible ways to assign plus and minus signs to the four differences.

	cAMP (pmol/oocyte)		
Frog	Control	Progesterone	d
1	6.01	5.23	0.78
2	2.28	1.21	1.07
3	1.51	1.40	0.11
4	2.12	1.38	0.74
Mean	2.98	2.31	0.675
SD	2.05	1.95	0.40

(a) For each of the 16 possible results, find the sum of the four differences and the mean difference.

(b) How many of the 16 results yield a mean difference as far from zero as 0.675, the mean difference for the observed data?

(c) Consider testing the null hypothesis that progesterone has no effect on cAMP against the alternative that progesterone affects cAMP. Considering your results from part (b), what is the P-value for this randomization test, and what conclusions would you draw if $\alpha = 0.10$?

8.2 The Paired-Sample *t* Test and Confidence Interval

In this section we discuss the use of Student's *t* distribution to obtain tests and confidence intervals for paired data.

ANALYZING DIFFERENCES

In Chapter 7 we considered how to analyze data from two independent samples. When we have paired data, we make a simple shift of viewpoint: Instead of considering Y_1 and Y_2 separately, we consider the *difference D*, defined as

$$D = Y_1 - Y_2$$

Note that it is often natural to consider a difference as the response variable of interest in a study. For example, if we were studying the growth rates of plants, we might grow plants under control conditions for a while at the beginning of a study and then apply a treatment for one week. We would measure the growth that takes place during the week after the treatment is introduced as $D = Y_1 - Y_2$, where Y_1 = height one week after applying the treatment and Y_2 = height before the treatment is applied.* Sometimes data are paired in a way that is less obvious, but whenever we have paired data, it is the observed differences that we wish to analyze.

Let us denote the mean of sample D's as $\bar{D}$. The quantity $\bar{D}$ is related to the individual sample means as follows:

$$\bar{D} = (\bar{Y}_1 - \bar{Y}_2)$$

The relationship between population means is analogous:

$$\mu_D = \mu_1 - \mu_2$$

Thus, we may say that *the mean of the difference is equal to the difference of the means.* Because of this simple relationship, a comparison of two paired means can be carried out by concentrating entirely on the D's.

The standard error for $\bar{D}$ is easy to calculate. Because $\bar{D}$ is just the mean of a single sample, we can apply the SE formula of Chapter 6 to obtain the following formula:

$$\text{SE}_{\bar{D}} = \frac{s_D}{\sqrt{n_D}}$$

where s_D is the standard deviation of the D's and n_D is the number of D's. The following example illustrates the calculation.

Example 8.2.1 **Hunger Rating** The hunger rating data of Example 8.1.1 are shown in Table 8.1.1, which gives the mean for each condition as well as the mean of the differences. Note that the mean of the difference is equal to the difference of the means:

$$\bar{d} = -29.6 = 55.3 - 84.9$$

Figure 8.2.1 shows the distribution of the nine sample differences.

While the mean of the difference is the same as the difference of the means, note that the SD of the difference is *not* the difference of the SDs. Thus the standard error of the mean difference must be calculated using the SD of the differences.

*Exercises 7.2.11 and 7.2.12 both involve such "before versus after" data.

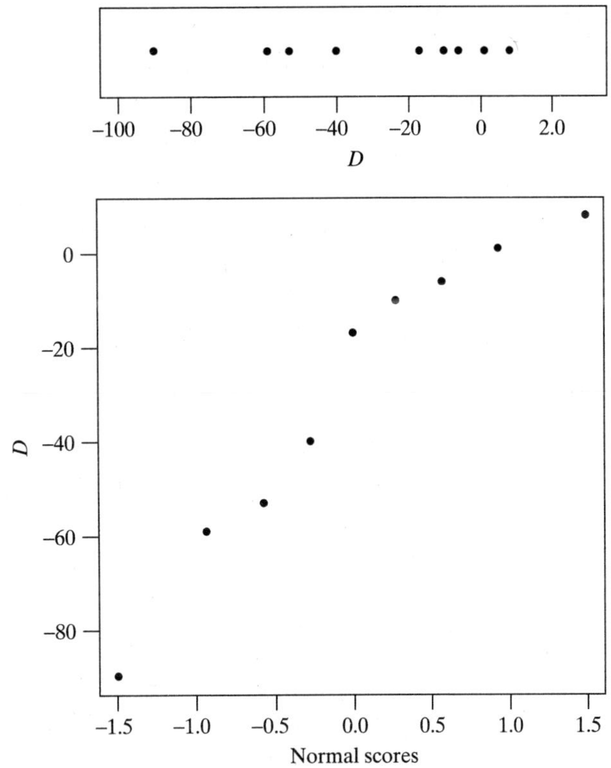

Figure 8.2.1 Dotplot of differences in hunger rating between mCPP and placebo, along with a normal quantile plot of the data

We calculate the standard error of the mean difference as follows:

$$s_D = 32.8$$

$$n_D = 9$$

$$\text{SE}_{\bar{D}} = \frac{32.8}{\sqrt{9}} = 10.9$$

CONFIDENCE INTERVAL AND TEST OF HYPOTHESIS

The standard error described previously is the basis for the **paired-sample *t* method** of analysis, which can take the form of a confidence interval or a test of hypothesis.

A 95% confidence interval for μ_D is constructed as

$$\bar{d} \pm t_{n_D-1,\, 0.025}\text{SE}_{\bar{D}}$$

where the constant $t_{n_D-1,\, 0.025}$ is determined from Student's *t* distribution with

$$\text{df} = n_D - 1$$

Intervals with other confidence coefficients (e.g., 90%, 99%) are constructed analogously (using $t_{0.05}$, $t_{0.005}$, etc.). The following example illustrates the confidence interval.

Example 8.2.2

Hunger Rating For the hunger rating data, we have df = 9 – 1 = 8. From Table 4 we find that $t_{8,\, 0.025} = 2.306$; thus, the 95% confidence interval for μ_D is

$$-29.6 \pm (2.306)\left(\frac{32.8}{\sqrt{9}}\right)$$

or

$$-29.6 \pm 25.2$$

or

$$(-54.8, -4.4)$$ ■

We can also conduct a t test. To test the null hypothesis

$$H_0: \mu_D = 0$$

we use the test statistic

$$t_s = \frac{\bar{d} - 0}{\mathrm{SE}_{\bar{D}}}$$

Critical values are obtained from Student's t distribution (Table 4) with df $= n_D - 1$. The following example illustrates the t test.

Example 8.2.3 **Hunger Rating** For the hunger rating data, let us formulate the null hypothesis and nondirectional alternative:

H_0: Mean hunger rating is the same after taking mCPP as it is after taking a placebo.

H_A: Mean hunger rating is different after taking mCPP than after taking a placebo.

or, in symbols,

$$H_0: \mu_D = 0$$
$$H_A: \mu_D \neq 0$$

Let us test H_0 against H_A at significance level $\alpha = 0.05$. The test statistic is

$$t_s = \frac{-29.6 - 0}{32.8/\sqrt{9}} = -2.71$$

From Table 4, $t_{8,\,0.02} = 2.449$ and $t_{8,\,0.01} = 2.896$. We reject H_0 and find that there is sufficient evidence $(0.02 < P < 0.04)$ to conclude that mean hunger rating is reduced more by mCPP than by a placebo.* (Using a computer gives the P-value as $P = 0.027$.) (Note that we can conclude that mCPP *caused* a greater decrease in hunger because this was an experiment and not an observational study.) ■

RESULT OF IGNORING PAIRING

Suppose that a study is conducted using a paired design, but that the pairing is ignored in the analysis of the data. Such an analysis is not valid because it assumes that the samples are independent when in fact they are not. The incorrect analysis can be misleading, as the following example illustrates.

Example 8.2.4 **Cheese Gumminess** A food scientist developed a new low-fat cheddar cheese by altering part of the ripening process. One outcome was the gumminess of the cheese, measured in milliNewtons. Ten wheels of cheese were measured for gumminess after 7 days and again after 30 days. The results are shown in Table 8.2.1.[3] Note that there is considerable variation from cheese to cheese. For example, cheese 4 had low gumminess on both days, but cheese 7 had high readings both times.

*Because the alternative hypothesis was non-directional, some others would say "We conclude that the mean hunger rating is affected by mCPP," rather than saying mCPP reduces hunger rating.

Table 8.2.1 Cheese gumminess (mN)

	Gumminess (mN)		
Cheese	Day 7 y_1	Day 30 y_2	Difference $d = y_1 - y_2$
1	7,296	5,544	1,752
2	6,325	6,120	205
3	8,003	5,720	2,283
4	5,013	2,508	2,505
5	4,637	3,743	894
6	8,525	5,272	3,253
7	9,445	7,189	2,256
8	8,794	6,794	2,000
9	5,213	4,409	804
10	3,399	4,083	–684
Mean	6,665	5,138	1,527
SD	2,045	1,451	1,196

For the cheese gumminess data, the SE for the mean difference is

$$\mathrm{SE}_{\overline{D}} = \frac{1196}{\sqrt{10}} = 378$$

Figure 8.2.2 shows the distribution of the 10 sample differences.

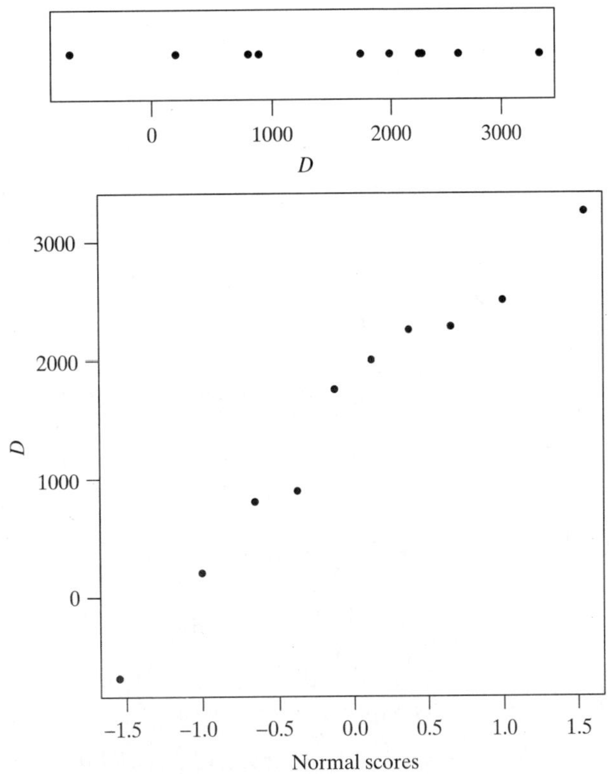

Figure 8.2.2 Dotplot of differences in gumminess for Day 7 and for Day 30, along with a normal quantile plot of the data.

A test of

$$H_0: \mu_d = 0$$

versus

$$H_A: \mu_d \neq 0$$

gives a test statistic of

$$t_s = \frac{1527 - 0}{378} = 4.04$$

This test statistic has 9 degrees of freedom. Using a computer gives the P-value as $P = 0.003$.

Figure 8.2.3 displays the Day 7 and Day 30 data separately. There is considerable overlap in the two distributions. This plot does not show compelling evidence that the gumminess increases over time (as determined from the paired analysis above) because this plot does not take into account the paired nature of this data.

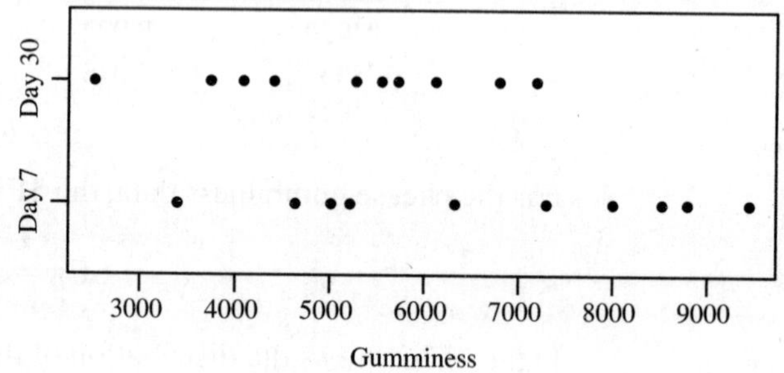

Figure 8.2.3 Parallel dotplots of cheese gumminess on days 7 and 30

Looking at the Day 7 and Day 30 data separately, the two sample SDs are $s_1 = 2045$ and $s_2 = 1451$. If we improperly proceed as if the samples were independent and apply the SE formula of Chapter 7, we obtain

$$\mathrm{SE}_{\bar{Y}_1 - \bar{Y}_2} = \sqrt{\frac{s_1^2}{n_1} + \frac{s_2^2}{n_2}}$$

$$= \sqrt{\frac{2045^2}{10} + \frac{1451^2}{10}} = 793$$

This SE is quite a bit larger than the value ($\mathrm{SE}_{\bar{D}} = 378$) that we calculated using the pairing.

Continuing to (wrongly) proceed as if the samples were independent, the test statistic is

$$t_s = \frac{6665 - 5138}{793} = 1.93.$$

The P-value for this test is 0.072, which is much greater than the P-value for the correct test, 0.003.

To further compare the paired and unpaired analysis, let us consider the 95% confidence interval for $(\mu_1 - \mu_2)$. For the unpaired analysis, formula (6.7.1) yields $16.3 \approx 16$ degrees of freedom; this gives a t multiplier of $t_{16,0.025} = 2.120$ and yields a confidence interval of

$$1527 \pm (2.262)(793) \quad \text{or} \quad 1527 \pm 1794 \quad \text{or} \quad (-267, 3321)$$

This erroneous confidence interval is wider than the correct confidence interval from a paired analysis. A paired analysis yields the narrower interval

$$(6665 - 5138) \pm (2.120)(378) \quad \text{or} \quad 1527 \pm 801 \quad \text{or} \quad (726, 2328)$$

The paired-sample interval is narrower because it uses a smaller SE; this effect is slightly offset by a larger value of $t_{0.025}$ (2.262 vs. 2.120).

Why is the paired-sample SE smaller than the independent-samples SE calculated from the same data (SE = 378 vs. SE = 793)? Table 8.2.1 reveals the reason. The data show that there is large variation from one cheese wheel to the next. For instance, cheese 4 has low gumminess (both for Day 7 and for Day 30), and cheese 7 has high values. The independent-samples SE formula incorporates all of this variation (expressed through s_1 and s_2); in the paired-sample approach, intercheese variation in gumminess has no influence on the calculations because only the D's are used. By using each cheese as its own control, the experimenter has increased the precision of the experiment. But if the pairing is ignored in the analysis, the extra precision is wasted. ■

The preceding example illustrates the gain in precision that can result from a paired design coupled with a paired analysis. The choice between a paired and an unpaired design will be discussed in Section 8.3.

CONDITIONS FOR VALIDITY OF STUDENT'S *t* ANALYSIS

The conditions for validity of the paired-sample t test and confidence interval are as follows:

1. It must be reasonable to regard the *differences* (the D's) as a random sample from some large population.
2. The population distribution of the D's must be normal. The methods are approximately valid if the population distribution is approximately normal or if the sample size (n_D) is large.

The preceding conditions are the same as those given in Chapter 6; in the present case, the conditions apply to the D's because the analysis is based on the D's. Verification of the conditions can proceed as described in Chapter 6. First, the design should be checked to assure that the D's are independent of each other, and especially that there is no hierarchical structure within the D's. (Note, however, that the Y_1's are not independent of the Y_2's because of the pairing.) Second, a histogram or dotplot of the D's can provide a rough check for approximate normality. A normal quantile plot can also be used to assess normality.

Notice that normality of the Y_1's and Y_2's is not required, because the analysis depends only on the D's. The following example shows a case in which the Y_1's and Y_2's are not normally distributed, but the D's are.

Example 8.2.5

Squirrels If you walk toward a squirrel that is on the ground, it will eventually run to the nearest tree for safety. A researcher wondered whether he could get closer to the squirrel than the squirrel was to the nearest tree before the squirrel would start to run. He made 11 observations, which are given in Table 8.2.2. Figure 8.2.4 shows that the distribution of distances from squirrel to person appear to be reasonably normal, but that the distances from squirrel to tree are far from being normally distributed. However, panel (c) of Figure 8.2.4 shows that the 11 differences do meet the normality condition. Since a paired t test analyzes the differences, a t test (or confidence interval) is valid here.[4] ■

Table 8.2.2 Distances (in inches) from person and from tree when squirrel started to run

Squirrel	From person y_1	From tree y_2	Difference $d = y_1 - y_2$
1	81	137	−56
2	178	34	144
3	202	51	151
4	325	50	275
5	238	54	184
6	134	236	−102
7	240	45	195
8	326	293	33
9	60	277	−217
10	119	83	36
11	189	41	148
Mean	190	118	72
SD	89	101	148

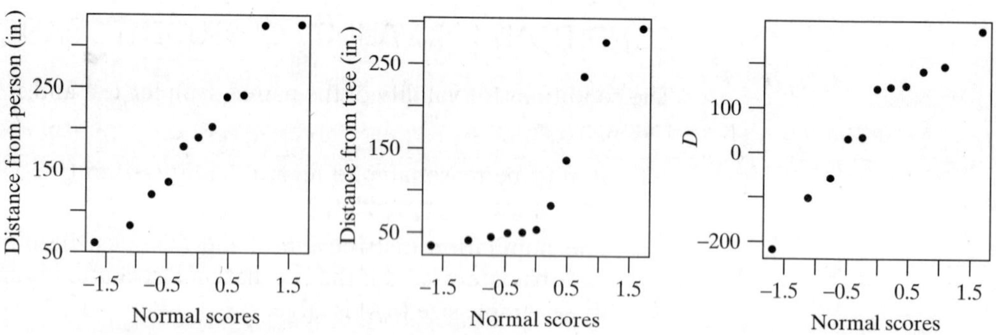

Figure 8.2.4 Normal quantile plots of distance from squirrel to person, from squirrel to tree, and their difference.

SUMMARY OF FORMULAS

For convenient reference, we summarize the formulas for the paired-sample methods based on Student's *t*.

Standard Error of $\overline{D}$

$$SE_{\overline{D}} = \frac{s_D}{\sqrt{n_D}}$$

***t* Test**

$$H_0: \mu_D = 0$$

$$t_s = \frac{\overline{d} - 0}{SE_{\overline{D}}}$$

95% Confidence Interval for μ_d

$$\overline{d} \pm t_{0.025} SE_{\overline{D}}$$

Intervals with other confidence levels (e.g., 90%, 99%) are constructed analogously (e.g., using $t_{0.05}$, $t_{0.005}$).

Exercises 8.2.1–8.2.11

8.2.1 In an agronomic field experiment, blocks of land were subdivided into two plots of 346 square feet each. Each block provided two paired observations: one for each of the varieties of wheat. The plot yields (lb) of wheat are given in the table.[5]

(a) Calculate the standard error of the mean difference between the varieties.

(b) Test for a difference between the varieties using a paired t test at $\alpha = 0.05$. Use a nondirectional alternative.

(c) Test for a difference between the varieties the wrong way, using an independent-samples test. Compare with the result of part (b).

	Variety		
Block	1	2	Difference
1	32.1	34.5	−2.4
2	30.6	32.6	2.0
3	33.7	34.6	−0.9
4	29.7	31.0	−1.3
Mean	31.53	33.18	−1.65
SD	1.76	1.72	0.68

8.2.2 In an experiment to compare two diets for fattening beef steers, nine pairs of animals were chosen from the herd; members of each pair were matched as closely as possible with respect to hereditary factors. The members of each pair were randomly allocated, one to each diet. The following table shows the weight gains (lb) of the animals over a 140-day test period on diet 1 (Y_1) and on diet 2 (Y_2).[6]

Pair	Diet 1	Diet 2	Difference
1	596	498	98
2	422	460	−38
3	524	468	56
4	454	458	−4
5	538	530	8
6	552	482	70
7	478	528	−50
8	564	598	−34
9	556	456	100
Mean	520.4	497.6	22.9
SD	57.1	47.3	59.3

(a) Calculate the standard error of the mean difference.

(b) Test for a difference between the diets using a paired t test at $\alpha = 0.10$. Use a nondirectional alternative.

(c) Construct a 90% confidence interval for μ_D.

(d) Interpret the confidence interval from part (c) in the context of this setting.

8.2.3 Refer to the cyclic adenosine monophosphate (cAMP) data in Exercise 8.1.1. Use a t test to investigate the effect of progesterone on cAMP. Let H_A be nondirectional and let $\alpha = 0.10$.

	cAMP (pmol/oocyte)		
Frog	Control	Progesterone	d
1	6.01	5.23	0.78
2	2.28	1.21	1.07
3	1.51	1.40	0.11
4	2.12	1.38	0.74
Mean	2.98	2.31	0.68
SD	2.05	1.95	0.40

8.2.4 During a weight loss study, each of nine subjects was given (1) the active drug m-chlorophenylpiperazine (mCPP) for 2 weeks and then a placebo for another 2 weeks, or (2) the placebo for the first 2 weeks and then mCPP for the second 2 weeks. The following table shows the amount of weight loss (kg) for the nine subjects when taking the drug mCPP and when taking a placebo.[1] (Note that if a subject gained weight, then the recorded weight loss is negative, as is the case for subject 2, who gained 0.3 kg when on the placebo.) Use a t test to investigate the claim that mCPP affects weight loss. Let H_A be nondirectional, and let $\alpha = 0.01$.

	Weight change		
Subject	mCPP	Placebo	Difference
1	1.1	0.0	1.1
2	1.3	−0.3	1.6
3	1.0	0.6	0.4
4	1.7	0.3	1.4
5	1.4	−0.7	2.1
6	0.1	−0.2	0.3
7	0.5	0.6	−0.1
8	1.6	0.9	0.7
9	−0.5	−2.0	1.5
Mean	0.91	−0.09	1.00
SD	0.74	0.88	0.72

(a) What is the value of the t test statistic for assessing whether weight change when taking mCPP is different from a placebo?

(b) In the context of this study, state the null and alternative hypotheses.

(c) The P-value for the t test is 0.003. If $\alpha = 0.10$, what is your conclusion regarding the hypotheses in (b)?

8.2.5 Refer to Exercise 8.2.4.

(a) Construct a 99% confidence interval for μ_D.

(b) Interpret the confidence interval from part (a) in the context of this setting.

8.2.6 Under certain conditions, electrical stimulation of a beef carcass will improve the tenderness of the meat. In one study of this effect, beef carcasses were split in half; one side (half) was subjected to a brief electrical current, and the other side was an untreated control. For each side, a steak was cut and tested in various ways for tenderness. In one test, the experimenter obtained a specimen of connective tissue (collagen) from the steak and determined the temperature at which the tissue would shrink; a tender piece of meat tends to yield a low collagen shrinkage temperature. The data are given in the following table.[7]

(a) Construct a 95% confidence interval for the mean difference between the treated side and the control side.

(b) Construct a 95% confidence interval the wrong way, using the independent-samples method. How does this interval differ from the one you obtained in part (a)?

	Collagen shrinkage temperature (°C)		
Carcass	Treated side	Control side	Difference
1	69.50	70.00	−0.50
2	67.00	69.00	−2.00
3	70.75	69.50	1.25
4	68.50	69.25	−0.75
5	66.75	67.75	−1.00
6	68.50	66.50	2.00
7	69.50	68.75	0.75
8	69.00	70.00	−1.00
9	66.75	66.75	0.00
10	69.00	68.50	0.50
11	69.50	69.00	0.50
12	69.00	69.75	−0.75
13	70.50	70.25	0.25
14	68.00	66.25	1.75
15	69.00	68.25	0.75
Mean	68.750	68.633	0.117
SD	1.217	1.302	1.118

8.2.7 Refer to Exercise 8.2.6. Use a t test to test the null hypothesis of no effect against the alternative hypothesis that the electrical treatment tends to reduce the collagen shrinkage temperature. Let $\alpha = 0.10$.

8.2.8 Trichotillomania is a psychiatric illness that causes its victims to have an irresistible compulsion to pull their own hair. Two drugs were compared as treatments for trichotillomania in a study involving 13 women. Each woman took clomipramine during one time period and desipramine during another time period in a double-blind experiment. Scores on a trichotillomania-impairment scale, in which high scores indicate greater impairment, were measured on each woman during each time period. The average of the 13 measurements for clomipramine was 6.2; the average of the 13 measurements for desipramine was 4.2.[8] A paired t test gave a value of $t_s = 2.47$ and a two-tailed P-value of 0.03. Interpret the result of the t test. That is, what does the test indicate about clomipramine, desipramine, and hair pulling?

8.2.9 A scientist conducted a study of how often her pet parakeet chirps. She recorded the number of distinct chirps the parakeet made in a 30-minute period, sometimes when the room was silent and sometimes when music was playing. The data are shown in the following table.[9] Construct a 95% confidence interval for the mean increase in chirps (per 30 minutes) when music is playing over when music is not playing.

	Chirps in 30 minutes		
Day	With music	Without music	Difference
1	12	3	9
2	14	1	13
3	11	2	9
4	13	1	12
5	20	5	15
6	14	3	11
7	10	0	10
8	12	2	10
9	8	6	2
10	13	3	10
11	14	2	12
12	15	4	11
13	12	3	9
14	13	2	11
15	8	0	8
16	18	5	13
17	15	3	12
18	12	2	10
19	17	2	15
20	15	4	11
21	11	3	8
22	22	4	18
23	14	2	12
24	18	4	14
25	15	5	10
26	8	1	7
27	13	2	11
28	16	3	13
Mean	13.7	2.8	10.9
SD	3.4	1.5	3.0

8.2.10 Consider the data in Exercise 8.2.9. There are two outliers among the 28 differences: the smallest value, which is 2, and the largest value, which is 18. Delete these two observations and construct a 95% confidence interval for the mean increase, using the remaining 26 observations. Do the outliers have much of an effect on the confidence interval?

8.2.11 Invent a paired data set, consisting of five pairs of observations, for which $\bar{y}_1$ and $\bar{y}_2$ are not equal, and $SE_{\bar{Y}_1} > 0$ and $SE_{\bar{Y}_2} > 0$, but $SE_{\bar{D}} = 0$.

8.3 The Paired Design

Ideally, in a paired design the members of a pair are relatively similar to each other—that is, more similar to each other than to members of other pairs—with respect to extraneous variables. The advantage of this arrangement is that, when members of a pair are compared, the comparison is free of the extraneous variation that originates in between-pair differences. We will expand on this theme after giving some examples.

EXAMPLES OF PAIRED DESIGNS

Paired designs can arise in a variety of ways, including the following:

Experiments in which similar experimental units form pairs
Observational studies of identical twins
Repeated measurements on the same individual at two different times
Pairing by time

Experiments with Pairs of Units Often, researchers who wish to compare two treatments will first form pairs of experimental units (pairs of animals, pairs of plots of land, etc.) that are similar (e.g., animals of the same age and sex or plots of land with the same type of soil and exposure to wind, rain, and sun). Then one member of a pair is randomly chosen to receive the first treatment and the other member is given the second treatment. The following is an example.

Example 8.3.1 **Fertilizers for Eggplants** In a greenhouse experiment to compare two fertilizer treatments for eggplants, individually potted plants are arranged on the greenhouse bench in pairs, such that two plants in the same pair are subject to the same amount of sunlight, the same temperature, and so on. Within each pair, one (randomly chosen) plant will receive treatment 1, and the other will receive treatment 2. ■

Observational Studies As noted in Section 7.4, randomized experiments are preferred over observational studies, due to the many confounding variables that can arise within an observational study. An observational study may tell us that X and Y are *associated,* but only an experiment can address the question of whether X *causes* Y. If no experiment is possible and an observational study must be carried out, then it is preferable (although rarely possible) to study identical twins as the observational units. For example, in a study of the effect of "secondhand smoke" it would be ideal to enroll several sets of nonsmoking twins for which, in each pair, one of the twins lived with a smoker and the other twin did not. Because sets of twins are rarely, if ever, available, **matched-pair designs,** in which two groups are matched with respect to various extraneous variables, are often used.[10] Here is an example.*

*We also discussed this type of design in Example 1.2.10.

Example 8.3.2 **Smoking and Lung Cancer** In a case-control study of lung cancer, 100 lung cancer patients were identified. For each case, a control was chosen who was individually matched to the case with respect to age, sex, and education level. The smoking habits of the cases and controls were compared. ■

Repeated Measurements Many biological investigations involve repeated measurements made on the same individual under different conditions. These include studies of growth and development, studies of biological processes, and studies in which measurements are made before and after application of a certain treatment. When only two conditions are involved, the measurements are paired, as in Example 8.1.1. The following is another example.

Example 8.3.3 **Working Memory** Many neuroscientists believe that the capacity of a person's working memory is seven items, give or take two. How much does memorizing something new crowd out something previously memorized? In one study, subjects were given 1 minute to memorize a list of 10 words (each 8–10 letters long), after which they wrote down as many of the words as they could remember; this number is called "Recall 1." The subjects then repeated this process with a second list of 10 words, after which they were also asked to write down as many words as they could remember from the *first* list. This number is "Recall 2." The results are shown in Table 8.3.1.[11] Note that there is considerable variation from person to person. For example, subject 4 had good recall twice, but subject 5 did poorly both times. ■

Table 8.3.1 Recall of 10 words

Subject	Recall 1	Recall 2
1	7	2
2	4	2
3	9	3
4	7	5
5	3	0
6	7	3
7	4	0
8	7	1

Pairing by Time In some situations, pairs are formed implicitly when replicate measurements are made at different times. The following is an example.

Example 8.3.4 **Stream Coliform Contamination** Pollutants in a stream may accumulate or attenuate as water flows down the stream. In a study to monitor the accumulation and attenuation of fecal contamination in a stream running through cattle rangeland, monthly water specimens were collected at two locations along the stream over a period of 21 months. The data are shown in Table 8.3.2.[12] Each number represents the total coliform count (MPN/100ml) for a water specimen.

It is natural to expect that during some parts of the season contamination levels across the entire stream will be high and at other times low, thus making the up- and downstream data linked or paired. This association can be seen in the data in Table 8.3.2. Note that between October 2008 and April 2009 the coliform levels were low at both the upstream and downstream locations, while at other times both locations

Table 8.3.2 Stream coliform count (MPN/100ml)

Collection date	Upstream	Downstream
Jan-08	2909	2359
Feb-08	2382	3873
Mar-08	2046	1725
Apr-08	1483	771
May-08	4611	1529
Jun-08	3448	2909
Jul-08	4106	2014
Aug-08	2755	1872
Sep-08	3448	2481
Oct-08	2098	613
Nov-08	789	733
Dec-08	988	960
Jan-09	1187	833
Feb-09	638	388
Mar-09	2481	1421
Apr-09	2481	933
May-09	3654	1076
Jun-09	3130	1374
Jul-09	4611	1553
Aug-09	2755	1198
Sep-09	3076	1300
$\bar{y}$	2622.7	1519.8

tended to be high. This tendency to co-vary arises from variations in the environmental conditions that affect the entire stream such as the presence of cattle, rainfall, solar exposure, and so forth. The variability in these conditions cannot be prevented but can be accounted for when the data are properly analyzed as pairs. ■

Examples 8.3.3 and 8.3.4 both involve two measurements on the same unit. But notice that the pairing structure in the two examples is different. In Example 8.3.3 the members of a pair are measurements on the same individual under two conditions (i.e., before and after reading a second list of words), whereas in Example 8.3.4 the members of a pair are measurements on the same stream at the same time. Nevertheless, in both examples the principle of pairing is the same: Members of a pair are similar to each other with respect to extraneous variables. In Example 8.3.4 time is an extraneous variable, whereas in Example 8.3.3 the comparison between two times (before and after reading a second list of words) is of primary interest and interperson variation is extraneous.

PURPOSES OF PAIRING

Pairing in an experimental design can serve to reduce bias, to increase precision, or both. Usually the primary purpose of pairing is to increase precision.

We noted in Section 7.4 that pairing or matching can reduce bias by controlling variation due to extraneous variables. The variables used in the matching are necessarily balanced in the two groups to be compared and therefore cannot distort the comparison. For instance, if two groups are composed of age-matched pairs of people, then a comparison between the two groups is free of any bias due to a difference in age distribution.

In randomized experiments, where bias can be controlled by randomized allocation, a major reason for pairing is to increase precision. Effective pairing increases precision by increasing the information available in an experiment. An appropriate analysis, which extracts this extra information, leads to more powerful tests and narrower confidence intervals. Thus, an effectively paired experiment is more efficient; it yields more information than an unpaired experiment with the same number of observations.

We saw an instance of effective pairing in the cheese springiness data of Example 8.2.4. The pairing was effective because much of the variation in the measurements was due to variation between wheels of cheese, which did not enter the comparison between the times. As a result, the experiment yielded more precise information about the difference as the cheese aged than would a comparable unpaired study—that is, a study that would compare the springiness of 10 wheels of cheese at the start of aging (7 days old) to the springiness of 10 different wheels of aged cheese (90 days old).

The effectiveness of a given pairing can be displayed visually in a scatterplot of Y_2 against Y_1; each point in the scatterplot represents a single pair (Y_1, Y_2). Figure 8.3.1 shows a scatterplot for the stream contamination data of Example 8.3.4, together with a boxplot of the differences; each point in the scatterplot represents a single run. Notice that the points in the scatterplot show a definite upward trend. This upward trend indicates the effectiveness of the pairing: Measurements taken on the same day have more in common than measurements on different days so that a day with a relatively high value of Y_1 tends to have a relatively high value of Y_2, and similarly for low values.

Figure 8.3.1 Scatterplot for the stream contamination data, with boxplot of the differences

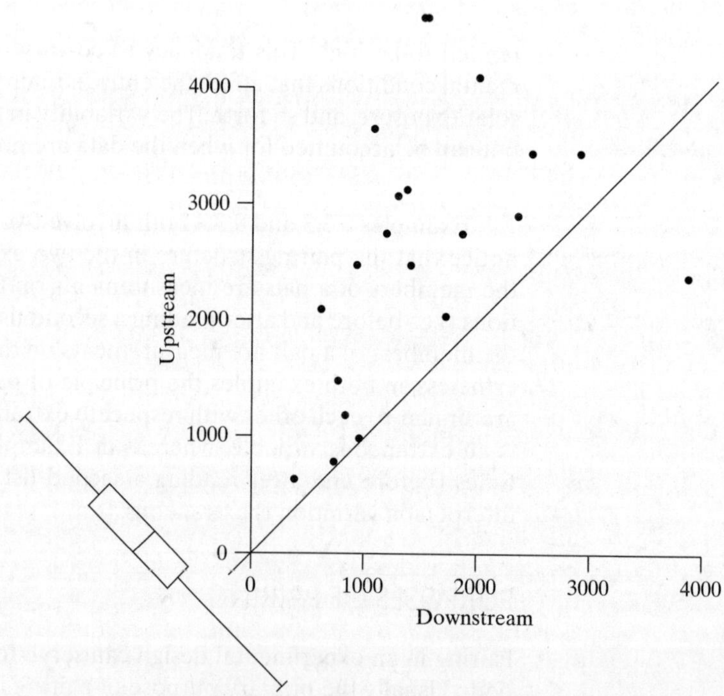

Note that pairing is a strategy of *design,* not of analysis, and is therefore carried out *before* the Y's are observed. It is not correct to use the observations themselves to form pairs. Such a data manipulation could severely distort the experimental results and could be considered scientific fraud.

RANDOMIZED PAIRS DESIGN VERSUS COMPLETELY RANDOMIZED DESIGN

In planning a randomized experiment, the experimenter may need to decide between a paired design and a design that uses random assignment without any pairing, called a completely randomized design. We have said that effective pairing can greatly enhance the precision of an experiment. On the other hand, pairing in an experiment may not be effective, if the observed variable Y is not related to the factors used in the pairing. For instance, suppose pairs were matched on age only, but in fact Y turned out not to be age related. It can be shown that ineffective pairing can actually yield less precision than no pairing at all. For instance, in relation to a t test, ineffective pairing would not tend to reduce the SE, but it would reduce the degrees of freedom, and the net result would be a loss of power.

The choice of whether to use a paired design depends on practical considerations (pairing may be expensive or unwieldy) and on precision considerations. With respect to precision, the choice depends on how effective the pairing is expected to be. The following example illustrates this issue.

Example 8.3.5 **Fertilizers for Eggplants** A horticulturist is planning a greenhouse experiment with individually potted eggplants. Two fertilizer treatments are to be compared, and the observed variable is to be Y = yield of eggplants (pounds). The experimenter knows that Y is influenced by such factors as light and temperature, which vary somewhat from place to place on the greenhouse bench. The allocation of pots to positions on the bench could be carried out according to a completely randomized design, or according to a paired design, as in Example 8.3.1. In deciding between these options, the experimenter must use her knowledge of how effective the pairing would be—that is, whether two pots sitting adjacent on the bench would be more similar in yield than pots farther apart. If she judges that the pairing would not be very effective, she may opt for the completely randomized design. ■

Note that effective pairing is *not* the same as simply holding experimental conditions constant. Pairing is a way of *organizing* the unavoidable variation that still remains after experimental conditions have been made as constant as possible. The ideal pairing organizes the variation in such a way that the variation within each pair is minimal and the variation between pairs is maximal.

CHOICE OF ANALYSIS

The analysis of data should fit the design of the study. If the design is paired, a paired-sample analysis should be used; if the design is unpaired, an independent-samples analysis (as in Chapter 7) should be used.

Note that the extra information made available by an effectively paired design is *entirely wasted* if an unpaired analysis is used. (We saw an illustration of this in Example 8.2.4.) Thus, the paired design does not increase efficiency unless it is accompanied by a paired analysis.

Exercises 8.3.1–8.3.5

8.3.1 (Sampling exercise) This exercise illustrates the application of a matched-pairs design to the population of 100 ellipses (shown with Exercise 1.3.6). The accompanying table shows a grouping of the 100 ellipses into 50 pairs.

Pair	Ellipse ID numbers		Pair	Ellipse ID numbers		Pair	Ellipse ID numbers	
01	20	45	18	11	46	35	16	66
02	03	49	19	09	29	36	18	58
03	07	27	20	19	39	37	30	50
04	42	82	21	00	10	38	76	86
05	81	91	22	40	55	39	17	83
06	38	72	23	21	56	40	04	52
07	60	70	24	08	62	41	12	64
08	31	61	25	24	78	42	23	57
09	77	89	26	67	93	43	98	99
10	01	41	27	35	80	44	36	96
11	14	48	28	74	88	45	44	84
12	59	87	29	94	97	46	06	51
13	22	68	30	02	28	47	85	90
14	47	79	31	26	71	48	37	63
15	05	95	32	25	65	49	43	69
16	53	73	33	15	75	50	34	54
17	13	33	34	32	92			

To better appreciate this exercise, imagine the following experimental setting. We want to investigate the effect of a certain treatment, T, on the organism *C. ellipticus*. We will observe the variable Y = length. We can measure each individual only once, and so we will compare n treated individuals with n untreated controls. We know that the individuals available for the experiment are of various ages, and we know that age is related to length, so we have formed 50 age-matched pairs, some of which will be used in the experiment. The purpose of the pairing is to increase the power of the experiment by eliminating the random variation due to age. (Of course, the ellipses do not actually have ages, but the pairing shown in the table has been constructed in a way that *simulates* age matching.)

(a) Use random digits (from Table 1 or your calculator) to choose a random sample of five pairs from the list.

(b) For each pair, use random digits (or toss a coin) to randomly allocate one member to treatment (T) and the other to control (C).

(c) Measure the lengths of all 10 ellipses. Then, to simulate a treatment effect, add 6 mm to each length in the T group.

(d) Apply a paired-sample t test to the data. Use a nondirectional alternative and let $\alpha = 0.05$.

(e) Did the analysis of part (d) lead you to a Type II error?

8.3.2 (Continuation of Exercise 8.3.1) Apply an independent-samples t test to your data. Use a nondirectional alternative and let $\alpha = 0.05$. Does this analysis lead you to a Type II error?

8.3.3 (Sampling exercise) Refer to Exercise 8.3.1. Imagine that a matched-pairs experiment is not practical (perhaps because the ages of the individuals cannot be measured), so we decide to use a completely randomized design to evaluate the treatment T.

(a) Use random digits (from Table 1 or your calculator) to choose a random sample of 10 individuals from the ellipse population (shown with Exercise 3.1.1). From these 10, randomly allocate 5 to T and 5 to C. (Or, equivalently, just randomly select 5 from the population to receive T and 5 to receive C.)

(b) Measure the lengths of all 10 ellipses. Then, to simulate a treatment effect, add 6 mm to each length in the T group.

(c) Apply an independent-samples t test to the data. Use a nondirectional alternative and let $\alpha = 0.05$.

(d) Did the analysis of part (c) lead you to a Type II error?

8.3.4 Refer to each of the following exercises. Construct a scatterplot of the data. Does the appearance of the scatterplot indicate that the pairing was effective?

(a) Exercise 8.2.1

(b) Exercise 8.2.2

(c) Exercise 8.2.6

8.3.5 Does temperature affect heart rate in humans? One way to address this question is to conduct an experiment. Take 20 test subjects and randomly split them into two groups of size 10 each: 10 to be placed in a cool (65°F) room and 10 to be placed in a warm (90°F) room. After 15 minutes, measure the pulse of each of the 20 individuals and compare the rates.

(a) Describe how one could conduct a similar study with the same 20 subjects, but using a paired design.

(b) What is the primary advantage of the paired design in this example?

8.4 The Sign Test

The **sign test** is a nonparametric test that can be used to compare two paired samples. It is not particularly powerful, but it is very flexible in application and is especially simple to use and understand—a blunt but handy tool.

METHOD

Like the paired-sample t test, the sign test is based on the differences

$$D = Y_1 - Y_2$$

The only information used by the sign test is the *sign* (positive or negative) of each difference. If the differences are preponderantly of one sign, this is taken as evidence for the alternative hypothesis (provided that the data deviate in a manner consistent with the alternative hypothesis, in the case of a directional alternative). The following examples illustrate the sign test.

Example 8.4.1 **Skin Grafts** Skin from cadavers can be used to provide temporary skin grafts for severely burned patients. The longer such a graft survives before its inevitable rejection by the immune system, the more the patient benefits. A medical team investigated the usefulness of matching graft to patient with respect to the HL-A (Human Leukocyte Antigen) antigen system. Each patient received two grafts, one with close HL-A compatibility and the other with poor compatibility. The survival times (in days) of the skin grafts are shown in the Table 8.4.1.[13]

Table 8.4.1 Skin graft survival times

	HL-A Compatibility		
Patient	Close y_1	Poor y_2	Sign of $d = y_1 - y_2$
1	37	29	+
2	19	13	+
3	57+	15	+
4	93	26	+
5	16	11	+
6	23	18	+
7	20	26	−
8	63	43	+
9	29	18	+
10	60+	42	+
11	18	19	−

Notice that a t test could not be applied here because two of the observations are incomplete; patient 3 died with a graft still surviving and the observation on patient 10 was incomplete for an unspecified reason. Nonetheless, we can proceed with a sign test, since the sign test depends only on the sign of the difference for each patient and we know that $Y_1 - Y_2$ is positive for both of these patients.

Let us carry out a sign test to compare the survival times of the two sets of skin grafts using $\alpha = 0.05$. A directional research (alternative) hypothesis is appropriate for this experiment:

H_A: Skin grafts tend to last longer when the HL-A compatibility is close.

The null hypothesis is

H_0: The survival time distribution is the same for close compatibility as it is for poor compatibility.

The first step is to determine the following counts:

$$N_+ = \text{Number of positive differences}$$
$$N_- = \text{Number of negative differences}$$

Because H_A is directional and it predicts that most of the differences will be positive, the test statistic B_s is

$$B_s = N_+$$

For the present data, we have

$$N_+ = 9$$
$$N_- = 2$$
$$B_s = 9$$

The next step is to find the P-value. We use the letter B in labeling the test statistic B_s because the distribution of B_s is based on the binomial distribution. Let p represent the probability that a difference will be positive. If the null hypothesis is true, then $p = 0.5$. Thus, the null distribution of B_s is a binomial with $n = 11$ and $p = 0.5$. That is, the null hypothesis implies that the sign of each difference is like the result of a coin toss, with heads corresponding to a positive difference and tails to a negative difference.

For the skin graft data, the P-value for the test is the probability of getting 9 or more positive differences in 11 patients if $p = 0.5$. This is the probability that a binomial random variable with $n = 11$ and $p = 0.5$ will be greater than or equal to 9. Using the binomial formula from Chapter 3, or a computer, we find that this probability is 0.03272.*

Because the P-value is less than α, we find significant evidence for H_A that skin grafts tend to last longer when the HL-A compatibility is close than when it is poor. ■

Example 8.4.2 **Stream Coliform Contamination** Table 8.4.2 shows the stream coliform data of Example 8.3.4, together with the signs of the differences.

Let's carry out a sign test to compare the coliform levels at the two locations, using $\alpha = 0.01$. The null hypothesis and nondirectional alternative are

H_0: The coliform level is the same at the two locations.

H_A: One of the locations has a higher coliform level than the other.

For these data,

$$N_+ = 1$$
$$N_- = 20$$

When the alternative is nondirectional, B_s is defined as

$$B_s = \text{Larger of } N_+ \text{ and } N_-$$

*Later in this section we shall learn how to use a table to compute these P-values; however, if you have covered the optional section on the binomial distribution, you can compute this probability using the binomial formula

$${}_{11}C_9(0.5)^9(0.5)^2 + {}_{11}C_{10}(0.5)^{10}(0.5)^1 + {}_{11}C_{11}(0.5)^{11} = 0.02686 + 0.00537 + 0.00049 = 0.03272$$

Table 8.4.2 Stream coliform count (MPN/100ml)

Collection date	Upstream	Downstream	Sign of $d = y_1 - y_2$
Jan-08	2,909	2,359	$-$
Feb-08	2,382	3,873	$+$
Mar-08	2,046	1,725	$-$
Apr-08	1,483	771	$-$
May-08	4,611	1,529	$-$
Jun-08	3,448	2,909	$-$
Jul-08	4,106	2,014	$-$
Aug-08	2,755	1,872	$-$
Sep-08	3,448	2,481	$-$
Oct-08	2,098	613	$-$
Nov-08	789	733	$-$
Dec-08	988	960	$-$
Jan-09	1,187	833	$-$
Feb-09	638	388	$-$
Mar-09	2,481	1,421	$-$
Apr-09	2,481	933	$-$
May-09	3,654	1,076	$-$
Jun-09	3,130	1,374	$-$
Jul-09	4,611	1,553	$-$
Aug-09	2,755	1,198	$-$
Sep-09	3,076	1,300	$-$

so for the virus growth data,

$$B_s = 20$$

The P-value for the test is the probability of getting 20 or more successes, plus the probability of getting 1 or fewer successes, in a binomial experiment with $n = 21$. We could use the binomial formula to calculate the P-value. As an alternative, critical values and P-values for the sign test are given in Table 7 (at the end of the book). Using Table 7 with $n_D = 21$, we obtain the critical values and corresponding P-values shown in Table 8.4.3:

Table 8.4.3 Critical values and P-values for the sign test when $n_D = 21$

n_D	0.20	0.10	0.05	0.02	0.01	0.002	0.001
21	**14** *0.189*	**15** *0.078*	**16** *0.027*	**17** *0.007*	**17** *0.007*	**18** *0.0015*	**19** *0.0002*

From the table we see that for $B_s = 20$ the P-value is less than 0.0002 since the B_s value exceeds those listed (**19** *0.0002*) in the table, so there is significant evidence for H_A. That is, we reject H_0 and conclude that the data provide significant evidence that the coliform levels are not the same at the two locations on the stream. ■

Bracketing the *P*-Value Like the Wilcoxon-Mann-Whitney test, the sign test has a discrete null distribution. Certain critical value entries in Table 7 are blank, for in some cases the most extreme data possible do not lead to a small *P*-value. Table 7 has another peculiarity that is not shared by the Wilcoxon-Mann-Whitney test: Some critical values appear more than once in the same row due to the discreteness of the null distribution.

Directional Alternative To use Table 7 if the alternative hypothesis is directional, we proceed with the familiar two-step procedure:

Step 1. Check directionality (see if the data deviate from H_0 in the direction specified by H_A).

(a) If not, the *P*-value is greater than 0.50.

(b) If so, proceed to step 2.

Step 2. The *P*-value is half what it would be if H_A were nondirectional.

Caution Note that Table 7, for the sign test, and Table 4, for the *t* test, are organized differently: Table 7 is entered with n_D, while Table 4 is entered with df$(= n_D - 1)$.

Treatment of Zeros It may happen that some of the differences $(Y_1 - Y_2)$ are equal to zero. Should these be counted as positive or negative in determining B_s? A recommended procedure is to drop the corresponding pairs from the analysis and reduce the sample size n_D accordingly. In other words, each pair whose difference is zero is ignored entirely; such pairs are regarded as providing no evidence against H_0 in either direction. Notice that this procedure has no parallel in the *t* test; the *t* test treats differences of zero the same as any other value.

Example 8.4.3

Null Distribution Consider an experiment with 10 pairs so that $n_D = 10$. If H_0 is true, then the probability distribution of N_+ is a binomial distribution with $n = 10$ and $p = 0.5$. Figure 8.4.1(a) shows this binomial distribution, together with the associated values of N_+ and N_-. Figure 8.4.1(b) shows the null distribution of B_s, which is a "folded" version of Figure 8.4.1(a). (We saw a similar relationship between parts (a) and (b) of Figure 7.10.4.)

If N_+ is 7 and H_A is directional (and predicts that positive differences are more likely than negative differences), then the *P*-value is the probability of 7 or more (+) signs in 10 trials. Using the binomial formula from Chapter 3, or a computer, we find

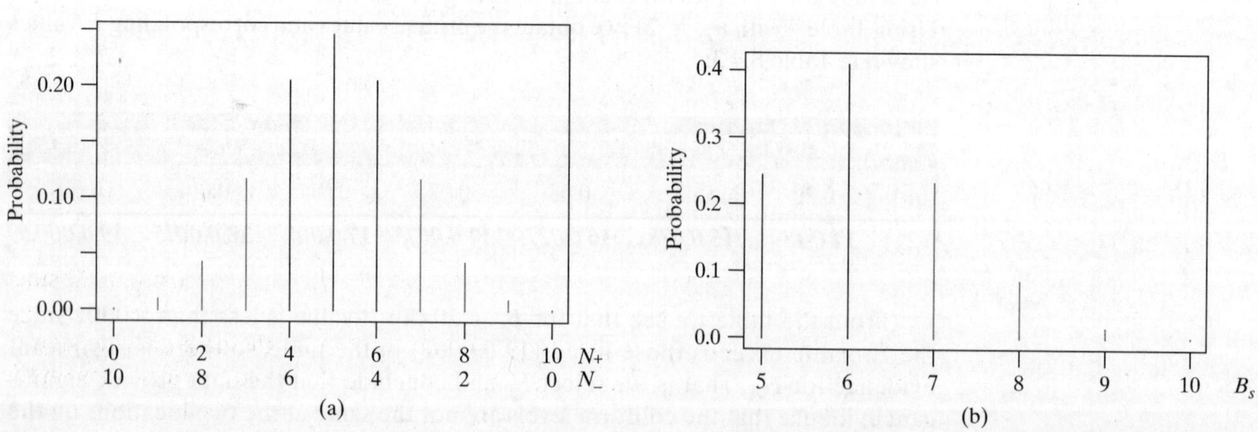

Figure 8.4.1 Null distributions for the sign test when $n_d = 10$. (a) Distribution of N_+ and N_- and (b) distribution of B_s.

that this probability is 0.17188.* This value (0.171876) is the sum of the shaded bars in the right-hand tail in Figure 8.4.1(a). If H_A is nondirectional, then the P-value is the sum of the shaded bars in the left-hand tail and of the right-hand tail of Figure 8.4.1(a). The two shaded areas are both equal to 0.17188; consequently, the total shaded area, which is the P-value, is

$$P = 2(0.171876) = 0.343752 \approx 0.34$$

In terms of the null distribution of B_s, the P-value is an upper-tail probability; thus, the sum of the shaded bars in Figure 8.4.1(b) is equal to 0.34. ■

How Table 7 Is Calculated Throughout your study of statistics you are asked to take on faith the critical values given in various tables. Table 7 is an exception; the following example shows how you could (if you wished to) calculate the critical values yourself. Understanding the example will help you to appreciate how the other tables of critical values have been obtained.

Example 8.4.4 Suppose $n_D = 10$. We saw in Example 8.4.3 that

If $B_s = 7$, the P-value of the data is 0.34376.

Similar calculations using the binomial formula show that

If $B_s = 8$, the P-value of the data is 0.10938.
If $B_s = 9$, the P-value of the data is 0.02148.
If $B_s = 10$, the P-value of the data is 0.00195.

For $n_D = 10$, the critical values from Table 7 are reproduced in Table 8.4.4.

Table 8.4.4 Critical values and P-values for the sign test when $n_D = 10$

n_D	0.20	0.10	0.05	0.02	0.01	0.002	0.001
10	**8** *0.109*	**9** *0.021*	**9** *0.021*	**10** *0.002*	**10** *0.002*	**10** *0.0020*	

The smallest value of B_s that gives a P-value less than 0.20 is $B_s = 8$, so this is the entry in the 0.20 column. For $\alpha = 0.10$ or $\alpha = 0.05$, $B_s = 9$ is needed. The most extreme possibility, $B_s = 10$, gives a P-value of 0.00195, which is rounded to 0.0020 in the table. It is not possible to obtain a nondirectional P-value as small as 0.001, so that entry is left blank. ■

APPLICABILITY OF THE SIGN TEST

The sign test is valid in any situation where the D's are independent of each other and the null hypothesis can be appropriately translated as

$$H_0\text{: Pr}\{D \text{ is positive}\} = 0.5$$

Thus, the sign test is distribution free; its validity does not depend on any conditions about the form of the population distribution of the D's. This broad validity is bought at a price: If the population distribution of the D's is indeed normal, then the sign test is much less powerful than the t test.

*Applying the binomial formula we have

$$ {}_{10}C_7(0.5)^7(0.5)^3 + {}_{10}C_8(0.5)^8(0.5)^2 + {}_{10}C_9(0.5)^9(0.5)^1 + {}_{10}C_{10}(0.5)^{10} $$
$$ = 0.117188 + 0.043945 + 0.009766 + 0.000977 = 0.171876 $$

The sign test is useful because it can be applied quickly and in a wide variety of settings. In fact, sometimes the sign test can be applied to data that do not permit a t test at all, as was shown in Example 8.4.1. There is another test for paired data, the Wilcoxon signed-ranks test, which is presented in Section 8.5, that is generally more powerful than the sign test and yet is distribution free. However, the Wilcoxon signed-ranks test is more difficult to carry out than the sign test and, like the t test, there are situations in which it cannot be conducted. The following is another example in which only a sign test is possible.

Example 8.4.5 **THC and Chemotherapy** Chemotherapy for cancer often produces nausea and vomiting. The effectiveness of THC (tetrahydrocannabinol—the active ingredient of marijuana) in preventing these side effects was compared with the standard drug Compazine. Of the 46 patients who tried both drugs (but were not told which was which), 21 expressed no preference, while 20 preferred THC and 5 preferred Compazine. Since "preference" indicates a sign for the difference, but not a magnitude, a t test is impossible in this situation. For a sign test, we have $n_d = 25$ and $B_s = 20$, so the P-value is 0.004; even at $\alpha = 0.005$ we would reject H_0 and find that the data provide sufficient evidence to conclude that THC is preferred to Compazine.[14] ■

Exercises 8.4.1–8.4.11

8.4.1 Use Table 7 to find the P-value for a sign test (against a nondirectional alternative), assuming that $n_D = 9$ and

(a) $B_s = 6$

(b) $B_s = 7$

(c) $B_s = 8$

(d) $B_s = 9$

8.4.2 Use Table 7 to find the P-value for a sign test (against a nondirectional alternative), assuming that $n_D = 15$ and

(a) $B_s = 10$

(b) $B_s = 11$

(c) $B_s = 12$

(d) $B_s = 13$

(e) $B_s = 14$

(f) $B_s = 15$

8.4.3 A group of 30 postmenopausal women were given oral conjugated estrogen for one month. Plasma levels of plasminogen-activator inhibitor type 1 (PAI-1) went down for 22 of the women, but went up for 8 women.[15] The P-value for a sign test, with a nondirectional alternative, is 0.016.

(a) State the null and alternative hypotheses in context.

(b) If $\alpha = 0.10$, what is your conclusion regarding H_0?

8.4.4 Can mental exercise build "mental muscle"? In one study of this question, 12 littermate pairs of young male rats were used; one member of each pair, chosen at random, was raised in an "enriched" environment with toys and companions, while its littermate was raised alone in an "impoverished" environment. After 80 days, the animals were sacrificed and their brains were dissected by a researcher who did not know which treatment each rat had received. One variable of interest was the weight of the cerebral cortex, expressed relative to total brain weight. For 10 of the 12 pairs, the relative cortex weight was greater for the "enriched" rat than for his "impoverished" littermate; in the other 2 pairs, the "impoverished" rat had the larger cortex. Use a sign test to compare the environments at $\alpha = 0.05$; let the alternative hypothesis be that environmental enrichment tends to increase the relative size of the cortex.[16]

8.4.5 Twenty institutionalized epileptic patients participated in a study of a new anticonvulsant drug, valproate. Ten of the patients (chosen at random) were started on daily valproate, and the remaining 10 received an identical placebo pill. During an 8-week observation period, the numbers of major and minor epileptic seizures were counted for each patient. After this, all patients were "crossed over" to the other treatment, and seizure counts were made during a second 8-week observation period. The numbers of minor seizures are given in the accompanying table.[17] Test for efficacy of valproate using the sign test at $\alpha = 0.05$. Use a directional alternative. (Note that this analysis ignores the possible effect of time—that is, first versus second observation period.)

Patient number	Placebo period	Valproate period	Patient number	Placebo period	Valproate period
1	37	5	11	7	8
2	52	22	12	9	8
3	63	41	13	65	30
4	2	4	14	52	22
5	25	32	15	6	11
6	29	20	16	17	1
7	15	10	17	54	31
8	52	25	18	27	15
9	19	17	19	36	13
10	12	14	20	5	5

8.4.6 An ecological researcher studied the interaction between birds of two subspecies, the Carolina Junco and the Northern Junco. He placed a Carolina male and a Northern male, matched by size, together in an aviary and observed their behavior for 45 minutes beginning at dawn. This was repeated on different days with different pairs of birds. The table shows counts of the episodes in which one bird displayed dominance over the other—for instance, by chasing it or displacing it from its perch.[18] Consider a sign test, using a nondirectional alternative, to compare the subspecies.

	Number of episodes in which	
Pair	Northern was dominant	Carolina was dominant
1	0	9
2	0	6
3	0	22
4	2	16
5	0	17
6	2	33
7	1	24
8	0	40

(a) What is the value of the test statistic?

(b) The P-value for the sign test is 0.008. If $\alpha = 0.01$, what is your conclusion?

8.4.7

(a) Suppose a paired data set has $n_D = 4$ and $B_s = 4$. Calculate the exact P-value of the data as analyzed by the sign test (against a nondirectional alternative).

(b) Explain why, in Table 7 with $n_D = 3$, no critical values are given in any column.

8.4.8 Suppose a paired data set has $n_D = 15$. Calculate the exact P-value of the data as analyzed by the sign test (against a nondirectional alternative) if $B_s = 15$.

8.4.9 The study described in Example 8.1.1, involving the compound mCPP, included a group of men. The men were asked to rate how hungry they were at the end of each 2-week period and differences were computed (hunger rating when taking mCPP minus hunger rating when taking the placebo). The distribution of the differences was not normal. Nonetheless, a sign can be conducted using the following information: Out of eight men who recorded hunger ratings, three reported greater hunger on mCPP than on the placebo and five reported lower hunger on mCPP than on the placebo.[1] Conduct a sign test at the $\alpha = 0.10$ level; use a nondirectional alternative.

8.4.10 Refer to Exercise 8.4.9. Calculate the exact P-value of the data as analyzed by the sign test. (*Note:* H_A is nondirectional.)

8.4.11 (Power) A researcher is planning to conduct an experiment to compare two treatments in which matched pairs of subjects will be given the treatments and a sign test will be used, with a nondirectional alternative, to analyze the difference in responses.

Suppose the researcher believes that one treatment will always do better than the other. How many pairs does he need to have in the experiment if he wants to be able to reject H_0 when $\alpha = 0.05$? If one treatment "wins" in every pair, what will be the P-value from the resulting test?

8.5 The Wilcoxon Signed-Rank Test

The **Wilcoxon signed-rank test,** like the sign test, is a nonparametric method that can be used to compare paired samples. Conducting a Wilcoxon signed-rank test is somewhat more complicated than conducting a sign test, but the Wilcoxon test is more powerful than the sign test. Like the sign test, the Wilcoxon signed-rank test does *not* require that the data be a sample from a normally distributed population.

The Wilcoxon signed-rank test is based on the set of differences, $D = Y_1 - Y_2$. It combines the main idea of the sign test—"look at the signs of the differences"—with the main idea of the paired t test—"look at the magnitudes of the differences."

METHOD

The Wilcoxon signed-rank test proceeds in several steps, which we present here in the context of an example.

Example 8.5.1

Nerve Cell Density For each of nine horses, a veterinary anatomist measured the density of nerve cells at specified sites in the intestine. The results for site I (midregion of jejunum) and site II (mesenteric region of jejunum) are given in the accompanying table.[19] Each density value is the average of counts of nerve cells in five equal sections of tissue. The null hypothesis of interest is that in the population of all horses there is no difference between the two sites.

1. The first step in the Wilcoxon signed-rank test is to calculate the differences, as shown in Table 8.5.1.

Table 8.5.1 Nerve cell density at each of two sites

Animal	Site I	Site II	Difference
1	50.6	38.0	12.6
2	39.2	18.6	20.6
3	35.2	23.2	12.0
4	17.0	19.0	−2.0
5	11.2	6.6	4.6
6	14.2	16.4	−2.2
7	24.2	14.4	9.8
8	37.4	37.6	−0.2
9	35.2	24.4	10.8

2. Next we find the absolute value of each difference.
3. We then rank these absolute values, from smallest to largest, as shown in Table 8.5.2.

Table 8.5.2 Absolute values of the difference of new cell density at each of two sites

Animal	Difference, d	$\|d\|$	Rank of $\|d\|$
1	12.6	12.6	8
2	20.6	20.6	9
3	12.0	12.0	7
4	−2.0	2.0	2
5	4.6	4.6	4
6	−2.2	2.2	3
7	9.8	9.8	5
8	−0.2	0.2	1
9	10.8	10.8	6

4. Next we restore the + and − signs to the ranks of the absolute differences to produce signed ranks, as shown in Table 8.5.3.

Table 8.5.3 Signed ranks for the difference of new cell density at each of two sites

Animal	Difference, d	Rank of $\|d\|$	Signed rank
1	12.6	8	8
2	20.6	9	9
3	12.0	7	7
4	−2.0	2	−2
5	4.6	4	4
6	−2.2	3	−3
7	9.8	5	5
8	−0.2	1	−1
9	10.8	6	6

5. We sum the positive signed ranks to get W_+; we sum the absolute values of the negative signed ranks to get W_-. For the nerve cell data, $W_+ = 8 + 9 + 7 + 4 + 5 + 6 = 39$ and $W_- = 2 + 3 + 1 = 6$. The test statistic, W_s is defined as

$$W_s = \text{Larger of } W_+ \text{ and } W_-$$

For the nerve cell data, $W_s = 39$.

6. To find the P-value, we consult Table 8 (at the end of the book). Part of Table 8 is reproduced in Table 8.5.4.

Table 8.5.4 Critical values for the Wilcoxon signed-rank test when $n_D = 9$

n	0.20	0.10	0.05	0.02	0.01	0.002	0.001
9	**35** *0.164*	**37** *0.098*	**40** *0.039*	**42** *0.020*	**44** *0.0078*		

From Table 8.5.4, we see that for $W_s = 37$ the P-value is 0.098 and for $W_s = 40$ the P-value is 0.039. There is weak but suggestive evidence ($0.039 < P\text{-value} < 0.098$) that there is a difference in nerve cell density in the two regions. (We reject H_0 if α is 0.10 or larger.) ■

Bracketing the P-Value Like the sign test, the Wilcoxon signed-rank test has a discrete null distribution. Certain critical value entries in Table 8 are blank; this situation is familiar from our study of the Wilcoxon-Mann-Whitney test and the sign test. For example, if $n_D = 9$, then the strongest possible evidence against H_0 occurs when all 9 differences are positive (or when all 9 differences are negative), in which case $W_s = 45$. But the chance that W_s will equal 45 when H_0 is true is $(1/2)^9 + (1/2)^9$, which is approximately 0.0039. Thus, it is not possible to have a two-tailed P-value smaller than 0.002, let alone 0.001. This is why the last two entries are blank in the $n_D = 9$ row of Table 8. Also note that if $W_s = 34$, for example, then the table only tells us that $P > 0.20$.

Directional Alternative To use Table 8 if the alternative hypothesis is directional, we proceed with the familiar two-step procedure:

Step 1. Check directionality (see if the data deviate from H_0 in the direction specified by H_A).

(a) If not, the P-value is greater than 0.50.

(b) If so, proceed to step 2.

Step 2. The P-value is half what it would be if H_A were nondirectional.

Treatment of Zeros If any of the differences ($Y_1 - Y_2$) are zero, then those data points are deleted and the sample size is reduced accordingly. For example, if one of the 9 differences in Example 8.5.1 had been zero, we would have deleted that point when conducting the Wilcoxon test, so that the sample size would have become 8.

Treatment of Ties If there are ties among the absolute values of the differences (in step 3) we average the ranks of the tied values. If there are ties, then the *P*-value given by the Wilcoxon signed-rank test is only approximate.

APPLICABILITY OF THE WILCOXON SIGNED-RANK TEST

The Wilcoxon signed-rank test can be used in any situation in which the *D*'s are independent of each other and come from a symmetric distribution; the distribution need not be normal.* The null hypothesis of "no treatment effect" or "no difference between populations" can be stated as

$$H_0\colon \mu_D = 0$$

Sometimes the Wilcoxon signed-rank test can be carried out even with incomplete information. For example, a Wilcoxon test is possible for the skin graft data of Example 8.4.1. It is true that an exact value of d cannot be calculated for two of the patients, but for both of these patients the difference is positive and is larger than either of the negative differences. The data in Table 8.5.5 show that there only are two negative differences. The smaller of these is −1, for patient 11. This is the smallest difference in absolute value, so it has signed rank −1. The only other negative signed rank is for patient 7; all of the other signed ranks are positive. (The rest of this example is left as an exercise.)

Table 8.5.5 Skin graft survival times

	HL-A compatibility		
	Close	Poor	
Patient	y_1	y_2	$d = y_1 - y_2$
1	37	29	8
2	19	13	6
3	57+	15	42+
4	93	26	67
5	16	11	5
6	23	18	5
7	20	26	−6
8	63	43	20
9	29	18	11
10	60+	42	18+
11	18	19	−1

As with the Wilcoxon-Mann-Whitney test for independent samples, there is a procedure associated with the Wilcoxon signed-rank test that can be used to construct a confidence interval for μ_D. The procedure is beyond the scope of this book.

*Strictly speaking, the distribution must be continuous, which means that the probability of a tie is zero.

In summary, when dealing with paired data we have three inference procedures: the paired t test, the Wilcoxon signed-rank test, and the sign test. The t test requires that the data come from a normally distributed population; if this condition is met then the t test is recommended, as it is more powerful than the Wilcoxon test or sign test. The Wilcoxon test does not require normality but does require that the differences come from a symmetric distribution and that they can be ranked; it has more power than the sign test. The sign test is the least powerful of the three methods, but the most widely applicable, since it only requires that we determine whether each difference is positive or negative.

Exercises 8.5.1–8.5.7

8.5.1 Use Table 8 to find the P-value for a Wilcoxon signed-rank test (against a nondirectional alternative), assuming that $n_D = 7$ and

(a) $W_s = 22$

(b) $W_s = 25$

(c) $W_s = 26$

(d) $W_s = 28$

8.5.2 Use Table 8 to find the P-value for a Wilcoxon signed-rank test (against a nondirectional alternative), assuming that $n_D = 12$ and

(a) $W_s = 55$

(b) $W_s = 65$

(c) $W_s = 71$

(d) $W_s = 73$

8.5.3 The study described in Example 8.1.1, involving the compound mCPP, included a group of nine men. The men were asked to rate how hungry they were at the end of each 2-week period and differences were computed (hunger rating when taking mCPP—hunger rating when taking the placebo). Data for one of the subjects are not available; the data for the other eight subjects are given in the accompanying table.[1] Analyze these data with a Wilcoxon signed-rank test at the $\alpha = 0.10$ level; use a nondirectional alternative.

	Hunger rating		
	mCPP	Placebo	Difference
Subject	y_1	y_2	$d = y_1 - y_2$
1	64	69	−5
2	119	112	7
3	0	28	−28
4	48	95	−47
5	65	145	−80
6	119	112	7
7	149	141	8
8	NA	NA	NA
9	99	119	−20

8.5.4 As part of the study described in Example 8.1.1 (and in Exercise 8.5.3), involving the compound mCPP, weight change was measured for nine men. For each man two measurements were made: weight change when taking mCPP and weight change when taking the placebo. The data are given in the accompanying table.[1]

	Weight change		
	mCPP	Placebo	Difference
Subject	y_1	y_2	$d = y_1 - y_2$
1	0.0	−1.1	1.1
2	−1.1	0.5	−1.6
3	−1.6	0.5	−2.1
4	−0.3	0.0	−0.3
5	−1.1	−0.5	−0.6
6	−0.9	1.3	−2.2
7	−0.5	−1.4	0.9
8	0.7	0.0	0.7
9	−1.2	−0.8	−0.4

(a) What is the null hypothesis for a Wilcoxon signed-rank test here?

(b) Here is computer output for a Wilcoxon signed-rank test. Explain what the P-value means in the context of this study.

```
Wilcoxon signed rank test
data: mCpp and Placebo   p-value = 0.4258
alternative hypothesis: true location
shift is not equal to 0
```

(c) If $\alpha = 0.05$, what is your conclusion regarding H_0? State your conclusion in the context of the research.

8.5.5 Consider the skin graft data of Example 8.4.1. Table 8.5.5, at the end of Section 8.5, shows the first steps in conducting a Wilcoxon signed-rank test of the null hypothesis that HL-A compatibility has no effect on graft survival time. Complete this test. Use $\alpha = 0.05$ and use the directional alternative that survival time tends to be greater when compatibility score is close.

8.5.6 In an investigation of possible brain damage due to alcoholism, an X-ray procedure known as a computerized tomography (CT) scan was used to measure brain densities in 11 chronic alcoholics. For each alcoholic, a nonalcoholic control was selected who matched the alcoholic on age, sex, education, and other factors. The brain density measurements on the alcoholics and the matched controls are reported in the accompanying table.[20]

Pair	Alcoholic	Control	Difference
1	40.1	41.3	−1.2
2	38.5	40.2	−1.7
3	36.9	37.4	−0.5
4	41.4	46.1	−4.7
5	40.6	43.9	−3.3
6	42.3	41.9	0.4
7	37.2	39.9	−2.7
8	38.6	40.4	−1.8
9	38.5	38.6	−0.1
10	38.4	38.1	0.3
11	38.1	39.5	−1.4
Mean	39.15	40.66	−1.52
SD	1.72	2.56	1.58

(a) What is the value of the Wilcoxon signed-rank test statistic for comparing the means?

(b) The *P*-value for the Wilcoxon signed-rank test is 0.0049. If $\alpha = 0.01$, what is your conclusion regarding H_0?

8.5.7 In a study, on the effect of caffeine on myocardial blood flow, 10 subjects had their blood flow measured before and after consuming caffeine.[21] For this setting the differences do not follow a normal distribution, so a *t* test would not be valid. Use a Wilcoxon signed-rank test to test the null hypothesis of no difference against the alternative that caffeine has an effect on myocardial blood flow. Let $\alpha = 0.01$.

Subject	Baseline	Caffeine	Difference
1	3.43	2.72	0.71
2	3.08	2.94	0.14
3	3.07	1.76	1.31
4	2.65	2.16	0.49
5	2.49	2.00	0.49
6	2.33	2.37	−0.04
7	2.31	2.35	−0.04
8	2.24	2.26	−0.02
9	2.17	1.72	0.45
10	1.34	1.22	0.12
Mean	2.51	2.15	0.36
SD	0.59	0.50	0.43

8.6 Perspective

In this section we consider some limitations to the analysis of paired data.

BEFORE–AFTER STUDIES

Many studies in the life sciences compare measurements before and after some experimental intervention, which can present another limitation. These studies can be difficult to interpret, because the effect of the experimental intervention may be confounded with other changes over time. For example, in one study, doctors measured myocardial blood flow (MBF) during bicycle exercise before and after giving subjects a dose of caffeine. They found that MBF went down for each of the eight subjects in the study, but it is possible that blood flow would have decreased with the passage of time even if the subjects had not taken caffeine.[22] One way to protect against this difficulty is to use randomized concurrent controls, as in the following example.

Example 8.6.1 **Wrist Fracture and Pain** Adults who broke their wrists were randomly divided into two groups. The "intervention" group was given exercises to do under the direction of a physiotherapist. The "control" group did not receive such intervention. The patients were asked to rate how much pain they felt, on a scale of 0 to 100, at the beginning of the experiment and again after 6 weeks.[23]

Table 8.6.1 Pain rating

Group	n	Pain rating: At baseline	6 weeks later	Difference: Mean	SE
Intervention	27	49.0	23.2	25.8	3.8
Control	20	50.3	37.4	12.9	4.0

Let us analyze the before–after changes by paired t tests at $\alpha = 0.05$. In the intervention group, the mean pain rating fell by 25.8. To evaluate the statistical significance of this increase, the test statistic is

$$t_s = \frac{25.8 - 0}{3.8} = 6.8$$

which is highly significant (P-value < 0.0001). However, this result alone does not demonstrate the effectiveness of the intervention; the improvement in pain might be partly or entirely due to other factors, such as the special attention received by all the participants or simply the passage of time. Indeed, a paired t test applied to the control group gives

$$t_s = \frac{12.9 - 0}{4.0} = 3.2$$

with P-value $= 0.0048$.

Thus, the people who did *not* receive any physiotherapy intervention *also* experienced a statistically significant drop in pain.

To isolate the effect of the intervention, we can compare the experience of the two treatment groups, using an independent-samples t test *on the two samples of differences*. We again choose $\alpha = 0.05$. The difference between the mean changes in the two groups is

$$25.8 - 12.9 = 12.9$$

and the standard error of this difference is

$$\sqrt{3.8^2 + 4.0^2} = 5.5$$

Thus, the t statistic is

$$t_s = \frac{12.9}{5.5} = 2.34$$

This test provides strong evidence ($P = 0.0238$) that the intervention is effective. If the experimental design had not included the control group, then this last crucial comparison would not have been possible, and the support for efficacy of the intervention would have been shaky indeed. ■

In analyzing real data, it is wise to keep in mind that the statistical methods we have been considering address only limited questions.

The paired t test is limited in two ways:

1. It is limited to questions concerning $\overline{D}$.
2. It is limited to questions about *aggregate* differences.

The second limitation is very broad; it applies not only to the methods of this chapter but also to those of Chapter 7 and to many other elementary statistical techniques. We will discuss these two limitations separately.

LIMITATION OF $\overline{D}$

One limitation of the paired t test and confidence interval is simple, but too often overlooked: When some of the D's are positive and some are negative, the magnitude of $\overline{D}$ does not reflect the "typical" magnitude of the D's. The following example shows how misleading $\overline{D}$ can be.

Example 8.6.2 **Measuring Serum Cholesterol** Suppose a clinical chemist wants to compare two methods of measuring serum cholesterol; she is interested in how closely the two methods agree with each other. She takes blood specimens from 400 patients, splits each specimen in half, and assays one half by method A and the other by method B. Table 8.6.2 shows fictitious data, exaggerated to clarify the issue.

Table 8.6.2 Serum cholesterol (mg/dl)

Specimen	Method A	Method B	$d = A - B$
1	200	234	−34
2	284	272	+12
3	146	153	−7
4	263	250	+13
5	258	232	+26
⋮	⋮	⋮	⋮
400	176	190	−14
Mean	215.2	214.5	0.7
SD	45.6	59.8	18.8

In Table 8.6.2, the sample mean difference is small ($\bar{d} = 0.7$). Furthermore, the data indicate that the population mean difference is small (a 95% confidence interval is $-1.1 \text{ mg/dl} < \mu_D < 2.5 \text{ mg/dl}$). But such discussion of $\overline{D}$ or μ_D does not address the central question, which is: How closely do the methods agree? In fact, Table 8.6.2 indicates that the two methods do not agree well; the individual differences between method A and method B are not small *in magnitude.* The mean $\bar{d}$ is small because the positive and negative differences tend to cancel each other. A graph similar to Figure 8.3.1 would be very helpful in visually determining how well the methods agree. We would examine such a graph to see how closely the points cluster around the $y = x$ line as well as to see the spread in the boxplot of differences. To make a numerical assessment of agreement between the methods we should not focus on the mean difference, $\overline{D}$. It would be far more relevant to analyze the absolute (unsigned) magnitudes of the d's (i.e., 34, 12, 7, 13, 26, etc.). These magnitudes could be analyzed in various ways: We could average them, we could count how many are "large" (say, more than 10 mg/dl), and so on. ■

LIMITATION OF THE AGGREGATE VIEWPOINT

Consider a paired experiment in which two treatments, say A and B, are applied to the same person. If we apply a t test, a sign test, or a Wilcoxon signed-rank test, we are viewing the people as an ensemble rather than individually. This is appropriate if we are willing to assume that the difference (if any) between A and B is in a consistent direction for all people—or, at least, that the important features of the difference are preserved even when the people are viewed *en masse.* The following example illustrates the issue.

Example 8.6.3

Treatment of Acne Consider a clinical study to compare two medicated lotions for treating acne. Twenty patients participate. Each patient uses lotion A on one (randomly chosen) side of his face and lotion B on the other side. After 3 weeks, each side of the face is scored for total improvement.

First, suppose that the A side improves more than the B side in 10 patients, while in the other 10 the B side improves more. According to a sign test, this result is in perfect agreement with the null hypothesis. And yet, two very different interpretations are logically possible:

Interpretation 1: Treatments A and B are in fact completely equivalent; their action is indistinguishable. The observed differences between A and B sides of the face were entirely due to chance variation.

Interpretation 2: Treatments A and B are in fact completely different. For some people (about 50% of the population), treatment A is more effective than treatment B, whereas in the remaining half of the population treatment B is more effective. The observed differences between A and B sides of the face were biologically meaningful.*

The same ambiguity of interpretation arises if the results favor one treatment over another. For instance, suppose the A side improved more than the B side in 18 of the 20 cases, while B was favored in 2 patients. This result, which is statistically significant ($P < 0.001$), could again be interpreted in two ways. It could mean that treatment A is in fact superior to B for everybody, but chance variation obscured its superiority in two of the patients; or it could mean that A is superior to B for most people, but for about 10% of the population ($2/20 = 0.10$) B is superior to A. ■

The difficulty illustrated by Example 8.6.3 is not confined to experiments with randomized pairs. In fact, it is particularly clear in another type of paired experiment—the measurement of change over time. Consider, for instance, the wrist pain data of Example 8.6.1. Our discussion of that study hinged on an aggregate measure of wrist pain: the mean. If some patients' pain rose as a result of the intervention and others fell, these details were ignored in the analysis based on Student's t; only the average change was analyzed.

The difficulties described previously aren't only confined to human experiments either. Suppose, for instance, that two fertilizers, A and B, are to be compared in an agronomic field experiment using a paired design, with the data to be analyzed by a paired t test. If treatment A is superior to B on acid soils, but B is better than A on alkaline soils, this fact would be obscured in an experiment that included soils of both types.

The issue raised by the preceding examples is a very general one. Simple statistical methods such as the sign test and the t test are designed to evaluate treatment effects *in the aggregate*—that is, *collectively*—for a population of people, or of mice, or of plots of ground. The segregation of differential treatment effects in subpopulations requires more delicate handling, both in design and analysis.

This confinement to the aggregate point of view applies to Chapter 7 (independent samples) even more forcefully than to the present chapter. For instance, if

*This may seem farfetched, but phenomena of this kind do occur; as an obvious example, consider the response of patients to blood transfusions of type A or type B blood.

treatment A is given to one group of mice and treatment B to another, it is quite impossible to know how a mouse in group A would have responded *if* it had received treatment B; the only possible comparison is an aggregate one. In Section 7.9 we stated that the statistical comparison of independent samples depends on an "implicit assumption"; essentially, the assumption is that the phenomenon under study can be adequately perceived from an aggregate viewpoint.

In many, perhaps most, biological investigations the phenomena of interest are reasonably universal, so this issue of submerging the individual in the aggregate does not cause a serious problem. Nevertheless, one should not lose sight of the fact that aggregation may obscure important individual detail.

REPORTING OF DATA

In communicating experimental results, it is desirable to choose a form of reporting that conveys the extra information provided by pairing. With small samples, a graphical approach can be used, as in Figure 8.1.1, where the line segments gave clear visual evidence that blood flow decreased for each subject.

In published reports of biological research, the crucial information related to pairing is often omitted. For instance, a common practice is to report the means and standard deviations of Y_1 and Y_2 but to omit the standard deviation of the difference, D! This is a serious flaw. It is best to report some description of D, using either a display like Figure 8.1.1, a histogram of the D's, or at least the standard deviation of the D's.

Exercises 8.6.1–8.6.5

8.6.1 Thirty-three men with high serum cholesterol, all regular coffee drinkers, participated in a study to see whether abstaining from coffee would affect their cholesterol level. Twenty-five of the men (chosen at random) drank no coffee for 5 weeks, while the remaining 8 men drank coffee as usual. The accompanying table shows the serum cholesterol levels (in mg/dl) at baseline (at the beginning of the study) and the change from baseline after 5 weeks.[24]

	No coffee ($n = 25$)		Usual coffee ($n = 8$)	
	Mean	SD	Mean	SD
Baseline	341	37	331	30
Change from baseline	−35	27	+26	56

For the following t tests use nondirectional alternatives and let $\alpha = 0.05$.

(a) The no-coffee group experienced a 35 mg/dl drop in mean cholesterol level. Use a t test to assess the statistical significance of this drop.

(b) The usual-coffee group experienced a 26 mg/dl rise in mean cholesterol level. Use a t test to assess the statistical significance of this rise.

(c) Use a t test to compare the no-coffee mean change (−35) to the usual-coffee mean change (+26).

8.6.2 Eight young women participated in a study to investigate the relationship between the menstrual cycle and food intake. Dietary information was obtained every day by interview; the study was double-blind in the sense that the participants did not know its purpose and the interviewer did not know the timing of their menstrual cycles. The table shows, for each participant, the average caloric intake for the 10 days preceding and the 10 days following the onset of the menstrual period (these data are for one cycle only). For these data, prepare a display like that of Figure 8.1.1.[25]

	Food intake (CAL)	
Participant	Premenstrual	Postmenstrual
1	2,378	1,706
2	1,393	958
3	1,519	1,194
4	2,414	1,682
5	2,008	1,652
6	2,092	1,260
7	1,710	1,239
8	1,967	1,758

8.6.3 For each of 29 healthy dogs, a veterinarian measured the glucose concentration in the anterior chamber of the left eye and the right eye, with the results shown in the table.[26]

Animal number	Glucose (mg/dl) Right eye	Left eye	Animal number	Glucose (mg/dl) Right eye	Left eye
1	79	79	16	80	80
2	81	82	17	78	78
3	87	91	18	112	110
4	85	86	19	89	91
5	87	92	20	87	91
6	73	74	21	71	69
7	72	74	22	92	93
8	70	66	23	91	87
9	67	67	24	102	101
10	69	69	25	116	113
11	77	78	26	84	80
12	77	77	27	78	80
13	84	83	28	94	95
14	83	82	29	100	102
15	74	75			

Using the paired t method, a 95% confidence interval for the mean difference is $-1.1 \text{ mg/dl} < \mu_D < 0.7 \text{ mg/dl}$. Does this result suggest that, for the typical dog in the population, the difference in glucose concentration between the two eyes is less than 1.1 mg/dl? Explain.

8.6.4 Tobramycin is a powerful antibiotic. To minimize its toxic side effects, the dose can be individualized for each patient. Thirty patients participated in a study of the accuracy of this individualized dosing. For each patient, the predicted peak concentration of Tobramycin in the blood serum was calculated, based on the patient's age, sex, weight, and other characteristics. Then Tobramycin was administered and the actual peak concentration (μg/ml) was measured. The results were reported as in the table.[27]

	Predicted	Actual
Mean	4.52	4.40
SD	0.90	0.85
n	30	30

Does the reported summary give enough information for you to judge whether the individualized dosing is, on the whole, accurate in its prediction of peak concentration? If so, describe how you would make this judgment. If not, describe what additional information you would need and why.

8.6.5 Can keeping cows out of a creek on ranchland improve the water quality? To answer this question environmental researchers measured the coliform counts at a variety of locations along two creeks on grazed ranch land. Then, a fence was installed along one of the creeks to keep the cattle out. Several weeks later, coliform counts were measured again along the two creeks at the same locations (i.e., the locations formed pairs). The accompanying table summarizes the coliform data (MPN/100ml).[28]

Coliform (MPN/100ml)	Fenced creek ($n = 8$ locations) Mean	SD	Control creek ($n = 7$ locations) Mean	SD
Before fencing	256.1	57.8	284.3	91.4
After fencing	225.0	75.2	269.0	81.5
Difference	31.1	17.8	15.3	16.3

(a) Use an appropriate t test to test to assess the statistical significance of the reduction in coliform along the fenced creek after the installation of fencing. Use a nondirectional alternative and $\alpha = 0.10$.

(b) Briefly explain why considering the data for fenced creek data alone, one cannot determine if the fencing was effective.

(c) Use an appropriate t test to test to compare the reduction in coliform along the fenced and control creeks. Use a nondirectional alternative and $\alpha = 0.10$. (You may use $df \approx 12$).

(d) Is the answer to part (c) consistent with the hypothesis that the fencing is effective?

Supplementary Exercises 8.S.1–8.S.23

8.S.1 A volunteer working at an animal shelter conducted a study of the effect of catnip on cats at the shelter. She recorded the number of "negative interactions" each of 15 cats made in 15-minute periods before and after being given a teaspoon of catnip. The paired measurements were collected on the same day within 30 minutes of one another; the data are given in the accompanying table.[29]

(a) Construct a 95% confidence interval for the difference in mean number of negative interactions.

(b) Construct a 95% confidence interval the wrong way, using the independent-samples method. How does this interval differ from the one obtained in part (a)?

Cat	Before (Y_1)	After (Y_2)	Difference
Amelia	0	0	0
Bathsheba	3	6	−3
Boris	3	4	−1
Frank	0	1	−1
Jupiter	0	0	0
Lupine	4	5	−1
Madonna	1	3	−2
Michelangelo	2	1	1
Oregano	3	5	−2
Phantom	5	7	−2
Posh	1	0	1
Sawyer	0	1	−1
Scary	3	5	−2
Slater	0	2	−2
Tucker	2	2	0
Mean	1.8	2.8	−1
SD	1.66	2.37	1.20

8.S.2 Refer to Exercise 8.S.1. Compare the before and after populations using a t test at $\alpha = 0.05$. Use a positive directional alternative.

8.S.3 Refer to Exercise 8.S.1. Compare the before and after populations using a sign test at $\alpha = 0.01$. Use a nondirectional alternative.

8.S.4 Refer to Exercise 8.S.1. Construct a scatterplot of the data. Does the appearance of the scatterplot indicate that the pairing was effective? Explain.

8.S.5 As part of a study of the physiology of wheat maturation, an agronomist selected six wheat plants at random from a field plot. For each plant, she measured the moisture content in two batches of seeds: one batch from the "central" portion of the wheat head, and one batch from the "top" portion, with the results shown in the following table.[30] Construct a 95% confidence interval for the mean difference in moisture content of the two regions of the wheat head.

	Percentage moisture	
Plant	Central	Top
1	62.7	59.7
2	63.6	61.6
3	60.9	58.2
4	63.0	60.5
5	62.7	60.6
6	63.7	60.8

8.S.6 Biologists noticed that some stream fishes are most often found in pools, which are deep, slow-moving parts of the stream, while others prefer riffles, which are shallow, fast-moving regions. To investigate whether these two habitats support equal levels of diversity (i.e., equal numbers of species), they captured fish at 15 locations along a river. At each location, they recorded the number of species captured in a riffle and the number captured in an adjacent pool. The following table contains the data.[31] Construct a 90% confidence interval for the difference in mean diversity between the types of habitats.

Location	Pool	Riffle	Difference
1	6	3	3
2	6	3	3
3	3	3	0
4	8	4	4
5	5	2	3
6	2	2	0
7	6	2	4
8	7	2	5
9	1	2	−1
10	3	2	1
11	4	3	1
12	5	1	4
13	4	3	1
14	6	2	4
15	4	3	1
Mean	4.7	2.5	2.2
SD	1.91	0.74	1.86

8.S.7 Refer to Exercise 8.S.6. What conditions are necessary for the confidence interval to be valid? Are those conditions satisfied? How do you know?

8.S.8 Refer to Exercise 8.S.6. Compare the habitats using a t test at $\alpha = 0.05$. Use a negative directional alternative.

8.S.9 Refer to Exercise 8.S.6.

(a) Compare the habitats using a sign test at $\alpha = 0.10$. Use a nondirectional alternative.

(b) Use the binomial formula to calculate the exact P-value for part (a).

8.S.10 Refer to Exercise 8.S.6. Analyze these data using a Wilcoxon signed-rank test.

8.S.11 Refer to the Wilcoxon signed-rank test from Exercise 8.S.10. On what grounds could it be argued that the P-value found in this test might not be accurate? This is, why might it be argued that the Wilcoxon test P-value is not a completely accurate measure of the strength of the evidence against H_0 in this case?

8.S.12 In a study of the effect of caffeine on muscle metabolism, nine male volunteers underwent arm exercise tests on two separate occasions. On one occasion, the volunteer took a placebo capsule an hour before the test; on the other occasion he received a capsule containing pure caffeine. (The time order of the two occasions was randomly determined.) During each exercise test, the subject's respiratory exchange ratio (RER) was measured. The RER is the ratio of carbon dioxide produced to oxygen consumed and is an indicator of whether energy is being obtained from carbohydrates or from fats. The results are presented in the accompanying table.[32] Use a t test to assess the effect of caffeine. Use a nondirectional alternative and let $\alpha = 0.05$.

	RER(%)	
Subject	Placebo	Caffeine
1	105	96
2	119	99
3	92	89
4	97	95
5	96	88
6	101	95
7	94	88
8	95	93
9	98	88

8.S.13 For the data of Exercise 8.S.12, construct a display like that of Figure 8.1.1.

8.S.14 Refer to Exercise 8.S.12. Analyze these data using a t test for an independent sample. Use a nondirectional alternative and let at $\alpha = 0.01$.

8.S.15 Certain types of nerve cells have the ability to regenerate a part of the cell that has been amputated. In an early study of this process, measurements were made on the nerves in the spinal cord in rhesus monkeys. Nerves emanating from the left side of the cord were cut, while nerves from the right side were kept intact. During the regeneration process, the content of creatine phosphate (CP) was measured in the left and the right portion of the spinal cord. The following table shows the data for the right (control) side (Y_1), and for the left (regenerating) side (Y_2). The units of measurement are mg CP per 100 gm tissue.[33] Consider a t test to compare the two sides at $\alpha = 0.05$ using a nondirectional alternative.

Animal	Right side (Control)	Left side (Regenerating)	Difference
1	16.3	11.5	4.8
2	4.8	3.6	1.2
3	10.9	12.5	−1.6
4	14.2	6.3	7.9
5	16.3	15.2	1.1
6	9.9	8.1	1.8
7	29.2	16.6	12.6
8	22.4	13.1	9.3
Mean	15.50	10.86	4.64
SD	7.61	4.49	4.89

(a) What is the value of the t test statistic?

(b) In the context of this study, state the null and alternative hypotheses.

(c) The P-value for the t test is 0.031. If $\alpha = 0.05$, what is your conclusion regarding the hypotheses in (b)?

(d) Compute a 95% confidence interval for the mean difference in creatine phosphate between the cut and control sides.

(e) A 90% confidence interval for the mean difference in creatine phosphate between the cut and control sides is (1.36, 7.91) mg/100gm. Interpret this interval in the context of the research.

8.S.16 Aldosterone is a hormone involved in maintaining fluid balance in the body. In a veterinary study, six dogs with heart failure were treated with the drug Captopril, and plasma concentrations of aldosterone were measured before and after the treatment. The results are given in the table below.[34] Consider a t test, using a nondirectional alternative, to investigate the claim that Captopril affects aldosterone level.

Animal	Before	After	Difference
1	749	374	375
2	469	300	169
3	343	146	197
4	314	134	180
5	286	69	217
6	223	20	203
Mean	397.3	173.8	223.5
SD	190.5	136.4	76.1

(a) What is the value of the test statistic?

(b) If $\alpha = 0.10$, what is your conclusion?

8.S.17 Refer to Exercise 8.S.16. Analyze these data using a Wilcoxon signed-rank test.

8.S.18 Refer to Exercise 8.S.16. Note that the dogs in this study are not compared to a control group. How does this weaken any inference that might be made about the effectiveness of Captopril?

8.S.19 (Computer exercise) For an investigation of the mechanism of wound healing, a biologist chose a paired design, using the left and right hindlimbs of the salamander *Notophthalmus viridescens.* After amputating each limb, she made a small wound in the skin and then kept the limb for 4 hours in either a solution containing benzamil or a control solution. She theorized that the benzamil would impair the healing. The accompanying table shows the amount of healing, expressed as the area (mm^2) covered with new skin after 4 hours.[35]

Animal	Control limb	Benzamil limb	Animal	Control limb	Benzamil limb
1	0.55	0.14	10	0.42	0.21
2	0.15	0.08	11	0.49	0.11
3	0.00	0.00	12	0.08	0.03
4	0.13	0.13	13	0.32	0.14
5	0.26	0.10	14	0.18	0.37
6	0.07	0.08	15	0.35	0.25
7	0.20	0.11	16	0.03	0.05
8	0.16	0.00	17	0.24	0.16
9	0.03	0.05			

(a) Assess the effect of benzamil using a t test at $\alpha = 0.05$. Let the alternative hypothesis be that the researcher's expectation is correct.

(b) Proceed as in part (a) but use a sign test.

(c) Construct a 95% confidence interval for the mean effect of benzamil.

(d) Construct a scatterplot of the data. Does the appearance of the scatterplot indicate that the pairing was effective? Explain.

8.S.20 (Computer exercise) In a study of hypnotic suggestion, 16 male volunteers were randomly allocated to an experimental group and a control group. Each subject participated in a two-phase experimental session. In the first phase, respiration was measured while the subject was awake and at rest. (These measurements were also described in Exercises 7.5.6 and 7.10.4.) In the second phase, the subject was told to imagine that he was performing muscular work, and respiration was measured again.

For subjects in the experimental group, hypnosis was induced between the first and second phases; thus, the suggestion to imagine muscular work was "hypnotic suggestion" for experimental subjects and "waking suggestion" for control subjects. The accompanying table shows the measurements of total ventilation (liters of air per minute per square meter of body area) for all 16 subjects.[36]

(a) Use a t test to compare the mean resting values in the two groups. Use a nondirectional alternative and let $\alpha = 0.05$. This is the same as Exercise 7.5.6(a).

(b) Use suitable paired and unpaired t tests to investigate (i) the response of the experimental group to suggestion; (ii) the response of the control group to suggestion; (iii) the difference between the responses of the experimental and control groups. Use directional alternatives (suggestion increases ventilation, and hypnotic suggestion increases it more than waking suggestion) and let $\alpha = 0.05$ for each test.

Experimental group			Control group		
Subject	Rest	Work	Subject	Rest	Work
1	5.74	6.24	9	6.21	5.50
2	6.79	9.07	10	4.50	4.64
3	5.32	7.77	11	4.86	4.61
4	7.18	16.46	12	4.78	3.78
5	5.60	6.95	13	4.79	5.41
6	6.06	8.14	14	5.70	5.32
7	6.32	11.72	15	5.41	4.54
8	6.34	8.06	16	6.08	5.98

(c) Repeat the investigations of part (b) using suitable nonparametric tests (sign and Wilcoxon-Mann-Whitney tests).

(d) Use suitable graphs to investigate the reasonableness of the normality condition underlying the t tests of part (b). How does this investigation shed light on the discrepancies between the results of parts (b) and (c)?

8.S.21 Suppose we want to test whether an experimental drug reduces blood pressure more than does a placebo. We are planning to administer the drug or the placebo to some subjects and record how much their blood pressures are reduced. We have 20 subjects available.

(a) We could form 10 matched pairs, where we form a pair by matching subjects, as best we can, on the basis of age and sex, and then randomly assign one subject in each pair to the drug and the other subject in the pair to the placebo. Explain why using a matched pairs design might be a good idea.

(b) Briefly explain why a matched pairs design might *not* be a good idea. That is, how might such a design be inferior to a completely randomized design?

8.S.22 A group of 20 postmenopausal women were given transdermal estradiol for one month. Plasma levels of plasminogen-activator inhibitor type 1 (PAI-1) went down for 10 of the women and went up for the other 10 women.[37] Use a sign test to test the null hypothesis that transdermal estradiol has no effect on PAI-1 level. Use $\alpha = 0.10$ and use a nondirectional alternative.

8.S.23 Six patients with renal disease underwent plasmapheresis. Urinary protein excretion (grams of protein per gram of creatinine) was measured for each patient before and after plasmapheresis. The data are given in the following table.[38] Use these data to investigate whether or not plasmapheresis affects urinary protein excretion in patients with renal disease. (*Hint.* Graph the data and consider whether a t test is appropriate in the original scale.)

Patient	Before	After	Difference
1	20.3	0.8	19.5
2	9.3	0.1	9.2
3	7.6	3.0	4.6
4	6.1	0.6	5.5
5	5.8	0.9	4.9
6	4.0	0.2	3.8
Mean	8.9	0.9	7.9
SD	5.9	1.1	6.0

HIGHLIGHTS AND STUDY

(REFLECTIONS ON CHAPTERS 6, 7, AND 8)

Unit

II

In Chapters 6–8 we introduced the concept of statistical inference in the context of comparing the means of two populations (or treatment groups), first by studying confidence intervals and then by studying hypothesis testing. Here we review the major points of these two methods of statistical inference in the context of a nutrition study that investigated the relationship between acai berry consumption and blood serum lipids in rats.[1]

Sixteen rats were fed a hypercholesterolemic diet (i.e., high cholesterol diet). Eight of these rats also received supplementary acai pulp in their diet, while the other eight did not. After 6 weeks the total blood cholesterol (mmol/L) was measured for each rat. A summary of the data appears in Table II.1 below.

Table II.1 Total blood cholesterol (mmol/L)

	Acai	Control
Mean ± SE	5.42 ± 0.35	8.12 ± 0.82
n	8	8

Hypothesis Testing

The intention of the researchers was to investigate whether or not acai pulp would be beneficial at reducing the negative physiological effects of a high cholesterol diet. Their research intention can be stated as an alternative hypothesis: The mean blood cholesterol level for rats on a hypercholesterolemic diet is lower for rats that consume acai than for rats that do not. To test this hypothesis, we examine the data in the context of and in contrast to a null hypothesis, which represents the status quo, that there is no difference in the mean blood cholesterol levels for rats that consume acai and those that do not.

Symbolically, our null hypothesis, H_0, is that $\mu_1 - \mu_2 = 0$ and our alternative hypothesis, H_A, is that $\mu_1 - \mu_2 < 0$.* (We have arbitrarily labeled the acai group as group 1 and the control group as group 2 so that $\mu_1 - \mu_2 < 0$ indicates that the mean cholesterol levels are lower in the acai group.) If $\bar{y}_1 - \bar{y}_2$ is substantially far from zero (in the negative direction), we can infer that H_0 is false. That is, we could infer that there is significant evidence that H_A is true. How far is "substantial"? One method we

*This is an example of a directional hypothesis (lower tail, H_A: $\mu_1 - \mu_2 < 0$). H_A: $\mu_1 - \mu_2 \neq 0$ would be an example of a non-directional hypothesis where the direction of the difference is not specified; that there is a difference in mean total cholesterol for rats consuming acai compared to those who don't.

covered in Chapter 7 is the two-sample t-test in which we compare (as a ratio) $\bar{y}_1 - \bar{y}_2$ to the magnitude of the standard error of $\bar{Y}_1 - \bar{Y}_2$ to obtain a (calculated) t statistic.

$$t_s = \frac{\bar{y}_1 - \bar{y}_2}{SE_{\bar{Y}_1 - \bar{Y}_2}}$$

If H_0 were true, this ratio should be relatively close to zero because $\bar{Y}_1 - \bar{Y}_2$ should be close to zero for most samples. If the calculated t ratio is large in magnitude (and in this case negative since we wish to show acai lowers cholesterol), we would have evidence for H_A. For these data we have

$$t_s = \frac{5.42 - 8.12}{\sqrt{0.35^2 + 0.82^2}} = \frac{-2.70}{0.892} = -3.027$$

Thus $\bar{y}_1 - \bar{y}_2$, the difference in our sample means, is more than 3.027 SEs away from what we would expect if acai had no effect on cholesterol (i.e., if H_0 were true). Is this difference "large enough" to be statistically significant? In Chapter 7 we learned how to address this question by considering a related one: How unlikely is it to obtain a (standardized) difference in sample means as large as ours in this study if H_0 is true? In context, if acai has no effect, how unlikely is it for two sample means (control sample and acai sample) to be 3.027 SEs apart? We considered randomization distributions as one method to compute this probability and then considered the more common method in which we compare our t statistic to an appropriate t distribution to compute a tail area. The tail area is called the P-value for the test. If the P-value is smaller than a predetermined cut-off value of α, then we declare that the difference between the sample means is "statistically significant," meaning that a difference this extreme (or more extreme) is unlikely to arise by chance alone. While comparing the P-value to α is a common practice to assess statistical significance, we recommend that the P-value always be reported, which allows the reader of a statistical report to use his or her own α when deciding whether or not to believe H_A.

In our example, we compare $t_s = -3.027$ to a t distribution with approximately 9 degrees of freedom to obtain the one-sided P-value of 0.0072. (Without a computer, we can use Table 4 to find the bracketed P-value of $0.005 < P\text{-value} < 0.01$.) This value indicates that if acai really has no effect on total cholesterol, a difference as large as our observed difference in sample means would only occur about 0.72% of the time in studies similar to this one. Since this would be very rare, our data provide compelling evidence that acai lowers cholesterol. (*Note:* We can say that acai *affects* [in this case, lowers] cholesterol because this study was an experiment.) We declare this result statistically significant at the 5% level since the P-value is less than $\alpha = 0.05$.

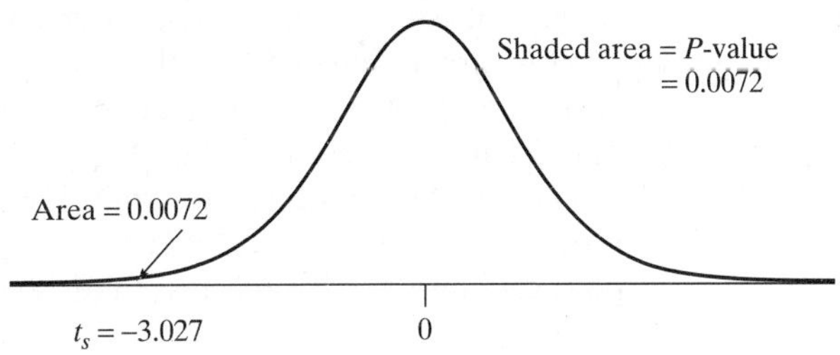

TYPE I AND TYPE II ERRORS

As discussed in Section 7.3, we recognize that conclusions from hypothesis tests aren't certain. By using $\alpha = 0.05$ as our level of significance, and abiding by the decision rule to reject H_0 when the P-value is less than $\alpha = 0.05$ we run the risk of wrongly rejecting H_0 (i.e., having statistically significant findings when there is no real difference between the populations studied) 5% of the time. We defined this type of error as a Type I error. The Type I error rate is determined by the choice of α.

To avoid Type I errors, we might choose a very small α, but doing so will increase Type II error rates, the chance of not detecting a difference between populations when there really is one (or more generally, not finding evidence for H_A when H_A is true). We learned in Section 7.3 that there are other factors that influence Type II error rates beyond the choice of α, namely sample sizes, the magnitude of the difference between the population means, and the population standard deviations. We noted that the most practical way to reduce Type II error is to increase sample sizes.

EFFECT SIZE AND CONFIDENCE INTERVALS

Even though the result was statistically significant (i.e., the difference in cholesterol was large enough to not be easily explained by chance variation), the reduction in cholesterol might not be large enough to have any practical utility. To determine practical utility, or importance, we compute a confidence interval for the difference (as covered in Chapter 6).

For these data, a 95% confidence interval for the difference in mean total cholesterol for the acai and control diets is (–4.72, –0.68) mmol/L as shown below.*

$$\bar{y}_1 - \bar{y}_2 \pm t_{0.025}SE_{\bar{Y}_1-\bar{Y}_2}$$

$$5.42 - 8.12 \pm 2.262 \times \sqrt{0.35^2 + 0.82^2}$$

$$-2.70 \pm 2.262 \times 0.8916$$

$$-2.70 \pm 2.017$$

$$(-4.717, -0.683) \approx (-4.72, -0.68) \text{ mmol/L}$$

In the context of this study, we can say that we are 95% confident that acai lowers† cholesterol by 0.68 to 4.72 mmol/L, on average. This interval is consistent with our above findings since the interval does not contain zero.‡ This interval is also important to a clinician who understands the biological significance of cholesterol. If changes in blood cholesterol as small as 0.5 mmol/L are biologically important, then the confidence interval indicates the practical significance of our findings. We are confident that the difference is larger than 0.5 mmol/L because both endpoints of the interval are above 0.5 mmol/L in magnitude. If, however, changes smaller than, say, 1.5 mmol/L are biologically *in*significant, the findings are of questionable utility, as our interval doesn't indicate with confidence that the difference in mean cholesterol is larger than 1.5 mmol/L (the population means could differ by as little as 0.68 mmol/L, which is biologically insignificant). Last, perhaps differences smaller than 5.0 mmol/L are not of biological importance. Then, while we can conclude acai has an effect (i.e., statistically significant at $\alpha = 0.05$ level), the effect is too small to be biologically relevant. The interval suggests the largest we believe the difference to be is 4.7 mmol/L, which is less than 5.0 mmol/L.

---

*Using as before.

†Again, since this was an experiment we can make strong cause–effect claims. If this were only an observational study, our result would be expressed in a more observational manner: that we are 95% confident that rats on an acai diet have mean cholesterol levels that are between 0.7 to 4.7 mmol/L lower than rats consuming a control diet.

‡Since our test was directional, a case could be made that a one-sided confidence interval should be used, but for simplicity we'll ignore this.

Another common method to communicate the practical importance of statistical findings is to report the effect size, as discussed in Section 7.6. The effect size summarizes the difference between the means relative to the variability of the individuals in the population (or sample). For our example, the estimated effect size is computed to be

$$\frac{|\bar{y}_1 - \bar{y}_2|}{s} = \frac{|5.42 - 8.12|}{2.32} = 1.16$$

Thus, it appears that acai lowers cholesterol about 1.16 standard deviations. (Note that the sample standard deviation for the acai and control groups can be computed from the reported standard error values in Table II.1. The standard deviations are $0.35 \times \sqrt{8} = 0.99$ and $0.82 \times \sqrt{8} = 2.32$, respectively. When computing the estimated effect size, it is customary and conservative to use the larger of the two standard deviation values, which is why the value 2.32 is used in the computation.)

NONSIGNIFICANT FINDINGS

In addition to comparing total cholesterol levels, this study also compared levels of blood triacylglycerol (mmol/L). The data appear in Table II.2. Does the acai diet affect triacylglycerol? Here, the alternative hypothesis, H_A, is nondirectional: $\mu_1 - \mu_2 \neq 0$; that the mean triacylglycerol levels are different for rats receiving the acai compared to those that are not. The null hypothesis, H_0, is $\mu_1 - \mu_2 = 0$ as before; that there is no difference in mean triacylglycerol for rats consuming the two diets.

Table II.2 Triacylglycerol (mmol/L)

	Acai	Control
Mean ± SE	0.36 ± 0.0212	0.31 ± 0.0141
n	8	8

For these data, $t_s = 1.964$. Referring to a t distribution with approximately 12 degrees of freedom, the two-tailed (nondirectional) P-value is 0.0731. This means that for similar studies, a difference in sample means as large as (or larger than) ours in this study happens about 7.31% of the time by chance alone (i.e., when acai has no effect on triacylglycerol). Thus, the data are somewhat consistent with the null hypothesis; we would not reject the null hypothesis using $\alpha = 0.05$. We learned in Chapter 7 to be very cautious with our interpretation in this case and to not say that we have evidence that the null hypothesis is true. There are other reasons for data to be consistent with the null hypothesis besides the null being true. For example, a Type II error could have occurred: The null hypothesis could be false, but the sample sizes or differences are too small to detect the difference, or we simply had an unlucky sample. Confidence intervals may be helpful in interpreting this situation further.

A 95% confidence interval for the difference in mean triacylglycerol ($\mu_{acai} - \mu_{control}$) is $(-0.006, 0.106)$ mmol/L. Consistent with our results above (not rejecting H_0 and not finding significant evidence for a difference in population means), the interval contains zero. Accounting for our uncertainty with the interval, we can say with 95% confidence that the mean triacylglycerol under the acai diet could be as much as 0.006 mmol/L lower to as much as 0.106 mmol/L higher than the mean triacylglycerol under the control diet. Viewing the interval in this way, we see that we cannot say we have evidence for a difference (for H_A). We can, however, say with confidence that if there is a difference, it is no more than 0.106 mmol/L in magnitude. A medical professional could determine whether or not this result is useful. If differences less than 0.106 mmol/L are of no clinical importance, then we could argue that we have evidence of no clinical difference between the two diets. If, however, subtle differences (say 0.075 mmol/L) could be of clinical importance, then we would want to investigate further with a larger study, as the results of this study would be inconclusive. Our interval indicates that the difference could be larger or smaller than 0.075 mmol/L in magnitude.

REQUIREMENTS

The conclusions drawn from both the hypothesis tests and confidence intervals in this case study are only valid provided that the study design and data meet the necessary conditions for the two-sample *t* test and interval. First, it must be reasonable to regard each of the two samples of rats as random samples from two populations. In this case, we can think of one population as the population of all rats who might eat a control diet, and the other as all rats who might eat an acai-fortified diet. Because little information is actually given about how the rats were sampled, we'll assume that the rats were chosen randomly. In practice, if we were responsible for the study, we would carefully report how the rats were sampled to inform readers that this condition is satisfied. The researchers do note in their article that the rats are all male Fischer rats, so one may need to be cautious when generalizing these results to other sexes, species of rats, or other animal species (e.g., humans) in general.

Because we are using independent sample *t* tests and intervals (as opposed to paired methods), a further requirement is that the two samples of rats must be independent of one another. In the case of this experiment, independence is achieved through the process of randomly assigning the rats to the two groups: control and acai.

Since the tests and intervals in this case study are based on the *t* distribution (as opposed to permutation tests or nonparametric tests like the Wilcoxon-Mann-Whitney) we require that the sampling distribution of $\overline{Y}_1 - \overline{Y}_2$ be normally distributed. Ideally, we would produce normal probability plots, conduct Shapiro-Wilk's tests of normality for the data from each sample, or both, but with such small samples it is unlikely that we would have any real power to detect anything but the strongest cases of nonnormality. With larger samples, we would have more power to detect nonnormality and, ironically, we would be less concerned about normality due to the Central Limit Theorem! In cases like this (very small samples), the most conservative approach is to use nonparametric or permutation-based tests unless there is a good historical reasoning (i.e., data from prior studies) that suggests the data are normally distributed. In the case of blood cholesterol levels, normality is likely a safe assumption.

OUTLIERS

Sometimes an outlier will be present in a data set, calling into question the result of a *t* test. It is not legitimate to simply delete the outlier. A sensible procedure is to conduct the analysis with the outlier included and then delete the outlier and repeat the analysis. If the conclusion is unchanged when the outlier is removed, then we can feel confident that no single observation is having undue influence on the inferences we draw from the data. If the conclusion changes when the outlier is removed, then we cannot be confident in the inferences we draw. For example, if the *P*-value for a test is small with the outlier present but large when the outlier is deleted, then we might state, "There is evidence that the populations differ from one another, but this evidence is largely due to a single observation." Such a statement warns the reader that not too much should be read into any differences that were observed between the samples.

Unit II Summary Exercises

II.1 Let Y denote the fruit weight of a nectarine. Suppose Nancy wants to know how weights in her orchard compared from this season to the last. In particular, suppose she is interested in the averages μ_1 and μ_2. You may assume that Nancy has taken several statistics courses and knows a lot about statistics, including how to interpret confidence intervals and hypothesis tests. She usually chooses to limit her type I error rate to 0.05. You have random samples of fruit from each season and are to analyze the data and write a report. You plan to report to Nancy the two sample means, but you aren't sure what to say about how they compare. You seek advice from four persons:

Rudd says, "Conduct an $\alpha = 0.05$ test of $H_0: \mu_1 = \mu_2$ versus $H_A: \mu_1 \neq \mu_2$ and tell Nancy whether or not you reject H_0 at the $\alpha = 0.05$ level."

Linda says, "Report a 95% confidence interval for $\mu_1 - \mu_2$."

Steve says, "Conduct a test of $H_0: \mu_1 = \mu_2$ versus $H_A: \mu_1 \neq \mu_2$ and report to Nancy the P-value from the test."

Gloria says, "Compare $\bar{y}_1$ to $\bar{y}_2$. If $\bar{y}_1 > \bar{y}_2$ then test $H_0: \mu_1 = \mu_2$ versus $H_A: \mu_1 > \mu_2$ using $\alpha = 0.05$ and tell Nancy whether or not you reject H_0. If $\bar{y}_1 < \bar{y}_2$ then test $H_0: \mu_1 = \mu_2$ versus $H_A: \mu_1 < \mu_2$ using $\alpha = 0.05$ and tell Nancy whether or not you reject H_0."

Rank the four pieces of advice from worst to best and explain why you rank them as you do. That is, explain what makes one better than another.

II.2 Each of 41 students at a college was asked to calculate their "ecological footprint"—the number of hectares required to support their existence, taking into account such things as the land needed to produce the food they eat and so forth. For 27 women, the average was 6.59, and the standard deviation was 3.89 hectares. For 14 men, the average was 3.96, and the standard deviation was 1.06 hectares.

(a) Use these data to conduct a t test of the null hypothesis that there is no difference between the population means for men and women. Use a two-sided alternative and use $\alpha = 0.02$. (*Note:* There are approximately 32 degrees of freedom here for the t test.) Provide all steps: Construct the value of the test statistic, give bounds on the P-value, and state your conclusion regarding H_0.

(b) In nontechnical language, explain to a nonstatistician what this means about the ecological footprints of men and women at this college. Be specific.

(c) Suppose differences in ecological footprints smaller than 0.7 hectares are considered to be not ecologically important. Compute a confidence interval for the difference in mean global footprint between men and women and discuss whether the difference can be considered "ecologically important."

II.3 After seeing the sample means and SDs from Question II.2, Norman became concerned that a t test might not be appropriate here, so he wants Rebecca to do a different analysis.

(a) Why is Norman concerned? That is, what is it about the data that would make someone question whether a t test would be valid?

(b) If Norman is right, then what should Rebecca do to analyze her data?

(c) Alana says that Norman shouldn't be so worried and should let Rebecca go ahead with a t test. On what grounds can Alana look at the information in Question II.2 and justify using a t test (with the original data)?

II.4 Suppose someone constructs a 95% confidence interval for $\mu_1 - \mu_2$ and gets (1.3,12.7). If we were to test $H_0: \mu_1 = \mu_2$ versus $H_A: \mu_1 \neq \mu_2$ with $\alpha = 0.10$, would we reject H_0? Or can we not tell from the information given? Explain your reasoning.

II.5 Nitric oxide is sometimes given to newborns who experience respiratory failure. In one experiment, nitric oxide was given to 114 infants. This group was compared to a control group of 121 infants. The length of hospitalization (in days) was recorded for each of the 235 infants. The mean in the nitric oxide sample was $\bar{y}_1 = 36.4$; the mean in the control sample was $\bar{y}_2 = 29.5$. A 95% confidence interval for $\mu_1 - \mu_2$ is $(-2.3, 16.1)$, where μ_1 is the population mean length of hospitalization for infants who get nitric oxide and μ_2 is the mean length of hospitalization for infants in the control population For each of the following, say whether the statement is true or false and explain why.

(a) 95% of infants who experience respiratory failure would have their hospital stays altered by between a 2.3-day decrease to a 16.1-day increase if they received nitric oxide.

(b) We are 95% confident that nitric oxide has no effect on the length of hospitalization.

(c) We are 95% confident that if nitric oxide affects the length of hospitalization, its effect is less than 16.1 days, on average.

II.6 Researchers took random samples of subjects from two populations and applied a two-sample t test to the data using $\alpha = 0.10$; the P-value for the test, using a nondirectional alternative, was 0.06. For each of the following, say whether the statement is true or false and explain why.

(a) There is a 6% chance that the two population distributions actually are the same.

(b) If the two population distributions actually are the same, then a difference between the two samples as extreme as the difference that these researchers observed would only happen 6% of the time.

(c) If a new study were done that compared the two populations, there is a 6% probability that H_0 would be rejected again.

(d) If $\alpha = 0.05$ and a directional alternative were used, and the data departed from H_0 in the direction specified by the alternative hypothesis, then H_0 would be rejected.

II.7 Researchers measured blood levels of the hormone Androstenedione (Andro) in each of 24 women. Among 12 women who had recently fallen in love, the sample mean was 2.1 ng/ml; the SD of these data was 0.7. Among 12 women who had *not* recently fallen in love, the mean was 1.9, and the SD was 0.7. Formula (6.7.1) yields 22 degrees of freedom.

(a) Conduct a two-sample t test, with nondirectional alternative; show all steps, including bounds on the P-value. If $\alpha = 0.05$, do you reject H_0? Why or why not?

(b) A 90% confidence interval for the difference in mean Andro levels for the two populations is $(-0.39, 0.79)$ ng/ml. Suppose a change of 0.4 ng/ml is medically important. Are these results medically important? Briefly explain your reasoning.

II.8 Consider the process of planning for adequate power when designing an experiment for which a two-sample t test will be used.

(a) Suppose the effect size is constant and the chosen alpha level is also constant (a) (e.g., at 0.05). If the sample sizes for the two groups go up from 12 to 18, how does this affect power? That is, does power go down, stay the same, or go up? Why? How is this related to the standard error?

(b) The graph below shows two normally distributed populations. Calculate the effect size here.

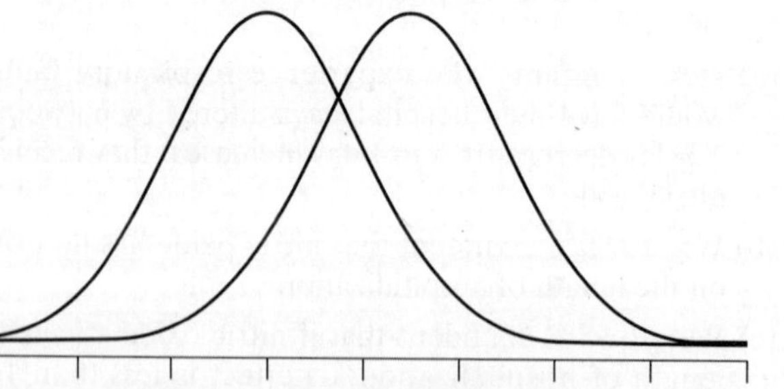

II.9 A researcher did a randomization test to compare men and women on a variable Y. The null hypothesis is that men and women are the same, and the alternative is that they are different. There are 28 possible randomizations of the data. The graph below shows the statistic "Difference in means" for each randomization. This includes the original assignment of observations to the male and female groups, which had a difference in means of 31.

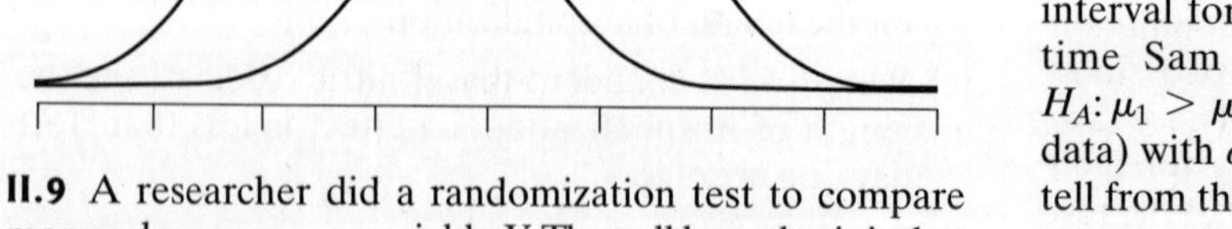

(a) What is the randomization test P-value?

(b) Using $\alpha = 0.10$, is there statistically significant evidence that men and women differ with respect to the mean of variable Y?

(c) The difference in sample means between men and women was 31 $(\bar{y}_{men} - \bar{y}_{women})$. What is the randomization test P-value for the test that considers the alternative hypothesis $\mu_{men} > \mu_{women}$?

II.10 Consider studying whether drug A and drug B are equally effective in lowering blood pressure. For each of (a), (b), and (c) answer two questions (you do not need to explain your answers):

(I) What kind of paired design is this (i.e., What is the nature of the pairing)?

(II) True or false: "It is legitimate to conduct a paired t test analysis on these differences (assuming that the A−B differences follow a normal distribution)."

(a) We get a sample of 20 patients and record their ages, ranking them from 1 to 20. We match patient 1 with patient 2, patient 3 with patient 4, and so forth. In each pair, one patient gets drug A, and one gets drug B (randomly). Then we record change in blood pressure and look at the differences within each pair.

(b) For each of 10 consecutive weeks, we randomly select two patients, give one of them drug A and the other drug B (randomly), and record change in blood pressure. We then look at the differences within the pair from the first week, the pair from the second week, and so forth.

(c) We get a sample of 20 patients and give 10 of them drug A and the other 10 drug B (randomly). Then, we measure change in blood pressure. We match the patient with the greatest change on A with the patient with the greatest change on B, the second greatest change on A with the second greatest change on B, and so forth. Then, we look at the differences within each pair.

II.11 Fred collects data from two populations. Maria uses those data to construct a (two-sided) 90% confidence interval for $\mu_1 - \mu_2$, and gets $(-3.4, 23.7)$. At the same time Sam uses the data to test $H_0\colon \mu_1 = \mu_2$ versus $H_A\colon \mu_1 > \mu_2$ (which he chose to test before seeing the data) with $\alpha = 0.05$. Does Sam reject H_0? Or can we not tell from the information given? Explain your reasoning.

II.12 [Based on Koh, et al. (1997). Effects of hormone-replacement therapy on fibrinolysis in postmenopausal women. *New England Journal of Medicine* **336**, 683–690, which is the basis of Exercise 8.S.22.] Researchers wanted to test the effect of oral conjugated estrogen on plasminogen-activator inhibitor type 1 (PAI-1). They took measurements on each of 30 women before taking conjugated estrogen and after taking conjugated estrogen. They then prepared to analyze the data from this paired design. One researcher, Smith, wanted to conduct a sign test. Another, Jones, advocated a paired *t* test on the grounds that the sign test uses only part of the information in the data (namely, whether the before—after difference is positive or negative) and thus is less powerful than the *t* test.

(a) What did Jones mean in saying that the sign test "is less powerful than the *t* test"?

(b) Assuming that Jones is correct, why would anyone ever conduct a sign test?

II.13 A random sample of 50 subjects received biofeedback training to reduce blood pressure. Researchers measured the decrease in blood pressure for each of them; the average decrease was 11.4, and the SD was 1.3. A second sample of 40 control subjects had an average decrease in blood pressure during the study of 5.0, with an SD of 1.4.

(a) Construct a 95% confidence interval for the population difference average in blood pressure decrease between treatment (biofeedback) and control patients.

(b) Consider just the treatment (biofeedback) group. True or false: We can estimate that roughly 95% of all persons given biofeedback training will experience a decrease in blood pressure in the range $11.4 \pm 2(1.3)$, that is, between 8.8 and 14.0. Explain your reasoning.

II.14 Suppose we want to test whether an experimental drug reduces blood pressure more than does a placebo. We are planning to administer the drug or placebo to some subjects and record how much their blood pressures are reduced. We have 20 subjects available.

(a) We could form 10 matched pairs, where we form a pair by matching subjects, as best we can, on the basis of age and sex. Briefly explain why using a matched pairs design might be a good idea.

(b) Briefly explain why a matched pairs design might *not* be a good idea. That is, how might such a design be inferior to a completely randomized design?

Background for II.15–II.19

Researchers wanted to assess whether or not training a cat can affect its acceptance of blood testing. Fourteen cats were selected to not be familiarized (trained) with having their blood drawn, while 17 other cats were selected to be trained by familiarizing them with being restrained and shaved for a blood draw. Blood cortisol levels (a stress hormone) was measured to compare the stress levels of the cats during their blood draw (μg/dL). The following table presents a summary of the data.[2]

	Cortisol (μg/dL)		
Group	Mean	SE	N
Trained	0.541	0.215	17
Untrained	2.324	0.239	14

II.15 Consider the research question: Can training *reduce* stress in cats when they are getting their blood drawn?

(a) In plain English, write the null and alternative hypotheses of interest to the researcher (regardless of whether or not this is truly testable with these data).

(b) Given the context of this study design, is this research question in (a) testable with these data? Briefly explain.

(c) Use an appropriate *t* test to test the hypothesis in part (a). You may use df $\approx$ 28 and $\alpha = 0.05$.

(d) Suppose that medically speaking, cortisol levels that differ by less 1.0 μg/dL are not treated as being different. Do you think we should regard the mean cortisol levels for trained and untrained cats as being different from a medical (i.e., practical) perspective? Use the data and an appropriate statistical procedure to support your answer.

II.16 Below are histograms and normal probability plots and Shapiro-Wilk test *P*-values of the cortisol levels of the 31 cats in the study.

Trained

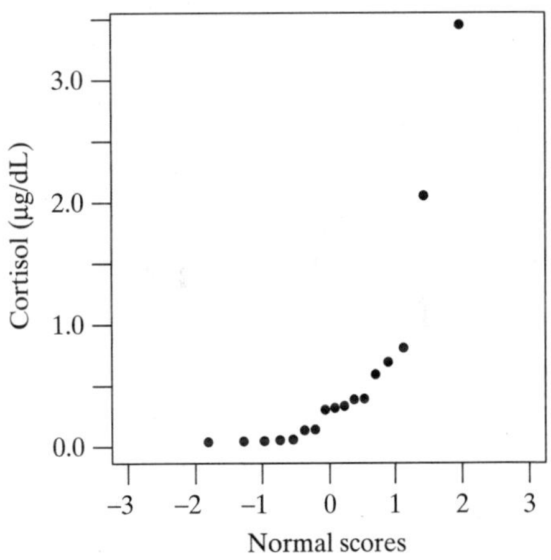

Shapiro-Wilk *P*-value = 0.0001

Untrained

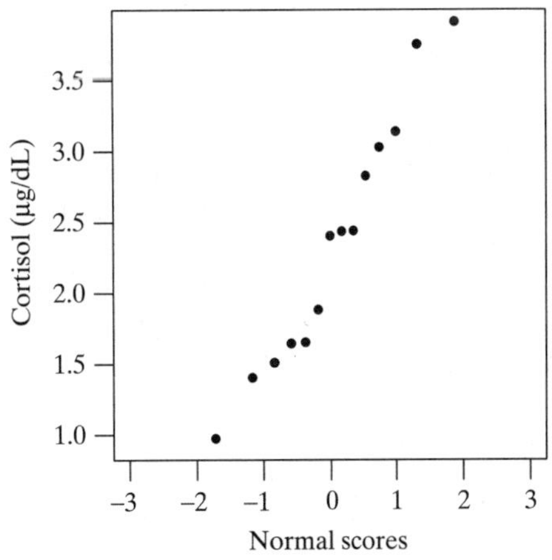

Shapiro-Wilk *P*-value = 0.6686

(a) Do these data meet the normality requirements for the two-sample t test?

(b) If we obtained larger samples, would we expect the resulting plots to appear more normal? Briefly explain.

(c) In a sentence or two, what would be the primary argument for using the Wilcoxon-Mann-Whitney test to compare cortisol levels for trained and untrained cats?

(d) What is the primary drawback of using the Wilcoxon-Mann-Whitney test over the two-sample t test when you don't "need" to use it?

II.17 For each of the following statements say whether they are true or false and explain why.

(a) If training is truly unrelated to stress (cortisol) levels, a larger study would have more power than a smaller one.

(b) Changing the value of α will affect the P-value of the test.

(c) Consider a test to compare the number of escape attempts for trained and untrained cats. If training is really associated with fewer escape attempts, we would have a better chance of detecting this relationship by choosing a small α rather than a large one.

(d) As in (c) above, consider a test to compare the mean number of escape attempts for trained and untrained cats. If trained cats try to escape less than untrained cats, on average, we would have a better chance of detecting this difference if we studied more cats.

(e) A two-sample t test comparing the number of escape attempts in this study had a P-value of 0.022. If $\alpha = 0.01$ it is possible that we could be making a Type I error with these data.

II.18 The researchers also measured the number of escape attempts during the medical procedure for each cat. A 95% confidence interval for the difference in mean number of escape attempts (per cat) was $(\mu_{\text{trained}} - \mu_{\text{untrained}})$ and was given to be $(-1.29, -0.17)$ attempts per cat. Suppose a nondirectional two-sample t test was conducted on these same data. State whether the following statements are true or false or cannot be determined without further computations. Explain why.

(a) The P-value for the two-sample t test would be greater than 0.05.

(b) There is statistically significant evidence (at the $\alpha = 0.05$ level) that the mean number of escape attempts per cat is not the same for trained and untrained cats.

(c) There is statistically significant evidence (at the $\alpha = 0.01$ level) that the mean number of escape attempts per cat is not the same for trained and untrained cats.

(d) There is statistically significant evidence (at the $\alpha = 0.10$ level) that the mean number of escape attempts per cat is not the same for trained and untrained cats.

II.19 Write a sentence interpreting the confidence interval reported in II.18 in the context of the problem. That is, use the interval to carefully describe to a reader what the study reveals about how often trained cats try to escape versus untrained cats.

II.20 As part of a study to determine the effect of flaxseed on blood thiocyanate concentration in rats, the authors reported the amount of hydrogen cyanide (HCN) in the flaxseed fed to the rats. This value was reported to be 255 ± 8.3 mg/kg seed. Based on the authors' intent to communicate that flaxseed source is fairly consistent in its HCN content, do you think these values represent mean $\pm$ SD or mean $\pm$ SE?

II.21 For each of the following research studies, identify whether independent samples (e.g., two-sample t test, Wilcoxon-Mann-Whitney) or paired samples (e.g., paired t test, sign test) would most likely be appropriate and explain why.

(a) To investigate whether or not there are differences in turbidity (i.e., suspended sediment) at the surface of a stream versus near the bottom, 15 stream locations were sampled, and turbidity measurements were made at the surface and depth of 0.5 meters.

(b) To investigate if a marine protection policy off the coast of California improves the health of rockfish, researchers measured catch per unit effort (a surrogate measure for fish abundance) at eight rockfish habitat locations inside a marine protected area and eight additional nearby habitat locations outside the protected area.

(c) To compare the therapeutic value of different milk-based supplements in malnourished children, researchers randomly assigned one of two common supplements (10% milk or 25% milk) to approximately 1,900 malnourished children in rural Malawi and measured their weights at the start of the study, and again after 8 weeks of therapy.

II.22 Consider the study described in II.21(b).

(a) Considering conducting a study of the same size to answer the same question as described in II.21(b), describe in detail a study design that would use paired samples analysis.

(b) Which design (independent or paired) is likely more powerful and why?

Categorical Data: One-Sample Distributions

Chapter 9

OBJECTIVES

In this chapter we study categorical data. We will

- explore sampling distributions for estimators that describe dichotomous populations.
- demonstrate how to make and interpret confidence intervals for proportions.
- provide a method for finding an optimal sample size for estimating a proportion.
- show how and when to conduct a chi-square goodness-of-fit test.

9.1 Dichotomous Observations

In Chapter 5 we worked with problems involving numeric variables and examined the sampling distribution of the sample mean. In Chapter 6 we used the sampling distribution to explain how the sample mean tends to vary from the population mean, and we constructed confidence intervals for the population mean. We begin this chapter by proceeding in a similar manner by first considering a simple dichotomous categorical variable (i.e., a categorical variable that has only two possible values) and the sampling distribution of the sample proportion. In Section 9.2 we will use the sampling distribution of the sample proportion to construct a confidence interval for a population proportion.

THE WILSON-ADJUSTED SAMPLE PROPORTION, $\tilde{P}$

When sampling from a large dichotomous population, a natural estimate of the population proportion, p, is the sample proportion, $\hat{p} = y/n$, where y is the number of observations in the sample with the attribute of interest and n is the sample size.

Example 9.1.1

Contaminated Soda At any given time, soft-drink dispensers may harbor bacteria such as *Chryseobacterium meningosepticum* that can cause illness.[1] To estimate the proportion of contaminated soft-drink dispensers in a community in Virginia, researchers randomly sampled 30 dispensers and found 5 to be contaminated with *Chryseobacterium meningosepticum*. Thus the sample proportion of contaminated dispensers is

$$\hat{p} = \frac{5}{30} = 0.167$$

■

The estimate, $\hat{p} = 0.167$, given in Example 9.1.1 is a good estimate of the population proportion of contaminated soda dispensers, but it is not the only possible estimate. The Wilson-adjusted sample proportion, $\tilde{p}$, is another estimate of the population proportion and is given by the formula in the following box.

Wilson-Adjusted Sample Proportion, $\tilde{p}$

$$\tilde{p} = \frac{y + 2}{n + 4}$$

Example 9.1.2 **Contaminated Soda** The Wilson-adjusted sample proportion of contaminated dispensers is

$$\tilde{p} = \frac{5 + 2}{30 + 4} = 0.206^*$$ ■

As the previous example illustrates, $\tilde{P}$ is equivalent to computing the ordinary sample proportion $\hat{P}$ on an augmented sample: one that includes four extra observations of soft-drink dispensers—two that are contaminated and two that are not. This augmentation has the effect of biasing the estimate towards the value 1/2. Generally speaking we would like to avoid biased estimates, but as we shall see in Section 9.2, confidence intervals based on this biased estimate, $\tilde{P}$, actually are more reliable than those based on $\hat{P}$.

THE SAMPLING DISTRIBUTION OF $\tilde{P}$

For random sampling from a large dichotomous population, we saw in Chapter 3 how to use the binomial distribution to calculate the probabilities of all the various possible sample compositions. These probabilities in turn determine the sampling distribution of $\tilde{P}$. An example follows.

Example 9.1.3 **Contaminated Soda** Suppose that in a certain region of the United States, 17% of all soft-drink dispensers are contaminated with *Chryseobacterium meningosepticum.* If we were to examine a random sample of two drink dispensers from this population of dispensers, then we will get either zero, one, or two contaminated machines. The probability that both dispensers are contaminated is $0.17 \times 0.17 = 0.0289$. The probability that neither are contaminated is $(1 - 0.17) \times (1 - 0.17) = 0.6889$. There are two ways to get a sample in which one machine is contaminated and one is not: The first could be contaminated, but not the second, or vice versa. Thus, the probability that exactly one machine is contaminated is

$$0.17 \times (1 - 0.17) + 0.17 \times (1 - 0.17) = 0.2822$$

If we let $\tilde{P}$ represent the Wilson-adjusted sample proportion of contaminated dispensers, then a sample that contains no contaminated dispensers has $\tilde{p} = \frac{0 + 2}{2 + 4} = 0.33$, which occurs with probability 0.6889. A sample that contains one contaminated machine has $\tilde{p} = \frac{1 + 2}{2 + 4} = 0.50$; this happens with probability 0.2822. Finally, a sample that contains two contaminated machines has $\tilde{p} = \frac{2 + 2}{2 + 4} = 0.67$, which occurs with probability 0.0289.[†] Thus, there is roughly a 69% chance that $\tilde{P}$ will equal 0.33, a 28% chance that $\tilde{P}$ will equal 0.50, and a 3% chance that $\tilde{P}$ will equal 0.67.

---

*In keeping with our convention, $\tilde{P}$ denotes a random variable, whereas $\tilde{p}$ denotes a particular number (e.g., 0.206 in this example).

[†]It is worth noting that with a small sample size ($n = 2$) the possible values of $\tilde{p}$ are 0.33, 0.50, and 0.67 while the possible values of $\hat{p}$ are 0.00, 0.50, and 1.00. This sheds some light as to why $\tilde{p}$ is a sensible estimator of the population proportion, particularly for small samples. With a small sample it is quite likely that one could obtain no contaminated machines even if a reasonable proportion of the population is contaminated. It would be unwise, with such a small sample, to assert that the population proportion of contaminated machines is 0.

This sampling distribution is given in Table 9.1.1 and Figure 9.1.1.

Table 9.1.1 Sampling distribution of Y (the number of contaminated dispensers) and of $\tilde{P}$ (the Wilson-adjusted proportion of contaminated dispensers) for samples of size $n = 2$ for a population with 17% of the dispensers contaminated

Y	$\tilde{P}$	Probability
0	0.33	0.6889
1	0.50	0.2822
2	0.67	0.0289

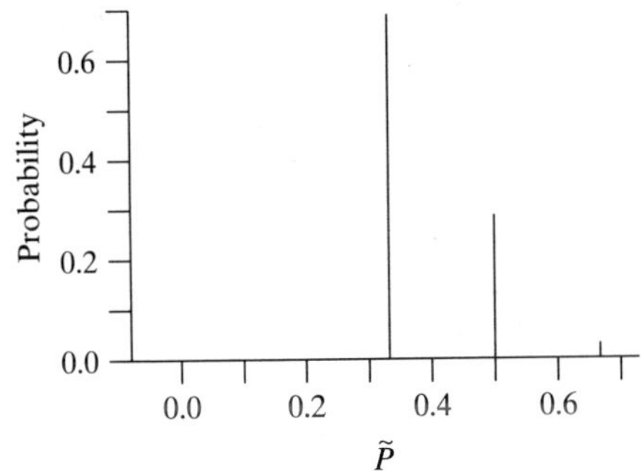

Figure 9.1.1 Sampling distribution of $\tilde{P}$ for $n = 2$ and $p = 0.17$

Example 9.1.4 **Contaminated Soda and a Larger Sample** Suppose we were to examine a sample of 20 dispensers from a population in which 17% are contaminated. How many contaminated dispensers might we expect to find in the sample? As was true in Example 9.1.3, this question can be answered in the language of probability. However, since $n = 20$ is rather large, we will not list each possible sample. Rather, we will make calculations using the binomial distribution with $n = 20$ and $p = 0.17$. For instance, let us calculate the probability that 5 dispensers in the sample would be contaminated and 15 would not:

$$\begin{aligned}\Pr\{5 \text{ contaminated, } 15 \text{ not contaminated}\} &= {}_{20}C_5(0.17)^5(0.83)^{15} \\ &= 15{,}504(0.17)^5(0.83)^{15} \\ &= 0.1345\end{aligned}$$

Letting $\tilde{P}$ represent the Wilson-adjusted sample proportion of contaminated dispensers, a sample that contains 5 contaminated dispensers has $\tilde{p} = \frac{5+2}{20+4} = 0.2917$. Thus, we have found that

$$\Pr\{\tilde{P} = 0.2917\} = 0.1345$$

The binomial distribution can be used to determine the entire sampling distribution of $\tilde{P}$. The distribution is displayed in Table 9.1.2 and as a probability histogram in Figure 9.1.2.

Table 9.1.2 Sampling distribution of Y, the number of successes, and of $\tilde{P}$, the Wilson-adjusted proportion of successes, when $n = 20$ and $p = 0.17$

Y	$\tilde{P}$	Probability	Y	$\tilde{P}$	Probability
0	0.0833	0.0241	11	0.5417	0.0001
1	0.1250	0.0986	12	0.5833	0.0000
2	0.1667	0.1919	13	0.6250	0.0000
3	0.2083	0.2358	14	0.6667	0.0000
4	0.2500	0.2053	15	0.7083	0.0000
5	0.2917	0.1345	16	0.7500	0.0000
6	0.3333	0.0689	17	0.7917	0.0000
7	0.3750	0.0282	18	0.8333	0.0000
8	0.4167	0.0094	19	0.8750	0.0000
9	0.4583	0.0026	20	0.9167	0.0000
10	0.5000	0.0006			

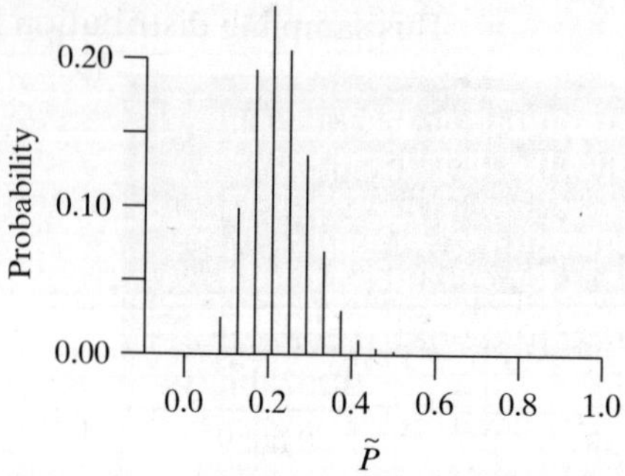

Figure 9.1.2 Sampling distribution of $\tilde{P}$ when $n = 20$ and $p = 0.17$

We can use this distribution to answer questions such as, "If we take a random sample of size $n = 20$, what is the probability that no more than 5 will be contaminated?" Notice that this question can be asked in two equivalent ways: "What is $\Pr\{Y \leq 5\}$?" and "What is $\Pr\{\tilde{P} \leq 0.2917\}$?" The answer to either question is found by adding the first six probabilities in Table 9.1.2:

$$\begin{aligned}\Pr\{Y \leq 5\} &= \Pr\{\tilde{P} \leq 0.2917\} \\ &= 0.0241 + 0.0986 + 0.1919 + 0.2358 + 0.2053 + 0.1345 \\ &= 0.8902\end{aligned}$$

■

RELATIONSHIP TO STATISTICAL INFERENCE

In making a statistical inference from a sample to the population, it is reasonable to use $\tilde{P}$ as our estimate of p. The sampling distribution of $\tilde{P}$ can be used to predict how much sampling error to expect in this estimate. For example, suppose we want to know whether the sampling error will be less than 5 percentage points, in other words, whether $\tilde{P}$ will be within ± 0.05 of p. We cannot predict for certain whether this event will occur, but we can find the probability of it happening, as illustrated in the following example.

Example 9.1.5 **Contaminated Soda** In the soda-dispenser example with $n = 20$, we see from Table 9.1.2 that

$$\begin{aligned}\Pr\{0.12 \leq \tilde{P} \leq 0.22\} &= 0.0986 + 0.1919 + 0.2358 \\ &= 0.5263 \approx 0.53\end{aligned}$$

Thus, there is a 53% chance that, for a sample of size 20, $\tilde{P}$ will be within $\pm$ 0.05 of p.

■

DEPENDENCE ON SAMPLE SIZE

Just as the sampling distribution of $\bar{Y}$ depends on n, so does the sampling distribution of $\tilde{P}$. The larger the value of n, then the more likely it is $\tilde{P}$ will be close to p.* The following example illustrates this effect.

Example 9.1.6 **Contaminated Soda** Figure 9.1.3 shows the sampling distribution of $\tilde{P}$, for three different values of n, for the soft-drink dispenser population of Example 9.1.1. (Each sampling distribution is determined by a binomial distribution with $p = 0.17$.) You

*This statement should not be interpreted too literally. As a function of n, the probability that $\tilde{P}$ is close to p has an overall increasing trend, but it can fluctuate somewhat.

Figure 9.1.3 Sampling distributions of $\widetilde{P}$ for $p = 0.17$ and various values of n

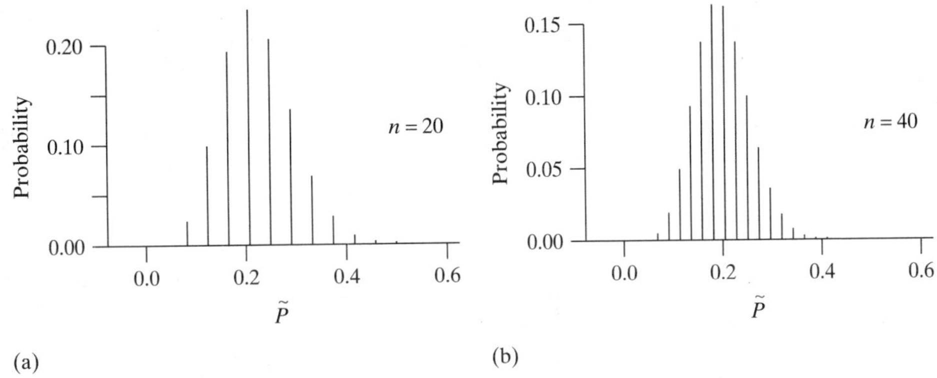

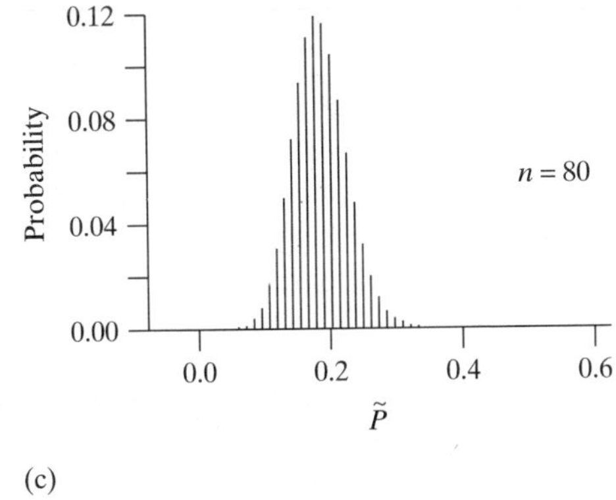

Table 9.1.3

n	$\Pr\{0.12 \le \widetilde{P} \le 0.22\}$
20	0.53
40	0.56
80	0.75
400	0.99

can see from the figure that as n increases, the sampling distribution becomes more compressed around the value $p = 0.17$; thus, the probability that $\widetilde{P}$ is close to p tends to increase as n increases. For example, consider the probability that $\widetilde{P}$ is within ± 5 percentage points of p. We saw in Example 9.1.5 that for $n = 20$ this probability is equal to 0.53; Table 9.1.3 and Figure 9.1.3 show how the probability depends on n.

Note: A larger sample improves the probability that $\widetilde{P}$ will be close to p. We should be mindful, however, that the probability that $\widetilde{P}$ is exactly *equal* to p is very small for large n. In fact,

$$\Pr\{\widetilde{P} = 0.17\} = 0.110 \text{ for } n = 80^*$$

The value $\Pr\{0.12 \le \widetilde{P} \le 0.22\} = 0.75$ is the sum of many small probabilities, the largest of which is 0.118; you can see this effect clearly in Figure 9.1.3(c). ■

*For $n = 80$, $\widetilde{p} = 0.1667$ when $y = 12$, is the closest possible value to 0.17.

Exercises 9.1.1–9.1.10

9.1.1 Consider taking a random sample of size 3 from a population of persons who smoke and recording how many of them, if any, have lung cancer. Let $\widetilde{P}$ represent the Wilson-adjusted proportion of persons in the sample with lung cancer. What are the possible values in the sampling distribution of $\widetilde{P}$?

9.1.2 Suppose we are to draw a random sample of three individuals from a large population in which 37% of the individuals are mutants (as in Example 3.6.4). Let $\widetilde{P}$ represent the Wilson-adjusted proportion of mutants in the sample. Calculate the probability that $\widetilde{P}$ will be equal to

(a) 2/7 (b) 3/7

Is it possible to obtain a sample of three individuals for which $\widetilde{P}$ is zero? Explain.

9.1.3 Suppose we are to draw a random sample of five individuals from a large population in which 37% of the individuals are mutants (as in Example 3.6.4). Let $\widetilde{P}$ represent the Wilson-adjusted proportion of mutants in the sample.

(a) Use the results in Table 3.6.3 to determine the probability that $\widetilde{P}$ will be equal to

(i) 2/9 (ii) 3/9 (iii) 4/9

(iv) 5/9 (v) 6/9 (vi) 7/9

(b) Display the sampling distribution of $\widetilde{P}$ in a graph similar to Figure 9.1.1.

9.1.4 A new treatment for acquired immune deficiency syndrome (AIDS) is to be tested in a small clinical trial on 15 patients. The Wilson-adjusted proportion $\tilde{P}$ who respond to the treatment will be used as an estimate of the proportion p of (potential) responders in the entire population of AIDS patients. If, in fact, $p = 0.2$, and if the 15 patients can be regarded as a random sample from the population, find the probability that

(a) $\tilde{P} = 5/19$ (b) $\tilde{P} = 2/19$

9.1.5 In a certain forest, 25% of the white pine trees are infected with blister rust. Suppose a random sample of four white pine trees is to be chosen, and let $\tilde{P}$ be the Wilson-adjusted sample proportion of infected trees.

(a) Compute the probability that $\tilde{P}$ will be equal to

(i) 2/8 (ii) 3/8 (iii) 4/8 (iv) 5/8 (v) 6/8

(b) Display the sampling distribution of $\tilde{P}$ in a graph similar to Figure 9.1.1.

9.1.6 Refer to Exercise 9.1.5.

(a) Determine the sampling distribution of $\tilde{P}$ for samples of size $n = 8$ white pine trees from the same forest.

(b) Construct graphs of the sampling distributions of $\tilde{P}$ for $n = 4$ and for $n = 8$, using the same horizontal and vertical scales for both. Compare the two distributions visually. How do they differ?

9.1.7 The shell of the land snail *Limocolaria marfensiana* has two possible color forms: streaked and pallid. In a certain population of these snails, 60% of the individuals have streaked shells (as in Exercise 3.6.4). Suppose a random sample of six snails is to be chosen from the population; let $\tilde{p}$ be the Wilson-adjusted sample proportion of streaked snails. Find

(a) $\Pr\{\tilde{P} = 0.5\}$ (b) $\Pr\{\tilde{P} = 0.6\}$

(c) $\Pr\{\tilde{P} = 0.7\}$ (d) $\Pr\{0.5 \le \tilde{P} \le 0.7\}$

(e) the percentage of samples for which $\tilde{P}$ is within ± 0.10 of p.

9.1.8 In a certain community, 17% of the soda dispensers are contaminated (as in Example 9.1.3). Suppose a random sample of five dispensers is to be chosen and the contamination observed. Let $\tilde{P}$ represent the Wilson-adjusted sample proportion contaminated dispensers.

(a) Compute the sampling distribution of $\tilde{P}$.

(b) Construct a histogram of the distribution found in part (a) and compare it visually with Figure 9.1.3. How do the two distributions differ?

9.1.9 Consider random sampling from a dichotomous population; let E be the event that $\tilde{P}$ is within ± 0.05 of p. In Example 9.1.5, we found that $\Pr\{E\} = 0.53$ for $n = 20$ and $p = 0.17$. Calculate $\Pr\{E\}$ for $n = 20$ and $p = 0.25$. (Perhaps surprisingly, the two probabilities are roughly equal.)

9.1.10 Consider taking a random sample of size 10 from the population of students at a certain college and asking each of the 10 students whether or not they smoke. In the context of this setting, explain what is meant by the sampling distribution of $\hat{P}$, the ordinary sample proportion.

9.2 Confidence Interval for a Population Proportion

In Section 6.3 we described confidence intervals when the observed variable is quantitative. Similar ideas can be used to construct confidence intervals in situations in which the variable is *categorical* and the parameter of interest is a population *proportion.* We assume that the data can be regarded as a random sample from some population. In this section we discuss construction of a confidence interval for a population proportion.

Consider a random sample of n categorical observations, and let us fix attention on one of the categories. For instance, suppose a geneticist observes n guinea pigs whose coat color can be either black, sepia, cream, or albino; let us fix attention on the category "black." Let p denote the population proportion of the category of interest, and let $\tilde{p}$ denote the Wilson-adjusted sample proportion (as in Section 9.1), which is our estimate of p. The situation is schematically represented in Figure 9.2.1.

Figure 9.2.1 Notation for population and sample proportion

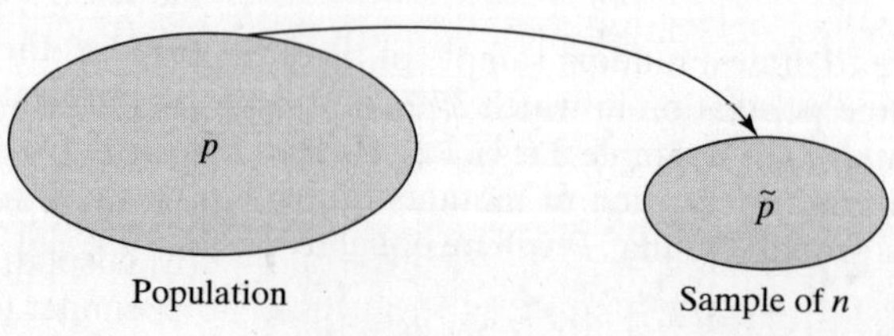

How close to p is $\tilde{P}$ likely to be? We saw in Section 9.1 that this question can be answered in terms of the sampling distribution of $\tilde{P}$ (which in turn is computed from the binomial distribution). As we shall see, by using properties of the sampling distribution of $\tilde{P}$, such as the standard error and $\tilde{P}$'s approximately normal behavior under certain situations, we will be able to construct confidence statements for p. To construct the intervals we will use the same rationale used for numeric data in Section 6.3 where we constructed confidence statements for μ based on the properties of the sampling distribution of $\overline{Y}$.

Although a confidence interval for p can be constructed directly from the binomial distribution, for many practical situations a simple approximate method can be used instead. When the sample size, n, is large, the sampling distribution of $\tilde{P}$ is approximately normal; this approximation is related to the Central Limit Theorem. If you review Figure 9.1.3, you will see that the sampling distributions resemble normal curves, especially the distribution with $n = 80$. (The approximation is described in detail in optional Section 5.4.) In Section 6.3 we stated that when the data come from a normal population, a 95% confidence interval for a population mean μ is constructed as

$$\bar{y} \pm t_{0.025}\mathrm{SE}_{\overline{Y}}$$

A confidence interval for a population proportion p is constructed analogously. We will use $\tilde{P}$ as the center of a 95% confidence interval for p. In order to proceed we need to calculate the standard error for $\tilde{P}$.

STANDARD ERROR OF $\tilde{P}$

The standard error of the estimate is found using the following formula.

Standard Error of $\tilde{p}$ (for a 95% Confidence Interval)

$$\mathrm{SE}_{\tilde{p}} = \sqrt{\frac{\tilde{p}(1-\tilde{p})}{n+4}}$$

This formula for the standard error of the estimate looks similar to the formula for the standard error of a mean, but with $\sqrt{\tilde{p}(1-\tilde{p})}$ playing the role of s and with $n + 4$ in place of n.

Example 9.2.1 **Smoking During Pregnancy** In the Pregnancy Risk Assessment Monitoring System survey, 999 women who had given birth were asked about their smoking habits.[2] Smoking during the last 3 months of pregnancy was reported by 125 of those sampled, which is 12.5%. Thus, $\tilde{p}$ is $\frac{125+2}{999+4} = \frac{127}{1003} = 0.127$; the standard error is $\sqrt{\frac{0.127(1-0.127)}{1003}} = 0.011$ or 1.1%. A sample value $\tilde{p}$ is typically within ± 2 standard errors of the population proportion p. Based on this standard error, we can expect that the proportion, p, of all women who smoked during the last 3 months of pregnancy is in the interval (0.105, 0.149) or (10.5%, 14.9%). A confidence interval for p makes this idea more precise. ■

95% CONFIDENCE INTERVAL FOR P

Once we have the standard error of $\tilde{P}$, we need to know how likely it is that $\tilde{P}$ will be close to p. The general process of constructing a confidence interval for a proportion is similar to that used in Section 6.3 to construct a confidence interval for a mean. However, when constructing a confidence interval for a mean, we multiplied the standard error by a t multiplier. This was based on having a sample from a normal distribution. When dealing with proportion data we know that the population is not normal—there only are two values in the population!—but the Central Limit Theorem tells us that the sampling distribution of $\tilde{P}$ is approximately normal if the sample size, n, is large. Moreover, it turns out that even for moderate or small samples, intervals based on $\tilde{P}$ and Z multipliers do a very good job of estimating the population proportion, p.[3]

For a 95% confidence interval, the appropriate Z multiplier is $z_{0.025} = 1.960$. Thus, the approximate 95% confidence interval for a population proportion p is constructed as shown in the following box.*

95% Confidence Interval for p

95% confidence interval: $\tilde{p} \pm 1.96\text{SE}_{\tilde{P}}$

Example 9.2.2 **Breast Cancer** *BRCA1* is a gene that has been linked to breast cancer. Researchers used DNA analysis to search for *BRCA1* mutations in 169 women with family histories of breast cancer. Of the 169 women tested, 27 (16%) had *BRCA1* mutations.[4] Let p denote the probability that a woman with a family history of breast cancer will have a *BRCA1* mutation. For these data, $\tilde{p} = \dfrac{27 + 2}{169 + 4} = 0.168$. The standard error for $\tilde{P}$ is $\sqrt{\dfrac{0.168(1 - 0.168)}{169 + 4}} = 0.028$. Thus, a 95% confidence interval for p is

$$0.168 \pm (1.96)(0.028)$$

or

$$0.168 \pm 0.055$$

or

$$0.113 < p < 0.223$$

Thus, we are 95% confident that the probability of a BRCA1 mutation in a woman with a family history of breast cancer is between 0.113 and 0.223 (i.e., between 11.3% and 22.3%). ■

Note that the size of the standard error is inversely proportional to $\sqrt{n}$, as illustrated in the following example.

Example 9.2.3 **Breast Cancer** Suppose, as in Example 9.2.2, that a sample of n women with family histories of breast cancer contains 16% with *BRCA1* mutations. Then $\tilde{p} \approx 0.168$ and

$$\text{SE}_{\tilde{P}} \approx \sqrt{\frac{0.168(0.832)}{n + 4}}$$

We saw in Example 9.2.2 that if $n = 169$, then

$$\text{SE}_{\tilde{P}} = 0.028$$

*Many statistics books present the confidence interval for a proportion as $\hat{p} \pm 1.96\sqrt{\dfrac{\hat{p}(1 - \hat{p})}{n}}$ where $\hat{p} = y/n$. This commonly used interval is similar to the interval we present, particularly if n is large. For small or moderate sample sizes, the interval we present is more likely to cover the population proportion p. A technical discussion of the Wilson-interval using $\tilde{P}$ is given in Appendix 9.1.

If $n = 4 \times 169 = 676$, then

$$SE_{\tilde{P}} = 0.014$$

Thus, a sample with the same composition (i.e., 16% with *BRCA1* mutations) but four times as large, would yield twice as much precision in the estimation of p. ■

The Wilson-adjusted sample proportion can be used to construct a confidence interval for p even when the sample size is small, as the following example illustrates.

Example 9.2.4

ECMO Extracorporeal membrane oxygenation (ECMO) is a potentially life-saving procedure that is used to treat newborn babies who suffer from severe respiratory failure. An experiment was conducted in which 11 babies were treated with ECMO; none of the 11 babies died.[5] Let p denote the probability of death for a baby treated with ECMO. The fact that none of the babies in the experiment died should not lead us to believe that the probability of death, p, is precisely zero—only that it is close to zero. The estimate given by $\tilde{p}$ is $2/15 = 0.133$. The standard error of $\tilde{p}$ is

$$\sqrt{\frac{0.133(0.867)}{15}} = 0.088^*$$

Thus, a 95% confidence interval for p is

$$0.133 \pm (1.96)(0.088)$$

or

$$0.133 \pm 0.172$$

or

$$-0.039 < p < 0.305$$

We know that p cannot be negative, so we state the confidence interval as (0, 0.305).

Thus, we are 95% confident that the probability of death in a newborn with severe respiratory failure who is treated with ECMO is between 0 and 0.305 (i.e., between 0% and 30.5%). ■

CONDITIONS FOR USE OF THE WILSON 95% CONFIDENCE INTERVAL FOR p

In order for the Wilson confidence interval to be applicable, it must be reasonable to regard the data as a random sample from some population. In particular, it is important that the observations are chosen independently and that all items in the population have the same chance of being sampled. The Wilson interval does not require large sample sizes to be valid.[3]

ONE-SIDED CONFIDENCE INTERVALS

Most confidence intervals are of the form "estimate $\pm$ margin of error"; these are known as two-sided intervals. However, it is possible to construct a one-sided confidence interval, which is appropriate when only a lower bound, or only an upper bound, is of interest. The following example provides an illustration.

*Note that if we used the commonly presented method of $\hat{p} \pm 1.96\sqrt{\frac{\hat{p}(1-\hat{p})}{n}}$ we would find that the standard error is zero, leading to a confidence interval of 0 ± 0. Such an interval would not seem to be very useful in practice!

Example 9.2.5

ECMO—One-Sided Consider the ECMO data from Example 9.2.4, which are used to estimate the probability of death, p, in a newborn with severe respiratory failure. We know that p cannot be less than zero, but we might want to know how large p might be. Whereas a two-sided confidence interval is based on capturing the middle 95% of a standard normal distribution and thus uses the Z multipliers of ± 1.96, a one-sided 95% (upper) confidence interval uses the fact that $\Pr(-\infty < Z < 1.645) = 0.95$. Thus, the upper limit of the confidence interval is $\tilde{p} + 1.645 \times SE_{\tilde{p}}$ and the lower limit of the interval is negative infinity. In this case we get

$$0.133 + (1.645)(0.088) = 0.133 + 0.145 = 0.278$$

as the upper limit. The resulting interval is $(-\infty, 0.278)$, but since p cannot be negative, we state the confidence interval as $(0, 0.278)$. That is, we are 95% confident that the probability of death is at most 27.8%. ■

PLANNING A STUDY TO ESTIMATE p

In Section 6.4 we discussed a method for choosing the sample size n so that a proposed study would have sufficient precision for its intended purpose. The approach depended on two elements: (1) a specification of the desired $SE_{\bar{Y}}$ and (2) a preliminary guess of the SD. In the present context, when the observed variable is categorical, a similar approach can be used. If a desired value of $SE_{\tilde{p}}$ is specified, and if a rough informed guess of $\tilde{p}$ is available, then the required sample size n can be determined from the following equation:

$$\text{Desired SE} = \sqrt{\frac{(\text{Guessed } \tilde{p})(1 - \text{Guessed } \tilde{p})}{n + 4}}$$

The following example illustrates the use of the method.

Example 9.2.6

Vegetarians In a survey of 136 students at a U.S. college, 19 of them said that they were vegetarians.[6]

The sample estimate of the proportion is

$$\tilde{p} = \frac{19 + 2}{136 + 4} = 0.15$$

Suppose we regard these data as a pilot study and we now wish to plan a study large enough to estimate p with a standard error of two percentage points, that is, 0.02. We choose n to satisfy the following relation:

$$\sqrt{\frac{0.15(0.85)}{n + 4}} \le 0.02$$

This equation is easily solved to give $n + 4 \ge 318.75$. We should plan a sample of 315 students. ■

Planning in Ignorance Suppose no preliminary informed guess of p is available. Remarkably, in this situation it is still possible to plan an experiment to achieve a desired value of $SE_{\tilde{p}}$.* Such a "blind" plan depends on the fact that the crucial quantity $\sqrt{\tilde{p}(1 - \tilde{p})}$ is *largest* when $\tilde{p} = 0.5$; you can see this in the graph of Figure 9.2.2.

*By contrast, it would not be possible if we were planning a study to estimate a population mean μ and we had no information whatsoever about the value of the SD.

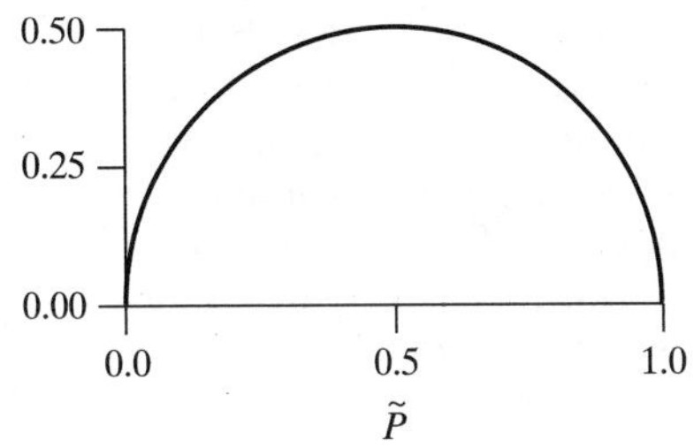

Figure 9.2.2 How $\sqrt{\tilde{p}(1-\tilde{p})}$ depends on $\tilde{p}$

It follows that a value of n calculated using "guessed $\tilde{p}$" = 0.5 will be *conservative*—that is, it will certainly be large enough. (Of course, it will be much larger than necessary if $\tilde{p}$ is really very different from 0.5.) The following example shows how such "worst-case" planning is used.

Example 9.2.7 **Vegetarians** Suppose, as in Example 9.2.6, that we are planning a study of vegetarianism and that we want $SE_{\tilde{P}}$ to be 0.02, but suppose that we have no preliminary information whatsoever. We can proceed as in Example 9.2.6, but using a guessed value of $\tilde{p}$ of 0.5. Then we have

$$\sqrt{\frac{0.5(0.5)}{n+4}} \leq 0.02$$

which means that $n + 4 \geq 625$, so we need $n = 621$. Thus, a sample of 621 students would be adequate to estimate p with a standard error of 0.02, regardless of the actual value of p. (Of course, if $p = 0.15$, this value of n is much larger than is necessary.) ■

Exercises 9.2.1–9.2.17

9.2.1 A series of patients with bacterial wound infections were treated with the antibiotic Cefotaxime. Bacteriologic response (disappearance of the bacteria from the wound) was considered "satisfactory" in 84% of the patients.[7] Determine the standard error of $\tilde{P}$, the Wilson-adjusted observed proportion of "satisfactory" responses, if the series contained

(a) 50 patients of whom 42 were considered "satisfactory."

(b) 200 patients of whom 168 were considered "satisfactory."

9.2.2 In an experiment with a certain mutation in the fruitfly *Drosophila*, n individuals were examined; of these, 20% were found to be mutants. Determine the standard error of $\tilde{P}$ if

(a) $n = 100$ (20 mutants).

(b) $n = 400$ (80 mutants).

9.2.3 Refer to Exercise 9.2.2. In each case ($n = 100$ and $n = 400$) construct a 95% confidence interval for the population proportion of mutants.

9.2.4 In a natural population of mice (*Mus musculus*) near Ann Arbor, Michigan, the coats of some individuals are white spotted on the belly. In a sample of 580 mice from the population, 28 individuals were found to have white-spotted bellies.[8] Construct a 95% confidence interval for the population proportion of this trait.

9.2.5 To evaluate the policy of routine vaccination of infants for whooping cough, adverse reactions were monitored in 339 infants who received their first injection of vaccine. Reactions were noted in 69 of the infants.[9]

(a) Construct a 95% confidence interval for the probability of an adverse reaction to the vaccine.

(b) Interpret the confidence interval from part (a). What does the interval say about whooping cough vaccinations?

(c) Using your interval from part (a), can we be confident that the probability of an adverse reaction to the vaccine is less than 0.25?

(d) What level of confidence is associated with your answer to part (c)? (*Hint*: What is the associated one-sided interval confidence level?)

9.2.6 In a study of human blood types in nonhuman primates, a sample of 71 orangutans were tested, and 14 were found to be blood type B.[10] Construct a 95% confidence interval for the relative frequency of blood type B in the orangutan population.

9.2.7 In populations of the snail *Cepaea*, the shells of some individuals have dark bands, while other individuals have unbanded shells.[11] Suppose that a biologist is planning a study to estimate the percentage of banded individuals in a certain natural population, and that she wants to estimate the percentage—which she anticipates will be in the neighborhood of 60%—with a standard error not to exceed 4 percentage points. How many snails should she plan to collect?

9.2.8 (Continuation of Exercise 9.2.7) What would the answer be if the anticipated percentage of banded snails were 50% rather than 60%?

9.2.9 The ability to taste the compound phenylthiocarbamide (PTC) is a genetically controlled trait in humans. In Europe and Asia, about 70% of people are "tasters."[12] Suppose a study is being planned to estimate the relative frequency of tasters in a certain Asian population, and it is desired that the standard error of the estimated relative frequency should be 0.01. How many people should be included in the study?

9.2.10 Refer to Exercise 9.2.9. Suppose a study is being planned for a part of the world for which the percentage of tasters is completely unknown, so the 70% figure used in Exercise 9.2.9 is not applicable. What sample size is needed so that the standard error will be no larger than 0.01?

9.2.11 Refer to Exercise 9.2.9. Suppose the SE requirement is relaxed by a factor of 2—from 0.01 to 0.02. Would this reduce the required sample size by a factor of 2? Explain.

9.2.12 The "Luso" variety of wheat is resistant to the Hessian fly. In order to understand the genetic mechanism controlling this resistance, an agronomist plans to examine the progeny of a certain cross involving Luso and a nonresistant variety. Each progeny plant will be classified as resistant or susceptible and the agronomist will estimate the proportion of progeny that are resistant.[13] How many progeny does he need to classify in order to guarantee that the standard error of his estimate of this proportion will not exceed 0.05?

9.2.13 (Continuation of Exercise 9.2.12) Suppose the agronomist is considering two possible genetic mechanisms for the inheritance of resistance; the population ratio of resistant to susceptible progeny would be 1:1 under one mechanism and 3:1 under the other. If the agronomist uses the sample size determined in Exercise 9.2.12, can he be sure that a 95% confidence interval will exclude at least one of the mechanisms? That is, can he be sure that the confidence interval will *not* contain both 0.50 and 0.75? Explain.

9.2.14 In a study of *in vitro* fertilization, 264 women ages 40–44 underwent a procedure known as elective single-embryo transfer (eSET) to attempt to get pregnant. Sixty of these women successfully became pregnant and gave birth.[14] Use these data to construct a 95% confidence interval for the probability of success using eSET for a woman ages 40–44.

9.2.15 Consider the data from Exercise 9.2.14. Suppose a new study of the effectiveness of eSET is being considered for women ages 35–39, and it is desired to have a standard error no greater than 0.04. How large a sample would be needed

(a) if the data from Exercise 9.2.14 are used to get a preliminary estimate of $\tilde{p}$?

(b) if we don't trust that data from women ages 40–44 are useful and thus no estimate of $\tilde{p}$ is available (i.e., if we wish to guard against the worst case)?

9.2.16 A researcher found that 54 out of 66 trees sampled in an arboretum were native species to the area. Using these data, he calculated a 95% confidence interval as (0.71, 0.89).[15]

(a) The confidence interval is an estimate of what quantity?

(b) Write a sentence that interprets the confidence interval in the context of this setting. (See Example 9.2.2.)

9.2.17 In a study of hand hygiene compliance in hospital surgical wards, researchers observed 375 instances for which hand washing would have been appropriate. Among the 375 observations, hands were washed only 161 times.[16]

(a) Using these data, a 95% confidence interval is (0.381, 0.481). Based on this interval, is it reasonable to conclude that the overall hand-washing rate at this surgical ward is less than 50%? Briefly explain.

(b) Over the period of observation of hand-washing opportunities in this study it is unlikely that more than 40 different doctors, nurses, and technicians were in the surgical ward. How might this information cast doubt regarding the validity of the confidence interval as an estimate of the proportion of workers who wash their hands?

9.3 Other Confidence Levels (Optional)

The procedure outlined in Section 9.2 can be used to construct 95% confidence intervals. In order to construct intervals with other confidence coefficients, some modifications to the procedure are needed. The first modification concerns $\tilde{p}$. For a

95% confidence interval we defined $\tilde{p}$ to be $\frac{y+2}{n+4}$. In general, for a confidence interval of level $100(1 - \alpha)\%$, $\tilde{p}$ is defined as

$$\tilde{p} = \frac{y + 0.5(z^2_{\alpha/2})}{n + z^2_{\alpha/2}}$$

For a 95% confidence interval $z_{\alpha/2}$ is 1.96, so $\tilde{p} = \frac{y + 0.5(1.96^2)}{n + 1.96^2}$. This is equal to $\frac{y + 1.92}{n + 3.84}$, which we rounded off as $\frac{y+2}{n+4}$. However, any confidence level can be used. As an example, for a 90% confidence interval, $\tilde{p} = \frac{y + 0.5(1.645^2)}{n + 1.645^2}$; this is equal to $\frac{y + 1.35}{n + 2.7}$.

The second modification concerns the standard error. For a 95% confidence interval we used $\sqrt{\frac{\tilde{p}(1-\tilde{p})}{n+4}}$ as the standard error term. In general, we use $\sqrt{\frac{\tilde{p}(1-\tilde{p})}{n+z^2_{\alpha/2}}}$ as the standard error term.

Finally, the Z multiplier must match the confidence level (1.645 for a 90% confidence interval, etc.). These can be found most easily from Table 4 with df $= \infty$. (Recall from Section 6.3 that the t distribution with df $= \infty$ is a normal (Z) distribution.) The following example illustrates these modifications.

Example 9.3.1 **Vegetarians** In Example 9.2.6 we considered a survey of college students in which 19 of 136 students were found to be vegetarians. Let us construct a 90% confidence interval for the proportion, p, of vegetarians in the population.[6]

The sample estimate of the proportion is

$$\tilde{p} = \frac{19 + 0.5(1.645^2)}{136 + 1.645^2} = \frac{19 + 1.35}{136 + 2.7} \approx 0.147$$

and the SE is

$$\sqrt{\frac{0.147(0.853)}{138.7}} = 0.030$$

A 90% confidence interval for p is

$$0.147 \pm (1.645)(0.030)$$

or

$$0.098 < p < 0.196$$

Thus, we are 90% confident that between 9.8% and 19.6% of the population that was sampled are vegetarians. ■

Exercises 9.3.1–9.3.4

9.3.1 In a sample of 848 children ages 3 to 5 it was found that 3.7% of them had iron deficiency.[17] Use these data to construct a 90% confidence interval for the proportion of all 3- to 5-year-old children with iron deficiency.

9.3.2 Researchers tested patients with cardiac pacemakers to see if use of a cellular telephone interferes with the operation of the pacemaker. There were 959 tests conducted for one type of cellular telephone; interference with the pacemaker (detected with electrocardiographic monitoring) was found in 15.7% of these tests.[18]

(a) Use these data to construct an appropriate 90% confidence interval.

(b) The confidence interval from part (a) is a confidence interval for what quantity? Answer in the context of the setting.

9.3.3 Gene mutations have been found in patients with muscular dystrophy. In one study, it was found that there were defects in the gene coding of sarcoglycan proteins in 23 of 180 patients with limb-girdle muscular dystrophy.[19] Use these data to construct a 99% confidence interval for the corresponding population proportion.

9.3.4 In an ecological study of the Carolina Junco, 53 birds were captured from a certain population; of these, 40 were male.[20] Use these data to construct a 90% confidence interval for the proportion of male birds in the Carolina Junco population.

9.4 Inference for Proportions: The Chi-Square Goodness-of-Fit Test

In Section 9.2 we described methods for constructing confidence intervals when the observed variable is categorical. We now turn our attention to hypothesis testing for categorical data. We will begin by considering analysis of a single sample of categorical data. We assume that the data can be regarded as a random sample from some population and we will test a null hypothesis, H_0, that specifies the population proportions, or probabilities, of the various categories. Here is an example.

Example 9.4.1 **Deer Habitat and Fire** Does fire affect deer behavior? Six months after a fire burned 730 acres of homogenous deer habitat, researchers surveyed a 3,000-acre parcel surrounding the area, which they divided into four regions: the region near the heat of the burn (1), the inside edge of the burn (2), the outside edge of the burn (3), and the area outside of the burned area (4); see Figure 9.4.1 and Table 9.4.1.[21] The null hypothesis is that that deer show no preference to any particular type of burned/unburned habitat—that they are randomly distributed over the 3,000 acres. The alternative hypothesis is that the deer do show a preference for some of the regions—that they are not randomly distributed across all 3,000 acres.

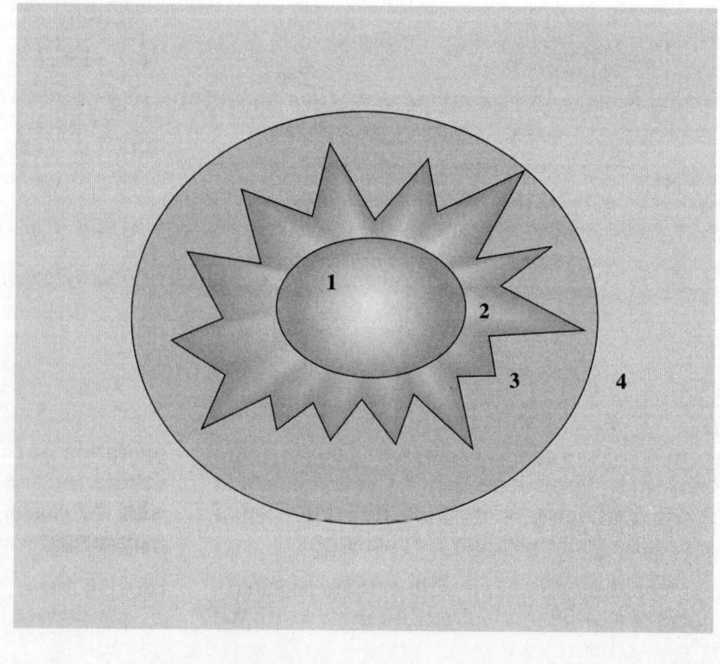

Figure 9.4.1 Schematic of 3,000-acre parcel with an interior 730-acre fire (not to scale)

Table 9.4.1 Deer distribution

Region	Acres	Proportion
1. Inner burn	520	0.173
2. Inner edge	210	0.070
3. Outer edge	240	0.080
4. Outer unburned	2,030	0.677
	3,000	1.000

Under the null hypothesis, if deer were randomly distributed over the 3,000 acres, then we would expect the counts of deer in the regions to be in proportion to the sizes of the regions. Expressing the null hypothesis numerically we have the following probabilities of sighting deer:

$$H_0\text{: } \Pr\{\text{inner burn}\} = \frac{520}{3{,}000} = 0.173$$

$$\Pr\{\text{inner edge}\} = \frac{210}{3{,}000} = 0.070$$

$$\Pr\{\text{outer edge}\} = \frac{240}{3{,}000} = 0.080$$

$$\Pr\{\text{outer unburned}\} = \frac{2{,}030}{3{,}000} = 0.677$$

Because the alternative hypothesis is not specific (it only states that the deer prefer some regions over others but doesn't indicate the nature of the preference), there is no simple symbolic way to express the alternative hypothesis. Thus, typically we do not use a symbolic representation. If we chose to express the alternative symbolically we could write:

$$H_A\text{: } \Pr\{\text{inner burn}\} \neq 0.173, \text{ and/or } \Pr\{\text{inner edge}\} \neq 0.070, \text{ and/or}$$
$$\Pr\{\text{outer edge}\} \neq 0.080, \text{ and/or } \Pr\{\text{outer unburned}\} \neq 0.677$$ ■

Given a random sample of n categorical observations, how can one judge whether they provide evidence against a null hypothesis H_0 that specifies the probabilities of the categories? There are two complementary approaches to this question: The first considers an examination of the observed relative frequencies of each category while the second examines the frequencies directly. Considering the first method, the observed relative frequencies serve as estimates of the probabilities of the categories. The following notation for relative frequencies is useful: When a probability $\Pr\{E\}$ is estimated from observed data, the estimate is denoted by a hat ("^"); thus,

$$\hat{\Pr}\{E\} = \text{the estimated probability of event } E$$

Example 9.4.2

Deer Habitat and Fire Researchers observed a total of 75 deer in the 3,000-acre parcel described in Example 9.4.1: Two were in the region near the heat of the burn (Region 1), 12 were on the inside edge of the burn (Region 2), 18 were on the outside edge of the burn (Region 3), and 43 were outside of the burned area (Region 4).

These data are shown in Figure 9.4.2.

Figure 9.4.2 Bar chart of deer distribution data

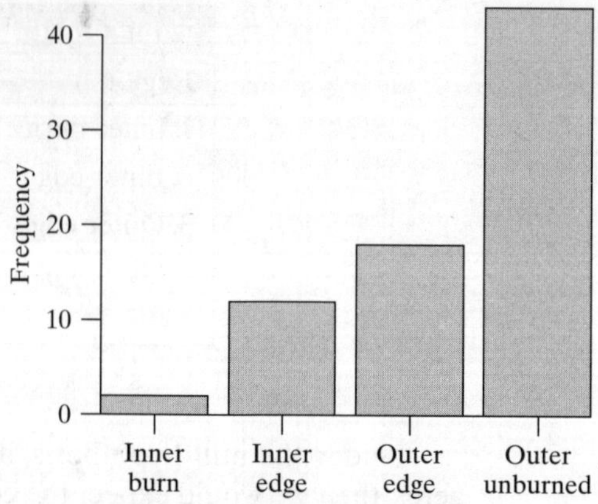

The estimated category probabilities are

$$\hat{\Pr}\{\text{inner burn}\} = \frac{2}{75} = 0.027$$

$$\hat{\Pr}\{\text{inner edge}\} = \frac{12}{75} = 0.160$$

$$\hat{\Pr}\{\text{outer edge}\} = \frac{18}{75} = 0.240$$

$$\hat{\Pr}\{\text{outer unburned}\} = \frac{43}{75} = 0.573$$

These estimated probabilities differ quite a bit from those in the model that is specified by H_0. Figure 9.4.3 shows stacked bar charts of the observed and hypothesized proportions. ■

Figure 9.4.3 Stacked bar charts of the deer proportions

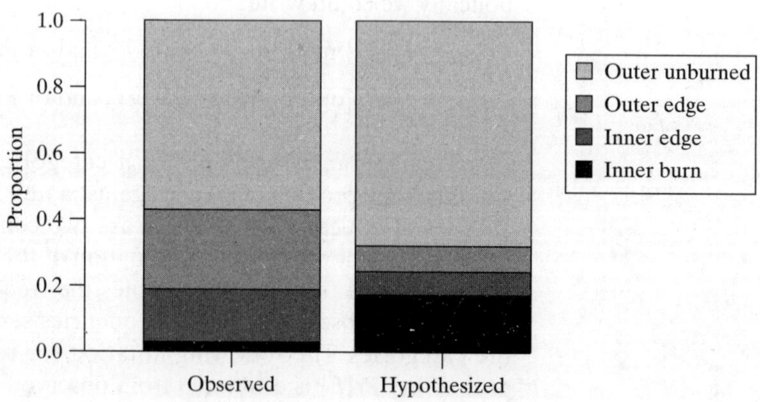

THE CHI-SQUARE STATISTIC

The second approach, which considers the actual frequencies, is to use a statistical test, called a **goodness-of-fit test,** to assess the compatibility of the data with H_0. The most widely used goodness-of-fit test is the **chi-square test** or χ^2 test (χ is the Greek letter "chi").

The calculation of the chi-square test statistic is done in terms of the absolute, rather than the relative, frequencies of the categories. For each category level, i, let

o_i represent the **observed frequency** of the category and let e_i represent the **expected frequency**—that is, the frequency that would be expected according to H_0. The e_i's are calculated by multiplying each probability specified in H_0 by n, as shown in Example 9.4.3.

Example 9.4.3

Deer Habitat and Fire Consider the null hypothesis specified in Example 9.4.1 and the data from Example 9.4.2. If the null hypothesis is true, then we expect 17.3% of the 75 deer to be in the inner burn region; 17.3% of 75 is 13.0:

$$\text{Inner burn: } e_1 = (0.173)(75) = 13.00$$

The corresponding expected frequencies for the other regions are

$$\text{Inner edge: } e_2 = (0.070)(75) = 5.25$$

$$\text{Outer edge: } e_3 = (0.080)(75) = 6.00$$

$$\text{Outer unburned: } e_4 = (0.677)(75) = 50.75$$

The test statistic for the chi-square goodness-of-fit test is then calculated from the o_i's and the e_i's using the formula given in the accompanying box with k equal to the number of category levels. Example 9.4.4 illustrates the calculation of the chi-square statistic.

The Chi-Square Statistic

$$\chi_s^2 = \sum_{i=1}^{k} \frac{(o_i - e_i)^2}{e_i}$$

where the summation is over all k categories.

Example 9.4.4

Deer Habitat and Fire The observed frequencies of 75 deer locations are

Region	Inner Burn	Inner Edge	Outer Edge	Outer Unburned	Total
Observed (o_i)	2	12	18	43	75

The expected frequencies are

Region	Inner Burn	Inner Edge	Outer Edge	Outer Unburned	Total
Expected (e_i)	13	5.25	6	50.75	75

Note that the sum of the expected frequencies is the same as the sum of the observed frequencies (75). The χ^2 statistic is

$$\chi_s^2 = \frac{(2-13)^2}{13} + \frac{(12-5.25)^2}{5.25} + \frac{(18-6)^2}{6} + \frac{(43-50.75)^2}{50.75}$$
$$= 43.2$$

Computational Note: In calculating a chi-square statistic the o_i's must be *absolute*, rather than relative, frequencies.

THE χ^2 DISTRIBUTION

From the way in which χ_s^2 is defined, it is clear that small values of χ_s^2 would indicate that the data agree with H_0, while large values of χ_s^2 would indicate disagreement. In order to base a statistical test on this agreement or disagreement, we need to know how much χ_s^2 may be affected by sampling variation.

We consider the null distribution of χ_s^2—that is, the sampling distribution that χ_s^2 follows if H_0 is true. It can be shown (using the methods of mathematical statistics) that, if the sample size is large enough, then the null distribution of χ_s^2 can be approximated by a distribution known as a **χ^2 distribution.** The form of a χ^2 distribution depends on a parameter called "degrees of freedom" (df). Figure 9.4.4 shows the χ^2 distribution with df = 5.

Figure 9.4.4 The χ^2 distribution with df = 5

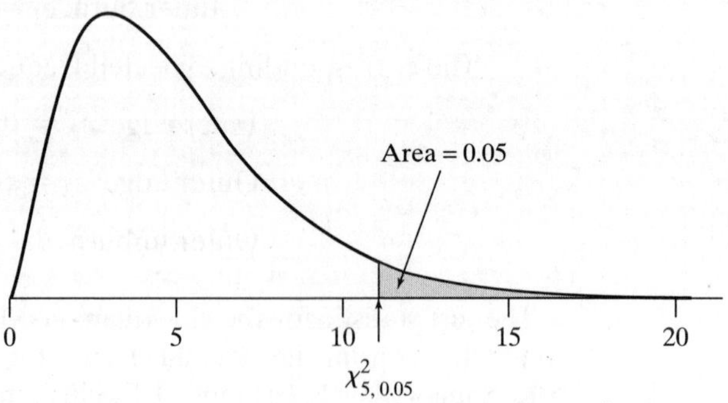

Table 9 gives critical values for the χ^2 distribution. For instance, for df = 5, the 5% critical value is $\chi^2_{5, 0.05} = 11.07$. This critical value corresponds to an area of 0.05 in the upper tail of the χ^2 distribution, as shown in Figure 9.4.4.

THE GOODNESS-OF-FIT TEST

For the chi-square goodness-of-fit test we have presented, the null distribution of χ_s^2 is approximately a χ^2 distribution with*

$$\text{df} = k - 1, \text{ where } k \text{ equals the number of categories}$$

For example, for the setting presented in Example 9.4.4 there are four categories so $k = 4$. The null hypothesis specifies the probabilities for each of the four categories. However, once the first three probabilities are specified, the last one is determined, since the four probabilities must sum to 1. There are four categories, but only three of them are "free"; the last one is constrained by the first three.

The test of H_0 is carried out using critical values from Table 9, as illustrated in the following example.

Example 9.4.5

Deer Habitat and Fire For the deer habitat data of Example 9.4.4, the observed chi-square statistic was $\chi_s^2 = 43.2$. Because there are four categories, the degrees of freedom for the null distribution are calculated as

$$\text{df} = 4 - 1 = 3$$

From Table 9 with df = 3 we find that $\chi^2_{3, 0.0001} = 21.11$. Since $\chi_s^2 = 43.2$ is greater than 21.11, the upper tail area beyond 43.2 is less than 0.0001. Thus, the

*The chi-square test can be extended to more general situations in which parameters are estimated from the data before the expected frequencies are calculated. In general, the degrees of freedom for the test are (number of categories) − (number of parameters estimated) −1. We are considering only the case in which there are no parameters to be estimated from the data.

P-value is less than 0.0001 and we have strong evidence against H_0 and in favor of the alternative hypothesis that the deer show preference for some areas over others. Upon comparing the observed and expected frequencies (or equivalently the hypothesized and estimated probabilities), we note that deer moved away from the burned and unburned regions (1) and (4) to be near the edge regions (2) and (3) (where there is likely to be new growth of vegetation yet proximity to old-growth shelter). ■

The chi-square test can be used with any number of categories. In Example 9.4.6 the test is applied to a variable with six categories.

Example 9.4.6 **Flax Seeds** Researchers studied a mutant type of flax seed that they hoped would produce oil for use in margarine and shortening. The amount of palmitic acid in the flax seed was an important factor in this research; a related factor was whether the seed was brown or was variegated. The seeds were classified into six combinations of palmitic acid and color, as shown in Table 9.4.2.[22] According to a hypothesized (Mendelian) genetic model, the six combinations should occur in a 3:6:3:1:2:1 ratio. That is, brown and low acid level should occur with probability 3/16, brown and intermediate acid level should occur with probability 6/16, and so on. The null hypothesis is that the model is correct; the alternative hypothesis is that the model is incorrect. The χ^2 statistic is

$$\chi_s^2 = \frac{(15-13.5)^2}{13.5} + \frac{(26-27)^2}{27} + \frac{(15-13.5)^2}{13.5} + \frac{(0-4.5)^2}{4.5} + \frac{(8-9)^2}{9} + \frac{(8-4.5)^2}{4.5}$$
$$= 7.7$$

Table 9.4.2 Flax seed distribution

Color	Acid level	Observed (o_i)	Expected (e_i)
Brown	Low	15	13.5
Brown	Intermediate	26	27
Brown	High	15	13.5
Variegated	Low	0	4.5
Variegated	Intermediate	8	9
Variegated	High	8	4.5
Total		72	72

The χ^2 test has $6 - 1 = 5$ degrees of freedom. From Table 9 with df $= 5$, we find that $\chi^2_{5,\,0.20} = 7.29$ and $\chi^2_{5,\,0.10} = 9.24$. Thus, the P-value is bracketed as $0.10 < P\text{-value} < 0.20$. (Computer software computes P-value $= 0.173$.) If the level of α chosen for the test is 0.10 or smaller, then the P-value is larger than α and we would not reject H_0. We conclude that there is no significant evidence that the data are inconsistent with the Mendelian model. (Note that we have not necessarily demonstrated that the Mendelian model is correct, only that we cannot reject this model.) ■

Note that the critical values for the chi-square test do not depend on the sample size, n. However, the test procedure *is* affected by n, through the value of the chi-square statistic. If we change the size of a sample while keeping its percentage composition fixed, then χ_s^2 varies directly as the sample size, n. For instance, imagine appending a replicate of a sample to the sample itself. Then the expanded sample would have twice as many observations as the original, but they would be in the same relative proportions. The value of each o_i would be doubled, the value of each e_i would be doubled, and so the value of χ^2 would be doubled [because in each term of χ_s^2 the numerator $(o_i - e_i)^2$ would be multiplied by 4, and the denominator e_i would be multiplied by 2]. That is, the value of χ_s^2 would go up by a factor of 2, despite the fact that the pattern in the data stayed the same! In this way, an increased sample size magnifies any discrepancy between what is observed and what is expected under the null hypothesis.

COMPOUND HYPOTHESES AND DIRECTIONALITY

Let us examine the goodness-of-fit null hypothesis more closely. In a two-sample comparison such as a t test, the null hypothesis contains exactly one assertion—for instance, that two population means are equal. By contrast, a goodness-of-fit null hypothesis can contain more than one assertion. Such a null hypothesis may be called a **compound null hypothesis.** An example follows.

Example 9.4.7 **Deer Habitat and Fire** The null hypothesis of Example 9.4.1 is

$$H_0\text{: Pr\{inner burn\}} = 0.173, \text{Pr\{inner edge\}} = 0.070, \text{Pr\{outer edge\}} = 0.080, \text{Pr\{outer unburned\}} = 0.677$$

This is a compound hypothesis because it makes three independent assertions, namely

$$\text{Pr\{inner burn\}} = 0.173,\ \text{Pr\{inner edge\}} = 0.070, \text{and}\ \text{Pr\{outer edge\}} = 0.080$$

Note that the fourth assertion (Pr{outer unburned} = 0.677) is not an independent assertion because it follows from the other three. ■

When the null hypothesis is compound, the chi-square test has two special features. First, the alternative hypothesis is necessarily nondirectional. Second, if H_0 is rejected, the test does not yield a directional conclusion. (However, if H_0 is rejected, then an examination of the observed proportions will sometimes show an interesting pattern of departure from H_0, as in Example 9.4.5.)

When H_0 is compound, the chi-square test is nondirectional in nature (perhaps "omnidirectional" would be a better term) because the chi-square statistic measures deviations from H_0 in all directions. Statistical methods are available that do yield directional conclusions and that can handle directional alternatives, but such methods are beyond the scope of this book.

DICHOTOMOUS VARIABLES

If the categorical variable analyzed by a goodness-of-fit test is dichotomous, then the null hypothesis is not compound, and directional alternatives and directional conclusions do not pose any particular difficulty.*

*When the data are dichotomous, there is an alternative to the goodness-of-fit test that is known as the Z test for a single proportion. The calculations used in the Z test look quite different from those of the goodness-of-fit test but, in fact, the two tests are mathematically equivalent. However, unlike the goodness-of-fit test, which can handle any number of categories, the Z test can be used only when the data are limited to two categories. Thus, we do not present it here.

Directional Conclusion The following example illustrates the directional conclusion.

Example 9.4.8

Deer Habitat, Fire, and Two Regions Suppose that the deer habitat data of Example 9.4.1 had been presented as being from only two regions, A and B, where region A is the area at the edge of the fire, which combines regions (2) and (3), and region B is the remainder of the parcel, combining regions (1) and (4). There were 30 deer seen in region A and 45 deer seen in region B. Is this evidence that deer prefer one region over the other?

An appropriate null hypothesis is

$$H_0\text{: Pr\{region A\}} = \frac{450}{3{,}000} = 0.15, \ \text{Pr\{region B\}} = \frac{2{,}550}{3{,}000} = 0.85$$

This hypothesis is not compound because it contains only one independent assertion. (Note that the second assertion—Pr{region B} = 0.85 —is redundant; it follows from the first.)

Let us test H_0 against the nondirectional alternative

$$H_A\text{: Pr\{region A\}} \neq 0.15$$

The observed and expected frequencies are shown in Table 9.4.3.

Table 9.4.3 Deer habitat data for two regions

	A	B	Total
Observed	30	45	75
Expected	11.25	63.75	75

The data yield $\chi_s^2 = 36.8$ and from Table 9 we find that $P < 0.0001$. Even at $\alpha = 0.0001$ we would reject H_0 and find that there is sufficient evidence to conclude that the population of deer prefers one region over the other. Comparing the observed and expected counts we observe that they prefer region A over region B. ■

To recapitulate, the directional conclusion in Example 9.4.8 is legitimate because we know that if H_0 is false, then necessarily either Pr{region A} < 0.15 or Pr{region A} > 0.15. By contrast, in Example 9.4.7 H_0 may be false but Pr{outer unburned} may still be equal to 0.677; the chi-square analysis does not determine which of the probabilities specified by H_0 are wrong.

Directional Alternative A chi-square goodness-of-fit test against a directional alternative (when the observed variable is dichotomous) uses the familiar two-step procedure:

Step 1. Check directionality (see if the data deviate from H_0 in the direction specified by H_A).

(a) If not, the P-value is greater than 0.50.

(b) If so, proceed to step 2.

Step 2. The P-value is half what it would be if H_A were nondirectional.

The following example illustrates the procedure.

Example 9.4.9

Harvest Moon Festival Can people who are close to death postpone dying until after a symbolically meaningful occasion? Researchers studied death from natural causes among elderly Chinese women (over age 75) living in California. They chose to study the time around the Harvest Moon Festival because (1) the date of the traditional Chinese festival changes somewhat from year to year, making it less likely that a time-of-year effect would be confounded with the effect they were studying and (2) it is a festival in which the role of the oldest woman in the family is very important.

Previous research had suggested that there might be a decrease in the mortality rate among elderly Chinese women immediately prior to the festival, with a corresponding increase afterward. The researchers found that over a period of several years there were 33 deaths in the group in the week preceding the Harvest Moon Festival and 70 deaths in the week following the festival.[23] How strongly does this support the interpretation that people can prolong life until a symbolically meaningful event?

We may formulate null and alternative hypotheses as follows:

H_0: Given that an elderly Chinese woman dies within one week of the Harvest Moon Festival, she is equally likely to die before the festival or after the festival.

H_A: Given that an elderly Chinese woman dies within one week of the Harvest Moon Festival, she is more likely to die after the festival than before the festival.

These hypotheses can be translated as

$$H_0\text{: Pr\{die after festival\}} = \frac{1}{2}$$

$$H_A\text{: Pr\{die after festival\}} > \frac{1}{2}$$

where it is understood that Pr{die after festival} is the probability of death after the festival, given that the woman dies within one week before or after the festival. The observed and expected frequencies are shown in Table 9.4.4.

Table 9.4.4 Harvest moon festival data

	Before	After	Total
Observed	33	70	103
Expected	51.5	51.5	103

From the data on the 103 deaths, we first note that the data do, indeed, deviate from H_0 in the direction specified by H_A, because the observed relative frequency of deaths after the festival is 70/103, which is greater than 1/2. The value of the chi-square statistic is $\chi_s^2 = 13.3$; from Table 9 we see that the P-value would have been bracketed between 0.0001 and 0.001 had H_A been nondirectional. However, for the directional alternative hypothesis specified in this test, we bracket the P-value as $0.00005 < P\text{-value} < 0.0005$. We conclude that the evidence is very strong that the death rate among elderly Chinese women goes up after the festival.* ■

*Based on these results, one might jump to the conclusion that this festival should be canceled to protect elderly Chinese women. As this study is only observational, however, we must not jump to causal conclusions!

Exercises 9.4.1–9.4.13

9.4.1 A cross between white and yellow summer squash gave progeny of the following colors[24]:

Color	White	Yellow	Green
Number of progeny	155	40	10

Are these data consistent with the 12:3:1 ratio predicted by a certain genetic model? Use a chi-square test at $\alpha = 0.10$.

9.4.2 Refer to Exercise 9.4.1. Suppose the sample had the same composition but was 10 times as large: 1,550 white, 400 yellow, and 100 green progeny. Would the data be consistent with the 12:3:1 model?

9.4.3 How do bees recognize flowers? As part of a study of this question, researchers used the following two artificial "flowers"[25]:

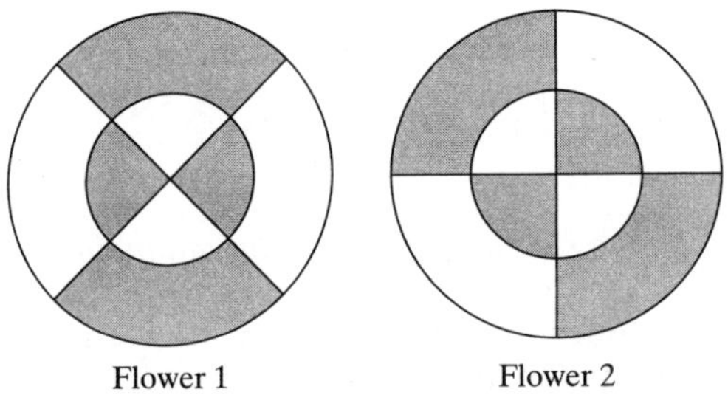

The experiment was conducted as a series of trials on individual bees; each trial consisted of presenting a bee with both flowers and observing which flower it landed on first. (Flower 1 was sometimes on the left and sometimes on the right.) During the "training" trials, flower 1 contained a sucrose solution, and flower 2 did not; thus, the bee was trained to prefer flower 1. During the testing trials, neither flower contained sucrose. In 25 testing trials with a particular bee, the bee chose flower 1 twenty times and flower 2 five times.

Use a goodness-of-fit test to assess the evidence that the bee could remember and distinguish the flower patterns. Use a directional alternative and let $\alpha = 0.05$.

9.4.4 At a Midwestern hospital there were a total of 932 births in 20 consecutive weeks. Of these births, 216 occurred on weekends.[26] Do these data reveal more than chance deviation from random timing of the births? Consider a goodness of fit test with two categories of births—weekday and weekend—using a nondirectional alternative.

(a) What is the value of the χ^2 test statistic?

(b) In the context of this study, state the null and alternative hypotheses.

(c) The P-value for the test is 0.0003. If $\alpha = 0.05$, what is your conclusion regarding the hypotheses in (b)?

9.4.5 In a breeding experiment, white chickens with small combs were mated and produced 190 offspring of the types shown in the accompanying table.[27] Are these data consistent with the Mendelian expected ratios of 9:3:3:1 for the four types? Use a chi-square test at $\alpha = 0.10$.

Type	Number of offspring
White feathers, small comb	111
White feathers, large comb	37
Dark feathers, small comb	34
Dark feathers, large comb	8
Total	190

9.4.6 Among n babies born in a certain city, 51% were boys.[28] Suppose we want to test the hypothesis that the true probability of a boy is $\frac{1}{2}$. Calculate the value of χ_s^2, and bracket the P-value for testing against a nondirectional alternative, if

(a) $n = 1{,}000$

(b) $n = 5{,}000$

(c) $n = 10{,}000$

9.4.7 In an agronomy experiment, peanuts with shriveled seeds were crossed with normal peanuts. The genetic model that the agronomists were considering predicted that the ratio of normal to shriveled progeny would be 3:1. They obtained 95 normal and 54 shriveled progeny.[29] Do these data support the hypothesized model?

Here is computer output for a chi-square test using a nondirectional alternative.

```
X-squared = 10.0425, df = 1, p-value = 0.00153
```

(a) State the null and alternative hypotheses in context.

(b) Compute the expected frequencies.

(c) If $\alpha = 0.01$, what is your conclusion regarding H_0?

9.4.8 An experimental design using litter-matching was employed to test a certain drug for cancer-causing potential. From each of 50 litters of rats, three females were selected; one of these three, chosen at random, received the test drug, and the other two were kept as controls. During a 2-year observation period, the time of occurrence of a tumor, and/or death from various causes, was recorded for each animal. One way to analyze the data is to note simply which rat (in each triplet) developed a tumor first. Some triplets were uninformative on this point because either (a) none of the three littermates developed a tumor, or (b) a rat developed a tumor after its littermate had died from some other cause. The results for the 50 triplets are shown in the table.[30] Use a goodness-of-fit test to evaluate the evidence that the drug causes cancer. Use a directional alternative and let $\alpha = 0.01$. State your conclusion from part (a) in the context of this setting. (*Hint*: Use only the 20 triplets that provide complete information.)

	Number of triplets
Tumor first in the treated rat	12
Tumor first in one of the two control rats	8
No tumor	23
Death from another cause	7
Total	50

9.4.9 A study of color vision in squirrels used an apparatus containing three small translucent panels that could be separately illuminated. The animals were trained to choose, by pressing a lever, the panel that appeared different from the other two. (During these "training" trials, the panels differed in brightness, rather than color.) Then the animals were tested for their ability to discriminate between various colors. In one series of "testing" trials on a single animal, one of the panels was red and the other two were white; the location of the red panel was varied randomly from trial to trial. In 75 trials, the animal chose correctly 45 times and incorrectly 30 times.[31] How strongly does this support the interpretation that the animal can discriminate between the two colors?

(a) Test the null hypothesis that the animal cannot discriminate red from white. Use a directional alternative and let $\alpha = 0.02$.

(b) Why is a directional alternative appropriate in this case?

9.4.10 Scientists have used Mongolian gerbils when conducting neurological research. A certain breed of these gerbils were crossed and gave progeny of the following colors[32]:

Color	Black	Brown	White
Number of progeny	40	59	42

(a) What is the value of the chi-square test statistic for investigating whether these data are consistent with the 1:2:1 ratio predicted by a certain genetic model?

(b) The P-value for the chi-square test is 0.149. If $\alpha = 0.05$, what is your conclusion regarding H_0?

9.4.11 Each of 36 men was asked to touch the foreheads of three women, one of whom was their romantic partner, while blindfolded. The two "decoy" women were the same age, height, and weight as the man's partner. Of the 36 men tested, 18 were able to correctly identify their partner.[33] Do the data provide sufficient evidence to conclude that men can do better than they would do by merely guessing?

Conduct an appropriate test.

9.4.12 Geneticists studying the inheritance pattern of cowpea plants classified the plants in one experiment according to the nature of their leaves. The data follow[34]:

Type	I	II	III
Number	179	44	23

Test the null hypothesis that the three types occur with probabilities 12/16, 3/16, and 1/16. Use a chi-square test with $\alpha = 0.10$.

9.4.13 In the snapdragon (*Antirrhinum majus*), individual plants can be red flowered, pink flowered, or white flowered. According to a certain Mendelian genetic model, self-pollination of pink-flowered plants should produce progeny that are red, pink, and white in the ratio 1:2:1. A geneticist self-pollinated pink-flowered snapdragon plants and produced 234 progeny with the following colors[35]:

Type	Red	Pink	White
Number	54	122	58

Test the null hypothesis that the three colors occur with probabilities 1/4, 1/2, and 1/4. Use a chi-square test with $\alpha = 0.10$.

9.5 Perspective and Summary

In this chapter we have discussed inference for categorical data, including confidence intervals and hypothesis tests. The procedures that we have developed, which are summarized next, can be applied if (1) the data can be regarded as a random sample from a large population and (2) the observations are independent.

Summary of Inference Methods for Categorical Data

95% Confidence interval for p

$$\tilde{p} \pm 1.96 \times SE_{\tilde{p}}$$

where

$$\tilde{p} = \frac{y + 2}{n + 4}$$

and

$$\text{SE}_{\tilde{p}} = \sqrt{\frac{\tilde{p}(1 - \tilde{p})}{n + 4}}$$

General confidence interval for *p*

$$\tilde{p} \pm z_{\alpha/2} \times \text{SE}_{\tilde{p}}$$

where

$$\tilde{p} = \frac{y + 0.5(z^2_{\alpha/2})}{n + z^2_{\alpha/2}}$$

$$\text{SE}_{\tilde{p}} = \sqrt{\frac{\tilde{p}(1 - \tilde{p})}{n + z^2_{\alpha/2}}}$$

Goodness-of-fit test

Data:

$$o_i = \text{the observed frequency of category } i$$

Null hypothesis:

H_0 specifies the probability of each category.*

Calculation of expected frequencies:

$$e_i = n \times \text{Probability specified for category } i \text{ by } H_0$$

Test statistic:

$$\chi^2_s = \sum_{i=1}^{k} \frac{(o_i - e_i)^2}{e_i}$$

Null distribution (approximate):

$$\chi^2 \text{ distribution with df} = k - 1$$

where k = the number of categories

This approximation is adequate if $e_i \geq 5$ for every category.

*A slightly modified form of the goodness-of-fit test can be used to test a hypothesis that merely constrains the probabilities rather than specifying them exactly. An example would be testing the fit of a binomial distribution to data (see optional Section 3.7). The details of this test are beyond the scope of this text.

Supplementary Exercises 9.S.1–9.S.22

9.S.1 In a certain population, 83% of the people have Rh-positive blood type.[36] Suppose a random sample of $n = 12$ people is to be chosen from the population and let $\tilde{P}$ represent the Wilson-adjusted proportion of Rh-positive people in the sample. Find

(a) $\Pr\{\tilde{P} = 0.614\}$

(b) $\Pr\{\tilde{P} = 0.686\}$

9.S.2 In a population of flatworms (*Planaria*) living in a certain pond, one in five individuals is adult and four are juvenile.[37] An ecologist plans to count the adults in a random sample of 20 flatworms from the pond; she will then use $\tilde{P}$, the Wilson-adjusted sample proportion of adults in the sample, as her estimate of p, the proportion of adults in the pond population. Find

(a) $\Pr\{\tilde{P} = p\}$

(b) $\Pr\{p - 0.05 \le \tilde{P} \le p + 0.05\}$

9.S.3 In a study of environmental effects upon reproduction, 123 female adult white-tailed deer from the central Adirondack area were captured, and 97 were found to be pregnant.[38] Construct a 95% confidence interval for the proportion of females pregnant in this deer population.

9.S.4 Refer to Exercise 9.S.3. Which of the conditions for validity of the confidence interval might have been violated in this study?

9.S.5 A sample of 32 breastfed infants found that 2 of them developed iron deficiency by age 5.5 months.[39]

(a) Use these data to construct an appropriate 95% confidence interval.

(b) What conditions are necessary for the confidence interval from part (a) to be valid?

(c) Interpret your confidence interval from part (a) in the context of this setting. That is, what do the numbers in the confidence interval tell us about iron deficiency in breastfed infants?

9.S.6 A certain California winery produces 720,000 bottles of wine each year. Suppose you want to estimate the proportion of those bottles that have cork taint (i.e., the wine is spoiled due to a failure of the cork). Suppose that 4% of all corked wine has cork taint. Using this as a preliminary guess of p, how many bottles of wine would need to be included in a random sample if you want the standard error of your estimate to be less than or equal to 1 percentage point?[40]

9.S.7 Refer to Exercise 9.S.6. Suppose you do not trust that the 4% taint rate for wines in general is a useful guess for this particular winery.

(a) Suppose that, based on previous years of data at this winery, about 10% of the wines have had cork taint. How many bottles would need to be included in a random sample if you want the standard error of your estimate to be less than or equal to 1 percentage point?

(b) How many bottles would need to be included in a random sample if you want the standard error of your estimate to be less than or equal to 1 percentage point, no matter what the value of p is?

9.S.8 When male mice are grouped, one of them usually becomes dominant over the others. In order to see how a parasitic infection might affect the competition for dominance, male mice were housed in groups, three mice to a cage; two mice in each cage received a mild dose of the parasitic worm *H. polygyrus.* Two weeks later, criteria such as the relative absence of tail wounds were used to identify the dominant mouse in each cage. It was found that the uninfected mouse had become dominant in 15 of 30 cages.[41] Is this evidence that the parasitic infection tends to inhibit the development of dominant behavior? Use a goodness-of-fit test against a directional alternative. Let $\alpha = 0.05$. (*Hint*: The observational unit in this experiment is not an individual mouse, but a cage of three mice.)

9.S.9 Are mice right-handed or left-handed? In a study of this question, 320 mice of a highly inbred strain were tested for paw preference by observing which forepaw—right or left—they used to retrieve food from a narrow tube. Each animal was tested 50 times, for a total of $320 \times 50 = 16{,}000$ observations. The results were as follows[42]:

	Right	Left
Number of observations	7,871	8,129

Suppose we assign an expected frequency of 8,000 to each category and perform a goodness-of-fit test; we find that $\chi_s^2 = 4.16$, so at $\alpha = 0.05$ we would reject the hypothesis of a 1:1 ratio and find that there is sufficient evidence to conclude that mice of this strain are (slightly) biased toward use of the left paw. This analysis contains a fatal flaw. What is it?

9.S.10 As part of the study of the inheritance pattern of cowpea plants, geneticists classified the plants in one experiment according to whether the plants had one leaf or three. The data are shown below[43]:

Number of leaves	1	3
Number of plants	74	61

(a) What is the value of the chi-square statistic for testing the null hypothesis that the two types of plants occur with equal probabilities?

(b) The P-value for the chi-square test (with a nondirectional alternative) is 0.26. If $\alpha = 0.05$, what is your conclusion regarding H_0? State your conclusions in the context of the problem.

9.S.11 People who harvest wild mushrooms sometimes accidentally eat the toxic "death cap" mushroom, *Amanita phalloides*. In reviewing 205 European cases of death-cap poisoning from 1971 through 1980, researchers found that 45 of the victims had died.[44] Consider a test to compare this mortality to the 30% mortality that was recorded before 1970. Let the alternative hypothesis be that mortality has decreased with time.

Here is computer output for a chi-square test *that used a nondirectional alternative.*

```
X-squared = 6.324, df = 1, p-value = 0.01191
```

If $\alpha = 0.01$ and the alternative hypothesis is directional, what is your conclusion regarding H_0? State your conclusions in the context of the problem.

9.S.12 The appearance of leaf pigment glands in the seedling stage of cotton plants is genetically controlled.

According to one theory of the control mechanism, the population ratio of glandular to glandless plants resulting from a certain cross should be 11:5; according to another theory it should be 13:3. In one experiment, the cross produced 89 glandular and 36 glandless plants.[45] Use goodness-of-fit tests (at $\alpha = 0.10$) to determine whether these data are consistent with

(a) the 11:5 theory

(b) the 13:3 theory

9.S.13 (Continuation of 9.S.12)

(a) If the 11:5 and 13:3 ratios are the only two reasonable theories to consider, would you have compelling evidence that the theory you selected in Exercise 9.S.12 is the correct theory? Explain.

(b) If there are also other possible theoretical ratios that weren't considered, would you have compelling evidence that the theory you selected in Exercise 9.S.12 is the correct theory? Explain.

9.S.14 When fleeing a predator, the minnow *Fundulus notti* will often head for shore and jump onto the bank. In a study of spatial orientation in this fish, individuals were caught at various locations and later tested in an artificial pool to see which direction they would choose when released: Would they swim in a direction which, at their place of capture, would have led toward shore? The following are the directional choices ($\pm 45°$) of 50 fish tested under cloudy skies[46]:

Toward shore	18
Away from shore	12
Along shore to the right	13
Along shore to the left	7

Use chi-square tests at $\alpha = 0.05$ to test the hypothesis that directional choice under cloudy skies is random,

(a) using the four categories listed in the table.

(b) collapsing to two categories—"toward shore" and "away from or along shore"—and using a directional H_A.

(*Note:* Although the chi-square test is valid in this setting, it should be noted that more powerful tests are available for analysis of orientation data.)[47]

9.S.15 Refer to the cortex-weight data of Exercise 8.4.4.

(a) Use a goodness-of-fit test to test the hypothesis that the environmental manipulation has no effect. As in Exercise 8.4.4, use a directional alternative and let $\alpha = 0.05$. (This exercise shows how, by a shift of viewpoint, the sign test can be reinterpreted as a goodness-of-fit test. Of course, the chi-square goodness-of-fit test described in this chapter can be used only if the number of observations is large enough.)

(b) Is the number of observations large enough for the test in part (a) to be valid?

9.S.16 A biologist wanted to know if the cowpea weevil has a preference for one type of bean over others as a place to lay eggs. She put equal amounts of four types of seeds into a jar and added adult cowpea weevils. After a few days she observed the following data[48]:

Type of bean	Number of eggs
Pinto	167
Cowpea	176
Navy beans	174
Northern beans	194

Do these data provide evidence of a preference for some types of beans over others? That is, are the data consistent with the claim that the eggs are distributed randomly among the four types of bean?

9.S.17 An experiment was conducted in which two types of acorn squash were crossed. According to a genetic model, 1/2 of the resulting plants should have dark stems and dark fruit, 1/4 should have light stems and light fruit, and 1/4 should have light stems and plain fruit. The actual data were 220, 129, and 105 for these three categories.[49] Do these data refute this model? Consider a chi-square test.

(a) What is the value of the chi-square test statistic for investigating whether these data are consistent with the 1/2, 1/4, 1/4 probabilities model?

(b) The P-value for the chi-square test is 0.23. If $\alpha = 0.10$, what is your conclusion regarding H_0?

9.S.18 Each of 36 men was asked to touch the backs of the hands of three women, one of whom was the man's romantic partner, while blindfolded. The two "decoy" women were the same age, height, and weight as the man's partner.[33] Of the 36 men tested, 16 were able to correctly identify their partner. Do the data provide sufficient evidence that the men are able to sense their partners better than guessing would predict? Conduct a goodness-of-fit test of the data, using $\alpha = 0.05$.

9.S.19 In a study of resistance to a certain soybean virus, biologists cross-fertilized two soybean cultivars. They expected to get a 3:1 ratio of resistant to susceptible plants. The observed data were 58 resistant and 26 susceptible plants.[50] Are these data significantly inconsistent with the expected 3:1 ratio? Consider a test, using $\alpha = 0.10$ and a nondirectional alternative.

(a) What are the expected counts for the two categories under the null hypothesis?

(b) The P-value for the chi-square test (with a nondirectional alternative) is 0.21. If $\alpha = 0.10$, what is your conclusion regarding H_0?

(c) Do these results confirm the 3:1 ratio expected by the researchers?

9.S.20 A group of 1,438 sexually active patients were counseled on condom use and the risk of contracting a sexually transmitted disease (STD). After 6 months, 103 of the patients had new STDs.[51] Construct a 95% confidence interval for the probability of contracting an STD within 6 months after being part of a counseling program like the one used in this study.

9.S.21 (Continuation of 9.S.20) Suppose that for (uncounseled) sexually active individuals the probability of acquiring an STD in a 6-month period is 10%.

(a) Using your interval computed in Exercise 9.S.21, is there compelling evidence that the 6-month STD probability is different for those who receive counseling?

(b) Using the data from Exercise 9.S.21, conduct a nondirectional chi-square test to determine if the 6-month STD rate is different for counseled individuals compared to the uncounseled population.

(c) Do your answers to parts (a) and (b) agree? Explain.

Chapter 10

CATEGORICAL DATA: RELATIONSHIPS

OBJECTIVES

In this chapter we extend our study of categorical data to several populations. We will

- discuss independence and association for categorical variables.
- describe a chi-square test to assess the independence between two categorical variables.
- consider the conditions under which a chi-square test is valid.
- describe Fisher's exact test of independence between two categorical variables.
- present McNemar's test to analyze paired categorical data.
- calculate relative risk, the odds ratio, and its associated confidence interval.

10.1 Introduction

In Chapter 9 we considered the analysis of a single sample of categorical data. The basic techniques we employed were estimation of category probabilities and comparison of observed category frequencies with frequencies "expected" according to a null hypothesis. In this chapter we will extend these basic techniques to more complicated situations. To set the stage, here are two examples, the first of which presents an experiment; the second, an observational study.

Example 10.1.1

Migraine Headache Patients who suffered from moderate to severe migraine headache took part in a double-blind clinical trial to assess an experimental surgery. A group of 75 patients were randomly assigned to receive either the real surgery on migraine trigger sites ($n = 49$) or a sham surgery ($n = 26$) in which an incision was made but no further procedure was performed. The surgeons hoped that patients would experience "a substantial reduction* in migraine headaches," which we will label as "success." Table 10.1.1 shows the results of the experiment.[1]

Table 10.1.1 Response to migraine surgery

		Surgery	
		Real	Sham
Substantial reduction in migraine headaches?	Success	41	15
	No success	8	11
	Total	49	26

A natural way to express the results is in terms of percentages, as follows:

Of the real surgeries, $\frac{41}{49}$ or 83.7% were successful.

Of sham surgeries, $\frac{15}{26}$ or 57.7% were successful.

*"Substantial reduction" means at least a 50 percent reduction in migraine headache frequency, intensity, or duration when compared with baseline (presurgery) values.

In this study successful reduction in migraine headache was more common among patients who received the real surgery than among those who received the sham surgery—83.7% versus 57.7%. Table 10.1.2 provides a summary of the data; Figure 10.1.1 is a bar chart showing the percentages of successful surgeries for the two groups. ■

Table 10.1.2 Response to migraine surgery

	Surgery	
	Real	Sham
n	49	26
Success	41	15
Percent	83.7%	57.7%

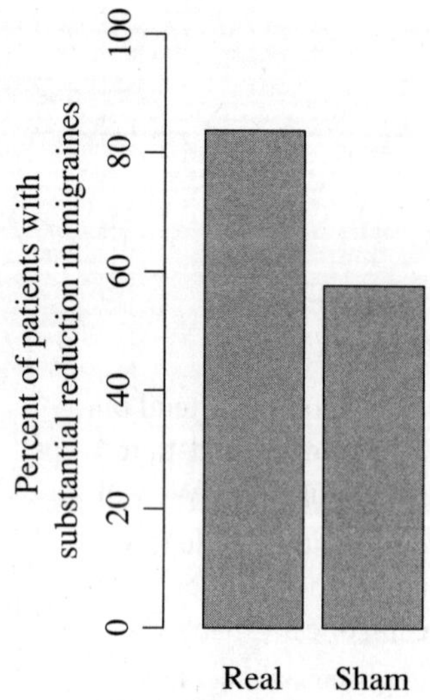

Figure 10.1.1 Bar chart of migraine surgery data

Example 10.1.2 **HIV Testing** A random sample of 120 college students found that 9 of the 61 women in the sample had taken an HIV test, compared to 8 of the 59 men.[2] These data are shown in Table 10.1.3.

Table 10.1.3 HIV testing data

	Female	Male
HIV test	9	8
No HIV test	52	51
Total	61	59

Of the women $\frac{9}{61} = 0.148$ or 14.8% had been tested for HIV. Of the men $\frac{8}{59} = 0.136$ or 13.6% had been tested for HIV.

These two percentages are nearly identical. ■

Tables such as Tables 10.1.1 and 10.1.3 are called **contingency tables.** The focus of interest in a contingency table is the dependence or association between the column variable and the row variable—for instance, between treatment and response in Tables 10.1.1 and 10.1.3. (The word *contingent* means "dependent.") In particular, Tables 10.1.1 and 10.1.3 are called **2 × 2** ("two-by-two") **contingency tables,** because they consist of two rows (excluding the "total" row) and two columns. Each category in the contingency table is called a **cell;** thus, a 2 × 2 contingency table has four cells.

We will consider the analysis and interpretation of 2 × 2 contingency tables before extending the discussion to larger tables. When analyzing a 2 × 2 contingency table it is natural to think of the probability of an event under either of two conditions being compared. We will find it is useful to extend the language of probability to include a new concept: conditional probability.*

CONDITIONAL PROBABILITY

Recall that the probability of an event predicts how often the event will occur. A **conditional probability** predicts how often an event will occur under specified conditions. The notation for a conditional probability is

$$\Pr\{E \mid C\}$$

which is read "probability of E, given C." When a conditional probability is estimated from observed data, the estimate is denoted by a hat ("^"); thus,

$$\hat{\Pr}\{E \mid C\}$$

The following example illustrates these ideas.

Example 10.1.3 **Migraine Headache** Consider the migraine headache data from Example 10.1.1. The conditional probabilities of interest are as follows:

Pr{substantial reduction in migraines | real surgery} = Pr{Success | Real}
= probability that a patient will have a substantial reduction in headache if given the real surgery

Pr{substantial reduction in migraines | sham surgery} = Pr{Success | Sham}
= probability that a patient will have a substantial reduction in headache if given the sham surgery

The estimates of these conditional probabilities from the data of Table 10.1.1 are

$$\hat{\Pr}\{\text{Success} \mid \text{Real}\} = \frac{41}{49} = 0.837$$

and

$$\hat{\Pr}\{\text{Success} \mid \text{Sham}\} = \frac{15}{26} = 0.577$$

■

A RANDOMIZATION TEST

Example 10.1.4 **ECMO** Extracorporeal membrane oxygenation (ECMO) is a potentially life-saving procedure that is used to treat newborn babies who suffer from severe respiratory failure. An experiment was conducted in which 29 babies were treated with ECMO and 10 babies were treated with conventional medical therapy (CMT). The data are shown in Table 10.1.4.[3]

The data in Table 10.1.4 show that 34 of the 39 babies survived while 5 of them died. The death rate was 40% for those given CMT and was 3.4% for those given ECMO. However, the sample sizes here are quite small. Is it possible that the difference in death rates happened simply by chance?

*Conditional probability is also discussed in optional Section 3.3.

Table 10.1.4 ECMO experiment data

		Treatment		
		CMT	ECMO	Total
Outcome	Died	4	1	5
	Lived	6	28	34
	Total	10	29	39

The null hypothesis of interest is that outcome (lived or died) is independent of treatment (CMT or ECMO). If the null hypothesis is true, then we can think of the data in the following way: The two column headings of "CMT" and "ECMO" are arbitrary labels. Five of the babies would have died no matter which treatment group they were in; one of these babies ended up in the ECMO group (and four in the CMT group) by chance.

Prior to this experiment, there was evidence to suggest that ECMO was better than CMT. Thus, a directional alternative hypothesis asserting that the probability of death is greater for CMT than for ECMO is appropriate here.

A question of interest is this: "If the null hypothesis is true, how likely is it to get a table of data like Table 10.1.4 or an even more extreme table, such as Table 10.1.5, which shows a possible result that is even more lopsided in favor of ECMO than the data from the actual experiment?" We can conduct a randomization test to find the probability that either of these two tables would arise by chance, given that the marginal totals—5 deaths and 34 survivors, 10 given CMT and 29 given ECMO—are fixed. To make this more concrete, suppose the null hypothesis is true and another experiment is conducted, with 10 babies given CMT and 29 given ECMO. Further, suppose that 5 of these 39 babies are going to die no matter which group they are in. That is, there are 5 babies who are so seriously ill that neither treatment would be able to save them. What is the probability that at most 1 of them will be assigned to the ECMO group (and either 4 or 5 will be assigned to CMT)?

Table 10.1.5 A more extreme outcome for the ECMO experiment

		Treatment		
		CMT	ECMO	Total
Outcome	Died	5	0	5
	Lived	5	29	34
	Total	10	29	39

We can estimate this probability by (1) taking 39 blank cards and writing Died on 5 of them and Lived on the other 34; (2) shuffling the cards and dealing out 29 into a pile that we call "ECMO" and the other 10 to a pile that we call "CMT"; and (3) counting the number of cards in the "ECMO" pile that say Died. We can then repeat steps (1)–(3) many times and find the proportion of times that the ECMO pile has 1 or 0 deaths; that is, we can find the proportion of tables for which ECMO/Died cell has a 1 or a 0 under the null hypothesis.

With the aid of a computer we can do this very efficiently. In one computer simulation for which steps (1)–(3) were repeated 50,000 times, there were 529 results for which there was 1 ECMO death and 24 results for which there were 0 ECMO deaths. Thus the randomization P-value is $(529 + 24)/50{,}000$ or $553/50{,}000$, which is 0.011. ■

10.2 The Chi-Square Test for the 2 × 2 Contingency Table

We often want to test the hypothesis that the conditional probabilities associated with a 2 × 2 table are equal, which is to say that the probability of an event E does not depend on whether the first condition, C, is present or the second condition, "not C," is present.

$$H_0: \Pr\{E \mid C\} = \Pr\{E \mid \text{not } C\}$$

The following example illustrates this null hypothesis.

Example 10.2.1 **Migraine Headache** For the migraine study of Example 10.1.1, the null hypothesis is

$$H_0: \Pr\{\text{Success} \mid \text{Real}\} = \Pr\{\text{Success} \mid \text{Sham}\}$$

or equivalently

$$H_0: \Pr\{\text{Success} \mid \text{Real}\} = \Pr\{\text{Success} \mid \text{not Real}\}$$ ■

THE CHI-SQUARE STATISTIC

Clearly, a natural way to test the preceding null hypothesis would be to reject H_0 if $\hat{\Pr}\{E \mid C\}$ and $\hat{\Pr}\{E \mid \text{not } C\}$ are different by a sufficient amount. We describe a test procedure that compares $\hat{\Pr}\{E \mid C\}$ and $\hat{\Pr}\{E \mid \text{not } C\}$ indirectly, rather than directly. The procedure is a chi-square test, based on the test statistic χ_s^2 that was introduced in Section 9.4:

$$\chi_s^2 = \sum_{i=1}^{4} \frac{(o_i - e_i)^2}{e_i}$$

In the formula, the sum is taken over all four cells in the contingency table. Each o represents an observed frequency, and each e represents the corresponding expected frequency according to H_0. We now describe how to calculate the e's.

The first step in determining the e's for a contingency table is to calculate the row and column total frequencies (these are called the **marginal frequencies**) and also the grand total of all the cell frequencies. The e's then follow from a simple rationale, as illustrated in Example 10.2.2.

Example 10.2.2 **Migraine Headache** Table 10.2.1 shows the migraine data of Example 10.1.1, together with the marginal frequencies.

Table 10.2.1 Observed frequencies for migraine study

	Surgery		
	Real	Sham	Total
Success	41	15	56
No success	8	11	19
Total	49	26	75

The e's should agree exactly with the null hypothesis. Because H_0 asserts that the probability of success does not depend on the treatment, we can generate an estimate of this probability by pooling the two treatment groups; from Table 10.2.1, the pooled estimate, based on the marginal totals, is $\frac{56}{75}$. That is, if H_0 is true, then the two columns "Real" and "Sham" are equivalent, and we can pool them together. Our best estimate of Pr{successful outcome} is then the pooled estimate $\frac{56}{75}$. We can then apply this estimate to each treatment group to yield the number of successful outcomes expected according to H_0, as follows:

Real surgery group: $\frac{56}{75} \times 49 = 36.59$ successful outcomes expected

Sham surgery group: $\frac{56}{75} \times 26 = 19.41$ successful outcomes expected

Likewise, the pooled estimate of Pr{a surgery will *not* be successful} is $\frac{19}{75}$. Applying this probability to the two treatment groups gives

Real surgery group: $\frac{19}{75} \times 49 = 12.41$ unsuccessful outcomes expected

Sham surgery group: $\frac{19}{75} \times 26 = 6.59$ unsuccessful outcomes expected

The expected frequencies are shown in parentheses in Table 10.2.2. Note that the marginal totals for the *e*'s are the same as for the *o*'s. ■

Table 10.2.2 Observed and expected frequencies for migraine study

	Surgery		
	Real	Sham	Total
Success	41 (36.59)	15 (19.41)	56
No success	8 (12.41)	11 (6.59)	19
Total	49	26	75

Notice that the preceding calculations for the expected frequencies only involve the marginal total frequencies; that is, the row, column, and grand totals. Hence, the procedure for calculating the *e*'s can be condensed into a simple formula. The expected frequency for each cell is calculated from the marginal total frequencies for the same row and column, as follows:

Expected Frequencies in a Contingency Table

$$e = \frac{(\text{Row total}) \times (\text{Column total})}{\text{Grand total}}$$

The formula produces the same calculation as does the rationale given in Example 10.2.2, as the following example shows.

Example 10.2.3 **Migraine Headache** We will apply the preceding formula to the migraine data of Example 10.1.1. The expected frequency of successful outcomes for the real surgery is calculated from the marginal totals as

$$e = \frac{56 \times 49}{75} = 36.59$$

Note that this is the same answer obtained in Example 10.2.1. Proceeding similarly for each cell in the contingency table, we would obtain all the *e*'s shown in Table 10.2.2. ■

Note: Although the formula for χ_s^2 for contingency tables is the same as given for goodness-of-fit tests in Section 9.4, the method of calculating the *e*'s is quite different for contingency tables because the null hypothesis is different.

THE TEST PROCEDURE

Other than the differences noted previously when computing expected counts, the chi-square test for a contingency table is carried out similarly to the chi-square goodness-of-fit test. Large values of χ_s^2 indicate evidence against H_0. Critical values are determined from Table 9 or *P*-values may be obtained using software; the number of degrees of freedom for a 2 × 2 contingency table is

$$\text{df} = 1$$

The chi-square test for a 2 × 2 table has 1 degree of freedom because, in a sense, there only is one free cell in the table. Table 10.2.2 has four cells, but once we have determined that the expected cell frequency for the top-left cell is 36.59, the expected frequency for the top-right cell is constrained to be 19.41, since the top row adds across to a total of 56. Likewise, the bottom-left cell is constrained to be 12.41, since the left column adds down to a total of 49. Once these three cells are determined, the remaining cell, on the bottom right, is constrained as well. Thus, there are four cells in the table, but only one of them is "free"; once we have used the null hypothesis to determine the expected frequency for one of the cells, the other cells are constrained.

For a 2 × 2 contingency table, the alternative hypothesis can be directional or nondirectional. Directional alternatives are handled by the familiar two-step procedure, cutting the nondirectional *P*-value in half if the data deviate from H_0 in the direction specified by H_A (or reporting that the *P*-value is >0.50 if the data deviate from H_0 in the direction opposite to the direction specified by H_A). Note that χ_s^2 itself does not express directionality; to determine the directionality of the data, one must calculate and compare the estimated probabilities.

The following example illustrates the chi-square test.

Example 10.2.4 **Migraine Headache** For the migraine experiment of Example 10.1.1, let us apply a chi-square test. Given that the experiment involves cranial surgery, and that a Type I error would be quite serious, a conservative choice of α is called for; we will use $\alpha = 0.01$. We may state the null hypothesis and a directional alternative informally as follows:

H_0: The real surgery is no better than the sham surgery for reducing migraine headache.

H_A: The real surgery is better than the sham surgery for reducing migraine headache.

Using the notation of conditional probability, the statements are

$$H_0: \Pr\{\text{Success} \mid \text{Real}\} = \Pr\{\text{Success} \mid \text{Sham}\}$$
$$H_A: \Pr\{\text{Success} \mid \text{Real}\} > \Pr\{\text{Success} \mid \text{Sham}\}$$

To check the directionality of the data, we calculate the estimated probabilities of response:

$$\hat{\Pr}\{\text{Success} \mid \text{Real}\} = \frac{41}{49} = 0.837$$
$$\hat{\Pr}\{\text{Success} \mid \text{Sham}\} = \frac{15}{26} = 0.577$$

and we note that

$$\hat{\Pr}\{\text{Success} \mid \text{Real}\} > \hat{\Pr}\{\text{Success} \mid \text{Sham}\}$$

Thus, the data do deviate from H_0 in the direction specified by H_A. We proceed to calculate the chi-square statistic from Table 10.2.2 as

$$\chi_s^2 = \frac{(41 - 36.59)^2}{36.59} + \frac{(15 - 19.41)^2}{19.41} + \frac{(8 - 12.41)^2}{12.41} + \frac{(11 - 6.59)^2}{6.59}$$
$$= 6.06$$

From Table 9 with df $= 1$, we find that $\chi^2_{1,0.02} = 5.41$ and $\chi^2_{1,0.01} = 6.63$, and so we have $0.005 < P\text{-value} < 0.01$. (Computer software gives a nondirectional P-value of 0.0138. Since our test was directional, we multiply this value by 1/2 to obtain the directional P-value of 0.0069.) Thus, we reject H_0 and find that the data provide sufficient evidence to conclude that the real surgery is better than the sham surgery for reducing migraine headache.

Note that, even though $\hat{\Pr}\{\text{Reduction} \mid \text{Real}\}$ and $\hat{\Pr}\{\text{Reduction} \mid \text{Sham}\}$ do not enter into the calculation of χ_s^2, the calculation of $\hat{\Pr}\{\text{Reduction} \mid \text{Real}\}$ and $\hat{\Pr}\{\text{Reduction} \mid \text{Sham}\}$ is an important part of the test procedure; the information provided by the quantities $\hat{\Pr}\{\text{Success} \mid \text{Real}\}$ and $\hat{\Pr}\{\text{Success} \mid \text{Sham}\}$ is essential for meaningful interpretation of the results.* ■

Computational Notes The following tips are helpful in analyzing a 2 × 2 contingency table:

1. The contingency table format is convenient for computations. For presenting the data in a report, however, it is usually better to use a more readable form of display such as Table 10.1.2; some additional examples are shown in the exercises.
2. For calculating χ_s^2, the observed frequencies (o's) must be *absolute*, rather than relative, frequencies; also, *the table must contain all four cells* so that the sum of the o's is equal to the total number of observations.

ILLUSTRATION OF THE NULL HYPOTHESIS

The chi-square statistic measures discrepancy between the data and the null hypothesis in an indirect way; the sample conditional probabilities are involved indirectly in the calculation of the expected frequencies. If sample conditional probabilities are equal, then the value of χ_s^2 is zero. Here is an example.

Example 10.2.5 **Fictitious Migraine Study** Table 10.2.3 shows fictitious data for a migraine study similar to that described in Example 10.1.1.

*It is natural to wonder why we do not use a more direct comparison of $\hat{\Pr}\{E \mid C\}$ and $\hat{\Pr}\{E \mid \text{not } C\}$. In fact, there is a test procedure based on a t-type statistic, calculated by dividing ($\hat{\Pr}\{E \mid C\} - \hat{\Pr}\{E \mid \text{not } C\}$) by its standard error. This t-type procedure is equivalent to the chi-square test. We have chosen to present the chi-square test instead, for two reasons: (1) It can be extended to contingency tables larger than 2 × 2; (2) in certain applications the chi-square statistic is more natural than the t-type statistic; some of these applications appear in Section 10.3.

For the data of Table 10.2.3, the estimated probabilities of successful surgery are *equal*:

$$\hat{\Pr}\{\text{Success}\,|\,\text{Real}\} = \frac{30}{150} = 0.20$$
$$\hat{\Pr}\{\text{Success}\,|\,\text{Sham}\} = \frac{20}{100} = 0.20$$

You can easily verify that, for Table 10.2.3, the expected frequencies are equal to observed frequencies, so the value of χ_s^2 is zero. Also notice that the columns of the table are proportional to each other:

$$\frac{30}{120} = \frac{20}{80}$$

Table 10.2.3 Fictitious data for migraine study

	Surgery		
	Real	Sham	Total
Success	30	20	50
No success	120	80	200
Total	150	100	250

As the preceding example suggests, an "eyeball" analysis of a contingency table is based on checking for proportionality of the columns. If the columns are nearly proportional, then the data agree fairly well with H_0; if they are highly nonproportional, then the data disagree with H_0. The following example shows a case in which the data agree quite well with the expected frequencies under H_0.

Example 10.2.6 **HIV Testing** The data from Example 10.1.2 show similar percentages of men and women who had been tested for HIV. The natural null hypothesis is that $\Pr\{\text{HIV test}\,|\,\text{Female}\} = \Pr\{\text{HIV test}\,|\,\text{Male}\}$ and that the sample proportions differ only due to chance error in the sampling process. The expected frequencies are shown in parentheses in Table 10.2.4. The chi-square test statistic is $\chi_s^2 = 0.035$. From Table 9 with df $= 1$, we find that $\chi^2_{1,0.20} = 1.64$. Thus, the P-value is greater than 0.20 (using a computer yields P-value $= 0.85$), and we do not reject the null hypothesis. Our conclusion is that the data provide no significant evidence that there is a difference in the rates at which men and women (at the college where the study was conducted) have been tested for HIV.

Table 10.2.4 Observed and expected frequencies for HIV study

	Female	Male	Total
HIV test	9 (8.64)	8 (8.36)	17
No HIV test	52 (52.36)	51 (50.64)	103
Total	61	59	120

Note that the actual value of χ_s^2 depends on the sample sizes as well as the degree of nonproportionality; as discussed in Section 9.4, the value of χ_s^2 varies directly with the number of observations if the percentage composition of the data is kept fixed and the number of observations is varied. This reflects the fact that a given percentage deviation from H_0 is less likely to occur by chance with a larger number of observations.

Exercises 10.2.1–10.2.15

10.2.1 The accompanying partially complete contingency table shows the responses to two treatments:

		Treatment	
		1	2
Response	Success	70	
	Failure		
	Total	100	200

(a) Invent a fictitious data set that agrees with the table and for which $\chi_s^2 = 0$.

(b) Calculate the estimated probabilities of success $\hat{\Pr}\{\text{Success} \mid \text{Treatment 1}\}$ and $\hat{\Pr}\{\text{Success} \mid \text{Treatment 2}\}$ for your data set. Are they equal?

10.2.2 Proceed as in Exercise 10.2.1 for the following contingency table:

		Treatment	
		1	2
Response	Success	30	
	Failure		
	Total	300	100

10.2.3 Proceed as in Exercise 10.2.1 for the following contingency table:

		Treatment	
		1	2
Response	Success	5	20
	Failure	10	

10.2.4 Most salamanders of the species *P. cinereus* are red striped, but some individuals are all red. The all-red form is thought to be a mimic of the salamander *N. viridescens,* which is toxic to birds. In order to test whether the mimic form actually survives more successfully, 163 striped and 41 red individuals of *P. cinereus* were exposed to predation by a natural bird population. After 2 hours, 65 of the striped and 23 of the red individuals were still alive.[4] Use a chi-square test to assess the evidence that the mimic form survives more successfully. Use a directional alternative and let $\alpha = 0.05$.

(a) State the null hypothesis in words.

(b) State the null hypothesis in symbols.

(c) Compute the sample survival proportions for each group and display the values in a table similar to Table 10.1.2.

(d) Find the value of the test statistic and the *P*-value.

(e) State the conclusion of the test in the context of this setting.

10.2.5 Can attack of a plant by one organism induce resistance to subsequent attack by a different organism? In a study of this question, individually potted cotton (*Gossypium*) plants were randomly allocated to two groups. Each plant in one group received an infestation of spider mites (*Tetranychus*); the other group were kept as controls. After 2 weeks the mites were removed, and all plants were inoculated with *Verticillium,* a fungus that causes wilt disease. The accompanying table shows the numbers of plants that developed symptoms of wilt disease.[5] Do the data provide sufficient evidence to conclude that infestation with mites induces resistance to wilt disease? Use a chi-square test against a directional alternative following the five steps (a–e) in Exercise 10.2.4. Let $\alpha = 0.01$.

		Treatment	
		Mites	No mites
Response	Wilt disease	11	17
	No wilt disease	15	4
	Total	26	21

10.2.6 It has been suspected that prolonged use of a cellular telephone increases the chance of developing brain cancer due to the microwave-frequency signal that is transmitted by the cell phone. According to this theory, if a cell phone is repeatedly held near one side of the head, then brain tumors are more likely to develop on that side of the head. To investigate this, a group of patients were studied who had used cell phones for at least 6 months prior to developing brain tumors. The patients were asked whether they routinely held the cell phone to a certain ear and, if so, which ear. The 88 responses (from those who preferred one side over the other) are shown in the following table.[6] Do the data provide sufficient evidence to conclude that use of cellular telephones leads to an increase in brain tumors on that side of the head? Use a chi-square test against a directional alternative following the five steps (a–e) in Exercise 10.2.4. Let $\alpha = 0.05$.

		Phone holding side	
		Left	Right
Brain tumor side	Left	14	28
	Right	19	27
	Total	33	55

10.2.7 Phenytoin is a standard anticonvulsant drug that unfortunately has many toxic side effects. A study was undertaken to compare phenytoin with valproate, another drug in the treatment of epilepsy. Patients were randomly allocated to receive either phenytoin or valproate for 12 months. Of 20 patients receiving valproate, 6 were free of seizures for the 12 months, while 6 of 17 patients receiving phenytoin were seizure free.[7]

(a) Consider a chi-square test to compare the seizure-free response rates for the two drugs using a nondirectional alternative.

(i) State the null and alternative hypotheses in symbols.

(ii) What is the value of the test statistic?

(iii) The *P*-value for the test is 0.73. If $\alpha = 0.10$, what is your conclusion regarding the hypotheses in (ii)?

(b) Do your conclusions in part (a)(iii) provide evidence that valproate and phenytoin are equally effective in preventing seizures? Discuss.

10.2.8 Estrus synchronization products are used to bring cows into heat at a predictable time so that they can be reliably impregnated by artificial insemination. In a study of two estrus synchronization products, 42 mature cows (ages 4–8 years) were randomly allocated to receive either product A or product B, and then all cows were bred by artificial insemination. The table shows how many of the inseminations resulted in pregnancy.[8] Consider a chi-square test to compare the effectiveness of the two products in producing pregnancy.

	Treatment	
	Product A	Product B
Total number of cows	21	21
Number of cows pregnant	8	15

Here is computer output for a chi-square test using a nondirectional alternative.

```
X-squared = 4.71, df = 1, p-value = 0.03
```

(a) State the null and alternative hypotheses in context.

(b) Compute the expected frequencies.

(c) If $\alpha = 0.05$, what is your conclusion regarding H_0?

(d) Suppose that prior to conducting the study, the researchers were expecting product A to be superior to product B. Using $\alpha = 0.05$ and a directional alternative, what is your conclusion?

10.2.9 Experimental studies of cancer often use strains of animals that have a naturally high incidence of tumors. In one such experiment, tumor-prone mice were kept in a sterile environment; one group of mice was maintained entirely germ free, while another group was exposed to the intestinal bacterium *Escherichia coli*. The accompanying table shows the incidence of liver tumors.[9]

		Mice with liver tumors	
Treatment	Total number of mice	Number	Percent
Germ free	49	19	39%
E. coli	13	8	62%

(a) How strong is the evidence that tumor incidence is higher in mice exposed to *E. coli*? Use a chi-square test against a directional alternative following the five steps (a–e) in Exercise 10.2.4. Let $\alpha = 0.05$.

(b) How would the result of part (a) change if the percentages (39% and 62%) of mice with tumors were the same, but the sample sizes were (i) doubled (98 and 26)? (ii) tripled (147 and 39)? [*Hint*: Part (b) requires almost no calculation.]

10.2.10 In a randomized clinical trial to determine the most effective timing of administration of chemotherapeutic drugs to lung cancer patients, 16 patients were given four drugs simultaneously and 11 patients were given the same drugs sequentially. Objective response to the treatment (defined as shrinkage of the tumor by at least 50%) was observed in 11 of the patients treated simultaneously and in 3 of the patients treated sequentially.[10] Do the data provide evidence as to which timing is superior? Use a chi-square test against a nondirectional alternative following the five steps (a–e) in Exercise 10.2.4. Let $\alpha = 0.05$.

10.2.11 Physicians conducted an experiment to investigate the effectiveness of external hip protectors in preventing hip fractures in elderly people. They randomly assigned some people to get hip protectors and others to be the control group. They recorded the number of hip fractures in each group.[11] Do the data in the following table provide sufficient evidence to conclude that hip protectors reduce the likelihood of fracture? Use a chi-square test against a directional alternative following the five steps (a–e) in Exercise 10.2.4. Let $\alpha = 0.01$.

		Treatment	
		Hip protector	Control
Response	Hip fracture	13	67
	No hip fracture	640	1081
	Total	653	1148

10.2.12 A sample of 276 healthy adult volunteers were asked about the variety of social networks that they were in (e.g., relationships with parents, close neighbors, workmates, etc.). They were then given nasal drops containing a rhinovirus and were quarantined for 5 days. Of the 123 subjects who were in five or fewer types of social relationships 57 (46.3%) developed colds. Of 153 who were in at least six types of social relationships 52 (34.0%) developed colds.[12] Thus, the data suggest that having more types of social relationships helps one develop resistance to the common cold. Determine whether this difference is statistically significant. That is, use a chi-square test to test the null hypothesis that the probability of getting a cold does not depend on the number of social relationships a person is in following the five steps (a–e) in Exercise 10.2.4. Use a nondirectional alternative and let $\alpha = 0.05$.

10.2.13 The drug ancrod was tested in a double-blind clinical trial in which subjects who had strokes were randomly assigned to get either ancrod or a placebo. One response variable in the study was whether or not a subject experienced intracranial hemorrhaging.[13] The data are provided in the following table. Use a chi-square test to determine whether the difference in hemorrhaging rates is statistically significant following the five steps (a–e) in Exercise 10.2.4. Use a nondirectional alternative and let $\alpha = 0.05$.

		Treatment	
		Ancrod	Placebo
Hemorrhage?	Yes	13	5
	No	235	247
	Total	248	252

10.2.14 Do women respond to men's solicitations more readily during the fertile phase of their menstrual cycles? In a study of this question, each of two hundred 18- to 25-year-old women who were walking alone in a city were approached by an attractive 20-year-old man who solicited the woman's telephone number. Previous research suggested that during the fertile phase of her menstrual cycle a woman would be more receptive to this kind of request than at other times. Of 60 women who were in the fertile phase of their cycles, 13 gave out their phone numbers and 47 refused. The corresponding numbers for the 140 women not in the fertile phase of their cycles were 11 and 129.[14] The data are summarized in the following table. Consider a chi-square test to determine whether the difference in success rates provides significant evidence in favor of an appropriate directional alternative.

Here is computer output for a chi-square test *that used a nondirectional alternative.*

```
X-squared = 7.585, df = 1, p-value = 0.0059
```

(a) State the null and appropriate directional alternative hypotheses in context.

(b) Compute the sample proportions and the expected frequencies.

(c) If $\alpha = 0.02$, what is your conclusion regarding H_0?

		Phase	
		Fertile	Not
Success?	Yes	13	11
	No	47	129
	Total	60	140

10.2.15 To compare the impacts of roads in two adjacent ecoregions in Brazil researchers sampled stretches of roadway over a period of 7 years in each region and recorded the species of animals killed by vehicles. In the Atlantic Forest habitat there were 178 roadkills observed, of which 51 were six-banded armadillos (*Euphractus sexcinctus*). In the adjacent Cerrado habitat, there were 318 roadkills observed, of which 66 were six-banded armadillos.[15] Use a chi-square test to assess the evidence that six-banded armadillos make up different proportions of species killed for the two regions.

(a) State the null hypothesis in words.

(b) State the null hypothesis in symbols.

(c) Compute the sample proportion of six-banded armadillos killed among roadkills in each ecosystem and display the results in a table similar to Table 10.1.2.

(d) Create a table displaying the observed and expected counts.

(e) The value of $\chi_s^2 = 3.948$ and the nondirectional P-value is 0.047. If $\alpha = 0.05$, what are your conclusions?

(f) Based on these findings, would it be reasonable to surmise that six-banded armadillos are in more danger of being hit at one location than the other, or is further information needed? If so, what information would be helpful?

10.3 Independence and Association in the 2 × 2 Contingency Table

The 2 × 2 contingency table is deceptively simple. In this section we explore further the relationships that it can express.

TWO CONTEXTS FOR CONTINGENCY TABLES

A 2 × 2 contingency table can arise in two contexts, namely:

1. Two independent samples with a dichotomous observed variable
2. One sample with two dichotomous observed variables

The first context is illustrated by the migraine data of Example 10.1.1, which can be viewed as two independent samples—the real surgery group and the sham surgery group—of sizes $n_1 = 49$ and $n_2 = 26$. The observed variable is success (or failure) of the surgery. Any study involving a dichotomous observed variable and completely randomized allocation to two treatments can be viewed this way. The second context is illustrated by the HIV data of Example 10.1.2, which can be viewed as a single sample of $n = 120$ students, observed with respect to two dichotomous variables—sex (male or female) and HIV test status (whether or not the student had been tested for HIV).

The two contexts—two samples with one variable or one sample with two variables—are not always sharply differentiated. For instance, the HIV data of Example 10.1.2 could have been collected in two samples—61 women and 59 men—observed with respect to one dichotomous variable (HIV test status).

The arithmetic of the chi-square test is the same in both contexts, but the statement and interpretation of hypotheses and conclusions can be very different.

INDEPENDENCE AND ASSOCIATION

In many contingency tables, the columns of the table play a different role than the rows. For instance, in the migraine data of Example 10.1.1, the columns represent treatments and the rows represent responses. Also, in Example 10.1.2 it seems more natural to define the columnwise conditional probabilities Pr{HIV test | F} and Pr{HIV test | M} rather than the rowwise conditional probabilities Pr{F | HIV test} and Pr{M | HIV test}.

On the other hand, in some cases it is natural to think of the rows and the columns of the contingency table as playing interchangeable roles. In such a case, conditional probabilities may be calculated either rowwise or columnwise, and the null hypothesis for the chi-square test may be expressed either rowwise or columnwise. The following is an example.

Example 10.3.1

Hair Color and Eye Color To study the relationship between hair color and eye color in a German population, an anthropologist observed a sample of 6,800 men, with the results shown in Table 10.3.1.[16]

Table 10.3.1 Hair color and eye color

		Hair color		
		Dark	Light	Total
Eye color	Dark	726	131	857
	Light	3,129	2,814	5,943
	Total	3,855	2,945	6,800

The data of Table 10.3.1 would be naturally viewed as a single sample of size $n = 6{,}800$ with two dichotomous observed variables—hair color and eye color. To describe the data, let us denote dark and light eyes by DE and LE, and dark and light hair by DH and LH. We may calculate estimated columnwise conditional probabilities as follows:

$$\hat{\Pr}\{\text{DE} \mid \text{DH}\} = \frac{726}{3855} \approx 0.19$$

$$\hat{\Pr}\{\text{DE} \mid \text{LH}\} = \frac{131}{2945} \approx 0.04$$

A natural way to analyze the data is to compare these values: 0.19 versus 0.04. On the other hand, it is just as natural to calculate and compare estimated rowwise conditional probabilities:

$$\hat{\Pr}\{\mathrm{DH}|\mathrm{DE}\} = \frac{726}{857} \approx 0.85$$

$$\hat{\Pr}\{\mathrm{DH}|\mathrm{LE}\} = \frac{3129}{5943} \approx 0.53$$

Corresponding to these two views of the contingency table, the null hypothesis for the chi-square test can be stated columnwise as

$$H_0\text{: } \Pr\{\mathrm{DE}|\mathrm{DH}\} = \Pr\{\mathrm{DE}|\mathrm{LH}\}$$

or rowwise as

$$H_0\text{: } \Pr\{\mathrm{DH}|\mathrm{DE}\} = \Pr\{\mathrm{DH}|\mathrm{LE}\}$$

As we shall see, these two hypotheses are equivalent—that is, any population that satisfies one of them also satisfies the other. ■

When a data set is viewed as a single sample with two observed variables, the relationship expressed by H_0 is called **statistical independence** of the row variable and the column variable. Variables that are not independent are called **dependent** or **associated.** Thus, the chi-square test is sometimes called a "test of independence" or a "test for association."

Example 10.3.2 **Hair Color and Eye Color** The null hypothesis of Example 10.3.1 can be stated verbally as

H_0: Eye color is independent of hair color

or

H_0: Hair color is independent of eye color

or, more symmetrically,

H_0: Hair color and eye color are independent ■

The null hypothesis of independence can be stated generically as follows. Two groups, G_1 and G_2, are to be compared with respect to the probability of a characteristic C. The null hypothesis is

$$H_0\text{: } \Pr\{C|G_1\} = \Pr\{C|G_2\}$$

Note that each of the two statements of H_0 in Example 10.3.1 is of this form.

To further clarify the meaning of the null hypothesis of independence, in the following example we examine a data set that agrees *exactly* with H_0.

Example 10.3.3 **Plant Height and Disease Resistance** Consider a (fictitious) species of plant that can be categorized as short (S) or tall (T) and as resistant (R) or nonresistant (NR) to a certain disease. Consider the following null hypothesis:

H_0: Plant height and disease resistance are independent

Each of the following is a valid statement of H_0:

1. H_0: $\Pr\{\mathrm{R}|\mathrm{S}\} = \Pr\{\mathrm{R}|\mathrm{T}\}$
2. H_0: $\Pr\{\mathrm{NR}|\mathrm{S}\} = \Pr\{\mathrm{NR}|\mathrm{T}\}$

3. H_0: $\Pr\{S|R\} = \Pr\{S|NR\}$
4. H_0: $\Pr\{T|R\} = \Pr\{T|NR\}$

The following is not a statement of H_0:

5. H_0: $\Pr\{R|S\} = \Pr\{NR|S\}$

Note the difference between statements 5 and 1. Statement 1 compares two groups (short and tall plants) with respect to disease resistance, whereas statement 5 is a statement about the distribution of disease resistance in only *one* group (short plants); statement 5 merely asserts that half (50%) of short plants are resistant and half are nonresistant.

Suppose, now, that we choose a random sample of 100 plants from the population and we obtain the data in Table 10.3.2.

Table 10.3.2 Plant height and disease resistance

		Height		
		S	T	Total
Resistance	R	12	18	30
	NR	28	42	70
	Total	40	60	100

The data in Table 10.3.2 agree exactly with H_0; this agreement can be checked in four different ways, corresponding to the four different symbolic statements of H_0:

1. $\hat{\Pr}\{R|S\} = \hat{\Pr}\{R|T\}$

$$\frac{12}{40} = 0.30 = \frac{18}{60}$$

2. $\hat{\Pr}\{NR|S\} = \hat{\Pr}\{NR|T\}$

$$\frac{28}{40} = 0.70 = \frac{42}{60}$$

3. $\hat{\Pr}\{S|R\} = \hat{\Pr}\{S|NR\}$

$$\frac{12}{30} = 0.40 = \frac{28}{70}$$

4. $\hat{\Pr}\{T|R\} = \hat{\Pr}\{T|NR\}$

$$\frac{18}{30} = 0.60 = \frac{42}{70}$$

Note that the data in Table 10.3.2 do *not* agree with statement 5:

$$\hat{\Pr}\{R|S\} = \frac{12}{40} = 0.30 \text{ and } \hat{\Pr}\{NR|S\} = \frac{28}{40} = 0.70$$

$$0.30 \neq 0.70$$

■

FACTS ABOUT ROWS AND COLUMNS

The data in Table 10.3.2 display independence whether viewed rowwise or columnwise. This is no accident, as the following fact shows.

Fact 10.3.1 The columns of a 2 × 2 table are proportional if and only if the rows are proportional. Specifically, suppose that a, b, c, and d are any positive numbers, arranged as in Table 10.3.3.

Table 10.3.3 A general 2 × 2 contingency table

			Total
	a	b	$a + b$
	c	d	$c + d$
Total	$a + c$	$b + d$	

Then

$$\frac{a}{c} = \frac{b}{d} \text{ if and only if } \frac{a}{b} = \frac{c}{d}$$

Another way to express this is

$$\frac{a}{a+c} = \frac{b}{b+d} \text{ if and only if } \frac{a}{a+b} = \frac{c}{c+d}$$

You can easily show that Fact 10.3.1 is true; just use simple algebra. Because of Fact 10.3.1, the relationship of independence in a 2 × 2 contingency table is the same whether the table is viewed rowwise or columnwise. Note also that the expected frequencies, and therefore the value of χ_s^2, would remain the same if the rows and columns of the contingency table were interchanged. The following fact shows that the *direction* of dependence is also the same whether viewed rowwise or columnwise.

Fact 10.3.2 Suppose that a, b, c, and d are any positive numbers, arranged as in Table 10.3.3. Then

$$\frac{a}{a+c} > \frac{b}{b+d} \text{ if and only if } \frac{a}{a+b} > \frac{c}{c+d}$$

Also

$$\frac{a}{a+c} < \frac{b}{b+d} \text{ if and only if } \frac{a}{a+b} < \frac{c}{c+d}$$

Note: For more discussion of conditional probability and independence, see optional Section 3.3.

VERBAL DESCRIPTION OF ASSOCIATION

Ideas of logical implication are expressed in everyday English in subtle ways. The following excerpt is from *Alice in Wonderland*, by Lewis Carroll:

> "... you should say what you mean," the March Hare went on.
>
> "I do," Alice hastily replied; "at least—at least I mean what I say—that's the same thing, you know."
>
> "Not the same thing a bit!" said the Hatter. "Why, you might just as well say that 'I see what I eat' is the same thing as 'I eat what I see'!"
>
> ... "You might just as well say," added the Dormouse ..., "That 'I breathe when I sleep' is the same thing as 'I sleep when I breathe'!"
>
> "It *is* the same thing with you," said the Hatter ...

We also use ordinary language to express ideas of probability, conditional probability, and association. For instance, consider the following four statements:

Color-blindness is more common among males than among females.

Maleness is more common among color-blind people than femaleness.

Most color-blind people are male.

Most males are color-blind.

The first three statements are all true; they are actually just different ways of saying the same thing. However, the last statement is false.[17]

In interpreting contingency tables, it is often necessary to describe probabilistic relationships in words. This can be quite a challenge. If you become fluent in such description, then you can always "say what you mean" *and* "mean what you say." The following two examples illustrate some of the issues.

Example 10.3.4

Plant Height and Disease Resistance For the plant height and disease resistance study of Example 10.3.3, we considered the null hypothesis

H_0: Height and resistance are independent.

This hypothesis could also be expressed verbally in various other ways, such as

H_0: Short and tall plants are equally likely to be resistant.

H_0: Resistant and nonresistant plants are equally likely to be tall.

H_0: Resistance is equally common among short and tall plants. ■

Example 10.3.5

Hair Color and Eye Color Let us consider the interpretation of Table 10.3.1. The chi-square statistic is $\chi_s^2 = 314$; from Table 9 we see that the P-value is tiny, so the null hypothesis of independence is overwhelmingly rejected. We might state our conclusion in various ways. For instance, suppose we focus on the incidence of dark eyes. From the data we found that

$$\hat{\Pr}\{\text{DE}\,|\,\text{DH}\} > \hat{\Pr}\{\text{DE}\,|\,\text{LH}\}$$

that is,

$$\frac{726}{3855} = 0.19 > \frac{131}{2945} = 0.04$$

A natural conclusion from this comparison would be

Conclusion 1: There is sufficient evidence to conclude that dark-haired men have a greater tendency to be dark-eyed than do light-haired men.

This statement is carefully phrased, because the statement

"Dark-haired men have a greater tendency to be dark-eyed."

is ambiguous by itself; it could mean

"Dark-haired men have a greater tendency to be dark-eyed than do light-haired men."

or

"Dark-haired men have a greater tendency to be dark-eyed than to be light-eyed."

The first of these statements says that

$$\hat{\Pr}\{\text{DE}\,|\,\text{DH}\} > \hat{\Pr}\{\text{DE}\,|\,\text{LH}\}$$

whereas the second says that

$$\hat{\Pr}\{DE|DH\} > \hat{\Pr}\{LE|DH\}$$

The second statement asserts that more than half of dark-haired men have dark eyes. Note that the data do not support this assertion; of the 3,855 dark-haired men, only 19% have dark eyes.

Conclusion 1 is only one of several possible wordings of the conclusion from the contingency table analysis. For instance, one might focus on dark hair and find

Conclusion 2: There is sufficient evidence to conclude that dark-eyed men have a greater tendency to be dark-haired than do light-eyed men.

A more symmetrical phrasing would be

Conclusion 3: There is sufficient evidence to conclude that dark hair is associated with dark eyes.

However, the phrasing in conclusion 3 is easily misinterpreted; it may suggest something like

"There is sufficient evidence to conclude that most dark-haired men are dark-eyed."

which is not a correct interpretation. ■

We emphasize once again the principle that we stated in Section 10.2: *The calculation and comparison of appropriate conditional probabilities or* $\hat{\Pr}$*'s is an essential part of the chi-square test.* Example 10.3.5 provides ample illustration of this point.

Exercises 10.3.1–10.3.12

10.3.1 Consider a fictitious population of mice. Each animal's coat is either black (B) or gray (G) in color and is either wavy (W) or smooth (S) in texture. Express each of the following relationships in terms of probabilities or conditional probabilities relating to the population of animals.

(a) Smooth coats are more common among black mice than among gray mice.

(b) Smooth coats are more common among black mice than wavy coats are.

(c) Smooth coats are more often black than are wavy coats.

(d) Smooth coats are more often black than gray.

(e) Smooth coats are more common than wavy coats.

10.3.2 Consider a fictitious population of mice in which each animal's coat is either black (B) or gray (G) in color and is either wavy (W) or smooth (S) in texture (as in Exercise 10.3.1). Suppose a random sample of mice is selected from the population and the coat color and texture are observed; consider the accompanying partially complete contingency table for the data.

		Height	
		B	G
Texture	W		50
	S		
	Total	60	150

(a) Invent fictitious data sets that agree with the table and for which

(i) $\hat{\Pr}\{W|B\} > \hat{\Pr}\{W|G\}$;

(ii) $\hat{\Pr}\{W|B\} = \hat{\Pr}\{W|G\}$

In each case, verify your answer by calculating the estimated conditional probabilities.

(b) For each of the two data sets you invented in part (a), calculate $\hat{\Pr}\{B|W\}$ and $\hat{\Pr}\{B|S\}$.

(c) Which of the data sets of part (a) has $\hat{\Pr}\{B|W\} > \hat{\Pr}\{B|S\}$? Can you invent a data set for which

$$\hat{\Pr}\{W|B\} > \hat{\Pr}\{W|G\} \text{ but } \hat{\Pr}\{B|W\} < \hat{\Pr}\{B|S\}$$

If so, do it. If not, explain why not.

10.3.3 Men with prostate cancer were randomly assigned to undergo surgery ($n = 347$) or "watchful waiting" (no surgery, $n = 348$). Over the next several years there were 83 deaths in the first group and 106 deaths in the second group. The results are given in the table.[18]

		Treatment		
		Surgery	WW	Total
Survival	Died	83	106	189
	Alive	264	242	506
	Total	347	348	695

(a) Let D and A represent died and alive, respectively, and let S and WW represent surgery and watchful waiting. Calculate Pr{D | S} and Pr{D | WW}.

(b) The value of the contingency-table chi-square statistic for these data is $\chi_s^2 = 3.75$. Test for a relationship between the treatment and survival. Use a nondirectional alternative and let $\alpha = 0.05$.

10.3.4 In a study of behavioral asymmetries, 2,391 women were asked which hand they preferred to use (e.g., to write) and which foot they preferred to use (for instance, to kick a ball). The results are reported in the table.[19]

Preferred hand	Preferred foot	Number of women
Right	Right	2,012
Right	Left	142
Left	Right	121
Left	Left	116
	Total	2,391

(a) Estimate the conditional probability that a woman is right-footed, given that she is right-handed.

(b) Estimate the conditional probability that a woman is right-footed, given that she is left-handed.

(c) Suppose we want to test the null hypothesis that hand preference and foot preference are independent. Calculate the chi-square statistic for this hypothesis.

10.3.5 Consider a study to investigate a certain suspected disease-causing agent. One thousand people are to be chosen at random from the population; each individual is to be classified as diseased or not diseased and as exposed or not exposed to the agent. The results are to be cast in the following contingency table:

		Exposure	
		Yes	No
Disease	Yes		
	No		

Let EY and EN denote exposure and nonexposure and let DY and DN denote presence and absence of the disease. Express each of the following statements in terms of conditional probabilities. (Note that "a majority" means "more than half.")

(a) The disease is more common among exposed than among nonexposed people.

(b) Exposure is more common among diseased people than among nondiseased people.

(c) Exposure is more common among diseased people than is nonexposure.

(d) A majority of diseased people are exposed.

(e) A majority of exposed people are diseased.

(f) Exposed people are more likely to be diseased than are nonexposed people.

(g) Exposed people are more likely to be diseased than to be nondiseased.

10.3.6 Refer to Exercise 10.3.5. Which of the statements express the assertion that occurrence of the disease is associated with exposure to the agent? (There may be more than one.)

10.3.7 Refer to Exercise 10.3.5. Invent fictitious data sets as specified, and verify your answer by calculating appropriate estimated conditional probabilities. (Your data need not be statistically significant.)

(a) Invent a data set for which

$$\hat{\Pr}\{DY|EY\} > \hat{\Pr}\{DY|EN\} \text{ but}$$
$$\hat{\Pr}\{EY|DY\} < \hat{\Pr}\{EN|DY\}$$

or explain why it is not possible.

(b) Invent a data set that agrees with statement (a) of Exercise 10.3.5 but with neither (d) nor (e); or, explain why it is not possible.

(c) Invent a data set for which

$$\hat{\Pr}\{DY|EY\} > \hat{\Pr}\{DY|EN\} \text{ but}$$
$$\hat{\Pr}\{EY|DY\} < \hat{\Pr}\{EY|DN\}$$

or explain why it is not possible.

10.3.8 An ecologist studied the spatial distribution of tree species in a wooded area. From a total area of 21 acres, he randomly selected 144 quadrats (plots), each 38 feet square, and noted the presence or absence of maples and hickories in each quadrat. The results are shown in the table.[20]

		Maples	
		Present	Absent
Hickories	Present	26	63
	Absent	29	26

The value of the chi-square statistic for this contingency table is $\chi_s^2 = 7.96$. Test the null hypothesis that the two species are distributed independently of each other. Use a nondirectional alternative and let $\alpha = 0.01$. In stating your conclusion, indicate whether the data suggest attraction between the species or repulsion. Support your interpretation with estimated conditional probabilities from the data.

10.3.9 Refer to Exercise 10.3.8. Suppose the data for fictitious tree species, A and B, were as presented in the accompanying table. The value of the chi-square statistic for this contingency table is $\chi_s^2 = 9.07$. As in Exercise

10.3.8, test the null hypothesis of independence and interpret your conclusion in terms of attraction or repulsion between the species.

		Species A	
		Present	Absent
Species B	Present	30	10
	Absent	49	55

10.3.10 A randomized experiment was conducted in which patients with coronary artery disease either had angioplasty or bypass surgery. The accompanying table shows the incidence of angina (chest pain) among the patients 5 years after treatment.[21]

		Treatment		
		Angioplasty	Bypass	Total
Angina?	Yes	111	74	185
	No	402	441	843
	Total	513	515	1,028

Let A represent angioplasty and B represent bypass.

(a) Calculate $\hat{\Pr}\{\text{Yes}|\text{A}\}$ and $\hat{\Pr}\{\text{Yes}|\text{B}\}$.

(b) Calculate $\hat{\Pr}\{\text{A}|\text{Yes}\}$ and $\hat{\Pr}\{\text{A}|\text{No}\}$.

10.3.11 Refer to Exercise 10.3.10. Invent a fictitious data set on coronary treatment and angina for 1,000 patients, for which $\hat{\Pr}\{\text{Yes}|\text{A}\}$ is twice as great as $\hat{\Pr}\{\text{Yes}|\text{B}\}$, but nevertheless the majority of patients who have angina also had bypass surgery (as opposed to angioplasty).

10.3.12 Suppose pairs of fraternal twins are examined and the handedness of each twin is determined; assume that all the twins are brother–sister pairs. Suppose data are collected for 1,000 twin pairs, with the results shown in the following table.[22] State whether each of the following statements is true or false.

(a) Most of the brothers have the same handedness as their sisters.

(b) Most of the sisters have the same handedness as their brothers.

(c) Most of the twin pairs are either both right-handed or both left-handed.

(d) Handedness of twin sister is independent of handedness of twin brother.

(e) Most left-handed sisters have right-handed brothers.

		Sister		
		Left	Right	Total
Brother	Left	15	85	100
	Right	135	765	900
	Total	150	850	1,000

10.4 Fisher's Exact Test (Optional)

In this optional section we consider an alternative to the chi-square test for 2×2 contingency tables. This procedure, known as **Fisher's exact test,** is particularly appropriate when dealing with small samples. In the ECMO data from Example 10.1.4 (shown again in Table 10.4.1) we used a randomization test to evaluate whether or not ECMO improves the survival outcomes for newborn babies suffering from severe respiratory failure. The randomization test used simulation to estimate the *P*-value—the probability of obtaining data that is as favorable or more favorable towards ECMO than the observed data from the experiment under the assumption that ECMO is no better than standard therapy. In the following example, we consider the same data and research question using Fisher's exact test.

Table 10.4.1 ECMO experiment data

		Treatment		
		CMT	ECMO	Total
Outcome	Die	4	1	5
	Live	6	28	34
	Total	10	29	39

Example 10.4.1 **ECMO** The ECMO data from Example 10.1.4 show that 34 out of 39 babies survived but 5 of the babies died. The death rate was 40% for those given CMT (4 out of 10) and was 3.4% for those given ECMO (1 out of 29). This difference casts doubt

on the null hypothesis that outcome (lived or died) is independent of treatment (CMT or ECMO).

In conducting Fisher's exact test we find the probability that the observed table, Table 10.1.5, would arise by chance, given that the marginal totals—5 deaths and 34 survivors, 10 given CMT and 29 given ECMO—are fixed.

To find this probability, we need to determine the following:

1. The number of ways of assigning exactly 4 of the 5 babies who are fated to die to the CMT group,
2. The number of ways of assigning exactly 6 of the 34 babies who are going to survive to the CMT group,

and

3. The number of ways of assigning 10 of the 39 babies to the CMT group.

The product of (1) and (2), divided by (3), gives the probability in question. ■

COMBINATIONS

In Section 3.6 we presented the binomial distribution formula. Part of that formula is the quantity ${}_nC_j$ (which in Section 3.6 we called a binomial coefficient). The quantity ${}_nC_j$ is the number of ways in which j objects can be chosen out of a set of n objects. For instance, the number of ways that a group of 4 babies can be chosen out of 5 babies is ${}_5C_4$. The numerical value of ${}_nC_j$ is given by formula (10.4.1):

$${}_nC_j = \frac{n!}{j!(n-j)!} \tag{10.4.1}$$

where $n!$ ("n factorial") is defined for any positive integer n as

$$n! = n(n-1)(n-2)\ldots(2)(1)$$

and $0! = 1$.

For example, if $j = 1$ then we have ${}_nC_1 = \dfrac{n!}{1!(n-1)!} = n$, which makes sense: There are n ways to choose 1 object from a set of n objects. If $j = n$ then we have ${}_nC_n = \dfrac{n!}{n!0!} = 1$, since there is only one way to choose all n objects from a set of size n.

Example 10.4.2 **ECMO** We can apply formula (10.4.1) as follows.

1. The number of ways of assigning 4 babies to the CMT group from among the 5 who are fated to die is ${}_5C_4 = \dfrac{5!}{4!1!} = 5$.
2. The number of ways of assigning 6 babies to the CMT group from among the 34 who are going to survive is ${}_{34}C_6 = \dfrac{34!}{6!28!} = 1{,}344{,}904$.
3. The number of ways of assigning 10 babies to the CMT group from among the 39 total babies is ${}_{39}C_{10} = \dfrac{39!}{10!29!} = 635{,}745{,}396$.*

*It is evident from this example that a computer or a graphing calculator is a very handy tool when conducting Fisher's exact test. This is a statistical procedure that is almost never carried out without the use of technology.

Thus, the probability of getting the same data as those in Table 10.4.1, given that the marginal totals are fixed, is $\frac{{}_5C_4 \times {}_{34}C_6}{{}_{39}C_{10}} = \frac{5 \times 1344904}{635745396} = 0.01058.$ ■

When conducting Fisher's exact test of a null hypothesis against a directional alternative, we need to find the probabilities of all tables of data (having the same margins as the observed table) that provide evidence at least as strongly against H_0, in the direction predicted by H_A, as the observed table.

Example 10.4.3 **ECMO** Prior to this experiment described in Example 10.4.1, there was evidence that suggested that ECMO is better than CMT. Hence, a directional alternative hypothesis is appropriate:

$$H_A\text{: Pr}\{\text{death}\,|\,\text{ECMO}\} < \text{Pr}\{\text{death}\,|\,\text{CMT}\}$$

The data in the observed table, Table 10.4.1, support H_A. There is one other possible table, shown as Table 10.4.2, that has the same margins as Table 10.4.1 but which is even more extreme in supporting H_A. Given that 5 of 39 babies died and that 10 babies were assigned to CMT, the most extreme possible result supporting the alternative hypothesis (that ECMO is better than CMT) is the table in which none of the ECMO babies die and all 5 deaths occur in the CMT group.

Table 10.4.2 A more extreme table that could have resulted from the ECMO experiment

		Treatment		
		CMT	ECMO	Total
Outcome	Die	5	0	5
	Live	5	29	34
	Total	10	29	39

The probability of Table 10.4.2 occurring, if H_0 is true, is $\frac{{}_5C_5 \times {}_{34}C_5}{{}_{39}C_{10}} = \frac{1 \times 278256}{635745396} = 0.00044$. The P-value is the probability of obtaining data at least as extreme as those observed, if H_0 is true. In this case, the P-value is the probability of obtaining either the data in Table 10.4.1 or in Table 10.4.2, if H_0 is true. Thus, P-value $= 0.01058 + 0.00044 = 0.01102$. This P-value is quite small, so the experiment provided strong evidence that H_0 is false and that ECMO really is better than CMT. ■

COMPARISON TO THE CHI-SQUARE TEST

The chi-square test presented in Section 10.2 is often used for analyzing 2 × 2 contingency tables. One advantage of the chi-square test is that it can be extended to 2 × 3 tables and other tables of larger dimension, as will be shown in Section 10.6. The P-value for the chi-square test is based on the chi-square distribution, as the name implies. It can be shown that as the sample size becomes large, this distribution provides a good approximation to the theoretical sampling distribution of the chi-square test statistic χ^2_s. If the sample size is small, however, then the approximation can be poor and the P-value from the chi-square test can be misleading.

Fisher's exact test is called an "exact" test because the P-value is determined exactly, using calculations such as those shown in Example 10.4.2, rather than being based on an asymptotic approximation. Example 10.4.4 shows how the exact test and the chi-square test compare for the ECMO data.

Example 10.4.4

ECMO Conducting a chi-square test on the ECMO experiment data in Table 10.4.1 gives a test statistic of

$$\chi_s^2 = \frac{(4 - 1.28)^2}{1.28} + \frac{(1 - 3.72)^2}{3.72} + \frac{(6 - 8.72)^2}{8.72} + \frac{(28 - 25.28)^2}{25.28}$$
$$= 8.89$$

The P-value (using a directional alternative) is 0.0014. This is quite a bit smaller than the P-value found with the exact test of 0.01102.* ■

NONDIRECTIONAL ALTERNATIVES AND THE EXACT TEST

Typically, the difference between a directional and a nondirectional test is that the P-value for the nondirectional test is twice the P-value for the directional test (assuming that the data deviate from H_0 in the direction specified by H_A). For Fisher's exact test this is not true. The P-value when H_A is nondirectional is not found by simply doubling the P-value from the directional test. Rather, a generally accepted procedure is to find the probabilities of all tables that are as likely or less likely than the observed table. These probabilities are added together to get the P-value for the nondirectional test.† Example 10.4.5 illustrates this idea.

Example 10.4.5

Flu Shots A random sample of college students found that 13 of them had gotten a flu shot at the beginning of the winter and 28 had not. Of the 13 who had a flu shot, 3 got the flu during the winter. Of the 28 who did not get a flu shot, 15 got the flu.[23] These data are shown in Table 10.4.3. Consider the null hypothesis that the probability of getting the flu is the same whether or not one gets a flu shot. The probability of the data in Table 10.4.3, given that the margins are fixed, is $\frac{{}_{18}C_3 \times {}_{23}C_{10}}{{}_{41}C_{13}} = 0.05298.$

Table 10.4.3 Flu shot data

		No shot	Flu shot	Total
Flu?	Yes	15	3	18
	No	13	10	23
	Total	28	13	41

A natural directional alternative would be that those who get flu shots have a lower chance of getting the flu. Figure 10.4.1 shows the obtained data (from Table 10.4.3) along with tables of possible outcomes that more strongly support H_A. The probability of each table is given in Figure 10.4.1, as well.

Table	Probability
15 3 13 10	0.05298
16 2 12 11	0.01174
17 1 11 12	0.00138
18 0 10 13	0.00006

Figure 10.4.1

*We expect the chi-square test to not be trustworthy, given that the first two expected counts are each smaller than 5.

†There is not universal agreement on this process. The P-value can be taken to be the sum of the probabilities of all "extreme" tables, but there are several ways to define "extreme." One alternative to the method presented here is to order tables according to the values of χ_s^2 and to count a table as extreme if it has a value of χ_s^2 that is at least as large as the χ_s^2 found from the observed table. Another approach is to order the tables according to $|\hat{p}_1 - \hat{p}_2|$. These methods will sometimes lead to a different P-value than the P-value being presented here.

Table		Probability
5	13	
23	0	0.00000
6	12	
22	1	0.00002
7	11	
21	2	0.00046
8	10	
20	3	0.00440
9	9	
19	4	0.02443
10	8	
18	5	0.08356

Figure 10.4.2

The P-value for the directional test is the sum of the probabilities of these tables: P-value $= 0.05298 + 0.01174 + 0.00138 + 0.00006 = 0.06616$.

A nondirectional alternative states that the probability of getting the flu depends on whether or not one gets a flu shot but does not state whether a flu shot is associated with an increase or decrease the probability. (Some people might get the flu *because* of the shot, so it is plausible that the overall flu rate is higher among people who get the shot than among those who don't—although public health officials certainly hope otherwise!)

Figure 10.4.2 shows tables of possible outcomes for which the flu rate is higher among those who got the shot than among those who didn't. The probability of each table is given, as well. The first five tables all have probabilities less than 0.05298, which is the probability of the observed data in Table 10.4.3, but the probability of the sixth table is greater than 0.05298. Thus, the contribution to the P-value from this set of tables is the sum of the first five probabilities: $0.00000 + 0.00002 + 0.00046 + 0.00440 + 0.02443 = 0.02931$. Adding this to the P-value for the directional test of 0.06616 gives the P-value for the nondirectional test: P-value $= 0.06616 + 0.02931 = 0.09547$.

As this example shows, the calculation of a P-value for Fisher's exact test is quite cumbersome, particularly when the alternative is nondirectional. It is highly recommended that statistics software be used to carry out the test. ■

Exercises 10.4.1–10.4.9

10.4.1 Consider conducting Fisher's exact test with the following fictitious table of data. Let the null hypothesis be that treatment and response are independent, and let the alternative be the directional hypothesis that treatment B is better than treatment A. List the tables of possible outcomes that more strongly support H_A.

		Treatment		
		A	B	Total
Outcome	Die	4	2	6
	Live	10	14	24
	Total	14	16	30

10.4.2 Repeat Exercise 10.4.1 with the following table of data.

		Treatment		
		A	B	Total
Outcome	Die	5	3	8
	Live	12	13	25
	Total	17	16	33

10.4.3 In a randomized, double-blind clinical trial, 156 subjects were given an antidepressant medication to help them stop smoking; a second group of 153 subjects were given a placebo. Insomnia was more common in the antidepressant group than in the placebo group; Fisher's exact test of the insomnia data gave a P-value of 0.008.[24] Interpret this P-value in the context of the clinical trial.

10.4.4 A randomized trial of clinically obese women examined the efficacy and safety of a partial meal replacement diet using nutritionally fortified meal-replacement shakes. Fourteen clinically obese women received the nutritional shakes and healthy diet counseling during their pregnancy. Another nine clinically obese women only received healthy diet counseling (control). Among the meal-replacement group, two women had preterm deliveries (i.e., before 37 weeks' gestation), while none in the control group did.[25]

(a) Does the use of the meal-replacement shakes cause excessive risk of preterm birth? Fisher's exact test with a directional alternative gives a P-value of 0.3597. Interpret the conclusions of this test.

(b) Given the results above, do the results provide compelling evidence that the meal-replacements are safe with respect to causing preterm birth?

(c) Provide an argument for why a directional test could be preferred over a nondirectional test given the context of this research

(d) Provide an argument for why a nondirectional test could be preferred over a directional test given the context of this research

10.4.5 (Computer exercise) A random sample of 99 students in a Conservatory of Music found that 9 of the 48 women sampled had "perfect pitch" (the ability to identify, without error, the pitch of a musical note), but only 1 of the 51 men sampled had perfect pitch.[26] Conduct Fisher's exact test of the null hypothesis that having perfect pitch is independent of sex. Use a directional alternative and let $\alpha = 0.05$. Do you reject H_0? Why or why not?

10.4.6 Consider the data from Exercise 10.4.5. Conduct a chi-square test and compare the results of the chi-square test to the results of Fisher's exact test.

10.4.7 (Computer exercise) The growth factor pleiotrophin is associated with cancer progression in humans. In an attempt to monitor the growth of tumors, doctors measured serum pleiotrophin levels in patients with pancreatic cancer and in a control group of patients. They found that only 2 of 28 control patients had serum levels more than two standard deviations above the control group mean, whereas 20 of 41 cancer patients had serum levels this high.[27] Use Fisher's exact test to determine whether a discrepancy this large (2 of 28 versus 20 of 41) is likely to happen by chance. Use a directional alternative and let $\alpha = 0.05$.

10.4.8 (Computer exercise) A group of 225 men with benign prostatic hyperplasia were randomly assigned to take saw palmetto extract or a placebo in a double-blind trial. One year into the experiment 45 of the 112 men in the saw palmetto group (40%) thought they were taking saw palmetto, compared to 52 of the 113 men in the placebo group (46%).[28] Is this difference consistent with chance variation? Conduct Fisher's exact test using a nondirectional alternative.

10.4.9 (Computer exercise) An experiment involving subjects with schizophrenia compared "personal therapy" to "family therapy." Only 2 out of 23 subjects assigned to the personal therapy group suffered psychotic relapses in the first year of the study, compared to 8 of the 24 subjects assigned to the family therapy group.[29] Is this sufficient evidence to conclude, at the 0.05 level of significance, that the two types of therapies are not equally effective? Conduct Fisher's exact test using a nondirectional alternative.

10.5 The $r \times k$ Contingency Table

The ideas of Sections 10.2 and 10.3 extend readily to contingency tables that are larger than 2×2. We now consider a contingency table with r rows and k columns, which is termed an **$r \times k$ contingency table.** Here is an example.

Example 10.5.1

Plover Nesting Wildlife ecologists monitored the breeding habitats of mountain plovers for 3 years and made note of where the plovers nested. They found 66 nests on agricultural fields (AF), 67 nests in shortgrass prairie dog habitat (PD), and 20 nests on other grassland (G). The nesting choices varied across the years for these 153 sampled plover broods; Table 10.5.1 shows the data.[30]

Table 10.5.1 Plover nest locations across 3 years

	Year			
Location	2004	2005	2006	Total
Agricultural field (AF)	21	19	26	66
Prairie dog habitat (PD)	17	38	12	67
Grassland (G)	5	6	9	20
Total	43	63	47	153

To compare the distributions in the three locations, we can calculate the columnwise percentages, as displayed in Table 10.5.2. (For instance, in the 2004 sample $\frac{21}{43}$, or 48.8%, of the nests were on agricultural fields.) Inspection of Table 10.5.2 shows some clear differences among the three percentage distributions (columns), with prairie dog habitat being much more common in 2005 than in the other years. ■

Table 10.5.2 Percentage distributions of plover nests by year

Location	Year 2004	2005	2006
Agricultural field (AF)	48.8	30.2	55.3
Prairie dog habitat (PD)	39.5	60.3	25.5
Grassland (G)	11.6	9.5	19.1
Total	99.9*	100.0	99.9*

*The sums of the 2004 and 2006 percentages differ from 100% due to rounding.

Figure 10.5.1 is a bar chart of the data, which gives a visual impression of the distributions.

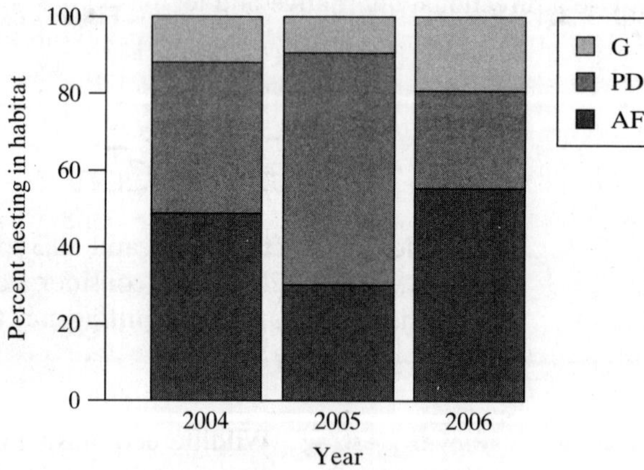

Figure 10.5.1 Stacked relative frequency (percentage) chart of plover nest locations by year

THE CHI-SQUARE TEST FOR THE $r \times k$ TABLE

The goal of statistical analysis of an $r \times k$ contingency table is to investigate the relationship between the row variable and the column variable. Such an investigation can begin with an inspection of the columnwise or rowwise percentages, as in Table 10.5.2. One route to further analysis is to ask whether the discrepancies in percentages are too large to be explained as sampling error. This question can be answered by a chi-square test. The chi-square statistic is calculated from the familiar formula

$$\sum_{\text{all cells}} \frac{(o_i - e_i)^2}{e_i}$$

where the sum is over all $I = r \times k$ cells of the contingency table, and the expected frequencies (e's) are calculated as

$$e = \frac{(\text{Row total}) \times (\text{Column total})}{\text{Grand total}}$$

This method of calculating the e's can be justified by a simple extension of the rationale given in Section 10.2. Critical values for the chi-square test are obtained from Table 9 with

$$\text{df} = (r - 1)(k - 1)$$

The following example illustrates the chi-square test.

Example 10.5.2

Plover Nesting Let us apply the chi-square test to the plover nesting data of Example 10.5.1. The null hypothesis is

H_0: The population distributions of nest locations are the same in the 3 years.

This hypothesis can be stated symbolically in conditional probability notation as follows:

$$H_0: \begin{Bmatrix} \Pr\{AF|2004\} = \Pr\{AF|2005\} = \Pr\{AF|2006\} \\ \Pr\{PD|2004\} = \Pr\{PD|2005\} = \Pr\{PD|2006\} \\ \Pr\{G|2004\} = \Pr\{G|2005\} = \Pr\{G|2006\} \end{Bmatrix}$$

Note that the percentages in Table 10.5.2 are the estimated conditional probabilities; that is,

$$\hat{\Pr}\{AF|2004\} = 0.488$$

$$\hat{\Pr}\{AF|2005\} = 0.302$$

and so on. We test H_0 against the nondirectional alternative hypothesis

H_A: The population distributions of nest locations are not the same in all 3 years.

Table 10.5.3 shows the observed and expected frequencies.

Table 10.5.3 Observed and expected frequencies of plover nests

	Year			
Location	2004	2005	2006	Total
Agricultural field (AF)	21 (18.55)	19 (21.18)	26 (20.27)	66
Prairie dog habitat (PD)	17 (18.83)	38 (27.59)	12 (20.58)	67
Grassland (G)	5 (5.62)	6 (8.24)	9 (6.14)	20
Total	43	63	47	153

From Table 10.5.3, we can calculate the test statistic as

$$\chi_s^2 = \frac{(21 - 18.55)^2}{18.55} + \frac{(19 - 21.18)^2}{21.18} + \cdots + \frac{(9 - 6.14)^2}{6.14}$$

$$= 14.09$$

For these data, $r = 3$ and $k = 3$, so

$$df = (3 - 1)(3 - 1) = 4$$

From Table 9 with df $= 4$, we find that $\chi^2_{4,0.01} = 13.28$ and $\chi^2_{4,0.001} = 18.47$, and so we have $0.001 < P\text{-value} < 0.01$ (computer software gives a P-value $= 0.0070$). Thus, the chi-square test shows that there is significant evidence that the nesting location preferences differed across the 3 years. ■

Note that H_0 in Example 10.5.2 is a compound null hypothesis in the sense defined in Section 9.4—that is, H_0 contains more than one independent assertion. This will always be true for contingency tables larger than 2×2, and consequently for such tables the alternative hypothesis for the chi-square test will always be nondirectional and the conclusion, if H_0 is rejected, will be nondirectional. Thus, the chi-square test will often not represent a complete analysis of an $r \times k$ contingency table.

TWO CONTEXTS FOR $r \times k$ CONTINGENCY TABLES

We noted in Section 10.3 that a 2×2 contingency table can arise in two different contexts. Similarly, an $r \times k$ contingency table can arise in the following two contexts:

1. k independent samples; a categorical observed variable with r categories
2. One sample; two categorical observed variables—one with k categories and one with r categories

As with the 2×2 table, the calculation of the chi-square statistic is the same for both contexts, but the statement of hypotheses and conclusions can differ. The following example illustrates the second context.

Example 10.5.3 **Hair Color and Eye Color** Table 10.5.4 shows the relationship between hair color and eye color for 6,800 German men.[31] (This is the same study as in Example 10.3.2.)

Let us use a chi-square test to test the hypothesis

H_0: Hair color and eye color are independent.

Table 10.5.4 Hair color and eye color

		Hair color			
		Brown	Black	Fair	Red
Eye Color	Brown	438	288	115	16
	Gray or Green	1,387	746	946	53
	Blue	807	189	1,768	47

For the data of Table 10.5.4, one can calculate $\chi_s^2 = 1{,}074$. The degrees of freedom for the test are df $= (3-1)(4-1) = 6$. From Table 9 we find $\chi^2_{6,0.0001} = 27.86$. Thus, H_0 is overwhelmingly rejected, and we conclude that there is extremely strong evidence that hair color and eye color are associated. ■

Exercises 10.5.1–10.5.11

10.5.1 Patients with painful knee osteoarthritis were randomly assigned in a clinical trial to one of five treatments: glucosamine, chondroitin, both, placebo, or Celebrex, the standard therapy. One outcome recorded was whether or not each patient experienced substantial improvement in pain or in ability to function. The data are given in the following table.[32]

		Successful outcome	
Treatment	Sample size	Number	Percent
Glucosamine	317	192	60.6
Chondroitin	318	202	63.5
Both	317	208	65.6
Placebo	313	178	56.9
Celebrex	318	214	67.3

(a) Use a chi-square test to compare the success rates at $\alpha = 0.05$. (The value of the chi-square statistic is $\chi_s^2 = 9.29$.)

(b) Verify the value of χ_s^2 given in part (a).

10.5.2 For a study of free-living populations of the fruitfly *Drosophila subobscura,* researchers placed baited traps in two woodland sites and one open-ground area. The numbers of male and female flies trapped in a single day are given in the table.[33]

	Woodland site I	Woodland site II	Open ground
Males	89	34	74
Females	31	20	136
Total	120	54	210

Here is computer output for a chi-square test to compare the sex ratios at the three sites.

```
X-squared = 49.7, df = 2, p-value = 1.58e-11
```

(a) State the null and alternative hypotheses in context.

(b) Compute the expected frequencies.

(c) If $\alpha = 0.05$, what is your conclusion regarding H_0? [*Note:* 1.58e-11 means $1.58*10^{-11}$].

(d) Construct a table that displays the data in a more readable format, such as the one in Exercise 10.5.1.

10.5.3 In a classic study of peptic ulcer, blood types were determined for 1,655 ulcer patients. The accompanying table shows the data for these patients and for an independently chosen group of 10,000 healthy controls from the same city.[34]

Blood type	Ulcer patients	Controls
O	911	4,578
A	579	4,219
B	124	890
AB	41	313
Total	1,655	10,000

(a) The value of the chi-square statistic for this contingency table is $\chi_s^2 = 49.0$. Carry out the chi-square test at $\alpha = 0.01$.

(b) Construct a table showing the percentage distributions of blood type for patients and for controls.

(c) Verify the value of χ_s^2 given in part (a).

10.5.4 The two claws of the lobster *(Homarus americanus)* are identical in the juvenile stages. By adulthood, however, the two claws normally have differentiated into a stout claw called a "crusher" and a slender claw called a "cutter." In a study of the differentiation process, 26 juvenile lobsters were reared in smooth plastic trays and 18 were reared in trays containing oyster chips (which they could use to exercise their claws). Another 23 lobsters were reared in trays containing only one oyster chip. The claw configurations of all the lobsters as adults are summarized in the table.[35]

	Claw configuration		
Treatment	Right crusher, Left cutter	Right cutter, Left crusher	Right cutter, Left cutter
Oyster chips	8	9	1
Smooth plastic	2	4	20
One oyster chip	7	9	7

(a) The value of the contingency-table chi-square statistic for these data is $\chi_s^2 = 24.35$. Carry out the chi-square test at $\alpha = 0.01$.

(b) Verify the value of χ_s^2 given in part (a).

(c) Construct a table showing the percentage distribution of claw configurations for each of the three treatments.

(d) Interpret the table from part (c): In what way is claw configuration related to treatment? (For example, if you wanted a lobster with two cutter claws, which treatment would you choose and why?)

10.5.5 A randomized, double-blind, placebo-controlled experiment was conducted in which patients with Alzheimer's disease were given either extract of Ginkgo biloba (EGb) or a placebo for one year. The change in each patient's Alzheimer's Disease Assessment Scale—Cognitive subscale (ADAS-Cog) score was measured. The results are given in the table.[36] (*Note*: If the ADAS-Cog went down, then the patient improved.)

	Change in ADAS-Cog score				
	−4 or Better	−3 to −2	−1 to +1	+2 to +3	+4 or Worse
EGb	22	18	12	7	16
Placebo	10	11	19	11	24

(a) Use a chi-square test to compare the prevalence rates at $\alpha = 0.05$. (The value of the chi-square statistic is $\chi_s^2 = 10.26$.)

(b) Verify the value of χ_s^2 given in part (a).

10.5.6 (Computer exercise) Marine biologists have noticed that the color of the outermost growth band on a clam tends to be related to the time of the year in which the clam dies. A biologist conducted a small investigation of whether this is true for the species *Protothaca staminea.* She collected a sample of 78 clam shells from this species and cross-classified them according to (1) month when the clam died and (2) color of the outermost growth band. The data are shown in the following table.[37]

	Color		
	Clear	Dark	Unreadable
February	9	26	9
March	6	25	3
Total	15	51	12

(a) Create a stacked bar chart similar to Figure 10.5.1 to compare the color distribution across the 2 months.

(b) Create a summary table similar to Table 10.5.2 to compare the percentage distributions of colors across the 2 months.

(c) Use a chi-square test statistic for investigating whether color distribution is independent of month. Let $\alpha = 0.05$, What is your conclusion regarding H_0?

10.5.7 A group of patients with a binge-eating disorder were randomly assigned to take either the experimental drug fluvoxamine or a placebo in a 9-week long double-blind clinical trial. At the end of the trial the condition of each patient was classified into one of four categories: no response, moderate response, marked response, or remission. The following table shows a cross classification of the data.[38] Is there statistically significant evidence, at the 0.10 level, to conclude that there is an association between treatment group (fluvoxamine versus placebo) and condition?

	No response	Moderate response	Marked response	Remission	Total
Fluvoxamine	15	7	3	15	40
Placebo	22	7	3	11	43
Total	37	14	6	26	

10.5.8 Patients with coronary artery disease were randomly assigned to either receive angioplasty plus medical therapy ($n = 1149$) or medical therapy alone ($n = 1138$) in a clinical trial. Over the next several years 85 angioplasty and 95 medical therapy patients died, with cause of death categorized as cardiac, other, or unknown. The following table shows a cross classification of the data.[39] Is there statistically significant evidence, at the 0.10 level, to conclude that there is an association between treatment group (angioplasty versus medical therapy) and outcome?

(a) State the null and alternative hypotheses in context.

(b) How many degrees of freedom are there for a chi-square test?

(c) The P-value for the chi-square test is 0.87. If $\alpha = 0.10$, what is your conclusion regarding H_0?

	Cardiac death	Other death	Unknown cause of death	Alive	Total
Angioplasty	23	45	17	1,064	1,149
Medical therapy	25	51	19	1,043	1,138
Total	48	96	36	2,107	2,287

10.5.9 Patients with pleural infections (fluid buildup in the chest) were randomly assigned to placebo, the treatment "tPA" (tissue plasminogen activator), the treatment "DNase" (deoxyribonuclease), or a combination of tPA and DNase in a double-blind clinical trial. Some of the patients needed to be given penicillin. The following table shows a cross-classification of the data.[40] Is there evidence of an association between treatment group and use of penicillin?

	Placebo	tPA	DNase	tPA+DNase	Total
Penicillin	4	2	3	3	12
No penicillin	51	50	48	49	198
Total	55	52	51	52	210

Here is computer output for a chi-square test to compare the four groups.

```
X-squared = 0.59, df = 3, p-value = 0.90
```

(a) State the null and alternative hypotheses in context.

(b) Compute the expected frequency of the Penicillin/Placebo cell under the null hypothesis.

(c) If $\alpha = 0.10$, what is your conclusion regarding H_0?

10.5.10 Is acupuncture effective on back pain? Patients with chronic low back pain were randomly assigned to one of four groups. Some patients received standardized acupuncture, while others received acupuncture that was individualized. A third group was given simulated acupuncture in which a toothpick was used but the skin was not penetrated. A fourth group received medical care that did not include acupuncture. The researchers recorded whether or not each patient improved by at least 3 points on the Roland-Morris Disability Questionnaire scale. The following table shows the data.[41] Is there evidence of an association between treatment group and improvement?

Here is computer output for a chi-square test to compare the four groups.

```
X-squared = 8.93, df = 3, p-value = 0.0302
```

(a) State the null and alternative hypotheses in context.

(b) Compute the observed "Yes" percentages for each of the four groups.

(c) If $\alpha = 0.05$, what is your conclusion regarding H_0?

		Individualized	Standardized	Simulated	Usual care	Total
Improved?	Yes	92	96	90	72	350
	No	49	51	62	71	233
	Total	141	147	152	143	

10.5.11 Consider the data from Exercise 10.5.10. Suppose the Usual Care group is excluded and comparison is made only between the first three treatments: the two types of real acupuncture and the simulated acupuncture. Here is computer output for a chi-square test to compare the three groups.

```
X-squared = 1.57, df = 2, p-value = 0.4551
```

(a) If $\alpha = 0.05$, what is your conclusion regarding H_0?

(b) Write a sentence that combines the conclusion from 10.5.10 and that from 10.5.11(a).

10.6 Applicability of Methods

In this section we discuss guidelines for deciding when to use a chi-square test.

CONDITIONS FOR VALIDITY

A chi-square test is valid under the following conditions:

1. *Design conditions* For the contingency-table chi-square test, it must be appropriate to view the data in one of the following ways:
   (a) As two or more independent random samples, observed with respect to a categorical variable; or
   (b) As one random sample, observed with respect to two categorical variables.

   For either type of chi-square test, the observations within a sample must be independent of each other.
2. *Sample size conditions* The sample size must be large enough. The critical values given in Table 9 are only approximately correct for determining the P-value associated with χ_s^2. As a rule of thumb, the approximation is considered adequate if each expected frequency (e) is at least equal to 5.* (If the expected frequencies are small and the data form a 2×2 contingency table, then Fisher's exact test might be appropriate—see optional Section 10.4.)
3. *Form of H_0* A generic form of the null hypothesis for the contingency-table chi-square test may be stated as follows:

   H_0: The row variable and the column variable are independent.
4. *Scope of inference* As with other statistical tests, if the data arise from an experiment with random assignment of treatments, as in Example 10.1.1, then we can draw a causal inference; if the experimental units were drawn at random from a population, then we can extend the causal inference to that population. However, if the data arise from an observational study, as in Example 10.1.2, then a small P-value only allows us to infer that the observed association is not duc to chancc, but wc cannot rule out other explanations.

VERIFICATION OF DESIGN CONDITIONS

To verify the design conditions, we need to identify a population from which the data may be viewed as a random sample. If the data consist of several samples [situation 1(a)], then the samples are required to be independent of each other. Failure to observe this restriction may result in a loss of power. If the design includes any pairing or matching of experimental units, then the samples would not be independent. A method of analysis for dependent samples is described in Section 10.8.

---

*For an $r \times k$ table with more than 2 rows and columns, the approximation is adequate if the average expected frequency is at least 5, and no expected frequency is less than 1.

As always, bias in the sampling procedure must be ruled out. Moreover, chi-square methods are not appropriate when complex random sampling schemes such as cluster sampling or stratified random sampling are used. Finally, there must be no dependency or hierarchical structure in the design. Failure to observe this restriction can result in a vastly inflated chance of Type I error (which is usually much more serious than a loss of power). The following examples show the relevance of checking for dependency in the observations.

Example 10.6.1 **Food Choice by Insect Larvae** In a behavioral study of the clover root curculio *Sitona hispidulus*, 20 larvae were released into each of six petri dishes. Each dish contained nodulated and nonnodulated alfalfa roots, arranged in a symmetric pattern. (This experiment was more fully described in Example 1.1.5.) After 24 hours the location of each larva was noted, with the results shown in Table 10.6.1.[42]

Table 10.6.1 Food choice by *Sitona* larvae

	Number of larvae		
Dish	Nodulated roots	Nonnodulated roots	Other (died, lost, etc.)
1	5	3	12
2	9	1	10
3	6	3	11
4	7	1	12
5	5	1	14
6	14	3	3
Total	46	12	62

Suppose the following analysis is proposed. A total of 58 larvae made a choice; the observed frequencies of choosing nodulated and nonnodulated roots were 46 and 12, and the corresponding expected frequencies (assuming random choice) would be 29 and 29; these data yield $\chi_s^2 = 19.93$, from which (using a directional alternative) we find from Table 9 that P-value < 0.00005. The validity of this proposed analysis is highly doubtful because it depends on the assumption that all the observations in a given dish are independent of each other; this assumption would certainly be false if (as is biologically plausible) the larvae tend to follow each other in their search for food.

How, then, should the data be analyzed? One approach is to make the reasonable assumption that the observations in one dish are independent of those in another dish. Under this assumption one could use a paired analysis on the six dishes ($n_d = 6$); a paired t test yields P-value ≈ 0.005 and a sign test yields P-value ≈ 0.02. Note that the questionable assumption of independence within dishes led to a P-value that was much too small. ■

Example 10.6.2 **Pollination of Flowers** A study was conducted to determine the adaptive significance of flower color in the scarlet gilia *(Ipomopsis aggregata)*. Six red-flowered plants and six white-flowered plants were chosen for observation in field conditions; hummingbirds were permitted to visit the flowers, but the other major pollinator, a moth, was excluded by covering the plants at night. Table 10.6.2 shows, for each plant, the total number of flowers at the end of the season and the number that had set fruit.[43]

The question of interest is whether the percentage of fruit set is different for red-flowered than for white-flowered plants. Suppose this question is approached by regarding the individual flower as the observational unit; then the data could be cast in the contingency table format of Table 10.6.3.

Table 10.6.2 Fruit set in scarlet gilia flowers

	Red-flowered plants			White-flowered plants		
	Number of flowers	Number setting fruit	Percentage setting fruit	Number of flowers	Number setting fruit	Percentage setting fruit
	140	26	19	125	21	17
	116	11	9	134	17	13
	34	0	0	273	81	30
	79	9	11	146	38	26
	185	28	15	103	17	17
	106	11	10	82	24	29
Sum	660	85		863	198	

Table 10.6.3 Fruit set in scarlet gilia flowers

		Flower color	
		Red	White
Fruit set	Yes	85	198
	No	575	665
Total		660	863
	Percentage setting fruit	13	23

Table 10.6.3 yields $\chi_s^2 = 25.0$, for which Table 9 gives P-value < 0.0001. However, this analysis is not correct, because the observations on flowers on the same plant are not independent of each other; they are dependent because the pollinator (the hummingbird) tends to visit flowers in groups, and perhaps also because the flowers on the same plant are physiologically and genetically related. The chi-square test is invalidated by the hierarchical structure in the data.

A better approach would be to treat the entire plant as the observational unit. For instance, one could take the "Percentage Setting Fruit" column of Table 10.6.2 as the basic observations; applying a t test to the values yields $t_s = 2.88$ (with $0.01 < P$-value < 0.02), and applying a Wilcoxon-Mann-Whitney test yields $U_s = 32$ (with $0.02 < P$-value < 0.05). Thus, the P-value from the inappropriate chi-square analysis is much too small. ■

POWER CONSIDERATIONS

In many studies the chi-square test is valid but is not as powerful as a more appropriate test. Specifically, consider a situation in which the rows or the columns (or both) of the contingency table correspond to a *rankable* categorical variable with more than two categories. The following is an example.

Example 10.6.3

Physiotherapy A randomized clinical trial was conducted to determine whether the addition of Saturday physiotherapy sessions (the "treatment") to the usual Monday–Friday sessions (the "control") would benefit patients undergoing rehabilitation in a hospital. One outcome measure was the destination of a patient upon being discharged, with the categories being home, low-level residential care (LLRC), high-level residential care (HLRC), or acute hospital transfer (AHT). The results are shown in Table 10.6.4.[44]

Table 10.6.4 Discharge destination for physiotherapy patients

		Group	
		Treatment	Control
Discharge destination	Home	107	103
	LLRC	10	15
	HLRC	6	1
	AHT	7	13
	Total	130	132

A contingency-table chi-square test would be valid to compare treatment and control, but the test would lack power because it does not use the information contained in the *ordering* of the discharge destination categories (home is preferred to LLRC, which is preferred to HLRC, which is preferred to AHT). A related weakness of the chi-square test is that, even if H_0 is rejected, the test does not yield a directional conclusion such as "the treatment leads to better discharge destinations than does the control." ■

Methods are available to analyze contingency tables with rankable row and/or column variables; such methods, however, are beyond the scope of this book.

Exercises 10.6.1–10.6.3

10.6.1 Refer to the chemotherapy data of Exercise 10.2.10. Are the sample sizes large enough for the approximate validity of the chi-square test?

10.6.2 In a study of prenatal influences on susceptibility to seizures in mice, pregnant females were randomly allocated to a control group or a "handled" group. Handled mice were given sham injections three times during gestation, while control mice were not touched. The offspring were tested for their susceptibility to seizures induced by a loud noise. The investigators noted that the response varied considerably from litter to litter. The accompanying table summarizes the results.[45]

If these data are analyzed as a 2 × 3 contingency table, the chi-square statistic is $\chi_s^2 = 8.45$ and Table 9 gives $0.01 < P\text{-value} < 0.02$. Is this an appropriate analysis for this experiment? Explain. (*Hint:* Does the design meet the conditions for validity of the chi-square test?)

Treatment	Number of litters	Number of mice	Response to loud noise: No response	Wild running	Seizure
Handled	19	104	23	10	71
Control	20	120	47	13	60

10.6.3 In control of diabetes it is important to know how blood glucose levels change after eating various foods. Ten volunteers participated in a study to compare the effects of two foods—a sugar and a starch. A blood specimen was drawn before each volunteer consumed a measured amount of food; then additional blood specimens were drawn at 11 times during the next 4 hours. Each volunteer repeated the entire test on another occasion with the other food. Of particular concern were blood glucose levels that dropped below the initial level; the accompanying table shows the number of such values.[46]

Food	No. of values less than initial value	Total number of observations
Sugar	26	110
Starch	14	110

Suppose we analyze the given data as a contingency table. The test statistic would be

$$\chi_s^2 = \frac{(26-20)^2}{20} + \frac{(14-20)^2}{20} + \frac{(84-90)^2}{90} + \frac{(96-90)^2}{90} = 4.40$$

At $\alpha = 0.05$ we would reject H_0 and find that there is sufficient evidence to conclude that blood glucose values below the initial value occur more often after ingestion of sugar than after ingestion of starch. This analysis contains two flaws. What are they? (*Hint*: Are the conditions for validity of the test satisfied?)

10.7 Confidence Interval for Difference Between Probabilities

The chi-square test for a 2 × 2 contingency table answers only a limited question: Do the estimated probabilities—call them $\hat{p}_1$ and $\hat{p}_2$—differ enough to conclude that the true probabilities—call them p_1 and p_2—are not equal? A complementary mode of analysis is to construct a confidence interval for the magnitude of the difference, $(p_1 - p_2)$.

When we discussed constructing a confidence interval for a single proportion, p, in Section 9.2, we defined an estimate $\tilde{p}$, based on the idea of "adding 2 successes and 2 failures to the data." Making this adjustment to the data resulted in a confidence interval procedure that has good coverage properties. Likewise, when constructing a confidence interval for the difference in two proportions, we will define new estimates that are based on the idea of adding 1 observation to each cell of the table (so that a *total* of 2 successes and 2 failures are added to the data).

Consider a 2 × 2 contingency table that can be viewed as a comparison of two samples, of sizes n_1 and n_2, with respect to a dichotomous response variable. Let the 2 × 2 table be given as

Sample 1	Sample 2
y_1	y_2
$n_1 - y_1$	$n_2 - y_2$
n_1	n_2

We define

$$\tilde{p}_1 = \frac{y_1 + 1}{n_1 + 2}$$

and

$$\tilde{p}_2 = \frac{y_2 + 1}{n_2 + 2}$$

We will use the difference in the new values, $(\tilde{p}_1 - \tilde{p}_2)$, to construct a confidence interval for $(p_1 - p_2)$. Like all quantities calculated from samples, the quantity $(\tilde{P}_1 - \tilde{P}_2)$ is subject to sampling error. The magnitude of the sampling error can be expressed by the standard error of $(\tilde{P}_1 - \tilde{P}_2)$, which is calculated from the following formula:

$$\mathrm{SE}_{(\tilde{P}_1 - \tilde{P}_2)} = \sqrt{\frac{\tilde{p}_1(1 - \tilde{p}_1)}{n_1 + 2} + \frac{\tilde{p}_2(1 - \tilde{p}_2)}{n_2 + 2}}$$

Note that $\mathrm{SE}_{(\tilde{P}_1 - \tilde{P}_2)}$ is analogous to $\mathrm{SE}_{(\bar{Y}_1 - \bar{Y}_2)}$ as described in Section 6.6.

An approximate confidence interval can be based on $\mathrm{SE}_{(\tilde{P}_1 - \tilde{P}_2)}$; for instance, a 95% confidence interval is

$$(\tilde{p}_1 - \tilde{p}_2) \pm (1.96)\,\mathrm{SE}_{(\tilde{P}_1 - \tilde{P}_2)}$$

Confidence intervals constructed this way have good coverage properties (i.e., approximately 95% of all 95% confidence intervals cover the true difference $p_1 - p_2$) for almost any sample sizes n_1 and n_2.[47] The following example illustrates the construction of the confidence interval.*

*In Section 9.3 we presented a general version of the "add 2 successes and 2 failures" idea, in which the formula for $\tilde{p}$ depends on the confidence level (95%, 90%, etc.). When constructing a confidence interval for a difference in two proportions, the coverage properties of the interval are best when 1 is added to each cell in the 2 × 2 table, no matter what confidence level is being used.[48]

Example 10.7.1 **Migraine Headache** For the migraine headache data of Example 10.1.1, the sample sizes are $n_1 = 49$ and $n_2 = 26$, and the estimated probabilities of substantial reduction in migraines are

$$\tilde{p}_1 = \frac{42}{51} = 0.824$$
$$\tilde{p}_2 = \frac{16}{28} = 0.571$$

The difference between these is

$$\begin{aligned}\tilde{p}_1 - \tilde{p}_2 &= 0.824 - 0.571\\ &= 0.253\\ &\approx 0.25\end{aligned}$$

Thus, we estimate that the real surgery increases the probability of substantial reduction in migraines by 0.25, compared to the sham surgery. To set confidence limits on this estimate, we calculate the standard error as

$$\begin{aligned}\text{SE}_{(\tilde{P}_1-\tilde{P}_2)} &= \sqrt{\frac{0.824(0.176)}{51} + \frac{0.571(0.429)}{28}}\\ &= 0.1077\end{aligned}$$

The 95% confidence interval is

$$0.253 \pm (1.96)(0.1077)$$
$$0.253 \pm 0.211$$
$$0.042 < p_1 - p_2 < 0.464$$

We are 95% confident that the probability of substantial reduction in migraines is between 0.042 and 0.464 higher with the real surgery than with the sham surgery. ■

Relationship to Test The chi-square test for a 2 × 2 contingency table (Section 10.2) is approximately, but not exactly, equivalent to checking whether a confidence interval for $(p_1 - p_2)$ includes zero. [Recall from Section 7.3 that there is an exact equivalence between a t test and a confidence interval for $(\mu_1 - \mu_2)$.]

Exercises 10.7.1–10.7.7

10.7.1 Elderly patients who had suffered hip fractures were randomly assigned to receive either a placebo ($n = 1{,}062$) or zolendronic acid ($n = 1{,}065$) in a double-blind clinical trial. During the trial 139 placebo patients and 92 zolendronic acid patients had new fractures.[49] Let p_1 and p_2 represent the probabilities of fracture on placebo and zolendronic acid, respectively. Construct a 95% confidence interval for $(p_1 - p_2)$.

10.7.2 Refer to the liver tumor data of Exercise 10.2.9.

(a) Construct a 95% confidence interval for (Pr{liver tumor | germ-free} – Pr{liver tumor | *E. coli*}).

(b) Interpret the confidence interval from part (a). That is, explain what the interval tells you about tumor probabilities.

10.7.3 For women who are pregnant with twins, complete bed rest in late pregnancy is commonly prescribed in order to reduce the risk of premature delivery. To test the value of this practice, 212 women with twin pregnancies were randomly allocated to a bed-rest group or a control group. The accompanying table shows the incidence of preterm delivery (less than 37 weeks of gestation).[50]

	Bed rest	Controls
No. of preterm deliveries	32	20
No. of women	105	107

Construct a 95% confidence interval for (Pr{preterm | bed rest} – Pr{preterm | control}). Does the confidence interval suggest that bed rest is beneficial?

10.7.4 Refer to Exercise 10.7.3. The numbers of infants with low birthweight (2,500 gm or less) born to the women are shown in the table.

	Bed rest	Controls
No. of low-birthweight babies	76	92
Total no. of babies	210	214

Let p_1 and p_2 represent the probabilities of a low-birthweight baby in the two conditions. Explain why the above information is not sufficient to construct a confidence interval for $(p_1 - p_2)$.

10.7.5 Refer to the blood type data of Exercise 10.5.3. Let p_1 and p_2 represent the probabilities of type O blood in the patient population and the control population, respectively.

(a) Construct a 95% confidence interval for $(p_1 - p_2)$.

(b) Interpret the confidence interval from part (a). That is, explain what the interval tells you about the difference in probabilities of type O blood.

10.7.6 In an experiment to treat patients with "generalized anxiety disorder," the drug hydroxyzine was given to 71 patients, and 30 of them improved. A group of 70 patients were given a placebo, and 20 of them improved.[51] Let p_1 and p_2 represent the probabilities of improvement using hydroxyzine and the placebo, respectively. Construct a 95% confidence interval for $(p_1 - p_2)$.

10.7.7 Children with juvenile arthritis were randomly assigned to receive the drug tocilizumab or a placebo for 12 weeks. In the tocilizumab group, 64 out of 75 patients showed marked improvement versus 9 out of 37 for the placebo group.[52] An appropriate 95% confidence interval is (0.43, 0.75). Write a sentence that interprets this confidence interval, in context. Use cause–effect language if appropriate or say why no causal statement can be made.

10.8 Paired Data and 2 × 2 Tables (Optional)

In Chapter 8 we considered paired data when the response variable is continuous. In this section we consider the analysis of paired categorical data.

Example 10.8.1 **HIV Transmission to Children** A study was conducted to determine a woman's risk of transmitting HIV to her unborn child. A sample of 114 HIV-infected women who gave birth to two children found that HIV infection occurred in 19 of the 114 older siblings and in 20 of the 114 younger siblings.[53] These data are shown in Table 10.8.1.

Table 10.8.1 HIV infection data

		Older sibling	Younger sibling
HIV?	Yes	19	20
	No	95	94
	Total	114	114

At first glance, it might appear that a regular chi-square test could be used to test the null hypothesis that the probability of HIV infection is the same for older siblings as for younger siblings. However, as we stated in Section 10.6, for the chi-square test to be valid the two samples—of 114 older siblings and of 114 younger siblings—must be independent of each other. In this case the samples are clearly dependent. Indeed, these are paired data, with a family generating the pair (older sibling, younger sibling).

Table 10.8.2 presents the data in a different format. This format helps focus attention on the relevant part of the data.*

Table 10.8.2 HIV infection data shown by pairs

		Younger sibling HIV?	
		Yes	No
Older sibling	Yes	2	17
HIV?	No	18	77

*Note that Table 10.8.2 cannot be derived from Table 10.8.1.

From Table 10.8.2 we can see that there are 79 pairs in which both siblings have the same HIV status: Two are "yes/yes" pairs, and 77 are "no/no" pairs. These 79 pairs, which are called **concordant pairs,** do not help us determine whether HIV infection is more likely for younger siblings than for older siblings. The remaining 35 pairs—17 "yes/no" pairs and 18 "no/yes" pairs—do provide information on the relative likelihood of HIV infection for older and younger siblings. These pairs are called **discordant pairs;** we will focus on these 35 pairs in our analysis.

If the chance of HIV infection is the same for older siblings as it is for younger siblings, then the two kinds of pairs—"yes/no" and "no/yes"—are equally likely. Thus, the null hypothesis

H_0: the probability of HIV infection is the same for older siblings as it is for younger siblings

is equivalent to

$$H_0\text{: among discordant pairs, } \Pr\{\text{"yes/no"}\} = \Pr\{\text{"no/yes"}\} = \frac{1}{2}$$ ■

MCNEMAR'S TEST

The hypothesis that discordant pairs are equally likely to be "yes/no" or "no/yes" can be tested with the chi-square goodness-of-fit test developed in Section 9.4. This application of the goodness-of-fit test is known as **McNemar's test** and has a particularly simple form.* Let n_{11} denote the number of "yes/yes" pairs, n_{12} the number of "yes/no" pairs, n_{21} the number of "no/yes" pairs, and n_{22} the number of "no/no" pairs, as shown in Table 10.8.3. If H_0 is true, the expected number of "yes/no" pairs is $\frac{n_{12} + n_{21}}{2}$, as is the expected number of "no/yes" pairs. Thus, the test statistic is

$$\chi_s^2 = \frac{\left(n_{12} - \frac{(n_{12} + n_{21})}{2}\right)^2}{\frac{(n_{12} + n_{21})}{2}} + \frac{\left(n_{21} - \frac{(n_{12} + n_{21})}{2}\right)^2}{\frac{(n_{12} + n_{21})}{2}}$$

which simplifies to

$$\chi_s^2 = \frac{(n_{12} - n_{21})^2}{n_{12} + n_{21}}$$

The distribution of χ_s^2 under the null hypothesis is approximately a χ^2 distribution with 1 degree of freedom.

Table 10.8.3 A general table of paired proportion data

	Yes	No
Yes	n_{11}	n_{12}
No	n_{21}	n_{22}

*The null hypothesis tested by McNemar's test can also be tested by using the binomial distribution. The null hypothesis states that among discordant pairs, Pr{"yes/no"} = Pr{"no/yes"} = 1/2. Thus, under the null hypothesis, the number of "yes/no" pairs has a binomial distribution with n = the number of discordant pairs and $p = 0.5$.

Example 10.8.2 **HIV Transmission to Children** For the data given in Example 10.8.1, $n_{12} = 17$ and $n_{21} = 18$. Thus,

$$\chi_s^2 = \frac{(17 - 18)^2}{17 + 18} = 0.0286$$

From Table 9 we see that the P-value is greater than 0.20. (Using a computer gives P-value = 0.87.) The data are very much consistent with the null hypothesis that the probability of HIV infection is the same for older siblings as it is for younger siblings. ■

Exercises 10.8.1–10.8.4

10.8.1 As part of a study of risk factors for stroke, 155 women who had experienced a hemorrhagic stroke (cases) were interviewed. For each case, a control was chosen who had not experienced a stroke; the control was matched to the case by neighborhood of residence, age, and race. Each woman was asked whether she used oral contraceptives. The data for the 155 pairs are displayed in the table. "Yes" and "No" refer to use of oral contraceptives.[54]

		Case	
		No	Yes
Control	No	107	30
	Yes	13	5

To test for association between oral contraceptive use and stroke, consider only the 43 discordant pairs (pairs who answered differently) and test the hypothesis that a discordant pair is equally likely to be "yes/no" or "no/yes." Use McNemar's test to test the hypothesis that having a stroke is independent of use of oral contraceptives against a nondirectional alternative at $\alpha = 0.05$.

10.8.2 Example 10.8.1 referred to a sample of HIV-infected women who gave birth to two children. One of the outcomes that was studied was whether the gestational age of the child was less than 38 weeks; this information was recorded for 106 of the families. The data for this variable are shown in the table below. Consider using McNemar's test with a nondirectional alternative.

		Younger sibling < 38 weeks?	
		Yes	No
Older sibling	Yes	26	5
< 38 weeks?	No	21	54

(a) State the null hypothesis in context.

(b) What is the value of the test statistic?

(c) The P-value for the test is 0.0017. If $\alpha = 0.10$, what is your conclusion regarding the hypotheses in (ii)?

10.8.3 A study of 85 patients with Hodgkin's disease found that 41 had had their tonsils removed. Each patient was matched with a sibling of the same sex. Only 33 of the siblings had undergone tonsillectomy. The data are shown in the following table.[55] Use McNemar's test to test the hypothesis that "yes/no" and "no/yes" pairs are equally likely. Previous research had suggested that having a tonsillectomy is associated with an increased risk of Hodgkin's disease; thus, use a directional alternative. Let $\alpha = 0.05$.

		Sibling Tonsillectomy?	
		Yes	No
Hodgkin's patient	Yes	26	15
tonsillectomy	No	7	37

10.8.4 In a study of the mating behavior of *Gryllus campestris*, pairs of female crickets were placed in a plexiglass arena with a single male cricket. There were 54 cases in which the females fought; these resulted in 42 cases in which the winning female copulated with the male, 8 cases in which the losing female copulated with the male, and 4 cases that ended with no copulation. The data are summarized in the following table.[56] Use McNemar's test to test the hypothesis that winners and losers are equally likely to copulate. Use an appropriate directional alternative and let $\alpha = 0.05$.

		Winners	
	Copulate	Yes	No
Losers	Yes	0	8
	No	42	4

10.9 Relative Risk and the Odds Ratio (Optional)

It is quite common to test the null hypothesis that two population proportions, p_1 and p_2, are equal. A chi-square test, based on a 2×2 table, is often used for this purpose. A confidence interval for $(p_1 - p_2)$ provides information about the magnitude of the difference between p_1 and p_2. In this section we consider two other measures of dependence: the relative risk and the odds ratio.

RELATIVE RISK

Sometimes researchers prefer to compare probabilities in terms of their *ratio,* rather than their difference. When the outcome event is deleterious (e.g., having a heart attack, getting cancer) the ratio of probabilities is called the **relative risk,** or the risk ratio. The relative risk is defined as p_1/p_2. This measure is widely used in studies of human health. The following is an example.

Example 10.9.1 **Smoking and Lung Cancer** The health histories of 11,900 middle-age male smokers or former smokers were tracked over many years. During the study 126 of the men developed lung cancer, including 89 men who were smokers and 37 men who were former smokers. Table 10.9.1 shows the data.[57]

Table 10.9.1 Incidence of lung cancer and smoking status

		Smoking history	
		Smoker	Former smoker
Lung cancer?	Yes	89	37
	No	6,063	5,711
	Total	6,152	5,748

The probabilities of primary interest are the columnwise conditional probabilities:

$$p_1 = \Pr\{\text{lung cancer} \mid \text{smoker}\}$$

$$p_2 = \Pr\{\text{lung cancer} \mid \text{former smoker}\}$$

The estimates of these from the data are

$$\hat{p}_1 = \frac{89}{6152} = 0.01447 \approx 0.014$$

$$\hat{p}_2 = \frac{37}{5748} = 0.00644 \approx 0.006$$

The estimated relative risk is

$$\frac{\hat{p}_1}{\hat{p}_2} = \frac{0.01447}{0.00644} = 2.247 \approx 2.2$$

Thus, we estimate that the risk (i.e., the conditional probability) of developing lung cancer is about 2.2 times as great for smokers as for former smokers. (Of course, because this is an observational study, we would not be justified in concluding that smoking *causes* lung cancer.) ■

THE ODDS RATIO

Another way to compare two probabilities is in terms of **odds.** The odds of an event E is defined to be the ratio of the probability that E occurs to the probability that E does not occur:

$$\text{odds of } E = \frac{\Pr\{E\}}{1 - \Pr\{E\}}$$

For instance, if the probability of an event is 1/4, then the odds of the event are $\frac{1/4}{3/4} = 1/3$ or 1:3. As another example, if the probability of an event is 1/2, then the odds of the event are $\frac{1/2}{1/2} = 1$ or 1:1.

The **odds ratio** is simply the ratio of odds under two conditions. Specifically, suppose that p_1 and p_2 are the conditional probabilities of an event under two different conditions. Then the odds ratio, which we will denote by θ ("theta"), is defined as follows:

$$\theta = \frac{\dfrac{p_1}{1 - p_1}}{\dfrac{p_2}{1 - p_2}}$$

If the estimated probabilities $\hat{p}_1$ and $\hat{p}_2$ are calculated from a 2×2 contingency table, the corresponding estimated odds ratio, denoted $\hat{\theta}$, is calculated as

$$\hat{\theta} = \frac{\dfrac{\hat{p}_1}{1 - \hat{p}_1}}{\dfrac{\hat{p}_2}{1 - \hat{p}_2}}$$

We illustrate with an example.

Example 10.9.2 **Smoking and Lung Cancer** From the data of Example 10.9.1, we estimate the odds of developing lung cancer as follows:

$$\widehat{\text{odds}} = \frac{0.01447}{1 - 0.01447} = 0.01468 \text{ among smokers}$$

$$\widehat{\text{odds}} = \frac{0.00644}{1 - 0.00644} = 0.00648 \text{ among former smokers}$$

The estimated odds ratio is

$$\hat{\theta} = \frac{0.01468}{0.00648} = 2.265 \approx 2.3$$

Thus, we estimate that the odds of developing lung cancer are about 2.3 times as great for smokers as for former smokers. ■

ODDS RATIO AND RELATIVE RISK

The odds ratio measures association in an unfamiliar way; the relative risk is a more natural measure. Fortunately, in many applications the two measures are approximately equal. In general the relationship between the odds ratio and the relative risk is given by

$$\text{odds ratio} = \text{relative risk} \times \frac{1 - p_2}{1 - p_1}$$

Notice that if p_1 and p_2 are small, then the relative risk is approximately equal to the odds ratio. We illustrate with the smoking and lung cancer data.

Example 10.9.3 **Smoking and Lung Cancer** For the data in Table 10.9.1 we found that the estimated relative risk of lung cancer is

$$\text{estimated relative risk} = 2.247$$

and the estimated odds ratio is

$$\hat{\theta} = 2.265$$

These are approximately equal because the outcome of interest (developing lung cancer) is rare, so $\hat{p}_1$ and $\hat{p}_2$ are small. ■

ADVANTAGE OF THE ODDS RATIO

Both the relative risk p_1/p_2 and the difference $(p_1 - p_2)$ are easier to interpret than the odds ratio. Why, then, is the odds ratio used at all? One important advantage of the odds ratio is that, in certain kinds of studies, the odds ratio can be estimated even though p_1 and p_2 *cannot* be estimated. To explain this property, we must first discuss the question of estimability of conditional probabilities in contingency tables.

In a 2×2 contingency table, the conditional probabilities can be defined by rows or by columns. Whether or not these probabilities can be estimated from the observed data depends on the study design. The following example illustrates this point.

Example 10.9.4 **Smoking and Lung Cancer** In studying the relationship between smoking and lung cancer, the conditional probabilities of primary interest are

$$p_1 = \Pr\{\text{lung cancer} \mid \text{smoker}\}$$

and

$$p_2 = \Pr\{\text{lung cancer} \mid \text{former smoker}\}$$

These are columnwise probabilities in a table like Table 10.9.1. One could, however, also consider the following rowwise conditional probabilities:

$$p_1^* = \Pr\{\text{smoker} \mid \text{lung cancer}\}$$

and

$$p_2^* = \Pr\{\text{smoker} \mid \text{no lung cancer}\}$$

(Of course, p_1^* and p_2^* are not particularly meaningful biologically.) From the study described in Example 10.9.1, that is, a single sample of size $n = 11{,}900$ observed with respect to smoking status and lung cancer, one can estimate not only p_1 and p_2 but also p_1^* and p_2^*. However, there are other important study designs that do not provide enough information to estimate all these conditional probabilities. For example, suppose that a study is conducted by choosing a group of 500 smokers and a group of 500 former smokers and then observing how many of them develop lung cancer. This kind of study is called a prospective study or **cohort study.** Such a study might produce the fictitious but realistic data of Table 10.9.2.

Table 10.9.2 Fictitious data for cohort study of smoking and lung cancer mortality

		Smoking history	
		Smoker	Former smoker
Lung cancer?	Yes	7	3
	No	493	497
	Total	500	500

The data of Table 10.9.2 can be viewed as two independent samples. From the data we can estimate the conditional probabilities of lung cancer in the two populations (smokers and former smokers):

$$\hat{p}_1 = \frac{7}{500} = 0.014 \quad \hat{p}_2 = \frac{3}{500} = 0.006$$

By contrast, the rowwise probabilities p_1^* and p_2^* cannot be estimated from Table 10.9.2. Because the relative numbers of smokers and former smokers were predetermined by the design of the study ($n_1 = 500$ and $n_2 = 500$), the data contain no information about the prevalence of smoking, and therefore no information about the population values of

Pr{smoker | lung cancer} and Pr{smoker | no lung cancer}

Table 10.9.2 was generated by fixing the column totals and observing the row variable. Consider now the reverse sort of design. Suppose we choose 500 men who died from lung cancer and 500 men who did not die from lung cancer and we then determine the smoking histories of the men. This design is called a **case-control design.** Such a design might generate the fictitious but realistic data of Table 10.9.3.

Table 10.9.3 Fictitious data for case-control study of smoking and lung cancer mortality

		Smoking history		
		Smoker	Former smoker	Total
Lung cancer?	Yes	273	227	500
	No	173	327	500

From Table 10.9.3 we can estimate the rowwise conditional probabilities

$$\hat{p}_1^* = \frac{273}{500} = 0.546 \approx 0.55$$

$$\hat{p}_2^* = \frac{173}{500} = 0.346 \approx 0.35$$

However, from the data in Table 10.9.3 we cannot estimate the columnwise conditional probabilities p_1 and p_2: Because the row totals were predetermined by design, the data contain no information about Pr{lung cancer | smoker} and Pr{lung cancer | former smoker}. ■

The preceding example shows that, depending on the design, a study may not permit estimation of both columnwise probabilities p_1 and p_2 and rowwise probabilities p_1^* and p_2^*. Fortunately, the odds ratio is the same whether it is determined columnwise or rowwise. Specifically,

$$\theta = \frac{\dfrac{p_1}{1 - p_1}}{\dfrac{p_2}{1 - p_2}} = \frac{\dfrac{p_1^*}{1 - p_1^*}}{\dfrac{p_2^*}{1 - p_2^*}}$$

Because of this relationship, the odds ratio θ can be estimated by estimating p_1 and p_2 or by estimating p_1^* and p_2^*. This fact has important applications, especially for case-control studies, as illustrated by the following example.

Example 10.9.5 **Smoking and Lung Cancer** To characterize the relationship between smoking and lung cancer mortality, the columnwise probabilities p_1 and p_2 are more biologically meaningful than the rowwise probabilities p_1^* and p_2^*. If we investigate the relationship using a case-control design, neither p_1 nor p_2 can be estimated from the data. (See Example 10.9.4.) However, the odds ratio *can* be estimated from the data. For instance, from Table 10.9.3 we obtain

$$\hat{\theta} = \frac{\dfrac{\hat{p}_1^*}{1 - \hat{p}_1^*}}{\dfrac{\hat{p}_2^*}{1 - \hat{p}_2^*}}$$

$$= \frac{\dfrac{0.546}{1 - 0.546}}{\dfrac{0.346}{1 - 0.346}} = 2.273 \approx 2.27$$

We can interpret this odds ratio as follows: We know that the outcome event—developing lung cancer—is rare, and so we know that the odds ratio is approximately equal to the relative risk, p_1/p_2. We therefore estimate that the risk of lung cancer is about 2.3 times as great for smokers as for former smokers. ■

There is an easier way to compute the odds ratio for a 2 × 2 contingency table. For a general 2 × 2 table, let n_{11} denote the number of observations in the first row and the first column. Likewise, let n_{12} be the number of observations in the first row and second column, and so on. The general 2 × 2 table then has the form

n_{11}	n_{12}
n_{21}	n_{22}

The estimated odds ratio from the table is given in the box.*

Estimated Odds Ratio ($\hat{\theta}$)

$$\hat{\theta} = \frac{n_{11}n_{22}}{n_{12}n_{21}}$$

*If any of the observed frequencies are zero, 0.5 can be added to all observed cell counts to avoid unreasonable or undefined estimates.

Example 10.9.6

Smoking and Lung Cancer From the data in Table 10.9.1, we can calculate the estimated odds ratio as

$$\hat{\theta} = \frac{89 \times 5{,}711}{37 \times 6{,}063} = 2.265 \approx 2.27$$

■

The case-control design is often the most efficient design for investigating rare outcome events, such as rare diseases. Although Table 10.9.3 was constructed assuming that the two samples, cases and controls, were chosen independently, a more common design is to incorporate matching of cases and controls with respect to potential confounding factors (e.g., age). As we have seen, by taking advantage of the odds ratio, one can estimate the relative risk from a case-control study of a rare event even though one cannot estimate the risks p_1 and p_2 separately.

If the odds ratio (or the relative risk) is equal to 1.0, then the odds (or the risk) are the same for both of the groups being compared. In the smoking and lung cancer data of Table 10.9.1 the calculated odds ratio was *greater* than 1.0, indicating that the odds of lung cancer are greater for smokers than for former smokers. Notice that we could have focused attention on the odds of not getting lung cancer. In this case, the odds ratio would be *less* than 1.0, as shown in Example 10.9.7.

Example 10.9.7

Smoking and Lung Cancer Suppose we rearrange the data in Table 10.9.1 by putting lung cancer in the second row and not getting lung cancer in the first row (see Table 10.9.4):

Table 10.9.4 Rearranged smoking and lung cancer data

		Smoking history	
		Smoker	Former smoker
Lung cancer?	No	6,063	5,711
	Yes	89	37
	Total	6,152	5,748

In this case the odds ratio is the odds of not getting lung cancer for a smoker divided by the odds of not getting lung cancer for a former smoker. We can calculate the estimated odds ratio as

$$\hat{\theta} = \frac{6{,}063 \times 37}{5{,}711 \times 89} = 0.44$$

This is the reciprocal of the odds ratio calculated in Example 10.9.6: $\frac{1}{2.27} = 0.44$. The fact that the odds ratio is less than 1.0 means that the event (being free of lung cancer) is less likely for smokers than for former smokers. ■

CONFIDENCE INTERVAL FOR THE ODDS RATIO

In Chapter 6 we discussed confidence intervals for proportions, which are of the form $\tilde{p} \pm z_{\alpha/2}\text{SE}_{\tilde{p}}$, where $\tilde{p} = \frac{y + 2}{n + 4}$. In particular, a 95% confidence interval for p is given by $\tilde{p} \pm z_{0.025}\text{SE}_{\tilde{p}}$. Such confidence intervals are based on the fact that for large samples the sampling distribution of $\tilde{p}$ is approximately normal (according to the Central Limit Theorem).

In a similar way, we can construct a confidence interval for an odds ratio. One problem is that the sampling distribution of $\hat{\theta}$ is not normal. However, if we take the natural logarithm of $\hat{\theta}$, then we have a distribution that is approximately normal. Hence, we construct a confidence interval for θ by first finding a confidence interval for $\ln(\theta)$ and then transforming the endpoints back to the original scale.

In order to construct a confidence interval for $\ln(\theta)$, we need the standard error of $\ln(\hat{\theta})$. The formula for the standard error of $\ln(\hat{\theta})$ is given in the box.*

Standard Error of $\ln(\hat{\theta})$

$$SE_{\ln(\hat{\theta})} = \sqrt{\frac{1}{n_{11}} + \frac{1}{n_{12}} + \frac{1}{n_{21}} + \frac{1}{n_{22}}}$$

A 95% confidence interval for $\ln(\theta)$ is given by $\ln(\hat{\theta}) \pm (1.96)\, SE_{\ln(\hat{\theta})}$. We then exponentiate the two endpoints of the interval to get a 95% confidence interval for θ. Intervals with other confidence coefficients are constructed analogously; for instance, for a 90% confidence interval one would use $z_{0.05}$ (1.645) instead of $z_{0.025}$ (1.960). The process for finding a confidence interval for θ is summarized in the following box.†

Confidence Interval for θ

To construct a 95% confidence interval for θ,

1. Calculate $\ln(\hat{\theta})$.
2. Construct a confidence interval for $\ln(\theta)$ using the formula $\ln(\hat{\theta}) \pm (1.96)\, SE_{\ln(\hat{\theta})}$.
3. Exponentiate the endpoints to get a confidence interval for θ.

This process is illustrated in the following examples.

Example 10.9.8 **Smoking and Lung Cancer** From the data in Table 10.9.1, the estimated odds ratio is

$$\hat{\theta} = \frac{89 \times 5{,}711}{37 \times 6{,}063} = 2.27$$

Thus, $\ln(\hat{\theta}) = \ln(2.27) = 0.820$.

The standard error is given by $SE_{\ln(\hat{\theta})} = \sqrt{\frac{1}{89} + \frac{1}{37} + \frac{1}{6{,}063} + \frac{1}{5{,}711}} = 0.1965$.

A 95% confidence interval for $\ln(\theta)$ is $0.820 \pm (1.96)(0.1965)$ or 0.820 ± 0.385. This interval is (0.435, 1.205).

To get a 95% confidence interval for θ, we evaluate $e^{0.435} = 1.54$ and $e^{1.205} = 3.24$. Thus, we are 95% confident that the population value of the odds ratio is between 1.54 and 3.24. ■

*If any observed frequencies are zero, the standard error will be undefined. To obtain a reasonable approximation, if there are observed counts of zero, 0.5 may be added to each observed count before computing the standard error.

†A confidence interval for the relative risk can be found in a suitably modified manner for those situations in which the relative risk can be estimated from the data.

Example 10.9.9 **Heart Attacks and Aspirin** During the Physician's Health Study, 11,037 physicians were randomly assigned to take 325 mg of aspirin every other day; 104 of them had heart attacks during the study. Another 11,034 physicians were randomly assigned to take a placebo; 189 of them had heart attacks. These data are shown in Table 10.9.5.[58] The odds ratio for comparing the heart attack rate on aspirin to the heart attack rate on placebo is

$$\hat{\theta} = \frac{189 \times 10{,}933}{104 \times 10{,}845} = 1.832$$

Thus, $\ln(\hat{\theta}) = \ln(1.832) = 0.605$.
The standard error is

$$SE_{\ln(\hat{\theta})} = \sqrt{\frac{1}{189} + \frac{1}{104} + \frac{1}{10{,}845} + \frac{1}{10{,}933}} = 0.123.$$

A 95% confidence interval for $\ln(\theta)$ is $0.605 \pm (1.96)(0.123)$ or 0.605 ± 0.241. This interval is (0.364, 0.846).

To get a 95% confidence interval for θ, we evaluate $e^{0.364} = 1.44$ and $e^{0.846} = 2.33$. Thus, we are 95% confident that the population value of the odds ratio is between 1.44 and 2.33. Because heart attacks are relatively rare in this data set, the relative risk is nearly equal to the odds ratio. Thus, we can say that we are 95% confident that the probability of a heart attack is about 1.44 to 2.33 times greater when taking the placebo than when taking aspirin. ■

Table 10.9.5 Heart attacks on placebo and on aspirin

	Placebo	Aspirin
Heart attack	189	104
No heart attack	10,845	10,933
Total	11,034	11,037

Exercises 10.9.1–10.9.9

10.9.1 For each of the following tables, calculate (i) the relative risk and (ii) the odds ratio.

(a)

25	23
492	614

(b)

12	8
93	84

10.9.2 For each of the following tables, calculate (i) the relative risk and (ii) the odds ratio.

(a)

14	16
322	412

(b)

15	7
338	82

10.9.3 Hip dysplasia is a hip socket abnormality that affects many large breed dogs. A review of medical records of dogs seen at 27 veterinary medical teaching hospitals found that hip dysplasia was more common in Golden Retrievers than in Border Collies; the data are shown in the following table.[59] Calculate the relative risk of hip dysplasia for Golden Retrievers compared to Border Collies.

		Golden retriever	Border collie
Hip dysplasia?	Yes	3,995	221
	No	42,946	5,007
	Total	46,941	5,228

10.9.4 Consider the data from Exercise 10.9.3.

(a) Calculate the sample value of the odds ratio.

(b) Construct a 95% confidence interval for the population value of the odds ratio.

(c) Interpret the confidence interval from part (b) in the context of this setting.

10.9.5 As part of the National Health Interview Survey, occupational injury data were collected on thousands of American workers. The table below summarizes part of these data.[60]

		Self-employed	Employed by others
Injured?	Yes	210	4,391
	No	33,724	421,502
	Total	33,934	425,893

(a) Calculate the sample value of the odds ratio.

(b) According to the odds ratio, are self-employed workers more likely, or less likely, to be injured than persons who work for others?

(c) Construct a 95% confidence interval for the population value of the odds ratio.

(d) Interpret the confidence interval from part (b) in the context of this setting.

10.9.6 Many over-the-counter decongestants and appetite suppressants contain the ingredient phenylpropanolamine. A study was conducted to investigate whether this ingredient is associated with strokes. The study found that 6 of 702 stroke victims had used an appetite suppressant containing phenylpropanolamine, compared to only 1 of 1,376 subjects in a control group. The following table summarizes these data.[61]

		Stroke	No stroke
Appetite suppressant?	Yes	6	1
	No	696	1,375
	Total	702	1,376

(a) Calculate the sample value of the odds ratio.

(b) Construct a 95% confidence interval for the population value of the odds ratio.

(c) Upon hearing of these data, some scientists called the study "inconclusive" because the numbers of users of appetite suppressants containing phenylpropanolamine (7 total: 6 in one group and 1 in the other) are so small. What is your response to these scientists?

10.9.7 Two treatments, heparin and enoxaparin, were compared in a double-blind, randomized clinical trial of patients with coronary artery disease. The subjects can be classified as having a positive or negative response to treatment; the data are given in the following table.[62]

		Heparin	Enoxaparin
Outcome	Negative	309	266
	Positive	1,255	1,341
	Total	1,564	1,607

(a) Calculate the sample value of the odds ratio.

(b) Construct a 95% confidence interval for the population value of the odds ratio.

(c) Interpret the confidence interval from part (b) in the context of this setting.

10.9.8 Consider the data from Exercise 10.7.1. Given that there were 139 hip fractures in 1,062 placebo patients and 92 hip fractures in 1,065 zolendronic acid patients, construct a 95% confidence interval for the population value of the odds ratio.[49]

10.9.9 Researchers in Norway found that of 61,042 children born to women who took folic acid while pregnant there were 64 who developed autism.[63] This compares to 50 out of 24,134 for women who did not take folic acid.

(a) Calculate the sample value of the odds ratio.

(b) Construct a 95% confidence interval for the population value of the odds ratio.

(c) Interpret the confidence interval from part (b) in the context of this setting.

10.10 Summary of Chi-Square Test

The chi-square test is often applied to contingency tables; it is summarized here.

Summary of Chi-Square Test for a Contingency Table

Null hypothesis:

H_0: Row variable and column variable are independent

Calculation of expected frequencies:

$$e_i = \frac{(\text{Row Total}) \times (\text{Column Total})}{\text{Grand Total}}$$

Test statistic:

$$\chi_s^2 = \sum_{\text{all cells}} \frac{(o_i - e_i)^2}{e_i}$$

Null distribution (approximate):

$$\chi^2 \text{ distribution with df} = (r - 1)(k - 1)$$

where r is the number of rows and k is the number of columns in the contingency table. This approximation is adequate if $e_i \geq 5$ for every cell. If r and k are large, the condition that $e_i \geq 5$ is less critical and the χ^2 approximation is adequate if the average expected frequency is at least 5, and no expected frequency is less than 1.

The observations must be independent of one another. If paired data are collected for a 2 × 2 table, then McNemar's test is appropriate (Section 10.8).

Supplementary Exercises 10.S.1–10.S.21

(Note: Exercises preceded by an asterisk refer to optional sections.)

10.S.1 In the Women's Health Initiative Dietary Modification Trial women were randomly assigned to an intervention or a control group. The intervention included counseling sessions designed to reduce fat intake and to increase consumption of fruits and vegetables. Over 6 years data were collected on coronary heart disease (CHD); results are shown in the table.[64] Do the data provide evidence that the intervention makes a difference? The value of the chi-square statistic for this contingency table is $\chi_s^2 = 0.69$.

		Group		
		Intervention	Control	
CHD?	Yes	1,000	1,549	2,549
	No	18,541	27,745	46,286
	Total	19,541	29,294	48,835

(a) State the null and alternative hypotheses in symbols.

(b) The (nondirectional) P-value for the test is 0.407. If $\alpha = 0.10$, what is your conclusion regarding the hypotheses in (a)?

10.S.2 Use the data from Exercise 10.S.1 to construct a 99% confidence interval for $(\Pr\{\text{CHD} \mid \text{intervention}\} - \Pr\{\text{CHD} \mid \text{control}\})$.

10.S.3 As part of a study of environmental influences on sex determination in the fish *Menidia,* eggs from a single mating were divided into two groups and raised in either a warm or a cold environment. It was found that 73 of 141 offspring in the warm environment and 107 of 169 offspring in the cold environment were females.[65] In each of the following chi-square tests, use a nondirectional alternative and let $\alpha = 0.05$.

(a) Test the hypothesis that the population sex ratio is 1:1 in the warm environment.

(b) Test the hypothesis that the population sex ratio is 1:1 in the cold environment.

(c) Test the hypothesis that the population sex ratio is the same in the warm as in the cold environment.

(d) Define the population to which the conclusions reached in parts (a)–(c) apply. (Is it the entire genus *Menidia?*)

10.S.4 The cilia are hairlike structures that line the nose and help to protect the respiratory tract from dust and foreign particles. A medical team obtained specimens of nasal tissue from nursery school children who had viral upper respiratory infections, and also from healthy children in the same classroom. The tissue was sectioned, and the cilia were examined with a microscope for specific defects, with the results shown in the accompanying table.[66] The data show that the percentage of defective cilia was much higher in the tissue from infected children (15.7% versus 3.1%). Would it be valid to apply a chi-square test to compare these percentages? If so, do it. If not, explain why not.

	Number of children	Total number of cilia counted	Cilia with defects	
			Number	Percent
Control	7	556	17	3.1
Respiratory infection	22	1,493	235	15.7

10.S.5 A group of mountain climbers participated in a trial to investigate the usefulness of the drug acetazolamide in preventing altitude sickness. The climbers were randomly assigned to receive either drug or placebo during an ascent of Mt. Rainier. The experiment was supposed to be double-blind, but the question arose whether some of the climbers might have received clues (perhaps from the presence or absence of side effects or from a perceived therapeutic effect or lack of it) as to which treatment they were receiving. To investigate this possibility, the climbers were asked (after the trial was over) to guess which treatment they had received.[67] The results can be cast in the following contingency table, for which $\chi_s^2 = 5.07$:

		Treatment received	
		Drug	Placebo
Guess	Correct	20	12
	Incorrect	11	21

Alternatively the same results can be rearranged in the following contingency table, for which $\chi_s^2 = 0.01$:

		Treatment received	
		Drug	Placebo
Guess	Drug	20	21
	Placebo	11	12

Consider the null hypothesis

H_0: The blinding was perfect (the climbers received no clues).

Carry out the chi-square test of H_0 against the alternative that the climbers did receive clues. Let $\alpha = 0.05$. (You must decide which contingency table is relevant to this question.) (*Hint*: To clarify the issue for yourself, try inventing a fictitious data set in which most of the climbers *have* received strong clues, so we would expect a large value of χ_s^2; then arrange your fictitious data in each of the two contingency table formats and note which table would yield a larger value of χ_s^2.)

***10.S.6** Desert lizards *(Dipsosaurus dorsalis)* regulate their body temperature by basking in the sun or moving into the shade, as required. Normally the lizards will maintain a daytime temperature of about 38°C. When they are sick, however, they maintain a temperature about 2° to 4° higher—that is, a "fever." In an experiment to see whether this fever might be beneficial, lizards were given a bacterial infection; then 36 of the animals were prevented from developing a fever by keeping them in a 38° enclosure, while 12 animals were kept at a temperature of 40°. The following table describes the mortality after 24 hours.[68] How strongly do these results support the hypothesis that fever has survival value? Use Fisher's exact test against a directional alternative. Let $\alpha = 0.05$.

	38°	40°
Died	18	2
Survived	18	10
Total	36	12

10.S.7 Consider the data from Exercise 10.S.6. Compute expected cell frequencies for a chi-square test.

10.S.8 In a randomized clinical trial, 154 women with breast cancer were assigned to receive chemotherapy. Another 164 women were assigned to receive chemotherapy combined with radiation therapy. Survival data after 15 years are given in the following table.[69] Use these data to conduct a test of the null hypothesis that type of treatment does not affect survival rate.

	Chemotherapy only	Chemotherapy and radiation therapy
Died	78	66
Survived	76	98
Total	154	164

(a) State the alternative hypothesis.

(b) How many degrees of freedom are there?

(c) Compute the chi-square statistic.

(d) Draw your conclusion at $\alpha = 0.010$.

***10.S.9** Refer to the data in Exercise 10.S.8.

(a) Calculate the sample odds ratio.

(b) Find a 95% confidence interval for the population value of the odds ratio.

10.S.10 Two drugs, zidovudine and didanosine, were tested for their effectiveness in preventing progression of HIV disease in children. In a double-blind clinical trial, 276 children with HIV were given zidovudine, 281 were given didanosine, and 274) were given zidovudine plus didanosine. The following table shows the survival data for the three groups.[70] Use these data to conduct a test of the null hypothesis that survival and treatment are independent. Let $\alpha = 0.05$.

	Zidovudine	Didanosine	Zidovudine and didanosine
Died	17	7	10
Survived	259	274	264
Total	276	281	274

10.S.11 The blood types of malaria patients at a clinic in India were compared with those obtained in a sample of visitors to a nearby hospital. The data are shown in the following table.[71] Use these data to conduct a test of the null hypothesis that blood type is independent of contracting malaria.

	A	B	O	AB	Total
Malaria cases	138	199	106	33	476
Controls	229	535	428	96	1,288

Here is computer output for a chi-square test.

```
X-squared = 34.929, p-value = 1.26e-07
```

(a) If the null hypothesis were true, how many malaria cases would we expect to see in individuals with type A blood in similar studies of this size?

(b) How many degrees of freedom are there?

(c) If $\alpha = 0.05$, what is your conclusion regarding H_0?

10.S.12 The habitat selection behavior of the fruitfly *Drosophila subobscura* was studied by capturing flies from two different habitat sites. The flies were marked with colored fluorescent dust to indicate the site of capture and then released at a point midway between the original sites. On the following 2 days, flies were recaptured at the two sites. The results are summarized in the table.[72] The value of the chi-square statistic for this contingency table is $\chi_s^2 = 10.44$. Test the null hypothesis of independence against the alternative that the flies preferentially tend to return to their site of capture. Let $\alpha = 0.01$.

		Site of recapture	
		I	II
Site of original capture	I	78	56
	II	33	58

10.S.13 In the garden pea *Pisum sativum*, seed color can be yellow (Y) or green (G), and seed shape can be round (R) or wrinkled (W). Consider the following three hypotheses describing a population of plants:

$$H_0^{(1)}\text{: Pr}\{Y\} = \frac{1}{2}$$

$$H_0^{(2)}\text{: Pr}\{R\} = \frac{3}{4}$$

$$H_0^{(3)}\text{: Pr}\{R \mid Y\} = \text{Pr}\{R \mid G\}$$

The first hypothesis asserts that yellow and green plants occur in a 1:1 ratio; the second hypothesis asserts that round and wrinkled plants occur in a 3:1 ratio, and the third hypothesis asserts that color and shape are independent.

Suppose a random sample of 600 plants is to be observed, with the data to be arranged in the following contingency table:

		Color	
		Y	G
Shape	R		
	W		
			600

Invent fictitious data sets as specified, and verify each answer by calculating the estimated conditional probabilities. (*Hint*: In each case, begin with the marginal frequencies.)

(a) A data set that agrees perfectly with $H_0^{(1)}$, $H_0^{(2)}$, and $H_0^{(3)}$.

(b) A data set that agrees perfectly with $H_0^{(1)}$ and $H_0^{(2)}$ but not with $H_0^{(3)}$.

(c) A data set that agrees perfectly with $H_0^{(3)}$ but not with $H_0^{(1)}$ or $H_0^{(2)}$.

***10.S.14** A study of 36,080 persons who had heart attacks found that men were more likely to survive than were women. The following table shows some of the data collected in the study.[73]

		Men	Women
Survived at least 24 hours?	Yes	25,339	8,914
	No	1,141	686
	Total	26,480	9,600

(a) Calculate the odds ratio for comparing survival of men to survival of women.

(b) Calculate a 95% confidence interval for the population value of the odds ratio.

(c) Does the odds ratio give a good approximation to the relative risk for these data? Why or why not?

***10.S.15** In the study described in Exercise 10.9.6, one of the variables measured was whether the subjects had used any products containing phenylpropanolamine. The odds ratio was calculated to be 1.49, with stroke victims more likely than the control subjects to have used a product containing phenylpropanolamine.[61] Compute a 99% confidence interval for the population value of the odds ratio. Interpret this confidence interval in the context of this setting.

10.S.16 (Computer exercise) In a study of the effects of smoking cigarettes during pregnancy, researchers examined the placenta from each of 58 women after childbirth. They noted the presence or absence (P or A) of a particular placental abnormality—atrophied villi. In addition, each woman was categorized as a nonsmoker (N), moderate smoker (M), or heavy smoker (H). The following table shows, for each woman, an ID number (#) and the results for smoking (S) and atrophied villi (V).[74]

#	S	V	#	S	V	#	S	V	#	S	V
1	N	A	16	H	P	31	M	A	46	M	A
2	M	A	17	H	P	32	M	A	47	H	P
3	N	A	18	N	A	33	N	A	48	H	P
4	M	A	19	M	P	34	N	A	49	H	A
5	M	A	20	N	P	35	N	A	50	N	P
6	M	P	21	M	A	36	H	P	51	N	A
7	H	P	22	H	A	37	N	A	52	M	P
8	N	A	23	M	P	38	H	P	53	M	A
9	N	A	24	N	A	39	H	P	54	H	P
10	M	P	25	N	P	40	N	A	55	H	A
11	N	A	26	N	A	41	M	A	56	M	P
12	N	P	27	N	A	42	N	A	57	H	P
13	H	P	28	M	P	43	H	A	58	H	P
14	M	A	29	N	A	44	M	A			
15	M	P	30	N	A	45	M	P			

(a) Test for a relationship between smoking status and atrophied villi. Use a chi-square test at $\alpha = 0.05$.

(b) Prepare a table that shows the total number of women in each smoking category, and the number and percentage in each category who had atrophied villi.

(c) What pattern appears in the table of part (b) that is not used by the test of part (a)?

***10.S.17** Researchers studied the cellular telephone records of 699 persons who had automobile accidents. They determined that 170 of the 699 had made a cellular telephone call during the 10-minute period prior to their accident; this period is called the hazard interval. There were 37 persons who had made a call during a corresponding 10-minute period on the day before their accident; this period is called the control interval. Finally, there were 13 who made calls both during the hazard interval and the control interval.[75] Do these data indicate that use of a cellular telephone is associated with an increase in accident rate? Analyze these data using McNemar's test. Use a directional alternative and let $\alpha = 0.01$.

		Call during control interval?	
		Yes	No
Call during hazard interval?	Yes	13	157
	No	24	505

10.S.18 Prior to an influenza season subjects were randomly assigned to receive either a flu vaccine or a placebo. During that season there were 28 cases of the flu among 813 vaccine recipients and 35 cases of the flu among the 325 subjects who were given the placebo.[76] Do these data indicate that the vaccine was effective? Conduct an appropriate test using a directional alternative with $\alpha = 0.05$.

***10.S.19** Refer to the data in Exercise 10.S.18.

(a) Calculate the sample odds ratio.

(b) Find a 95% confidence interval for the population value of the odds ratio.

10.S.20 Consider Exercise 9.S.18. The romantic partners of the 36 men discussed in Exercise 9.S.18 were also tested, in the same manner as the men (i.e., they were blindfolded and asked to identify their partner by touching the backs of the hands of three men, one of whom was their partner). Among the women, 25 were successful and 11 were not. Are these data significant evidence for the hypothesis that men and women differ in their ability to identify their partners?

Consider an appropriate test that uses a nondirectional alternative.

Here is computer output for a chi-square test.

```
X-squared = 4.59, df = 1, p-value = 0.03219
```

(a) State the null and appropriate directional alternative hypotheses in context.

(b) Compute the sample proportions and the expected frequencies.

(c) If $\alpha = 0.05$, what is your conclusion regarding H_0?

10.S.21 A researcher found that 54 out of 66 randomly selected trees in an arboretum in northern Ohio were native species.[77] This compared to 42 out of 72 randomly chosen trees in "managed" land in the same general area. Is there significant evidence of a difference in proportions? If so, how strong is that evidence? Compute a 95% confidence interval for the difference between these two proportions.

Highlights and Study

(Reflections on Chapters 9 and 10)

Unit

III

While Chapters 6–8 emphasized statistical inference for numeric data, Chapters 9 and 10 introduced inference for categorical data.[1] In Chapter 9 we examined one-sample methods and focused attention on describing a single population proportion or assessing a particular population model. Chapter 10 considered methods to compare qualitative properties of two populations or relate two qualitative variables from a single population. The following related studies will showcase many of the methods of these two chapters.

Researchers were interested in whether or not acupuncture administered on the day of embryo transfer could improve the efficacy of *in vitro* fertilization in humans. To investigate, they studied 465 patients at a private infertility clinic in Washington who received embryo transfers (ET). Of these women, 188 had also elected to have acupuncture at the time of transfer, while the remaining 277 did not receive acupuncture. Of the 188 who had acupuncture, 106 (56.4%) had pregnancies leading to a live birth; for the 277 not receiving acupuncture, 126 (45.5%) led to a live birth. The following table more succinctly summarizes the results for easy comparison. Note that in this summary table it is only necessary to present the number of live births or the birth rates, as one could easily be inferred from the other; however, for the purpose of simple comparisons the proportion or percentage of live births (i.e., birth rates) are more useful for comparison because of the unequal sample sizes.

Table III.1 Number of live births

	Number (percentage) of live births	Sample size
Acupuncture	106 (56.4)	188
No acupuncture	126 (45.5)	277

Chi-Square Test of Independence

Symbolically, the researcher's null hypothesis is H_0: Pr(live birth | Acupuncture) = Pr(live birth | No Acupuncture). In other words, the probability of a live birth is the same whether or not acupuncture is used. The researcher's alternative hypothesis was directional, that acupuncture increases (improves) the live birth rate, or H_A: Pr(live birth | Acupuncture) > Pr(live birth | No Acupuncture). For the purpose of this example, we'll consider this directional hypothesis; however, it is good to consider why a nondirectional test might be ethically more appropriate. A nondirectional test could provide useful information for either the efficacy *or* harm of this procedure with respect to the outcome of a live birth. We might learn that acupuncture is harmful and actually reduces the chances of a live birth.

In Chapter 10 we learned that we can test the relationship between two categorical variables using a chi-square test of independence. Table III.2 displays the observed counts along with the expected counts in parentheses. Recall that the expected counts for a chi-square test of independence are computed as

$$e = \frac{(\text{Row Total}) \times (\text{Column Total})}{\text{Grand Total}}$$

To illustrate, for the acupuncture group we have $e = \frac{188 \times 232}{465} = 93.8$. Unlike the observed counts, the expected counts may be non-integer since they represent the average count that would be expected if the experiment were run many times and there were no relationship between the two variables (Acupuncture and Live Birth). It is worth noting that all the expected counts in this study are greater than or equal to 5, a requirement for validity in this analysis.

Table III.2 Observed and expected birth outcomes

	Live birth	Not live birth	Total
Acupuncture	106 (93.8)	82 (94.2)	188
No acupuncture	126 (138.2)	151 (138.8)	277
Total	232	233	465

For these data we have

$$\begin{aligned} \chi_s^2 &= \frac{(106 - 93.8)^2}{93.8} + \frac{(82 - 94.2)^2}{94.2} + \frac{(126 - 138.2)^2}{138.2} + \frac{(151 - 138.8)^2}{138.8} \\ &= 5.318 \end{aligned}$$

Referring to Table 9 with 1 degree of freedom we find the directional *P*-value to be $\frac{0.02}{2} < P\text{-value} < \frac{0.05}{2}$ or $0.01 < P\text{-value} < 0.025$. (Using software, the exact directional *P*-value is reported as 0.0106.) This value indicates that if there is no relationship between live birth rate and acupuncture, obtaining a difference in birth rates (i.e., sample proportions) as different as ours would only occur between 1% and 2.5% of the time. Assuming the common choice of $\alpha = 0.05$, we would declare this result to be statistically significant.

Drawing Conclusions Based on the Study Design

Considering the study design, could we say that there is statistically significant evidence that acupuncture increases live birth rate? No, not in this study. If the researchers had randomly assigned subjects to the acupuncture and no acupuncture groups, then a cause–effect claim could be made since the study would have been an experiment. In this study, however, the subjects chose their own groups. Thus, there could be many other reasons for the apparent difference between the birth rates; for example, perhaps women who choose acupuncture are more health conscious than those who would not choose acupuncture. Of course, this does not mean the study has no value. Because the *P*-value was so small, we can still say that the difference between the live birth rates was not well explained by chance variation, so we have evidence that there is something causing a difference between the two groups. We just don't know if acupuncture is the cause, or perhaps something else that is associated with electing to have acupuncture.

It is also worth noting that this study was conducted at a private infertility clinic in Washington State. For the results to be generalizable to a larger population, we must think about who constitutes that population and whether or not we can regard our observational units as a random sample from that population. A natural population to consider for this type of study would be the population of women who are seeking *in vitro* fertilization. In our example, the researchers are studying women who are residents of Washington State and wealthy enough to attend a private clinic. Thus, not all women in the population can be considered to be equally likely to participate in this study, and we must be cautious in generalizing the results.

Chi-Square Goodness-of-Fit Test

In a different study of single embryo transfer for older women, researchers examined *in vitro* fertilization outcomes for 264 women between 40 and 44 years of age.[2] In this study there were three possible outcomes recorded: live birth (LB), miscarriage (M), and no pregnancy (NP). The results are summarized in Table III.3 below. How do these rates compare to the greater population at large? Do older women have rates that are different from the general population?

Table III.3 Number and (proportion) of each type of pregnancy outcome for the women in the study and population proportions

	Live births	Miscarriages	No pregnancy	Total
Sample counts and proportions	36 (0.1364)	24 (0.0909)	204 (0.7727)	264 (1.000)
Null population proportions	0.224	0.070	0.706	1.000

According to Web MD*, in 2009 the *in vitro* live birth rate in the United States was 22.4%; the miscarriage rate, 7.0%; and, the no pregnancy rate, 70.6%. So, to address our research question, we consider the null hypothesis H_0: Pr[LB] $= 0.224$, Pr[M] $=$ 0.070, and Pr[NP] $= 0.706$ and the alternative hypothesis that the probabilities are not all equal to those listed. We then test the hypotheses with a chi-square goodness-of-fit test.

Table III.4 displays the observed and expected counts for all three categories of outcomes in this study. Each observed count is computed as

$$e_i = np_i$$

where p_i is the population proportion for the ith category specified in the null hypothesis. For example, the expected number of live births among the 264 women

*www.webmd.com/infertility-and-reproduction/guide/in-vitro-fertilization?page=2

Table III.4 Observed and expected counts for the chi-square goodness of fit test

	Live births	Miscarriages	No pregnancy	Total
Observed counts	36	24	204	264
Expected counts	59.1	18.5	186.4	264

in this study, if the null hypothesis were true, is $264 \times 0.224 = 59.1$. That is, in the long run, if the null hypothesis were true and this study were repeated over and over, we would expect an average of 59.1 live births per study.

We compute the value of the chi-square statistic to be

$$\chi_s^2 = \frac{(36 - 59.1)^2}{59.1} + \frac{(24 - 18.5)^2}{18.5} + \frac{(204 - 186.4)^2}{186.4} = 12.37$$

Referring to Table 9 with df = number of categories $-1 = 3 - 1 = 2$ we obtain an (omni-directional) bracketed P-value of $0.001 < P\text{-value} < 0.01$. (Recall that with three categories, there is no sensible directional test possible unless we reduce the problem first to only consider two categories.) With $\alpha = 0.05$, there is statistically significant evidence that the proportions of live births, miscarriages, and nonpregnancies for older women differs from the population proportions. To clarify, we certainly don't expect this sample's proportions to match the null population proportions exactly. We can, however, say that the observed proportions for these sampled older women are not easily explained by chance variation, if the underlying null population proportions are accurate. Observing 264 women with proportions that differ as much as or more extremely than the null population proportions listed in Table III.3 would occur less than 1% of the time if only chance variation were at play.

Wilson Adjusted Confidence Interval for a Proportion

If it is reasonable to regard the 264 older women as a random sample from a larger population of older women seeking *in vitro* fertilization, we can compute a 95% confidence interval for the population proportion of live births for this population. The Wilson-adjusted proportion for the 95% confidence interval is computed to be

$$\tilde{p} = \frac{36 + 2}{264 + 4} = \frac{38}{268} = 0.1418$$

with standard error

$$SE_{\tilde{p}} = \sqrt{\frac{(0.1418)(1 - 0.1418)}{264 + 4}} = 0.0213$$

The estimate and standard error can be combined to produce the following 95% confidence interval

$$0.1418 \pm 1.96 \times 0.0213$$
$$(0.1000, 0.1835)$$

Interpreting the interval, we are 95% confident that the proportion of live births among Washington women seeking *in vitro* fertilization is between 0.100 and 0.184. It is interesting to note that these values are lower than the U.S. rate of 0.224 suggesting that older women are less likely to have a successful live birth from *in vitro* fertilization than the general population of all women.

Unit III Summary Exercises

III.1 Randomly chosen college students were asked if they could touch their nose with their tongue. Six out of 53 men said yes, compared to 15 out of 56 women.

(a) Use these data to construct a 95% confidence interval (CI) for the difference in population proportions.

(b) Use your CI from part (a) to test, using $\alpha = 0.05$, H_0: $p_{\text{men}} = p_{\text{women}}$ against H_a: $p_{\text{men}} \neq p_{\text{women}}$. Do you reject H_0? Why or why not?

(c) State the conclusion from your hypothesis test in part (b) in everyday language *in the context of this problem.*

III.2 A random sample of 46 college students were asked if they were doing community service work during a particular semester. The following table gives the data, comparing men and women. Do the data provide sufficient evidence to conclude that men are less likely than women to be involved in community service?

Community service			
Sex	Yes	No	Total
Male	5	18	23
Female	11	12	23
Total	16	30	46

(a) Consider a chi-square test against a directional alternative with $\alpha = 0.05$. The value of the test statistic is 3.45. Give bounds for the *P*-value and state the conclusion that you reach in the context of the problem.

(b) Now consider Fisher's Exact Test, with a directional alternative, which involves calculating the probabilities of several possible 2×2 tables. Don't calculate any probabilities. Instead, state the number of tables for which the probability would need to be calculated if you were to carry out the test.

III.3 A researcher planted various kinds of seeds and recorded whether or not they germinated within 5 weeks. The following table gives the data.

Type of seed				
Germinate?	Okra	Sunflower	Eggplant	Total
Yes	9	12	11	32
No	11	6	8	25
Total	20	18	19	57

(a) Consider the null hypothesis that type of seed is independent germination. The value of the expected cell count for the "Okra/Yes" cell is $32 \times 20/57 = 11.23$. Why is this correct? That is, explain the statistical reasoning that leads to (row total) $\times$ (column total)/(grand total) being the formula for an expected cell count.

(b) The value of the test statistic is 1.84. Conduct a test of the null hypothesis that type of seed is independent germination. Use $\alpha = 0.05$. Provide all steps, including degrees of freedom, *P*-value, and conclusion.

III.4 In a study of the effect of oral contraceptives on thromboembolic disease (blood clots) 175 pairs of women were studied.[3] The pairs were matched on age, race, and other variables. Within each pair one woman (the "case") had the disease but the other woman (the "control") did not. The following table shows the data on oral contraceptive use.

Oral contraceptives used by	Number of pairs
Both women in the pair	10
Only the case	57
Only the control	13
Neither woman in the pair	95
Total	175

Conduct a test of the hypothesis that presence of the disease is independent of oral contraceptive use. Use a nondirectional alternative and let $\alpha = 0.01$. (*Hint*: You might first want to display the data in a different format.)

(a) State the null hypothesis in symbols.

(b) Do you reject H_0? Why or why not?

(c) State your conclusion from part (b) in the context of this setting.

III.5 You think that approximately 25% of the members of a certain population smoke, but you want to take a sample in order to make an estimate. You want the standard error of $\tilde{p}$ to be no greater than 0.06.

(a) Using 0.25 as your guess of $\tilde{p}$, how large does your sample need to be?

(b) Suppose you have no guess for $\tilde{p}$. How large does your sample need to be so that the SE is no greater than 0.06 no matter what $\tilde{p}$ is?

Background for III.6–III.9

The following questions are motivated by the February 28, 2013 article: "A Brain-to-Brain Interface for Real-Time Sharing of Sensorimotor Information" (*Scientific Reports* **3**, 1319). The primary research question was to determine if the brain could assimilate signals from sensors from a different body. Here is a description of the study found on the Duke University Website (www.dukehealth.org/health_library/news/brain-to-brain-interface-allows-transmission-of-tactile-and-motor-information-between-rats).[4]

> To test this hypothesis, the researchers first trained pairs of rats to solve a simple problem: to press the correct lever when an indicator light above the lever switched on, which rewarded the rats with a sip of water. They next connected the two animals' brains via arrays of microelectrodes inserted into the area of the cortex that processes motor information.
>
> One of the two rodents was designated as the "encoder" animal. This animal received a visual cue that showed it which lever to press in exchange for a water reward. Once this "encoder" rat pressed the right lever, a sample of its brain activity that coded its behavioral decision was translated into a pattern of electrical stimulation that was delivered directly into the brain of the second rat, known as the "decoder" animal.
>
> The decoder rat had the same types of levers in its chamber, but it did not receive any visual cue indicating which lever it should press to obtain a reward. Therefore, to press the correct lever and receive the reward it craved, the decoder rat would have to rely on the cue transmitted from the encoder via the brain-to-brain interface.

III.6 Before connecting the encoder and decoder rats, the researchers wanted to verify that the decoder rats were sufficiently trained to choose the correct lever when seeing a visual cue. After training, one of the decoder rats was tested 25 times on his ability to select the "correct" lever. He selected the correct lever 19 times.

(a) Compute a 95% Wilson-adjusted confidence interval for the accuracy rate for this rat *and* interpret the interval in the context of the question.

(b) Using your interval computed in part (a) to support your answer, are you convinced that this rat's accuracy rate is better than simple guessing (i.e., 50% accuracy)?

(c) Using your interval computed in part (a) to support your answer, are you convinced that this rat's accuracy rate is better than 70%?

(d) Use a χ^2 goodness-of-fit test, to test whether or not the accuracy rate exceeds 70%. Use $\alpha = 0.05$.

III.7 Once a brain-to-brain connection was established, a pair of rats would undergo 50 lever-pressing trials. At each trial, the researchers would record whether or not the encoder rat pressed the "correct" lever and whether or not the decoder rat also pressed the "correct" lever. Following is a summary of the data.

	Decoder correct	Decoder wrong	Total
Encoder correct	27	11	38
Encoder wrong	6	6	12
Total	33	17	50

(a) Consider the question: Is decoder accuracy related to encoder accuracy? That is, does the decoder accuracy depend on whether or not the encoder pressed the right lever? Express the null hypothesis related to this research question symbolically. You may use the notation EC, EW, DC, and DW as needed to describe the groups (e.g., EC = Encoder correct).

(b) Complete the table of expected counts used for the χ^2 test of independence.

(c) The value of $\chi^2_s = 1.801$ for these data. Is there statistically significant evidence ($\alpha = 0.05$) that the decoder is *more* likely to press the correct lever if the encoder also presses the correct lever than if the encoder does not? Justify your answer.

(d) There is one condition for the χ^2-test of independence that is clearly violated. Which condition is violated?

III.8 Consider the research question: Does the decoder rat do what the encoder rat does? That is, when the encoder is correct, is the decoder also correct, and when the encoder is wrong, is the decoder also wrong? Thinking about the data in this way, the data in Exercise III.7 can be reduced as shown below.

	Agreement in response	Disagreement in response	Total
Count	33	17	50

If the rats do not have a "mental link," then the probability of agreeing would equal the probability of disagreeing or H_0: Pr[agree] = Pr[disagree].

(a) What is the alternative hypothesis for this research question?

(b) Use a χ^2 goodness of fit test to test the hypothesis in (a).

III.9 The study described above was actually carried out on four pairs of connected rats. The table below summarizes the results from the 200 (50 × 4) trials.

	Decoder correct	Decoder wrong	Total
Encoder correct	112	42	154
Encoder wrong	21	25	46
Total	133	67	200

The value of $\chi^2_s = 11.66$ and the corresponding P-value is 0.001, suggesting there is very strong evidence that the accuracy of the decoder is linked to the accuracy of the encoder. Briefly explain why the validity of this χ^2 test is questionable.

III.10 State whether the following statements are true or false and explain why.

(a) It is more difficult to estimate a population proportion precisely if its value is near 0.50.

(b) A χ^2 goodness-of-fit test has the null hypothesis that category proportions are all equal.

(c) A χ^2 goodness-of-fit test is used for observational studies whereas a χ^2 test of independence is used for experiments.

(d) $\chi_s^2 = 0$ when the observed data are in perfect agreement with the null hypothesis.

III.11 Consider the study of behavioral asymmetries in Exercise 10.3.4. Suppose we want to test the null hypothesis that righthanded women are equally likely to be right-footed or left-footed.

(a) Calculate the chi-square statistic for this hypothesis.

(b) What conclusions can be drawn? Let $\alpha = 0.05$.

Chapter 11

COMPARING THE MEANS OF MANY INDEPENDENT SAMPLES

OBJECTIVES

In this chapter we study analysis of variance (ANOVA). We will

- discuss when and why an analysis of variance may be conducted.
- develop the intuition behind the ANOVA model.
- demonstrate how ANOVA calculations are carried out.
- describe and examine the conditions under which ANOVA is valid.
- see how blocking is used and how to conduct randomized blocks ANOVA.
- describe interactions and main effects in factorial ANOVA models.
- construct contrasts and other linear combinations of means.
- introduce and compare several methods for dealing with multiple comparisons.

11.1 Introduction

In Chapter 7 we considered the comparison of two independent samples with respect to a quantitative variable *Y*. The classical techniques for comparing the two sample means $\bar{Y}_1$ and $\bar{Y}_2$ are the test and the confidence interval based on Student's *t* distribution. In this chapter we consider the comparison of the means of *I* independent samples, where *I* may be greater than 2. The following example illustrates an experiment with $I = 5$.

Example 11.1.1

Sweet Corn When growing sweet corn, can organic methods be used successfully to control harmful insects and limit their effect on the corn? In a study of this question researchers compared the weights of ears of corn under five conditions in an experiment in which sweet corn was grown using organic methods. In one plot of corn a beneficial soil nematode was introduced. In a second plot a parasitic wasp was used. A third plot was treated with both the nematode and the wasp. In a fourth plot a bacterium was used. Finally, a fifth plot of corn acted as a control; no special treatment was applied here. Thus, the treatments were

Treatment 1: Nematodes

Treatment 2: Wasps

Treatment 3: Nematodes and wasps

Treatment 4: Bacteria

Treatment 5: Control

Ears of corn were randomly sampled from each plot and weighed. The results are given in Table 11.1.1 and plotted in Figure 11.1.1.[1] Note that in addition to the differences between the treatment means, there is also considerable variation within each treatment group.

A RANDOMIZATION TEST

One way of analyzing data from *I* independent samples is to extend the ideas of Section 7.1 by creating a randomization test that compares all *I* groups at once. We illustrate that idea now.

Table 11.1.1 Weights (ounces) of ears of sweet corn

	Treatment				
	1	2	3	4	5
	16.5	11.0	8.5	16.0	13.0
	15.0	15.0	13.0	14.5	10.5
	11.5	9.0	12.0	15.0	11.0
	12.0	9.0	10.0	9.0	10.0
	12.5	11.5	12.5	10.5	14.0
	9.0	11.0	8.5	14.0	12.0
	16.0	9.0	9.5	12.5	11.0
	6.5	10.0	7.0	9.0	9.5
	8.0	9.0	10.5	9.0	18.5
	14.5	8.0	10.5	9.0	17.0
	7.0	8.0	13.0	6.5	10.0
	10.5	5.0	9.0	8.5	11.0
Mean	11.58	9.63	10.33	11.13	12.29
SD	3.47	2.42	1.96	3.12	2.87
n	12	12	12	12	12

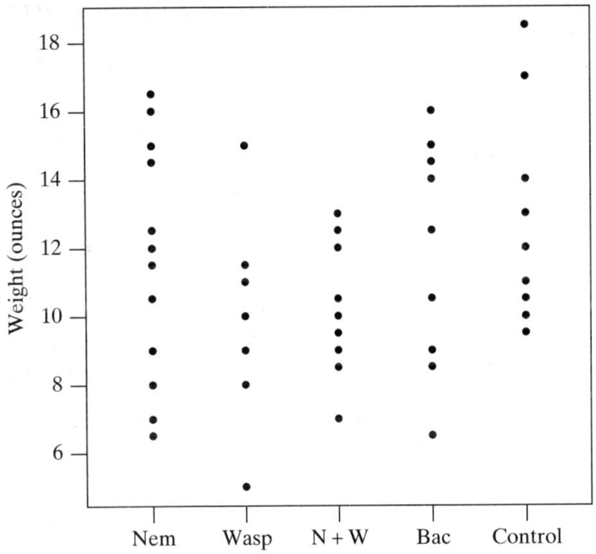

Figure 11.1.1 Weights of ears of corn receiving five different treatments

Example 11.1.2

Sweet Corn Suppose the five treatments for the sweet corn experiment were identical so that the labels of Nem, Wasp, N + W, Bac, and Control in Figure 11.1.1 were arbitrary. Then, we could think of each of the 60 ears of corn as having landed in one of the five groups at random. If that were the case, then how likely would it be that one of the five sample means would be as large as 12.29 (for the 12 ears that ended up in the Control group) and one of the sample means would be as small as 9.36 (for the 12 ears that ended up in the Wasp group)? Even if the population means are the same, we would not expect the five sample means to be identical. But would it be surprising to see a difference between the largest sample mean and the smallest sample mean as great as $12.29 - 9.63 = 2.66$, as happened here?

Consider doing the following. (1) Write the 60 ear weights on each of 60 cards; (2) shuffle the cards; (3) randomly deal out the cards into five piles of a dozen cards each, and call the groups Nem, Wasp, N + W, Bac, and Control; (4) compute the sample mean for each of the five groups; (5) calculate the difference between the largest and smallest of the five sample means; (6) record whether that difference is at least 2.66; (7) repeat steps (1)–(6) many times.

With the aid of a computer, we can conduct a simulation to learn how often a difference as great as 2.7 happens just by chance. In one simulation of 10,000 trials, there were 1,626 trials for which the difference was at least as large as 2.66, which means that the randomization P-value is 0.163. The differences among the five group means that were observed in the experiment could easily have happened by chance if there were no true differences among the five treatments. ■

The classical method of analyzing data from I independent samples is called an **analysis of variance**, or **ANOVA.** In applying analysis of variance, the data are regarded as random samples from I populations. We denote the means of these populations as $\mu_1, \mu_2, \ldots, \mu_I$ and the standard deviations as $\sigma_1, \sigma_2, \ldots, \sigma_I$. We test a null hypothesis of equality among all I of the population means.

WHY NOT REPEATED *t* TESTS?

It is natural to wonder why the comparison of the means of I samples requires any new methods. For instance, why not just use a two-sample t test on each pair of samples? There are three reasons why this is not a good idea.

1. *The problem of multiple comparisons* The most serious difficulty with a naive "repeated t tests" procedure concerns Type I error: The probability of false rejection of a null hypothesis may be much higher than it appears to be. For instance, suppose $I = 4$ and consider the null hypothesis that all four population means are equal (H_0: $\mu_1 = \mu_2 = \mu_3 = \mu_4$) versus the alternative hypothesis that the four means are not all equal.* Among four means there are six possible pairs to compare. The pairings are displayed in Figure 11.1.2. The six resulting hypotheses are

$$H_0: \mu_1 = \mu_2 \quad H_0: \mu_1 = \mu_3 \quad H_0: \mu_1 = \mu_4$$
$$H_0: \mu_2 = \mu_3 \quad H_0: \mu_2 = \mu_4 \quad H_0: \mu_3 = \mu_4$$

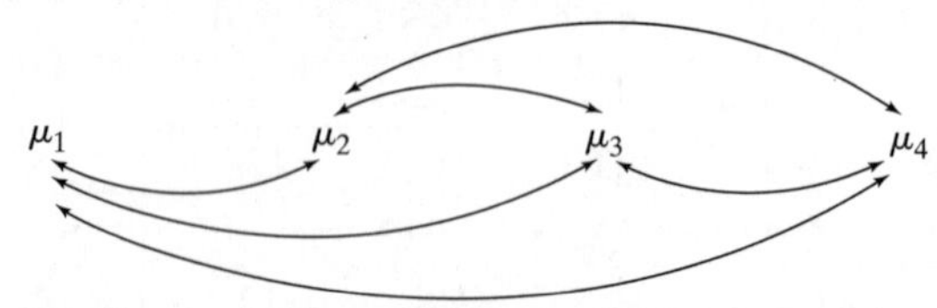

Figure 11.1.2 Comparing four population means requires six comparisons

Let's consider the risk of a Type I error for testing our primary null hypothesis that all four means are equal by conducting six separate t tests. If *any* of the six t tests finds a significant difference between a pair of means, we would reject our primary null hypothesis that all four means are equal. A Type I error would occur if *any* of the six t tests found a significant difference between a pair of means when in fact all four means are equal. Thus, using $\alpha = 0.05$ for each of the individual t tests carries an overall risk of a Type I error that is greater than 5%.

Our intuition might suggest that the risk of an overall Type I error in the preceding example should be $6 \times 0.05 = 0.3 = 30\%$ (in each of six tests we had a 5% chance of wrongly finding evidence for a difference), but this is not the case. The computation of this overall Type I error rate is more complex. Table 11.1.2 displays the overall risk of Type I error,† that is,

Table 11.1.2 Overall risk of type I error in using repeated t tests at $\alpha = 0.05$

I	Overall risk
2	0.05
3	0.12
4	0.20
6	0.37
8	0.51
10	0.63

*In Section 11.2 we will elaborate more on the form of this alternative hypothesis.

†Table 11.1.2 was computed assuming that the sample sizes are large and equal and that the population distributions are normal with equal standard deviations.

Overall Type I error risk $=$ Probability that at least one of the t tests will reject its null hypothesis, when in fact $\mu_1 = \mu_2 = \mu_3 = \cdots = \mu_I$.

If $I = 2$, then the overall risk is 0.05, as it should be, but with larger I the risk increases rapidly; for $I = 6$ it is 0.37. It is clear from Table 11.1.2 that the researcher who uses repeated t tests is highly vulnerable to Type I error unless I is quite small.

The difficulties illustrated by Table 11.1.2 are due to **multiple comparisons**—that is, many comparisons on the same set of data. These difficulties can be reduced when the comparison of several groups is approached through ANOVA.

2. *Estimation of the standard deviation.* The ANOVA technique combines information on variability from all the samples simultaneously. This global sharing of information can yield improved precision in the analysis.
3. *Structure in the groups.* In many studies the logical structure of the treatments or groups to be compared may inspire questions that cannot be answered by simple pairwise comparisons. For example, we may wish to study the effects of two experimental factors simultaneously. ANOVA can be used to analyze data in such settings (see Sections 11.6, 11.7, and 11.8).

A GRAPHICAL PERSPECTIVE ON ANOVA

When data are analyzed by analysis of variance, the usual first step is to test the following global null hypothesis:

$$H_0\colon \mu_1 = \mu_2 = \mu_3 = \cdots = \mu_I$$

which asserts that all the population means are equal. A statistical test of H_0 will be described in Section 11.4. However, we will first consider analysis of variance from a graphical perspective.

Consider the dotplots shown in Figure 11.1.3(a). These dotplots were generated in a setting in which H_0 is true. The sample means, which are shown as lines on the graph, differ from one another only as a result of chance error. For the data shown in Figure 11.1.3(b), H_0 is false. The sample means are quite different—there is substantial variability between the group means, which provides evidence that the corresponding population means (μ_1, μ_2, μ_3, and μ_4) are not all equal. In this particular case, it appears that μ_1 and μ_2 differ from μ_3 and μ_4.

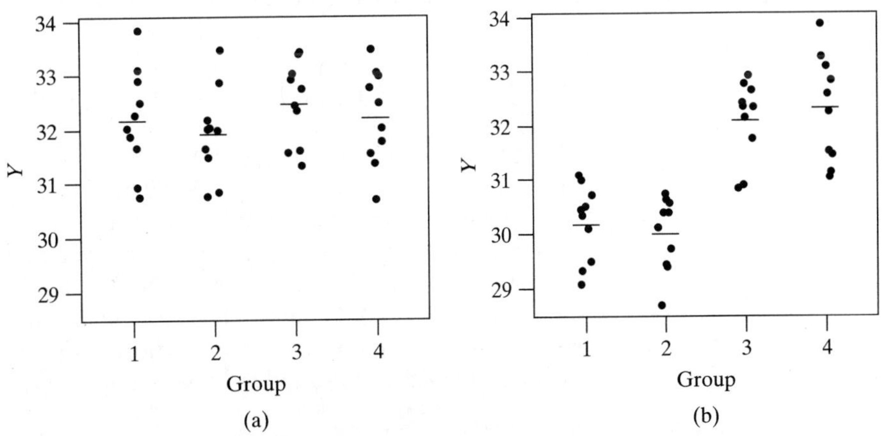

Figure 11.1.3 (a) H_0 true, (b) H_0 false, with small SDs for the groups

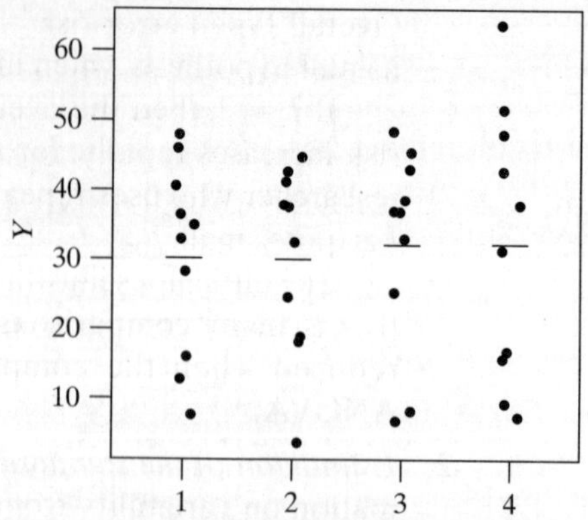

Figure 11.1.4 H_0 false, with large SDs for the groups

Figure 11.1.4 shows a situation that is less clear. In fact, H_0 is false here—the means in Figure 11.1.4 are identical to those in Figure 11.1.3(b). However, the individual group standard deviations are quite large, which makes it hard to tell that the population means differ.*

We need to know how much inherent variability there is in the data before we can judge whether a difference in sample means is fairly small and attributable to chance or whether it is too large to be due to chance alone. As Figures 11.1.3 and 11.1.4 illustrate, in order to find compelling evidence for a difference in population means, not only must there be (1) variation among the group means, but it must be large *relative* to (2) the inherent variability in the groups. It is through comparing the relative magnitudes of these two kinds of *variability*—this "analysis of variance"—that we are able to make an inference about *means.*

A LOOK AHEAD

If the global null hypothesis that $\mu_1 = \mu_2 = \mu_3 = \cdots = \mu_I$ is rejected, then the data provide sufficient evidence to conclude that at least *some* of the μ's are unequal; the researcher would usually proceed to detailed comparisons to determine the *pattern* of differences among the μ's. If there is a lack of evidence against the global null hypothesis, then the researcher might choose to construct one or more confidence intervals to characterize the lack of significant differences among the μ's.

All the statistical procedures of this chapter—the test of the global null hypothesis and various methods of making detailed comparisons among the means—depend on the same basic calculations. These calculations are presented in Section 11.2.

11.2 The Basic One-Way Analysis of Variance

The ANOVA model presented in Section 11.1 that compares the means of three or more groups is called a **one-way ANOVA.** The term "one-way" refers to the fact that there is one variable that defines the groups or treatments (e.g., in the sweet corn example the treatments were based on the type of harmful insect/bacteria). Later in this chapter we will examine other ANOVA models such as the randomized complete block ANOVA (Section 11.6) and the two-way ANOVA model (Section 11.7), which consider the impact of having more than one variable defining the groups or how treatments are assigned to experimental units.

*Note the change in scale on the vertical axis in Figure 11.1.4.

In this section we present the basic one-way ANOVA calculations that are used to describe the data and to facilitate further analysis. In the previous section we noted that if the between-group mean variability is large relative to within-group variability, we will take this as evidence against the null hypothesis that the population means are all equal. Hence, the analysis of variance of I samples, or groups, begins with the calculation of quantities that describe the variability of the data *between* the groups and *within* the groups.* (For clarity, in this chapter we will often refer to the samples as "groups" of observations.)

NOTATION

To describe several groups of quantitative observations, we will use two subscripts: one to keep track of group membership and the other to keep track of observations within the groups. Thus, we will denote observation j in group i as

$$y_{ij} = \text{observation } j \text{ in group } i$$

Thus, the first observation in the first group is y_{11}, the second observation in the first group is y_{12}, the third observation in the second group is y_{23}, and so on.

We will also use the following notation:

$$\begin{aligned} I &= \text{number of groups} \\ n_i &= \text{number of observations in group } i \\ \bar{y}_i &= \text{mean for group } i \\ s_i &= \text{standard deviation for group } i \end{aligned}$$

The total number of observations is

$$n_{\bullet} = \sum_{i=1}^{I} n_i$$

Finally, the **grand mean**—the mean of all the observations—is

$$\bar{\bar{y}} = \frac{\sum_{i=1}^{I}\sum_{j=1}^{n_i} y_{ij}}{n_{\bullet}}$$

Equivalently we can express $\bar{\bar{y}}$ as a weighted average of the group means

$$\bar{\bar{y}} = \frac{\sum_{i=1}^{I} n_i\bar{y}_i}{\sum_{i}^{I} n_i} = \frac{\sum_{i=1}^{I} n_i\bar{y}_i}{n_{\bullet}}$$

The following example illustrates this notation.

Example 11.2.1 **Weight Gain of Lambs** Table 11.2.1 shows the weight gains (in 2 weeks) of young lambs on three different diets. (These data are fictitious, but are realistic in all respects except for the fact that the group means are whole numbers.)[2]

The total number of observations is

$$n_{\bullet} = 3 + 5 + 4 = 12$$

*Grammatically speaking, the word *among* should be used rather than *between* when referring to three or more groups; however, we will use "between" because it more clearly suggests that the groups are being compared against each other.

Table 11.2.1 Weight gains of lambs (lb)*

	Diet 1	Diet 2	Diet 3
	8	9	15
	16	16	10
	9	21	17
		11	6
		18	
n_i	3	5	4
Sum $= \sum_{j=1}^{n_i} y_{ij}$	33	75	48
Mean $= \bar{y}_i$	11.000	15.000	12.000
SD $= s_i$	4.359	4.950	4.967

*Extra digits are reported for accuracy of subsequent calculations.

and the total of all the observations is

$$\sum_{i=1}^{I}\sum_{j=1}^{n_i} y_{ij} = 33 + 75 + 48 = 156 \text{ or, equivalently } 3 \times 11 + 5 \times 15 + 4 \times 12 = 156$$

The grand mean is

$$\bar{\bar{y}} = \frac{156}{12} = 13 \text{ lb}$$

■

If the sample sizes (n_i's) are all equal, then the grand mean $\bar{\bar{y}}$ is just the ordinary average (i.e., mean) of the group means (the $\bar{y}_i$'s); but if the sample sizes are unequal, this is not the case. For instance, in Example 11.2.1 note that

$$\frac{11 + 15 + 12}{3} \neq 13$$

MEASURING VARIATION WITHIN GROUPS

A combined measure of variation within the I groups is the pooled standard deviation s_{pooled}, often simply denoted as just s, which is computed as follows.*

Pooled Standard Deviation

$$s_{\text{pooled}} = s = \sqrt{\frac{\sum_{i=1}^{I}(n_i - 1)s_i^2}{\sum_{i=1}^{I}(n_i - 1)}} = \sqrt{\frac{\sum_{i=1}^{I}(n_i - 1)s_i^2}{n. - I}}$$

*There is no ambiguity in this notation since s_i (i.e., s with a subscript) denotes an individual group sample standard deviation.

We call $s^2_{\text{pooled}} = s^2$ the pooled variance*

$$s^2_{\text{pooled}} = s^2 = \frac{\sum_{i=1}^{I}(n_i - 1)s_i^2}{\sum_{i=1}^{I}(n_i - 1)}$$

Examining the formula we can see that the pooled variance is a weighted average of the group sample variances, and thus the pooled standard deviation can be very loosely interpreted as a weighted average of the group standard deviations.

The following example illustrates the computation of the pooled standard deviation, s.

Example 11.2.2

Weight Gain of Lambs Table 11.2.1 shows the group sample sizes and standard deviations for the lamb weight-gain data. The pooled variance and standard deviation are calculated as

$$s^2 = \frac{(3 - 1)4.359^2 + (5 - 1)4.950^2 + (4 - 1)4.967^2}{12 - 3} = \frac{210.025}{9} = 23.336$$

$$s = \sqrt{23.336} = 4.831$$

Observe that the pooled standard deviation, 4.831 lb, is a sensible representative value for the three group standard deviations, 4.359, 4.950, and 4.967 lb. If we assume that the population standard deviation of weight gains is the same for all three diets, then we would estimate this common value to be 4.83 lb. This estimate depends only on the variability within the groups and not on their mean values. Figure 11.2.1(a) displays the data from Table 11.2.1 while Figure 11.2.1(b) displays a modified version of the data for which 7 has been added to each Diet 2 observation and 5 has been subtracted from each Diet 3 observation. We see that while the group means are different for these two data sets, the pooled standard deviation—the inherent variability in each group—is the same.

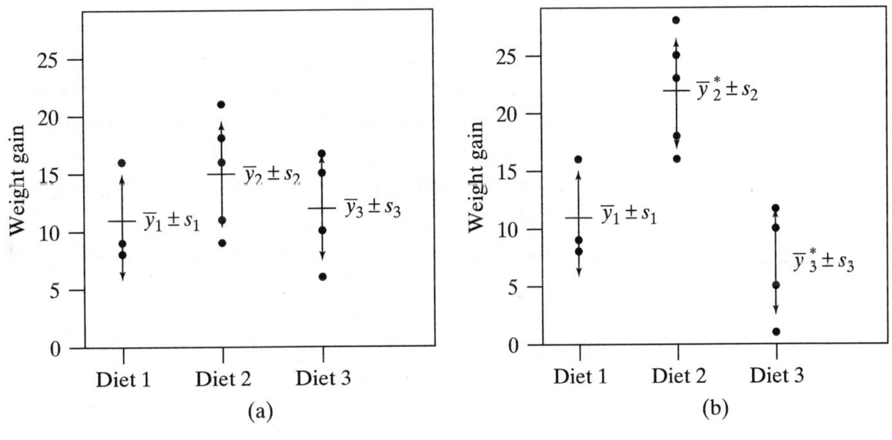

Figure 11.2.1 Examining within-group standard deviations. Plot (a) displays the weight gain data from Table 11.2.1 with $s = 4.831$. Plot (b) displays modified data with the same individual group standard deviations, and thus the same pooled standard deviation $s = 4.831$

ANOVA NOTATION

While our preceding formulas use familiar notation and terms, we will find it convenient to decompose the pooled variance into parts and subsequently define new terms to be used in the context of ANOVA.

*Recall from Chapter 2 that the variance is simply the standard deviation squared.

The numerator of the pooled variance is known as the **sum of squares within groups, SS(within),** while the denominator is known as the **degrees of freedom within groups, df(within).** The formulas for these are displayed in the following box.*

Sum of Squares and df Within Groups

$$\text{SS(within)} = \sum_{i=1}^{I}(n_i - 1)s_i^2$$

$$\text{df(within)} = n. - I$$

Their ratio is defined as the **mean square within groups,** or **MS(within).** Note that MS(within) is just another name for the pooled variance.

Mean Square Within Groups

$$\text{MS(within)} = \frac{\text{SS(within)}}{\text{df(within)}}$$

Hence, the quantity MS(within) measures the variability within the groups.†

The following example illustrates the calculation of SS(within), df(within), and MS(within).

Example 11.2.3 **Weight Gain of Lambs** In Example 11.2.2 when computing the pooled variance, we found

$$s^2 = \frac{(3-1)4.359^2 + (5-1)4.950^2 + (4-1)4.967^2}{12-3} = \frac{210.025}{9} = 23.336$$

Thus, SS(within) = 210.025, df(within) = 9, and MS(within) = 23.336. ■

VARIATION BETWEEN GROUPS

For two groups, the difference between the groups is simply described by $(\bar{y}_1 - \bar{y}_2)$. How can we describe between-group variability for more than two groups? One naive idea is to simply compute the sample variance of the group means. The **mean square between groups,** or **MS(between)** is motivated by this idea. In fact, were it not for the n_i in the numerator of the following expression (to adjust for the sample sizes of the groups), the MS(between) would indeed be the sample variance of the group means.

Mean Square Between Groups

$$\text{MS(between)} = \frac{\sum_{i=1}^{I} n_i(\bar{y}_i - \bar{\bar{y}})^2}{I - 1}$$

*A popular but less intuitive formula for SS(within) is given by SS(within) $= \sum_{i=1}^{I}\sum_{j=1}^{n_i}(y_{ij} - \bar{y}_i)^2$.

†If there were only one group, with n observations, then df(within) would be $n - 1$ and the SS(within) would be $(n-1)s^2$. MS(within) would then simply be $\frac{(n-1)s^2}{(n-1)} = s^2$, the sample variance.

As with the measures used for the within-group variation, MS(within), it is convenient to define the numerator of MS(between) as the **sum of squares between groups** or **SS(between)** and the denominator as the **degrees of freedom between groups** or **df(between)** so that

$$\text{MS(between)} = \frac{\text{SS(between)}}{\text{df(between)}}$$

where SS(between) and df(between) are explicitly defined as follows.

Sum of Squares and df Between Groups

$$\text{SS(between)} = \sum_{i=1}^{I} n_i(\bar{y}_i - \bar{\bar{y}})^2$$

$$\text{df(between)} = I - 1$$

The following example illustrates these definitions.

Example 11.2.4 **Weight Gain of Lambs** For the data of Example 11.2.1, the quantities that enter SS(between) are shown in Table 11.2.2.

Table 11.2.2 Calculation of SS(between) for lamb weight gains

	Diet 1	Diet 2	Diet 3
Mean: $\bar{y}_i$	11	15	12
n_i	3	5	4
	Grand mean $\bar{\bar{y}} = 13$		

From Table 11.2.2 we calculate

$$\text{SS(between)} = 3(11 - 13)^2 + 5(15 - 13)^2 + 4(12 - 13)^2 = 36$$

Since $I = 3$, we have

$$\text{df(between)} = 3 - 1 = 2$$

so that

$$\text{MS(between)} = \frac{36}{2} = 18$$

■

The SS(between) and MS(between) measure the variability between the sample means of the groups. This variability is shown graphically in Figure 11.2.2.

A FUNDAMENTAL RELATIONSHIP OF ANOVA

The name *analysis of variance* derives from a fundamental relationship involving SS(between) and SS(within). Consider an individual observation y_{ij}. It is obviously true that

$$y_{ij} - \bar{\bar{y}} = (y_{ij} - \bar{y}_i) + (\bar{y}_i - \bar{\bar{y}})$$

This equation expresses the deviation of an observation from the grand mean as the sum of two parts: a within-group deviation $(y_{ij} - \bar{y}_i)$ and a between-group deviation

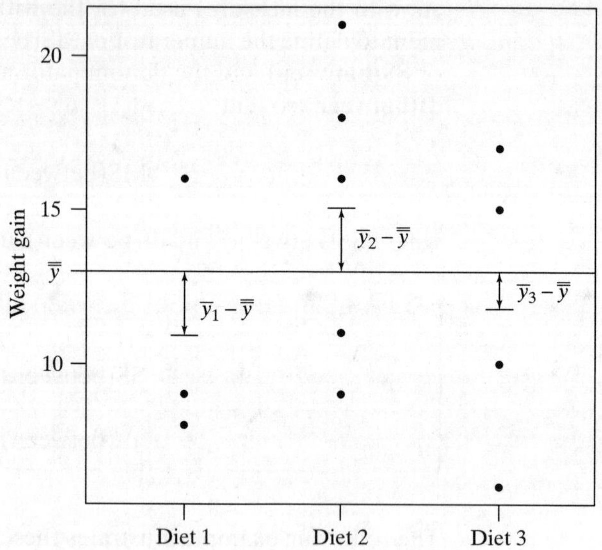

Figure 11.2.2 Measuring the differences between group means

$(\bar{y}_i - \bar{\bar{y}})$. It is also true (but not at all obvious) that the analogous relationship holds for the corresponding sums of squares; that is,

$$\sum_{i=1}^{I}\sum_{j=1}^{n_i}(y_{ij} - \bar{\bar{y}})^2 = \sum_{i=1}^{I}\sum_{j=1}^{n_i}(y_{ij} - \bar{y}_i)^2 + \sum_{i=1}^{I}\sum_{j=1}^{n_i}(\bar{y}_i - \bar{\bar{y}})^2 \qquad (11.2.1)$$

which, by rewriting each of the sums on the right-hand side, can be expressed as

$$\sum_{i=1}^{I}\sum_{j=1}^{n_i}(y_{ij} - \bar{\bar{y}})^2 = \sum_{i=1}^{I}(n_i - 1)s_i^2 + \sum_{i=1}^{I}n_i(\bar{y}_i - \bar{\bar{y}})^2$$
$$= \text{SS(within)} + \text{SS(between)}$$

The quantity on the left-hand side of formula (11.2.1) is called the **total sum of squares,** or **SS(total)**:

Definition of Total Sum of Squares

$$\text{SS(total)} = \sum_{i=1}^{I}\sum_{j=1}^{n_i}(y_{ij} - \bar{\bar{y}})^2$$

Note that SS(total) measures variability among all n observations in the I groups. The relationship [formula (11.2.1)] can be written as

Relationship between Sums of Squares

$$\text{SS(total)} = \text{SS(between)} + \text{SS(within)}$$

The preceding fundamental relationship shows how the total variation in the data set can be analyzed, or broken down, into two interpretable components: between-sample variation and within-sample variation. This partition is an analysis of variance.

The **total degrees of freedom, or df(total),** is defined as follows:

Total df

$$\text{df(total)} = n. - 1$$

With this definition, the degrees of freedom add, just as the sums of squares do; that is,

$$\text{df(total)} = \text{df(within)} + \text{df(between)}$$
$$n. - 1 = (n. - I) + (I - 1)$$

Notice that, if we were to consider all $n.$ observations as a single sample, then the SS for that sample (i.e., the numerator of the variance) would be SS(total) and the associated df (i.e., the denominator of the variance) would be df(total). Consequently, $\sqrt{\frac{\text{SS(total)}}{\text{df(total)}}}$ is the standard deviation of the entire data set when group membership is ignored.

The following example illustrates the fundamental relationships between the sums of squares and degrees of freedom.

Example 11.2.5

Weight Gain of Lambs For the data of Table 11.2.1, we found $\bar{\bar{y}} = 13$; we calculate SS(total) as

$$\begin{aligned}\text{SS(total)} &= \sum_{i=1}^{I}\sum_{j=1}^{n_i}(y_{ij} - \bar{\bar{y}})^2 \\ &= \left[(8-13)^2 + (16-13)^2 + (9-13)^2\right] \\ &\quad + \left[(9-13)^2 + (16-13)^2 + (21-13)^2 + (11-13)^2 + (18-13)^2\right] \\ &\quad + \left[(15-13)^2 + (10-13)^2 + (17-13)^2 + (6-13)^2\right] \\ &= 246\end{aligned}$$

For these data, we found that SS(between) = 36 and SS(within) = 210. We verify that

$$246 = 36 + 210$$

Also, we found that df(within) = 9 and df(between) = 2. We verify that

$$\text{df(total)} = 12 - 1 = 11 = 9 + 2$$

■

THE ANOVA TABLE

When working with the ANOVA quantities, it is customary to arrange them in a table. The following example shows a typical format for the ANOVA table.

Example 11.2.6

Weight Gain of Lambs Table 11.2.3 shows the ANOVA for the lamb weight-gain data. Notice that the ANOVA table clearly shows the additivity of the sums of squares and the degrees of freedom. ■

Comment on Terminology While the terms "between-groups" and "within-groups" are not technical terms, they are useful in describing and understanding the ANOVA model. Computer software and other texts commonly refer to these sources of variability as **treatment** (between groups) and **error** (within groups).

Table 11.2.3 ANOVA table for lamb weight gains

Source	df	SS	MS
Between diets	2	36	18.00
Within diets	9	210	23.33
Total	11	246	

SUMMARY OF FORMULAS

For convenient reference, we display in the box the definitional formulas for the basic ANOVA quantities.

ANOVA Quantities with Formulas

Source	df	SS (Sum of Squares)	MS (Mean Square)
Between groups	$I - 1$	$\sum_{i=1}^{I} n_i(\bar{y}_i - \bar{\bar{y}})^2$	SS/df
Within groups	$n. - I$	$\sum_{i=1}^{I} (n_i - 1)s_i^2$	SS/df
Total	$n. - 1$	$\sum_{i=1}^{I}\sum_{j=1}^{n_i} (y_{ij} - \bar{\bar{y}})^2$	

Exercises 11.2.1–11.2.7

11.2.1 The accompanying table shows fictitious data for three samples.

	Sample 1	2	3
	48	40	39
	39	48	30
	42	44	32
	43		35
Mean	43.00	44.00	34.00
SD	3.74	4.00	3.92

(a) Compute SS(between) and SS(within).

(b) Compute SS(total), and verify the relationship between SS(between), SS(within), and SS(total).

(c) Compute MS(between), MS(within), and s_{pooled}.

11.2.2 Proceed as in Exercise 11.2.1 for the following data:

	Sample 1	2	3
	23	18	20
	29	12	16
	25	15	17
	23		23
			19
Mean	25.00	15.00	19.00
SD	2.83	3.00	2.74

11.2.3 For the following data, SS(within) = 116 and SS(total) = 338.769.

Sample		
1	2	3
31	30	39
34	26	45
39	35	39
32	29	37
	30	

(a) Find SS(between).
(b) Compute MS(between), MS(within), and s_{pooled}.

11.2.4 The following ANOVA table is only partially completed.

Source	df	SS	MS
Between groups	3		45
Within groups	12	337	
Total		472	

(a) Complete the table.
(b) How many groups were there in the study?
(c) How many total observations were there in the study?

11.2.5 The following ANOVA table is only partially completed.

Source	df	SS	MS
Between groups	4		
Within groups		964	
Total	53	1123	

(a) Complete the table.
(b) How many groups were there in the study?
(c) How many total observations were there in the study?

11.2.6 The following ANOVA table is only partially completed.

Source	df	SS	MS
Between groups		258	
Within groups	26		
Total	29	898	

(a) Complete the table.
(b) How many groups were there in the study?
(c) How many total observations were there in the study?

11.2.7 Invent examples of data with
(a) SS(between) = 0 and SS(within) > 0
(b) SS(between) > 0 and SS(within) = 0
For each example, use three samples, each of size 5.

11.3 The Analysis of Variance Model

In Section 11.2 we introduced the notation y_{ij} for the jth observation in group i. We think of y_{ij} as a random observation from group i, where the population mean of group i is μ_i. We use analysis of variance to investigate the null hypothesis that $\mu_1 = \mu_2 = \cdots = \mu_I$. It can be helpful to think of ANOVA in terms of the following model:

$$y_{ij} = \mu + \tau_i + \varepsilon_{ij}$$

In this model, μ represents the grand population mean—the population mean when all the groups are combined. If the null hypothesis is true, then μ is the common population mean. If the null hypothesis is false, then at least some of the μ_i's differ from the grand population mean of μ.

The term τ_i represents the effect of group i—that is, the difference between the population mean for group i, μ_i, and the grand population mean, μ. (τ is the Greek letter "tau.") Thus,

$$\tau_i = \mu_i - \mu$$

The null hypothesis

$$H_0\text{: } \mu_1 = \mu_2 = \cdots = \mu_I$$

is equivalent to

$$H_0\colon \tau_1 = \tau_2 = \cdots = \tau_I = 0$$

If H_0 is false, then at least some of the groups differ from the others. If τ_i is positive, then observations from group i tend to be greater than the overall average; if τ_i is negative, then data from group i tend to be less than the overall average.

The term ε_{ij} in the model represents random error associated with observation j in group i. Thus, the model

$$y_{ij} = \mu + \tau_i + \varepsilon_{ij}$$

can be stated in words as

$$\text{observation} = \text{overall average} + \text{group effect} + \text{random error}$$

We estimate the overall average, μ, with the grand mean of the data:

$$\hat{\mu} = \bar{\bar{y}}$$

Likewise, we estimate the population average for group i with the sample average for group i:

$$\hat{\mu}_i = \bar{y}_i$$

Since the group effect is

$$\tau_i = \mu_i - \mu$$

we estimate τ_i as

$$\hat{\tau}_i = \bar{y}_i - \bar{\bar{y}}$$

Finally, we estimate the random error, ε_{ij}, for observation y_{ij} as

$$\hat{\varepsilon}_{ij} = y_{ij} - \bar{y}_i$$

Putting these estimates together, we have

$$y_{ij} = \bar{\bar{y}} + (\bar{y}_i - \bar{\bar{y}}) + (y_{ij} - \bar{y}_i)$$

or

$$y_{ij} = \hat{\mu} + \hat{\tau}_i + \hat{\varepsilon}_{ij}$$

Note. Some authors (and most computer software) use the terminology SS(error) for what we have called SS(within). This is due to the fact that the within-groups component $y_{ij} - \bar{y}_i$ estimates the random error term in the ANOVA model.

Example 11.3.1 **Weight Gain of Lambs** For the data of Example 11.2.1, the estimate of the grand population mean is $\hat{\mu} = 13$. The estimated group effects are

$$\hat{\tau}_1 = \bar{y}_1 - \bar{\bar{y}} = 11 - 13 = -2$$

$$\hat{\tau}_2 = 15 - 13 = 2$$

and

$$\hat{\tau}_3 = 12 - 13 = -1$$

Thus, we estimate that Diet 2 increases weight gain by 2 lb on average (when compared to the average of the three diets), Diet 1 decreases weight gain by an average of 2 lb, and Diet 3 decreases weight gain by 1 lb, on average. ■

When we conduct an analysis of variance, we are comparing the sizes of the sample group effects, the $\hat{\tau}_i$'s, to the sizes of the random errors in the data, the $\hat{\varepsilon}_{ij}$'s. We can see that

$$\text{SS(between)} = \sum_{i=1}^{I} n_i \hat{\tau}_i^2$$

and

$$\text{SS(within)} = \sum_{i=1}^{I}\sum_{j=1}^{n_i} \hat{\varepsilon}_{ij}^2$$

11.4 The Global F Test

The global null hypothesis is

$$H_0\text{: } \mu_1 = \mu_2 = \cdots = \mu_I$$

We consider testing H_0 against the nondirectional (or omnidirectional) alternative hypothesis

$$H_A\text{: The } \mu_i\text{'s are not all equal}$$

Note that H_0 is compound (unless $I = 2$), and so rejection of H_0 does not specify *which* μ_i's are different. If we reject H_0, then we conduct a further analysis to make detailed comparisons among the μ_i's. Testing the global null hypothesis may be likened to looking at a microscope slide through a low-power lens to see if there is anything on it; if we find something, we switch to a greater magnification to examine its fine structure.

THE F DISTRIBUTIONS

The ***F* distributions,** named after the statistician and geneticist R. A. Fisher, are probability distributions that are used in many kinds of statistical analysis. The form of an F distribution depends on two parameters: the **numerator degrees of freedom** and the **denominator degrees of freedom.** Figure 11.4.1 shows an F distribution with numerator df $= 4$ and denominator df $= 20$. Critical values for the F distribution are given in Table 10 at the end of this book. Note that Table 10 occupies 10 pages, each page having a different value of the numerator df. As a specific example, for numerator df $= 4$ and denominator df $= 20$, we find in Table 10 that $F(4, 20)_{0.05} = 2.87$; this value is shown in Figure 11.4.1.

Figure 11.4.1 The F distribution with numerator df $= 4$ and denominator df $= 20$

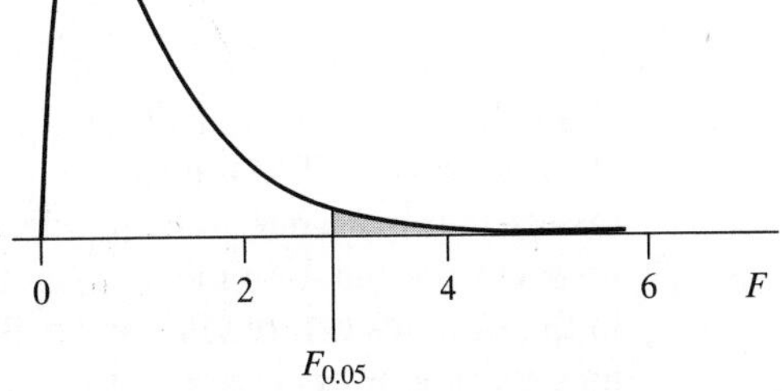

THE *F* TEST

The ***F* test** is a classical test of the global null hypothesis. The test statistic, the ***F* statistic,** is calculated as follows:

$$F_s = \frac{\text{MS(between)}}{\text{MS(within)}}$$

From the definitions of the mean squares (Section 11.2), it is clear that F_s will be large if the discrepancies among the group means ($\bar{Y}_i$'s) are large relative to the variability within the groups. Thus, large values of F_s tend to provide evidence against H_0—evidence for a difference among the group means.

To carry out the *F* test of the global null hypothesis, critical values are obtained from an *F* distribution (Table 10) with

$$\text{Numerator df} = \text{df(between)}$$

and

$$\text{Denominator df} = \text{df(within)}$$

It can be shown that (when suitable conditions for validity are met) the null distribution of F_s is an *F* distribution with df as given above.

The following example illustrates the global *F* test.

Example 11.4.1 **Weight Gain of Lambs** For the lamb feeding experiment of Example 11.2.1, the global null hypothesis and alternative can be stated verbally as

H_0: Mean weight gain is the same on all three diets.

H_A: Mean weight gain is not the same on all three diets.

or symbolically as

H_0: $\mu_1 = \mu_2 = \mu_3$

H_A: The μ_i's are not all equal

We saw in Figure 11.2.2 that the three sample means do not differ much when compared to the variability within the groups. Let us confirm this visual impression by carrying out the *F* test at $\alpha = 0.05$ and assess whether or not this difference could easily be explained by chance variation by obtaining the *P*-value for this test with these data. From the ANOVA table (Table 11.2.3) we find

$$F_s = \frac{18.00}{23.33} = 0.77$$

The degrees of freedom can also be read from the ANOVA table as

$$\text{Numerator df} = 2$$

$$\text{Denominator df} = 9$$

From Table 10 we find $F(2, 9)_{0.20} = 1.93$, so $P > 0.20$. (Computer software gives *P*-value $= 0.4907$.) Thus, there is a lack of significant evidence against H_0; there is insufficient evidence to conclude that there is any difference among the diets with respect to population mean weight gain. The observed differences in the mean gains in the samples can readily be attributed to chance variation. Because this study was an experiment (as opposed to an observational study), we can even make a slightly stronger summary of the results: There is insufficient evidence to conclude that among these three diets, diet *affects* weight gain.

RELATIONSHIP BETWEEN *F* TEST AND *t* TEST

Suppose only two groups are to be compared ($I = 2$). Then one could test H_0: $\mu_1 = \mu_2$ against H_A: $\mu_1 \neq \mu_2$ using either the F test or the t test. The t test from Chapter 7 can be modified slightly by replacing each sample standard deviation by s_{pooled}, as defined in Section 11.2, before calculating the standard error of $(\bar{Y}_1 - \bar{Y}_2)$. It can be shown that the F test and this "pooled" t test are actually equivalent procedures. The relationship between the test statistics is $t_s^2 = F_s$; that is, the value of the F statistic for any set of data is necessarily equal to the square of the value of the (pooled) t statistic. The corresponding relationship between the critical values is $t_{0.025}^2 = F_{0.05}$, $t_{0.005}^2 = F_{0.01}$, and so on. For example, suppose $n_1 = 10$ and $n_2 = 7$. Then the appropriate t distribution has df $= n_1 + n_2 - 2 = 15$, and $t_{15,0.025} = 2.131$, whereas the F distribution has numerator df $= I - 1 = 1$ and denominator df $= n. - I = 15$ so that $F(1, 15)_{0.05} = 4.54$; note that $(2.131)^2 = 4.54$. Because of the equivalence of the tests, the application of the F test to compare the means of two samples will always give exactly the same P-value as the pooled t test applied to the same data.

Exercises 11.4.1–11.4.9

11.4.1 Monoamine oxidase (MAO) is an enzyme that is thought to play a role in the regulation of behavior. To see whether different categories of patients with schizophrenia have different levels of MAO activity, researchers collected blood specimens from 42 patients and measured the MAO activity in the platelets. The results are partially summarized in the accompanying tables. (Values are expressed as nmol benzylaldehyde product/10^8 platelets/hour.)[3]

	Mao activity		
Diagnosis	Mean	SD	No. of patients
Chronic undifferentiated schizophrenia	9.81	3.62	18
Undifferentiated with paranoid features	6.28	2.88	16
Paranoid schizophrenia	5.97	3.19	8

Source	df	Sum of squares	Mean square
Between	XX	136.12	68.06
Within	XX	418.25	10.72
Total	XX	554.37	

(a) Dotplots of these data follow. Based on this graphical display, does it appear that the null hypothesis is true? Why or why not?

(b) How many degrees of freedom are there for the ANOVA F test?

(c) Compute the value of F_s.

(d) The P-value for the test is 0.004. If $\alpha = 0.05$, what is your conclusion regarding the global null hypothesis?

(e) Calculate the pooled standard deviation s_{pooled}.

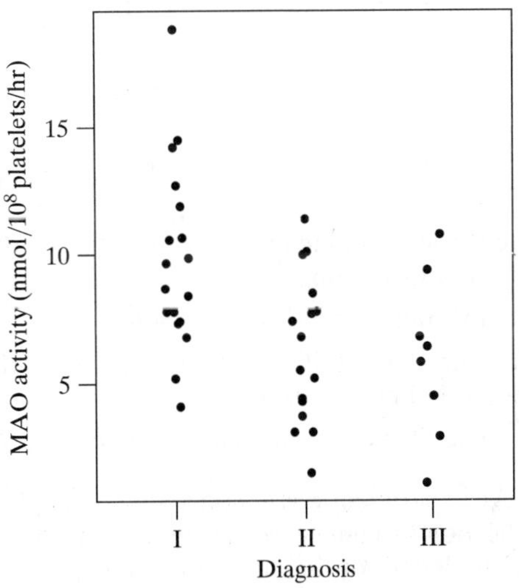

11.4.2 It is thought that stress may increase susceptibility to illness through suppression of the immune system. In an experiment to investigate this theory, 48 rats were randomly allocated to four treatment groups: no stress, mild stress, moderate stress, and high stress. The stress conditions involved various amounts of restraint and electric shock. The concentration of lymphocytes (cells/ml $\times 10^{-6}$) in the peripheral blood was measured for each rat, with the results given in the accompanying table.[4] Calculations based on the raw data yielded SS(between) = 89.036 and SS(within) = 340.24.

	No stress	Mild stress	Moderate stress	High stress
$\bar{y}$	6.64	4.84	3.98	2.92
s	2.77	2.42	3.91	1.45
n	12	12	12	12

(a) Construct the ANOVA table and test the global null hypothesis at $\alpha = 0.05$.

(b) Calculate the pooled standard deviation, s_{pooled}.

11.4.3 Human beta-endorphin (HBE) is a hormone secreted by the pituitary gland under conditions of stress. An exercise physiologist measured the resting (unstressed) blood concentration of HBE in three groups of men: 15 who had just entered a physical fitness program, 11 who had been jogging regularly for some time, and 10 sedentary people. The HBE levels (pg/ml) are shown in the following table.[5] Calculations based on the raw data yielded SS(between) = 240.69 and SS(within) = 6,887.6.

	Fitness program entrants	Joggers	Sedentary
Mean	38.7	35.7	42.5
SD	16.1	13.4	12.8
n	15	11	10

(a) State the appropriate null hypothesis in words, in the context of this setting.

(b) State the null hypothesis in symbols.

(c) Construct the ANOVA table and test the null hypothesis. Let $\alpha = 0.05$.

(d) Calculate the pooled standard deviation, s_{pooled}.

11.4.4 An experiment was conducted in which the antiviral medication zanamivir was given to patients who had the flu. The length of time until the alleviation of major flu symptoms was measured for three groups: 85 patients who were given inhaled zanamivir, 88 patients who were given inhaled and intranasal zanamivir, and 89 patients who were given a placebo. Summary statistics are given in the following table.[6] The ANOVA SS(between) is 53.67 and the SS(within) is 2034.52.

	Inhaled zanamivir	Inhaled and intranasal zanamivir	Placebo
Mean	5.4	5.3	6.3
SD	2.7	2.8	2.9
n	85	88	89

(a) State the appropriate null hypothesis in words, in the context of this setting.

(b) State the null hypothesis in symbols.

(c) How many degrees of freedom are there for the ANOVA F test?

(d) The P-value for the test is 0.034. If $\alpha = 0.05$, what is your conclusion regarding the global null hypothesis?

(e) Calculate the pooled standard deviation, s_{pooled}.

11.4.5 A researcher collected daffodils from four sides of a building and from an open area nearby. She wondered whether the average stem length of a daffodil depends on the side of the building on which it is growing. Summary statistics are given in the following table.[7] The ANOVA SS(between) is 871.408 and the SS(within) is 3588.54.

	North	East	South	West	Open
Mean	41.4	43.8	46.5	43.2	35.5
SD	9.3	6.1	6.6	10.4	4.7
n	13	13	13	13	13

(a) Dotplots of these data follow. Based on the dotplots, does it appear that the null hypothesis is true? Why or why not?

(b) State the null hypothesis in symbols.

(c) Construct the ANOVA table and test the null hypothesis. Let $\alpha = 0.10$.

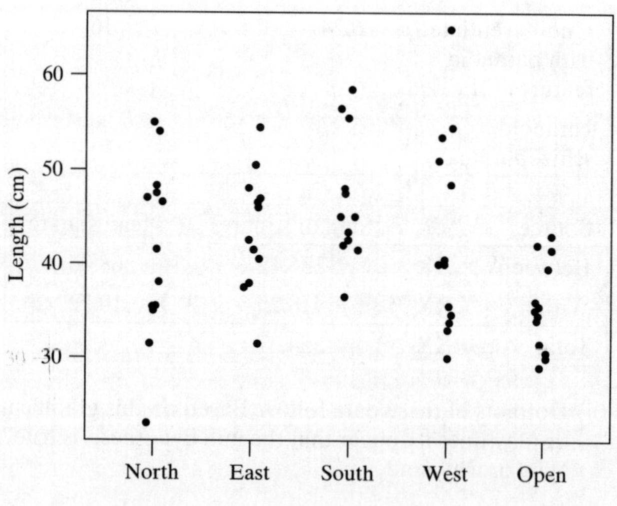

11.4.6 A researcher studied the flexibility of 10 women in an aerobic exercise class, 10 women in a modern dance class, and a control group of 9 women. One measurement she made on each woman was spinal extension, which is a measure of how far the woman could bend her back. Measurements were made before and after a 16-week training period. The change in spinal extension was recorded for each woman. Summary statistics are given in the following table.[8] The ANOVA SS(between) is 7.04 and the SS(within) is 15.08.

	Aerobics	Modern dance	Control
Mean	−0.18	0.98	0.13
SD	0.80	0.86	0.57
n	10	10	9

(a) Dotplots of these data are shown below. Based on the dotplots, does it appear that the null hypothesis is true? Why or why not?

(b) State the null hypothesis in symbols.

(c) Construct the ANOVA table and test the null hypothesis. Let $\alpha = 0.01$.

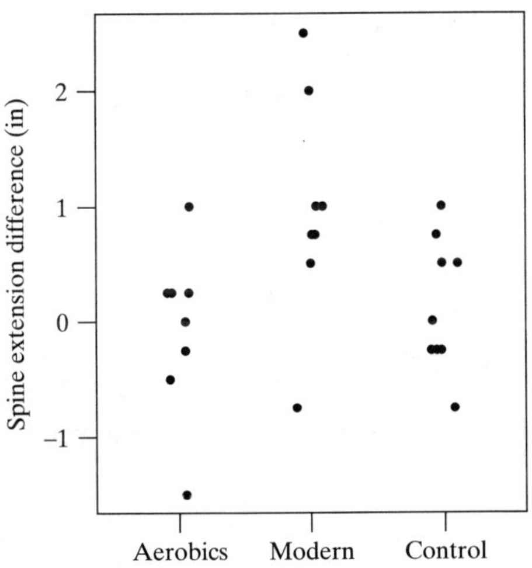

11.4.7 The following computer output is for an analysis of variance in which yields (bu/acre) of different varieties of oats were compared.[9]

Source	df	Sums of squares	Mean square	F ratio	Prob
Group	2	76.8950	38.4475	0.40245	0.6801
Error	9	859.808	95.5342		
Total	11	936.703			

(a) How many varieties (groups) were in the experiment?

(b) State the conclusion of the ANOVA.

(c) What is the pooled standard deviation, s_{pooled}?

11.4.8 Consider using an F test to test the null hypothesis of equality of population means for the example data in the following graphs. For each pair of plots, identify which would yield the larger value of F_s and briefly justify your answer.

(a)

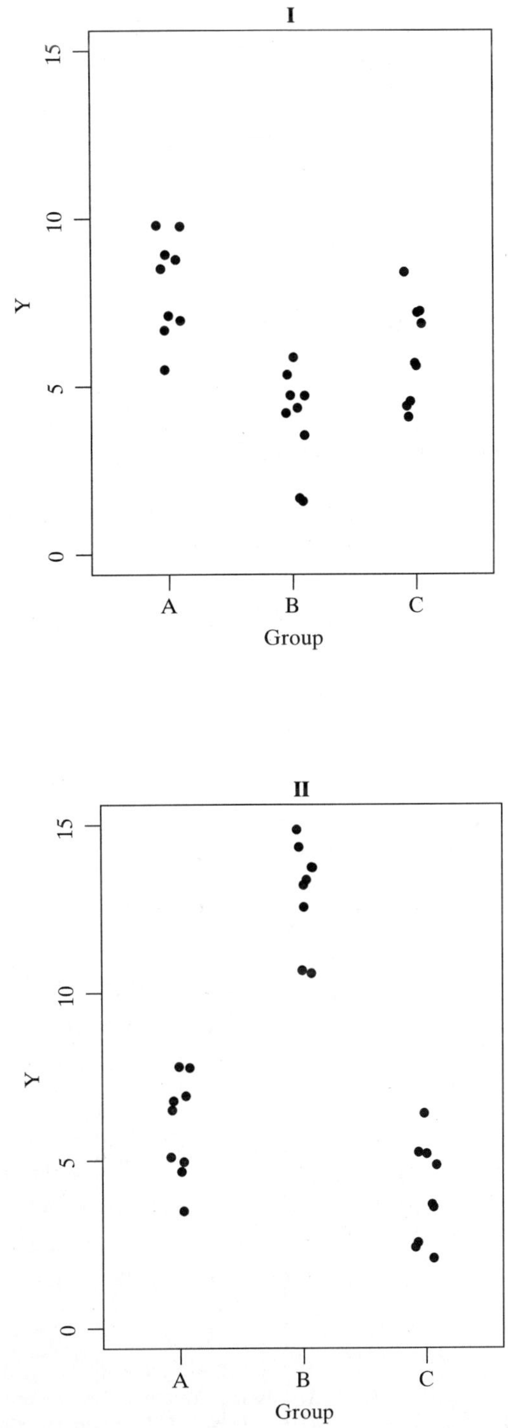

(b)

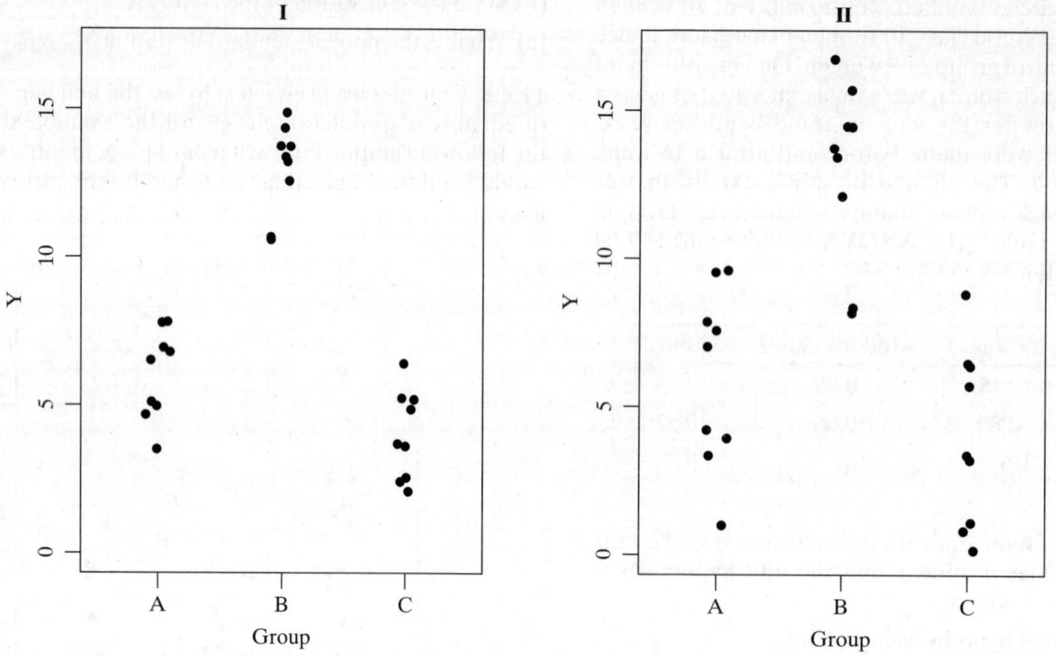

11.4.9 Radish seedlings were randomly distributed and then grown under four conditions: natural lighting, darkness, regular (incandescent) light bulb, and full spectrum light bulb (providing ultraviolet lighting). Heights of the seedlings were measured after 9 days. The data are summarized below.[10]

	Natural	Dark	Regular bulb	Full spectrum
Mean	5.72	9.06	1.40	5.16
SD	2.19	4.09	1.87	2.44
n	20	25	20	24

(a) State in words the null hypothesis that is tested using ANOVA here.

(b) The *P*-value for the *F* test is 2.9×10^{-12}. What is your conclusion regarding the null hypothesis?

(c) Is it proper to draw a cause–effect conclusion from these data? Why or why not?

11.5 Applicability of Methods

Like all other methods of statistical inference, the calculations and interpretations of ANOVA are based on certain conditions.

STANDARD CONDITIONS

The ANOVA techniques described in this chapter, including the global *F* test, are valid if the following conditions hold.

1. *Design conditions*
   (a) It must be reasonable to regard the groups of observations as random samples from their respective populations.
   (b) The I samples must be independent of each other.
2. *Population conditions* The I population distributions must be (approximately) normal with equal standard deviations:*

$$\sigma_1 = \sigma_2 = \cdots = \sigma_I$$

*There is a variation on standard ANOVA that does not require equal standard deviations. Most of the examples presented in this chapter analyze experimental data, for which it is usually reasonable to expect equal SDs. For observational data, the condition of equal SDs is more questionable, and so-called "Welch ANOVA" may be needed.

These conditions are extensions of the conditions given in Chapter 7 for the independent-samples t test with the added condition that the standard deviations be equal. The condition of normal populations with equal standard deviations is less crucial if the sample sizes (n_i) are large and approximately equal.

VERIFICATION OF CONDITIONS

The design conditions may be verified as for the independent-samples t test. To check condition 1(a), one looks for biases or hierarchical structure in the collection of the data. A completely randomized design assures independence of the samples [condition 1(b)]. If units have been allocated to treatment groups in a nonrandom manner (e.g., by a randomized blocks design to be discussed in Section 11.6), or if observations on the same experimental unit appear in different samples (e.g., for $I = 2$, paired data as seen in Chapter 9), then the samples are not independent.

As with the independent-samples t test, the population conditions can be roughly checked from the data. To check normality, a separate histogram or normal quantile plot can be made for each sample. Another option is to make a single histogram or normal quantile plot of the deviations ($y_{ij} - \bar{y}_i$) from all the samples combined. In the context of ANOVA we call these deviations from the group means **residuals.** Thus, a residual measures how far a data value falls from its respective group mean.

Equality of the population SDs is checked by comparing the sample SDs; one useful trick is to plot the SDs against the means ($\bar{y}_i$'s) to check for a trend. Another approach is to make a plot of the residuals ($y_{ij} - \bar{y}_i$) against the means ($\bar{y}_i$'s). As a rule of thumb, we would like the largest sample SD divided by the smallest sample SD to be less than 2 or so. If this ratio is much larger than 2, then we cannot be confident in the P-value from the ANOVA, particularly if the sample sizes are small and unequal. In particular, if the sample sizes are unequal and the sample SD from a small sample is quite a bit larger than the other SDs, then the P-value can be quite inaccurate.

Example 11.5.1 **Weight Gain of Lambs** Consider the lamb feeding experiment of Example 11.2.2. Figure 11.2.1 (in Section 11.2) shows that the variability within groups is nearly equal across the three diets: The three sample SDs are 4.36, 4.95, and 4.97. Figure 11.5.1 is

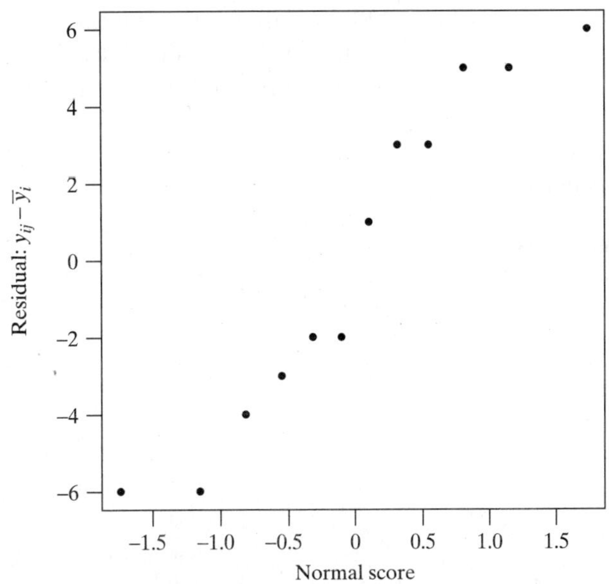

Figure 11.5.1 Normal quantile plot of residuals ($y_{ij} - \bar{y}_i$) in weight-gain data

a normal quantile plot of the 12 residuals $(y_{ij} - \bar{y}_i)$ (3 from Diet 1, 5 from Diet 2, and 4 from Diet 3). This plot is close to linear, which provides no evidence to cast doubt on the normality condition. ■

Example 11.5.2

Sweet Corn Consider the sweet corn data of Example 11.1.1. Figure 11.5.2(a) shows the data with each group receiving its own plotting symbol. Using those same plotting symbols for each group, Figure 11.5.2(b) displays the residuals $(y_{ij} - \bar{y}_i)$ plotted against each group's mean $(\bar{y}_i)$ (also known as a **fitted value** in the context of ANOVA). This second graph shows that the variability (as measured visually by the vertical spread) does not appreciably change as the mean changes (which is good—if the variability increased as the mean increased, then condition 2 would be violated). ■

Figure 11.5.2 Plot of residuals versus sample mean for the sweet corn data

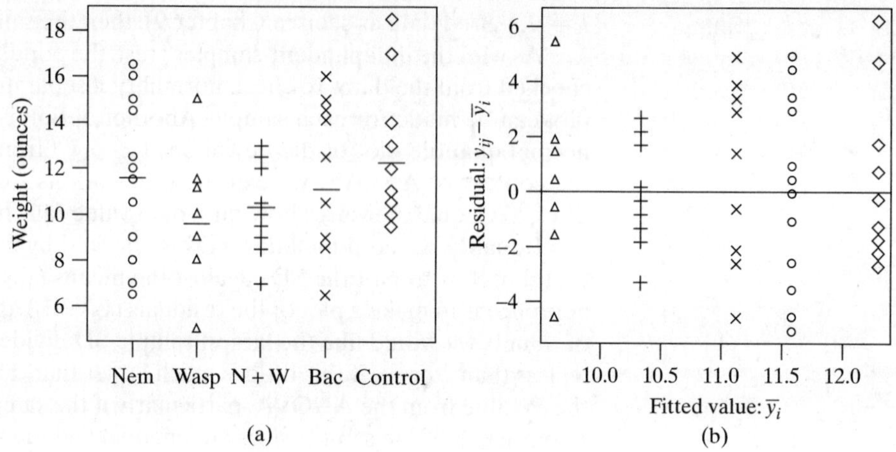

While one could look at a basic plot of the data, as in Figure 11.5.2(a), to visually inspect that the SDs are similar across all groups, plotting the data as in Figure 11.5.2(b) provides some visual advantages. First, by examining the residuals [Figure 11.5.2(b)] and not the raw data [Figure 11.5.2(a)], one can scan the graph from left to right allowing the eyes to more clearly compare the variability among the groups without being distracted by the changing means. Furthermore, a common violation of the equal SD requirement is that the group SDs grow with the means. To illustrate this violation, consider the fictitious data graphed in Figure 11.5.3(a) consisting of five treatment groups and seven observations per group. Clearly the variability is not the

Figure 11.5.3 Plot of residuals versus sample means for a fictitious data set for which the standard deviation increases with the mean

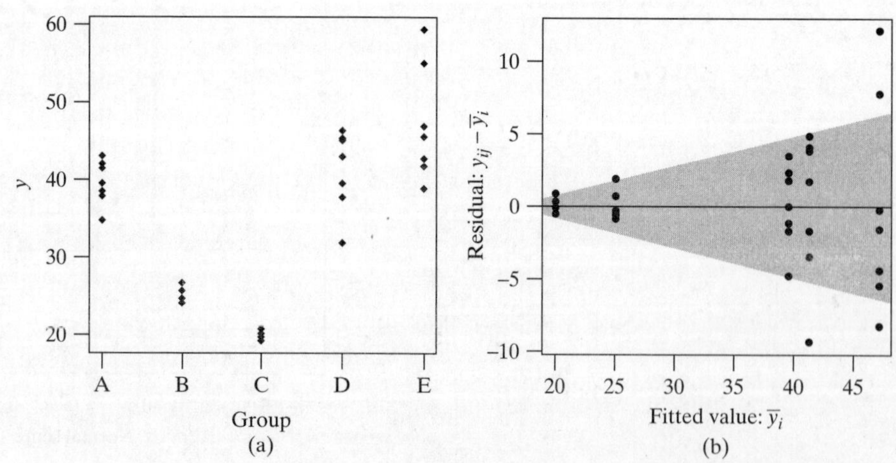

same in all five groups. The plot of the residuals versus means in Figure 11.5.3(b) exposes this problem more clearly and shows that the SD (represented by vertical spread) increases with the mean. We often describe this as *funnel* or *horn* shape in the residuals.

Example 11.5.3

Sweet Corn Again considering the sweet corn data of Example 11.2.1, we examine the normality of the groups through examination of the residuals. Figure 11.5.4 contains a histogram and a normal quantile plot of the 60 residuals $(y_{ij} - \bar{y}_i)$. The bell-shape nature of plot (a) and linearity of plot (b) cast little doubt upon the normality condition. ■

Figure 11.5.4 Histogram and normal quantile plot of deviations $(y_{ij} - \bar{y}_i)$ in sweet corn data

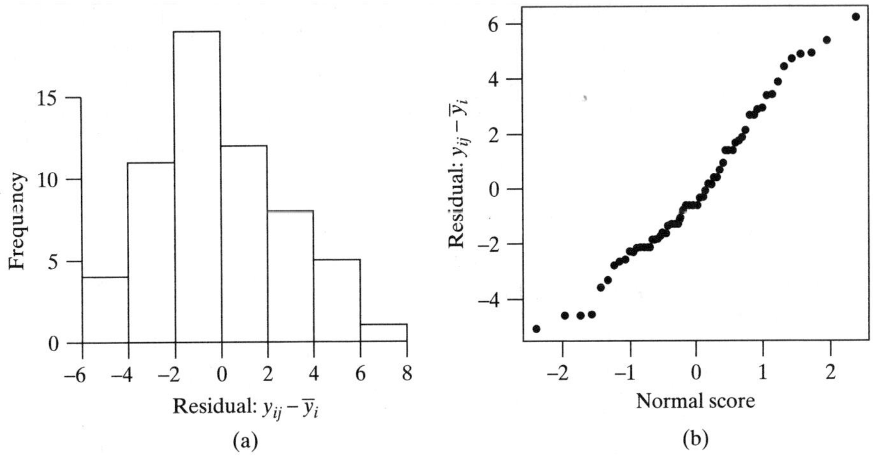

FURTHER ANALYSIS

In addition to their relevance to the F test, the standard conditions underlie many classical methods for further analysis of the data.

If the I populations have the same SD, then a pooled estimate of that SD from the data is

$$s_{\text{pooled}} = \sqrt{\text{MS(within)}}$$

from the ANOVA. This pooled standard deviation s_{pooled} is a better estimate than any individual sample SD because s_{pooled} is based on more observations.

A simple way to see the advantage of s_{pooled} is to consider the standard error of an individual sample mean, which can be calculated as

$$\text{SE}_{\bar{Y}} = \frac{s_{\text{pooled}}}{\sqrt{n}}$$

where n is the size of the individual sample. The df associated with this standard error is df(within), which is the sum of the degrees of freedom of all the samples. By contrast, if the individual SD were used in calculating $\text{SE}_{\bar{Y}}$, it would have only $(n - 1)$ df. When the SE is used for inference, larger df yield smaller critical values (see Table 4), which in turn lead to improved power and narrower confidence intervals.

In optional Sections 11.7 and 11.8 we will consider methods for detailed analysis of the group means $\bar{Y}_1, \bar{Y}_2, \ldots, \bar{Y}_I$. Like the F test, these methods were designed for independent samples from normal populations with equal standard deviations. The methods use standard errors based on the pooled standard deviation estimate s_{pooled}.

Exercises 11.5.1–11.5.4

11.5.1 Refer to the lymphocyte data of Exercise 11.4.2. The global F test is based on certain conditions concerning the population distributions.

(a) State the conditions.

(b) Which features of the data suggest that the conditions may be doubtful in this case?

11.5.2 Refer to the lymphocyte data of Exercise 11.4.2.

(a) Suppose the lymphocyte concentration for one of the 12 rats in the mild stress group was 5.21. What is the value of the residual for this rat?

(b) Suppose one of the rats in the high stress group had a residual of -2.10 cells/ml $\times$ 10^{-6}. Interpret this residual in the context of the problem. That is, what does the residual tell us about this particular rat?

11.5.3 Patients with advanced cancers of the stomach, bronchus, colon, ovary, or breast were treated with ascorbate. The purpose of the study was to determine if the survival times differ with respect to the organ affected by the cancer. The variable of interest is survival time (in days).[11] Here are parallel dotplots of the raw data.

An ANOVA was done after a square root transformation was applied to the raw data. There were two (related) reasons that the data were transformed. What were those two reasons?

11.5.4 Refer to the lymphocyte data of Exercise 11.4.2.

(a) Looking at the summary data (means, SDs, ns) for the four groups, why might the conclusion of the ANOVA F test be questioned?

(b) Despite the concern raised in part (a), why might we trust the ANOVA F test?

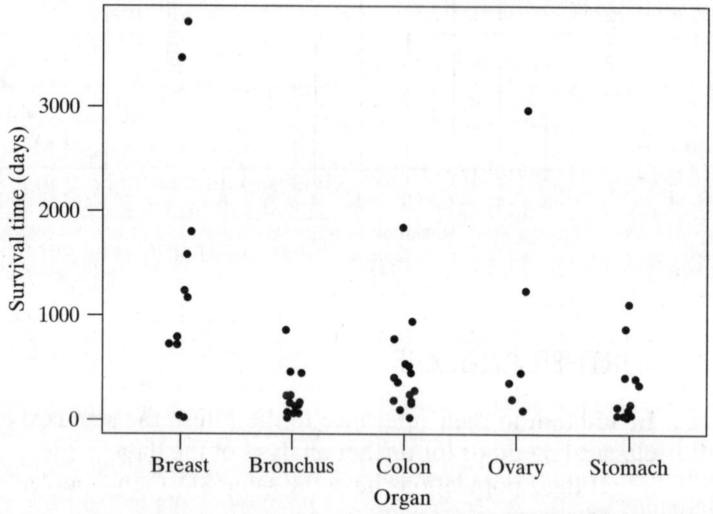

11.6 One-Way Randomized Blocks Design

The completely randomized design makes no distinctions among the experimental units. Often an experiment can be improved by a more refined approach, one that takes advantage of known patterns of variability in the experimental units.

In a **randomized blocks design,** we first group the experimental units into sets, or **blocks,** of relatively similar units and then we randomly allocate treatments within each block. Here is an example.

Example 11.6.1 **Alfalfa and Acid Rain** Researchers were interested in the effect that acid has on the growth rate of alfalfa plants. They created three treatment groups in an experiment: low acid, high acid, and control. The response variable in their experiment was the height of the alfalfa plants in a Styrofoam cup after 5 days of growth.* They had 5 cups for each of the 3 treatments, for a total of 15 observations. However, the

*More precisely, the response variable was the average height of plants within a cup, so the observational unit was a cup, rather than individual plants.

Figure 11.6.1 Design of the alfalfa experiment

Window	Block 1	Block 2	Block 3	Block 4	Block 5
	high	control	control	control	high
	control	low	high	low	low
	low	high	low	high	control

Organization of blocks for alfalfa experiment

cups were arranged near a window and they wanted to account for the effect of differing amounts of sunlight. Thus, they created 5 blocks—each block was a fixed distance away from the window (block 1 being the closest through block 5, the farthest). Within each block the three treatments were randomly assigned, as shown in Figure 11.6.1.[12] ■

Example 11.6.1 is an illustration of a randomized blocks design. To carry out a randomized blocks design, the experimenter creates or identifies suitable blocks of experimental units and then randomly assigns treatments within each block in such a way that each treatment appears in each block.* In Example 11.6.1, the rows of cups at each of the five distances from the window serve as blocks. In general, we create blocks in order to reduce or eliminate variability caused by extraneous variables so that the precision of the experiment is increased. We want the experimental units within a block to be homogenous; we want the extraneous variability to occur *between* the blocks. Here are more examples of randomized blocks designs in biological experiments.

Example 11.6.2

Blocking by Litter How does experience affect the anatomy of the brain? In a typical experiment to study this question, young rats are placed in one of three environments for 80 days:

T_1: *Standard environment.* The rat is housed with a single companion in a standard lab cage.

T_2: *Enriched environment.* The rat is housed with several companions in a large cage, furnished with various playthings.

T_3: *Impoverished environment.* The rat lives alone in a standard lab cage.

At the end of the 80-day experience, various anatomical measurements are made on the rats' brains.

Suppose a researcher plans to conduct the above experiment using 30 rats. To minimize variation in response, all 30 animals will be male, of the same age and strain. To reduce variation even further, the researcher can take advantage of the similarity of animals from the same litter. In this approach, the researcher would obtain three male rats from each of 10 litters. The three littermates from each litter would be assigned at random: one to T_1, one to T_2, and one to T_3.[13] ■

Another way to visualize the experimental design is in tabular form, as shown in Table 11.6.1. Each "Y" in the table represents an observation on one rat (Y_{ij} represents the measurement of rat i who received treatment j). Using the layout of Table 11.6.1, the experimenter can compare the responses of rats that received *different* treatments but are in the *same* litter. Such comparisons are not affected by any difference (genetic and other) that may exist between one litter and another.

*Strictly speaking, the design we discuss is termed a *randomized complete blocks design* because every treatment appears in every block. In an *incomplete blocks design*, each block contains some, but not necessarily all, of the treatments.

Table 11.6.1 Format for rat brain data

	Treatment		
	T_1	T_2	T_3
Litter 1	$Y_{1,1}$	$Y_{1,2}$	$Y_{1,3}$
Litter 2	$Y_{2,1}$	$Y_{2,2}$	$Y_{2,3}$
Litter 3	$Y_{3,1}$	$Y_{3,2}$	$Y_{3,3}$
·	·	·	·
·	·	·	·
·	·	·	·
Litter 10	$Y_{10,1}$	$Y_{10,2}$	$Y_{10,3}$

Example 11.6.3

Within-Subject Blocking (Pairing) A dermatologist is planning a study to compare two medicated lotions for their effectiveness in treating acne. Twenty patients are to participate in the study. Each patient will use lotion A on one side of his or her face and lotion B on the other; the dermatologist will observe the improvement on each side during a 3-month period. For each patient, the side of the face to receive lotion A is randomly selected; the other side receives lotion B. The bottles of medication have coded labels so that neither the patient nor the physician knows which bottle contains A and which contains B—that is, in addition to blocking, the experiment also makes use of blinding.[14] This example, with blocks of size 2, is an example of pairing: The left side of the face is paired with the right side of the face. We have considered the analysis of paired data in Chapter 8. ■

Example 11.6.4

Blocking in an Agricultural Field Study When comparing several varieties of grain, an agronomist will generally plant many field plots of each variety and measure the yield of each plot. Differences in yields may reflect not only genuine differences among the varieties, but also differences among the plots in soil fertility, pH, water-holding capacity, and so on. Consequently, the spatial arrangement of the plots in the field is important. An efficient way to use the available field area is to divide the field into large regions—the blocks—and to subdivide each block into several plots. Within each block the various varieties of grain are then randomly allocated to the plots, with a separate randomization done for each block. For instance, suppose we want to test four varieties of barley. Then each block would contain four plots. The resulting randomized allocation might look like Figure 11.6.2, which is a schematic map of the field. The "treatments" T_1, T_2, T_3, and T_4 are the four varieties of barley. ■

Figure 11.6.2 Layout of an agricultural randomized blocks design

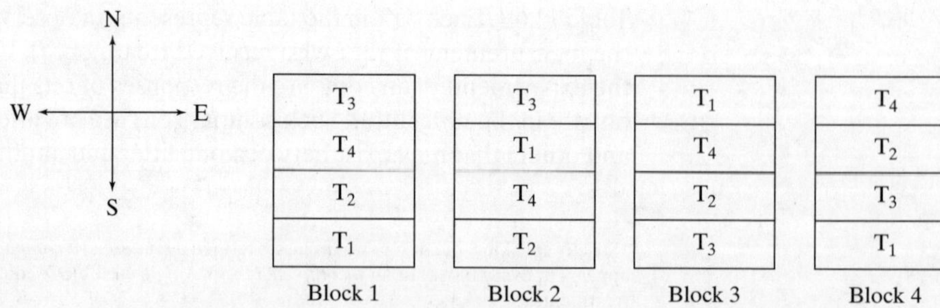

CREATING THE BLOCKS

As the preceding examples show, blocking is a way of *organizing* the inherent variation that exists among experimental units. Ideally, the blocking should be arranged so as to increase the information available from the experiment. To achieve this goal, *the experimenter should try to create blocks that are as homogeneous within themselves as possible so that the inherent variation between experimental units becomes, as far as possible, variation between blocks rather than within blocks.* This principle was illustrated in the preceding examples (e.g., in Example 11.6.2, where blocking by litter exploits the fact that littermates are more similar to each other than to nonlittermates). The following is another illustration.

Example 11.6.5 **Agricultural Field Study** For the barley experiment of Example 11.6.4, how would agronomists determine the best arrangement or layout of blocks in a field? They would design the blocks to take advantage of any prior knowledge they may have of fertility patterns in the field. For instance, if they know that an east–west fertility gradient exists in the field (perhaps the field slopes from east to west, with the result that the west end has a thicker layer of good soil or receives better irrigation), then they might choose blocks as in Figure 11.6.2; the layout maximizes soil differences between the blocks and minimizes differences between plots within each block. (But even if a field appears to be uniform, blocking is usually used in agronomic experiments, because plots closer together in the field are generally more similar than plots farther apart.) ■

To add solidity to this example, let us look at a set of data from a randomized blocks experiment on barley. Each entry in Table 11.6.2 shows the yield (bushels of barley per acre) of a plot 3.5 ft wide by 80 ft long.[15]

Table 11.6.2 Yield (lb) of barley

	Block 1	Block 2	Block 3	Block 4	Variety mean
Variety 1	93.5	66.6	50.5	42.4	63.3
Variety 2	102.9	53.2	47.4	43.8	61.8
Variety 3	67.0	54.7	50.0	40.1	53.0
Variety 4	86.3	61.3	50.7	46.4	61.2
Block mean	87.4	59.0	49.7	43.2	

It appears from Table 11.6.2 that the yield potential of the blocks varies greatly; the data clearly indicate a fertility gradient from block 1 to block 4. Because of the blocked design, comparison of the varieties is relatively unaffected by the fertility gradient. Of course, there also appears to be substantial variation within blocks. [You might find it an interesting exercise to peruse the data and ask yourself whether the observed differences between varieties are large enough to conclude that, for example, variety 1 is superior (in mean yield) to variety 3; use your intuition rather than a formal statistical analysis. The truth is revealed in Note 15.]

THE RANDOMIZATION PROCEDURE

Once the blocks have been created, the blocked allocation of experimental units is straightforward: It is as if a mini-experiment is conducted within each block. Randomization is carried out for each block separately, as illustrated in the following example.

Example 11.6.6 **Agricultural Field Study** Consider the agricultural field experiment of Example 11.6.4. In block 1, let us label the plots 1, 2, 3, 4, from north to south (see Figure 11.6.2); we will allocate one plot to each variety. The allocation proceeds as for the completely randomized design, by choosing plots at random from the four, and assigning the first plot chosen to T_1, the second to T_2, and so on. For instance, using a computer to randomly permute the numbers 1 through 4 (or even shuffled cards numbered 1 through 4) we might obtain the sequence 4, 3, 1, 2, which would lead to the following treatment allocation.

Block 1

T_1: Plot 4

T_2: Plot 3

T_3: Plot 1

T_4: Plot 2

This is in fact the assignment shown in Figure 11.6.2 for block 1. We can then repeat this procedure for blocks 2, 3, and so on. ■

ANALYZING DATA FROM A RANDOMIZED BLOCK EXPERIMENT

In the same way we cannot use a two-sample t test when data are paired, when an experiment has been blocked, we no longer can use our ANOVA methods of Section 11.4. Instead, we will use a **randomized blocks ANOVA** model. We will illustrate the concepts as we reconsider the alfalfa and acid rain experiment of Example 11.6.1 in which the researchers blocked the experiment based on rows of cups placed parallel to a window so that each block has roughly the same light exposure. The data are given in Table 11.6.3 and are graphed in Figure 11.6.3.

Table 11.6.3 Alfalfa plant height after 5 days (cm)

	High acid	Low acid	Control	Block mean
Block 1	1.30	1.78	2.67	1.917
Block 2	1.15	1.25	2.25	1.550
Block 3	0.50	1.27	1.46	1.077
Block 4	0.30	0.55	1.66	0.837
Block 5	1.30	0.80	0.80	0.967
Treatment mean = $\bar{y}_i$	0.910	1.130	1.768	
n	5	5	5	

Our usual ANOVA null hypothesis for comparing I populations or treatments is

$$H_0: \mu_1 = \mu_2 = \cdots = \mu_I$$

Example 11.6.7 **Alfalfa and Acid Rain** The null hypothesis for the alfalfa growth experiment is that acid has no effect on 5-day growth. (We can make a strong causal claim like this because this was an experiment.) More directly, the null hypothesis is that the mean 5-day growth is the same for all three treatments (high acid, low acid, and control).

$$H_0: \mu_1 = \mu_2 = \mu_3$$

■

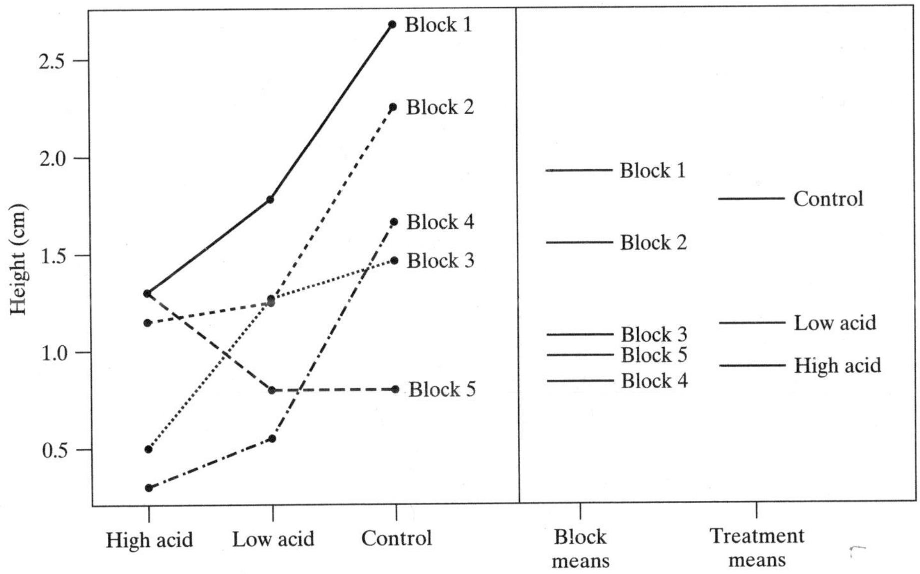

Figure 11.6.3 Dotplots of the alfalfa growth data with a summary of block and treatment means

This hypothesis can be tested with an analysis of variance *F* test, but first we want to remove the variability in the data that is due to differences between the blocks. To do this, we extend the ANOVA model presented in Section 11.3 to the following model:

$$y_{ijk} = \mu + \tau_i + \beta_j + \varepsilon_{ijk}$$

In this model y_{ijk} is the *k*th observation when treatment *i* is applied in block *j*. (In Example 11.6.1 there is only one observation for each treatment in each block, but in general there might be more than one.) Here, as before, μ represents the grand population mean and the term τ_i represents the effect of group *i* (i.e., treatment *i*). The new term in the model is β_j, which represents the effect of the *j*th block.

VISUALIZING THE BLOCK EFFECTS

To visualize how blocking affects our ANOVA, we can think of our model in a slightly different way:

$$(y_{ijk} - \tau_i) = \mu + \beta_j + \varepsilon_{ijk}$$

The left-hand side of the equation describes the data after treatment effects have been removed. With our data we estimate this left-hand side as

$$y_{ijk} - \hat{\tau}_i = y_{ijk} - (\bar{y}_{i.} - \bar{\bar{y}})$$

That is, within each treatment group, the treatment mean is subtracted from each data value.* We've seen this before—in the context of a one-way ANOVA (Section 11.2), we called these deviations or residuals. Figure 11.6.4 is a plot of the deviations from the treatment means for the alfalfa data broken down by block. We can see that there is still a lot of structure in the data: The mean deviations in blocks 1 and 2 are greater than zero while blocks 3, 4, and 5 are below zero (corresponding to above average growth near the window and below average growth farther from the

*Here we write $\bar{y}_{i\bullet}$ rather than $\bar{y}_i$ to distinguish the treatment means from the block means $\bar{y}_{\bullet j}$.

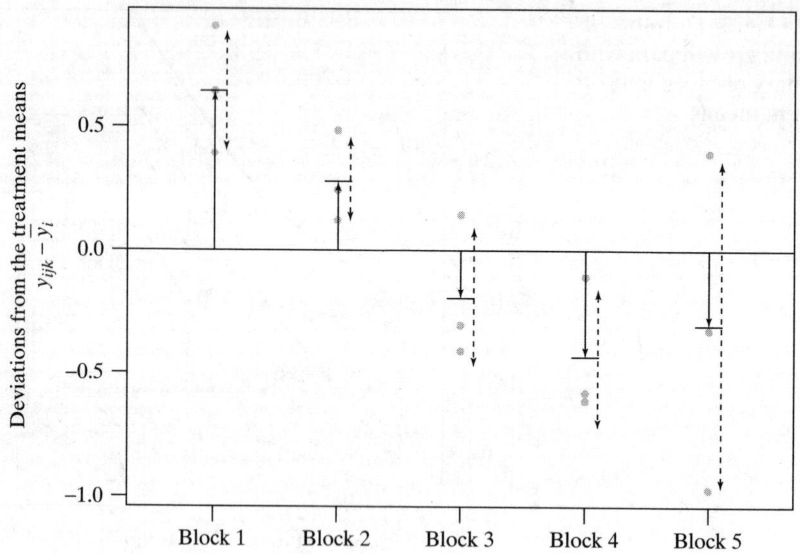

Figure 11.6.4 Deviations from the treatment means for the alfalfa growth data by blocks

window). The fact that these mean deviations are not all zero is a consequence of the variability due to the blocks. Soon we shall describe how to measure the variability of these mean deviations for the blocks through the **mean squares for blocks,** or **MS(blocks).**

To visualize how accounting for this block-to-block variation improves our ability to detect treatment effects, consider the alfalfa and acid rain data graphed in Figure 11.6.5. Figure 11.6.5(a) displays the growth data for each treatment group and simply ignores the blocks entirely while Figure 11.6.5(b) displays the growth data after adjusting for the estimated block effects.* While the variability among the

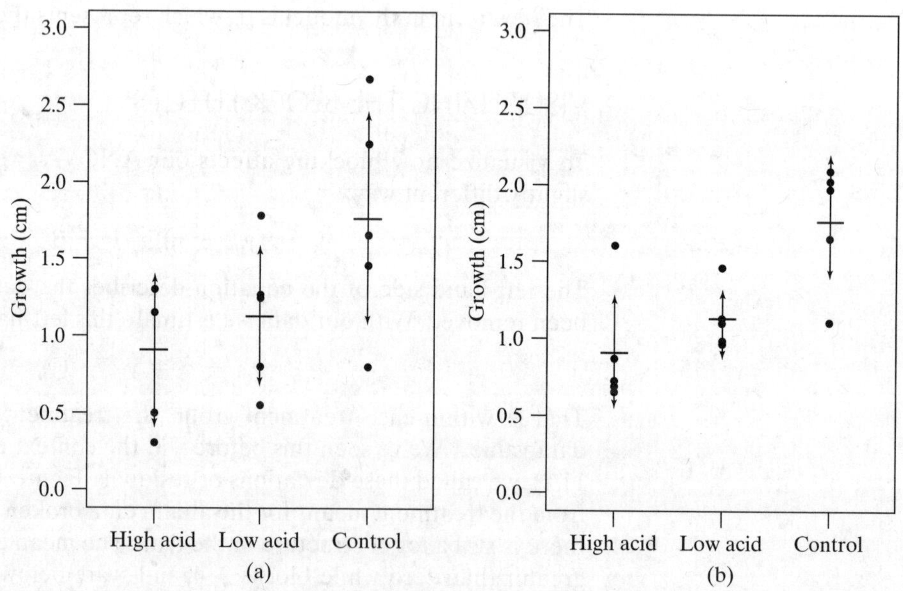

Figure 11.6.5 Visualizing the effect of blocking when comparing mean growth under the three acid treatments in the alfalfa experiment. Plot (a) displays the raw growth data while (b) displays the growth data after adjusting for the estimated block effects. Treatment means are indicated by horizontal lines and within-group standard deviations by arrows

*To account for the blocking, the adjusted growth data on the y-axis for each treatment group is computed as $y_{ijk} - \bar{y}_{\cdot} + \bar{y}$.

treatment means is unchanged between the plots, we observe that the variability within the treatment groups is much smaller after accounting for the blocks and thus the differences among the treatments are more pronounced.

THE ONE-WAY RANDOMIZED COMPLETE BLOCK *F* TEST

Recall that the ANOVA F test is a ratio that compares the variability among the treatment means to the within-group variability. As seen in Figure 11.6.5, accounting for the blocks has reduced the within-group variability and will thus increase the F statistic value. We now briefly discuss the computations involved in computing the ANOVA table for the randomized complete block F test.

In Section 11.2 for a one-way ANOVA, we discussed how the total sum of squares, SS(total), is broken down into SS(between), which measures variability attributed to differences among the treatment means, and SS(within), which measures unexplained random variation in the data. For a randomized blocks experiment, we write SS(treatments) rather than SS(between) to describe the variability between treatment means to be clear that we're measuring variability between treatments and not blocks. For a randomized blocks experiment we also split the one-way ANOVA SS(within) into two parts: SS(blocks), which measures variability among the block means, and SS(within), which measures the remaining unexplained variation in the data. Thus, we have

One-way ANOVA: SS(total) = SS(within) + SS(between)

One-way ANOVA with blocks: SS(total) = SS(within) + SS(blocks) + SS(treatments)

Usually we are not interested in testing a hypothesis about the blocks, but nonetheless we want to take into consideration the effect that blocking has on the response variable. Refining the one-way ANOVA by calculating SS(blocks) accomplishes this goal and furthermore, if blocks are chosen wisely, can lead to more powerful tests.

Computing the sums of squares is typically left to a computer and rarely performed by hand. Nonetheless, the formulas are worth noting as they mathematically reveal how the blocks are being accounted for.

The **mean squares between blocks** is calculated in a manner similar to our computation of MS(between) from the one-way ANOVA of Section 11.2. Roughly speaking, we compute a sort of weighted variance of the block means in which we weight the differences between a block mean and the overall mean by the block sample size. If we define the average of the observations in block j to be $\bar{y}_{\cdot j}$ and we let m_j denote the number of observations in block j, then the mean squares due to blocks is defined as follows:

Mean Squares Between Blocks

$$\text{MS(blocks)} = \frac{\sum_{j=1}^{J} m_j(\bar{y}_{\cdot j} - \bar{\bar{y}})^2}{J - 1}$$

Analogous to our formulas in Section 11.2 we define SS(blocks) and df(blocks) as the numerator and denominator of MS(blocks) as follows:

Sum of Squares and df Between Blocks

$$\text{SS(blocks)} = \sum_{j=1}^{J} m_j(\bar{y}_{\bullet j} - \bar{\bar{y}})^2$$

$$\text{df(blocks)} = J - 1$$

As noted previously, the blocking reduces MS(within). To compute MS(within) for the randomized complete block experiment we compute

$$\text{SS(within)} = \text{SS(total)} - \text{SS(treatment)} - \text{SS(blocks)}$$

where SS(treatment) and SS(total) are computed as in Section 11.2. As sums of squares are always nonnegative values, the preceding formula shows directly how the blocks reduce the within-group variability.

Similarly, to compute df(within) for the randomized complete block experiment, we have

$$\begin{aligned} \text{df(within)} &= \text{df(total)} - \text{df(treatment)} - \text{df(blocks)} \\ &= (n_{\bullet} - 1) - (I - 1) - (J - 1) \\ &= n_{\bullet} - I - J + 1 \end{aligned}$$

Example 11.6.8 **Alfalfa and Acid Rain** For the alfalfa growth data in Table 11.6.2, the total of all the observations is $1.30 + 1.15 + \cdots + 0.80 = 19.04$ and the grand mean is

$$\bar{\bar{y}} = \frac{19.04}{15} = 1.269$$

We calculate

$$\text{SS(treatments)} = 5(0.910 - 1.269)^2 + 5(1.130 - 1.269)^2 + 5(1.768 - 1.269)^2 = 1.986$$

Since $I = 3$, we have

$$\text{df(treatments)} = 3 - 1 = 2$$

so that

$$\text{MS(treatments)} = \frac{1.986}{2} = 0.993$$

We calculate

$$\begin{aligned} \text{SS(blocks)} &= 3(1.917 - 1.269)^2 + 3(1.550 - 1.269)^2 \\ &\quad + 3(1.077 - 1.269)^2 + 3(0.837 - 1.269)^2 \\ &\quad + 3(0.967 - 1.269)^2 \\ &= 2.441 \end{aligned}$$

Since $J = 5$, we have

$$\text{df(blocks)} = 5 - 1 = 4$$

and

$$\text{MS(blocks)} = \frac{2.441}{4} = 0.610$$

The total sum of squares is found as $(1.30 - 1.269)^2 + \cdots + (0.80 - 1.269)^2 = 5.879$.

By subtraction, we compute SS(within):

$$\begin{aligned}\text{SS(within)} &= \text{SS(total)} - \text{SS(treatments)} - \text{SS(blocks)} \\ &= 5.879 - 1.986 - 2.441 = 1.452\end{aligned}$$

Similarly, we compute df(within) as

$$\text{df(within)} = \text{df(total)} - \text{df(treatments)} - \text{df(blocks)}$$

which in this case gives us $14 - 2 - 4 = 8$.

Thus, $\text{MS(within)} = \dfrac{1.452}{8} = 0.182$. ■

The sums of squares, degrees of freedom, and resulting mean squares are collected in an expanded ANOVA table, which includes a line for the effect of the blocks.

To test the null hypothesis, we calculate

$$F_s = \frac{\text{MS(treatments)}}{\text{MS(within)}}$$

and reject H_0 if the P-value is sufficiently small.

Example 11.6.9 **Alfalfa and Acid Rain** For the alfalfa growth data of Example 11.6.1, the ANOVA summary is given in Table 11.6.4. The F statistic is $0.993/0.182 = 5.47$, with degrees of freedom 2 for the numerator and 8 for the denominator. From Table 10 we bracket the P-value as $0.02 < P\text{-value} < 0.05$. (Using a computer gives $P\text{-value} = 0.0318$.) The P-value is small, indicating that the differences between the three sample means are greater than would be expected by chance alone. There is significant evidence that acid affects the growth of alfalfa plants. (It is worth noting that if we ignore the blocks and conduct an erroneous one-way ANOVA, we would find $P\text{-value} = 0.0842$, which would not provide significant evidence for an acid effect at $\alpha = 0.05$.) ■

Table 11.6.4 ANOVA table for alfalfa experiment

Source	df	SS	MS	F ratio
Between treatments	2	1.986	0.993	5.47
Between blocks	4	2.441	0.610	
Within groups	8	1.452	0.182	
Total	14	5.879		

Exercises 11.6.1–11.6.12

(*Note:* In several of these exercises you are asked to prepare a randomized allocation. For this purpose you can use either Table 1, random digits from your calculator, or a computer.)

11.6.1 In an experiment to compare six different fertilizers for tomatoes, 36 individually potted seedlings are to be used, 6 to receive each fertilizer. The tomato plants will be grown in a greenhouse, and the total yield of tomatoes will be observed for each plant. The experimenter has decided to use a randomized blocks design: The pots are to be arranged in six blocks of 6 plants each on the greenhouse bench. Two possible arrangements of the blocks are shown in the accompanying figure.

Arrangement I:

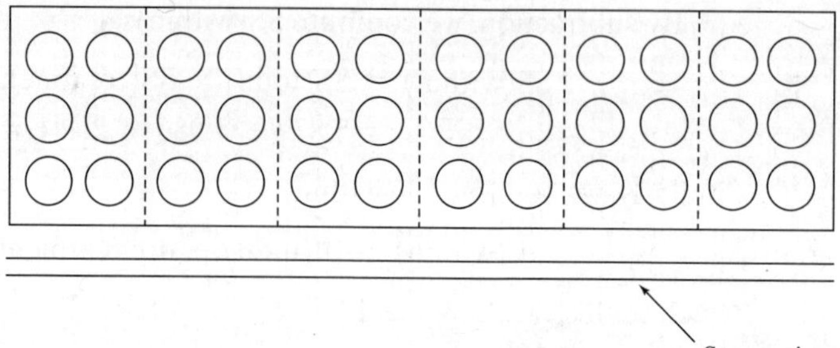

Arrangement II:

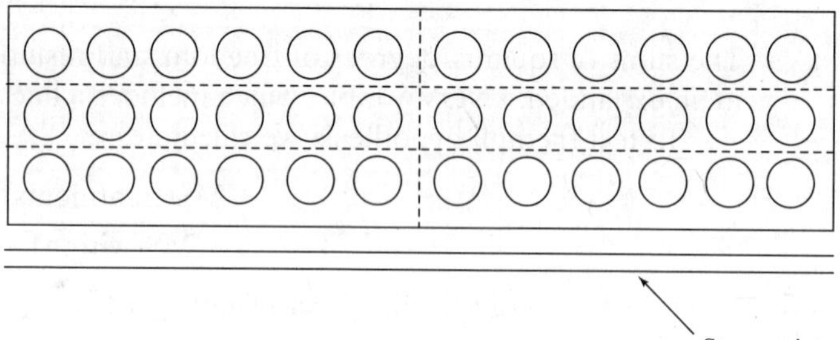

One factor that affects tomato yield is temperature, which cannot be held exactly constant throughout the greenhouse. In fact, a temperature gradient across the bench is likely. Heat for the greenhouse is provided by a steam pipe that runs lengthwise under one edge of the bench, so the side of the bench near the steam pipe is likely to be warmer.

(a) Which arrangement of blocks (I or II) is better? Why?

(b) Prepare a randomized allocation of treatments to the pots within each block. (Refer to Example 11.6.4 as a guide; assume that the assignments of seedlings to pots and of pots to positions within the block have already been made.)

11.6.2 An experiment on vitamin supplements is to be conducted on young piglets, using litters as blocks in a randomized blocks design. There will be five treatments: four types of supplement and a control. Thus, five piglets from each litter will be used. The experiment will include five litters. Prepare a randomized blocks allocation of piglets to treatments. (Refer to Example 11.6.4 as a guide.)

11.6.3 Refer to the vitamin experiment of Exercise 11.6.2. Suppose a colleague of the experimenter proposes an alternative design: All pigs in a given litter are to receive the same treatment, with the five litters being randomly allocated to the five treatments. He points out that his proposal would save labor and greatly simplify the record keeping. If you were the experimenter, how would you reply to this proposal?

11.6.4 In a pharmacological experiment on eating behavior in rats, 18 rats are to be randomly allocated to three treatment groups: T_1, T_2, and T_3. While under observation, the animals will be kept in individual cages in a rack. The rack has three tiers with six cages per tier. In spite of efforts to keep the lighting uniform, the lighting conditions vary somewhat from one tier to another (the bottom tier is darkest), and the experimenter is concerned about this because lighting is thought to influence eating behavior in rats. The following three plans are proposed for allocating the rats to positions in the rack (to be done after the allocation of rats to treatment groups):

Plan I. Randomly allocate the 18 rats to the 18 positions in the rack.

Plan II. Put all T_1 rats on the first tier, all T_2 rats on the second, and all T_3 rats on the third tier.

Plan III. On each tier, put two T_1 rats, two T_2 rats, and two T_3 rats.

Put these three plans in order, from best to worst. Explain your reasoning.

11.6.5 An experimenter is planning an agricultural field experiment to compare the yields of 25 varieties of corn. She will use a randomized blocks design with six blocks; thus, there will be 150 plots, and the yield of each plot must be measured. The experimenter realizes that the time required to harvest and weigh all the plots is so long that rain might interrupt the operation. If rain should intervene, there could be a yield difference between the harvests before and after the rain. The experimenter is considering the following plans.

Plan I. Harvest all plots of variety 1 first, all of variety 2 next, and so on.

Plan II. Harvest all plots of block 1 first, all of block 2 next, and so on.

Which plan is better? Why?

11.6.6 For an experiment to compare two methods of artificial insemination in cattle, the following cows are available:

Heifers (14–15 months old): 8 animals
Young cows (2–3 years old): 8 animals
Mature cows (4–8 years old): 10 animals

The animals are to be randomly allocated to the two treatment groups, using the three age groups as blocks. Prepare a suitable allocation, randomly dividing each stratum into two equal groups.

11.6.7 True or false (and say why): The primary reason for using a randomized blocks design in an experiment is to reduce bias.

11.6.8 In an experiment to understand the impact of fish grazing on invertebrate populations in streams, researchers established nine observation channels in three streams (three channels per stream). Each of the three channels within a stream received one of three treatments: No fish were added, Galaxias fish were added, or Trout fish were added. (The channels were constructed with mesh to prevent fish from entering or leaving.) Twelve days after establishing the channels, the number of *Deleatidium* mayfly nymphs present in a specified region in the center of the channel were counted. The number of nymphs for each treatment in each creek follows.[16]

		Creek		
		A	B	C
Treatment	No fish	11	8	7
	Galaxias	9	4	4
	Trout	6	4	0

(a) Identify the blocking, treatment (i.e., the explanatory variable of interest), and response variables in this study.

(b) In the context of this problem, explain to someone who has never taken a statistics course how blocking may help better identify treatment differences should they exist.

11.6.9 *(Continuation of 11.6.8)*

(a) The accompanying table is an (improper) ANOVA table for the data in Exercise 11.6.8. This analysis does not account for the blocking that was performed in the experiment. Based on this analysis, is there evidence that fish affect the number of mayfly nymphs present in the channels? Use $\alpha = 0.05$.

	df	Sum sq	Mean sq	*F* value
Between groups	2	42.889	21.444	2.924
Within groups	6	44.000	7.333	
Total	8	86.889		

(b) The proper ANOVA table for the data, which accounts for blocking, follows. Based on this proper analysis, is there evidence that fish affect the number of mayfly nymphs present in the channels? Use $\alpha = 0.05$.

	df	Sum sq	Mean sq	*F* value
Between groups	2	42.889	21.444	16.783
Between blocks	2	38.889	19.444	15.217
Within groups	4	5.111	1.278	
Total	8	86.889		

(c) Compute and compare s_{pooled} using the ANOVA table from parts (a) and (b). Why is one estimate larger than the other? What is s_{pooled} measuring in part (a)? In part (b)?

11.6.10 Consider the experiment described in Exercise 11.6.8. In addition to measuring the number of mayfly nymphs at the end of 12 days, stones of the same size were removed from each channel, and the algal ash free dry mass (mg/cm^2) was measured for each of nine stones. These data produced SS(blocks) = 0.889, SS(within) = 0.444, and SS(total) = 2.889.

(a) Construct an ANOVA table similar to Table 11.6.4 to summarize these data.

(b) State the null and alternative hypotheses in words and in symbols.

(c) The *P*-value for the test is 0.0493. If $\alpha = 0.05$, what is your conclusion regarding the null hypothesis?

(d) Can a causal conclusion be drawn from the analysis performed in part (b) based on these data? If so, what causal conclusion can be made? If not, explain why no causal conclusion is appropriate.

11.6.11 In an experiment to investigate the effects of different soil nutrient amendments on pomegranate fruit yield (number of fruits per tree) researchers first selected 24 crop rows within a pomegranate farm to follow the irrigation lines. Nine trees in each row were selected and were randomly assigned one of three soil nutrients (three trees per nutrient): potassium, KNO_3; magnesium, $MgSO_4$; or zinc, $ZnSO_4$. All of the fruit from each the treated trees were counted. In all 216 trees (24 × 9) were observed.[17]

(a) Identify the blocking, treatment, and response variables in this study.

(b) Why do you think the blocks were chosen in the way they were in this study?

(c) What were the researchers hoping to accomplish by using blocks?

(d) If it is found that there are statistically significant differences between the mean number of fruits produced across the treatments, will the design of this study allow the researchers to conclude that the nutrients contribute to this difference?

11.6.12 *(Continuation of 11.6.11)*

Below is the ANOVA summary of the experiment described in Exercise 11.6.11.

Source	df	Sum sq	Mean sq	F value
Between treatments	2	7,918.36	3,959.18	0.9651
Between blocks	23	170,443.33	7,410.58	1.8065
Within groups	190	779,407.64	4,102.15	
Total	215	957,769.33		

Is there evidence that the type of nutrient affects the total number of fruits per tree? Use $\alpha = 0.05$.

11.7 Two-Way ANOVA

FACTORIAL ANOVA

In a typical analysis of variance application there is a single explanatory variable or **factor** under study. For example, in the weight gain setting of Example 11.2.1, the factor is "type of diet," which takes on three **levels**: diet 1, diet 2, and diet 3. However, some analysis of variance settings involve the simultaneous study of two or more factors. The following is an example.

Example 11.7.1 **Growth of Soybeans** A plant physiologist investigated the effect of mechanical stress on the growth of soybean plants. Individually potted seedlings were randomly allocated to four treatment groups of 13 seedlings each. Seedlings in two groups were stressed by shaking for 20 minutes twice daily, while two control groups were not stressed. Thus, the first factor in the experiment was presence or absence of stress, with two levels: control or stress. Also, plants were grown in either low or moderate light. Thus, the second factor was amount of light, with two levels: low light or moderate light. This experiment is an example of a 2 × 2 *factorial experiment*; it includes four treatments:

Treatment 1: Control, low light

Treatment 2: Stress, low light

Treatment 3: Control, moderate light

Treatment 4: Stress, moderate light

After 16 days of growth, the plants were harvested, and the total leaf area (cm^2) of each plant was measured. The results are given in Table 11.7.1 and plotted in Figure 11.7.1.[18] There is evidence in Figure 11.7.1 that stress reduces leaf area. This is true under low light and under moderate light. Likewise, moderate light increases leaf area, whether or not the seedlings are stressed. ■

Table 11.7.1 Leaf area (cm^2) of soybean plants

	Treatment			
	Control, low light	Stress, low light	Control, moderate light	Stress, moderate light
	264	235	314	283
	200	188	320	312
	225	195	310	291
	268	205	340	259
	215	212	299	216
	241	214	268	201
	232	182	345	267
	256	215	271	326
	229	272	285	241
	288	163	309	291
	253	230	337	269
	288	255	282	282
	230	202	273	257
Mean	245.3	212.9	304.1	268.8
SD	27.0	29.7	26.9	35.2
n	13	13	13	13

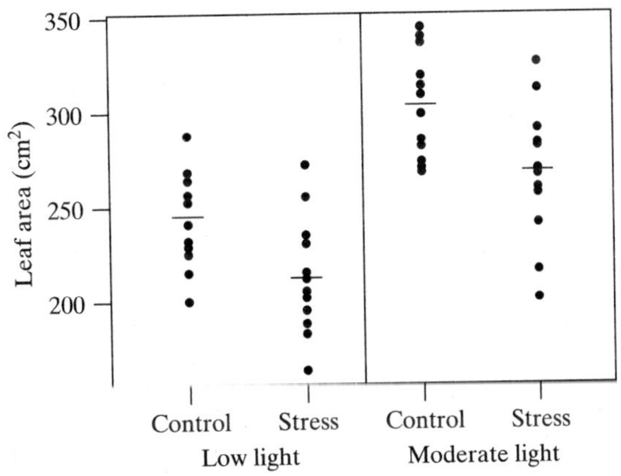

Figure 11.7.1 Leaf area of soybean plants receiving four different treatments. Group means indicated by (−)

A model for this setting is

$$y_{ijk} = \mu + \tau_i + \beta_j + \varepsilon_{ijk}$$

where y_{ijk} is the kth observation of level i of the first factor and level j of the second factor. The term τ_i represents the effect of level i of the first factor (stress condition in Example 11.7.1) and now the term β_j represents the effect of level j of the second factor (light condition in Example 11.7.1).

When studying two factors within a single experiment it helps to organize the sample means in a table that reflects the structure of the experiment and to present the means in a graph that features this structure.

Example 11.7.2

Growth of Soybeans Table 11.7.2 summarizes the data of Example 11.7.1. For example, when the first factor is at its first level (control) and the second factor is at its first level (low light), the sample mean is $\bar{y}_{11} = 245.3$. The format of this table permits us easily to consider the two factors—stress condition and light condition—separately and together. The last column shows the effect of light at each stress level. The numbers in this column confirm the visual impression of Figure 11.7.1: Moderate light increases average leaf area by roughly the same amount when the seedlings are stressed as it does when they are not stressed. Likewise, the last row (−32.4 versus −35.3) shows that the effect of stress is roughly the same at each level of light. ■

Table 11.7.2 Mean leaf areas for soybean experiment

		Light condition		
		Low light	Moderate light	Difference
Shaking condition	Control	245.3	304.1	58.8
	Stress	212.9	268.8	55.9
	Difference	−32.4	−35.3	

If the joint influence of two factors is equal to the sum of their separate influences, the two factors are said to be **additive** in their effects. For instance, consider the soybean experiment of Example 11.7.1. If stress reduces mean leaf area by the same amount in either light condition, then the effect of stress (a negative effect in this case) is *added* to the effect of light. To visualize this additivity of effects, consider Figure 11.7.2, which shows the data with the four treatment means. The solid lines connecting treatment means are almost parallel because the data display a pattern of nearly perfect additivity.*

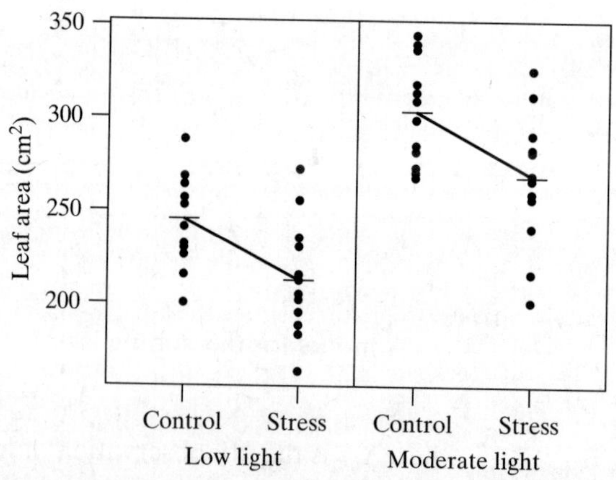

Figure 11.7.2 Data and treatment means for soybean experiment

*The difference between the mean leaf area for stress under low light (212.9) and the mean leaf area for control under low light of (245.3) is called the **simple effect** of shaking under low light. Thus, the simple effect of shaking under low light is $212.9 - 245.3 = -32.4$. Likewise, the simple effect of shaking under moderate light is $268.8 - 304.1 = -35.3$. A **main effect** is an average of simple effects. For example, the main effect of shaking is $(-32.4 + -35.3)/2 = -33.85$. The main effect of light is $(58.8 + 55.9)/2 = 57.35$.

When the effects of factors are additive, we say that there is no **interaction** between the factors. A graph that displays only the treatment means is often called an interaction graph. Figure 11.7.3, which is a summary version of Figure 11.7.2, is an interaction graph highlighting the effect of stress on mean leaf area for the two light conditions. Analogous graphs can be made to draw the focus to comparing the effect of light on mean leaf area for the two stress conditions.

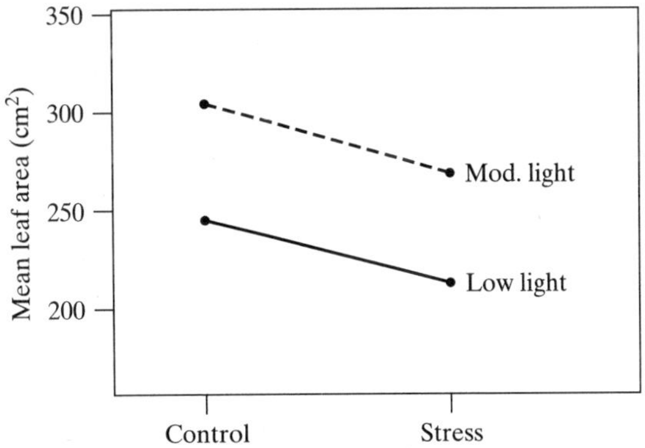

Figure 11.7.3 Interaction graph for soybean experiment

Sometimes the effect that one factor has on a response variable depends on the level of a second factor. When this happens we say that the two factors interact in their effect on the response. The following is an example.

Example 11.7.3

Iron Supplements in Milk-Based Fruit Beverages Iron and zinc fortification of milk-based fruit drinks are common practice. To better understand the effects of drink fortification on the cellular retention of iron, researchers conducted an experiment by fortifying milk-based fruit drinks with low and high levels of iron (Fe) and zinc (Zn). The drinks were digested in a simulated gastrointestinal tract, and cellular iron retention was measured (μg Fe/mg cell protein). Table 11.7.3 summarizes the data, which included eight observations for each combination of Fe and Zn supplementation levels.[19] Figure 11.7.4 is an interaction graph showing the four means. Note that when the Zn supplementation level is low, the effect of the Fe supplementation on cellular retention is much smaller than when the Zn supplementation level is high (i.e., the slopes of the two lines differ—the lines are not parallel). Thus, the effect of Fe supplementation on mean cellular retention depends on the amount of Zn supplementation used. We say that Fe and Zn interact in their effects on cellular retention. ■

Table 11.7.3 Mean iron retention (μg Fe/mg cell protein) for drink supplement experiment

		Zn level		
		Low	High	Difference
Fe	Low	0.707	0.215	−0.492
level	High	0.994	1.412	0.418
	Difference	0.287	1.197	

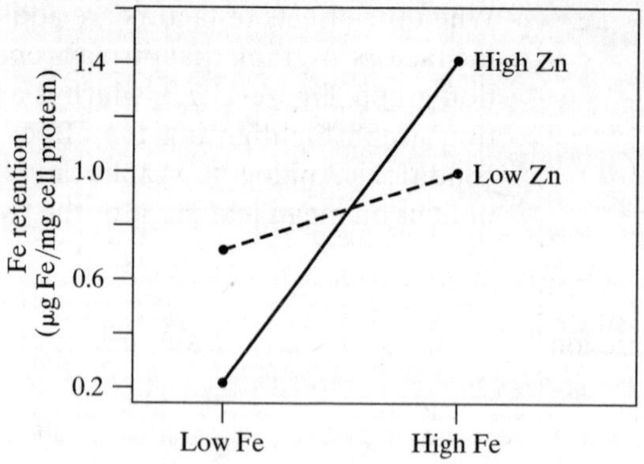

Figure 11.7.4 Interaction graph for drink supplementation experiment

When we suspect that two factors interact in an ANOVA setting, we can extend our model by adding an interaction term:

$$y_{ijk} = \mu + \tau_i + \beta_j + \gamma_{ij} + \varepsilon_{ijk}$$

Here the term γ_{ij} is the effect of the interaction between level i of the first factor and level j of the second factor. As before, if there are $n.$ total observations, then df(total) $= n. - 1$. If there are I levels of the first factor, then it has $I - 1$ degrees of freedom. Likewise, if there are J levels of the second factor, then it has $J - 1$ degrees of freedom. There are $(I - 1) \times (J - 1)$ interaction degrees of freedom. With I levels of the first factor and J levels of the second factor there are IJ treatment combinations. Thus, df(within) $= n. - IJ$.*

A null hypothesis of interest is that all interaction terms are zero:

$$H_0\colon \gamma_{11} = \gamma_{12} = \cdots = \gamma_{IJ} = 0$$

To test this null hypothesis we calculate

$$F_s = \frac{\text{MS(interaction)}}{\text{MS(within)}}$$

and reject H_0 if the P-value is too small.

Example 11.7.4 **Iron Supplements in Milk-Based Fruit Beverages** Table 11.7.4 shows the ANOVA results for the drink supplement experiment of Example 11.7.3. This table includes a line for the interaction term.† There were eight observations at each combination of Fe and Ze supplementation level; thus $n. = 32$ and df(total) $= 31$. In this example $I = J = 2$, so df(Fe levels) = df(Zn levels) = df(interaction) = 1. We can find df(within) by subtraction: df(within) $= 31 - 1 - 1 - 1 = 28$. (This agrees with the formula df(within) $= n. - IJ = 32 - 2 \times 2$.)

To test whether Fe and Zn supplementation levels interact we use the F ratio $1.6555/0.0019 = 871.3$, which has degrees of freedom 1 for the numerator and 28

*This is analogous to the definition of df(within) $= n. - I$ for one-way ANOVA from Section 11.2. In each setting df(within) = total number of observations − number of treatments.

†The ANOVA formulas that are used to calculate the sum of squares due to interaction are rather messy and aren't presented here. In particular, it matters whether or not the design is "balanced." The drink supplementation experiment is balanced in that there are eight observations in each of the four combinations of factor levels shown in Table 11.7.3. However, unbalanced designs, which lead to complicated calculations and analyses, are possible. We rely here on computer software to calculate the necessary sums of squares.

Table 11.7.4 ANOVA table for drink supplement experiment

Source	df	SS	MS	*F* ratio
Between Fe levels	1	4.4023	4.4023	2317.0
Between Zn levels	1	0.0109	0.0109	5.74
Interaction	1	1.6555	1.6555	871.3
Within groups	28	0.0523	0.0019	
Total	31	6.1210		

for the denominator. From Table 10 we bracket the *P*-value as *P*-value < 0.0001. The *P*-value is extremely small, indicating that the interaction pattern seen in Figure 11.7.4 is more pronounced than would be expected by chance alone. Thus, we reject H_0. ■

The concept of interaction occurs throughout biology. The terms "synergism" and "antagonism" describe interactions between biological agents. The term "epistasis" describes interaction between genes at two loci.

When interactions are present, as in Example 11.7.3, the main effects of factors don't have their usual interpretations. Regarding Example 11.7.3, it is difficult to state the independent effect of Fe because the nature and magnitude of the effect depends on the particular level of Zn supplementation. Because of this, we usually test for the presence of interactions first. If interactions are present, as in the drink supplementation example, then we often stop the analysis at this stage. If no evidence for an interaction effect is found (i.e., if we do not reject H_0), then we proceed to testing the main effects of the individual factors. The following example illustrates this process.

Example 11.7.5 **Growth of Soybeans** Table 11.7.5 is an ANOVA table for the soybean growth data of Example 11.7.1. The null hypothesis

$$H_0\text{: } \gamma_{11} = \gamma_{12} = \gamma_{21} = \gamma_{22} = 0$$

is tested with the *F* ratio

$$F_s = \frac{\text{MS(interaction)}}{\text{MS(within)}} = \frac{26.3}{895.34} = 0.029$$

Looking in Table 10 with degrees of freedom 1 and 48, we see that the *P*-value is greater than 0.20 (a computer gives *P*-value $= 0.8655$); thus there is no significant evidence for an interaction, and we do not reject H_0.

Table 11.7.5 ANOVA table for soybean growth experiment

Source	df	SS	MS	*F* ratio
Between stress levels	1	14858.5	14858.5	16.60
Between light levels	1	42751.6	42751.6	47.75
Interaction	1	26.3	26.3	0.029
Within groups	48	42976.3	895.34	
Total	51	100612.7		

Since there is no evidence of interactions, we test the main effect of stress level. Here the F ratio is

$$F_s = \frac{\text{MS(between stress levels)}}{\text{MS(within)}} = \frac{14858.5}{895.34} = 16.6$$

This is highly significant (i.e., the P-value is very small; P-value $= 0.0002$) and we reject H_0.

Likewise, the test for the main effect of light levels has an F ratio of

$$F_s = \frac{\text{MS(between light levels)}}{\text{MS(within)}} = \frac{42751.6}{895.34} = 47.75$$

Again, this is highly significant (P-value < 0.0001), and we reject H_0. ■

Interaction graphs can be used when there are more than two levels for a factor, as in the next example.

Example 11.7.6 **Toads** Researchers studied the effect that exposure to ultraviolet-B radiation has on the survival of embryos of the western toad *Bufo boreas.* They conducted an experiment in which several *B. borea* embryos were placed at one of three water depths—10 cm, 50 cm, or 100 cm—and one of two radiation settings—exposed to UV-B radiation or shielded. The response variable was the percentage of embryos surviving to hatching. Table 11.7.6 summarizes the data, which included four observations at each combination of depth and UV-B exposure. Figure 11.7.5 is an interaction graph showing the six means. The presence of interactions here is readily apparent. Table 11.7.7 summarizes the ANOVA.[20] ■

Table 11.7.6 Percentage embryos surviving for toads experiment

		UV-B		
		Exposed	Shielded	Difference
Water	10 cm	0.425	0.759	0.334
depth	50 cm	0.729	0.748	0.019
	100 cm	0.785	0.766	−0.019

Figure 11.7.5 Interaction graph for toad experiment

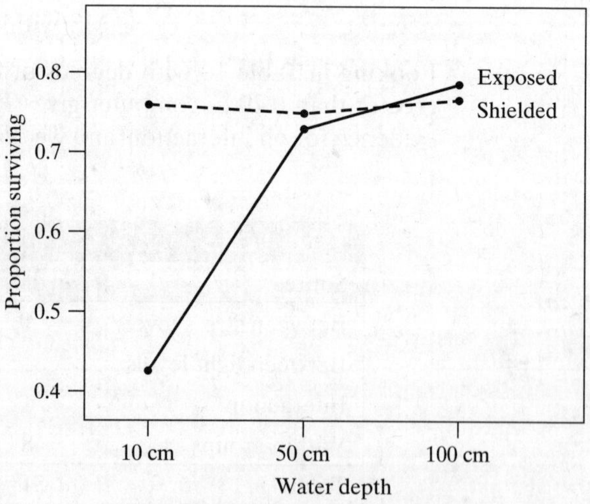

Table 11.7.7 ANOVA table for toad experiment

Source	df	SS	MS	F ratio
Between water depths	2	0.150676	0.075338	13.92
Between UV-B levels	1	0.074371	0.074371	13.74
Interaction	2	0.150185	0.075093	13.88
Within groups	18	0.097401	0.005411	
Total	23	0.472633		

The topic of interactions is also discussed in Section 11.8.

Exercises 11.7.1–11.7.9

11.7.1 A plant physiologist investigated the effect of flooding on root metabolism in two tree species: flood-tolerant river birch and the intolerant European birch. Four seedlings of each species were flooded for one day, and four were used as controls. The concentration of adenosine triphosphate (ATP) in the roots of each plant was measured. The data (nmol ATP per mg tissue) are shown in the table.[21]

	River birch		European birch	
	Flooded	Control	Flooded	Control
	1.45	1.70	0.21	1.34
	1.19	2.04	0.58	0.99
	1.05	1.49	0.11	1.17
	1.07	1.91	0.27	1.30
Mean	1.19	1.785	0.2925	1.20

Prepare an interaction graph (like Figure 11.7.3).

11.7.2 Consider the data from Exercise 11.7.1. For these data, SS(species of birch) = 2.19781, SS(flooding) = 2.25751, SS(interaction) = 0.097656, and SS(within) = 0.47438.

(a) Construct the ANOVA table.

(b) How many degrees of freedom are there for the ANOVA F test for interactions?

(c) The P-value for the test is 0.142. If $\alpha = 0.05$, what is your conclusion regarding the null hypothesis?

(d) What is the value of the F test statistic for testing that species has no effect on ATP concentration?

(e) The P-value for the testing that species has no effect on ATP concentration is 0.000008. If $\alpha = 0.01$, what is your conclusion regarding this null hypothesis?

(f) Assuming that each of the four populations has the same standard deviation, use the data to calculate an estimate of that standard deviation.

11.7.3 A completely randomized double-blind clinical trial was conducted to compare two drugs, ticrynafen (T) and hydrochlorothiazide (H), for effectiveness in treatment of high blood pressure. Each drug was given at either a low or a high dosage level for 6 weeks. The accompanying table shows the results for the drop (baseline minus final value) in systolic blood pressure (mm Hg).[22]

	Ticrynafen (T)		Hydrochlorothiazide (H)	
	Low dose	High dose	Low dose	High dose
Mean	13.9	17.1	15.8	17.5
No. of patients	53	57	55	58

Prepare an interaction graph (like Figure 11.7.3).

11.7.4 Consider the data from Exercise 11.7.3. The difference in response between T and H appears to be larger for the low dose than for the high dose.

(a) Carry out an F test for interactions to assess whether this pattern can be ascribed to chance variation. Let $\alpha = 0.10$. For these data SS(interaction) = 31.33 and SS(within) = 30648.81.

(b) Based on your results in part (a), is it sensible to examine and interpret the main effects of drug and of dose?

11.7.5 Consider the data from Exercise 11.7.3. For these data, SS(drug) = 69.22, SS(dose) = 330.00, SS(interaction) = 31.33, and SS(within) = 30648.81.

(a) Construct the ANOVA table.

(b) What is the value of the F test statistic for testing that the effects of the two drugs are equal?

(c) How many degrees of freedom are there for the test from part (b)?

(d) The P-value for the test is 0.485. If $\alpha = 0.05$, what is your conclusion regarding the null hypothesis?

11.7.6 In a study of lettuce growth, 36 seedlings were randomly allocated to receive either high or low light and to be grown in either a standard nutrient solution or one containing extra nitrogen. After 16 days of growth, the lettuce plants were harvested, and the dry weight of the leaves was determined for each plant. The accompanying table shows the mean leaf dry weight (gm) of the 9 plants in each treatment group.[23]

	Nutrient solution	
	Standard	Extra nitrogen
Low light	2.16	3.09
High light	3.26	4.48

For these data, SS(nutrient solution) = 10.4006, SS(light) = 13.95023, SS(interaction) = 0.18923, and SS(within) = 11.1392.

(a) Prepare an interaction graph (like Figure 11.7.3) and discuss any patterns.

(b) Construct the ANOVA table.

(c) Carry out an F test for interactions; use $\alpha = 0.05$.

(d) Test the null hypothesis that nutrient solution has no effect on weight. Use $\alpha = 0.01$.

11.7.7 Patients with pleural infections (fluid buildup in the chest) were randomly assigned to either placebo ($n = 55$), the treatment "tPA" (tissue plasminogen activator) ($n = 52$), the treatment "DNase" (deoxyribonuclease) ($n = 51$), or a combination of tPA and DNase ($n = 52$) in a double-blind clinical trial. Doctors used chest radiography to measure the percentage change, over 7 days, in the area of pleural fluid. The accompanying table shows the mean for each treatment group.[24] The Yes/Yes cell is for the patients who received both tPA and DNase; the No/No cell is for the placebo patients.

		DNase?	
		Yes	No
tPA?	Yes	−29.5	−17.2
	No	−14.7	−17.2

For these data, SS(DNase) = 1226, SS(tPA) = 2836, SS(interaction) = 2870, and SS(within) = 91827.

(a) Construct the ANOVA table.

(b) Carry out an F test for interactions; us $\alpha = 0.10$.

(c) Given the result of the test in part (b), is it appropriate to test the null hypothesis that tPA has no effect on change in area of pleural fluid? If so, then conduct that test using $\alpha = 0.05$. If not, then explain why not.

11.7.8 Forced expiratory volume (FEV) is a measure of the rate of airflow (L/min) during one deep exhalation. For each of 75 students, FEV was measured three times, and those results were averaged to give a summary FEV number. Some of the students were varsity soccer players. The table below shows mean summary values broken down by group.[25]

	Female, soccer no	Female, soccer yes	Male, soccer no	Male, soccer yes
Mean	396	472	560	653
n	14	17	20	24

For these data, SS(Sex) = 544,398; SS(Soccer) = 136,610; SS(interaction) = 1,207; and SS(within) = 662,408.

(a) Prepare an interaction graph (like Figure 11.7.3) to investigate whether the male versus female difference is the same for soccer players as for nonsoccer players.

(b) Does the interaction graph from part (a) suggest that there is an interaction present? If so, what does that mean? If not, what does that mean?

(c) Carry out an F test for interactions; use $\alpha = 0.10$.

(d) Given the result of the test in part (c), is it appropriate to test the null hypothesis that there is no soccer versus nonsoccer difference in mean FEV? If so, then conduct that test using $\alpha = 0.10$. If not, then explain why not.

11.7.9 A scientist recorded the weights of 2La malaria-vector mosquitoes and compared them to their counterpart 2L + mosquitoes. The table below shows summary values broken down by sex.[26]

	Female, 2La	Female, 2L+	Male, 2La	Male, 2L+
Mean	1.500	1.534	0.964	0.938
SD	0.273	0.277	0.138	0.295
N	20	20	20	20

(a) Prepare an interaction graph (like Figure 11.7.3) to investigate whether the male versus female difference is the same for 2La as for 2L + mosquitoes.

(b) Does the interaction graph from part (a) suggest that there is an interaction present? If so, what does that mean? If not, what does that mean?

(c) The P-value for the test of no 2La versus 2L + difference is 0.94. If $\alpha = 0.05$, state your conclusion regarding the null hypothesis, in the context of this setting.

(d) The P-value for the test of no female versus male difference is 1.9×10^{-15}. If $\alpha = 0.05$, state your conclusion regarding the null hypothesis, in the context of this setting.

11.8 Linear Combinations of Means (Optional)

In many studies, interesting questions can be addressed by considering linear combinations of the group means. A **linear combination** L is a quantity of the form

$$L = m_1\bar{y}_1 + m_2\bar{y}_2 + \cdots + m_I\bar{y}_I$$

where the m's are multipliers of the $\bar{y}_i$'s.

LINEAR COMBINATIONS FOR ADJUSTMENT

One use of linear combinations is to "adjust" for an extraneous variable, as illustrated by the following example.

Example 11.8.1

Forced Vital Capacity One measure of lung function is forced vital capacity (FVC), which is the maximal amount of air a person can expire in one breath. In a public health survey, researchers measured FVC in a large sample of people. The results for male ex-smokers, stratified by age, are shown in Table 11.8.1.[27]

Table 11.8.1 FVC in male ex-smokers

		FVC (liters)	
Age (years)	n	Mean	SD
25–34	83	5.29	0.76
35–44	102	5.05	0.77
45–54	126	4.51	0.74
55–64	97	4.24	0.80
65–74	73	3.58	0.82
25–74	481	4.56	

Suppose it is desired to calculate a summary value for FVC in male ex-smokers. One possibility would be simply to calculate the grand mean of the 481 observed values, which is 4.56 liters. But the grand mean has a serious drawback: It cannot be meaningfully compared with other populations that may have different age distributions. For instance, suppose we were to compare ex-smokers with nonsmokers; the observed difference in FVC would be distorted because ex-smokers as a group are (not surprisingly) older than nonsmokers. A summary measure that does not have this disadvantage is the "age-adjusted" mean, which is an estimate of the mean FVC value in a reference population with a specified age distribution. To illustrate, we will use the reference distribution in Table 11.8.2, which is (approximately) the distribution for the entire U.S. population.[28]

Table 11.8.2 Age distribution in reference population

Age	Relative frequency
25–34	0.22
35–44	0.22
45–54	0.24
55–64	0.20
65–74	0.12

The "age-adjusted" mean FVC value is the following linear combination:

$$L = 0.22\bar{y}_1 + 0.22\bar{y}_2 + 0.24\bar{y}_3 + 0.20\bar{y}_4 + 0.12\bar{y}_5$$

Note that the multipliers (m's) are the relative frequencies in the reference population. From Table 11.8.1, the value of L is

$$L = (0.22)(5.29) + (0.22)(5.05) + (0.24)(4.51) + (0.20)(4.24) + (0.12)(3.58)$$
$$= 4.63 \text{ liters}$$

This value is an estimate of the mean FVC in an idealized population of people who are biologically like male ex-smokers, but whose age distribution is that of the reference population. ■

CONTRASTS

A linear combination whose multipliers (m's) add to zero is called a **contrast.** The following example shows how contrasts can be used to describe the results of an experiment.

Example 11.8.2

Growth of Soybeans Table 11.8.3 shows the treatment means and sample sizes for the soybean growth experiment of Example 11.7.1. We can use contrasts to describe the effects of stress in the two temperature conditions.

Table 11.8.3 Soybean growth data

Treatment	Mean leaf area (cm^2)	n
1. Control, low light	245.3	13
2. Stress, low light	212.9	13
3. Control, moderate light	304.1	13
4. Stress, moderate light	268.8	13

(a) First, note that an ordinary pairwise difference is a contrast. For instance, to measure the effect of stress in low light we can consider the contrast

$$L = \bar{y}_1 - \bar{y}_2 = 245.3 - 212.9 = 32.4$$

For this contrast, the multipliers are $m_1 = 1, m_2 = -1, m_3 = 0, m_4 = 0$; note that they add to zero.

(b) To measure the effect of stress in moderate light we can consider the contrast

$$L = \bar{y}_3 - \bar{y}_4 = 304.1 - 268.8 = 35.3$$

For this contrast, the multipliers are $m_1 = 0, m_2 = 0, m_3 = 1, m_4 = -1$.

(c) To measure the overall effect of stress, we can average the contrasts in parts (a) and (b) to obtain the contrast

$$L = \frac{1}{2}(\bar{y}_1 - \bar{y}_2) + \frac{1}{2}(\bar{y}_3 - \bar{y}_4)$$
$$= \frac{1}{2}(32.4) + \frac{1}{2}(35.3) = 33.85$$

For this contrast, the multipliers are $m_1 = \frac{1}{2}, m_2 = -\frac{1}{2}, m_3 = \frac{1}{2}, m_4 = -\frac{1}{2}$. ■

STANDARD ERROR OF A LINEAR COMBINATION

Each linear combination L is an estimate, based on the $\bar{y}$'s, of the corresponding linear combination of the population means (μ's). As a basis for statistical inference, we need to consider the standard error of a linear combination, which is calculated as follows.

Standard Error of L

The standard error of the linear combination

$$L = m_1\bar{y}_1 + m_2\bar{y}_2 + \cdots + m_I\bar{y}_I$$

is

$$\text{SE}_L = s_{\text{pooled}}\sqrt{\sum_{i=1}^{I}\frac{m_i^2}{n_i}}$$

where $s_{\text{pooled}} = \sqrt{\text{MS(within)}}$ from the ANOVA.

The SE can be written explicitly as

$$\text{SE}_L = s_{\text{pooled}}\sqrt{\frac{m_1^2}{n_1} + \frac{m_2^2}{n_2} + \cdots + \frac{m_I^2}{n_I}}$$

If all the sample sizes (n_i) are equal, the SE can be written as

$$\text{SE}_L = s_{\text{pooled}}\sqrt{\frac{m_1^2 + m_2^2 + \cdots + m_I^2}{n}} = s_{\text{pooled}}\sqrt{\frac{1}{n}\sum_{i=1}^{I}m_i^2}$$

The following two examples illustrate the application of the standard error formula.

Example 11.8.3

Forced Vital Capacity For the linear combination L defined in Example 11.8.1, we find that

$$\sum_{i=1}^{I}\frac{m_i^2}{n_i} = \frac{0.23^2}{83} + \frac{0.22^2}{102} + \frac{0.24^2}{126} + \frac{0.22^2}{97} + \frac{0.09^2}{73}$$

$$= 0.0021789$$

The ANOVA for these data yields $s_{\text{pooled}} = \sqrt{0.59989} = 0.77453$. Thus, the standard error of L is

$$\text{SE}_L = 0.77453\sqrt{0.0021789} = 0.0362$$

■

Example 11.8.4

Growth of Soybeans For the linear combination L defined in Example 11.8.2(a), we find that

$$\sum_{i=1}^{I}m_i^2 = (1)^2 + (-1)^2 + (0)^2 + (0)^2 = 2$$

so that

$$\text{SE}_L = s_{\text{pooled}}\sqrt{\frac{2}{13}}$$

■

CONFIDENCE INTERVALS

Linear combinations of means can be used for testing hypotheses and for constructing confidence intervals. Critical values are obtained from Student's t distribution with

$$\text{df} = \text{df(within)}$$

from the ANOVA.* Confidence intervals are constructed using the familiar Student's t format. For instance, a 95% confidence interval is

$$L \pm t_{0.025}\text{SE}_L$$

The following example illustrates the construction of the confidence interval.

Example 11.8.5 **Growth of Soybeans** Consider the contrast defined in Example 11.8.2(c):

$$L = \frac{1}{2}(\bar{y}_1 - \bar{y}_2) + \frac{1}{2}(\bar{y}_3 - \bar{y}_4)$$

This contrast is an estimate of the quantity

$$\lambda = \frac{1}{2}(\mu_1 - \mu_2) + \frac{1}{2}(\mu_3 - \mu_4)$$

which can be described as the true (population) effect of stress, averaged over the light conditions. Let us construct a 95% confidence interval for this true difference.

We found in Example 11.8.2 that the value of L is

$$L = 33.85$$

To calculate SE_L, we first calculate

$$\sum_{i=1}^{I} \frac{m_i^2}{n_i} = \frac{(\frac{1}{2})^2}{13} + \frac{(-\frac{1}{2})^2}{13} + \frac{(\frac{1}{2})^2}{13} + \frac{(-\frac{1}{2})^2}{13} = \frac{1}{13}$$

From the ANOVA, which is shown in Table 11.8.4, we find that $s_{\text{pooled}} = \sqrt{895.34} = 29.922$; thus,

$$\text{SE}_L = s_{\text{pooled}}\sqrt{\sum_{i=1}^{I} \frac{m_i^2}{n_i}} = 29.922\sqrt{\frac{1}{13}} = 8.299$$

Table 11.8.4 ANOVA table for soybean growth experiment

Source	df	SS	MS	F ratio
Between stress depths	1	14858.5	14858.5	16.60
Between light levels	1	42751.6	42751.6	47.75
Interaction	1	26.3	26.3	0.029
Within groups	48	42976.3	895.34	
Total	51	100613		

*This method of determining critical values does not take account of multiple comparisons. See Section 11.9.

From Table 4 with df $= 40 \approx 48$, we find $t_{40,\,0.025} = 2.021$. The confidence interval is

$$33.85 \pm (2.021)(8.299)$$
$$33.85 \pm 16.77$$

or (17.1, 50.6).

We are 95% confident that the effect of stress, averaged over the light conditions, is to reduce the leaf area by an amount whose mean value is between 17.1 cm^2 and 50.6 cm^2. ■

t TESTS

To test the null hypothesis that the population value of a contrast is zero, the test statistic is calculated as

$$t_s = \frac{L}{SE_L}$$

and the *t* test is carried out in the usual way. The *t* test will be illustrated in Example 11.8.6.

CONTRASTS TO ASSESS INTERACTION

Sometimes an investigator wishes to study the separate and joint effects of two or more factors on a response variable Y. In Section 11.7 the concept of interaction between two factors was introduced. Linear contrasts provide another way to study such interactions. The following is an example.

Example 11.8.6

Growth of Soybeans In the soybean growth experiment (Example 11.7.1 and Example 11.8.2), the two factors of interest are stress condition and light level. Table 11.8.5 shows the treatment means, arranged in a new format that permits us easily to consider the factors separately and together.

Table 11.8.5 Mean leaf areas for soybean experiment

		Light condition		
		Low light	Moderate light	Difference
Shaking	Control	245.3 (1)	304.1 (3)	58.8
condition	Stress	212.9 (2)	268.8 (4)	55.9
	Difference	−32.4	−35.3	

At each light level, the mean effect of stress can be measured by a contrast:

Effect of stress in low light: $\bar{y}_2 - \bar{y}_1 = 212.9 - 245.3 = -32.4$

Effect of stress in moderate light: $\bar{y}_4 - \bar{y}_3 = 268.8 - 304.1 = -35.3$

Now consider the question: Is the reduction in leaf area due to stress the same in both light conditions? One way to address this question is to compare $(\bar{y}_2 - \bar{y}_1)$ versus $(\bar{y}_4 - \bar{y}_3)$; the difference between these two values is a contrast:

$$L = (\bar{y}_2 - \bar{y}_1) - (\bar{y}_4 - \bar{y}_3)$$
$$= -32.4 - (-35.3) = 2.9$$

This contrast L can be used as the basis for a confidence interval or a test of hypothesis. We illustrate the test. The null hypothesis is

$$H_0\text{: } (\mu_2 - \mu_1) = (\mu_4 - \mu_3)$$

or, in words,

H_0: The effect of stress is the same in the two light conditions.

For the preceding L, $\sum_{i=1}^{I} \frac{m_i^2}{n_i} = \frac{4}{13}$, and the standard error is

$$\text{SE}_L = s_{\text{pooled}}\sqrt{\sum_{i=1}^{I} \frac{m_i^2}{n_i}} = s_{\text{pooled}}\sqrt{\frac{4}{13}} = 29.922\sqrt{\frac{4}{13}} = 16.6$$

The test statistic is

$$t_s = \frac{2.9}{16.6} = 0.2$$

From Table 4 with df = 40 we find $t_{40,0.10} = 1.303$. The data provide virtually no evidence that the effect of stress is different in the two light conditions. This is consistent with the F test for interactions conducted in Example 11.7.5. ■

The statistical definition of interaction introduced in Section 11.7 and viewed through the lens of contrasts here is rather specialized. It is defined in terms of the observed variable rather than in terms of a biological mechanism. Further, interaction as measured by a contrast is defined by *differences* between means. In some applications the biologist might feel that ratios of means are more meaningful or relevant than differences. The following example shows that the two points of view can lead to different answers.

Example 11.8.7 **Chromosomal Aberrations** A research team investigated the separate and joint effects in mice of exposure to high temperature (35 °C) and injection with the cancer drug cyclophosphamide (CTX). A completely randomized design was used, with eight mice in each treatment group. For each animal, the researchers measured the incidence of a certain chromosomal aberration in the bone marrow; the result is expressed as the number of abnormal cells per 1,000 cells. The treatment means are shown in Table 11.8.6.[29]

Table 11.8.6 Mean incidence of chromosomal aberrations following various treatments

		Injection	
		CTX	None
Temperature	Room	23.5	2.7
	High	75.4	20.9

Is the observed effect of CTX greater at room temperature or at high temperature? The answer depends on whether "effect" is measured absolutely or relatively.

Measured as a difference, the effect of CTX is

Room temperature: 23.5 − 2.7 = 20.8

High temperature: 75.4 − 20.9 = 54.5

Thus, the absolute effect of CTX is greater at the high temperature. However, this relationship is reversed if we express the effect of CTX as a ratio rather than as a difference:

$$\textit{Room temperature: } \frac{23.5}{2.7} = 8.70$$

$$\textit{High temperature: } \frac{75.4}{20.9} = 3.61$$

At room temperature CTX produces almost a ninefold increase in chromosomal aberrations, whereas at high temperature the increase is less than fourfold; thus, in relative terms, the effect of CTX is much greater at room temperature. ■

If the phenomenon under study is thought to be multiplicative rather than additive, so that relative rather than absolute change is of primary interest, then ordinary contrasts should not be used. One simple approach in this situation is to use a logarithmic transformation—that is, to compute $Y' = \log(Y)$, and then analyze Y' using contrasts. The motivation for this approach is that relations of constant *relative* magnitude in the Y scale become relations of constant *absolute* magnitude in the Y' scale.

Exercises 11.8.1–11.8.10

11.8.1 Refer to the FVC data of Example 11.8.1.

(a) Verify that the grand mean of all 481 FVC values is 4.56.

(b) Taking into account the age distribution among the 481 subjects and the age distribution in the U.S. population, explain intuitively why the grand mean (4.56 liters) is smaller than the age-adjusted mean (4.67 liters).

11.8.2 To see if there is any relationship between blood pressure and childbearing, researchers examined data from a large health survey. The following table shows the data on systolic blood pressure (mm Hg) for random samples from two populations of women: women who had borne no children and women who had borne five or more children. The pooled standard deviation from all eight groups was $s_{\text{pooled}} = 18$ mm Hg.[30]

	No children		Five or more children	
Age	Mean blood pressure	No. of women	Mean blood pressure	No. of women
18–24	113	230	114	7
25–34	118	110	116	82
35–44	125	105	124	127
45–54	134	123	138	124
18–54	121	568	127	340

Carry out age adjustment, as directed, using the following reference distribution, which is the approximate distribution for U.S. women[31]:

Age	Relative frequency
18–24	0.17
25–34	0.29
35–44	0.31
45–54	0.23

(a) Calculate the age-adjusted mean blood pressure for women with no children.

(b) Calculate the age-adjusted mean blood pressure for women with five or more children.

(c) Calculate the difference between the values obtained in parts (a) and (b). Explain intuitively why the result is smaller than the unadjusted difference of $127 - 121 = 6$ mg Hg.

(d) Calculate the standard error of the value calculated in part (a).

(e) Calculate the standard error of the value calculated in part (c).

11.8.3 Refer to the ATP data of Exercise 11.7.1. The sample means and standard deviations are as follows:

	River birch		European birch	
	Flooded	Control	Flooded	Control
$\bar{y}$	1.19	1.78	0.29	1.20
s	0.18	0.24	0.20	0.16

Define linear combinations (i.e., specify the multipliers) to measure each of the following:

(a) The effect of flooding in river birch

(b) The effect of flooding in European birch

(c) The difference between river birch and European birch with respect to the effect of flooding (i.e., the interaction between flooding and species)

11.8.4 (*Continuation of Exercise 11.8.3*)

(a) Use a t test to investigate whether flooding has the same effect in river birch and in European birch. Use a nondirectional alternative and let $\alpha = 0.05$. (The pooled standard deviation is $s_{pooled} = 0.199$.)

(b) If the sample sizes were $n = 10$ rather than $n = 4$ for each group, but the means, standard deviations, and s_{pooled} remained the same, how would the result of part (a) change?

11.8.5 (*Continuation of Exercise 11.8.4*)

Consider the null hypothesis that flooding has no effect on ATP level in river birch. This hypothesis could be tested in two ways: as a contrast (using the method of Section 11.8), or with a two-sample t test (as in Exercise 7.2.11). Answer the following questions; do not actually carry out the tests.

(a) In what way or ways do the two test procedures differ?

(b) In what way or ways do the conditions for validity of the two procedures differ?

(c) One of the two procedures requires more conditions for its validity, but if the conditions are met, then this procedure has certain advantages over the other one. What are these advantages?

11.8.6 Consider the data from Exercise 11.7.3 in which the drugs ticrynafen (T) and hydrochlorothiazide (H) were compared. The data are summarized in the following table. The pooled standard deviation is $s_{pooled} = 11.83$ mm Hg.

	Ticrynafen (T)		Hydrochlorothiazide (H)	
	Low dose	High dose	Low dose	High dose
Mean	13.9	17.1	15.8	17.5
No. of patients	53	57	55	58

If the two drugs have equal effects on blood pressure, then T might be preferable because it has fewer side effects.

(a) Construct a 95% confidence interval for the difference between the drugs (with respect to mean blood pressure reduction), averaged over the two dosage levels.

(b) Interpret the confidence interval from part (a) in the context of this setting.

11.8.7 Consider the lettuce growth experiment described in Exercise 11.7.6. The accompanying table shows the mean leaf dry weight (gm) of the nine plants in each treatment group. MS(within) from the ANOVA was 0.3481.

	Nutrient solution	
	Standard	Extra nitrogen
Low light	2.16	3.09
High light	3.26	4.48

Construct a 95% confidence interval for the effect of extra nitrogen, averaged over the two light conditions.

11.8.8 Refer to the MAO data of Exercise 11.4.1.

(a) Define a contrast to compare the MAO activity for patients with schizophrenia without paranoid features versus the average of the two types with paranoid features.

(b) Calculate the value of the contrast in part (a) and its standard error.

(c) Apply a t test to the contrast in part (a). Let H_A be nondirectional and $\alpha = 0.05$.

11.8.9 Are the brains of left-handed people anatomically different? To investigate this question, a neuroscientist conducted postmortem brain examinations in 42 people. Each person had been evaluated before death for hand preference and categorized as consistently right-handed (CRH) or mixed-handed (MH). The table shows the results on the area of the anterior half of the corpus callosum (the structure that links the left and right hemispheres of the brain).[32] The MS(within) from the ANOVA was 2,498.

	Area (mm^2)		
Group	Mean	SD	n
1. Males: MH	423	48	5
2. Males: CRH	367	49	7
3. Females: MH	377	63	10
4. Females: CRH	345	43	20

(a) The difference between MH and CRH is 56 mm^2 for males and 32 mm^2 for females. Is this sufficient evidence to conclude that the corresponding population difference is greater for males than for females? Test an appropriate hypothesis. (Use a nondirectional alternative and let $\alpha = 0.10$.)

(b) As an overall measure of the difference between MH and CRH, one can consider the quantity $0.5(\mu_1 - \mu_2) + 0.5(\mu_3 - \mu_4)$. Construct a 95% confidence interval for this quantity. (This is a sex-adjusted comparison of MH and CRH, where the reference population is 50% male and 50% female.)

11.8.10 Consider the daffodil data of Exercise 11.4.5.

(a) Define a contrast to compare the stem length for daffodils from the open area versus the average of the north, south, east and west sides of the building.

(b) Consider applying a (nondirectional) t test to the contrast is part (a). What is the value of t_s and how many degrees of freedom are there?

(c) The P-value for the test is 0.0011. If $\alpha = 0.05$, state your conclusion regarding the contrast in the context of this setting.

11.9 Multiple Comparisons (Optional)

After conducting a global F test, we may find that there is significant evidence for a difference among the population means $\mu_1, \mu_2, \ldots, \mu_I$. In this situation, we are often interested in a detailed analysis of the sample means $\bar{Y}_1, \bar{Y}_2, \ldots, \bar{Y}_I$ considering all pairwise comparisons. That is, we wish to test all possible pairwise hypotheses:

$$H_0\colon \mu_1 = \mu_2$$
$$H_0\colon \mu_1 = \mu_3$$
$$H_0\colon \mu_2 = \mu_3$$

and so on.

We saw in Section 11.1 that using repeated t tests leads to an increased overall risk of Type I error (e.g., finding evidence for a difference in population means when, in fact, there is no difference). In fact, it was this increased risk of Type I error that motivated the global F test in the first place. In this section we describe three multiple comparison methods to control the overall risk of Type I error: Bonferroni's method, Fisher's Least Significant Difference, and Tukey's Honest Significant Difference. First, however, we must examine the different types of Type I error that arise when considering multiple comparisons.

EXPERIMENTWISE VERSUS COMPARISONWISE ERROR

Consider a study involving the comparison of four population means: μ_1, μ_2, μ_3, and μ_4. As noted in Section 11.1, there are six possible comparisons:

$$H_0\colon \mu_1 = \mu_2 \quad H_0\colon \mu_1 = \mu_3 \quad H_0\colon \mu_1 = \mu_4 \quad H_0\colon \mu_2 = \mu_3 \quad H_0\colon \mu_2 = \mu_4 \quad H_0\colon \mu_3 = \mu_4$$

When considering these six comparisons we can speak of the chance of a Type I error for a particular comparison, say H_0: $\mu_1 = \mu_2$, called the **comparisonwise Type I error rate (α_{cw})**, or we can speak of the chance of making a Type I error among *any* of the six comparisons, called the **experimentwise Type I error rate (α_{ew}).*** For example, Table 11.1.2 displays the experimentwise Type I error rates for comparing different numbers of groups when the comparisonwise Type I error rate is $\alpha_{cw} = 0.05$.

While the relationship between α_{cw} and α_{ew} may be complex, it is always true that

$$\alpha_{ew} \leq k \times \alpha_{cw}$$

where k is the number of comparisons. Thus, if six independent comparisons were made at the $\alpha_{cw} = 0.05$ level, the experimentwise Type I error rate (α_{ew}) is at most $6 \times 0.05 = 0.30$.

FISHER'S LEAST SIGNIFICANT DIFFERENCE

In optional Section 11.8 we described a procedure for estimating linear contrasts. Fisher's Least Significant Difference (LSD) uses this procedure to produce all pairwise confidence intervals for differences of population means using $\alpha_{cw} = \alpha$, the Type I error rate used in the ANOVA. Intervals that do not contain zero provide evidence for a significant difference between the compared population means.

An example of the procedure follows.

Example 11.1.1 **Oysters and Seagrass** In a study to investigate the effect of oyster density on seagrass biomass, researchers introduced oysters to thirty 1-m^2 plots of healthy seagrass. At the beginning of the study the seagrass was clipped short in all plots. Next, 10 randomly chosen plots received a high density of oysters; 10, an intermediate density; and 10, a low density. As a control, an additional 10 randomly chosen clipped 1-m^2 plots received no oysters. After 2 weeks, the belowground seagrass biomass was measured in each plot (g/m^2). Data from some plots are missing. A summary of the data (Table 11.9.1) as well as the ANOVA table (Table 11.9.2) follow.[33]

Table 11.9.1 Belowground seagrass biomass (g/m^2)

	Oyster density			
	None (1)	Low (2)	Intermediate (3)	High (4)
Mean	34.81	33.13	28.33	15.00
SD	13.44	17.36	17.11	10.97
n	9	10	8	10

*Although the term *experimentwise* contains the word experiment, this terminology pertains to both experiments and observational studies.

Table 11.9.2 ANOVA summary of belowground seagrass biomass (g/m^2)

	df	Sum of squares	Mean squares	F	P-value
Between	3	2365.5	788.51	3.5688	0.0243
Within	33	7291.1	220.94		
Total	36	9656.6			

The P-value for the ANOVA is 0.0243, indicating that there is significant evidence of a difference among the biomass means under these experimental conditions. Having evidence for a difference we proceed with comparisons.

Recall that for any linear contrast $L = m_1\bar{y}_1 + m_2\bar{y}_2 + \cdots + m_I\bar{y}_I$,

$$\text{SE}_L = s_{\text{pooled}}\sqrt{\sum_{i=1}^{I}\frac{m_i^2}{n_i}}$$

where

$$s_{\text{pooled}} = \sqrt{\text{MS(within)}}$$

Thus, to compare the no oyster condition (1) to the low oyster density condition (2) we define $D_{12} = \bar{Y}_1 - \bar{Y}_2$ so that as a linear contrast we have

$$\begin{aligned} d_{12} &= 1\bar{y}_1 + (-1)\bar{y}_2 + 0\bar{y}_3 + 0\bar{y}_4 \\ &= (1)(34.81) + (-1)(33.13) + (0)(28.33) + (0)(15.00) \\ &= 34.81 - 33.13 = 1.68 \end{aligned}$$

and, since $s_{\text{pooled}} = \sqrt{220.94} = 14.86$, we have

$$\begin{aligned} \text{SE}_{D_{12}} &= 14.86 \times \sqrt{\frac{1^2}{9} + \frac{(-1)^2}{10} + \frac{0^2}{8} + \frac{0^2}{10}} \\ &= 14.86 \times \sqrt{\frac{1}{9} + \frac{1}{10}} \\ &= 6.82 \end{aligned}$$

A 95% confidence interval for the population mean difference in belowground biomass for the no oyster condition compared to the low oyster density condition, $\mu_1 - \mu_2$, is given by

$$\begin{aligned} d_{12} \pm t_{33,0.025} \times \text{SE}_{D_{12}} &= 1.68 \pm 2.0345 \times 6.82 \\ &= 1.68 \pm 13.89 \\ &= (-12.21, 15.57) \end{aligned}$$

We are 95% confident that the mean belowground biomass when there are no oysters is between 12.21 g/m^2 lower to 15.57 g/m^2 higher than when there is a low density of oysters. Since this interval contains zero, there is no statistically significant evidence that the mean belowground biomass differs for these two conditions.

Repeating this process for the remaining five comparisons produces the intermediate computations and final intervals summarized in Table 11.9.3.

From Table 11.9.3 we observe that the only comparisons showing significant differences in mean biomass are the no- to high-oyster density and low- to high-oyster densities. ■

Table 11.9.3 Intermediate computations and 95% Fisher's LSD intervals comparing belowground biomass under different oyster density conditions*

Comparison	$d_{ab} = \bar{y}_a - \bar{y}_b$	$\sqrt{(1/n_a) + (1/n_b)}$	$SE_{D_{ab}} = s_{pooled} \times \sqrt{(1/n_a) + (1/n_b)}$	$t_{33,0.025} \times SE_{D_{ab}}$
None–low	1.68	0.459	6.828	13.891
None–intermediate	6.48	0.486	7.221	14.690
None–high	*19.81*	*0.459*	*6.828*	*13.891*
Low–intermediate	4.80	0.474	7.049	14.341
Low–high	*18.13*	*0.447*	*6.646*	*13.520*
Intermediate–high	13.33	0.474	7.049	14.341
Comparison	**Lower 95%**	**Upper 95%**		
None–low	−12.2	15.6		
None–intermediate	−8.2	21.2		
None–high	*5.9*	*33.7*		
Low–intermediate	−9.5	19.1		
Low–high	*4.6*	*31.7*		
Intermediate–high	−1.0	27.7		

*Intervals not containing zero (i.e., there is a statistically significant difference between the group means) are in italics. Note that an interval will not contain zero whenever $|D_{ab}| > t \times SE_{D_{ab}}$. (The value of $t_{33,0.025} = 2.0345$ was determined using a computer. Using Table 4 we would obtain very similar results using the value listed for 30 degrees of freedom, $t_{30,0.025} = 2.042$.)

A general formula for computing a $100(1 - \alpha)\%$ Fisher LSD interval for $(\mu_a - \mu_b)$ is given in the following box.

$100(1 - \alpha)\%$ Fisher LSD Interval for $(\mu_a - \mu_b)$

$$d_{ab} \pm t_{df,\alpha/2} \times SE_{D_{ab}}$$

where

$$d_{ab} = \bar{y}_a - \bar{y}_b$$

$$SE_{D_{ab}} = s_{pooled}\sqrt{\frac{1}{n_a} + \frac{1}{n_b}}$$

$$s_{pooled} = \sqrt{MS(within)}$$

and

$$df = df(within)$$

How does Fisher's LSD control the experimentwise Type I error rate? One should use Fisher's LSD comparisons only after rejecting the ANOVA global null hypothesis that all population means are equal: H_0: $\mu_1 = \mu_2 = \cdots = \mu_I$. The ANOVA global F test acts as a screening procedure for the multiple comparisons and thus offers control over α_{ew}.

DISPLAYING RESULTS

The presentation of all six Fisher LSD intervals for the seagrass example in Table 11.9.3 is a useful working summary but is not suitable for effective communication of results. To organize the results for presentation in a simple table we take the following steps.

Step 1. *Array of group labels.* Arrange the group labels in increasing order of their means.

Step 2. *Systematic comparison of means, underlining nonsignificant differences.*

(a) Begin by examining the interval comparing the largest and smallest means. If the interval contains zero, the difference in means is not statistically significant, and a line is drawn under the array of group labels to "connect" the groups with the largest and smallest means. If the interval does not contain zero, proceed to the next step.

(b) Ignore the group with the smallest mean and compare the remaining subarray of $I - 1$ means. As in step (2a), if the interval contains zero, the difference in means is not statistically significant, and a line is drawn under the array of group labels being compared to "connect" the groups. Next consider the other subarray of $I - 1$ means—the means that remain if the group with the largest mean is ignored. Again, underline this subarray if the interval contains zero.

(c) Repeat step (2b) by successively comparing all subarrays of size $I - 2, I - 3$, and so on, until an interval is produced that contains zero or no more comparisons are possible.

Important Notes: When sample sizes are similar, comparisons within any subarray that has already been underlined are unnecessary; these group means are automatically declared not statistically significantly different. Also, when underlining, use a separate line for each step; never join a line to one that has already been drawn.

Step 3. *Translate the underlines to a tabular summary.* Create a summary table of the data using superscript letters to indicate which groups are not statistically significantly different.

Example 11.9.2 **Oysters and Seagrass** In this example we will follow the preceding procedure to display the oyster and seagrass Fisher's LSD comparisons displayed in Table 11.9.3.

Step 1. We first arrange the labels in order of the means (shown in Table 11.9.1).

High Intermediate Low None

Step 2. We compare the groups with the smallest (high oyster density) and largest (no oysters) means: $\mu_{None} - \mu_{High} = (5.9, 33.7)$. This interval does not contain zero, so these means are significantly different and no underline is made.* We now proceed to the next set (step 2b), the comparisons of subarrays of three means. First, we compare Intermediate to None:

$$\mu_{None} - \mu_{Intermediate} = (-8.2, 21.2)$$

This interval contains zero, so an underline is drawn as shown.

High <u>Intermediate Low None</u>

*Intuitively, this interval should not contain zero since we have rejected the global F test null hypothesis, although there are some instances where the results of our multiple comparison procedure and global F test may not agree.

This underline indicates that these three groups do not have significantly different means. We now compare the next subarray of three means, High to Low: $\mu_{High} - \mu_{Low} = (4.6, 31.7)$. This interval does not contain zero, so no underlines are drawn. There is evidence for a difference in mean belowground biomass between the high and low oyster-density conditions.

Having compared all subarrays of three means, we continue with subarrays of two means. The only subarray of two means not already connected with an underline is the High–Intermediate comparison. This interval $\mu_{Intermediate} - \mu_{High} = (-1.0, 27.7)$ contains zero, so an underline is drawn as shown.

High Intermediate Low None

Step 3. Communicating these results, we give each line a letter and display these letters as superscripts in our table of group means as shown below and in Table 11.9.4. A graphical display is also possible and is displayed in Figure 11.9.1.

High Intermediate Low None
a
b

Table 11.9.4 Belowground seagrass biomass (g/m²) for different levels of oyster density

	Oyster density			
	None	Low	Intermediate	High
Mean	34.8^{a}	33.1^{a}	$28.3^{a,b}$	15.0^{b}
SD	13.4	17.4	17.1	11.0
n	9	10	8	10

*Groups sharing a common superscript have means that are not statistically significantly different based on Fisher's LSD comparisons with $\alpha_{cw} = 0.05$.

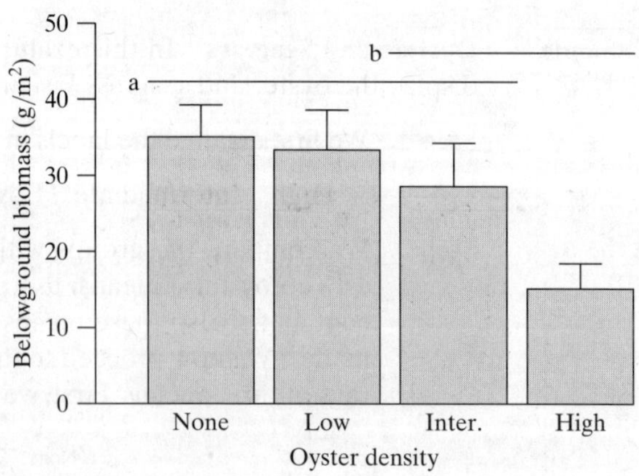

Figure 11.9.1 Belowground seagrass biomass (g/m²) for different levels of oyster density. Bars display means plus one standard error. Groups sharing a common overbar are not statistically significantly different based on Fisher's LSD comparisons with $\alpha_{cw} = 0.05$

THE BONFERRONI METHOD

The **Bonferroni method** is based on a very simple and general relationship: The probability that at least one of several events will occur cannot exceed the sum of the individual probabilities. For instance, suppose we conduct six tests of hypotheses, each at $\alpha_{cw} = 0.01$. Then the overall risk of Type I error α_{ew}—that is, the chance of rejecting at least one of the six hypotheses when in fact all of them are true—cannot exceed

$$0.01 + 0.01 + 0.01 + 0.01 + 0.01 + 0.01 = (6)(0.01) = 0.06$$

Turning this logic around, suppose an investigator plans to conduct six tests of hypotheses and wants the overall risk of Type I error not to exceed $\alpha_{ew} = 0.05$. A conservative approach is to conduct each of the separate tests at the significance level $\alpha_{cw} = 0.05/6 = 0.0083$; this is called a **Bonferroni adjustment.**

Note that the Bonferroni technique is very broadly applicable. The separate tests may relate to different response variables, different subsets, and so on; some may be t tests, some chi-square tests, and so on.

The Bonferroni approach can be used by a person reading a research report, if the author has included explicit P-values. For instance, if the report contains six P-values and the reader desires overall 5%-level protection against Type I error, then the reader will not regard a P-value as sufficient evidence of an effect unless it is smaller than $\alpha_{cw} = 0.0083$.

A Bonferroni adjustment can also be made for confidence intervals. For instance, suppose we wish to construct six confidence intervals and desire an overall probability of 95% that *all* the intervals contain their respective parameters ($\alpha_{ew} = 0.05$). Then this can be accomplished by constructing each interval at confidence level 99.17% (because $0.05/6 = 0.0083$ and $1 - 0.0083 = 0.9917$).

In general, to construct k Bonferonni-adjusted confidence intervals with an overall probability of $100(1 - \alpha_{ew})\%$ that *all* the intervals contain their respective parameters, we construct each interval at confidence level $100(1 - \alpha_{cw})\%$ where $\alpha_{cw} = \alpha_{ew}/k$. The mechanics of the computations are identical to those used for Fisher's LSD except the value of the t multiplier is modified: $t_{df,\alpha_{cw}/2}$. Note that the application of this idea requires unusual critical values, so standard tables are not sufficient. Table 11 (at the end of this book) provides Bonferroni multipliers for confidence intervals that are based on a t distribution. Software can also be used to produce appropriate multipliers. Example 11.9.3 illustrates this idea.

Example 11.9.3 **Oysters and Seagrass** To compute the Bonferroni adjusted experimentwise 95% ($\alpha_{ew} = 0.05$) confidence intervals for our oyster and seagrass example, we first recall that a total of six comparisons are required so that $\alpha_{cw} = 0.05/6 = 0.0083$ and $t_{30,0.0083/2} = 2.825$ [because not all values of df are listed in Table 11, we use df = 30, the closest value to df(within) = 33]. Table 11.9.5 summarizes the collection of intervals in a manner similar to the Fisher LSD intervals in Table 11.9.3.

Table 11.9.5 Intermediate computations and experimentwise 95% (99.17% comparisonwise) Bonferroni intervals comparing belowground biomass under different oyster density conditions

Comparison	$d_{ab} = \bar{y}_a - \bar{y}_b$	$SE_{D_{ab}}$	$t_{30,0.025/6} \times SE_{D_{ab}}$	Lower 99.17%	Upper 99.17%
None–low	1.68	6.828	19.289	−17.6	21.0
None–intermediate	6.48	7.221	20.399	−13.9	26.9
None–high	*19.81*	6.828	*18.775*	*0.5*	*39.1*
Low–intermediate	4.80	7.049	19.913	−15.1	24.7
Low–high	18.13	6.646	18.775	−0.6	36.9
Intermediate–high	13.33	7.049	19.913	−6.6	33.2

*Intervals not containing zero (i.e., where there is a statistically significant difference between the group means) are in italics. Note the first two columns (d_{ab} and $SE_{D_{ab}}$) are identical to those presented in Table 11.9.3.

Using the method of underlining to visualize the comparisons, we have

```
High   Intermediate   Low   None
      a___________________________
b___________________________
```

The underlines indicate that the only significant difference in mean belowground seagrass biomass is between the high oyster density and no oyster conditions. A summary of the results is presented in Table 11.9.6. ■

Table 11.9.6 Belowground seagrass biomass (g/m^2) for different levels of oyster density

	Oyster density			
	None	Low	Intermediate	High
Mean	34.8^{a}	$33.1^{a,b}$	$28.3^{a,b}$	15.0^{b}
SD	13.4	17.4	17.1	11.0
n	9	10	8	10

*Groups sharing a common superscript have means that are not statistically significantly different based on Bonferroni comparisons with $\alpha_{ew} = 0.05$.

Note that the Fisher LSD intervals and the Bonferroni intervals are not identical (the Bonferroni are wider due to the smaller value of α_{cw}). In addition, the conclusions differ as well. The Fisher LSD intervals indicate that there is evidence that the low and high oyster density conditions have different population means, while the Bonferroni intervals do not indicate a difference. This is because the Bonferroni intervals are less powerful and thus more conservative than the Fisher intervals. Unlike the Fisher intervals, the Bonferroni intervals are guaranteed to have α_{ew} less than or equal to the desired experimentwise Type I error rate.

Unfortunately, the Bonferroni intervals are often overly conservative so that the actual value of α_{ew} is much less than the desired experimentwise Type I error rate, and thus too much power is sacrificed for Type I error protection. A more complex procedure that (when sample sizes are equal) is able to achieve the desired experimentwise error exactly (and thus achieve higher power than Bonferroni) is Tukey's Honest Significant Difference.*

TUKEY'S HONEST SIGNIFICANT DIFFERENCE

Tukey's Honest Significant Difference (HSD) is very similar to the Fisher's LSD and Bonferonni adjusted intervals, but rather than using t multipliers in the confidence interval formulas, related values from a distribution known as the Studentized range distribution are used. Most computer packages will display all Tukey HSD pairwise intervals for any desired experimentwise Type I error rate, α_{ew}. As an example, Figure 11.9.2 displays the Tukey output from the statistical software package R using our oyster and seagrass data. Note that in addition to the intervals, most software also provides an "adjusted" P-value. Even though multiple comparisons are being made, if these "adjusted" P-values are compared to α_{ew}, an overall experimentwise Type I error rate of α_{ew} will still be maintained.

*There are many methods used to handle the "multiple comparisons/multiple testing" problem. One that has become popular in recent years is the False Discovery Rate, which is often used by researchers in genomics.[34]

Figure 11.9.2 R software output presenting experimentwise 95% Tukey HSD intervals for the oyster and seagrass example

	diff	lwr	upr	p-adj
int-high	13.33	−5.74	32.40	0.2515
low-high	18.13	0.15	36.11	0.0475
no-high	19.81	1.34	38.28	0.0318
low-int	4.80	−14.27	23.87	0.9037
zero-int	6.48	−13.06	26.02	0.8063
zero-low	1.68	−16.79	20.15	0.9947

The intervals in Figure 11.9.2 show that the conclusions drawn from the Tukey HSD intervals match those from the Fisher LSD intervals: the high and low oyster density as well as the high and no oyster density means differ significantly. The endpoints of the experimentwise 95% Tukey HSD intervals are, however, different from both the Fisher LSD and Bonferroni intervals.

CONDITIONS FOR VALIDITY

All three multiple comparison procedures as described require the same standard ANOVA conditions given in Section 11.5. In addition, the validity conditions for Fisher's LSD intervals also require that the procedure not be used unless the global null hypothesis of all means being equal is rejected. In contrast, Tukey's HSD and Bonferroni intervals do not require that the global F test be performed a priori (although the computation of s_{pooled} is still needed). To exactly achieve the desired experimentwise Type I error rate, Tukey's HSD requires that all samples be the same size. If the sample sizes are unequal, the actual error rate will be somewhat less than the nominal rate resulting in a loss of power.

An advantage of the Bonferroni method is that it is widely applicable and can easily be generalized to situations beyond ANOVA. One such example appears in the exercises.

Exercises 11.9.1–11.9.8

11.9.1 A botanist used a completely randomized design to allocate 45 individually potted eggplant plants to five different soil treatments. The observed variable was the total plant dry weight without roots (gm) after 31 days of growth. The treatment means were as shown in the following table.[35] The MS(within) was 0.2246. Use Fisher's LSD intervals to compare all pairs of means at $\alpha_{cw} = 0.05$. Present your results in a summary table similar to Table 11.9.4. (*Hint*: Take note that all sample sizes are equal; thus the calculated margin of error need only be calculated once for all comparisons. There are a total of 10 comparisons possible.)

Treatment	A	B	C	D	E
Mean	4.37	4.76	3.70	5.41	5.38
n	9	9	9	9	9

11.9.2 Repeat Exercise 11.9.1, but use Bonferroni intervals with $\alpha_{ew} = 0.05$.

11.9.3 In a study of the dietary treatment of anemia in cattle, researchers randomly divided 144 cows into four treatment groups. Group A was a control group, and groups B, C, and D received different regimens of dietary supplementation with selenium. After a year of treatment, blood samples were drawn and assayed for selenium. The accompanying table shows the mean selenium concentrations (μg/dl).[36] The MS(within) from the ANOVA was 2.071.

Group	Mean	n
A	0.8	36
B	5.4	36
C	6.2	36
D	5.0	36

(a) Compute three Bonferroni-adjusted intervals comparing diets B, C, and D to the control (diet A) using $\alpha_{ew} = 0.05$. (*Note*: This is an example of a situation for which the Bonferroni comparisons may be preferred over the Tukey HSD comparisons since not all comparisons are considered—we are only interested in comparing the control to each of the other three treatments.)

(b) In the context of the problem, interpret the Bonferroni interval computed in part (a) that compares the control (group A) to the group that is most different from it.

11.9.4 Consider the experiment and data in Exercise 11.9.3. The experimentwise 95% Tukey HSD intervals are displayed using the statistical software package R.

```
      diff    lwr    upr
B-A    4.6   3.72   5.48
C-A    5.4   4.52   6.28
D-A    4.2   3.32   5.08
C-B    0.8  -0.08   1.68
D-B   -0.4  -1.28   0.48
D-C   -1.2  -2.08  -0.32
```

(a) Using the preceding output to support your answer, is there evidence that each of the groups/diets B, C, and D, differs from the control, A?

(b) According to the preceding Tukey HSD intervals and summary of the data in Exercise 11.9.3, diet C yields the greatest mean selenium concentration and is significantly higher than the control. If the goal of the researchers is to find a diet that maximizes selenium concentration, is diet C the clear choice? That is, should we rule out diet B, diet D, or both? Refer to the Tukey HSD intervals to justify your answer.

11.9.5 Ten treatments were compared for their effect on the liver in mice. There were 13 animals in each treatment group. The ANOVA gave MS(within) = 0.5842. The mean liver weights are given in the table.[37]

Treatment	Mean liver weight (gm)	Treatment	Mean liver weight (gm)
1	2.59	6	2.84
2	2.28	7	2.29
3	2.34	8	2.45
4	2.07	9	2.76
5	2.40	10	2.37

(a) Use Fisher LSD intervals to compare all pairs of means with $\alpha_{cw} = 0.05$ and summarize the results in a table similar to Table 11.9.4. [*Time Saving Hints*: First note that the sample sizes are equal; hence the same margin of error ($t \times SE_{D_{ab}}$) can be used for all comparisons. Furthermore, since a summary table is desired, the actual intervals need not be computed: Simply check if $|d_{ab}| > t \times SE_{D_{ab}}$. If it is, then the computed interval would not contain zero, so the difference is significant. Finally, note that not all possible comparisons (there are 45) need to be checked: when using the method of underlining to summarize results, once a subarray of groups has been underlined all comparisons within the subarray are considered nonsignificant.]

(b) If Bonferroni's method is used with $\alpha_{ew} = 0.05$ instead of Fisher's LSD in part (a), are any pairs of means significantly different?

11.9.6 Consider the data from Example 11.2.1 on the weight gain of lambs. The MS(within) from the ANOVA for these data was 23.333. The sample mean of diet 2 was 15 and of diet 1 was 11.

(a) Use the Bonferroni method to construct a 95% confidence interval for the difference in population means of these two diets (assuming that intervals will also be computed for the other two possible comparisons as well).

(b) Suppose that the comparison in part (a) was the *only* comparison of interest (i.e., one comparison rather than three). How would the interval in part (a) change? Would it be wider, narrower, or stay the same? Explain.

11.9.7 As mentioned in this section, the Bonferroni procedure can be used in a variety of circumstances. Consider the plover nesting example from Section 10.5, which compares plover nest locations across 3 years. The percentage distribution appears in the following table.

	Year		
Location	2004	2005	2006
Agricultural field (AF)	48.8	30.2	55.3
Prairie dog habitat (PD)	39.5	60.3	25.5
Grassland (G)	11.6	9.5	19.1
Total	99.9*	100.0	99.9*

*The sums of the 2004 and 2006 percentages differ from 100% due to rounding.

The *P*-value for the chi-square test of these data was found to be 0.007, indicating a significant difference in the distribution of nesting locations across the 3 years with $\alpha = 0.10$. Considering reduced tables and using chi-square tests to compare nesting distributions for pairs of years, we obtain the following *P*-values:

Years compared	*P*-value
2004 to 2005	0.100
2004 to 2006	0.307
2005 to 2006	0.001

Using a Bonferonni adjustment to achieve $\alpha_{ew} = 0.10$, for which pair(s) of years is there evidence of a significant difference in nesting location distributions? Indicate the value of α_{cw} used.

11.9.8 Exercise 10.5.1 presented the following problem: Patients with painful knee osteoarthritis were randomly assigned in a clinical trial to one of five treatments: glucosamine, chondroitin, both, placebo, or Celebrex, the standard therapy. One outcome recorded was whether or not each patient experienced substantial improvement in pain or in ability to function. The data are given in the following table.

	Successful outcome		
Treatment	Sample size	Number	Percentage
Glucosamine	317	192	60.6
Chondroitin	318	202	63.5
Both	317	208	65.6
Placebo	313	178	56.9
Celebrex	318	214	67.3

(a) Suppose we wished to compare only the success rates of each of the treatments to the control (placebo) using four separate 2×2 chi-square tests. The P-values for these comparisons follow. Using a Bonferroni adjustment with $\alpha_{ew} = 0.05$, which treatments perform significantly different from the placebo? Indicate the value of α_{cw} used.

Treatments compared to placebo	P-value
Glucosamine	0.346
Chondroitin	0.088
Both	0.024
Celebrex	0.007

(b) The P-value of the chi-square test that considers the entire 5×2 table is 0.054, which provides insufficient evidence to demonstrate any difference among the success rates of the five treatments using $\alpha = 0.05$. Explain why this result does not contradict the results of part (a). [*Hint*: How many comparisons are being considered by this chi-square test as compared to the number of comparisons in part (a)? To achieve $\alpha_{ew} = 0.05$ using a Bonferroni adjustment, how large would α_{cw} need to be? How large was it in part (a)? How does conducting many tests with a Bonferroni adjustment affect the power of each test?]

11.10 Perspective

In Chapter 11 we have introduced some statistical issues that arise when analyzing data from more than two samples, and we have considered some classical methods of analysis. In this section we review these issues and briefly mention some alternative methods of analysis.

ADVANTAGES OF GLOBAL APPROACH

Let us recapitulate the advantages of analyzing I independent samples by a global approach rather than by viewing each pairwise comparison separately.

1. *Multiple comparisons* In Section 11.1 we saw that the use of repeated t tests can greatly inflate the overall risk of Type I error. Some control of Type I error can be gained by the simple device of beginning the data analysis with a global F test. For more stringent control of Type I error, other multiple comparison methods are available (e.g., Bonferroni and Tukey HSD) and are described in optional Section 11.9. (Note that the problem of multiple comparisons is not confined to an ANOVA setting.)
2. *Use of structure in the treatments or groups* Analysis of suitable combinations of group means can be very useful in interpreting data. Many of the relevant techniques are beyond the scope of this book. The discussion in optional Sections 11.7 and 11.8 gave a hint of the possibilities. In Chapter 12 we will discuss some ideas that are applicable when the treatments themselves are quantitative (e.g., doses).

3. *Use of a pooled SD* We have seen that pooling all of the within-sample variability into a single pooled SD leads to a better estimate of the common population SD and thus to a more precise analysis. This is particularly advantageous if the individual sample sizes (n's) are small, in which case the individual SD estimates are quite imprecise. Of course, using a pooled SD is proper only if the population SDs are equal. It sometimes happens that one cannot take advantage of pooling the SDs because the assumption of equal population SDs is not tenable. One approach that can be helpful in this case is to analyze a transformed variable, such as $\log(Y)$; the SDs may be more nearly equal in the transformed scale.

OTHER EXPERIMENTAL DESIGNS

The techniques of this chapter are valid only for independent samples. But the basic idea—partitioning variability within and between treatments into interpretable components—can be applied in many experimental designs. For instance, all the techniques discussed in this chapter can be adapted (by suitable modification of the SE calculation) to analysis of data from an experiment with more than two experimental factors or situations for which all or some experimental factors are numeric rather than categorical. These and related techniques belong to the large subject called *analysis of variance*, of which we have discussed only a small part.

NONPARAMETRIC APPROACHES

There are k-sample analogs of the Wilcoxon-Mann-Whitney test and other nonparametric tests (e.g., the Kruskal–Wallis test). These tests have the advantage of not assuming underlying normal distributions. However, many of the advantages of the parametric techniques—such as the use of linear combinations—do not easily carry over to the nonparametric setting.

Supplementary Exercises 11.S.1–11.S.20

(*Note*: Exercises preceded by an asterisk refer to optional sections.)

11.S.1 Consider the research described in Exercise 11.4.6 in which 10 women in an aerobic exercise class, 10 women in a modern dance class, and a control group of 9 women were studied. One measurement made on each woman was change in fat-free mass over the course of the 16-week training period. Summary statistics are given in the following table.[8] The ANOVA SS(between) is 2.465 and the SS(within) is 50.133.

	Aerobics	Modern dance	Control
Mean	0.00	0.44	0.71
SD	1.31	1.17	1.68
n	10	10	9

(a) State in words, in the context of this problem, the null hypothesis that is tested by the ANOVA.

(b) Construct the ANOVA table and test the null hypothesis. Let $\alpha = 0.05$.

11.S.2 Refer to Exercise 11.S.1. The F test is based on certain conditions concerning the population distributions.

(a) State the conditions.

(b) The following dotplots show the raw data. Based on these plots and on the information given in Exercise 11.S.1, does it appear that the F test conditions are met? Why or why not?

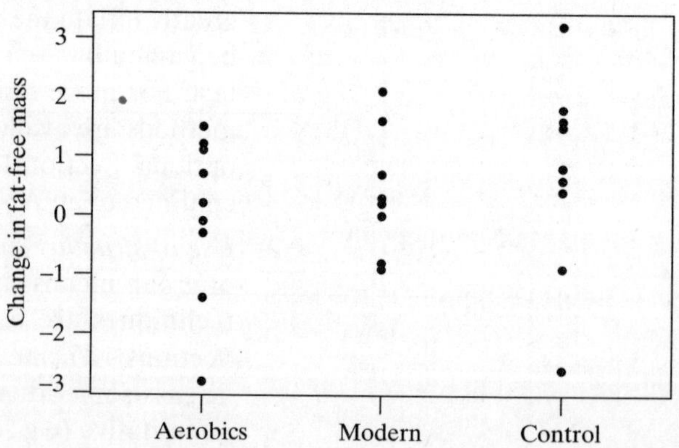

11.S.3 In a study of the eye disease retinitis pigmentosa (RP), 211 patients were classified into four groups according to the pattern of inheritance of their disease. Visual acuity (spherical refractive error, in diopters) was measured for each eye, and the two values were then averaged to give one observation per person. The accompanying table shows the number of persons in each group and the group mean refractive error.[38] The ANOVA of the 211 observations yields SS(between) = 129.49 and SS(within) = 2,506.8. Construct the ANOVA table and carry out the F test at $\alpha = 0.05$.

Group	Number of persons	Mean refractive error
Autosomal dominant RP	27	+0.07
Autosomal recessive RP	20	−0.83
Sex-linked RP	18	−3.30
Isolate RP	146	−0.84
Total	211	

11.S.4 (*Continuation of Exercise 11.S.3*) Another approach to the data analysis is to use the eye, rather than the person, as the observational unit. For the 211 persons there were 422 measurements of refractive error; the accompanying table summarizes these measurements. The ANOVA of the 422 observations yields SS(between) = 258.97 and SS(within) = 5,143.9.

Group	Number of eyes	Mean refractive error
Autosomal dominant RP	54	+0.07
Autosomal recessive RP	40	−0.83
Sex-linked RP	36	−3.30
Isolate RP	292	−0.84
Total	422	

(a) Construct the ANOVA table and bracket the P-value for the F test. Compare with the P-value obtained in Exercise 11.S.3. Which of the two P-values is of doubtful validity, and why?

(b) The mean refractive error for the sex-linked RP patients was −3.30. Calculate the standard error of this mean two ways: (i) regarding the person as the observational unit and using s_{pooled} from the ANOVA of Exercise 11.S.3; (ii) regarding the eye as the observational unit and using s_{pooled} from the ANOVA of this exercise. Which of these standard errors is of doubtful validity, and why?

***11.S.5** In a study of the mutual effects of the air pollutants ozone and sulfur dioxide, Blue Lake snap beans were grown in open-top field chambers. Some chambers were fumigated repeatedly with sulfur dioxide. The air in some chambers was carbon filtered to remove ambient ozone. There were three chambers per treatment combination, allocated at random. After one month of treatment, total yield (kg) of bean pods was recorded for each chamber, with results shown in the accompanying table.[39] For these data, SS(between) = 1.3538 and SS(within) = 0.27513.

	Ozone absent		Ozone present	
	Sulfur dioxide		Sulfur dioxide	
	Absent	Present	Absent	Present
	1.52	1.49	1.15	0.65
	1.85	1.55	1.30	0.76
	1.39	1.21	1.57	0.69
Mean	1.587	1.417	1.340	0.700
SD	0.237	0.181	0.213	0.056

(a) Construct the ANOVA table.

(b) The P-value for a one-way ANOVA F test is 0.0002. If $\alpha = 0.05$, what is your conclusion regarding the null hypothesis?

***11.S.6** Consider the data from Exercise 11.S.5. For these data, SS(ozone) = 0.696, SS(sulfur) = 0.492, SS(interaction) = 0.166, and SS(within) = 0.275.

(a) Prepare an interaction graph (like Figure 11.7.3).

(b) Construct the ANOVA table.

(c) What is the value of the F test statistic for interactions?

(d) The P-value for the interactions F test is 0.058. If $\alpha = 0.05$, what is your conclusion regarding the null hypothesis?

(e) The P-value for testing that ozone has no effect is 0.0019. If $\alpha = 0.05$, what is your conclusion regarding the null hypothesis?

***11.S.7** Refer to Exercise 11.S.5. Define contrasts to measure each effect specified, and calculate the value of each contrast.

(a) The effect of sulfur dioxide in the absence of ozone

(b) The effect of sulfur dioxide in the presence of ozone

(c) The interaction between sulfur dioxide and ozone

***11.S.8** (*Continuation of Exercises 11.S.6 and 11.S.7*) For the snap-bean data, use a t test to test the null hypothesis of no interaction against the alternative that sulfur dioxide is more harmful in the presence of ozone than in its absence. Let $\alpha = 0.05$. How does this compare with the F test of Exercise 11.S.6(b) (which has a nondirectional alternative)?

***11.S.9** (*Computer exercise*) Refer to the snap-bean data of Exercise 11.S.5. Apply a reciprocal transformation to the data. That is, for each yield value Y, calculate $Y' = 1/Y$.

(a) Calculate the ANOVA table for Y' and carry out the F test.

(b) It often happens that the SDs are more nearly equal for transformed data than for the original data. Is this true for the snap-bean data when a reciprocal transformation is used?

(c) Make a normal quantile plot of the residuals, $(y'_{ij} - \bar{y}'_i)$. Does this plot support the condition that the populations are normal?

***11.S.10** (*Computer exercise—continuation of Exercises 11.S.8 and 11.S.9*) Repeat the test in Exercise 11.S.7 using Y' instead of Y, and compare with the results of Exercise 11.S.7.

11.S.11 Suppose a drug for treating high blood pressure is to be compared to a standard blood pressure drug in a study of humans.

(a) Describe an experimental design for a study that makes use of blocking. Be careful to note which parts of the design involve randomness and which parts do not.

(b) Can the experiment you described in part (a) involve blinding? If so, explain how blinding could be used.

11.S.12 In a study of balloon angioplasty, patients with coronary artery disease were randomly assigned to one of four treatment groups: placebo, probucol (an experimental drug), multivitamins (a combination of beta carotene, vitamin E, and vitamin C), or probucol combined with multivitamins. Balloon angioplasty was performed on each of the patients. Later, "minimal luminal diameter" (a measurement of how well the angioplasty did in dilating the artery) was recorded for each of the patients. Summary statistics are given in the following table.[40]

	Placebo	Probucol	Multi-vitamins	Probucol and multi-vitamins
n	62	58	54	56
Mean	1.43	1.79	1.40	1.54
SD	0.58	0.45	0.55	0.61

(a) Complete the ANOVA table and bracket the P-value for the F test.

Source	df	SS	MS	F
Between treatments	____	5.4336	____	____
Within treatments	____	____	____	____
Total	229	73.9945	____	____

(b) What is the value of the F test statistic for testing that the effects of the four treatments are equal?

(c) How many degrees of freedom are there for the test from part (b)?

(d) The P-value for a one-way ANOVA F test is 0.0006. If $\alpha = 0.01$, what is your conclusion regarding the null hypothesis?

***11.S.13** Refer to Exercise 11.S.12. Define contrasts to measure each effect specified, and calculate the value of each contrast.

(a) The effect of probucol in the absence of multivitamins

(b) The effect of probucol in the presence of multivitamins

(c) The interaction between probucol and multivitamins

***11.S.14** Refer to Exercise 11.S.12. Construct a 95% confidence interval ($\alpha_{cw} = 0.05$) for the effect of probucol in the absence of multivitamins. That is, construct a 95% confidence interval for $\mu_{\text{probucol}} - \mu_{\text{placebo}}$.

***11.S.15** Refer to Exercise 11.S.12. Assuming all possible comparisons of group means will be computed, use the Bonferroni method to construct a 95% confidence interval for the effect of probucol in the absence of multivitamins. That is, construct a Bonferroni-adjusted 95% ($\alpha_{ew} = 0.05$) confidence interval for $\mu_{\text{probucol}} - \mu_{\text{placebo}}$.

***11.S.16** Three college students collected several pillbugs from a woodpile and used them in an experiment in which they measured the time, in seconds, that it took for a bug to move 6 inches within an apparatus they had created. There were three groups of bugs: one group was exposed to strong light, for one group the stimulus was moisture, and a third group served as a control. The data are shown in the following table.[41]

	Light	Moisture	Control
	23	170	229
	12	182	126
	29	286	140
	12	103	260
	5	330	330
	47	55	310
	18	49	45
	30	31	248
	8	132	280
	45	150	140
	36	165	160
	27	206	192
	29	200	159
	33	270	62
	24	298	180
	17	100	32
	11	162	54
	25	126	149
	6	229	201
	34	140	173
Mean	23.6	169.2	173.5
SD	12.3	83.5	86.0
n	20	20	20

Clearly the SDs show that the variability is not constant between groups, so a transformation is needed. Taking the natural logarithm of each observation results in the following dotplots and summary statistics.

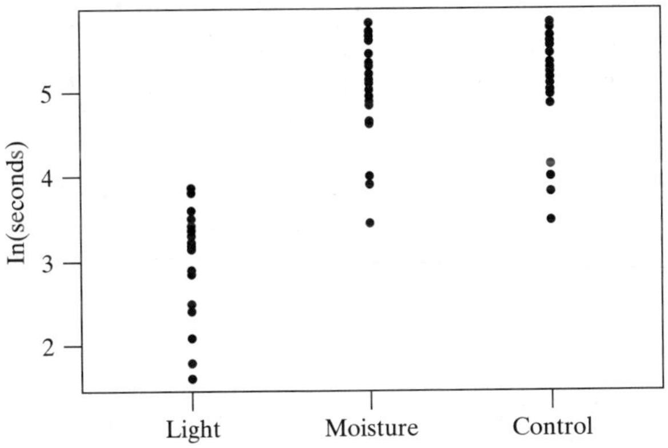

	Light	Moisture	Control
Mean	2.99	4.98	4.99
SD	0.65	0.62	0.66

For the transformed data, the ANOVA SS(between) is 53.1103 and the SS(within) is 23.5669.

(a) State the null hypothesis in symbols.

(b) Construct the ANOVA table and test the null hypothesis. Let $\alpha = 0.05$.

(c) Calculate the pooled standard deviation, s_{pooled}.

***11.S.17** Mountain climbers often experience several symptoms when they reach high altitudes during their climbs. Researchers studied the effects of exposure to high altitude on human skeletal muscle tissue. They set up a 2×2 factorial experiment in which subjects trained for 6 weeks on a bicycle. The first factor was whether subjects trained under hypoxic conditions (corresponding to an altitude of 3,850 m) or normal conditions. The second factor was whether subjects trained at a high level of energy expenditure or at a low level (25% less than the high level). There were either 7 or 8 subjects at each combination of factor levels. The accompanying table shows the results for the response variable "percentage change in vascular endothelial growth factor mRNA."[42]

	Hypoxic		Normal	
Energy	Low level	High level	Low level	High level
Mean	117.7	173.2	95.1	114.6
No. of patients	7	7	8	8

Prepare an interaction graph (like Figure 11.7.3).

***11.S.18** Consider the data from Exercise 11.S.17.

(a) Complete the following ANOVA table.

Source	df	SS	MS	F ratio
Between hypoxic and normal	1	12126.5	____	____
Between energy level	1	10035.7	____	____
Interaction	1	____	____	____
within groups	26	56076.0	____	____
Total	29	80738.7	____	____

(b) What is the value of the F test statistic used for testing for interactions?

(c) The P-value for the interactions F test is 0.29. If $\alpha = 0.05$, what is your conclusion regarding the null hypothesis?

(d) Based on your conclusion from part (c), is it sensible to examine the main effects of condition and of energy level?

(e) The P-value for testing the null hypothesis that energy level has no effect on the response is 0.04. If $\alpha = 0.05$, what is your conclusion regarding the null hypothesis?

(f) What is the value of the F test statistic used to test whether the effect on the response of hypoxic training is the same as the effect on the response of normal training?

(g) The P-value for the hypoxic training versus normal training F test is 0.025. If $\alpha = 0.05$, what is your conclusion regarding the null hypothesis?

***11.S.19** In a study to examine the utility of using ammonia gas to sanitize animal feeds, researchers inoculated corn silage with a strain of *Salmonella*. Next, two petri dishes of 5 g of contaminated feed were exposed to concentrated anhydrous ammonia gas and two control petri dishes of 5 g of contaminated feed were not treated with the gas. This experiment was repeated twice, for a total of three trials, as only two petri dishes could be placed in the pressurized gas chamber at any given time. Twenty-four hours after inoculation and gassing, the number of bacterial colonies (colony forming units or cfu) on each dish were counted. Because the data were highly skewed, the log(cfu) was analyzed.[43]

(a) Identify the blocking, treatment, and response variables in this problem.

(b) Complete the following ANOVA table for this blocked analysis.

	df	SS	MS	F ratio
Between treatments	1	1.141	1.141	7.107
Between trials	2	3.611	____	____
Within groups	8	____	____	
Total	11	6.036		

(c) Using the complete table from part (b), is there evidence that the ammonia gas treatment affects the contamination level (i.e., mean log cfu)? Use $\alpha = 0.05$.

(d) Do the preceding analysis and information allow you to infer that ammonia reduces contamination? If not, what other information would be necessary to make such a claim?

11.S.20 A biologist measured the post-orbital carapace length (mm) for specimens of two species of crayfish: Rusty crawfish and Sanborn crawfish. Summary statistics are given below.[44]

	Rusty, female	Rusty, male	Sanborn, female	Sanborn, male
Mean	16.78	21.62	16.64	15.69
SD	2.81	4.26	1.76	2.45
N	9	13	14	28

For these data, SS(Sex) = 11.5 SS(Species) = 198.8, SS(interaction) = 113.5, and SS(within) = 482.7.

(a) Prepare an interaction graph (like Figure 11.7.3) to investigate whether the male versus female difference is the same for Rusty as for Sanborn crayfish.

(b) Does the interaction graph from part (a) suggest that there is an interaction present? If so, what does that mean? If not, what does that mean?

(c) Carry out an F test for interactions; use $\alpha = 0.10$.

(d) Given the result of the test in part (c), is it appropriate to test the null hypothesis that species has no effect on carapace length? If so, then conduct that test using $\alpha = 0.05$. If not, then explain why not.

Chapter 12

Linear Regression and Correlation

OBJECTIVES

In this chapter we study correlation and regression. We will

- demonstrate how the correlation coefficient is calculated and interpreted.
- show how least-squares regression models are fit to data.
- examine the relationship between the regression line, sample correlation, and the prediction of means.
- show how to test whether a regression relationship is statistically significant.
- extend regression ideas to multiple regression, analysis of covariance, and logistic regression.

12.1 Introduction

In this chapter we discuss some methods for analyzing the relationship between two quantitative variables, X and Y. **Linear regression** and **correlation analysis** are techniques based on fitting a straight line to the data.

EXAMPLES

Data for regression and correlation analysis consist of pairs of observations (X, Y). Here are two examples.

Example 12.1.1

Amphetamine and Food Consumption Amphetamine is a drug that suppresses appetite. In a study of this effect, a pharmacologist randomly allocated 24 rats to three treatment groups to receive an injection of amphetamine at one of two dosage levels, or an injection of saline solution. She measured the amount of food consumed by each animal in the 3-hour period following injection. The results (gm of food consumed per kg body weight) are shown in Table 12.1.1.[1]

Figure 12.1.1 shows a **scatterplot** of

$$Y = \text{Food consumption}$$

against

$$X = \text{Dose of amphetamine}$$

The scatterplot suggests a definite dose-response relationship, with larger values of X tending to be associated with smaller values of Y.* ■

Example 12.1.2

Dissolved Oxygen The level of dissolved oxygen in a river is one measure of the overall health of the river. Researchers recorded water temperature (°C) and level

*In many dose–response relationships, the response depends linearly on log(dose) rather than on dose itself. We have chosen a linear portion of the dose–response curve to simplify the exposition.

Table 12.1.1 Food consumption (Y) of rats (gm/kg)

	X = Dose of amphetamine (mg/kg)		
	0	2.5	5.0
	112.6	73.3	38.5
	102.1	84.8	81.3
	90.2	67.3	57.1
	81.5	55.3	62.3
	105.6	80.7	51.5
	93.0	90.0	48.3
	106.6	75.5	42.7
	108.3	77.1	57.9
Mean	100.0	75.5	55.0
SD	10.7	10.7	13.3
No. of animals	8	8	8

Figure 12.1.1 Scatterplot of food consumption against dose of amphetamine

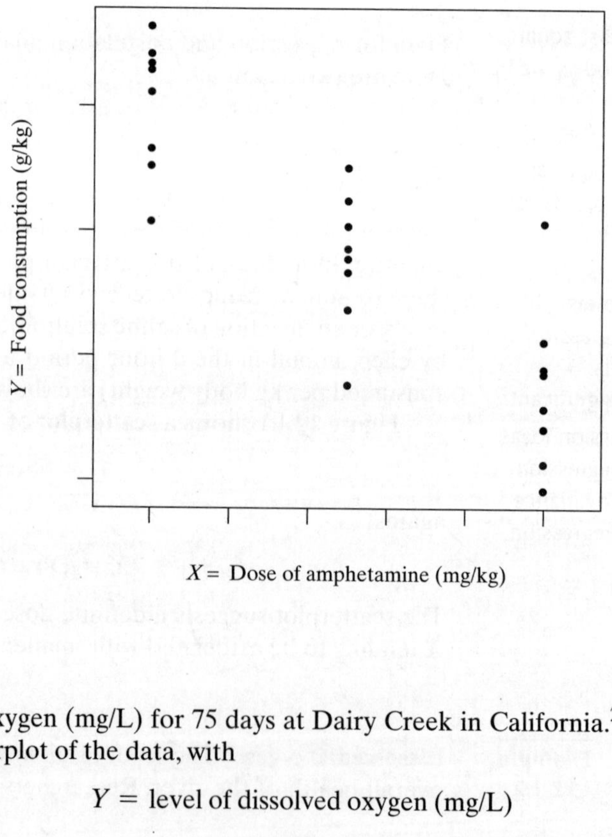

of dissolved oxygen (mg/L) for 75 days at Dairy Creek in California.[2] Figure 12.1.2 shows a scatterplot of the data, with

$$Y = \text{level of dissolved oxygen (mg/L)}$$

plotted against

$$X = \text{water temperature (°C)}$$

The scatterplot suggests that higher water temperatures (X) are associated with lower levels of dissolved oxygen (Y). ■

Figure 12.1.2 Scatterplot of dissolved oxygen level against water temperature

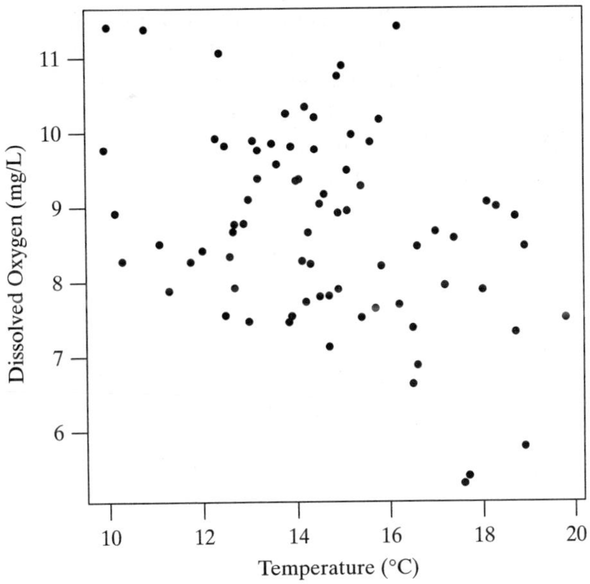

12.2 The Correlation Coefficient

Suppose we have a sample of n pairs for which each pair represents the measurements of two variables, X and Y. If a scatterplot of Y versus X shows a general linear trend, then it is natural to try to describe the strength of the linear association. In this section we will learn how to measure the strength of linear association using the **correlation coefficient.** The following example illustrates the kind of situation we wish to consider.

Example 12.2.1 **Length and Weight of Snakes** In a study of a free-living population of the snake *Vipera bertis,* researchers caught and measured nine adult females.[3] Their body lengths and weights are shown in Table 12.2.1 and are displayed as a scatterplot in Figure 12.2.1. The number of observations is $n = 9$. ■

Table 12.2.1

	Length X (cm)	Weight Y (g)
	60	136
	69	198
	66	194
	64	140
	54	93
	67	172
	59	116
	65	174
	63	145
Mean	63	152
SD	4.6	35.3

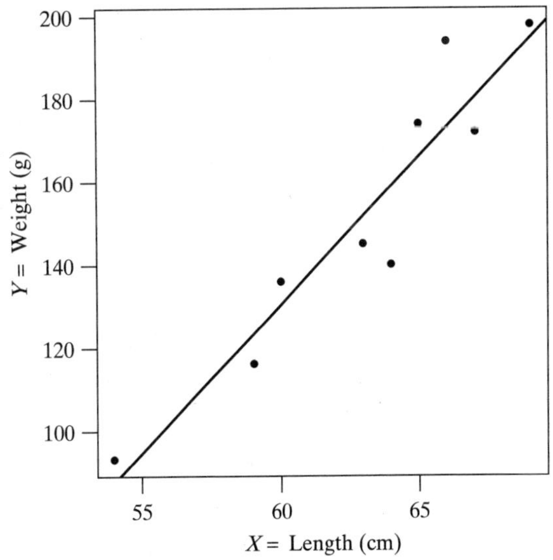

Figure 12.2.1 Body length and weight of nine snakes with fitted regression line

The scatterplot shown in Figure 12.2.1 shows a clear upward trend. We say that weight shows a **positive association** with length, indicating that greater lengths are associated with greater weights. Thus, snakes that are longer than the average length of $\bar{x} = 63$ tend to be heavier than the average weight of $\bar{y} = 152$. The line superimposed on the plot is called the **least-squares line** or **fitted regression line** of Y on X. We will learn how to compute and interpret the regression line in Section 12.3.

MEASURING STRENGTH OF LINEAR ASSOCIATION

How strong is the linear relationship between snake length and weight? Are the data points tightly clustered around the regression line, or is the scatter loose? To answer these questions we will compute the correlation coefficient, a scale-invariant numeric measure of the strength of linear association between two quantitative variables. Being scale invariant means that the correlation coefficient is unaffected by any changes in measurement scales. That is, the correlation between length and weight will be the same whether measured in centimeters and grams or inches and pounds. To understand how the correlation coefficient works, consider again the snake length and weight example. Rather than plotting the original data, Figure 12.2.2 plots the standardized data (z-scores) displayed in Table 12.2.2; note that this plot looks identical to our original plot except now our scales are unit-less.

Dividing the plot into quadrants based on the sign of the standardized score, we see that most of these points fall into the upper-right and lower-left quadrants. Points falling in these quadrants will have standardized scores whose *products* are positive. Likewise, points falling in the upper-left and lower-right quadrants will have standardized score products that are negative. Computing the sum of these products provides a numeric measure of where our points fall (i.e., which quadrants are dominant). In our case, since there is a positive association between length and weight, most points fall in the positive product quadrants; thus, the sum of the products of standardized scores is positive. If a negative relationship were present, most of the points would fall in the negative quadrants and the sum would be negative. And, if there were no *linear* relationship, the points would fall in evenly in all four quadrants so that the positive and negative products would balance and their sum would be zero.

Figure 12.2.2 Scatterplot of standardized weight versus standardized length

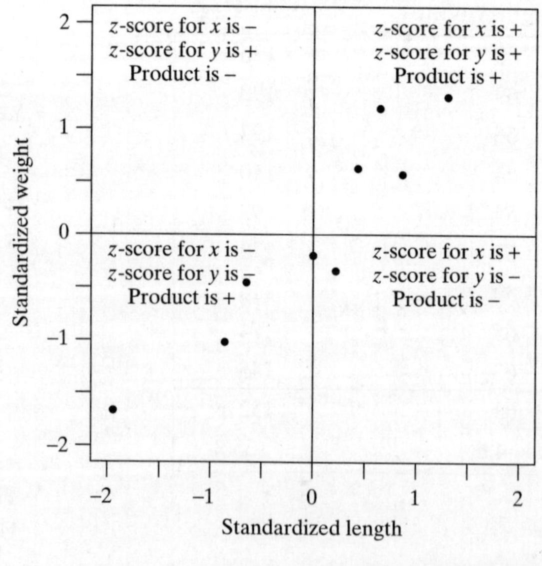

Table 12.2.2 Standardized snake weights, lengths, and their products

	Weight	Length	Standardized weight	Standardized length	Product of standardized values
	X	Y	$z_x = \frac{x - \bar{x}}{s_x}$	$z_y = \frac{y - \bar{y}}{s_y}$	$z_x z_y$
	60	136	−0.65 . . .	−0.45 . . .	0.29 . . .
	69	198	1.29 . . .	1.30 . . .	1.68 . . .
	66	194	0.65 . . .	1.19 . . .	0.77 . . .
	64	140	0.22 . . .	−0.34 . . .	−0.07 . . .
	54	93	−1.94 . . .	−1.67 . . .	3.24 . . .
	67	172	0.86 . . .	0.57 . . .	0.49 . . .
	59	116	−0.86 . . .	−1.02 . . .	0.88 . . .
	65	174	0.43 . . .	0.62 . . .	0.27 . . .
	63	145	0.00 . . .	−0.20 . . .	0.00 . . .
Sum	567	1368	0.00	0.00	7.5494
Mean	63.000	152.000	0.00	0.00	
SD	4.637	35.338	1.00	1.00	

Values in the table are truncated for ease of reading. Because the summary values will be used in subsequent calculations, they include more digits than one would typically report when following our rounding conventions.

The correlation coefficient is based on this sum. It is computed as the average product of standardized scores (using $n - 1$ rather than n to compute the average):*

The correlation coefficient, r

$$r = \frac{1}{n-1} \sum_{i=1}^{n} \left(\frac{x_i - \bar{x}}{s_x}\right)\left(\frac{y_i - \bar{y}}{s_y}\right)$$

From this formula it is clear that X and Y enter r symmetrically; therefore, if we were to interchange the labels X and Y of our variables, r would remain unchanged. In fact, this is one of the advantages of the correlation coefficient as a summary statistic: In interpreting r, it is not necessary to know (or to decide) which variable is labeled X and which is labeled Y.

INTERPRETING THE CORRELATION COEFFICIENT

Mathematically, the correlation coefficient is unit free and always between −1 and 1. The sign of the correlation indicates the sign of the relationship and matches the sign of the slope of the regression line: positive (increasing) or negative (decreasing). The

*By substituting $\sqrt{\sum_{i=1}^{n}(x - \bar{x})^2/(n-1)}$ for s_x and $\sqrt{\sum_{i=1}^{n}(y - \bar{y})^2/(n-1)}$ for s_y, the equation for the correlation coefficient can be rewritten as $r = \dfrac{\sum_{i=1}^{n}(x - \bar{x})(y - \bar{y})}{\sqrt{\sum_{i=1}^{n}(x - \bar{x})^2 \sum_{i=1}^{n}(y - \bar{y})^2}}$.

closer the correlation is to -1 or 1, the stronger the linear relationship between X and Y. A correlation equal to -1 or 1 indicates a perfect linear relationship between the two variables—a scatterplot of such data would display the data falling exactly on a straight line. Interestingly, a correlation of zero does not necessarily mean that there is no relationship between X and Y—it only means that there is no *linear* relationship between X and Y. The preceding computation of the correlation indicates that the sum of the products of standardized values will be zero whenever the positive and negative products balance; this can happen in many ways. Figure 12.2.3 displays several examples with a variety of correlation coefficient values.

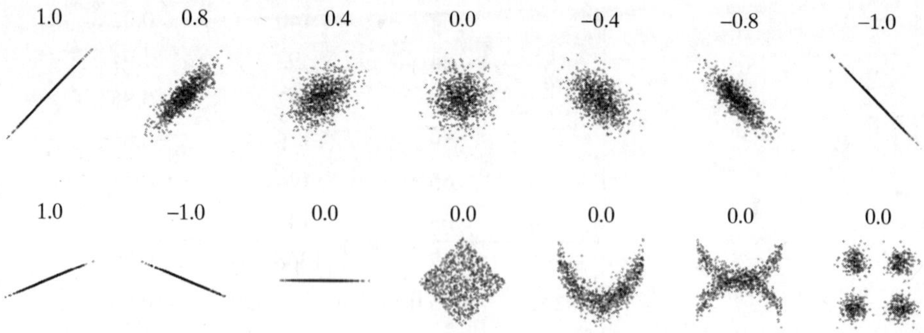

Figure 12.2.3 Scatterplots of data with a variety of sample correlation values

Example 12.2.2 **Length and Weight of Snakes** In Table 12.2.2 we showed that for the snake data the sum of the products of the standardized scores is 7.5494. Thus, the correlation coefficient for the lengths and weights of our sample of nine snakes is about 0.94.

$$r = \frac{1}{9-1} \times 7.5494 \approx 0.94$$

■

In this example we may also refer to the value 0.94 as the **sample correlation,** since the lengths and weights of these nine snakes comprise a sample from a larger population. The sample correlation is an estimate of the **population correlation** (often denoted by the Greek letter "rho," ρ)—in this case the correlation coefficient for the entire population of adult female *Vipera bertis* snakes. In order to regard the sample correlation coefficient r as an estimate of a population parameter, it must be reasonable to assume that both the X and the Y values were selected at random, as in the following **bivariate random sampling model**:

Bivariate Random Sampling Model:

We regard each pair (x_i, y_i) as having been sampled at random from a population of (x, y) pairs.

In the bivariate random sampling model, the observed X's are regarded as a random sample, and the observed Y's are also regarded as a random sample, so the marginal statistics $\bar{x}, \bar{y}, s_x$, and s_y are estimates of corresponding population values μ_x, μ_y, σ_x, and σ_y.

For many investigations the random sampling model is reasonable, but the additional assumption of a bivariate random sampling model is not. This is generally the case when the values of X are specified by the experimenter as in

Example 12.1.1 where the researchers assigned rats to one of three dosages of amphetamine. This type of sampling model is called the random subsampling model and is defined in Section 12.4. In these cases the sample correlation coefficient is not an appropriate estimate of the population correlation, though we shall see in Section 12.3 that it will still be a useful summary measure for identifying the line that best fits the data.

INFERENCE CONCERNING CORRELATION

We have described how the correlation coefficient describes a data set within the bivariate random sampling model. Now we shall consider statistical inference based on r for data from this model.

R TESTING THE HYPOTHESIS H_0: $\rho = 0$

In some investigations it is not a foregone conclusion that there is any relationship between X and Y. It then may be relevant to consider the possibility that any apparent trend in the data is illusory and reflects only sampling variability. In this situation it is natural to formulate the null hypothesis

H_0: X and Y are uncorrelated in the population

or, equivalently

H_0: There is no linear relationship between X and Y.

A RANDOMIZATION TEST

If the null hypothesis is true so that X and Y are uncorrelated in the population, it is still possible that a scatterplot of Y versus X will show a linear trend just by chance. Even if H_0 were true we would expect the sample correlation coefficient, r, to be at least *somewhat* different from zero. We can ask, "How likely is it that a correlation coefficient would be as far from zero as is our observed value of r, just by chance?" We can answer that question by creating a randomization test in which we break any link between X and Y in the data by holding the X values fixed while scrambling the Y values. If X and Y are unrelated in the population, then we can think of our observed correlation coefficient, r, as having arisen from such a process. We illustrate this idea now.

Example 12.2.3 **Blood Pressure and Platelet Calcium** It is suspected that calcium in blood platelets may be related to blood pressure. As part of a study of this relationship, researchers recruited 38 subjects whose blood pressure was normal (i.e., not abnormally elevated).[4] For each subject two measurements were made: pressure (average of systolic and diastolic measurements) and calcium concentration in the blood platelets. The data are shown in Figure 12.2.4. The sample size is $n = 38$, and the sample correlation is $r = 0.5832$.

Is there evidence that blood pressure and platelet calcium are linearly related? We will test the null hypothesis

$$H_0: \rho = 0$$

against the nondirectional alternative

$$H_A: \rho \neq 0$$

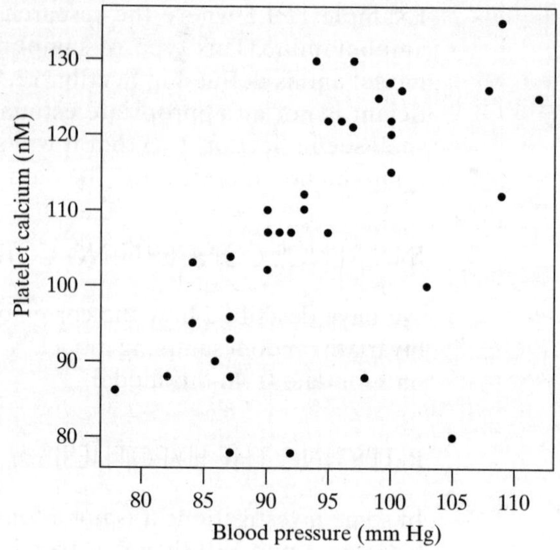

Figure 12.2.4 Blood pressure and platelet calcium for 38 persons with normal blood pressure

These hypotheses are translations of the verbal hypotheses

H_0: Platelet calcium is not linearly related to blood pressure.
H_A: Platelet calcium is linearly related blood pressure.

The scatterplot in Figure 12.2.4 suggests that high blood pressure levels are linked with high platelet calcium levels, but if H_0 were true then any of the $n = 38$ platelet calcium levels could have been associated with any of the 38 blood pressure levels. That is, knowing a person's blood pressure would not tell us anything about the person's platelet calcium level.

We can simulate this situation by doing the following. (1) Write down the 38 blood pressure levels in the order in which they appeared in the data set (i.e., create a fixed list of the 38 x-values); (2) write the 38 platelet calcium levels on each of 38 cards; (3) shuffle the cards; (4) randomly deal out the cards, placing one card (y) next to each blood pressure level (x); (5) compute the sample correlation coefficient for this random pairing of x with y; (6) record whether the absolute value of this sample correlation is at least 0.5832; (7) repeat steps (1)–(6) many times.

With the aid of a computer we can conduct a simulation to learn how often a correlation as extreme as 0.5832 happens just by chance. In one simulation of 100,000 trials there were only 12 trials for which the correlation was at least as large in magnitude as 0.5832, which means that the randomization P-value is 0.00012. This means that the sample correlation coefficient in the original data is not easily explained by chance. Instead, we have strong evidence for H_A. ■

A t TEST

A randomization test is one way to investigate the null hypothesis

H_0: X and Y are uncorrelated in the population

A more traditional approach is to use a t test that is based on the test statistic

$$t_s = r\sqrt{\frac{n-2}{1-r^2}}$$

Critical values are obtained from Student's t-distribution with

$$df = n - 2$$

The following example illustrates the application of this t test.

Example 12.2.4 **Blood Pressure and Platelet Calcium** Consider the blood pressure and blood platelet calcium data of Example 12.2.3. We wish to test the null hypothesis that there is no linear relationship between blood pressure and blood platelet calcium.

Let us choose $\alpha = 0.05$. The test statistic is

$$t_s = 0.5832\sqrt{\frac{38 - 2}{1 - 0.5832^2}} = 4.308$$

From Table 4 with $df = n - 2 = 36 \approx 40$, we find $t_{40,\,0.0005} = 3.551$. Thus, we find p-$value < 0.0005 \times 2 = 0.001$ (since H_A is nondirectional), and we reject H_0. The data provide strong evidence that platelet calcium is linearly related blood pressure ($t_s = 4.308$, $df = 36$, P-value < 0.001). ■

Why n − 2? The t statistic in the hypothesis test for the preceding population correlation coefficient has an associated df $= n - 2$. The origin of the $n - 2$ is easy to explain. Any two points determine a straight line, yet such a small data set ($n = 2$) provides no information about the inherent variability in the scatter of the points (or, equivalently, the strength of association between X and Y). It is not until we observe a third point that we are able to begin estimating the strength of any relationship. As in our earlier contexts related to t distributions and F-distributions (Chapters 6, 7, 8, and 11), the degrees of freedom is the number of pieces of information provided by the data about the "noise" from which the investigator wants to extract the "signal."

CONFIDENCE INTERVAL FOR ρ (OPTIONAL)

If the sample size is large, it is possible to construct a confidence interval for ρ. The sampling distribution of the sample correlation coefficient, r, is skewed, so in order to construct the confidence interval we apply what is known as the Fisher transformation of r:

$$z_r = \frac{1}{2}\ln\left[\frac{1 + r}{1 - r}\right]$$

where ln is the natural logarithm (base e). We can then construct a 95% confidence interval for $\frac{1}{2}\ln\left[\frac{1 + \rho}{1 - \rho}\right]$ as

$$z_r \pm 1.96\frac{1}{\sqrt{n - 3}}$$

Finally, we can convert the limits of the confidence interval for $\frac{1}{2}\ln\left[\frac{1 + \rho}{1 - \rho}\right]$ into a confidence interval for ρ by solving for ρ in the equations given by

$$\frac{1}{2}\ln\left[\frac{1 + \rho}{1 - \rho}\right] = z_r \pm 1.96\frac{1}{\sqrt{n - 3}}$$

Intervals with other confidence levels are constructed analogously. For example, to construct a 90% confidence interval, replace 1.96 with 1.645. The construction of a confidence interval for a correlation coefficient is illustrated in Example 12.2.5.

Example 12.2.5 **Blood Pressure and Platelet Calcium** For the data of Example 12.2.3 the sample size is $n = 38$ and the sample correlation is $r = 0.5832$. The Fisher transformation of r gives

$$z_r = \frac{1}{2}\ln\left[\frac{1+0.5832}{1-0.5832}\right] = \frac{1}{2}\ln\left[\frac{1.5832}{0.4168}\right] = 0.6673$$

A 95% confidence interval for $\frac{1}{2}\ln\left[\frac{1+\rho}{1-\rho}\right]$ is

$$0.6673 \pm 1.96\frac{1}{\sqrt{38-3}}$$

or 0.6673 ± 0.3313, which is (0.3360,0.9986).

Setting

$$\frac{1}{2}\ln\left[\frac{1+\rho}{1-\rho}\right] = 0.3360 \text{ gives } \rho = \frac{e^{2(0.3360)}-1}{e^{2(0.3360)}+1} = 0.32$$

Setting

$$\frac{1}{2}\ln\left[\frac{1+\rho}{1-\rho}\right] = 0.9986 \text{ gives } \rho = \frac{e^{2(0.9986)}-1}{e^{2(0.9986)}+1} = 0.76$$

We are 95% confident that the correlation between blood pressure and platelet calcium in the population is between 0.32 and 0.76. Thus, a 95% confidence interval for ρ is (0.32, 0.76). ■

CORRELATION AND CAUSATION

We have noted earlier that an observed association between two variables does not necessarily indicate any causal connection between them. It is important to remember this caution when interpreting correlation. The following example shows that even strongly correlated variables may be causally unrelated.

Example 12.2.6 **Reproduction of an Alga** Akinetes are sporelike reproductive structures produced by the green alga *Pithophora oedogonia.* In a study of the life cycle of the alga, researchers counted akinetes in specimens of alga obtained from an Indiana lake on 26 occasions over a 17-month period. Low counts indicated germination of the akinetes. The researchers also recorded the water temperature and the photoperiod (hours of daylight) on each of the 26 occasions. The data showed a rather strong negative correlation between akinete counts and photoperiod; the correlation coefficient was $r = -0.72$. The researchers, however, recognized that this observed correlation might not reflect a causal relationship. Longer days (increasing photoperiod) also tend to bring higher temperatures, and the akinetes might actually be responding to temperature rather than photoperiod. To resolve the question, the researchers conducted laboratory experiments in which temperature and photoperiod were varied independently; these experiments showed that temperature, not photoperiod, was the causal agent.[5] ■

As Example 12.2.6 shows, one way to establish causality is to conduct a controlled experiment in which the putative causal factor is varied and all other factors are either held constant or controlled by randomization. When such an experiment is not possible, indirect approaches using statistical analysis can shed some light on potentially causal relationships. (One such approach will be illustrated in Example 12.8.3.)

CAUTIONARY NOTES

To describe the results of testing a correlation coefficient, investigators often use the term *significant*, which can be misleading. For instance, a statement such as "A highly significant correlation was noted" is easily misunderstood. It is important to remember that statistical significance simply indicates rejection of a null hypothesis; it does not necessarily indicate a large or important effect. A "significant" correlation may in fact be quite a weak one; its "significance" means only that it cannot easily be explained away as a chance pattern. From the formula $t_s = r\sqrt{\frac{n-2}{1-r^2}}$ we can see that for a fixed value of r, t_s increases in magnitude as n increases. Thus, if the sample size is large enough, t_s will be large enough for the correlation to be "significant" no matter how close to zero r is (as long as r does not exactly equal zero). It is always wise to assess the practical significance of any result by considering a confidence interval for the population parameter of interest.

The correlation coefficient is highly sensitive to extreme points. For example, Figure 12.2.5(a) shows a scatterplot of 25 points with a correlation of $r = 0.2$; one of the points has been plotted as a blue dot. Figure 12.2.5(b) shows the same points, except that the point plotted as a blue dot has been changed. The change of that single point causes the correlation coefficient to climb from 0.2 to 0.6. Figure 12.2.5(c) shows a third version of the data. In this case $r = -0.1$. These three graphs illustrate how a single point can greatly influence the size of the correlation coefficient. It is important to always plot the data before using r (or any other statistic) to summarize the data.

Figure 12.2.5 The effect of outliers on the sample correlation coefficient

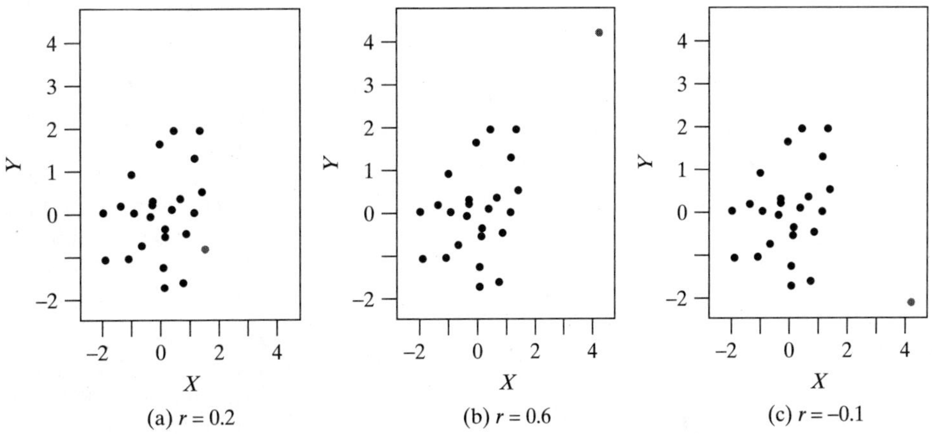

Exercises 12.2.1–12.2.12

12.2.1 Consider the following plots.

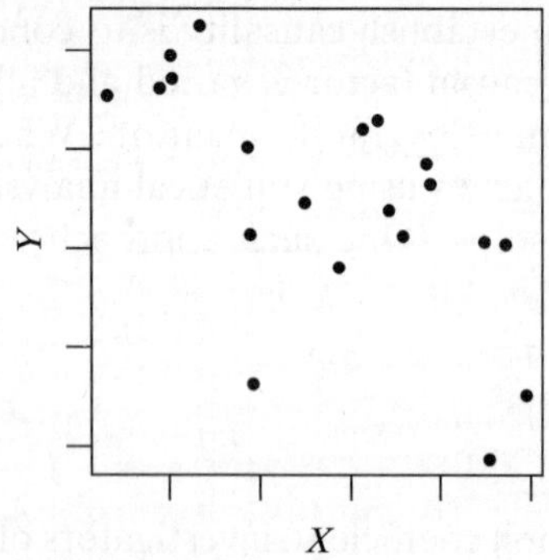
(i)

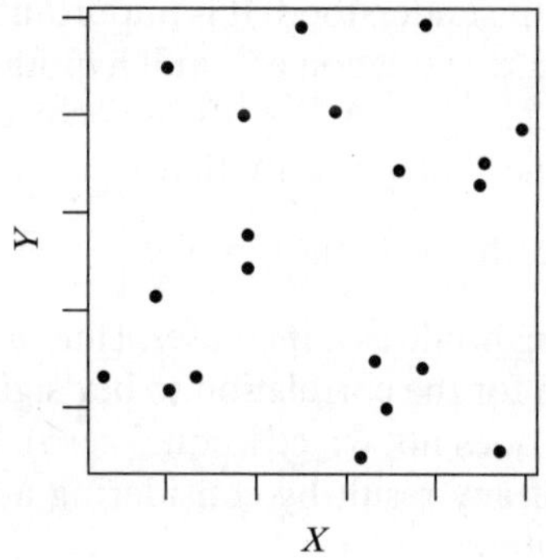
(ii)

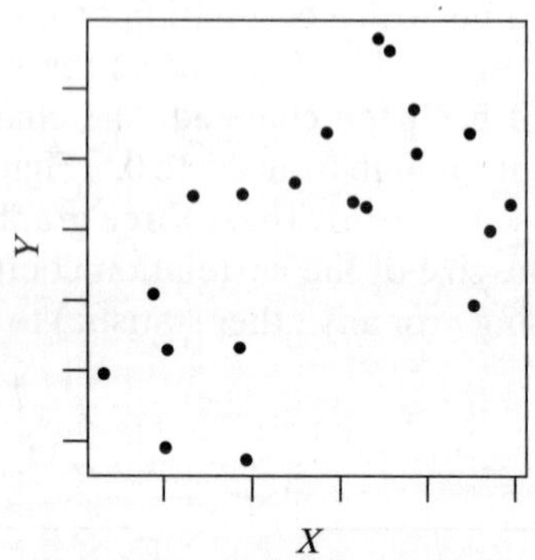
(iii)

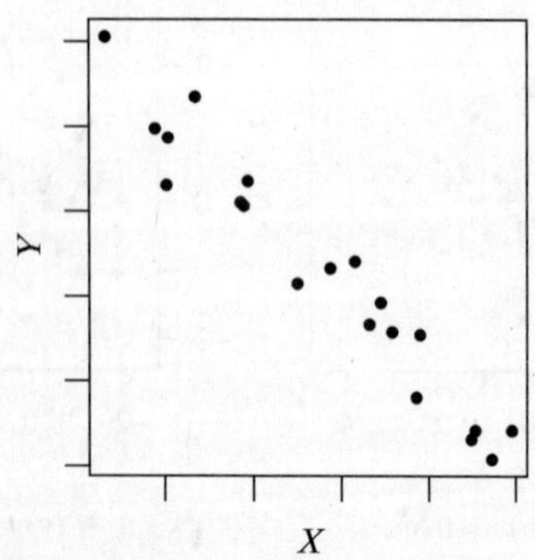
(iv)

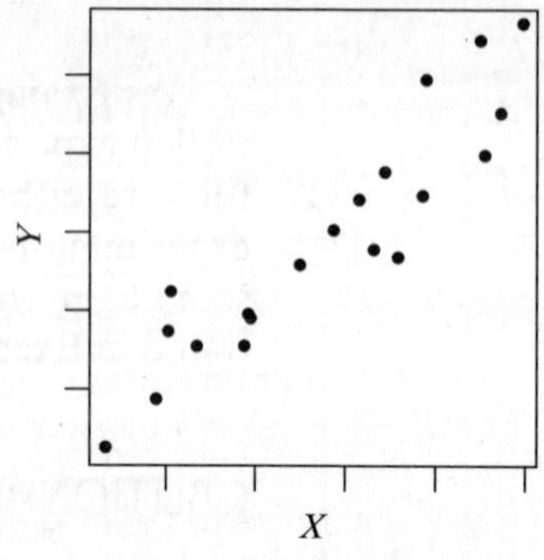
(v)

(a) Arrange the following plots in order of their correlations (from closest to −1 to closest to +1).

(b) Arrange plots (ii–iv) in order of their corresponding *P*-values (smallest to largest) for the test $H_0: \rho = 0$. *Note*: All of the plots display the same number of observations.

12.2.2 Consider the following data.

	X	*Y*
	6	6
	1	7
	3	3
	2	2
	5	14
Mean	3.4	6.4
SD	2.1	4.7

(a) Plot the data. Does there appear to be a relationship between *X* and *Y*? Is it linear or nonlinear? Weak or strong?

(b) Compute the sample correlation coefficient between *X* and *Y*.

(c) Is there significant evidence that *X* and *Y* are correlated? Conduct a test using $\alpha = 0.05$.

12.2.3 In a study of natural variation in blood chemistry, blood specimens were obtained from 284 healthy people. The concentrations of urea and of uric acid were measured for each specimen, and the correlation between these two concentrations was found to be $r = 0.2291$. Test the hypothesis that the population correlation coefficient is zero against the alternative that it is positive.[6] Let $\alpha = 0.05$.

12.2.4 Researchers measured the number of neurons in the CA1 region of the hippocampus in the brains of eight persons who had died of causes unrelated to brain function. They found that these data were negatively correlated with age. The sample value of *r* was −0.63.[7]

(a) Consider testing whether the population correlation is zero. State the null and alternative hypotheses in symbols.

(b) How many degrees of freedom are there for the test in part (a)?

(c) Compute the value of the test statistic.

(d) The P-value for the test is 0.094. If $\alpha = 0.10$, what is your conclusion regarding the null hypothesis?

12.2.5 (*Continuation of Exercise 12.2.4*)

Consider the study design and summary data presented in Exercise 12.2.4 to examine the correlation between age and the number of neurons in the CA1 region of the hippocampus.

(a) Can this study be used to gather evidence that aging is a cause for CA1 neuron loss? If not, what could be said? Briefly explain.

(b) Suppose the researchers had hypothesized that the number of neurons in the CA1 region increased with age and thus considered the directional test of H_A: $\rho > 0$. Considering all of the information above, what would be the resulting P-value and conclusions for this test?

12.2.6 Twenty plots, each 10 × 4 meters, were randomly chosen in a large field of corn. For each plot, the plant density (number of plants in the plot) and the mean cob weight (gm of grain per cob) were observed. The results are given in the table.[8]

Plant density X	Cob weight Y	Plant density X	Cob weight Y
137	212	173	194
107	241	124	241
132	215	157	196
135	225	184	193
115	250	112	224
103	241	80	257
102	237	165	200
65	282	160	190
149	206	157	208
85	246	119	224

Preliminary calculations yield the following results:

$$\bar{x} = 128.05 \qquad \bar{y} = 224.10$$
$$s_x = 32.61332 \qquad s_y = 24.95448$$
$$r = -0.94180$$

(a) Is there significant evidence for a linear relationship between cob weight and plant density? Carry out an appropriate test using $\alpha = 0.05$.

(b) Is this study an observational study or an experiment?

(c) Farmers are interested in whether manipulating plant density can alter cob weight. Could these data be used to answer this question? If not, what could be said? Briefly explain.

12.2.7 Laetisaric acid is a compound that holds promise for control of fungus diseases in crop plants. The accompanying data show the results of growing the fungus *Pythium ultimum* in various concentrations of laetisaric acid. Each growth value is the average of four radial measurements of a *P. ultimum* colony grown in a petri dish for 24 hours; there were two petri dishes at each concentration.[9]

(a) Consider testing whether the population correlation is zero. How many degrees of freedom are there for the test?

(b) Compute the value of the test statistic.

(c) The P-value for the test is 2.3×10^{-9}. If $\alpha = 0.05$, state your conclusion regarding the null hypothesis in the context of this setting.

	Laetisaric acid concentration X (μG/ml)	Fungus growth Y (mm)
	0	33.3
	0	31.0
	3	29.8
	3	27.8
	6	28.0
	6	29.0
	10	25.5
	10	23.8
	20	18.3
	20	15.5
	30	11.7
	30	10.0
Mean	11.500	23.642
SD	10.884	7.8471
	$r = -0.98754$	

12.2.8 (*Continuation of Exercise 12.2.7*)

Consider the study design and summary data presented in Exercise 12.2.7 to examine the relationship between fungus growth and latisaric acid concentration.

(a) Is this study an observational study or an experiment?

(b) It is suggested that acid could be used to retard fungus growth. Could these data be used to verify this claim? If not, what could be said? Briefly explain.

12.2.9 To investigate the dependence of energy expenditure on body build, researchers used underwater weighing techniques to determine the fat-free body mass for each of seven men. They also measured the total 24-hour energy expenditure for each man during conditions of quiet sedentary activity. The results are shown in the table.[10] (See also Exercise 12.5.5.)

Subject	Fat-free mass X (kg)	Energy expenditure Y (kcal)
1	49.3	1,894
2	59.3	2,050
3	68.3	2,353
4	48.1	1,838
5	57.6	1,948
6	78.1	2,528
7	76.1	2,568
Mean	62.400	2,168.429
SD	12.095	308.254
	$r = 0.98139$	

(a) The correlation between energy expenditure and fat-free mass is very large (near 1). It is 0.98139, but the sample size is quite small, only 7. Is there enough evidence to claim the correlation is different from zero? Carry out an appropriate test using $\alpha = 0.05$.

(b) Is this study an observational study or an experiment?

(c) Persons who exercise could increase their fat-free mass. Could these data be used to claim that their energy expenditure would also increase? If not, what could be said? Briefly explain.

12.2.10 Cellular ability to regulate homeostasis is measured by basal Ca pump activity. Deregulation of calcium homeostasis can trigger serious effects of cell functioning. Can maternal mercury exposure measured by mercury deposits in hair (μg/g) affect newborn's basal Ca pump activity (nmol/mg/hr)? The following data summaries and graph are from a human study involving a sample of 75 newborns and their mothers.[11]

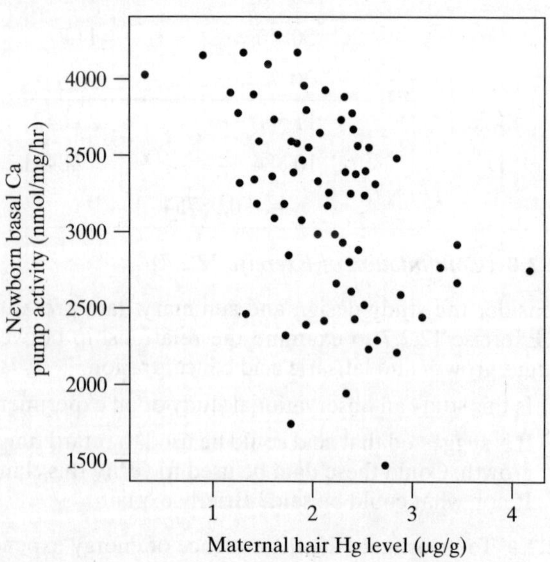

$$\bar{x} = 2.11183 \quad \bar{y} = 3196.8196$$

$$s_x = 0.61166 \quad s_y = 611.34876$$

$$r = -0.45289$$

(a) It is a good habit to always plot our data before analysis. Examining the preceding scatterplot, does there seem to be a linear trend in the data? Is it increasing or decreasing? Is it weak or strong?

(b) Examining the plot, we see there is a mother with a maternal hair level around 4.2 μg/g. If her child's basal Ca pump activity were changed from about 2800 to about 2000 nmol/mg/hr, would the sample correlation increase or decrease in magnitude?

(c) Is there evidence that newborn basal Ca pump activity linearly decreases with maternal hair level? Carry out an appropriate test using $\alpha = 0.05$.

(d) In part (c) you should have found that there is strong evidence for a linearly decreasing relationship between X and Y. Explain how the evidence can be so strong even though the graph displays substantial scatter and the sample correlation is not close to −1.

(e) Based on your answer to part (c) and the design of this study, what can we say regarding the primary research question: Is there statistical evidence that maternal mercury exposure measured by mercury deposits in hair (μg/g) *affects* newborn's basal Ca pump activity (nmol/mg/hr)?

12.2.11 For each of the following examples, explain whether or not it is reasonable to treat the sample correlation coefficient, r, as an estimate of a population correlation coefficient ρ. Briefly justify your answer.

(a) The blood chemistry data from Exercise 12.2.3.

(b) The CA1 neuron data from Exercise 12.2.4.

(c) The cob weight data from Exercise 12.2.6.

(d) The fungus growth data from Exercise 12.2.7.

(e) The basal Ca pump activity from Exercise 12.2.10.

12.2.12 (optional) For each of the following data sets, compute a 95% confidence interval for the population correlation coefficient.

(a) The blood chemistry data from Exercise 12.2.3.

(b) The cob weight data from Exercise 12.2.6.

(c) The energy expenditure data from Exercise 12.2.9.

12.3 The Fitted Regression Line

In Section 12.2 we learned how the correlation coefficient describes the strength of linear association between two numeric variables, X and Y. In this section we will learn how to find and interpret the line that best summarizes their linear relationship.

Example 12.3.1

Ocean Temperature Consider a data set for which there is a perfect linear relationship between X and Y, for example, temperature measured in X = Celsius and Y = Fahrenheit. Figure 12.3.1 displays 20 weekly ocean temperatures (in both °C and °F) for a coastal California city along with a line that perfectly describes the relationship:* $y = 32 + \frac{9}{5}x$. A summary of the data appears in Table 12.3.1.[12]

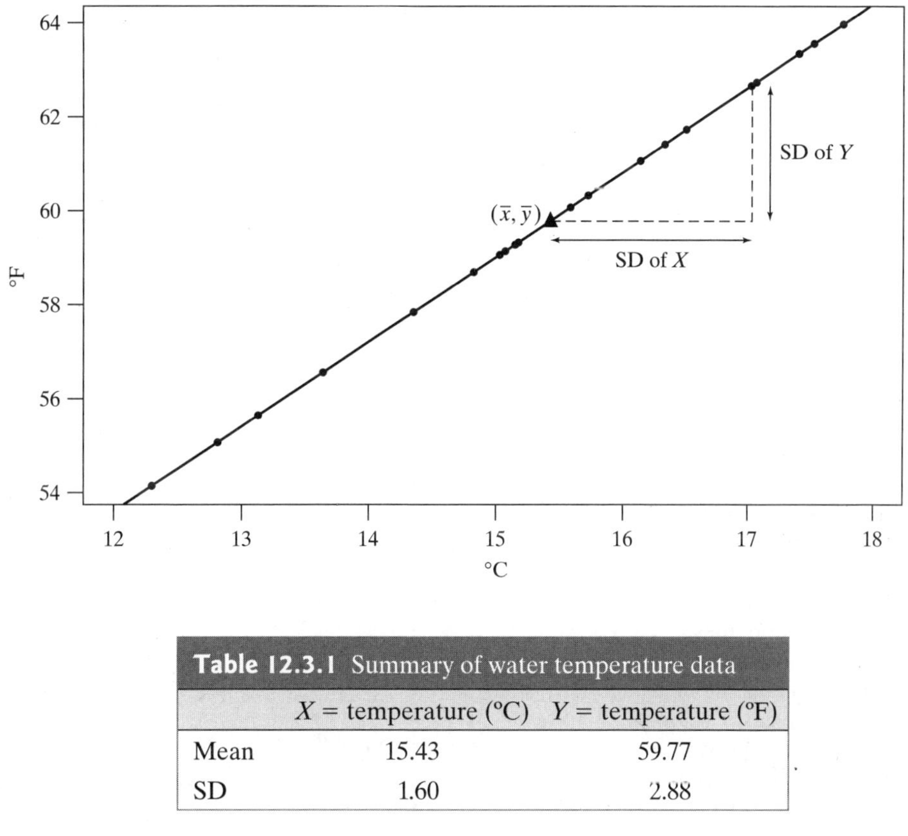

Figure 12.3.1 Scatterplot of Y = ocean temperature in °F versus X = ocean temperature in °C. The mean value $(\bar{x}, \bar{y})$ is denoted with a ▲

Table 12.3.1 Summary of water temperature data

	X = temperature (°C)	Y = temperature (°F)
Mean	15.43	59.77
SD	1.60	2.88

Because X and Y are measuring the same variable (temperature), it stands to reason that a water specimen that is 1 SD above average in °C ($s_x = 1.60$) will also be 1 SD above average in °F ($s_y = 2.88$). Combined, these values can describe the slope of the line that fits these data exactly:

$$\frac{\text{rise}}{\text{run}} = \frac{s_y}{s_x} = \frac{2.88}{1.60} = 1.80$$

In this example we also happen to know the equation of the line that describes the Celsius to Fahrenheit conversion. The slope of this line is $9/5 = 1.80$, the same value we found previously. ■

*This equation is the Celsius to Fahrenheit conversion formula.

THE SD LINE

In perfect linear relationships (i.e., when $r = \pm 1$) the line that fits the data exactly will have slope $\pm s_y/s_x$ (the sign of the slope matches the sign of the correlation coefficient) and passes through the point $(\bar{x}, \bar{y})$. This line is sometimes referred to as the **SD line.** Our previous temperature example displays this property. But what about situations in which r is not exactly ± 1, that is, when the relationship between X and Y is less than perfectly linear?

Example 12.3.2 **Dissolved Oxygen** In Section 12.1 we observed a scatterplot indicating that the amount of dissolved oxygen in a river and water temperature appear to be linearly related ($r = -0.391$). Figure 12.3.2 displays a scatterplot of these data along with the SD line (dashed line). At first glance the SD line appears to be a good fit to these data; however, further investigation suggests otherwise. Suppose we wanted to estimate the mean dissolved oxygen level when the water temperature is 11 °C. The SD line suggests an estimated mean dissolved oxygen level of approximately 10.8 mg/L. Another way to estimate this value would be to simply use the mean dissolved oxygen level for days in our sample that have water temperatures around 11 °C. The mean dissolved oxygen level for days with water temperatures between 10 and 12 °C is 9.2 mg/L (denoted by the ▲ on the graph), which is considerably less than the 10.8 mg/L value given by the SD line. Similarly, for days with water temperatures around 19 °C, the SD line indicates a dissolved oxygen level of about 6.2 mg/L, while the mean dissolved oxygen level for days with water temperatures between 18 and 20 °C in our sample is 8.0 mg/L, a much larger value. ■

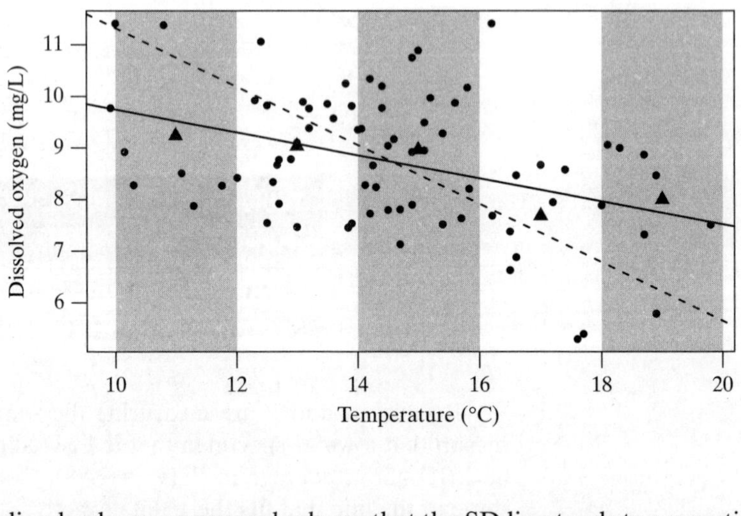

Figure 12.3.2 Levels of dissolved oxygen versus water temperature for 75 days. The dashed and solid lines are the SD and fitted regression lines, respectively. Each ▲ symbol indicates the mean dissolved oxygen level for a range of temperatures specified by the shading

The dissolved oxygen example shows that the SD line tends to overestimate the mean value of Y for below average X values and underestimate the mean value of Y for above average X values. Figure 12.3.3 shows an even more exaggerated example for a data set with a correlation even farther from ± 1; it is near zero ($r = -0.05$). Recall that a correlation of zero indicates no linear relationship between X and Y. This lack of linear relationship is demonstrated by the fact that the mean value of Y is about the same (≈ 17) regardless of the value of X (most of the ▲'s in the plot are near 17).

If the SD line can be such a poor summary, why bother studying it? Because it is an ideal starting place based on a perfect linear relationship. With a perfect

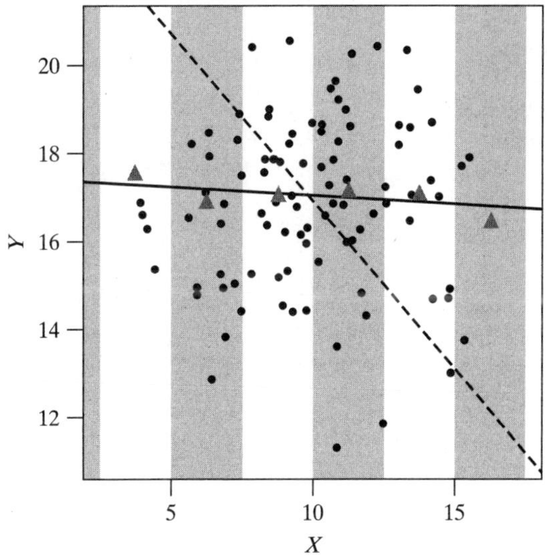

Figure 12.3.3 Scatterplot, SD line (dashed), and fitted regression line (solid) for a sample of 100 data (x, y) values with a correlation near zero. The ▲ symbols indicate the mean Y values for ranges of X values specified by the shading

(positive) linear relationship, the SD line is the best fitting line and has a slope of s_y/s_x. Our examples illustrate that if the relationship is not perfect, the relationship between the mean Y values and X values has a flatter slope. Mathematically, it can be shown that the line that is best suited to predicting Y (in a certain sense)—the so called **least-squares** or **fitted regression** line—has a slope equal to $r(s_y/s_x)$ and passes through the point $(\bar{x}, \bar{y})$. That is, for X values one standard deviation above average, the mean Y value will only be r standard deviations above average (assuming that r is positive; if r is negative, then for X values one standard deviation above average, the mean Y value will be r standard deviations below average).*

Example 12.3.3

Dissolved Oxygen A summary and scatterplot of our dissolved oxygen data appear in Table 12.3.2 and Figure 12.3.4. In this example we estimate that days with water temperatures that are $s_x = 2.30$ °C above average (i.e., one standard deviation above average) will have dissolved oxygen levels that are 0.51 mg/L lower than average ($r \times s_y = -0.391 \times 1.30 = -0.51$). Equivalently, the slope of the fitted regression line is

$$r(s_y/s_x) = -0.391 \times (1.30/2.30) = -0.22 \text{ (mg/L water)/(°C)}$$

meaning that each additional 1 °C increase in water temperature is associated with a 0.22 mg/L decrease in dissolved oxygen level, on average. ■

Table 12.3.2 Summary of dissolve oxygen data

	X = temperature (°C)	Y = dissolved oxygen (mg/L)
Mean	14.58	8.73
SD	2.30	1.30
	$r = -0.391$	

*This is called the *regression effect*.

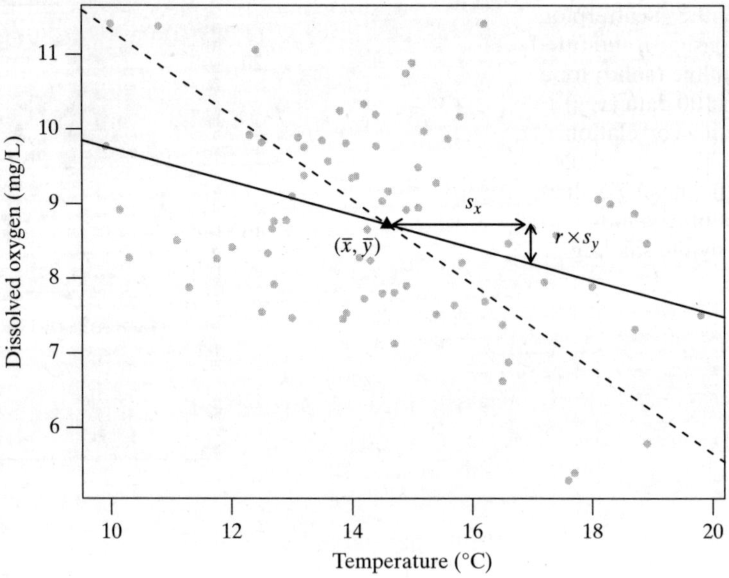

Figure 12.3.4 Levels of dissolved oxygen versus water temperature for 75 days with SD line (dashed) and fitted regression line (solid)

EQUATION OF THE FITTED REGRESSION LINE

The equation of a straight line can be written as

$$Y = b_0 + b_1X$$

where b_0 is the y-intercept and b_1 is the slope of the line. The slope b_1 is the rate of change of Y with respect to X.

The fitted regression line of Y on X is written $\hat{y} = b_0 + b_1x$. We write $\hat{y}$ (read "Y-hat") in place of Y to remind us that this line is providing only estimated or predicted Y values; unless the correlation is ± 1, we don't expect the data values to fall exactly on the line. The fitted regression line estimates the mean value of Y for any given value of X. We discuss this concept of the regression line as a *line of averages* in further detail below.

The slope and intercept of the least-squares* regression line are calculated from the data as follows:

Least-Squares Regression Line of Y on X

$$\text{Slope: } b_1 = r\left(\frac{s_y}{s_x}\right)$$

$$\text{Intercept: } b_0 = \bar{y} - b_1\bar{x}$$

Previously we saw the motivation for the formula for the slope, b_1. The formula for the intercept is also easy to motivate. We can rewrite the Y-intercept formula as

$$\bar{y} = b_0 + b_1\bar{x}$$

which shows that *regression line passes through the joint mean* $(\bar{x}, \bar{y})$ of our data.

We illustrate the use of these formulas by continuing our dissolved oxygen example.

*There are other methods of finding fitted regression lines. In this text, we consider only the least-squares regression line, which aims to minimize the sum of the squared vertical distances between the data values and the fitted line.

Example 12.3.4

Dissolved Oxygen Previously we found the slope of the regression line to be $r(s_y/s_x) = -0.391 \times (1.30/2.30) = -0.22$(mg/L water)/(°C). Using this value we find the Y-intercept,

$$b_0 = 8.73 - (-0.22) \times 14.58 = 11.94 \text{ mg/L}$$

Thus, our fitted regression line is $\hat{y} = 11.94 - 0.22x$ as previously displayed in Figure 12.3.4. ■

Note that the Y-intercept, the point $(0, b_0) = (0, 11.94)$, does not appear on the scatterplot in Figure 12.3.4 because the X-scale limits do not extend to zero; they range from about 10 to 20 to produce a plot for which the data fill the picture nicely.

GRAPH OF AVERAGES

If we have several observations of Y at a given level of X, we can estimate the population mean Y value for the given X value ($\mu_{Y|X}$) by simply using the sample average of Y, $\bar{y}$, for that given value of X; we can denote this sample average as $\bar{y}\,|\,X$.[†] Sometimes we are able to calculate a sample average, $\bar{y}$, for each of several X values. A graph of $\bar{y}\,|\,X$ is known as a **graph of averages,** since it shows the (observed) average of Y for different values of X.

Example 12.3.5

Amphetamine and Food Consumption Figure 12.3.5 is a graph of averages for the food consumption data in Table 12.1.1, showing the average y value for each of the three levels of X. Note that the three $\bar{y}$'s almost lie on a line. This supports the use of the linear model with these data. ■

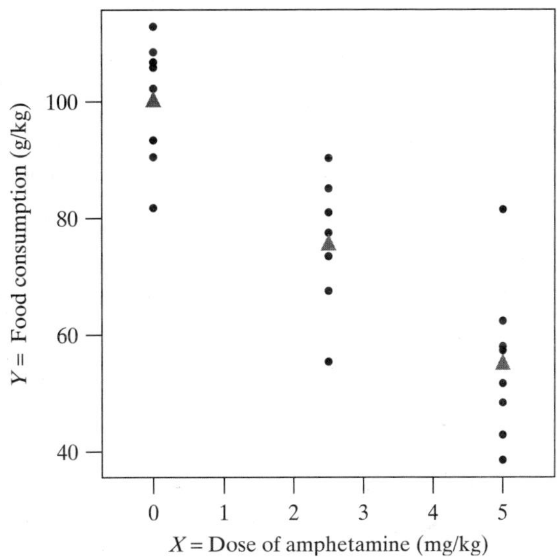

Figure 12.3.5 Graph of averages (▲) for food consumption data from Example 12.1.1 with the original data plotted as black dots

If the $\bar{y}$'s in a graph of averages fall exactly on a line, then that line is the regression line and $\mu_{Y|X}$ is estimated with $\bar{y}\,|\,X$. Usually, however, the $\bar{y}$'s are not perfectly collinear. In this case, the regression line is a *smoothed* version of the graph of averages, resulting in a fitted model in which all of the estimates of $\mu_{Y|X}$ fall on a line. By smoothing the graph of averages into a line, we use information from *all* the observations to estimate $\mu_{Y|X}$ at any level of X.

[†]A more detailed exposition of these "conditional means" appears in Section 12.4.

Example 12.3.6 **Amphetamine and Food Consumption** If we apply the preceding regression formulas to the food consumption data in Table 12.1.1, we obtain $b_0 = 99.3$ and $b_1 = -9.01$. Thus, the estimate of $\mu_{Y|X=0}$ is 99.3 g/kg. This estimate differs slightly from $\bar{y}|X = 0$, which is 100.0 g/kg. The estimate 99.3 makes use of (1) the 8 y values when $X = 0$ (which averaged to 100.0) and (2) the linear trend established by the other 16 data points, which showed higher food consumption associated with lower doses. Likewise, $\mu_{Y|X=2.5}$ is $99.3 - 9.01 \times 2.5 = 76.78$ g/kg, which differs slightly from $\bar{y}|X = 2.5$, which is 75.5 g/kg, and $\mu_{Y|X=5}$ is $99.3 - 9.01 \times 5 = 54.25$ g/kg, which differs slightly from $\bar{y}|X = 5$, which is 55.0 g/kg. ■

The idea of smoothing the graph of averages into a straight line carries over to the setting in which we have only a single observation at each level of X, as is the case with the dissolved oxygen example. When we draw a line through a set of (X, Y) data, we are expressing a belief that the underlying dependence of the mean value of Y on X is smooth, even though the data may show the relationship only roughly. Linear regression is one formal way of providing a smooth description of the data.

THE RESIDUAL SUM OF SQUARES

We now consider a statistic that describes the scatter of the points about the fitted regression line. The equation of the fitted line is $y = b_0 + b_1x$. Thus, for each observed x_i in our data there is a predicted Y value of

$$\hat{y}_i = b_0 + b_1x_i$$

Also associated with each observed pair (x_i, y_i) is a quantity called a **residual,** defined as

$$e_i = y_i - \hat{y}_i$$

Figure 12.3.6 shows $\hat{y}$ and the residual for a typical data point (x_i, y_i). It can be shown that the sum of the residuals, taking into account their signs, is always zero, because of "balancing" of data points above and below the fitted regression line. The *magnitude* (absolute value) of each residual is the vertical distance of the data point from the fitted line.

Figure 12.3.6 $\hat{y}$ and the residual for a typical data point (x, y)

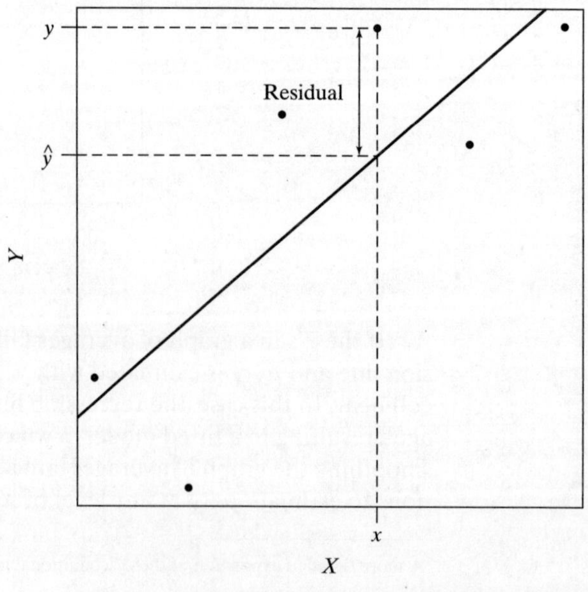

Note that a residual is calculated in terms of *vertical* distance. In using the regression model $\hat{y} = b_0 + b_1x$ we are thinking of the variable X as a predictor and the variable Y as a response that depends on X. We care primarily about how close each observed value, y_i, is to its predicted value, $\hat{y}_i$. Thus, we measure vertical distance from each point to the fitted line. A summary measure of the distances of the data points from the regression line is the **residual sum of squares,** or **SS(resid),** which is defined as follows:

Residual Sum of Squares

$$\text{SS(resid)} = \sum_{i=1}^{n}(y_i - \hat{y}_i)^2 = \sum_{i=1}^{n} e_i^2$$

It is clear from the definition that the residual sum of squares will be small if the data points all lie very close to the line.

The following example illustrates the computation of SS(resid).

Example 12.3.7 **Dissolved Oxygen** For the dissolved oxygen data, Table 12.3.3 indicates how SS(resid) would be calculated from its definition. The values displayed are abbreviated to improve readability. ■

Table 12.3.3 Calculation of SS(resid) for a portion of the dissolved oxygen data

Obs #	x	y	$\hat{y}$	$y - \hat{y}$	$(y - \hat{y})^2$
1	15.0	10.89	8.64 ...	2.25 ...	5.06 ...
2	15.7	7.64	8.48 ...	−0.84 ...	0.71 ...
3	17.7	5.40	8.04 ...	−2.64 ...	7.00 ...
4	12.7	7.92	9.14 ...	−1.22 ...	1.50 ...
5	17.2	7.95	8.15 ...	−0.20 ...	0.04 ...
6	18.9	5.79	7.78 ...	−1.99 ...	3.96 ...
7	14.2	10.34	8.81 ...	1.52 ...	2.32 ...
8	16.5	6.63	8.31 ...	−1.68 ...	2.82 ...
9	14.9	7.90	8.66 ...	−0.76 ...	0.58 ...
⋮	⋮	⋮	⋮	⋮	⋮
70	14.0	9.35	8.86 ...	0.49 ...	0.24 ...
71	15.1	9.49	8.61 ...	0.87 ...	0.76 ...
72	16.6	6.88	8.28 ...	−1.40 ...	1.98 ...
73	13.2	9.38	9.03 ...	0.34 ...	0.11 ...
74	12.0	8.42	9.29 ...	−0.87 ...	0.77 ...
75	18.1	9.06	7.95 ...	1.10 ...	1.21 ...
Sum				0.0	106.14 = SS(resid)

THE LEAST-SQUARES CRITERION

Many different criteria can be proposed to define the straight line that "best" fits a set of data points. The classical criterion is the least-squares criterion:

Least-Squares Criterion

The "best" straight line is the one that minimizes the residual sum of squares.

The formulas given for b_0 and b_1 were derived from the least-squares criterion by applying calculus to solve the minimization problem. (The derivation is given in Appendix 12.1.) The fitted regression line is also called the "least-squares line."

The least-squares criterion may seem arbitrary and even unnecessary. Why not fit a straight line by eye with a ruler? Actually, unless the data lie nearly on a straight line, it can be surprisingly difficult to fit a line by eye. The least-squares criterion provides an answer that does not rely on individual judgment and that (as we shall see in Sections 12.4 and 12.5) can be usefully interpreted in terms of estimating the distribution of Y values for each fixed X. Furthermore, we will see in Section 12.8 that the least-squares criterion is a versatile concept, with applications far beyond the simple fitting of straight lines.

THE RESIDUAL STANDARD DEVIATION

A summary of the results of the linear regression analysis should include a measure of the closeness of the data points to the fitted line. A measure derived from the residual sum of squares and easier to interpret is the **residual standard deviation,** denoted s_e, which is defined as follows:

Residual Standard Deviation

$$s_e = \sqrt{\frac{\sum_{i=1}^{n}(y_i - \hat{y}_i)^2}{n-2}} = \sqrt{\frac{\sum_{i=1}^{n}e_i^2}{n-2}} = \sqrt{\frac{\text{SS(resid)}}{n-2}}$$

The residual standard deviation tells how far above or below the regression line points tend to be. Thus, the residual standard deviation specifies how far off predictions made using the regression model tend to be. Notice the factor in the denominator $n - 2$, rather than the usual $n - 1$. The following example illustrates the calculation of s_e.

Example 12.3.8 **Dissolved Oxygen** For the dissolved oxygen data, we use SS(resid) from Example 12.3.7 to calculate

$$s_e = \sqrt{\frac{106.14}{75-2}} = \sqrt{1.454} = 1.21 \text{ mg/L}$$

Thus, predictions for the levels of dissolved oxygen based on the regression model tend to err by about 1.21 mg/L on average. ■

Note that the formula for s_e is closely analogous to the formula for s_y:

$$s_y = \sqrt{\frac{\sum_{i=1}^{n}(y_i - \bar{y})^2}{n-1}}$$

Both these SDs measure variability in Y, but the residual SD measures variability around the *regression line* and the ordinary SD measures variability around the mean, $\bar{y}$. Roughly speaking, s_e is a measure of the typical vertical distance of the data points from the regression line. (Notice that the unit of measurement of s_e is the same as that of Y—for instance, mg/L in the case of the dissolved oxygen data or grams in the case of the snake data from Example 12.2.1.) Figure 12.3.7 shows the scatterplot and regression line for the snake data from Example 12.2.1 with the residuals represented as vertical lines and the residual SD indicated as a vertical ruler line. Note that the residual SD roughly indicates the magnitude of a typical residual. Finding the equation of this line and the residual standard deviation appears as an exercise at the end of this section.

In many cases, s_e can be given a more definite quantitative interpretation. Recall from Section 2.6 that for a "nice" data set, we expect roughly 68% of the observations to be within 1 SD of the mean (and similarly for 95%, 2 SDs). Recall also that these rules work best if the data follow approximately a normal distribution. Similar interpretations hold for the residual SD: For "nice" data sets that are not too small, we expect roughly 68% of the observed Y's to be within $\pm 1s_e$ of the regression line. In other words, we expect roughly 68% of the data points to be within a vertical distance of s_e above and below the regression line (and similarly for 95%, $\pm 2s_e$). These rules work best if the residuals follow approximately a normal distribution. The dissolved oxygen data we've been working with are well-suited to illustrate the 68% rule.

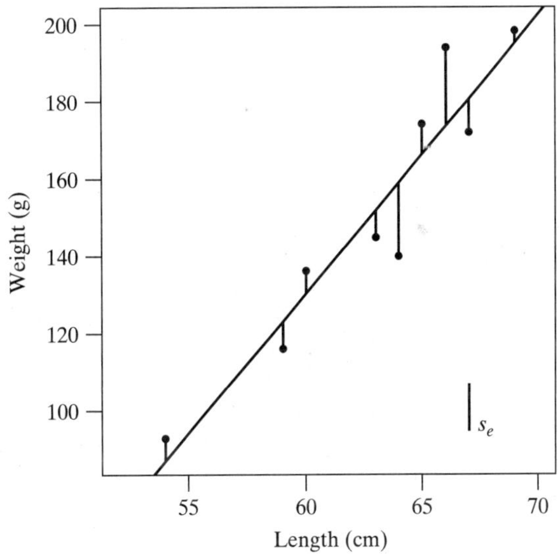

Figure 12.3.7 Weight versus length of nine snakes showing the residuals and a line segment denoting the magnitude of the residual SD

Example 12.3.9

Dissolved Oxygen For the dissolved oxygen data, the fitted regression line is $\hat{y} = 11.94 - 0.22x$ and the residual standard deviation is $s_e = 1.21$. Figure 12.3.8 shows the data and the regression line. The dashed lines are a vertical distance of s_e from the regression line. Of the 75 data points, 50 are within the dashed lines; thus, $50/75$ or $\approx 67\%$ of the observed Y's are within $\pm 1s_e$ of the regression line. ■

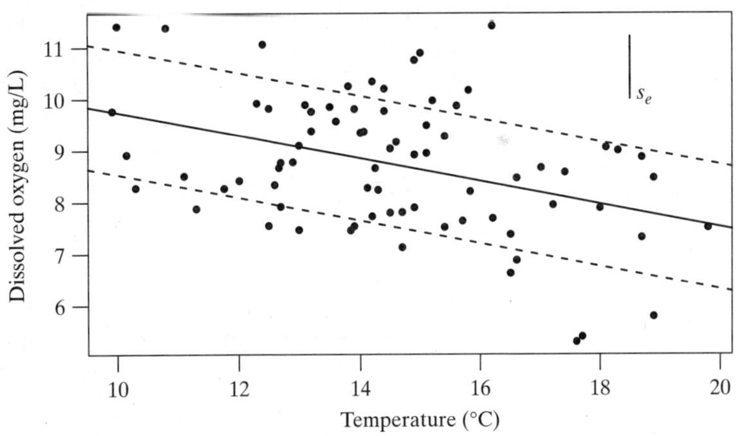

Figure 12.3.8 Levels of dissolved oxygen versus water temperature for 75 days. The dashed lines are a vertical distance of s_e from the regression line

THE COEFFICIENT OF DETERMINATION

We have said that the magnitude of r describes the tightness of the linear relationship between X and Y and have seen how its value is related to the slope of the regression line. When squared, it also provides an additional and very interpretable summary of the regression relationship. The **coefficient of determination,** r^2, describes the proportion of the variance in Y that is explained by the linear relationship between Y and X. This interpretation follows from the following fact (proved in Appendix 12.2).

Fact 12.3.1: Approximate Relationship of r to s_e and s_y

The correlation coefficient r obeys the following approximate relationship:

$$r^2 \approx \frac{s_y^2 - s_e^2}{s_y^2} = 1 - \frac{s_e^2}{s_y^2}$$

(The approximation in Fact 12.3.1 is best for large n, but it holds reasonably well even for n as small as 10.) The numerator, $s_y^2 - s_e^2$, can be roughly interpreted as the total variance in Y explained by the regression line: It is the difference between the variance in Y and the residual variance—the variance left over after fitting the regression line to the data. If the line fits the data very well, then s_e^2 will be close to zero so this numerator will be close to s_y^2; in this case r^2 will be close to 1. At the other extreme, if the line is a very poor fit, then s_e^2 will be close to s_y^2 and the numerator will be close to 0; in this case r^2 will be close to 0. The denominator, s_y^2, is the variance of Y; thus the ratio, r^2, is the proportion of the variance of Y that is explained by the regression relationship between Y and X. Note that because $-1 \leq r \leq 1, 0 \leq r^2 \leq 1$. The following examples illustrate the interpretation and an application of r^2 in context.

Example 12.3.10 **Dissolved Oxygen** For the dissolved oxygen data, we found $r = -0.391$, so $r^2 = 0.153$ or 15.3%. Thus, 15.3% of the variance in dissolved oxygen level is explained by the linear relationship between dissolved oxygen level and water temperature. ■

Example 12.3.11 **Amphetamine and Food Consumption** The standard deviation of food consumption for our entire sample of 24 rats (i.e., combining rats across all three doses of amphetamine) was $s_y = 21.84$ g/kg. Further, suppose r^2 was given to be 0.739. What is the estimated standard deviation of food consumption for rats given 4-mg/kg doses of amphetamine? That is, what is the value of $s_{Y|X=4}$?

To answer this question we first must recognize that the value of X is irrelevant; the residual standard deviation s_e describes the standard deviation of Y values for any given X value, and therefore for $X = 4$. Thus, we need to find the value of s_e. From Fact 12.3.1 we have

$$r^2 \approx 1 - \frac{s_e^2}{s_y^2}$$

After a little algebra, we find that the (approximate) standard deviation of food consumption for rats given 4-mg/kg doses of amphetamine is

$$s_e \approx s_y\sqrt{1 - r^2} = 21.84\sqrt{1 - 0.739} = 11.16 \text{ g/kg}$$ ■

Exercises 12.3.1–12.3.13

12.3.1 In a study of protein synthesis in the oocyte (developing egg cell) of the frog *Xenopus laevis,* a biologist injected individual oocytes with radioactively labeled leucine. At various times after injection, he made radioactivity measurements and calculated how much of the leucine had been incorporated into protein. The results are given in the accompanying table; each leucine value is the content of labeled leucine in two oocytes. All oocytes were from the same female.[13]

	Time	Leucine
	0	0.02
	10	0.25
	20	0.54
	30	0.69
	40	1.07
	50	1.50
	60	1.74
Mean	30.00	0.830
SD	21.60	0.637
	$r = 0.993$	
	SS(resid) = 0.035225	

(a) Plot the data. Does there appear to be a relationship between X and Y? Is it linear or nonlinear? Weak or strong?

(b) Use linear regression to estimate the rate of incorporation of the labeled leucine.

(c) Draw the regression line on your graph.

(d) Calculate the residual standard deviation.

12.3.2 (*Continuation of Exercise 12.3.1*)

Consider the leucine data and summaries presented in Exercise 12.3.1.

(a) Predict the amount of leucine incorporated at 45 minutes.

(b) Calculate the residual associated with data point (50, 1.50).

12.3.3 In an investigation of the physiological effects of alcohol (ethanol), 15 mice were randomly allocated to three treatment groups, each to receive a different oral dose of alcohol. The dosage levels were 1.5, 3.0, and 6.0 gm alcohol per kg body weight. The body temperature of each mouse was measured immediately before the alcohol was given and again 20 minutes afterward. The accompanying table shows the drop (before minus after) in body temperature for each mouse. (The negative value −0.1 refers to a mouse whose temperature rose rather than fell.)[14]

Alcohol		Drop in body temperature (°C)					
Dose (gm/kg)	Log(Dose) X	Individual values (Y)					Mean
1.5	0.176	0.2	1.9	−0.1	0.5	0.8	0.66
3.0	0.477	4.0	3.2	2.3	2.9	3.8	3.24
6.0	0.778	3.3	5.1	5.3	6.7	5.9	5.26

(a) Plot the mean drop in body temperature versus dose. Plot the mean drop in body temperature versus log(dose). Which plot appears more nearly linear?

(b) Plot the individual (x, y) data points [where $X = \log(\text{dose})$].

(c) For the regression of Y on $X = \log(\text{dose})$ preliminary calculations yield the following: $\bar{x} = 0.477$, $\bar{y} = 3.05333$, $s_x = 0.25439$, $s_y = 2.13437$, $r = 0.91074$. Calculate the fitted regression line and the (approximate) residual standard deviation.

(d) Draw the regression line on your graph.

(e) Is this study an example of an observational study or an experiment? How can you tell?

(f) Could data from this study be used to determine whether or not alcohol lowers body temperature? Briefly explain.

12.3.4 Consider the cob weight data from Exercise 12.2.6.

(a) Use the summaries in Exercise 12.2.6 to calculate the fitted regression line and approximate residual standard deviation.

(b) Interpret the value of the slope of the regression line, b_1, in the context of this setting.

(c) SS(resid) = 1337.3. Use this value to compute the residual standard deviation. How does it compare to the approximate value determined in part (a)?

(d) Interpret the value of s_e in the context of this setting.

(e) What proportion of the variation in cob weights is explained by the linear relationship between cob weight and density?

12.3.5 Consider the Fungus growth data from Exercise 12.2.7.

(a) Calculate the linear regression of Y on X.

(b) Plot the data and add the regression line to your graph. Does the line appear to fit the data well?

(c) SS(resid) = 16.7812. Use this to compute s_e. What are the units of s_e?

(d) Draw a ruler line on your graph to show the magnitude of s_e. (See Figure 12.3.8).

12.3.6 Consider the Energy Expenditure data from Exercise 12.2.9.

(a) Calculate the linear regression of Y on X.

(b) Plot the data and add the regression line to your graph. Does the line appear to fit the data well?

(c) Interpret the value of the slope of the regression line, b_1, in the context of this setting.

(d) SS(resid) = 21026.1. Use this to compute s_e. What are the units of s_e?

12.3.7 The rowan (*Sorbus aucuparia*) is a tree that grows in a wide range of altitudes. To study how the tree adapts to its varying habitats, researchers collected twigs with attached buds from 12 trees growing at various altitudes in North Angus, Scotland. The buds were brought back to the laboratory, and measurements were made of the dark respiration rate. The accompanying table shows the altitude of origin (in meters) of each batch of buds and the dark respiration rate (expressed as μl of oxygen per hour per mg dry weight of tissue).[15]

	Altitude of origin X (m)	Respiration rate Y (μl/hr × mg)
	90	0.11
	230	0.20
	240	0.13
	260	0.15
	330	0.18
	400	0.16
	410	0.23
	550	0.18
	590	0.23
	610	0.26
	700	0.32
	790	0.37
Mean	433.333	0.21000
SD	214.617	0.07711
	$r = 0.88665$	
	SS(resid) = 0.013986	

(a) Calculate the linear regression of Y on X.

(b) Plot the data and the regression line.

(c) Interpret the value of the slope of the regression line, b_1, in the context of this setting.

(d) Calculate the residual standard deviation.

12.3.8 Scientists studied the relationship between the length of the body of a bullfrog and how far it can jump. Eleven bullfrogs were included in the study. The results are given in the table.[16]

(a) Calculate the linear regression of Y on X.

(b) Interpret the value of the slope of the regression line, b_1, in the context of this setting.

(c) What proportion of the variation in maximum jump distances can be explained by the linear relationship between jump distance and frog length?

(d) Calculate the residual standard deviation and specify the units.

(e) Interpret the value of the residual standard deviation in the context of this setting.

Bullfrog	Length X (mm)	Maximum jump Y (cm)
1	155	71.0
2	127	70.0
3	136	100.0
4	135	120.0
5	158	103.3
6	145	116.0
7	136	109.2
8	172	105.0
9	158	112.5
10	162	114.0
11	162	122.9
Mean	149.6364	103.9909
SD	14.4725	17.9415
		$r = 0.28166$
		SS(resid) = 2,963.61

12.3.9 (*Continuation of Exercise 12.3.8*)

Consider the bullfrog jump data and summaries presented in Exercise 12.3.8.

(a) Predict the maximum jump for a frog that is 152 mm long.

(b) Assuming the residuals follow a normal model, would it be unusual for a 152 mm long frog to jump 145cm?

12.3.10 The peak flow rate of a person is the fastest rate at which the person can expel air after taking a deep breath. Peak flow rate is measured in units of liters per minute and gives an indication of the person's respiratory health. Researchers measured peak flow rate and height for each of a sample of 17 men. The results are given in the table.[17]

(a) Calculate the linear regression of Y on X.

(b) What proportion of the variation in flow rate is explained by the linear regression of flow rate on height?

(c) For each subject, calculate the predicted peak flow rate, using the regression equation from part (a).

(d) For each subject, calculate the residual, using the results from part (c).

Subject	Height X (cm)	Peak flow rate Y (l/min)
1	174	733
2	183	572
3	176	500
4	169	738
5	183	616
6	186	787
7	178	866
8	175	670
9	172	550
10	179	660
11	171	575
12	184	577
13	200	783
14	195	625
15	176	470
16	176	642
17	190	856
Mean	180.4118	660.0000
SD	8.5591	117.9952
	$r = 0.32725$	
	SS(resid) = 198,909	

(e) Calculate s_e and specify the units.

(f) What percentage of the data points are within $\pm s_e$ of the regression line? That is, what percentage of the 17 residuals are in the interval $(-s_e, s_e)$?

12.3.11 For each of the following data sets, prepare a plot like Figure 12.3.8, showing the data, the fitted regression line, and two lines whose vertical distance above and below the regression line is s_e. What percentage of the data points are within $\pm s_e$ of the regression line? What percentage of the data points do you expect to find within $\pm s_e$ of the regression line? How do these values compare?

(a) The body temperature data of Exercise 12.3.3.

(b) The corn yield data of Exercise 12.3.4.

12.3.12 Suppose a large sample of (x, y) pairs were used to fit the regression of Y on X. Now suppose we observed 100 further (x, y) pairs. About how many of these new observations would you expect to be farther than $2s_e$ from the regression line?

12.3.13 Forced expiratory volume (FEV) is a measure of the rate of airflow (L/min) during one deep exhalation. FEV was measured for each of 60 students between 65 and 73 inches tall. The scatterplot below shows FEV plotted against height. The triangles in the plot are the average FEV values for each height. The SD line and the regression line are shown. Which line—the dotted line or the dashed line—is the SD line, and which line is the regression line?[18]

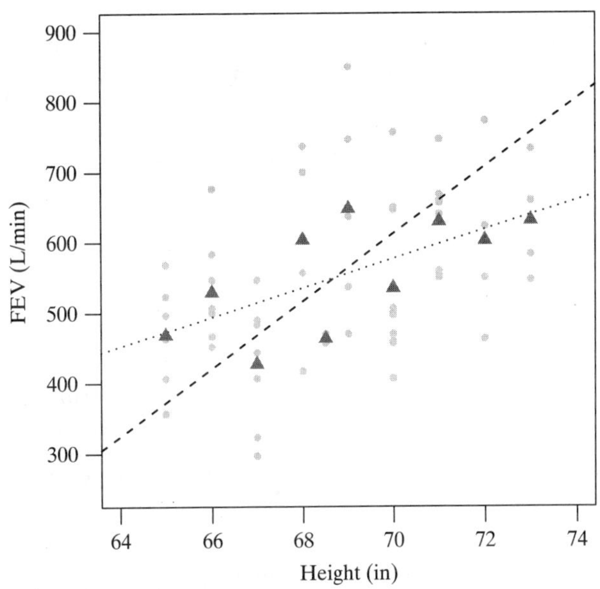

12.4 Parametric Interpretation of Regression: The Linear Model

One use of regression analysis is simply to provide a concise description of the data. The quantities b_0 and b_1 locate the regression line, and s_e describes the scatter of the points about the line.

For many purposes, however, data description is not enough. In this section we consider inference from the data to a larger population. In previous chapters we have spoken of one or several populations of Y values. Now, to encompass the X variable as well, we need to expand the notion of a population.

CONDITIONAL POPULATIONS AND CONDITIONAL DISTRIBUTIONS

A **conditional population** of Y values is a population of Y values associated with a fixed, or given, value of X. Within a conditional population we may speak of the **conditional distribution** of Y. The mean and standard deviation of a conditional population distribution are denoted as

$$\mu_{Y|X} = \text{Population mean } Y \text{ value for a given } X$$
$$\sigma_{Y|X} = \text{Population SD of } Y \text{ values for a given } X$$

(Note that the "given" symbol "|" is the same one used for conditional probability in Chapters 3 and 10.) The following example illustrates this notation.

Example 12.4.1 **Amphetamine and Food Consumption** In the rat experiment introduced in Example 12.1.1, the response variable Y was food consumption and the three values of X (dose) were $X = 0$, $X = 2.5$, and $X = 5$. In Example 12.3.5 we examined the graph of averages and considered the food consumption data as three independent samples (as for an ANOVA). In the ANOVA context we denote the three population means as μ_1, μ_2, and μ_3. In regression notation these means would be denoted as

$$\mu_{Y|X=0} \quad \mu_{Y|X=2.5} \quad \mu_{Y|X=5}$$

respectively. Similarly, the three population standard deviations, which would be denoted as σ_1, σ_2, and σ_3 in an ANOVA context, would be denoted as

$$\sigma_{Y|X=0} \quad \sigma_{Y|X=2.5} \quad \sigma_{Y|X=5}$$

respectively. In other words, the symbols

$$\mu_{Y|X} \text{ and } \sigma_{Y|X}$$

represent the mean and standard deviation of food consumption values for rats that are given dose X of amphetamine. ■

In observational studies, conditional distributions pertain to subpopulations rather than experimental treatment groups, as in the following example.

Example 12.4.2 **Height and Weight of Young Men** Consider the variables

$$X = \text{Height}$$

and

$$Y = \text{Weight}$$

for a population of young men. The conditional means and standard deviations are

$$\mu_{Y|X} = \text{Mean weight of men who are } X \text{ inches tall}$$
$$\sigma_{Y|X} = \text{SD of weights of men who are } X \text{ inches tall}$$

Thus, $\mu_{Y|X}$ and $\sigma_{Y|X}$ are the mean and standard deviation of weight in the *subpopulation* of men whose height is X. Of course, there is a different subpopulation for each value of X. ■

THE LINEAR MODEL

When we conduct a linear regression analysis, we think of Y as having a distribution that depends on X. The analysis can be given a parametric interpretation if two conditions are met. These conditions, which constitute the **linear model,** are given in the following box.

The Linear Model

1. *Linearity.* $Y = \mu_{Y|X} + \varepsilon$ where $\mu_{Y|X}$ is a linear function of X; that is

$$\mu_{Y|X} = \beta_0 + \beta_1 X$$

Thus, $Y = \beta_0 + \beta_1 X + \varepsilon$.

2. *Constancy of standard deviation.* $\sigma_{Y|X}$ does not depend on X. We denote this constant value as σ_ε.

In the linear model $Y = \beta_0 + \beta_1 X + \varepsilon$, the ε term represents **random error.** We include this term in the model to reflect the fact that Y varies, even when X is fixed. The variability of Y for a fixed value of X is measured by the conditional standard deviation of Y, $\sigma_{Y|X}$. But, because the linear model stipulates that this standard deviation is the same for every value of X, we commonly use the notation σ_ε to represent this standard deviation and refer to it as the standard deviation of the random error.

The following two examples show the meaning of the linear model.

Example 12.4.3 **Amphetamine and Food Consumption** For the rat food consumption experiment, the linear model asserts that (1) the population mean food consumption is a linear function of dose, and that (2) the population standard deviation of food consumption values is the same for all doses. Notice that the second condition is closely analogous to the condition in ANOVA that the population SDs are equal: $\sigma_1 = \sigma_2 = \sigma_3$. The linear model also allows for the fact that there is variability in Y when X is fixed. For example, there were eight observations for which $X = 5$. The eight y-values averaged 55.0, but none of the observations was equal to 55.0; there was substantial variability within the eight y-values. This variability is quantified by the SD of 13.3. ■

Example 12.4.4 **Height and Weight of Young Men** We consider an idealized fictitious population of young men whose joint height and weight distribution fits the linear model exactly. For our fictitious population we will assume that the conditional means and SDs of weight given height are as follows:

$$\mu_{Y|X} = -145 + 4.25X$$
$$\sigma_\varepsilon = 20$$

Thus, the regression parameters of the population are $\beta_0 = -145$ and $\beta_1 = 4.25$. (This fictitious population resembles that of U.S. 17-year-old males.)[19] Thus, the model is $Y = -145 + 4.25X + \varepsilon$.

Table 12.4.1 shows the conditional means and SDs of Y = weight for a few selected values of X = height. Figure 12.4.1 shows the conditional distributions of Y given X for these selected subpopulations.

Table 12.4.1 Conditional means and SDs of weight given height in a population of young men*

Height (in) X	Mean weight (lb) $\mu_{Y|X}$	Standard deviation of weights (lb) $\sigma_{Y|X}$
64	127	20
68	144	20
72	161	20
76	178	20

*Note that all values of $\sigma_{Y|X}$ are the same; they equal $\sigma_\varepsilon = 20$.

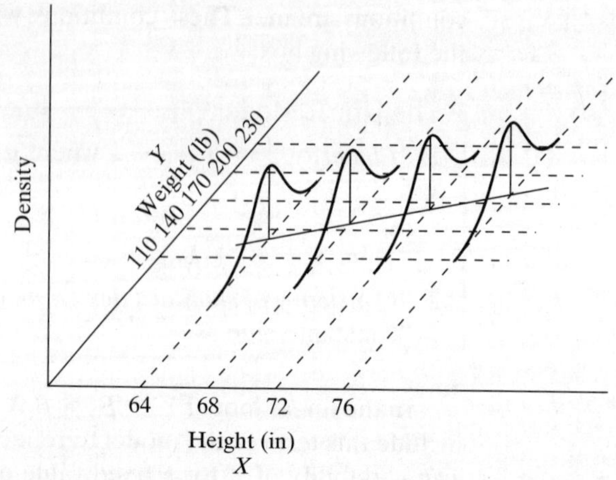

Figure 12.4.1 Conditional distributions of weight given height in a population of young men

Note, for example, that if height = 68 (in), then the mean weight is 144 (lb) and the SD of the weights is 20 (lb). For this subpopulation, $Y = 144 + \varepsilon$. If a particular young man who is 68 inches tall weighs 145 pounds, then $\varepsilon = 1$ for him. If another 68-inch-tall young man weighs 140 pounds, then $\varepsilon = -4$ in his case. Of course, β_0, β_1, and ε are generally not observable. This example is fictitious. ■

Remark Actually, the term *regression* is not confined to linear regression. In general, the relationship between $\mu_{Y|X}$ and X is called the *regression of Y on X*. The linearity assumption asserts that the regression of Y on X is linear rather than, for instance, a curvilinear function.

ESTIMATION IN THE LINEAR MODEL

Consider now the analysis of a set of (X, Y) data. Suppose we assume that the linear model is an adequate description of the true relationship of Y and X. Suppose further that we are willing to adopt the following **random subsampling model**:

> **Random Subsampling Model**
>
> For each observed pair (x, y), we regard the value y as having been sampled at random from the conditional population of Y values associated with the X value x.

Within the framework of the linear model and the random subsampling model, the quantities b_0, b_1, and s_e calculated from a regression analysis can be interpreted as estimates of population parameters:

b_0 is an estimate of β_0
b_1 is an estimate of β_1
s_e is an estimate of σ_ε

Example 12.4.5 **Length and Weight of Snakes** From the summaries of the snake data of Example 12.2.1 and 12.2.2, we can compute the following regression coefficients $b_0 = -301$, $b_1 = 7.19$, and $s_e = 12.5$ (computing these yourself from the provided summaries would be a good exercise). Thus,

-301 is our estimate of β_0
7.19 is our estimate of β_1
12.5 is our estimate of σ_ε

■

The application of the linear model to the snake data has yielded two benefits. First, the slope of the regression line, 7.19 gm/cm, is an estimate of a morphological parameter ("weight per unit change in length"), which is of potential biological interest in characterizing the population of snakes. Second, we have obtained an estimate (12.5 g) of the variability of weight among snakes of fixed length, even though no direct estimate of this variability was possible because no two of the observed snakes were the same length.

INTERPOLATION IN THE LINEAR MODEL

In Section 12.3 we regarded the regression line as a line of averages. The idea of smoothing the graph of averages into a straight line can be extended to the setting in which we have only a single observation at each level of X. When we draw a line through a set of (X, Y) data, we are expressing a belief that the underlying dependence of Y on X is smooth, even though the data may show the relationship only roughly. Linear regression is one formal way of providing a smooth description of the data as illustrated in the following example.

Example 12.4.6

Dissolved Oxygen What are the mean and standard deviation of dissolved oxygen levels in Dairy Creek for days when the water temperature is 17.5 °C? None of our observed days had a water temperature of 17.5 °C. If there were some observations at this temperature, we could average the associated dissolved oxygen levels to obtain one answer to our question, but because there is an apparent linear relationship between X and Y, we can use the line to obtain an even better estimate of the mean dissolved oxygen level that uses all of the data. In Examples 12.3.4 and 12.3.8 we found the regression equation to be $\hat{y} = 11.94 - 0.22x$ and $s_e = 1.21$. Thus the estimated mean dissolved oxygen level for days when the water temperature is 17.5 °C is $11.94 - 0.22 \times 17.5 = 8.1$ mg/L with a standard deviation of $s_e = 1.21$ mg/L. Figure 12.4.2 shows the interpolation graphically. ■

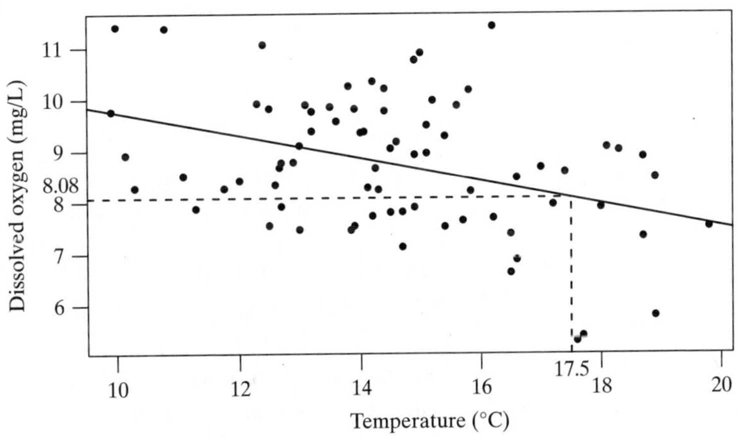

Figure 12.4.2 Levels of dissolved oxygen versus water temperature for 75 days

Note that estimation of the mean uses the linearity condition of the linear model, while estimation of the standard deviation uses the condition of constant standard deviation. In some situations only the linearity condition may be plausible, and then only the mean would be estimated.

Example 12.4.6 is an example of **interpolation,** because the X values we chose ($X = 17.5$ for the dissolved oxygen and 3.5 for the food consumption examples) were within the range of observed values of X. By contrast, **extrapolation** is the use of a regression line (or other curve) to predict Y for values of X that are outside the range of the data. Extrapolation should be avoided whenever possible, because there is usually no assurance that the relationship between $\mu_{Y|X}$ and X remains linear for X values outside the range of those observed. Many biological relationships are linear for only part of the possible range of X values. The following is an example.

Example 12.4.7 **Amphetamine and Food Consumption** The dose–response relationship for the rat food consumption experiment of Example 12.1.1 looks approximately like Figure 12.4.3.[20] The data cover only the linear portion of the relationship. Clearly it would be unwise to extrapolate the fitted line out to $X = 10$ or $X = 15$. ■

Figure 12.4.3 Dose–response curve (mean response versus dose) for rat food consumption experiment

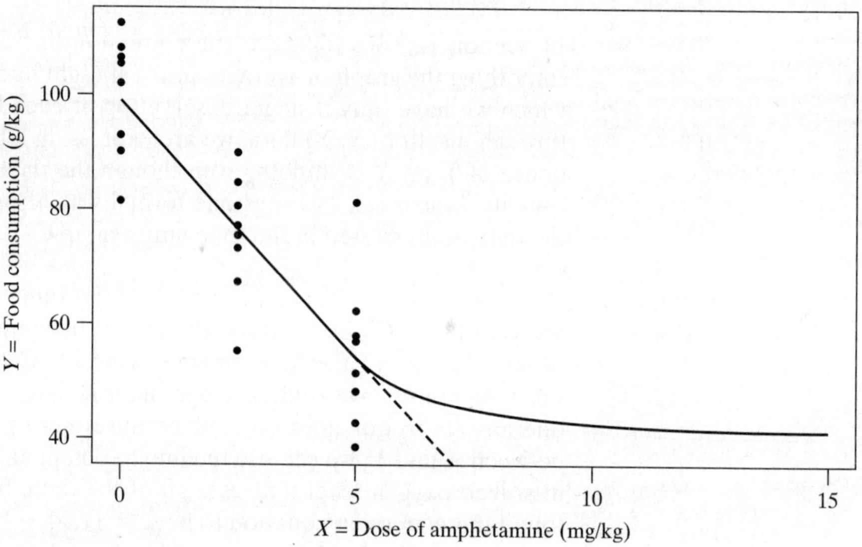

PREDICTION AND THE LINEAR MODEL

Consider the setting of using height, X, to predict weight, Y, for a large group of young men for whom the average weight is 150 pounds. Suppose a young man is chosen at random and we must predict his weight.

1. If we don't know anything about the height of the man, then the best estimate we can give of his weight is the overall average weight, $\bar{y} = 150$.
2. Suppose we learn that the man's height is 76 inches. If we know that the average weight of all 76-inch-tall men in the group is 180 pounds, then we can use this conditional average, $\bar{y}|x = 76$, as our prediction of the man's weight. We expect this prediction, which essentially is using the graph of averages (but without smoothing), to be more accurate than the one given in part 1.
3. Suppose we learn that the man's height is 76 inches and we also know that the least-squares regression equation is $Y = -140 + 4.3X$. Then we can use the value $x = 76$ to get a prediction, which would be $-140 + 4.3 \times 76 = 186.8$.

Is the prediction in 3 better than the prediction made in 2? Since using the regression equation amounts to smoothing the graph of averages, we expect prediction 3 to be better than prediction 2 *to the extent that we believe that there is a linear relationship between height and weight.* Prediction 3 has the advantage of using information from all the data points, not just those for which $x = 76$. Method 3 also has the advantage of allowing for predictions when the x value (the height) is not one that is in the original data set (as discussed in the preceding subsection "Interpolation in the Linear Model"), so $\bar{y}|x$ is not computable. However, method 3 will give poor predictions if the linear relationship does not hold. Thus it is very important to think about such relationships, and to explore them graphically, before using a regression model.

Exercises 12.4.1–12.4.9

12.4.1 For the data in Exercise 12.2.7 there were two observations for which $X = 0$. The average response (Y value) for these points is $\frac{33.3 + 31.0}{2} = 32.15$. However, the intercept of the regression line, b_0, is not 32.15. Why not? Why is b_0 a better estimate of the average fungus growth when laetisaric acid concentration is zero than 32.15?

12.4.2 Refer to the body temperature data of Exercise 12.3.3. Assuming that the linear model is applicable, estimate the mean and the standard deviation of the drop in body temperature that would be observed in mice given alcohol at a dose of 2 gm/kg. [*Tip*: Is the X variable dose or log(dose)?]

12.4.3 Refer to the cob weight data of Exercises 12.2.6 and 12.3.4. Assume that the linear model holds.

(a) Estimate the mean cob weight to be expected in a plot containing (i) 100 plants; (ii) 120 plants.

(b) Assume that each plant produces one cob. How much grain would we expect to get from a plot containing (i) 100 plants? (ii) 120 plants?

12.4.4 (*Continuation of Exercise 12.4.3*). For the cob weight data, SS(resid) = 1,337.3. Estimate the standard deviation of cob weight in plots containing (i) 100 plants; (ii) 120 plants.

12.4.5 Refer to the fungus growth data of Exercise 12.2.7. For these data, SS(resid) = 16.7812. Assuming that the linear model is applicable, find estimates of the mean and standard deviation of fungus growth at a laetisaric acid concentration of 15 μg/ml.

12.4.6 Refer to the energy expenditure data of Exercise 12.2.9. Assuming that the linear model is applicable, estimate the 24-hour energy expenditure of a man whose fat-free mass is 55 kg.

12.4.7 Refer to the Ca pump activity of Exercise 12.2.10. For these data SS(resid) = 21,984,623.

(a) Assuming that the linear model is applicable, estimate the mean and standard deviation basal Ca pump activity for children born to mothers with a hair Hg level of 3 μg/g.

(b) Using the values computed in part (a) to support your answer, would it be surprising for a mother with a hair Hg level of 3 μg/g to give birth to a child with a basal Ca pump activity above 4000 nmol/mg/hr?

12.4.8 Refer to the bullfrog data of Exercise 12.3.8. Assuming that the linear model is applicable, estimate the maximum jump length of a bullfrog whose body length is 150 mm.

12.4.9 Refer to the peak flow data of Exercise 12.3.10. Assuming that the linear model is applicable, find estimates of the mean and standard deviation of peak flow for men 180 cm tall.

12.5 Statistical Inference Concerning β_1

The linear model provides interpretations of b_0, b_1, and s_e that take them beyond data description into the domain of statistical inference. In this section we consider inference about the true slope β_1 of the regression line. The methods are based on the condition that the conditional population distribution of Y for each value of X is a normal distribution. This is equivalent to stating that in the linear model of $Y = \beta_0 + \beta_1 X + \varepsilon$, the ε values come from a normal distribution.

THE STANDARD ERROR OF b_1

Within the context of the linear model, b_1 is an estimate of β_1. Like all estimates calculated from data, b_1 is subject to sampling error. The standard error of b_1 is calculated as follows:

Standard Error of b_1

$$SE_{b_1} = \frac{s_e}{s_x\sqrt{n-1}}$$

The following example illustrates the calculation of SE_{b_1}.

Example 12.5.1 **Length and Weight of Snakes** For the snake data, we found in Table 12.2.2 that $n = 9, s_x = 4.637$, and in Example 12.4.5 that $s_e = 12.5$. The standard error of b_1 is

$$\mathrm{SE}_{b_1} = \frac{12.5}{4.637\sqrt{9-1}} = 0.9531$$

To summarize, the slope of the fitted regression line (from Example 12.4.5) is

$$b_1 = 7.19 \text{ gm/cm}$$

and the standard error of this slope is

$$\mathrm{SE}_{b_1} = 0.95 \text{ gm/cm}$$

■

Structure of the SE Let us see how the standard error of b_1 depends on various aspects of the data. In the same way that $\mathrm{SE}_{\bar{Y}}$ depends on the variability in the Y data (s_y) and the sample size (n), SE_{b_1} depends on the scatter of the data about the regression line (s_e) and the size of the sample (n). The formula for SE_{b_1} supports our intuition showing that data with less scatter about the regression line (smaller s_e) and larger sample sizes (larger n) produce more precise estimates of β_1 (i.e., a smaller SE_{b_1}). While variability in Y and sample size are the only two factors that affect our ability to estimate a population mean precisely ($\mathrm{SE}_{\bar{Y}}$), there is a third factor that is important for precise estimation of β_1: the variability of the X data. The more spread out our X values (larger s_x), the more precise our estimate of β_1 will be. The dependence on the spread in the X values is illustrated in Figure 12.5.1, which shows two data sets with the same value of s_e and the same value of n, but different values of s_x. Imagine using a ruler to fit a straight line by eye; it is intuitively clear that the data in case (b)—with the larger s_x—would determine the slope of the line more precisely.

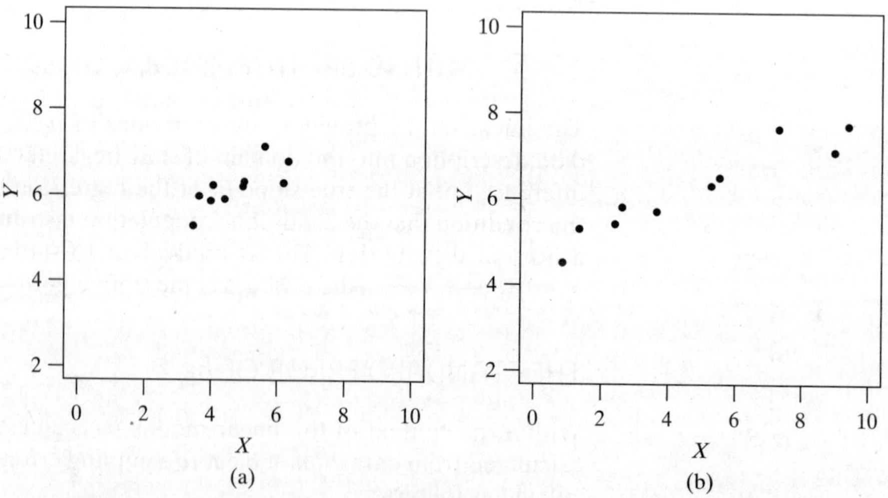

Figure 12.5.1 Two data sets with the same value of n and of s_e but different s_x: (a) smaller s_x and (b) larger s_x

As another way of thinking about this, imagine holding your arms out in front of you, extending the index finger on each hand, and balancing a meter stick on your two fingers. If you move your hands far apart from each other, balancing the meter stick is easy—this is like case (b). However, if you move your hands close together, balancing the meter stick becomes more difficult—this is like case (a). Having the base of support spread out increases stability. Likewise, having the x values spread out decreases the standard error of the slope.

Implications for Design The previous discussion implies that, for the purpose of gaining precise information about β_1, it is best to have the values of X as widely dispersed as possible. This fact can guide the experimenter when the design of the experiment includes choosing values of X. Other factors also play a role, however. For instance, if X is the dose of a drug, the criterion of widely dispersed X's would lead to using only two dosages, one very low and one very high. But in practice an experimenter would want to have at least a few observations at intermediate doses, to verify that the relation is actually linear within the range of the data.

CONFIDENCE INTERVAL FOR β_1

In many studies the quantity β_1 is a biologically meaningful parameter and a primary aim of the data analysis is to estimate β_1. A confidence interval for β_1 can be constructed by the familiar method based on the SE and Student's t distribution. For instance, a 95% confidence interval is constructed as

$$b_1 \pm t_{0.025}\,\mathrm{SE}_{b_1}$$

where the critical value $t_{0.025}$ is determined from Student's t distribution with

$$\mathrm{df} = n - 2$$

Intervals with other confidence coefficients are constructed analogously; for instance, for a 90% confidence interval one would use $t_{0.05}$.

Example 12.5.2 **Length and Weight of Snakes** Let us use the snake data to construct a 95% confidence interval for β_1. We found that $b_1 = 7.19186$ and $\mathrm{SE}_{b_1} = 0.9531$. There are $n = 9$ observations; we refer to Table 4 with df $= 9 - 2 = 7$, and obtain

$$t_{7,\,0.025} = 2.365$$

The confidence interval is

$$7.19186 \pm 2.365 \times 0.9531$$

or

$$4.94 \text{ gm/cm} < \beta_1 < 9.45 \text{ gm/cm}$$

We are 95% confident that the true slope of the regression of weight on length for this snake population is between 4.94 gm/cm and 9.45 gm/cm; this is a rather wide interval because the sample size is not very large. ■

TESTING THE HYPOTHESIS H_0: $\beta_1 = 0$

In some investigations it is not a foregone conclusion that there is any linear relationship between X and Y. It then may be relevant to consider the possibility that any apparent trend in the data is illusory and reflects only sampling variability. In this situation it is natural to formulate the null hypothesis

$$H_0\text{: } \mu_{Y|X} \text{ does not depend on } X$$

Within the linear model, this hypothesis can be translated as

$$H_0\text{: } \beta_1 = 0$$

A t test of H_0 is based on the test statistic*

$$t_s = \frac{b_1 - 0}{\text{SE}_{b_1}}$$

Critical values are obtained from Student's t distribution with

$$\text{df} = n - 2$$

The following example illustrates the application of this t test.

Example 12.5.3

Blood Pressure and Platelet Calcium The blood pressure and platelet calcium data from Example 12.2.3 are shown in Figure 12.5.2. Calculations from the data yield $\bar{x} = 94.50000, \bar{y} = 107.86840, s_x = 8.04968, s_y = 16.07780$, from which we can calculate†

$$b_0 = -2.2009 \text{ and } b_1 = 1.16475$$

The residual sum of squares is 6311.7618.

Thus,

$$s_e = \sqrt{\frac{6311.76}{38 - 2}} = 13.24 \text{ and SE}_{b_1} = \frac{13.24}{8.04968\sqrt{38 - 1}} = 0.2704$$

The values of b_0, b_1, SS(resid), and SE_{b_1} are generally found using computer software. The following computer output is typical:

```
The regression equation is
Platelet Calcium = -2.2 + 1.16 Blood Pressure
Predictor            Coef   SE Coef       T      P
Constant            -2.20     25.65   -0.09  0.932
Blood Pressure     1.1648    0.2704    4.31  0.000
S = 13.2411 R - Sq = 34.0% R - Sq(adj) = 32.2%

Analysis of Variance
Source          DF      SS      MS       F      P
Regression       1  3252.6  3252.6   18.55  0.000
Residual Error  36  6311.8   175.3
Total           37  9564.3
```

We will test the null hypothesis

$$H_0: \beta_1 = 0$$

against the nondirectional alternative

$$H_A: \beta_1 \neq 0$$

These hypotheses are translations, within the linear model, of the verbal hypotheses

H_0: Mean platelet calcium is not linearly related to blood pressure

H_A: Mean platelet calcium is linearly related to blood pressure

(*Note*: "Linearly related" does not necessarily refer to causal dependence as we have discussed in Section 12.2.)

---

*We include the "−0" in the numerator of the test statistic to remind us that we are comparing our estimated (observed) slope, b_1, to the slope we'd expect to observe if the null hypothesis were true. In the exercises we will consider a situation for which the hypothesized slope may be a value other than zero.

†As the following values are intermediate calculations used in the regression, we include more digits than one would typically display in a summary.

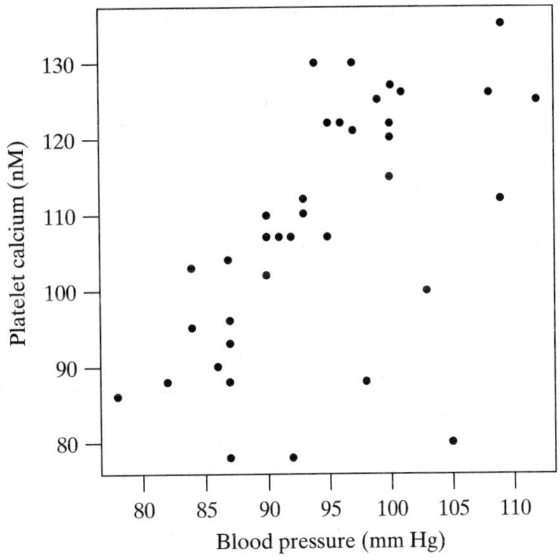

Figure 12.5.2 Blood pressure and platelet calcium for 38 persons with normal blood pressure

Let us choose $\alpha = 0.05$. The test statistic is

$$t_s = \frac{1.16475}{0.2704} = 4.308$$

From Table 4 with df $= n - 2 = 36 \approx 40$, we find $t_{40,0.0005} = 3.551$. Thus, we find P-value < 0.001 and we reject H_0. (Using a computer gives P-value $= 0.00012$) The data provide sufficient (and very strong) evidence to conclude that the true slope of the regression of platelet calcium on blood pressure in this population is positive (i.e., $\beta_1 > 0$). ■

While the forms of the test-statistic are quite different, testing H_0: $\beta_1 = 0$ is equivalent to testing H_0: $\rho = 0$. Recall that a population correlation of zero indicates that there is no linear relationship between X and Y. In this case, the slope that best summarizes "no linear relationship" is a slope of zero.

Note that the test on β_1 does not ask *whether* the relationship between $\mu_{Y|X}$ and X is linear. Rather, the test asks whether, *assuming* that the linear model holds, we can conclude that the slope is nonzero. It is therefore necessary to be careful in phrasing the conclusion from this test. For instance, the statement "There is a significant linear trend" could easily be misunderstood.*

As is the case with other hypothesis tests, if we wish to use a directional alternative hypothesis we follow the two-step procedure of (1) checking that the specified direction is correct (which in a regression setting means checking that the slope of the regression line has the correct + or − sign) and (2) cutting the nondirectional P-value in half if this condition is met.

A RANDOMIZATION TEST

The t test presented above is based on the condition that the population distribution of Y for each value of X is a normal distribution. If the normality condition is violated,

*There are tests that can (in some circumstances) test whether the true relationship is linear. Furthermore, there are tests that can test for a linear component of trend without assuming that the relationship is linear. These tests are beyond the scope of this book.

it is still possible to test the null hypothesis that the population value of the slope, β_1, is zero.

In Section 12.2 we presented a randomization test of the population correlation coefficient. The procedure described there can be adapted to testing the population slope. The following example illustrates the method.

Example 12.5.4 **Blood Pressure and Platelet Calcium** The sample slope for the blood pressure and platelet calcium data is $b_1 = 1.16475$. Suppose that H_0 is true and that $\beta_1 = 0$. How likely is it that the sample slope would be as far from zero as 1.16475? To conduct a randomization test we fix the 38 blood platelet values and scramble the 38 platelet calcium values to create a new "pseudo" data set. We then fit a regression model to these data and record whether or not the sample slope is as far from zero as 1.16475. After repeating this many times we have a measure of how unusual 1.16475 is, if H_0 is true.

In one computer simulation of 100,000 trials, there were only 13 trials for which the slope was at least as large in magnitude as 1.16475, which means that the randomization P-value is 0.00013. This means that the sample slope in the original data is not easily explained by chance. Instead, we have strong evidence for H_A. ■

Exercises 12.5.1–12.5.10

12.5.1 Refer to the leucine data given in Exercise 12.3.1.

(a) Construct a 95% confidence interval for β_1.

(b) Interpret the confidence interval from part (a) in the context of this setting.

12.5.2 Refer to the body temperature data of Exercise 12.3.3. For these data, $s_e = 0.91472$. Construct a 95% confidence interval for β_1.

12.5.3 Refer to the cob weight data of Exercise 12.2.6. For these data, SS(resid) = 1,337.3.

(a) Construct a 95% confidence interval for β_1.

(b) Interpret the confidence interval from part (a) in the context of this setting.

12.5.4 Refer to the fungus growth data of Exercise 12.2.7. For these data, SS(resid) = 16.7812.

(a) Calculate the standard error of the slope, SE_{b_1}.

(b) Consider the null hypothesis that laetisaric acid has no effect on growth of the fungus. Assuming that the linear model is applicable, state in symbols the null hypothesis about the true regression line and an alternative hypothesis that laetisaric acid inhibits growth of the fungus.

(c) How many degrees of freedom are there for the test in part (b)?

(d) The sample slope, b_1, is -0.7120. Compute the value of the test statistic.

(e) The P-value for the test is 1.16×10^{-9}. If $\alpha = 0.05$, state your conclusion regarding the null hypothesis in the context of this setting.

12.5.5 Refer to the energy expenditure data of Exercise 12.2.9. For these data, SS(resid) = 21,026.1.

(a) Construct a 95% confidence interval for β_1.

(b) Construct a 90% confidence interval for β_1.

12.5.6 Refer to the basal Ca pump data from Exercise 12.2.10. For these data, $s_e = 548.78$.

(a) Construct a 95% confidence interval for β_1.

(b) What do you think about a claim that β_1 is less than -800 (nmol/mg/hr)/(μg/g)? Use your interval from part (a) to support your answer.

(c) What do you think about a claim that β_1 is less than 800 (nmol/mg/hr)/(μg/g) in magnitude? Use your interval from part (a) to support your answer.

12.5.7 Refer to the respiration data of Exercise 12.3.7. Assuming that the linear model is applicable, test the null hypothesis of no relationship against the alternative that trees from higher altitudes tend to have higher respiration rates. Let $\alpha = 0.05$.

12.5.8 The following computer output is from fitting a regression model to the snake length data of Example 12.2.2.

```
The regression equation is
Weight = -301 + 7.19 Length

Predictor      Coef     Stdev   t-ratio      p
Constant    -301.09     60.19     -5.00  0.000
Length       7.1919    0.9531      7.55  0.000

s = 12.50 R-sq = 89.1% R-sq(adj) = 87.5%

Analysis of Variance

SOURCE       DF     SS       MS        F      p
Regression    1  8896.3   8896.3   56.94  0.000
Error         7  1093.7    156.2
Total         8  9990.0
```

(a) Use the output to construct a 95% confidence interval for β_1.

(b) Interpret the confidence interval from part (a) in the context of this setting.

12.5.9 Refer to the peak flow data of Exercise 12.3.10. Assume that the linear model is applicable.

(a) Test the null hypothesis of no relationship against the alternative that peak flow is related to height. Use a nondirectional alternative with $\alpha = 0.10$.

(b) Repeat the test from part (a), but this time use the directional alternative that peak flow tends to increase with height. Again let $\alpha = 0.10$.

12.5.10 Scientists recorded the dry weight (mg) and corolla diameter (cm) of 86 evening primrose (*O. harringtonii*) flowers growing in a natural habitat.[21]

(a) Use the following computer output, from fitting a regression model, to construct a 95% confidence interval for β_1.

(b) Interpret the confidence interval from part (a) in the context of this setting.

(c) Does the confidence interval from part (a) include zero? How is this related to the *P*-value given in the computer output?

```
             Estimate Std. Error t value  Pr(>|t|)
(Intercept)  -31.32      12.49     -2.51   0.014 *
corolla        2.29       0.23      9.97  <2e-16***
```

12.6 Guidelines for Interpreting Regression and Correlation

Any set of (X, Y) data can be submitted to a regression analysis and values of b_0, b_1, s_e, and r can be calculated. But these quantities require care in interpretation. In this section we discuss guidelines and cautions for interpretation of linear regression and correlation. We first consider the use of regression and correlation for purely descriptive purposes and then turn to inferential uses.

WHEN IS LINEAR REGRESSION DESCRIPTIVELY INADEQUATE?

Linear regression and correlation may provide inadequate description of a data set if any of the following features is present:

- curvilinearity
- outliers
- influential points

We briefly discuss each of these.

If the dependence of Y on X is actually curvilinear rather than linear, the application of linear regression and correlation can be very misleading. The following example shows how this can happen.

Example 12.6.1 **A Curvilinear Relationship with X** Figure 12.6.1 shows a set of fictitious data that obeys an exact relationship: $Y = -1 + 6X - X^2$. Nevertheless, X and Y are uncorrelated: $r = 0$ and $b_1 = 0$. The best straight line through the data would be a horizontal one, but of course the line would be a poor summary of the curvilinear relationship between X and Y. The residual SD is $s_e = 2.27$; however, since these data are nonrandom, s_e does not measure random variation, but rather measures deviation from linearity. ■

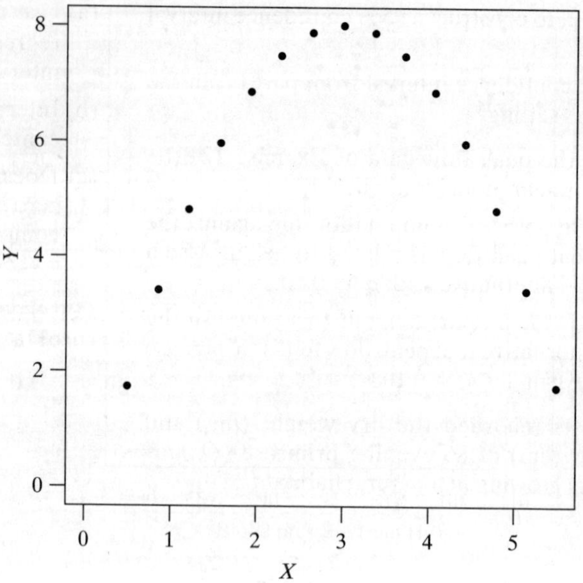

Figure 12.6.1 Data for which X and Y are uncorrelated but have a strong curvilinear relationship

Generally, the consequences of curvilinearity are that (1) the fitted line does not adequately represent the data; (2) the correlation is misleadingly small; (3) s_e is inflated. Of course, Example 12.6.1 is an extreme case of this distortion. A data set with mild, but still noticeable, curvilinearity is shown in Figure 12.6.2.

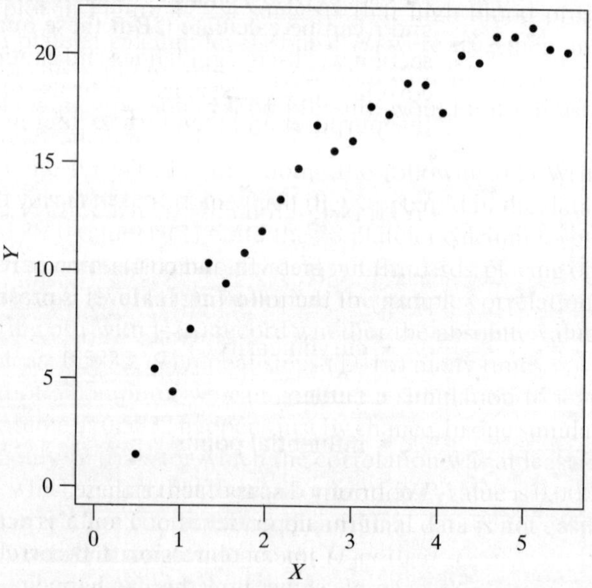

Figure 12.6.2 Data displaying mild curvilinearity

Outliers in a regression setting are data points that are unusually far from the linear trend formed by the data. Outliers can distort regression analysis in two ways: (1) by inflating s_e and reducing correlation; and (2) by unduly influencing the regression line. Note that a point can be an outlier in a scatterplot without being an outlier in either the distribution of X values or the distribution of Y values as we shall see in the following example.

Figure 12.6.3 displays a data set with a variety of outliers. Figure 12.6.3(a) displays a data set with no outliers, while (b) and (c) show data with regression outliers—they have points that fall far from the regression line. In plot (b) the outlying point

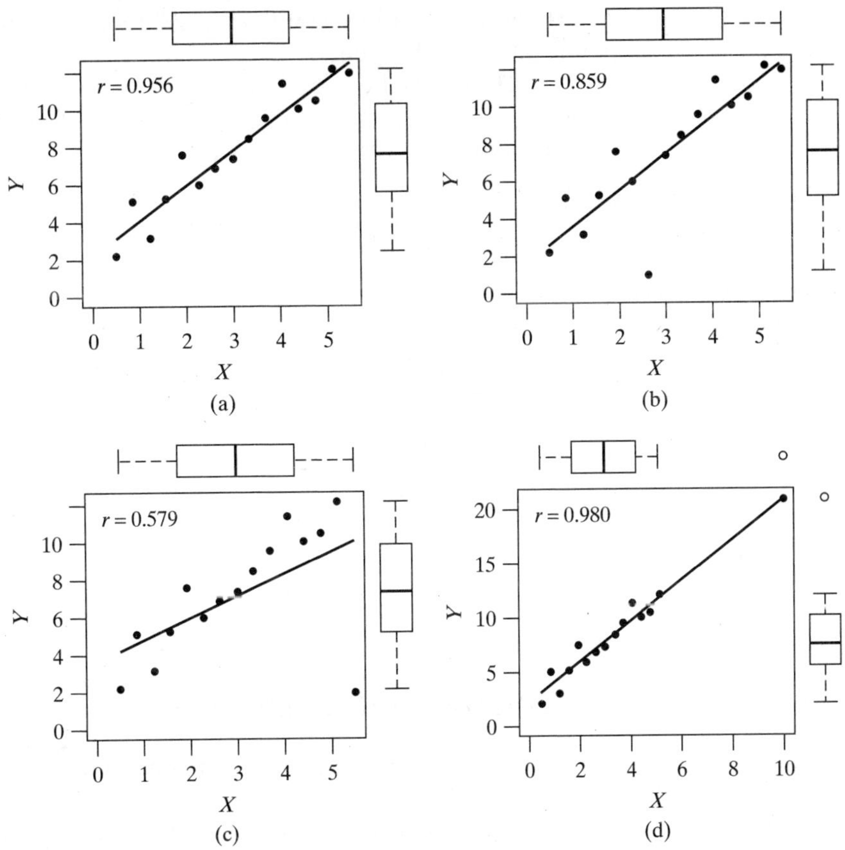

Figure 12.6.3 Different effects of outliers on the regression line. Boxplots of the X and Y data appear in the margins of each scatterplot. (a) A data set with no outliers; (b) the same data except for one outlier in the middle of the X values; (c) the same data except for one outlier at the high end of the X values (a point with leverage and influence); and (d) the same data except for one point that is an outlier with respect to the X (and Y) distribution, but not with respect to the regression line (a point with leverage, but little influence)

does not appear to affect the slope of the regression line very much, but it does increase the residual standard deviation, s_e, and reduce correlation. The outlying point in plot (c) appears to greatly affect the slope of the estimated regression line; it also increases s_e and reduces the correlation. While the unusual point in plot (d) is an outlier with respect to the X (and Y) distribution, it is not an outlier in the regression context as it does not fall far from the regression line.

Leverage points are points that have the potential to greatly influence the slope of the fitted regression model. The further a point's X value is from the center of the X distribution, the more leverage that point has on the overall regression model. *Having* and actually *exerting* leverage are two different things, however. Figure 12.6.3 plots (c) and (d) display examples of leverage points. In plot (c) the leverage point is shown to actually exert its leverage on the line, tipping the regression from the bulk of the data. A point that has a large effect on the regression model is called an **influential point.** Plot (d) shows a leverage point (because of the extreme X value) that is not influential because the regression line does not get pulled away from the trend in the bulk of the data. Note that the outlier in plot (b) is not considered a leverage point—its ability to affect the slope of the line is weak as its X value is near the center of the X distribution.

Influential points can also greatly affect (increase or decrease) the size of the correlation coefficient. In Figure 12.6.3, the influential point in (c) lowered the correlation from $r = 0.956$ in (a) to $r = 0.579$. Example 12.6.3 shows a situation for which the correlation is increased by the presence of an influential point.

Figure 12.6.4 (a) shows a data set and a regression line. Figure 12.6.4 (b) shows the same data set, but with an influential point added. Including the influential point in the data set changes the regression line noticeably. Although the influential point is an outlier in the X and Y distributions, it is not a regression outlier since the residual for this point is not very large.

The correlation coefficient for the data in Figure 12.6.4(a) is $r = 0.053$. Adding the influential point to the data set changes the correlation to $r = 0.759$ for the data in Figure 12.6.4(b).

Figure 12.6.4 The effect of an influential point on the regression line. (a) A data set; (b) the same data with an influential point added

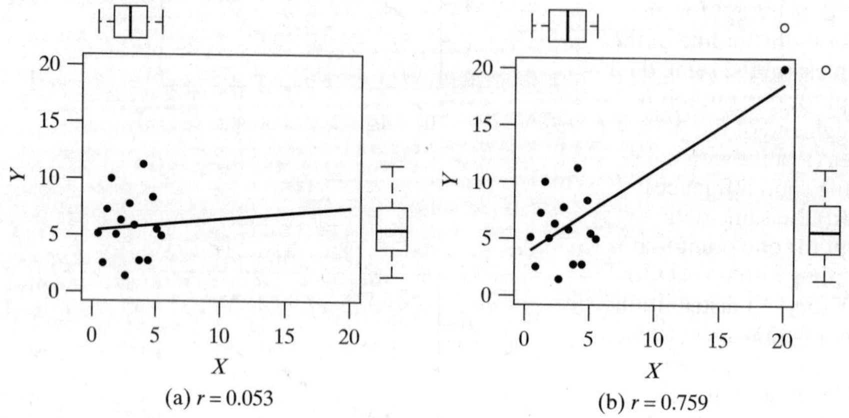

CONDITIONS FOR INFERENCE

The quantities b_0, b_1, s_e, and r can be used to describe a scatterplot that shows a linear trend. However, statistical inference based on these quantities depends on certain conditions concerning the design of the study, the parameters, and the conditional population distributions. We summarize these conditions and then discuss guidelines and cautions concerning them.

1. *Design conditions.* We have discussed two sampling models for regression and correlation:
   (a) Random subsampling model: For each observed X, the corresponding observed Y is viewed as randomly chosen from the conditional population distribution of Y values for that X.*
   (b) Bivariate random sampling model: Each observed pair (X, Y) is viewed as randomly chosen from the joint population distribution of bivariate pairs (X, Y).

   In either sampling model, each observed pair (X, Y) must be independent of the others. This means that the experimental design must not include any pairing, blocking, or hierarchical structure.
2. *Conditions concerning parameters.* The linear model states that
   (a) $\mu_{Y|X} = \beta_0 + \beta_1 X$.
   (b) σ_e does not depend on X.
3. *Condition concerning population distributions.* The confidence interval and t test are based on the conditional population distribution of Y for each fixed X having a normal distribution.

The random subsampling model is required if b_0, b_1, and s_e are to be viewed as estimates of the parameters β_0, β_1, and σ_ε mentioned in the linear model. The bivariate

*If the X variable includes measurement error, then X in the linear model must be interpreted as the measured value of X rather than some underlying "true" value of X. A linear model involving the "true" value of X leads to a different kind of regression analysis.

random sampling model is required if r is to be viewed as an estimate of a population parameter ρ. It can be shown that if the bivariate random sampling model is applicable, then the random subsampling model is also applicable. Thus, regression parameters can always be estimated if correlation can be estimated, but not vice versa.

GUIDELINES CONCERNING THE SAMPLING CONDITIONS

Departures from the sampling conditions not only affect the validity of formal techniques such as the confidence interval for β_1, but can also lead to faulty interpretation of the data even if no formal statistical analysis is performed. Two errors of interpretation that sometimes occur in practice are (1) failure to take into account dependency in the observations, and (2) insufficient caution in interpreting r when the X's do not represent a random sample.

The following two examples illustrate studies with dependent observations.

Example 12.6.2 **Serum Cholesterol and Serum Glucose** A data set consists of 20 pairs of measurements on serum cholesterol (X) and serum glucose (Y) in humans. However, the experiment included only two subjects; each subject was measured on 10 different occasions. Because of the dependency in the data, it is not correct to naively treat all 20 data points alike. Figure 12.6.5 illustrates the difficulty; the figure shows that there is no evidence of any correlation between X and Y, except for the modest fact that the subject who has larger X values happens also to have larger Y values. Clearly it would be impossible to properly interpret the scatterplot if all 20 points were plotted with the same symbol. By the same token, application of regression or correlation formulas to the 20 observations would be seriously misleading.[22] ■

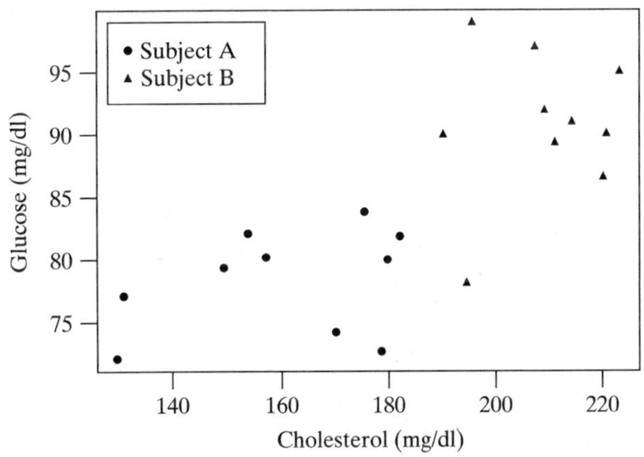

Figure 12.6.5 Twenty observations of X = serum cholesterol and Y = serum glucose in humans

Example 12.6.3 **Growth of Beef Steers** Figure 12.6.6 shows 20 pairs of measurements on the weight (Y) of beef steers at various times (X) during a feeding trial. The data represent four animals, each weighed at five different times; observations on the same animal are joined by lines in the figure. An ordinary regression analysis on the 20 data points would ignore the information carried in the lines and would yield inflated SEs and weak tests. Similarly, an ordinary scatterplot (without the lines) would be an inadequate representation of the data.[23] ■

In Example 12.6.2, ignoring the dependency in the observations would lead to *overinterpretation* of the data—that is, concluding that a relationship exists when there is actually very little evidence for it. By contrast, ignoring the dependency in

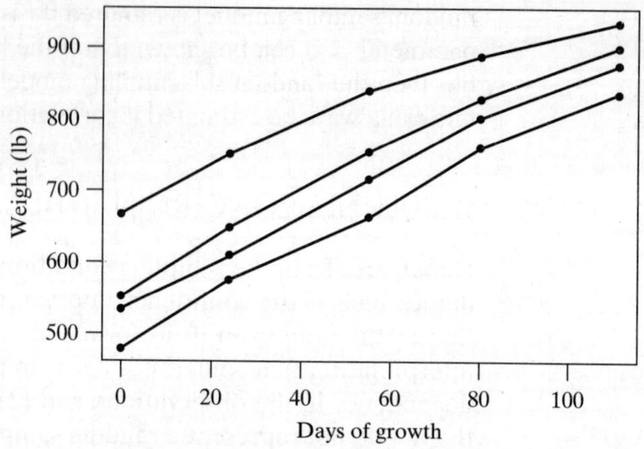

Figure 12.6.6 Twenty observations of X = days and Y = weight in steers. Data for individual animals are joined by lines

Example 12.6.3 would lead to *underinterpretation* of the data—that is, insufficiently extracting the "signal" from the "noise."

In interpreting the correlation coefficient r, one should recognize that r is influenced by the degree of spread in the values of X. If the regression quantities b_0, b_1, and s_e are unchanged, *more spread in the X values leads to a stronger correlation (larger magnitude of r)*. The following example shows how this happens.

Example 12.6.4 Figure 12.6.7 shows fictitious data that illustrate how r can be affected by the distribution of X. The data points in parts (a) and (b) have been plotted together in part (c). The regression line is nearly the same in all three scatterplots, but notice that X and Y appear more highly correlated in (c) than in either (a) or (b). The contrasting appearance of the scatterplots is reflected in the correlation coefficients; in fact, $r = 0.60$ for (a), $r = 0.58$ for (b), but $r = 0.85$ for (c). ■

The fact that r depends on the distribution of X does not mean that r is invalid as a descriptive statistic. But it does mean that, when the values of X cannot be viewed as a random sample, r must be interpreted cautiously. For instance, suppose two experimenters conduct separate studies of response (Y) to various doses (X) of a drug. Each of them could calculate r as a description of her or his own data, but they should *not* expect to obtain *similar* values of r unless they both use the same choice of doses (X values). By contrast, they might reasonably expect to obtain

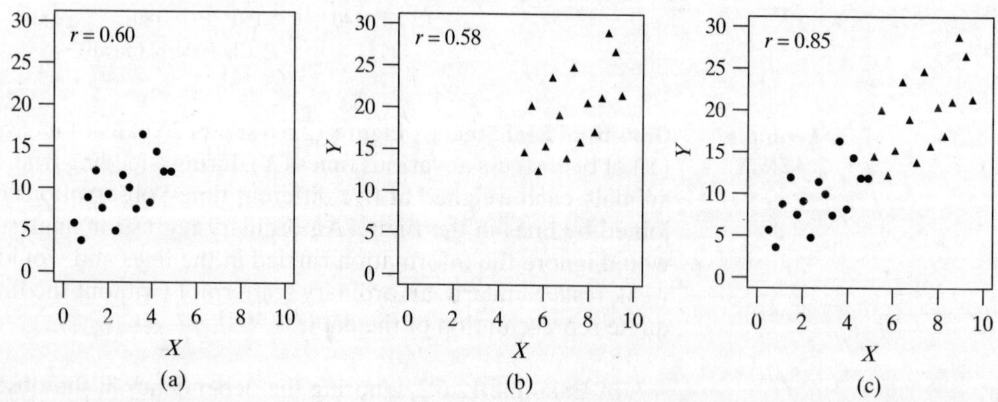

Figure 12.6.7 Dependence of r on the distribution of X. The data of (a) and (b) are plotted together in (c)

similar regression lines and similar residual standard deviations, regardless of their choice of X values, as long as the dose–response relationship remains the same throughout the range of doses used.

Labeling X and Y If the bivariate random sampling model is applicable, then the investigator is free to decide which variable to label X and which to label Y. Of course, for calculation of r the labeling does not matter. For regression calculations, the decision depends on the purpose of the analysis. The regression of Y on X yields (within the linear model) estimates of $\mu_{Y|X}$—that is, the population mean Y value for fixed X. Similarly, the regression of X on Y is aimed at estimating $\mu_{X|Y}$—that is, the mean X value for fixed Y. These approaches do not lead to the same regression line because they are directed at answering different questions. An intuitive example follows.

Example 12.6.5

Height and Weight of Young Men For the population of young men described in Example 12.4.4, the mean weight of young men 76″ (6′4″) tall is 178 lb. Now consider this question: What would be the mean height of young men who weigh 178 lb? There is no reason that the answer should be 76″. Intuition suggests that the answer should be less than 76″—and in fact it is about 71″. ■

GUIDELINES CONCERNING THE LINEAR MODEL AND NORMALITY CONDITION

The test and confidence interval for β_1 are based on the linear model and the condition of normality. The interpretation of these inferences can be seriously degraded if the linearity condition is not met; after all, we have seen earlier in this section that even the descriptive usefulness of regression is reduced if curvilinearity or outliers are present.

In addition to linearity, the linear model specifies that σ_ε is the same for all the observations. A common pattern of departure from this condition is a trend for larger means to be associated with larger SDs. Mild nonconstancy of the SDs does not seriously affect the interpretation of b_0, b_1, SE_{b_1}, and r (although it does invalidate the interpretation of s_e as a pooled estimate of a common SD).

RESIDUAL PLOTS

Formal statistical tests for curvilinearity, unequal standard deviations, nonnormality, and outliers are beyond the scope of this book. However, the single most useful instrument for detecting these features is the human eye, aided by scatterplots. For instance, notice how easily the eye detects the mild curvilinearity in Figure 12.6.2 and the outlier in Figure 12.6.3(b). Notice also in Figure 12.6.3(b) that examination of the marginal distributions of X and Y separately would not have revealed the outlier.

In addition to scatterplots of Y versus X, it is often useful to look at various displays of the residuals. A scatterplot of each residual $(y_i - \hat{y}_i)$ against $\hat{y}_i$ is called a **residual plot**. Residual plots are very useful for detecting curvature; they can also reveal trends in the conditional standard deviation. Figure 12.6.8 shows the data from Figure 12.6.2 together with a residual plot of those data.

A residual plot shows the data after the linear trend has been removed, which makes it easier to see nonlinear patterns in the data. The curvature in Figure 12.6.8(a) is apparent, but it is much more visible in the residual plot of Figure 12.6.8(b).

If the linear model holds, with no outliers, then the fitted regression line captures the trend in the data, leaving a random pattern in the residual plot. Thus, *we hope to*

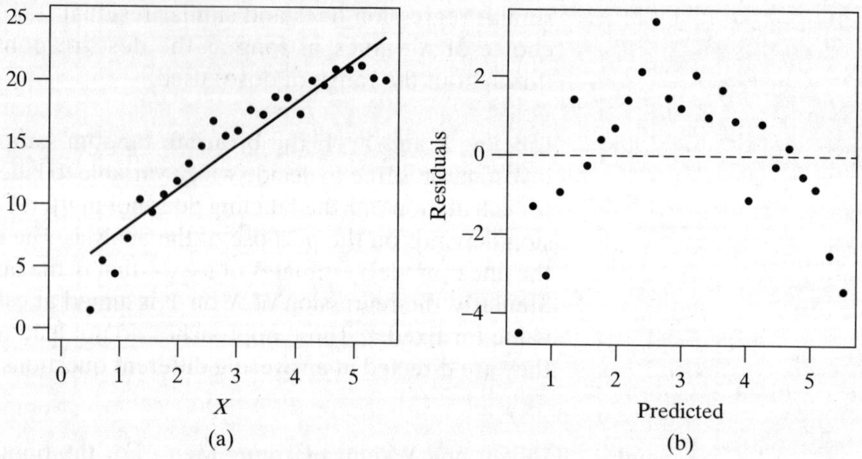

Figure 12.6.8 (a) Data displaying mild curvilinearity with linear regression line; (b) a residual plot of the data

see no striking pattern in a residual plot. For example, Figure 12.6.9 shows a residual plot of the snake data of Example 12.2.1. The lack of unusual features in this plot supports the use of a regression model for these data.

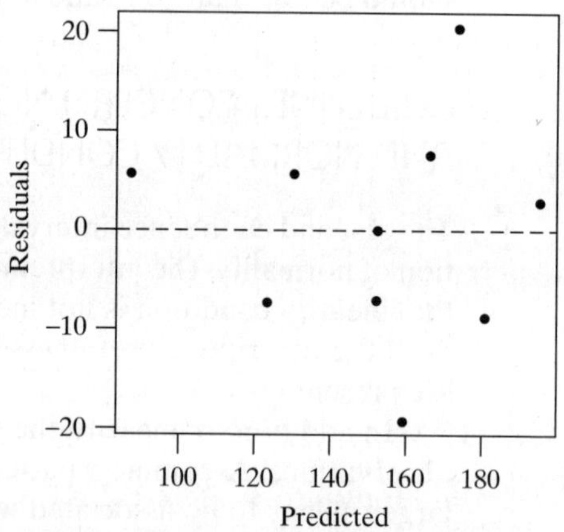

Figure 12.6.9 Residual plot of the snake data

If the condition of normality is met, then the distribution of the residuals should look roughly like a normal distribution.* A normal quantile plot of the residuals provides a useful check of the normality condition. The normal quantile plot of the snake data in Figure 12.6.10 is fairly linear, which supports the use of the t test and the confidence interval presented in Section 12.5.

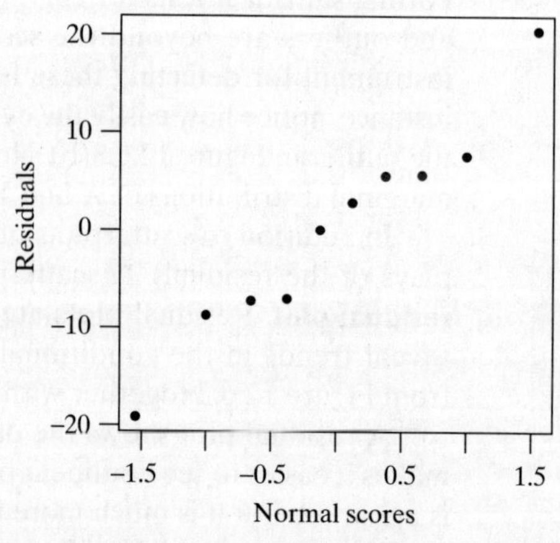

Figure 12.6.10 Normal quantile plot of the snake data

*This is the basis for the 68% and 95% interpretations of s_e given in Section 12.3.

THE USE OF TRANSFORMATIONS

If the conditions of linearity, constancy of standard deviation, and normality are not met, a remedy that is sometimes useful is to transform the scale of measurement of either Y, or X, or both. The following example illustrates the use of a logarithmic transformation.

Example 12.6.6

Growth of Soybeans A botanist placed 60 one-week-old soybean seedlings in individual pots. After 12 days of growth, she harvested, dried, and weighed 12 of the young soybean plants. She weighed another 12 plants after 23 days of growth, and groups of 12 plants each after 27 days, 31 days, and 34 days. Figure 12.6.11 shows the 60 plant weights plotted against days of growth; a smooth curve connects the group means. It is easy to see from Figure 12.6.11 that the relationship between mean plant weight and time is curvilinear rather than linear and that the conditional standard deviation is not constant but is strongly increasing.[24]

Figure 12.6.12 shows the logarithms (base 10) of the plant weights, plotted against days of growth together with the regression line. Notice that the logarithmic

Figure 12.6.11 Weight of soybean plants plotted against days of growth

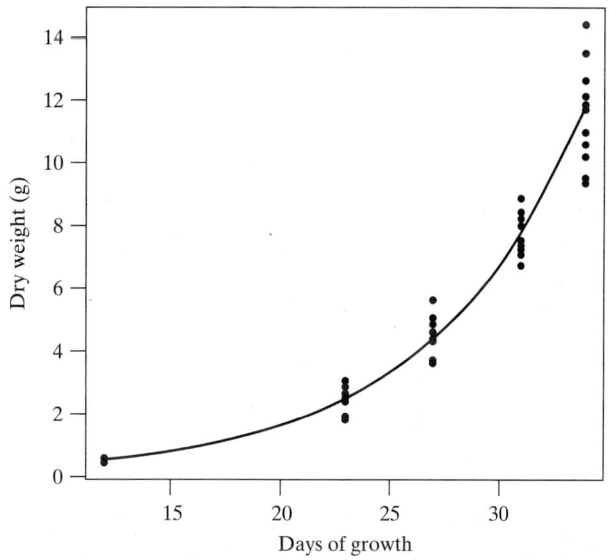

Figure 12.6.12 Log (weight) of soybean plants plotted against days of growth

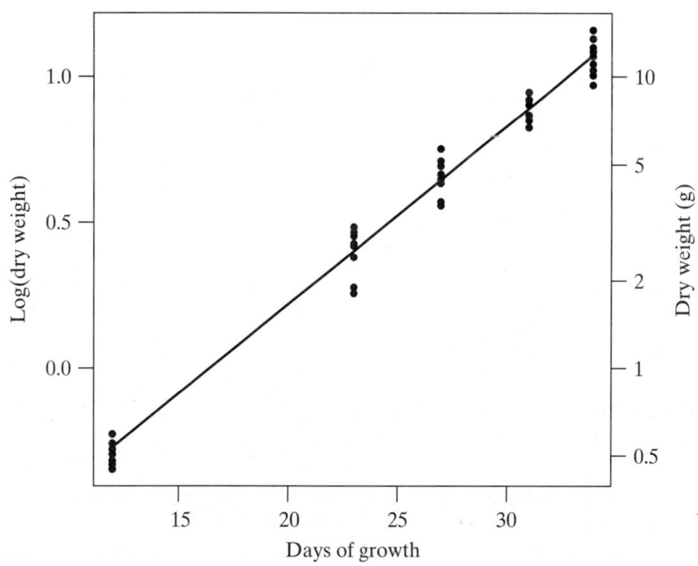

transformation has simultaneously straightened the curve and more nearly equalized the standard deviations. It would not be unreasonable to assume that the linear model is valid for the variables $Y = \log(\text{dry weight})$ and $X =$ days of growth. Table 12.6.1 shows the means and standard deviations before and after the logarithmic transformation. Note especially the effect of the transformation on the equality of the SDs. ■

Table 12.6.1 Summary of soybean growth data in original scale and after log transformation

		Dry weight (gm)		Log(dry weight)	
Days of growth	Number of plants	Mean	SD	Mean	SD
12	12	0.50	0.06	−0.31	0.055
23	12	2.63	0.37	0.42	0.062
27	12	4.67	0.70	0.67	0.066
31	12	7.57	1.19	0.87	0.069
34	12	11.20	1.62	1.04	0.064

Exercises 12.6.1–12.6.9

12.6.1 In a metabolic study, four male swine were tested three times: when they weighed 30 kg, again when they weighed 60 kg, and again when they weighed 90 kg. During each test, the experimenter analyzed feed intake and fecal and urinary output for 15 days, and from these data calculated the nitrogen balance, which is defined as the amount of nitrogen incorporated into body tissue per day. The results are shown in the accompanying table.[25]

	Nitrogen balance (gm/day)		
Animal number	Body weight 30 kg	60 kg	90 kg
1	15.8	21.3	16.5
2	16.4	20.8	18.2
3	17.3	23.8	17.8
4	16.4	22.1	17.5
Mean	16.48	22.00	17.50

Suppose these data are analyzed by linear regression. With $X =$ body weight and $Y =$ nitrogen balance, preliminary calculations yield $\bar{x} = 60$ and $\bar{y} = 18.7$. The slope is $b_1 = 0.017$, with standard error $SE_{b_1} = 0.032$. The t statistic is $t_s = 0.53$, which is not significant at any reasonable significance level. According to this analysis, there is insufficient evidence to conclude that nitrogen balance depends on body weight under the conditions of this study.

The above analysis is flawed in two ways. What are they? (*Hint*: Look for ways in which the conditions for inference are not met. There may be several minor departures from the conditions, but you are asked to find two major ones. No calculation is required.)

12.6.2 For measuring the digestibility of forage plants, two methods can be used: The plant material can be fermented with digestive fluids in a glass container, or it can be fed to an animal. In either case, digestibility is expressed as the percentage of total dry matter that is digested. Two investigators conducted separate studies to compare the methods by submitting various types of forage to both methods and comparing the results. Investigator A reported a correlation of $r = 0.8$ between the digestibility values obtained by the two methods, and investigator B reported $r = 0.3$. The apparent discrepancy between these results was resolved when it was noted that one of the investigators had tested only varieties of canary grass (whose digestibilities ranged from 56% to 65%), whereas the other investigator had used a much wider spectrum of plants, with digestibilities ranging from 35% for corn stalks to 72% for timothy hay.[26]

Which investigator (A or B) used only canary grass? How does the different choice of test material explain the discrepancy between the correlation coefficients?

12.6.3 Refer to the energy expenditure data of Exercise 12.2.9. Each subject's expenditure value (Y) is the average of two measurements made on different occasions. It might be proposed that it would be better to use the two measurements as separate data points, thus yielding 14 observations rather than 7. If this proposed approach were used, one of the conditions for inference would be highly doubtful. Which one, and why?

12.6.4 Refer to the fungus growth data of Exercise 12.2.7. In that exercise the investigator found $r = -0.98754$. Suppose a second investigator were to replicate the experiment, using concentrations of 0, 2, 4, 6, 8, and 10 mg, with two petri dishes at each concentration. Would you predict that the value of r calculated by this second investigator would be about the same as that found in Exercise 12.2.7, smaller in magnitude, or larger in magnitude? Explain.

12.6.5 In the following scatterplot of the Ca pump data of Exercise 12.2.10, one of the points is marked with an "×." In addition, there are two regression lines on the plot: The solid line includes all of the data and the dashed line omits the point marked "×."

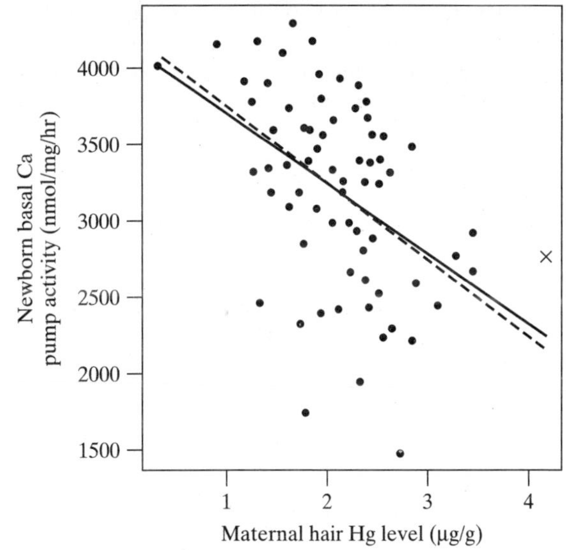

(a) Would we consider the point marked "×" an outlier? Explain.

(b) Would we consider the point marked "×" a leverage point? Explain.

(c) Noting the very small change in the slopes of the dashed and solid lines, would we consider the point marked "×" an influential observation? Explain.

12.6.6 The following three residual plots, (i), (ii), and (iii), were generated after fitting regression lines to the following three scatterplots, (a), (b), and (c). Which residual plot goes with which scatterplot? How do you know?

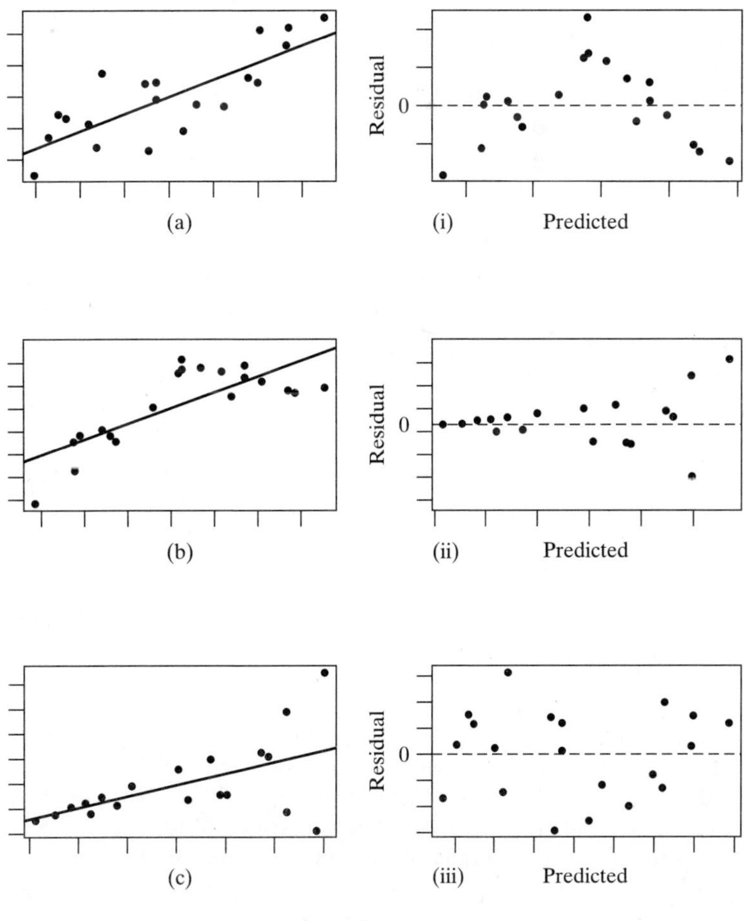

12.6.7 The following two residual plots, (i), and (ii), were generated after fitting regression lines to the two scatterplots (a) and (b). Which residual plot goes with which scatterplot? How do you know?

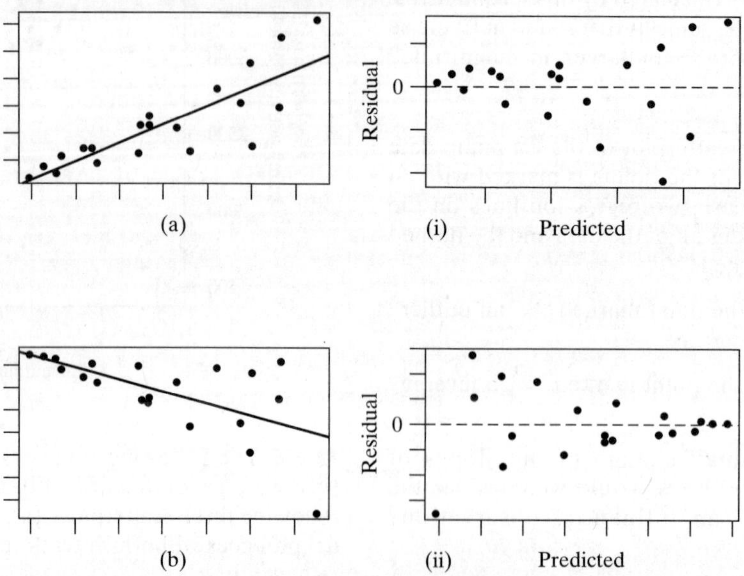

12.6.8 Sketch the residual plot that would be produced by fitting a regression line to the following scatterplot. One of the points is plotted with an "×." Indicate this point on the residual plot.

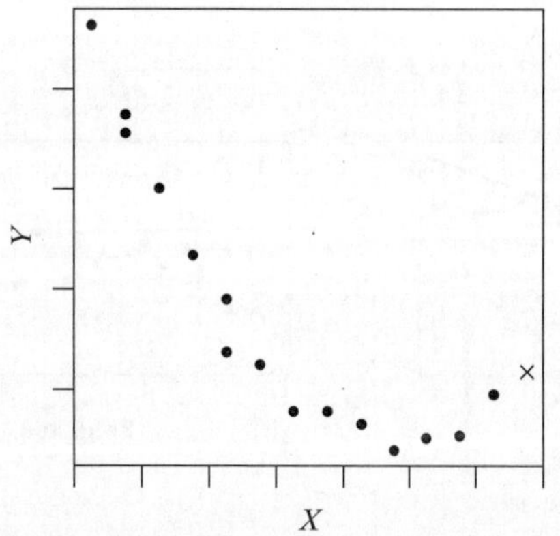

12.6.9 (Computer exercise) Researchers measured the diameters of 20 trees in a central Amazon rain forest and used ^{14}C-dating to determine the ages of these trees. The data are given in the following table.[27] Consider the use of diameter, X, as a predictor of age, Y.

Diameter (cm)	Age (yr)	Diameter (cm)	Age (yr)
180	1372	115	512
120	1167	140	512
100	895	180	455
225	842	112	352
140	722	100	352
142	657	118	249
139	582	82	249
150	562	130	227
110	562	97	227
150	552	110	172

(a) Make a scatterplot of Y = age versus X = diameter and fit a regression line to the data.

(b) Make a residual plot from the regression in part (a). Then make a normal quantile plot of the residuals. How do these plots call into question the use of a linear model and regression inference procedures?

(c) Take the logarithm of each value of age. Make a scatterplot of $Y = \log(\text{age})$ versus X = diameter and fit a regression line to the data.

(d) Make a residual plot from the regression in part (c). Next, make a normal quantile plot of the residuals. Based on these plots, does a regression model using a log scale, from part (c), seem appropriate?

12.7 Precision in Prediction (Optional)

In Section 12.4 we learned that one very practical use of regression is prediction. In this section we shall distinguish between the prediction of the *mean* Y value for a particular X value and the prediction of a *single* Y value for a particular X value. In particular, we will compare the precisions of these two very different types of predictions.

CONFIDENCE AND PREDICTION INTERVALS

In Example 12.4.6 we used a regression line to make a prediction: $\hat{y} = 11.94 - 0.22x$. Using this line again we could predict the *mean* dissolved oxygen level when the water temperature in Dairy Creek is 18 °C to be $\hat{y} = 11.94 - 0.22(18) = 7.98$ mg/L. What if instead of estimating the mean dissolved oxygen level for all days with this water temperature we wanted to predict *the* dissolved oxygen level on a particular day when the water temperature is 18 °C? Our estimate would still be the same, $\hat{y} = 7.98$ mg/L. That is, whether we are estimating the mean Y value or a single Y value for a particular value of X, we use the regression line in the same manner. However, the precisions of these estimates are very different.

Predicting a single Y value is much less precise than predicting the mean Y value because in addition to the uncertainty in the regression line (e.g., uncertainty in our estimates of the slope and intercept of the line), there is also uncertainty due to the inherent variability in Y values that have the same value of X. For example, there is variability among the dissolved oxygen levels for days when the water temperature is 18 °C (in fact, we estimate this variability to be s_e). The two graphs in Figure 12.7.1 illustrate the differences in our prediction precisions for the two types of estimates.

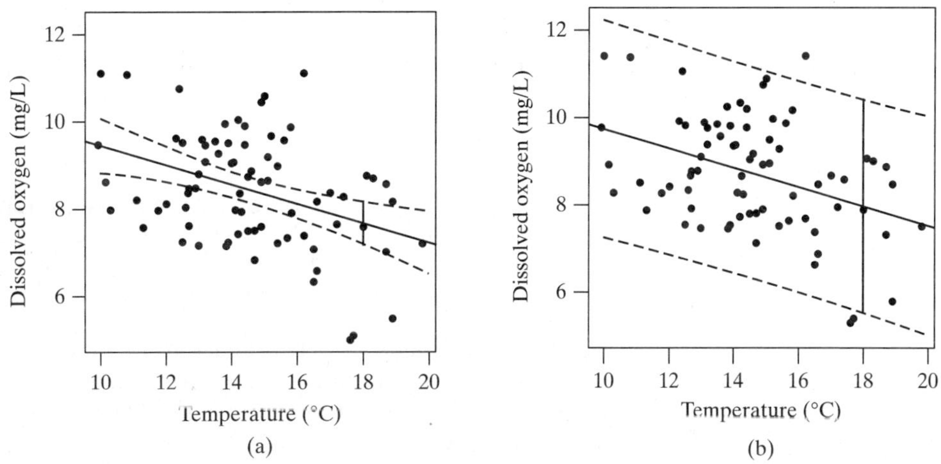

Figure 12.7.1 95% confidence and prediction bands for dissolved oxygen levels. Plot (a) shows a 95% confidence band for the predicted mean dissolved oxygen level and the 95% confidence interval for the predicted mean dissolved oxygen when the water temperature is 18 °C. Plot (b) shows a 95% prediction band for predicted dissolved oxygen levels and the 95% prediction interval for the predicted dissolved oxygen level when the water temperature is 18 °C

Figure 12.7.1 (a) displays a band representing all 95% confidence intervals for predicting mean dissolved oxygen levels as well as the specific interval for $X = 18$ °C marked by the vertical line. The confidence band reflects the uncertainty associated with estimating the slope and intercept of the regression line. Notice that the intervals are narrower (more precise) for water temperatures near the center of the data set and much wider near the extreme X values. We are 95% confident that the population regression line $\beta_0 + \beta_1 x$ lies within this band. The widening of the intervals on the end is a reflection of our uncertainty in our estimate of the slope of the regression line. The width of the band in the middle expresses our uncertainty of the overall height of the regression line (vis-à-vis b_0).

In contrast, Figure 12.7.1 (b) displays a band representing all 95% prediction intervals for predicting individual dissolved oxygen levels. The specific prediction interval for $X = 18$ is marked by the vertical line. Note how much wider this band is in (b) than in (a). Example 12.7.1 illustrates the use of confidence and prediction intervals for prediction in regression.

Example 12.7.1 **Dissolved Oxygen** Figure 12.7.1 shows that for days when the water temperature is 18 °C, the 95% confidence interval for the mean dissolved oxygen concentration is about 7.5 to 8.5 mg/L. In other words, we are 95% confident that the mean dissolved oxygen level on days when the water temperature is 18 °C is between 7.5 and 8.5 mg/L. On the other hand, using the prediction interval we estimate that 95% of days with water temperature of 18 °C will have dissolved oxygen levels between 5.5 and 10.4 mg/L. ■

Recall that the regression line can be interpreted as a "line of averages," and individuals will necessarily fall from this average. These graphs show us that we are much less certain about saying, "when the water temperature is X °C is there will be Y amount of dissolved oxygen" than we are about saying "when the water temperature is X °C there will, *on average,* be Y amount of dissolved oxygen in the water."

COMPUTING THE INTERVALS

Consider predicting $\mu_{Y|X=x^*}$ or $Y|X = x^*$; that is, predicting the mean or actual Y value when $X = x^*$. A 95% confidence interval for $\mu_{Y|X=x^*}$ is given by

$$\hat{y} \pm t_{0.025}\, s_e \sqrt{\frac{1}{n} + \frac{(x^* - \bar{x})^2}{(n-1)s_x^2}}$$

and a 95% prediction interval for $Y|X = x^*$ is given by

$$\hat{y} \pm t_{0.025}\, s_e \sqrt{1 + \frac{1}{n} + \frac{(x^* - \bar{x})^2}{(n-1)s_x^2}}$$

with the critical value $t_{0.025}$ determined from Student's t distribution with df $= n - 2$.

While these two formulas are very similar, note the extra "1" under the radical sign in the prediction interval formula. This "1" factors in the added variability associated with trying to make a prediction for an individual rather than for a population mean.

As we have seen in Figure 12.7.1, both confidence and prediction intervals are wider when we are making predictions far from the center of our data. Both formulas account for this additional uncertainty through the term $\frac{(x^* - \bar{x})^2}{(n-1)s_x^2}$. This term will be large when x^* is far from $\bar{x}$ and thus increases the width of the interval. Note that when $x^* = \bar{x}$ the confidence interval formula can reduce to a very familiar form: $\hat{y} \pm t_{0.025}\left(\frac{s_e}{\sqrt{n}}\right)$, which looks very similar to the formula for a confidence interval for a population mean from Chapter 6.

Most statistical software can compute and display confidence and prediction bands quite easily.

Exercises 12.7.1–12.7.4

12.7.1 In a study of heat stress on cows, researchers measured the rectal temperature (°C) for 1,280 lactating cows (Y) and relative humidity (%) (X).[28] The following graph displays the data and regression line (solid line). There are two other pairs of lines on this graph: dashed and dotted. One pair of lines shows the 95% confidence band, and the other shows the 95% prediction band.

(a) Which pair of lines shows the confidence band? What does this band tell us?

(b) Which pair of lines shows us the prediction band? What does this band tell us?

(c) If the data set were smaller, describe what would happen to these bands. Would we have narrower or wider bands around the regression line?

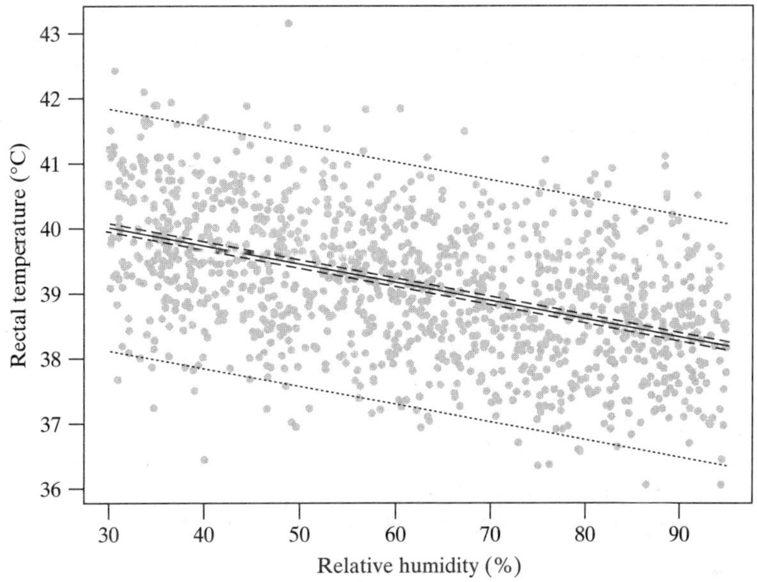

12.7.2 (*Continuation of 12.7.1*) Suppose 5,000 additional cows were included in the sample and a similar plot of the data, regression line, confidence and prediction bands were made of this new larger sample. Would the prediction band get much narrower? Explain your reasoning.

12.7.3 The following graph displays the regression line and 95% confidence and prediction bands for the peak respiration flow data from Exercise 12.3.10.

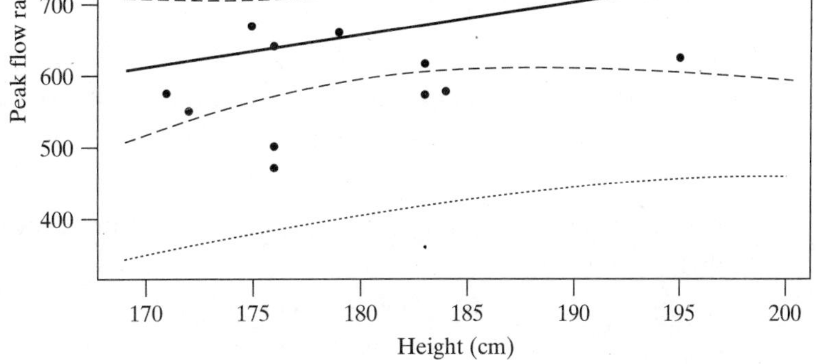

(a) Using the graph to justify your answer, would it be very surprising to find a 195-cm-tall individual with a peak flow rate above 900 l/min?

(b) Using the graph to justify your answer, would it be surprising to find a large group of 195-cm-tall individuals to have a mean peak flow rate above 900 l/min?

12.7.4 Biologists took a sample of 20 male toads and measured body length (*snout–vent length*, in mm) of each of them. They also recorded how deep the pitch was of each toad's croak (*call*, measured in Hz). Larger toads had deeper pitched croaks, as the scatterplot below shows. The graph shows a regression line that can be used to predict the size of a toad based on the pitch of its call. There are two other pairs of lines on this graph: dashed and dotted.[29]

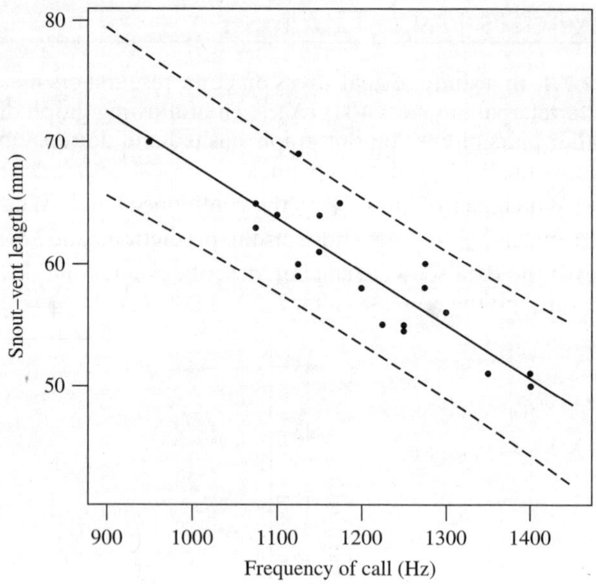

(a) Which pair of lines (dotted or dashed) shows the 95% confidence band, and which pair of lines shows the 95% prediction band?

(b) Suppose that the call of a toad is heard and it measures 1,300 Hz. Use the graph to estimate an interval that you can be 95% sure includes the snout–vent length of the toad.

12.8 Perspective

To put the methods of Chapter 12 in perspective, we will discuss their relationship to methods described in earlier chapters, and to methods that might be included in a second statistics course. We begin by relating regression to the methods of Chapters 7 and 11.

REGRESSION AND THE t TEST

When there are several Y values for each of two values of X, one could analyze the data with a two-sample t test or with a regression analysis. Each approach uses the data to estimate the conditional mean of Y for each fixed X; these parameters are estimated by the fitted line $b_0 + b_1 x$ in the regression approach and by the individual sample means $\bar{Y}$ in the t test approach. To test the null hypothesis of no dependence of Y on X, each approach translates the null hypothesis into its own terms. The following example illustrates the approaches.

Example 12.8.1 **Toluene and the Brain** In Chapter 7 we analyzed data on norepinephrine (NE) concentrations in the brains of six rats exposed to toluene and of five control rats. The data are reproduced in Table 12.8.1.

In Chapter 7 the null hypothesis

$$H_0: \mu_1 - \mu_2 = 0$$

was tested using the (unpooled) two-sample t test. The test statistic was

$$t_s = \frac{(540.83 - 444.20) - 0}{41.195} = 2.346$$

These data could be analyzed using a pooled t test (or, equivalently, with ANOVA). The pooled variance is

$$s^2_{\text{pooled}} = \frac{(6 - 1)66.12^2 + (5 - 1)69.64^2}{(6 + 5 - 2)} = 4584.24 = 67.71^2$$

Table 12.8.1 NE concentrations (ng/gm)

	Toluene	Control
	543	535
	523	385
	431	502
	635	412
	564	387
	549	
n	6	5
$\bar{y}$	540.83	444.20
s	66.12	69.64

and the pooled SE is

$$\text{SE}_{\text{pooled}} = 67.71\sqrt{\frac{1}{6} + \frac{1}{5}} = 41.00$$

This leads to a test statistic of

$$t_s = \frac{(540.83 - 444.20) - 0}{41.00} = 2.357$$

which is not much different than the unpooled t test result.

These data can also be analyzed with a regression model. To use regression, we define an **indicator variable**—a variable that indicates group membership—as follows. Let $X = 0$ for observations in the control group, and let $X = 1$ for observations in the toluene group. Then we can present the data graphically with a scatterplot, as in Figure 12.8.1.

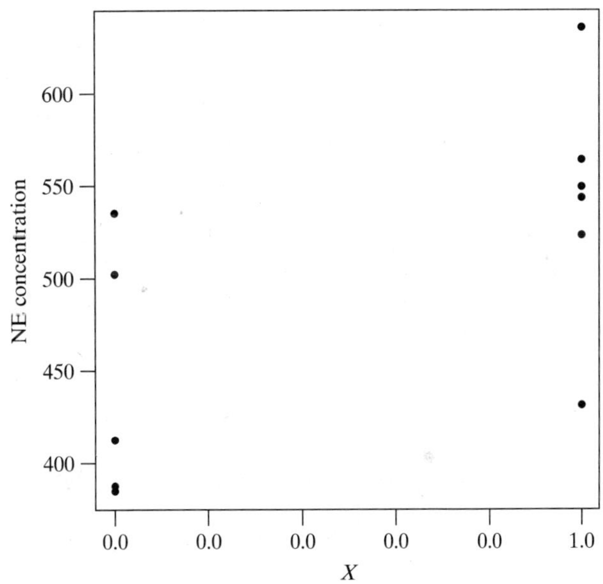

Figure 12.8.1 NE concentration data. $X = 0$ represents the control group; $X = 1$ represents the toluene group

We can analyze the data in the scatterplot with the linear model

$$Y = \beta_0 + \beta_1 X + \varepsilon$$

which states that $\mu_{Y|X} = \beta_0 + \beta_1 X$.

The linear model states that for rats in the control group, the (population) mean NE concentration is given by

$$\mu_{Y|X=0} = \beta_0 + \beta_1(0) = \beta_0$$

And, for rats in the toluene group, NE concentration is given by

$$\mu_{Y|X=1} = \beta_0 + \beta_1(1) = \beta_0 + \beta_1$$

The difference between the two group means is β_1. Thus, the null hypothesis

$$H_0\colon \mu_{Y|X=0} - \mu_{Y|X=1} = 0$$

is equivalent to the null hypothesis

$$H_0\colon \beta_1 = 0$$

The fitted regression line is $\hat{y} = 444.2 + 96.63\,x$. Note that when $X = 0$, the fitted regression line gives a value of $\hat{y} = 444.2$, which is the sample mean of the control group. When $X = 1$, the fitted regression line gives a value of $\hat{y} = 444.2 + 96.63 = 540.83$, which is the sample mean of the toluene group. That is, the sample value of the slope is equal to the change in the sample means when going from the control group ($X = 0$) to the toluene group ($X = 1$), as shown in Figure 12.8.2.

Figure 12.8.2 NE concentration data with regression line added

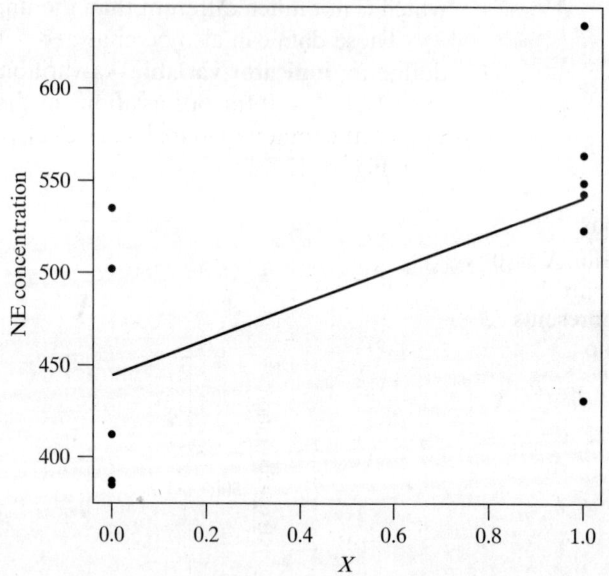

The test statistic for testing the hypothesis $H_0\colon \beta_1 = 0$ is

$$t_s = \frac{96.63}{41.0} = 2.36$$

This is identical to the previous pooled two-sample t test statistic. (Note that the regression analysis assumes that $\sigma_{Y|X} = \sigma_\varepsilon$ is constant. Thus, regression is similar to the pooled t test, rather than the unpooled t test.) The following computer output shows the coefficients for the fitted regression line as well as the t statistic.

```
The regression equation is
NE = 444 + 96.6X

Predictor      Coef  SE Coef      T      P
Constant     444.20    30.28  14.67  0.000
X             96.63    41.00   2.36  0.043
S = 67.7049 R-Sq = 38.2% R-Sq(adj) = 31.3%

Analysis of Variance

Source          DF     SS     MS     F      P

Regression       1  25467  25467  5.56  0.043
Residual Error   9  41256   4584
Total           10  66723
```

The following example compares the regression approach and the two-sample approach to a data set for which (unlike Example 12.8.1) X varies within as well as between the samples.

Example 12.8.2

Blood Pressure and Platelet Calcium In Example 12.5.3 we described blood pressure (X) and platelet calcium (Y) measurements on 38 subjects. Actually, the study included two groups of subjects: 38 volunteers with normal blood pressure, selected from hospital lab personnel and other nonpatients, and 45 patients with a diagnosis of high blood pressure. Table 12.8.2 summarizes the platelet calcium measurements in the two groups and Figure 12.8.3 shows the blood pressure and calcium measurements for all 83 subjects.[4]

Two ways to analyze the data are (1) as two independent samples and (2) by regression analysis. To test for a relationship between blood pressure and platelet calcium (1) a two-sample t test of H_0: $\mu_1 = \mu_2$ can be applied to Table 12.8.2; (2) a regression t test of H_0: $\beta_1 = 0$ can be applied to the data in Figure 12.8.3. The two-sample t statistic (unpooled) is $t_s = 11.2$ and the regression t statistic is $t_s = 20.8$. Both of these are highly significant, but the latter is more so because the regression analysis extracts more information from the data.

For these data, the regression approach is more enlightening and convincing than the two-sample approach. Figure 12.8.3 suggests that platelet calcium is correlated with blood pressure, not only between, but also within the two groups. Relevant regression analyses would include (1) testing for a correlation within each group separately (as in Examples 12.2.3 and 12.5.3); (2) testing for an overall correlation (as in the previous paragraph); (3) testing whether the regression lines in the two groups are identical (using methods not described in this book).

Formal testing aside, notice the advantage of the scatterplot as a tool for understanding the data and for communicating the results. Figure 12.8.3 provides eloquent

Table 12.8.2 Platelet calcium (nM) in two groups of subjects

	Normal blood pressure	High blood pressure
$\bar{y}$	107.9	168.2
s	16.1	31.7
n	38	45

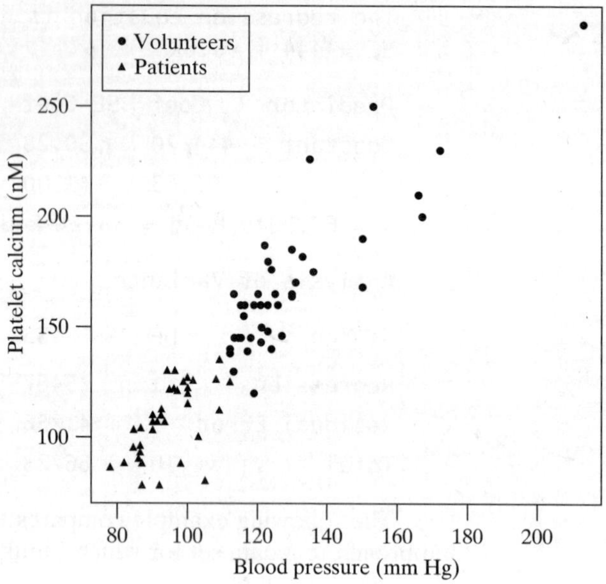

Figure 12.8.3 Blood pressure and platelet calcium for 83 subjects

testimony to the reality of the relationship between blood pressure and platelet calcium. (We emphasize once again, however, that a "real" relationship is not necessarily a causal relationship. Further, even if the relationship is causal, the data do not indicate the direction of causality—that is, whether high calcium causes high blood pressure or vice versa.*) ■

Example 12.8.2 illustrates a general principle: If quantitative information on a variable X is available, it is usually better to use that information than to ignore it.

EXTENSIONS OF LEAST SQUARES

We have seen that the classical method of fitting a straight line to data is based on the least-squares criterion. This versatile criterion can be applied to many other statistical problems. For instance, in **curvilinear regression,** the least-squares criterion is used to fit curvilinear relationships such as

$$Y = \beta_0 + \beta_1 X + \beta_2 X^2 + \varepsilon$$

Another application is **multiple regression and correlation,** in which the least-squares criterion is used to fit an equation relating Y to several X variables—X_1, X_2, and so on; for instance,

$$Y = \beta_0 + \beta_1 X_1 + \beta_2 X_2 + \varepsilon$$

The following example illustrates both curvilinear and multiple regression.

Example 12.8.3 **Serum Cholesterol and Blood Pressure** As part of a large health study, various measurements of blood pressure, blood chemistry, and physique were made on 2,599 men.[30] The researchers found a positive correlation between blood pressure and

*In fact, the authors of the study remark that "It remains possible . . . that an increased intracellular calcium concentration is a consequence rather than a cause of elevated blood pressure."

serum cholesterol ($r = 0.23$ for systolic blood pressure). But blood pressure and serum cholesterol are also related to age and physique. To untangle the relationships, the researchers used the method of least squares to fit the following equation:

$$Y = b_0 + b_1X_1 + c_1X_1^2 + b_2X_2 + b_3X_3 + b_4X_4$$

where

Y = Systolic blood pressure
X_1 = Age
X_2 = Serum cholesterol
X_3 = Blood glucose
X_4 = Ponderal index (height divided by the cube root of weight)

Note that the regression is curvilinear with respect to age (X_1) and linear in the other X variables.

By applying multiple regression and correlation analysis, the investigators determined that there is little or no correlation between blood pressure and serum cholesterol, after accounting for any relationship between blood pressure and age and ponderal index. They concluded that the observed correlation between serum cholesterol and blood pressure was an indirect consequence of the correlation of each of these with age and physique. ■

NONPARAMETRIC AND ROBUST REGRESSION AND CORRELATION

We have discussed the classical least-squares methods for regression and correlation analysis. There are also many excellent modern methods that are not based on the least-squares criterion. Some of these methods are *robust*—that is, they work well even if the conditional distributions of Y given X have long straggly tails or outliers. The nonparametric methods assume little or nothing about the form of dependence—linear or curvilinear—of Y on X, or about the form of the conditional distributions.

ANALYSIS OF COVARIANCE

Sometimes regression ideas can add greatly to the power of a data analysis, even if the relationship between X and Y is not of primary interest. The following is an example.

Example 12.8.4 **Caterpillar Head Size** Can diet affect the size of a caterpillar's head? Such an effect is plausible, because a caterpillar's chewing muscles occupy a large part of the head. To study the effect of diet, a biologist raised caterpillars (*Pseudaletia unipuncta*) on three different diets: diet 1, an artificial soft diet; diet 2, soft grasses; and diet 3, hard grasses. He measured the weight of the head and of the entire body in the final stage of larval development. The results are shown in Figure 12.8.4, where Y = ln(head weight) is plotted against X = ln(body weight), with different symbols for the three diets.[31] Note that the effect of diet is striking; there is virtually no overlap between the three groups of points. But if we were to ignore X and consider Y only, as displayed in Figure 12.8.5, the effect of diet would be much less pronounced. ■

Example 12.8.4 shows how comparison of several groups with respect to a variable Y can be strengthened by using information on an auxiliary variable X that is correlated

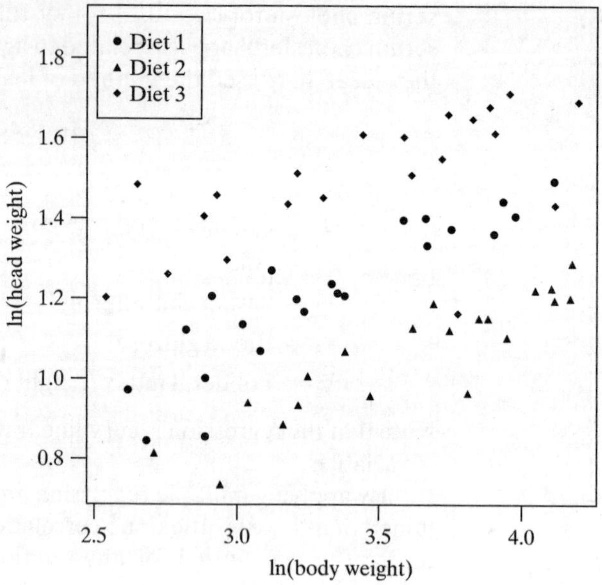

Figure 12.8.4 Head weight versus body weight (on logarithmic scales) for caterpillars on three different diets

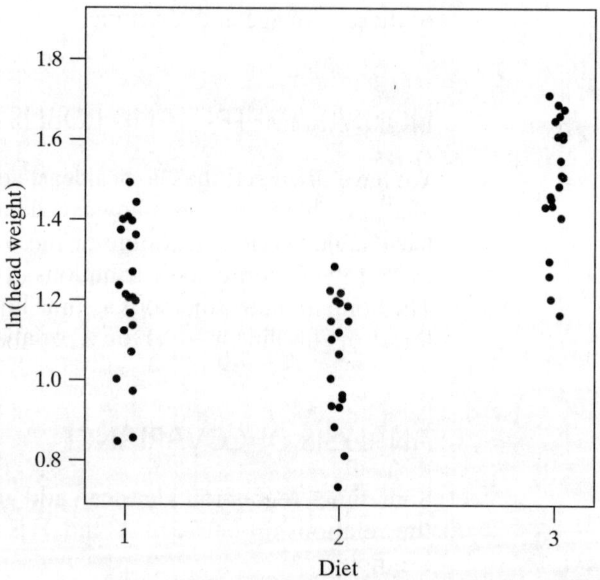

Figure 12.8.5 Head weight (on a logarithmic scale) for caterpillars on three different diets

with Y. A classical method of statistical analysis for such data is **analysis of covariance,** which proceeds by fitting regression lines to the (X, Y) data. But even without this formal technique, an investigator can often clarify the interpretation of data simply by constructing a scatterplot like Figure 12.8.4. Plotting the Y variable against X has the visual effect of removing that part of the variability in Y which is accounted for by X, causing the treatment effect to stand out more clearly against the residual background variation.

LOGISTIC REGRESSION

Regression and correlation are used to analyze the relationship between two quantitative variables, X and Y. Sometimes data arise in which a quantitative variable X is used to predict the response of a categorical variable Y. For example, we might wish to use $X =$ cholesterol level as a predictor of whether or not a person has heart disease. Here we could define a variable Y as 1 if a person has heart disease and 0 otherwise. We could then study how Y depends on X. When the response variable is dichotomous, as in this case, a technique known as **logistic regression** can be used to model the relationship. For example, logistic regression could be used to model how the probability of heart disease depends on blood pressure.

Example 12.8.5 provides a more detailed look at the use of logistic regression.

Example 12.8.5

Esophageal Cancer Esophageal cancer is a serious and very aggressive disease. Scientists conducted a study of 31 patients with esophageal cancer in which they studied the relationship between the size of the tumor that a patient had and whether or not the cancer had spread (metastasized) to the lymph nodes of the patient. In this study the response variable is dichotomous: $Y = 1$ if the cancer had spread to the lymph nodes and $Y = 0$ if not. The predictor variable is the size (recorded as the maximum dimension, in cm) of the tumor found in the esophagus. The data are given in Table 12.8.3 and plotted in Figure 12.8.6.[32]

Table 12.8.3 Esophageal cancer data

Patient number	Tumor size (cm), X	Lymph node metastasis, Y	Patient number	Tumor size (cm), X	Lymph node metastasis, Y
1	6.5	1	17	6.2	1
2	6.3	0	18	2.0	0
3	3.8	1	19	9.0	1
4	7.5	1	20	4.0	0
5	4.5	1	21	3.0	1
6	3.5	1	22	6.0	1
7	4.0	0	23	4.0	0
8	3.7	0	24	4.0	0
9	6.3	1	25	4.0	0
10	4.2	1	26	5.0	1
11	8.0	0	27	9.0	1
12	5.2	1	28	4.5	1
13	5.0	1	29	3.0	0
14	2.5	0	30	3.0	1
15	7.0	1	31	1.7	0
16	5.3	0			

The idea of logistic regression is to model the relationship between X and Y by fitting a response curve that is always between 0 and 1. With values bound between 0 and 1, the logistic regression model can be used to estimate the probability $Y = 1$ (e.g., metastasis) for a given value of X (e.g., tumor size). Thus, unlike linear regression, in which we model Y as a linear function of X (which does not remain between 0 and 1), with logistic regression we model the relationship between X and Y as having an "S" shape, as shown in Figure 12.8.7.

One way to begin understanding the data is to form groups on the basis of size, X, and calculate for each group the proportion of the Y values that are 1's. (This is somewhat analogous to finding the graph of averages described in Section 12.3, except that here we group together data points with differing X values.) Table 12.8.4

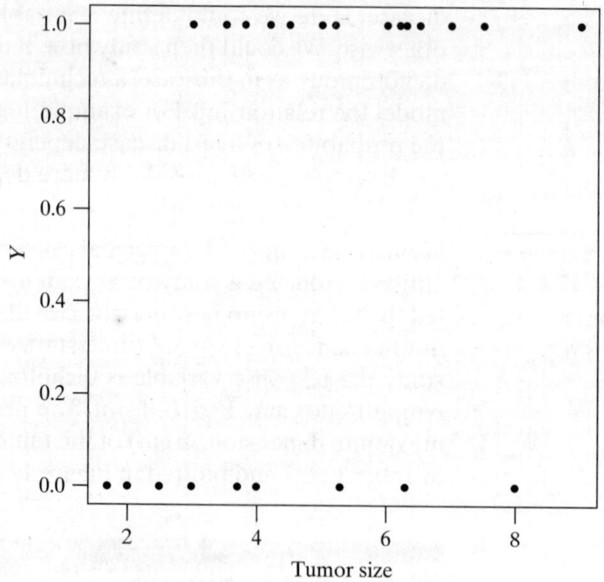

Figure 12.8.6 Lymph node metastasis, Y, as a function of tumor size, X

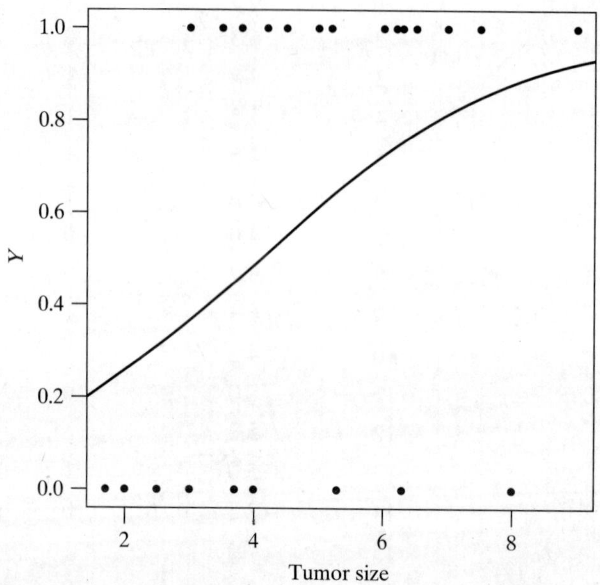

Figure 12.8.7 Lymph node metastasis, Y, as a function of tumor size, X, with smooth curve added

Table 12.8.4 Esophageal cancer data in groups

Size range	Points with $Y=1$	Points with $Y=0$	Fraction $Y=1$	Proportion $Y=1$
(1.5, 3.0]	2	4	2/6	0.33
(3.0, 4.5]	5	6	5/11	0.45
(4.5, 6.0]	4	1	4/5	0.80
(6.0, 7.5]	5	1	5/6	0.83
(7.5, 9.0]	2	1	2/3	0.67

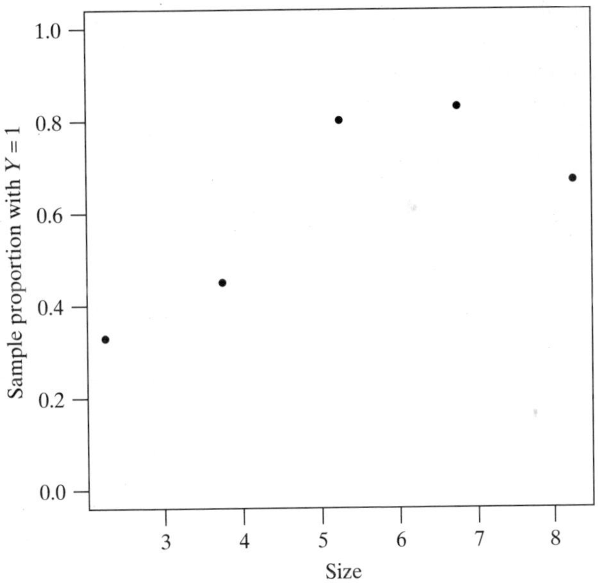

Figure 12.8.8 Sample proportion of patients with lymph node metastasis ($Y = 1$) for patients grouped by tumor size, X

provides such a summary, which is shown graphically in Figure 12.8.8. Note that the proportion of 1's (i.e., the proportion of patients for whom the cancer has metastasized) increases as tumor size increases (except for the last category of (7.5, 9], which has only three cases).

We can fit a smooth, continuous function to the data, to smooth out the percentages in the last column of Table 12.8.4. We can also impose the condition that the function be monotonically increasing, meaning that the probability of metastatis ($Y = 1$) strictly increases as tumor size increases. To do this, we use a computer to fit a **logistic response function.*** The fitted logistic response function for the esophageal cancer data is

$$\Pr\{Y = 1\} = \frac{e^{-2.086 + 0.5117 \times \text{size}}}{1 + e^{-2.086 + 0.5117 \times \text{size}}}$$

For example, suppose the size of a tumor is 4.0 cm. Then the predicted probability that the cancer has metastasized is

$$\frac{e^{-2.086 + 0.5117(4)}}{1 + e^{-2.086 + 0.5117(4)}} = \frac{e^{-0.0392}}{1 + e^{-0.0392}} = \frac{0.96156}{1 + 0.96156} = 0.49$$

On the other hand, suppose the size of a tumor is 8.0 cm. Then the predicted probability that the cancer has metastasized is

$$\frac{e^{-2.086 + 0.5117 \times 8}}{1 + e^{-2.086 + 0.5117 \times 8}} = \frac{e^{2.0076}}{1 + e^{2.0076}} = \frac{7.4454}{1 + 7.4454} = 0.88$$

We can calculate a predicted probability that $Y = 1$ for each value of X. Figure 12.8.9 shows a graph of such predictions, which have, generally speaking, an S shape. ■

The S shape of the logistic curve is easier to see if we extend the range of X, as shown in Figure 12.8.10. As X grows, the logistic curve approaches, but never exceeds, 1. Likewise, if we were to extend the curve into the region where X is less than zero

*Fitting a logistic model is quite a bit more complicated than is fitting a linear regression model. A technique known as maximum likelihood estimation is commonly used, with the help of a computer.

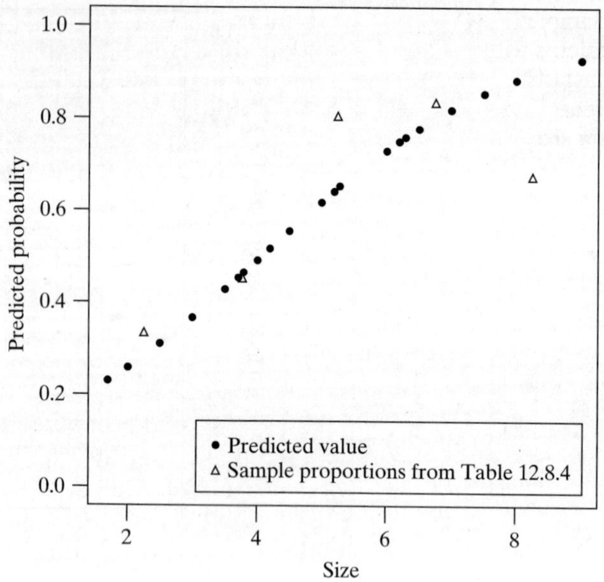

Figure 12.8.9 Predicted probability that $Y = 1$ as a function of tumor size, X with sample proportions from Table 12.8.4

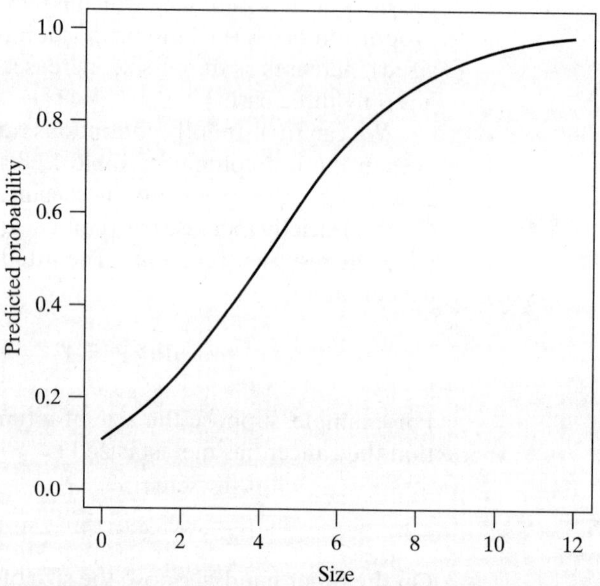

Figure 12.8.10 Logistic response function for the cancer data, shown over a larger range

we would see that as X gets smaller and smaller, the logistic curve approaches, but never drops below, 0. (Of course, in the setting of Example 12.8.5 it does not make sense to talk about tumor sizes that are negative. Thus, we only show the logistic curve for positive values of X.)

In general, if we have a logistic response function

$$\Pr\{Y = 1\} = \frac{e^{b_0 + b_1 x}}{1 + e^{b_0 + b_1 x}}$$

with b_1 positive, then as X grows, $\Pr\{Y = 1\}$ approaches one and as X gets smaller, $\Pr\{Y = 1\}$ approaches zero. Thus, unlike a linear regression model, a logistic curve stays between zero and one, which makes it appropriate for modeling a response probability.

12.9 Summary of Formulas

For convenient reference, we summarize the formulas presented in Chapter 12.

Correlation Coefficient

$$r = \frac{1}{n-1}\sum_{i=1}^{n}\left(\frac{x_i - \bar{x}}{s_x}\right)\left(\frac{y_i - \bar{y}}{s_y}\right)$$

Fact 12.3.1:

$$r^2 \approx \frac{s_y^2 - s_e^2}{s_y^2} = 1 - \frac{s_e^2}{s_y^2}$$

Fitted Regression Line

$$\hat{y} = b_0 + b_1 x$$

where

$$b_1 = r \times \left(\frac{s_y}{s_x}\right)$$

$$b_0 = \bar{y} - b_1\bar{x}$$

Residuals:

$$y_i - \hat{y}_i \quad \text{where} \quad \hat{y}_i = b_0 + b_1 x_i$$

Residual Sum of Squares:

$$\text{SS(resid)} = \sum (y_i - \hat{y}_i)^2$$

Residual Standard Deviation:

$$s_e = \sqrt{\frac{\text{SS(resid)}}{n-2}}$$

Inference

Standard Error of b_1:

$$\text{SE}_{b_1} = \frac{s_e}{s_x\sqrt{n-1}}$$

95% confidence interval for β_1:

$$b_1 \pm t_{0.025}\text{SE}_{b_1}$$

Test of H_0: $\beta_1 = 0$ or H_0: $\rho = 0$:

$$t_s = \frac{b_1}{\text{SE}_{b_1}} = r\sqrt{\frac{n-2}{1-r^2}}$$

Critical values for the test and confidence interval are determined from Student's t distribution with df $= n - 2$.

Prediction

A 95% confidence interval for $\mu_{Y|X=x^*}$ is given by

$$\hat{y} \pm t_{0.025}s_e\sqrt{\frac{1}{n} + \frac{(x^* - \bar{x})^2}{(n-1)s_x^2}}$$

A 95% prediction interval for $Y|X = x^*$ is given by

$$\hat{y} \pm t_{0.025}s_e\sqrt{1 + \frac{1}{n} + \frac{(x^* - \bar{x})^2}{(n-1)s_x^2}}$$

Critical values for intervals are determined from Student's t distribution with df $= n - 2$.

Supplementary Exercises 12.S.1–12.S.30

12.S.1 In a study of the Mormon cricket (*Anabrus simplex*), the correlation between female body weight and ovary weight was found to be $r = 0.836$. The standard deviation of the ovary weights of the crickets was 0.429 g. Assuming that the linear model is applicable, estimate the standard deviation of ovary weights of crickets whose body weight is 4 g.[33]

12.S.2 In a study of crop losses due to air pollution, plots of Blue Lake snap beans were grown in open-top field chambers, which were fumigated with various concentrations of sulfur dioxide. After a month of fumigation, the plants were harvested, and the total yield of bean pods was recorded for each chamber. The results are shown in the table.[34]

	X = Sulfur dioxide concentration (ppm)			
	0	0.06	0.12	0.30
	1.15	1.19	1.21	0.65
Y = yield (kg)	1.30	1.64	1.00	0.76
	1.57	1.13	1.11	0.69
Mean	1.34	1.32	1.11	0.70

Preliminary calculations yield the following results.

$\bar{x} = 0.12 \qquad \bar{y} = 1.117$

$s_X = 0.11724 \qquad s_Y = 0.31175$

$r = -0.8506 \qquad \text{SS(resid)} = 0.2955$

(a) Calculate the linear regression of Y on X.

(b) Plot the data and draw the regression line on your graph.

(c) Calculate s_e. What are the units of s_e?

12.S.3 Refer to Exercise 12.S.2.

(a) Estimate the intercept and slope of the regression line Y on X.

(b) Assuming that the linear model is applicable, find estimates of the mean and the standard deviation of yields of beans exposed to 0.4 ppm of sulphur dioxide.

(c) What is the unit of correlation coefficient for this set of data?

12.S.4 Refer to Exercise 12.S.2. Consider the null hypothesis that sulfur dioxide concentration has no effect on yield.

(a) Assuming that the linear model holds, formulate this as a hypothesis about the true regression line.

(b) Write a directional alternative, in symbols, that says increasing sulfur dioxide tends to decrease yield.

(c) How many degrees of freedom are there for the test that compares the hypotheses in (a) and (b)?

(d) The sample slope, b_1, is -2.262 and the standard error of the slope is 0.4421. Compute the value of the test statistic.

(e) The P-value for the test is 0.0002. If $\alpha = 0.05$, state your conclusion regarding the null hypothesis in the context of this setting.

12.S.5 Another way to analyze the data of Exercise 12.S.2 is to take each treatment mean as the observation Y; then the data would be summarized as in the accompanying table.

	Sulfur dioxide X (ppm)	Mean yield Y (kg)
	0.00	1.34
	0.06	1.32
	0.12	1.11
	0.30	0.70
Mean	0.1200	1.1175
SD	0.12961	0.29714

$r = -0.98666$

$\text{SS(resid)} = 0.007018$

(a) For the regression of mean yield on X, calculate the regression line and the residual standard deviation, and compare with the results of Exercise 12.S.2. Explain why the discrepancy is not surprising.

(b) What proportion of the variability in mean yield is explained by the linear relationship between mean yield and sulfur dioxide? Using the data in Exercise 12.S.2, what proportion of the variability in individual chamber yield is explained by the linear relationship between individual chamber yield and sulfur dioxide? Explain why the discrepancy is not surprising.

12.S.6 In a study of the tufted titmouse (*Parus bicolor*), an ecologist captured seven male birds, measured their wing lengths and other characteristics, and then marked and released them. During the ensuing winter, he repeatedly observed the marked birds as they foraged for insects and seeds on tree branches. He noted the branch diameter on each occasion, and calculated (from 50 observations) the average branch diameter for each bird. The results are shown in the table.[35]

Bird	Wing length X (mm)	Branch diameter Y (cm)
1	79.0	1.02
2	80.0	1.04
3	81.5	1.20
4	84.0	1.51
5	79.5	1.21
6	82.5	1.56
7	83.5	1.29
Mean	81.429	1.2614
SD	1.98806	0.21035
	$r = 0.80335$	
	SS(resid) $= 0.09415$	

(a) Calculate s_e and specify the units. Verify the approximate relationship between s_Y and s_e, and r.

(b) Do the data provide sufficient evidence to conclude that the diameter of the forage branches chosen by male titmice is correlated with their wing length? Test an appropriate hypothesis against a nondirectional alternative. Let $\alpha = 0.05$.

(c) The test in part (a) was based on 7 observations, but each branch diameter value was the mean of 50 observations. If we were to test the hypothesis of part (a) using the raw numbers, we would have 350 observations rather than only 7. Why would this approach not be valid?

12.S.7 (*Continuation of 12.S.6*) A scatterplot and fitted regression line of the data from Exercise 12.S.6 follow. The individual birds are labeled in the plot.

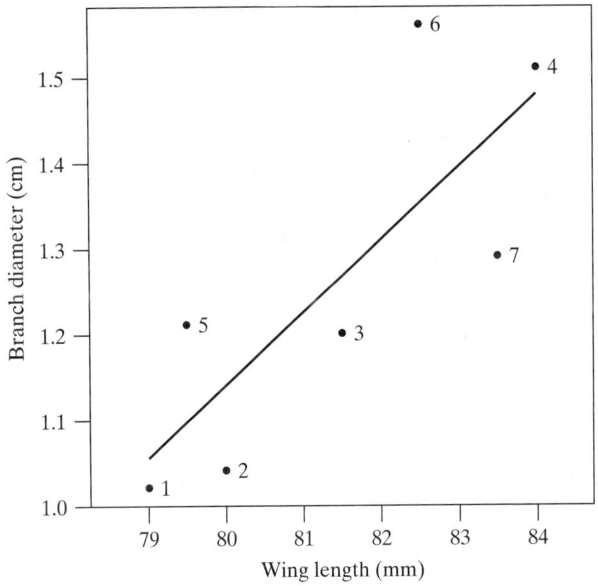

(a) Which bird/point has the largest regression residual?

(b) Which bird(s)/points(s) have the most leverage?

(c) Are there any birds/points that are influential?

(d) Invent your own bird observation of x = wing length and y = branch diameter that would be an example of a regression outlier, but not an influential observation.

(e) Invent your own bird observation of x = wing length and y = branch diameter that would be an example of a leverage point that is not influential.

(f) Invent your own bird observation of x = wing length and y = branch diameter that would an example of an influential point.

12.S.8 Exercise 12.3.7 deals with data on the relationship between body length and jumping distance of bullfrogs. A third variable that was measured in that study was the mass of each bullfrog. The following table shows these data.[16]

Bullfrog	Length X (mm)	Mass Y (g)
1	155	404
2	127	240
3	136	296
4	135	303
5	158	422
6	145	308
7	136	252
8	172	533.8
9	158	470
10	162	522.9
11	162	356
Mean	149.636	373.427
SD	14.4725	104.2922

Preliminary calculations yield the following results:

$$r = 0.90521 \quad \text{SS(resid)} = 19642$$

(a) Calculate the linear regression of Y on X.

(b) Interpret the value of the slope of the regression line, b_1, in the context of this setting.

(c) Calculate and interpret the value of s_e in the context of this setting.

(d) What percentage of the variation in length is explained by the linear relationship between mass and length?

12.S.9 (*Continuation of 12.S.8*). A residual plot and normal quantile plot from the linear regression of Y on X based on the bullfrog mass data in Exercise 12.S.8 follow.

Use these plots to comment on the required conditions for inference in regression. Is there any reason to substantially doubt that these conditions are met?

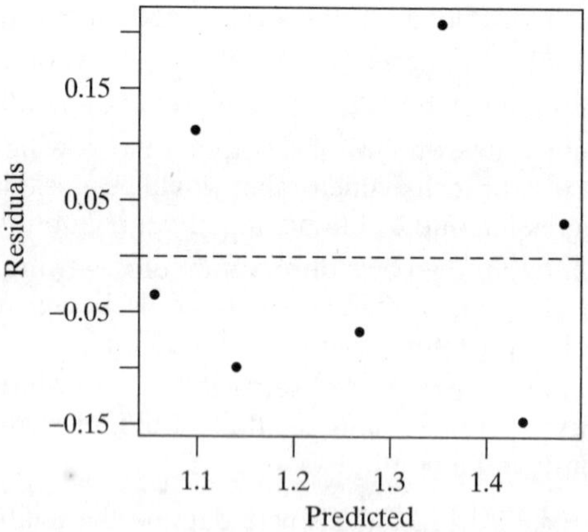

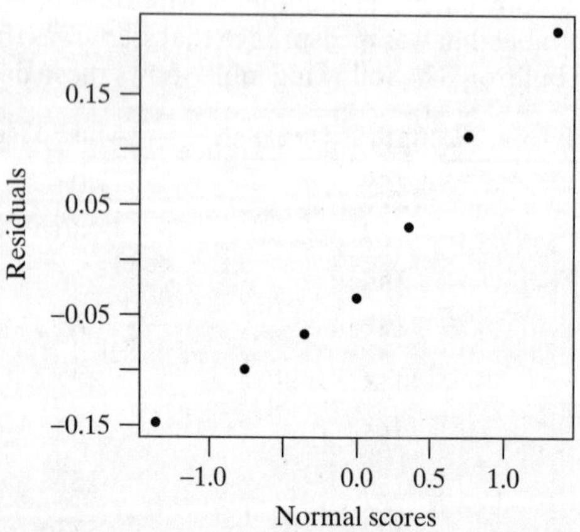

12.S.10 An exercise physiologist used skinfold measurements to estimate the total body fat, expressed as a percentage of body weight, for 19 participants in a physical fitness program. The body fat percentages and the body weights are shown in the table.[36]

Participant	Weight X (kg)	Fat Y (%)	Participant	Weight X (kg)	Fat Y (%)
1	89	28	11	57	29
2	88	27	12	68	32
3	66	24	13	69	35
4	59	23	14	59	31
5	93	29	15	62	29
6	73	25	16	59	26
7	82	29	17	56	28
8	77	25	18	66	33
9	100	30	19	72	33
10	67	23			

Actually, participants 1 to 10 are men, and participants 11 to 19 are women. A summary and graph of the data for men, women, and both sexes combined into a single sample follow.

Men (n = 10)	Women (n = 9)	Both sexes (n = 19)
$\bar{x}$ = 79.40	$\bar{x}$ = 63.1	$\bar{x}$ = 71.68
$\bar{y}$ = 26.30	$\bar{y}$ = 30.67	$\bar{y}$ = 28.37
s_X = 13.2430	s_X = 5.7975	s_X = 13.1320
s_Y = 2.6269	s_Y = 2.8723	s_Y = 3.4835
r = 0.9352	r = 0.8132	r = 0.0780

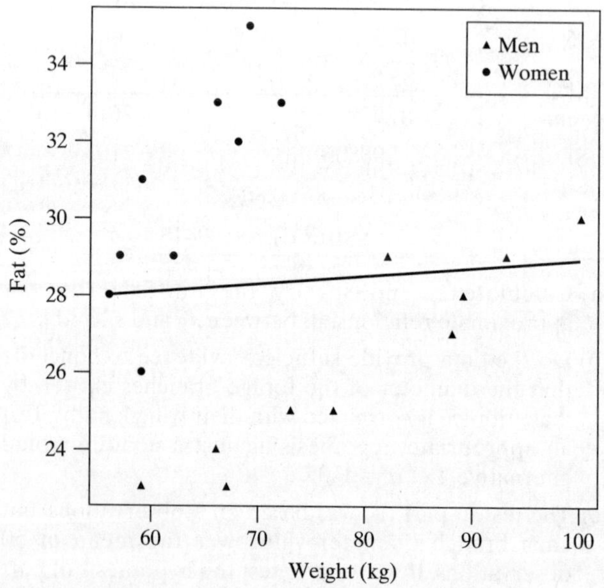

(a) Compute the regression equations for the males and females separately.

(b) The equation to the fitted regression line for both sexes combined, which is shown on the plot, is $\hat{y} = 26.88 + 0.021x$. How does the slope of this line compare to the slopes you computed in part (a)? Can you explain the discrepancy?

(c) Examine the correlation coefficients for (i) the males, (ii) the females, and (iii) both sexes combined. Do these values agree with your reasoning provided in part (b)?

12.S.11 Refer to the respiration rate data of Exercise 12.3.7. Construct a 99% confidence interval for β_1.

12.S.12 The following plot is a residual plot from fitting a regression model to some data. Make a sketch of the scatterplot of the data that led to this residual plot. (*Note*: There are two possible scatterplots—one in which b_1 is positive and one in which b_1 is negative.)

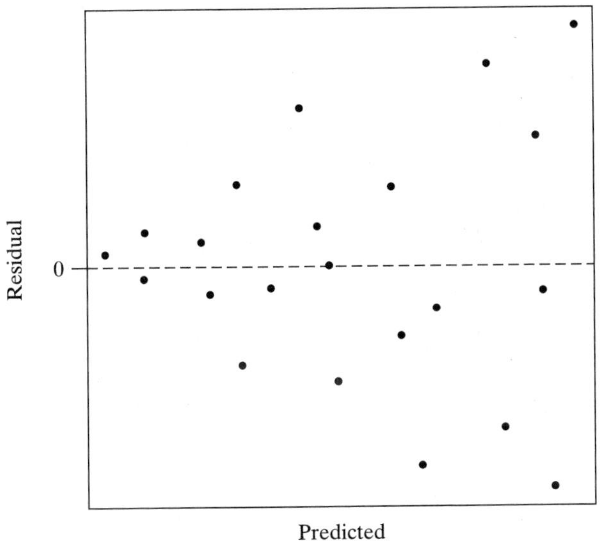

12.S.13 Biologists studied the relationship between embryonic heart rate and egg mass for 20 species of birds. They found that heart rate, Y, has a linear relationship with the logarithm of egg mass, X. The data are given in the following table.[37]

Species	Egg mass (g)	Log- (egg mass) X	Heart rate Y (beats/min)
Zebra finch	0.96	–0.018	335
Bengalese finch	1.10	0.041	404
Marsh tit	1.39	0.143	363
Bank swallow	1.42	0.152	298
Great tit	1.59	0.201	348
Varied tit	1.69	0.228	356
Tree sparrow	2.09	0.320	335
Budgerigar	2.19	0.340	314
House martin	2.25	0.352	357
Japanese bunting	2.56	0.408	370
Red-cheeked starling	4.14	0.617	358
Cockatiel	5.08	0.706	300
Brown-eared bulbul	6.40	0.806	333
Domestic pigeon	17.10	1.233	247
Fantail pigeon	19.70	1.294	267
Homing pigeon	19.80	1.297	230
Barn owl	20.10	1.303	219
Crow	20.50	1.312	297
Cattle egret	27.50	1.439	251
Lanner falcon	41.20	1.615	242
Mean	9.94	0.690	311

For these data the fitted regression equation is

$$\hat{y} = 368.06 - 82.452x$$

and

$$\text{SS(resid)} = 15748.6$$

(a) Interpret the value of the intercept of the regression line, b_0, in the context of this setting.

(b) Interpret the value of the slope of the regression line, b_1, in the context of this setting.

(c) Calculate s_e and specify the units.

(d) Interpret the value of s_e in the context of this setting.

12.S.14 An ornithologist measured the mass (g) and head length (the distance from the tip of the bill to the back of the skull, in mm) for a sample of 60 female blue jays. Here is a plot of the data and computer output:

```
Coefficients:

             Estimate Std. Error  t value    Pr(>|t|)
(Intercept) -50.6960    24.1843   -2.096    0.0404 *
Head          2.2052     0.4425    4.984 5.95e-06 ***
---
Signif. codes: 0 '***' 0.001 '**' 0.01 '*' 0.05 '.' 0.1
' ' 1 Residual standard error: 4.23 on 58 degrees of
freedom
Multiple R-squared: 0.2999,Adjusted R-squared: 0.2878
F-statistic: 24.84 on 1 and 58 DF, p-value: 5.954e-06
```

(a) Interpret the fitted slope of the regression model, in the context of this setting.

(b) The computer output says that the residual standard error (which we call the *residual standard deviation*) is 4.23. In the context of this setting, what does that mean? (Be sure to state the units for 4.23 as part of your answer.)

(c) The computer output says that R^2 is 0.2999. In the context of this setting, what does that mean?

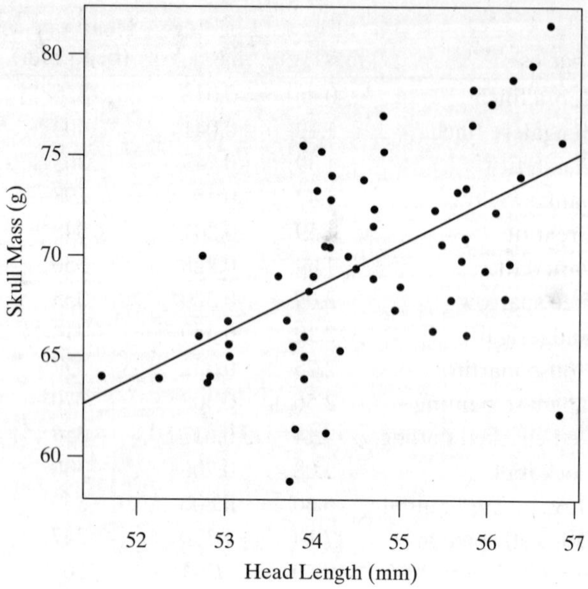

12.S.15 Consider the study and regression output in Exercise 12.S.14. The P-value given on the "Head" line is 0.00000595.

(a) What are the null hypothesis and alternative hypothesis for the regression coefficient of the 'Head' line? What is the test statistic?

(b) How would you take a decision based on the P-value? Based on the given P-value, what is your conclusion?

12.S.16 Consider the study and regression output in Exercise 12.S.14. Construct a 95% confidence interval for the population value of the slope.

12.S.17 Consider the study and regression output in Exercise 12.S.14. Use the regression model to predict the mass of a blue jay with head measurement of 56 mm.

12.S.18 Consider the study and regression output in Exercise 12.S.14. Use the regression model to predict the mean and SD of the masses for blue jays with heads that are 53 mm in length.

12.S.19 Consider the study and regression output in Exercise 12.S.14. Sadly, an ornithologist's cat brought in just the head of a blue jay. The head length was 47 mm. What would you predict the mass of the bird to have been? Is your prediction trustworthy? Explain.

12.S.20 (*Challenge question*) Consider the study and regression output in Exercise 12.S.14. Using only the numeric output to support your answer, would it be unusual for a female blue jay with a head length of 52 mm to weigh less than 53 g?

12.S.21 (*Computer exercise*) The accompanying table gives two data sets: (A) and (B). The values of X are the same for both data sets and are given only once.

	(A)	(B)		(A)	(B)
X	Y	Y	X	Y	Y
0.61	0.88	0.96	2.56	1.97	1.20
0.93	1.02	0.97	2.74	2.02	3.59
1.02	1.12	0.07	3.04	2.26	3.09
1.27	1.10	2.54	3.13	2.27	1.55
1.47	1.44	1.41	3.45	2.43	0.71
1.71	1.45	0.84	3.48	2.57	3.05
1.91	1.41	0.32	3.79	2.53	2.54
2.00	1.59	1.46	3.96	2.73	3.33
2.27	1.58	2.29	4.12	2.92	2.38
2.33	1.66	2.51	4.21	2.96	3.08

(a) Generate scatterplots of the two data sets.

(b) For each data set (i) estimate r visually and (ii) calculate r.

(c) For data set (a), multiply the values of X by 10, and multiply the values of Y by 3 and add 5. Recalculate r and compare with the value before the transformation. How is r affected by the linear transformation?

(d) Find the equations of the regression lines and verify that the regression lines for the two data sets are virtually identical (even though the correlation coefficients are very different).

(e) Draw the regression line on each scatterplot.

(f) Construct a scatterplot in which the two data sets are superimposed, using different plotting symbols for each data set.

12.S.22 (*Computer exercise*) This exercise shows the power of scatterplots to reveal features of the data that may not be apparent from the ordinary linear regression calculations. The accompanying table gives three fictitious data sets, A, B, and C. The values of X are the same for each data set, but the values of Y are different.[38]

Data set:	A	B	C
X	Y	Y	Y
10	8.04	9.14	7.46
8	6.95	8.14	6.77
13	7.58	8.74	12.74
9	8.81	8.77	7.11
11	8.33	9.26	7.81
14	9.96	8.10	8.84
6	7.24	6.13	6.08
4	4.26	3.10	5.39
12	10.84	9.13	8.15
7	4.82	7.26	6.42
5	5.68	4.74	5.73

(a) Verify that the fitted regression line is almost exactly the same for all three data sets. Are the residual standard deviations the same? Are the values of r the same?

(b) Construct a scatterplot for each of the data sets. What does each plot tell you about the appropriateness of linear regression for the data set?

(c) Plot the fitted regression line on each of the scatterplots.

12.S.23 (*Computer exercise*) In a pharmacological study, 12 rats were randomly allocated to receive an injection of amphetamine at one of two dosage levels or an injection of saline. Shown in the table is the water consumption of each animal (ml water per kg body weight) during the 24 hours following injection.[39]

Dose of amphetamine (ml/kg)		
0	1.25	2.5
122.9	118.4	134.5
162.1	124.4	65.1
184.1	169.4	99.6
154.9	105.3	89.0

(a) Calculate the regression line of water consumption on dose of amphetamine, and calculate the residual standard deviation.

(b) Construct a scatterplot of water consumption against dose.

(c) Draw the regression line on the scatterplot.

(d) Use linear regression to test the hypothesis that amphetamine has no effect on water consumption against the alternative that amphetamine tends to reduce water consumption. (Use $\alpha = 0.05$.)

(e) Use analysis of variance to test the hypothesis that amphetamine has no effect on water consumption. (Use $\alpha = 0.05$.) Compare with the result of part (d).

(f) What conditions are necessary for the validity of the test in part (d) but not for the test in part (e)?

(g) Calculate the pooled standard deviation from the ANOVA, and compare it with the residual standard deviation calculated in part (a).

12.S.24 (*Computer exercise*) Consider the Amazon tree data from Exercise 12.6.9. The researchers in this study were interested in how age, Y, is related to X = "growth rate," where growth rate is defined as diameter/age (i.e., cm of growth per year).

(a) Create the variable "growth rate" by dividing each diameter by the corresponding tree age.

(b) Make a scatterplot of Y = age versus X = growth rate and fit a regression line to the data.

(c) Make a residual plot from the regression in part (b). Then make a normal quantile plot of the residuals. How do these plots call into question the use of a linear model and regression inference procedures?

(d) Take the logarithm of each value of age and of each value of growth rate. Make a scatterplot of $Y = \log(\text{age})$ versus $X = \log(\text{growth rate})$ and fit a regression line to the data.

(e) Make a residual plot from the regression in part (d). Then make a normal quantile plot of the residuals. Based on these plots, does a regression model in log scale, from part (d), seem appropriate?

12.S.25 (*Computer exercise*) Researchers measured the blood pressures of 22 students in two situations: when the students were relaxed and when the students were taking an important examination. The table lists the systolic and diastolic pressures for each student in each situation.[40]

During exam		Relaxed	
Systolic pressure (mm Hg)	Diastolic pressure (mm Hg)	Systolic pressure (mm Hg)	Diastolic pressure (mm Hg)
132	75	110	70
124	170	90	75
110	65	90	65
110	65	110	80
125	65	100	55
105	70	90	60
120	70	120	80
125	80	110	60
135	80	110	70
105	80	110	70
110	70	85	65
110	70	100	60
110	70	120	80
130	75	105	75
130	70	110	70
130	70	120	80
120	75	95	60
130	70	110	65
120	70	100	65
120	80	95	65
120	70	90	60
130	80	120	70

(a) Compute the change in systolic pressure by subtracting systolic pressure when relaxed from systolic pressure during the exam; call this variable X.

(b) Repeat part (a) for diastolic pressure. Call the resulting variable Y.

(c) Make a scatterplot of Y versus X and fit a regression line to the data.

(d) Make a residual plot from the regression in part (c).

(e) Note the outlier in the residual plot [and on the scatterplot from part (c)]. Delete the outlier from the data set. Then repeat parts (c) and (d).

(f) What is the fitted regression model (after the outlier has been removed)?

12.S.26 (*Continuation of 12.S.25*) Consider the data from Exercise 12.S.25, part (f).

(a) Construct a 95% confidence interval for β_1.

(b) Interpret the confidence interval from part (a) in the context of this setting.

12.S.27 Selenium (Se) is an essential element that has been shown to play an important role in protecting marine mammals against the toxic effects of mercury (Hg) and other metals. It has been suggested that metal concentrations in marine mammal teeth can potentially be used as bioindicators for body burden. Twenty Belugas (*Delphinapterus leucas*) were harvested from the Mackenzie Delta, Northwest Territories, in 1996 and 2002, as part of an annual traditional Inuit hunt. Tooth and liver Se concentrations are reported in the table, summarized, and graphed.[41]

Whale	Liver Se (μg/g)	Tooth Se (ng/g)	Whale	Liver Se (μg/g)	Tooth Se (ng/g)
1	6.23	140.16	11	15.28	112.63
2	6.79	133.32	12	18.68	245.07
3	7.92	135.34	13	22.08	140.48
4	8.02	127.82	14	27.55	177.93
5	9.34	108.67	15	32.83	160.73
6	10.00	146.22	16	36.04	227.60
7	10.57	131.18	17	37.74	177.69
8	11.04	145.51	18	40.00	174.23
9	12.36	163.24	19	41.23	206.30
10	14.53	136.55	20	45.47	141.31

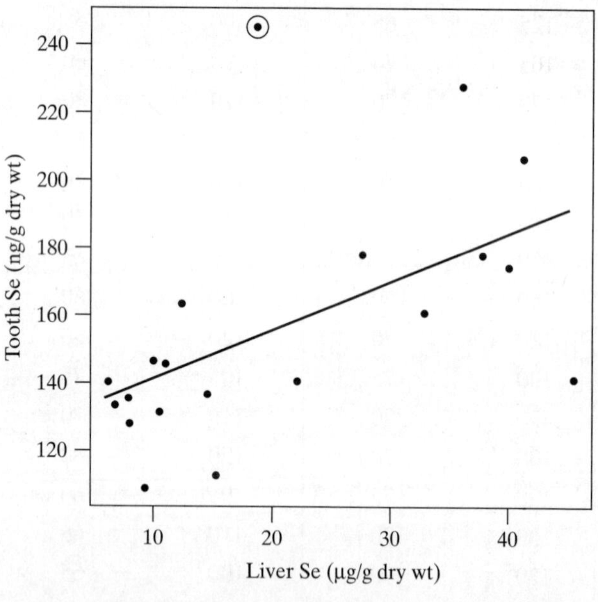

(a) Can we regard the sample correlation between Tooth (Y) and Liver (X) selenium, $r = 0.53726$, as an estimate of the population correlation coefficient? Briefly explain.

(b) If the circled point were removed from the data set, would the sample correlation listed in part (a) increase, decrease, or stay about the same?

(c) If the roles of X and Y were reversed (i.e., Y = Liver and X = Tooth selenium), would the sample correlation listed in part (a) increase, decrease, or stay about the same?

(d) Is the circled point on the plot a leverage and/or influential point? Explain briefly.

(e) Is the circled point on the plot an outlier?

12.S.28 (*Continuation of 12.S.27*) The following are summary statistics for the selenium data in Exercise 12.S.27.

$\bar{x} = 20.685$ $\quad\bar{y} = 156.599$

$s_X = 13.4491$ $\quad s_Y = 36.0595$

$r = 0.53729$ $\quad$ SS(resid) = 17,573.4

(a) Calculate the regression line of Liver selenium on Tooth selenium.

(b) Compute a 99% confidence interval for the slope of the regression line.

(c) Interpret the interval computed in part (b) in the context of the problem.

(d) Using the interval computed in part (b), is it reasonable to believe that the slope is as small as 0.15 (μg/g)/(ng/g)?

12.S.29 (*Continuation of 12.S.27 and 12.S.28*) Referring to the data plotted in Exercise 12.S.27, which of the following is a residual plot resulting from fitting the regression line in Exercise 12.S.28, part (a)? Justify your choice.

12.S.30 (*Continuation of 12.S.27*) The whales observed in this study were harvested during a traditional Inuit hunt in two particular years. What are we assuming about the captured whales to justify our analyses of these data in the preceding problems?

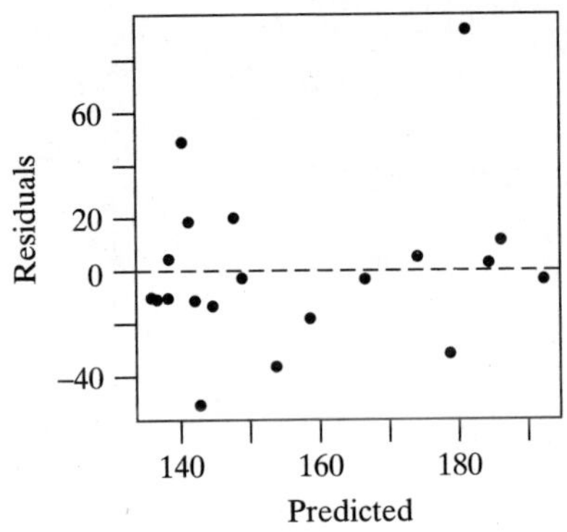
Residuals
60
20
0
–40
140
160
180
Predicted

(a)

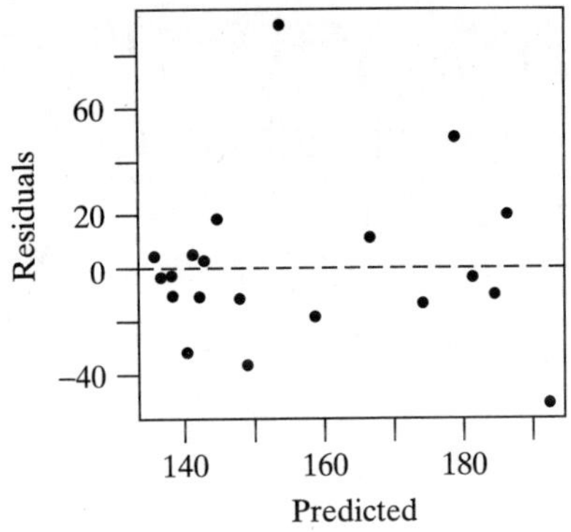
Residuals
60
20
0
–40
140
160
180
Predicted

(b)

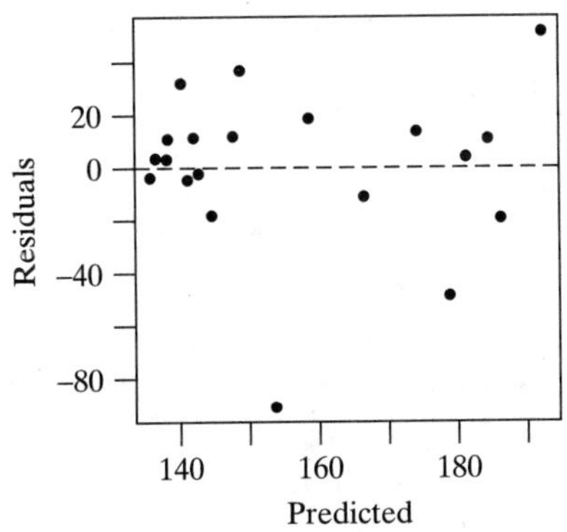
Residuals
20
0
–40
–80
140
160
180
Predicted

(c)

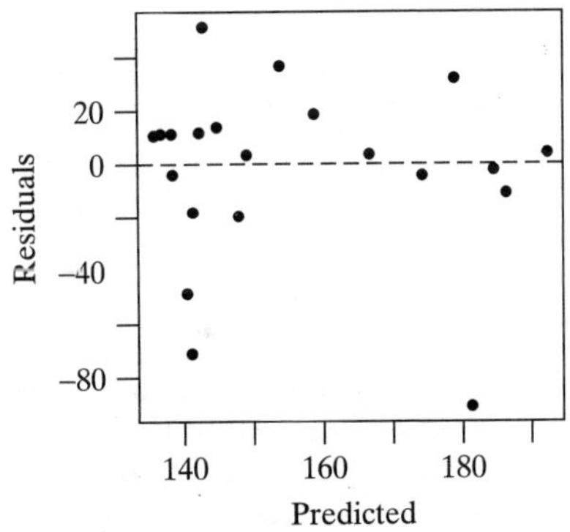
Residuals
20
0
–40
–80
140
160
180
Predicted

(d)

HIGHLIGHTS AND STUDY

(REFLECTIONS ON CHAPTERS 11 AND 12)

Unit

IV

In Chapters 11 and 12, we examined the relationship between a numeric response variable and an explanatory variable that was either categorical with three or more levels (ANOVA) or numeric (regression). In some situations, either tool may be used, such as when the explanatory variable takes on only a few discrete numeric values, as is the case in the following example.

To study the effects of a preservative foliar spray of varying dosages of abscisic acid on potted impatiens plants, researchers randomly sprayed 30 plants with either water (control) or a low dose (300 PPM) or medium dose (600 PPM) of abscisic acid.*,† After the initial treatment, the plants were examined daily to determine if they were in marketable condition. The response variable was then taken to be the total number of marketable days. The results are summarized in Table IV.1 and Figure IV.1.

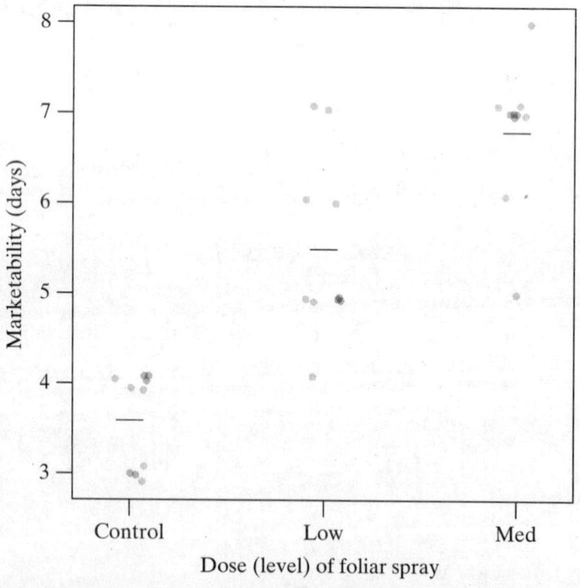

Figure IV.1 The number of marketable days of impatiens plants treated with varying doses of abscisic acid. The lines represent the sample means.

Table IV.1 Marketable days of impatiens plants treated with varying doses of abscisic acid

	Marketability (days)		
Treatment (dose)	Mean	SD	*N*
1. Control (0 PPM)	3.60	0.52	10
2. Low (300 PPM)	5.50	0.97	10
3. Med (600 PPM)	6.80	0.79	10

*Why is it important that the treatments were applied randomly? What might go wrong if the application was nonrandom?

†A high dose was also used, but is not included in this example.

Using analysis of variance and the Global F-test, we can test the null hypothesis that the mean number of marketable days is the same for all three treatments versus the alternative hypothesis that the mean number of marketable days is not the same (i.e., the treatment affects marketability).

$$H_0: \mu_1 = \mu_2 = \mu_3$$

$$H_A: \text{not all } \mu_i\text{'s are equal}$$

Before we begin the analysis, there are two points that are worth noting. The first regards the ability to make a causal claim with the alternative hypothesis. Since the researchers assigned the sprays to plants, this was an experiment, and thus determining cause–effect relationships is possible. The second point regards directionality, or more accurately, the lack of directionality. The researchers in this study most likely wish to show that the spray preserves the plants and increases marketability (a directional claim). At a minimum, they would be interested in showing that the two conditions using the foliar spray preserve the plants longer than the water control. Unfortunately, the Global F-test is an omni-directional test—it will only reveal if there is evidence for differences among the population means but will not indicate the nature of the differences. If the Global F-test yields significant results, we will have to carry out some post-hoc multiple comparisons (e.g., Fisher, Tukey, or Bonferroni intervals) as discussed in the optional Section 11.9.

Statistical software yields the ANOVA table displayed in Table IV.2 for these data.* As noted in Section 11.2, computer software often displays the between-group variability as "treatment" and within-group variability as "error." The ANOVA table in Table IV.2 also displays the P-value as < 0.0001 indicating that differences as extreme as those among the treatment means observed in these data are very unlikely to occur if the treatments have no effect. That is, there is very strong evidence that the treatment means are not all equal—that the foliar spray affects the mean number of marketable days.

Table IV.2 ANOVA table for impatiens preservation data

Source	df	Sum of squares	Mean square	F ratio	Prob $> F$
Treatment	2	51.800	25.900	42.382	<.0001
Error	27	16.500	0.611		
Total	29	68.300			

Multiple Comparisons (Optional)

Having very strong evidence for a difference among the treatments, it is now natural to investigate which groups differ from which as well as the magnitude of the differences (effect size) as considered in Section 11.9. Table IV.3 shows computer output displaying the conservative experimentwise 95% Bonferroni intervals for all pairwise comparisons (98.33% comparisonwise confidence level).

*It would be a good exercise to review the formulae behind the values listed in the table as well as to confirm their values.

Table IV.3 Experimentwise 95% Bonferroni confidence intervals comparing the mean number of marketable days for the three treatments

Comparison	$d_{ab} = \bar{y}_a - \bar{y}_b$	$SE_{D_{ab}}$	Lower 98.33%	Upper 98.33%
Med − Control: $\mu_3 - \mu_1$	3.20	0.35	2.31	4.09
Low − Control: $\mu_2 - \mu_1$	1.90	0.35	1.01	2.79
Med − Low: $\mu_3 - \mu_2$	1.30	0.35	0.41	2.19

The intervals listed in Table IV.3 provide compelling evidence that the foliar spray is effective when compared to a control, as both the confidence interval comparing the medium level to the control, $\mu_3 - \mu_1$, and low level to control, $\mu_2 - \mu_1$, are entirely positive. Furthermore, there is evidence that the efficacy increases with dosage, since the confidence interval comparing the medium to low level, $\mu_3 - \mu_2$, is also entirely positive. One might summarize these results in a publication in the following tabular (Table IV.4) or graphical forms (Figure IV.2). Because the focus is a comparison of means, the standard error of the mean, rather than the standard deviation of the sample, is listed for the reader's convenience.

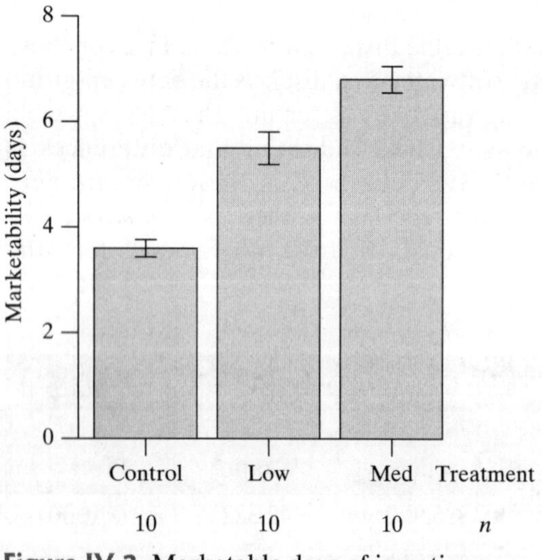

Figure IV.2 Marketable days of impatiens plants treated with varying dosages of abscisic acid. Error bars are ±SE.

Table IV.4 Marketable days of impatiens plants treated with varying dosages of abscisic acid. Means that do not share a common superscript are statistically significantly different

	Marketability (days)		
Treatment	Mean	SE	N
1. Control (0 PPM)	3.60[a]	0.16	10
2. Low (300 PPM)	5.50[b]	0.31	10
3. Med (600 PPM)	6.80[c]	0.25	10

Checking ANOVA Requirements

In order to trust our results, we should investigate whether or not our data meet the conditions for the ANOVA Global F-test as discussed in Section 11.5. We cannot ascertain from the raw data whether or not random sampling was used or if the samples are independent. These conditions must be verified by examining the process that was used to collect the data. In this case we know that the researchers randomly assigned the three treatments to the 30 potted plants; thus, the samples can be regarded as independent from one another. It would be good to know how the initial 30 plants were chosen. While they are probably not from a true random sample, is it reasonable to regard them as a random sample? For example, did the researchers choose the best plants to apply the sprays to? If so, these plants would not be representative of a larger population of interest.

By examining the residuals from the ANOVA model displayed in Figure IV.3, we can verify (or more accurately rule out gross violations of) the other two conditions: (1) that the several populations being compared are each normally distributed, and (2) that the populations have similar standard deviations. The normal quantile plot in Figure IV.3 shows data that are consistent with normality; the plot is fairly linear.* The horizontal banding is due to the measurements being taken discretely (days of marketability).

The residuals versus fitted value plot in Figure IV.3 (b) suggests that the population standard deviations may not be the same for all three groups (the dots on the left are not as spread out vertically as those in the center or right), and Table IV.1 shows that the sample standard deviation for the low dose is nearly double the standard deviation for the control. We also see mild evidence of the standard deviation growing with the mean, which is a common problem. (We would ordinarily expect to see 30 dots in the plot of the residuals versus fitted values, but because there were only 10 distinct values for the number of marketable days observed there appear to only be 10 observable values. If we were to jitter the points, we would observe that some of the dots on the plot are actually multiple observations.)

Figure IV.3 (a) Normal quantile plot of ANOVA residuals and (b) ANOVA residuals versus fitted values plot for the impatiens foliar spray study

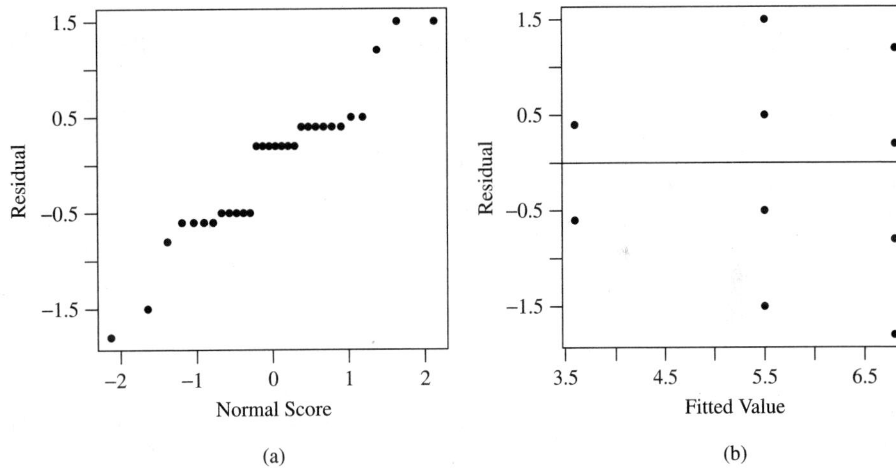

Regression

Looking back on Figures IV.1 and IV.2, there is a clear upward trend in preservation duration as the dose of abscisic acid is increased in the spray. If the relationship between these numeric variables is linear, we can investigate its statistical significance using regression.† Figure IV.4 shows a plot of days of marketability against dose (with some jittering to better reveal the data since both dose and days of marketability are discretely observed). A regression line is also included in the plot to draw attention to the positive linear trend. Table IV.5 provides numeric summaries of the data for a regression model, and Table IV.6 provides computer output from fitting a regression model.

---

*Some would argue that the discreteness of the bands indicates nonnormality because the normal distribution characterizes continuous, not discrete, random variables. Although this is true, the model still serves as a useful approximation.

†If the relationship were nonlinear, say quadratic, regression methods could still be used, but more complex models would be required.

Table IV.5 Summary of impatiens preservation data

	X = Dose (PPM)	Y = Marketability (days)
Mean	300.00	5.3000
SD	249.14	1.5345
	$r = 0.8658$	

Table IV.6 Computer output from regressing days of marketability on dose

```
Coefficients:
            Estimate Std. Error t value Pr(>|t|)
(Intercept) 3.7000000  0.2255945  16.401 6.86e-16 ***
Dose        0.0053333  0.0005825   9.156 6.49e-10 ***
---
Signif. codes:  0 `***' 0.001 `**' 0.01 `*' 0.05 `.' 0.1 ` ' 1
Residual standard error: 0.7815 on 28 degrees of freedom
Multiple R-squared:  0.7496,     Adjusted R-squared:  0.7407
F-statistic: 83.84 on 1 and 28 DF,  p-value: 6.492e-10
```

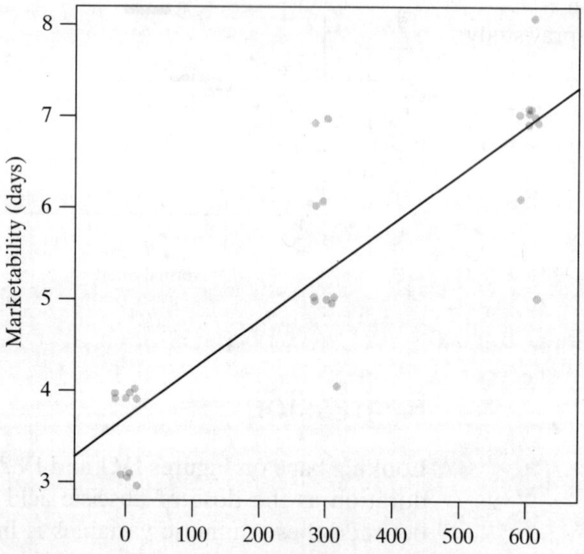

Figure IV.4 Jittered plot with regression line of the impatiens preservation data

From the output in Table IV.6, we can test the null hypothesis that there is no linear relationship between dose and marketability—that the slope of the regression line is zero ($\beta_1 = 0$)—against the alternative (directional) hypothesis that there is a positive linear relationship between dose and marketability ($\beta_1 > 0$). t_s is reported to be 9.156 with a P-value of 6.5×10^{-10}. This small P-value indicates that observing a linear association at least as strong as that observed in these data is extremely unlikely if dose and marketability are unrelated (i.e., results as extreme as these would occur less than 0.01% of the time due to chance). These data provide very strong evidence that increasing the dose provides greater preservation.

Using the output in Table IV.6, we obtain the equation of the regression line: $\hat{y} = 3.70 + 0.00533x$ indicating that for each PPM increase in the dose of abscisic acid, plants are preserved an additional 0.00533 days, on average. Equivalently, for each 100 PPM increase in the dose of abscisic acid, plants are preserved an additional 0.533 days on average. A 95% confidence interval for the slope of the regression line is $0.00533 \pm 2.0484 \times 0.0005825$ or (0.00414, 0.00652) days/PPM.

As discussed in Section 12.4, we must be careful to avoid extrapolating the results of regression models. The predicted increase in preservation is not likely to be valid for doses outside of the range of those studied, nor could we use the model to predict marketability duration for doses much above 600 PPM. If we were to use this model to estimate the mean marketability duration for plants sprayed with a 900 PPM dose, we would predict the plants to last on average $\hat{y} = 3.70 + 0.00533 \times 900 = 8.5$ days. It turns out, however, that for this experiment, the researcher actually did study this dose. At 900 PPM, the mean number of days of marketability was reported to be 6.7 days—much lower than our extrapolated value of 8.5 days. Clearly, the upward linear trend does not continue indefinitely as shown in Figure IV.5.

Figure IV.5 Jittered plot of the impatiens preservation data with regression model based only on doses 0 through 600 PPM (solid line) and loess curve based on all doses 0 through 900 PPM (dashed line)

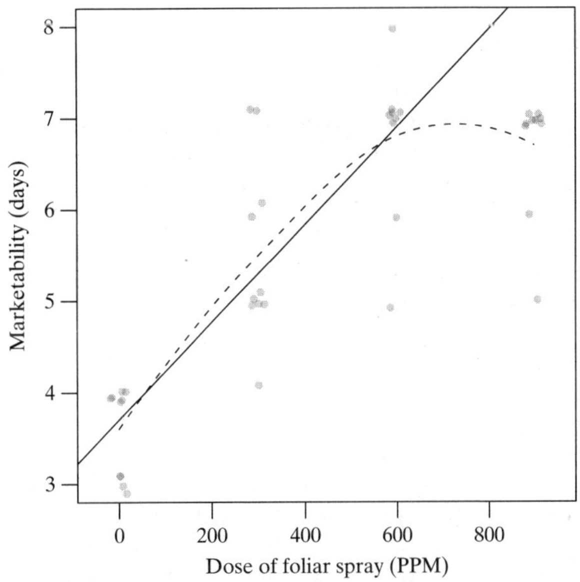

Checking Regression Requirements

The requirements for the linear regression model are very similar to the requirements of the ANOVA model: (1) random sampling (the random subsampling or bivariate random sampling model); (2) normal distribution of Y for each fixed X; and (3) constant standard deviation (σ_e does not depend on X); a further condition for regression (not ANOVA) is (4) linearity: $\mu_{Y|X} = \beta_0 + \beta_1 X$, that the mean value of Y changes linearly with X.

As discussed earlier in the context of ANOVA, we don't know if the 30 plants were truly chosen at random, but we'll regard them as such for this example. Because the researchers set the doses of spray and observed the days of marketability, these data fit the random subsampling model (X fixed, Y random). The normality condition may be evaluated by examining a normal quantile plot of the residuals as shown in Figure IV.6 (a). Because the ANOVA and regression models are slightly different, the ANOVA residuals in Figure IV.3 (a) are slightly different

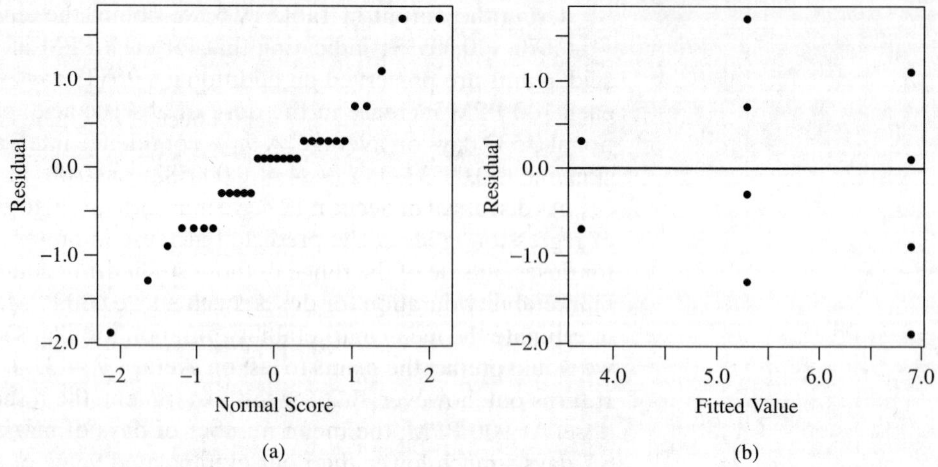

Figure IV.6 (a) Normal quantile plot of regression residuals and (b) regression residuals versus fitted values plot for the impatiens foliar spray study

from those in Figure IV.6(a). The linearity of this plot is consistent with normality of the residuals; the data appear to meet the normality condition. The linearity and constant SD condition are assessed using the residual versus fitted value plot in Figure IV.6(b). As observed earlier in the ANOVA context, the SD appears to be somewhat smaller for small fitted values, but overall the SD is fairly stable across the range. If the linearity condition were not met, we would see curved patterns in the residuals versus fitted value plot (as well as a curved pattern in the scatterplot of the original data as shown in Figure IV.4). No curved patterns are observed; thus, the data appear to meet the linearity condition.

Unit IV Summary Exercises

IV.1 The growth per day (in cm) of alfalfa sprouts was recorded for 50 sprouts kept in darkness and for 49 sprouts kept in light. The log of each observation was then taken to make the distributions reasonably normal. For the darkness sample, the average was 1.40, and the SD was 0.92. For the light sample, the average was 0.47, and the SD was 0.45.

George has all of the raw data and wants to test the null hypothesis that the two population means are equal. He wants to use ANOVA to do this test. Is this a good idea? Discuss (i) whether this is *possible* (saying why or why not, and if not, saying what you would do instead of ANOVA) and (ii) whether this is *wise* (saying why or why not, and if not, saying what you would do instead of ANOVA).

IV.2 Researchers measured initial weight, X, and weight gain, Y, of 15 rats on a high protein diet.[1] All weights are in grams. A scatterplot of the data shows a linear relationship. The fitted regression model is

$$\hat{y} = 54.95 + 1.06x$$

The sample correlation coefficient, r, is 0.489. The SE of b_1 is 0.526. Also, $s_e = 19.3$.

(a) Find r^2 and interpret r^2 in the context of this problem.

(b) Suppose that a rat initially weighs 60 g. What is the predicted weight gain for the rat?

(c) Interpret the value 1.06 from the fitted model *in the context of this problem*. (What does this 1.06 mean?)

(d) Construct a 95% confidence interval for β_1.

(e) Interpret the value of s_e *in the context of this problem*. That is, what does it mean to say that $s_e = 19.3$? How does this relate to your answer to part (b)?

IV.3 Researchers wanted to compare two drugs, formoterol and salbutamol, in aerosol solution, for the treatment of patients who suffer from exercise-induced asthma.[2] Patients were to take a drug, do some exercise, and then have their "forced expiratory volume" measured. There were 30 subjects available.

(a) Explain how to set up a randomized blocks design (RBD) here using age as the blocking variable and five blocks.

(b) How would an RBD be a helpful? That is, what is the main advantage of using a RBD in a setting like this?

IV.4 A confused researcher finds a dime on the sidewalk and wants to test $H_0\colon p = 0.5$ against $H_A\colon p \neq 0.5$ where $p = \Pr[\text{Heads}]$ when tossing the coin. This dime is an ordinary coin for which $p = 0.5$—but she doesn't know that. She tosses the coin 100 times, finds the P-value for a goodness-of-fit test, and compares it to $\alpha = 0.05$. However, if she retains H_0 (because the P-value is large), then she discards the first sample and gets a new sample by tossing the coin 100 more times and repeating the goodness-of-fit test with the new data. If she retains H_0 for this test, then she discards the data and collects a third sample and does another goodness-of-fit test, after which she stops no matter what. What is the probability that she will make a type I error?

IV.5 A researcher collected data on a random sample of 12 breakfast cereals. He recorded x = fiber (in grams/ounce) and y = price (in cents/ounce). A scatterplot of the data shows a linear relationship. The fitted regression model is

$$\hat{y} = 17.42 + 0.62x$$

The sample correlation coefficient, r, is 0.23. The SE of b_1 is 0.81. Also, $s_e = 3.1$.

(a) Find r^2 and interpret r^2 in the context of this problem.

(b) Suppose that a cereal has 2.63 grams of fiber/ounce and costs 17.3 cents/ounce. What is the residual for this cereal?

(c) Interpret the value of s_e *in the context of this problem.* That is, what does it mean to say that $s_e = 3.1$?

IV.6 Consider a regression setting in which we construct a scatterplot, fit the regression model $\hat{y}_i = b_0 + b_1x_i$, and generate a residual plot.

(a) Suppose the scatterplot of y versus x is as follows:

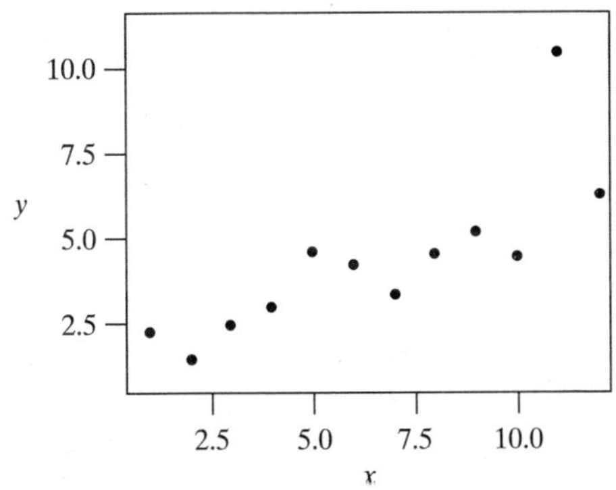

Draw a sketch of the resulting residual plot. Label the axes on your graph.

(b) Suppose we fit a regression line to a new set of data and the resulting residual plot is as follows:

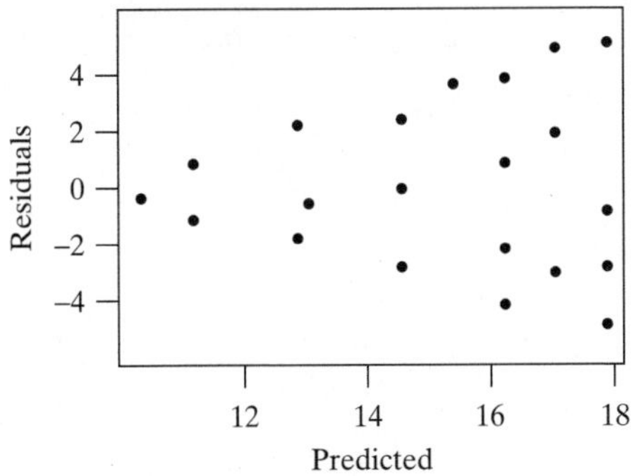

Draw a sketch of the scatterplot of y versus x. Label the axes on your graph.

IV.7 A researcher measured the number of tree species per 0.1 hectare plot along the Black, Huron, and Vermilion rivers. The data are summarized in the table below:

	Black	Huron	Vermilion
Mean	9.33	9.89	11.11
Median	10	11	11
SD	3.16	2.42	2.71
n	9	9	9

Here is a partial ANOVA table summarizing the results.

Source	Degree of freedom	Sum of squares	Mean square	F
Between groups		14.889		
Within groups	24	185.778		
Total		200.667		

(a) Find the value of the test statistic that is used to test H_0. (You do not need to complete the test.)

(b) We use ANOVA to test a null hypothesis, H_0. The P-value for the test is 0.397. State the conclusion regarding H_0 *in the context of this setting.*

(c) What conditions are necessary for the ANOVA to be valid?

IV.8 Consider the data from Exercise IV.7. We use $\sqrt{\text{MS(Within)}}$ as an estimate of what quantity? Give your answer in the context of the question.

IV.9 (a) Consider the data from Exercise IV.7. Suppose we want to compare the Vermilion River to the average of the other two rivers. Calculate the value of the contrast, L, to measure the difference between the mean number of species (per 0.1 hectare) along the Vermilion River and the mean number of species (per 0.1 hectare) along the other two rivers.

(b) Find the SE of the contrast L from part (a). (Don't make an interval or do a test.)

IV.10 Is there a relationship between wing length (mm) and wingbeat frequency (Hz) among hummingbirds? In one study, researchers measured the wing lengths and wingbeat frequencies of 12 hummingbirds.[3] The following are basic summaries and a plot of the 12 data values.

Variable	Mean	SD
Wing length (mm)	39.4034	11.0090
Wingbeat frequency (Hz)	41.6858	5.4176
	$r = -0.9061$	

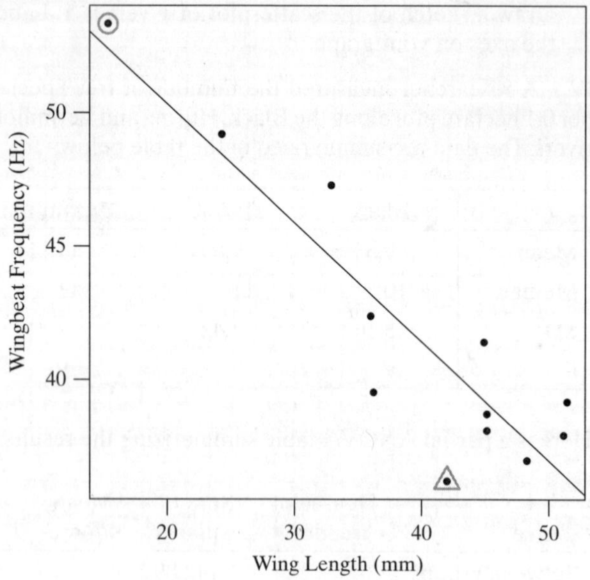

(a) If the circled point was removed from this dataset, would the value of the sample correlation increase in magnitude, decrease in magnitude, or stay about the same? Briefly explain.

(b) The residual SD for all these data (including the points in the circle and triangle) is $s_e = 2.40$ Hz. If the point in the triangle were removed, would the residual SD increase, decrease, or stay about the same? Briefly explain.

IV.11 Consider the hummingbird data in Exercise IV.10.

(a) What percentage of the variation in wingbeat frequency is explained by the relationship between wing length and wingbeat frequency?

(b) Formally speaking, an important adjective is missing from part (a) that describes the nature of the relationship being described. What is the missing adjective?

IV.12 Consider the hummingbird data and information provided in Exercise IV.10.

(a) Find the equation of the fitted regression line.

(b) Predict the wingbeat frequency for a hummingbird with 30 mm wings.

(c) Predict the mean wingbeat frequency for hummingbirds with 30 mm wings.

(d) Are the predictions of (b) and (c) interpolations or extrapolations? Briefly explain.

(e) Hummingbird B has wings that are 1mm longer than the wings of hummingbird A. How much faster or slower would you expect hummingbird B's wings to beat compared to A's?

IV.13 Consider the hummingbird data and information provided in Exercise IV.10.

(a) Do longer wings tend to beat more slowly (i.e., lower frequency) than shorter wings? In plain English, what are the null and alternative hypotheses to be tested?

(b) Compute the value of t_s used to test the hypothesis in part (a) for testing $H_A: \rho < 0$.

(c) Compute the value of t_s used to test the hypothesis in part (a) for testing $H_A: \beta < 0$.

(d) Explain why your answers to (b) and (c) agree.

(e) The P-value for the test of $H_A: \rho \neq 0$ is 0.004. Without using a table or computer, what is the P-value for the tests in parts (b) and (c)?

IV.14 The following is a plot of the residuals against the fitted values for the hummingbird data of Exercise IV.10.

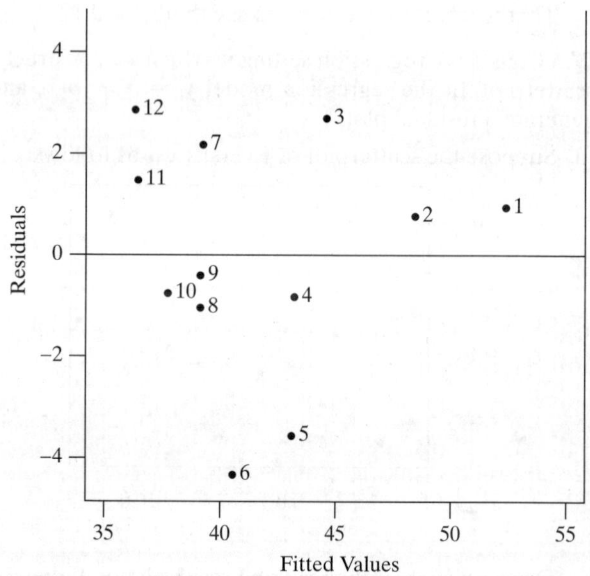

(a) Which point in the residual plot corresponds to the circled point in the data appearing in Exercise IV.10? How can you tell?

(b) Which point in the residual plot corresponds to the triangle enclosed point in the data appearing in Exercise IV.10? How can you tell?

A SUMMARY OF INFERENCE METHODS

Chapter

13

OBJECTIVES

In this chapter we summarize inference methods presented throughout the text. We will

- show the process of choosing an appropriate inference technique from among those presented in earlier chapters.
- consider several examples of choosing an inference method.

13.1 Introduction

In Chapters 2 and 6 through 12 we introduced many statistical methods for visually and numerically summarizing data and for making inferences. Statistics students are often overwhelmed by the number and variety of procedures that have been presented. What a statistician sees as a clearly arranged set of tools for analyzing data can appear as a blur to the novice. In this chapter we present a variety of examples that demonstrate the analysis process from exploration and summary to inference using some of the methods presented in earlier chapters. With the examples, we also provide some guidelines that are useful in deciding how to make an inference from a given set of data.

When presented with a set of data, it is useful to ask a series of questions:

1. *What question were the researchers attempting to answer when they collected these data?* Data analysis is done for a purpose: to extract information and to aid decision making. When looking at data, it helps to bear in mind the purpose for which the data were collected. For example, were the researchers trying to compare groups, perhaps patients given a new drug and patients given a placebo? Were they trying to see how two quantitative variables are related so that they can use one variable to make predictions of the other? Were they checking whether a hypothesized model gives accurate predictions of the probabilities associated with a categorical variable? A good understanding of why the data were collected often clarifies the next question:
2. *What is the response variable in the study?* For example, if the researchers were concerned with the effect of a medication on blood pressure, then the likely response variable is $Y =$ change in blood pressure of an individual (a continuous numeric variable). If they were concerned with whether or not a medication cures an illness, then the response variable is categorical with two levels: yes if a person is cured, no if a person is not cured, or maybe even categorical with three or more ordered levels: fully cured, improved, no improvement.
3. *What predictor variables, if any, were involved?* For example, if a new drug is being compared to a placebo, then the predictor variable is group membership: A patient is either in the group that gets the new drug or else the patient is in the placebo group. If height is used to predict weight, then height is the predictor (and weight is the response variable). Sometimes there is no predictor variable. For example, a researcher might be interested in the distribution of cholesterol levels in adults. In this case, the response variable is cholesterol

level, but there is no predictor variable. [One might argue that there is a predictor: whether or not someone is an adult. If we wished to compare cholesterol levels of adults to those of children, then whether or not someone is an adult would be a predictor. But if there is no comparison to be made, and everyone in the study is part of the same group (adults), then it is not accurate to speak of a predictor *variable*, since group membership does not vary from person to person.]

The answers to these questions help frame the analysis to be conducted. Sometimes the analysis will be entirely descriptive and will not include any statistical inference, such as when the data are not collected by way of a random sample. Even when a statistical inference is called for, there is generally more than one way to proceed. Two statisticians analyzing the same set of data may use somewhat different methods and draw different conclusions. However, there are commonly used statistical procedures in various situations. The flowchart given in Figure 13.1.1 helps to organize the inference methods that have been presented in this book.

To use this flowchart, we start by asking whether the response variable is quantitative or categorical. We then consider the type of predictor variables in the study and whether the samples collected are independent of one another or are dependent (e.g., matched pairs). Many of the methods, such as the confidence interval for a population mean presented in Chapter 6, depend on the data being from a population that has a normal distribution. (This condition is less important for large samples

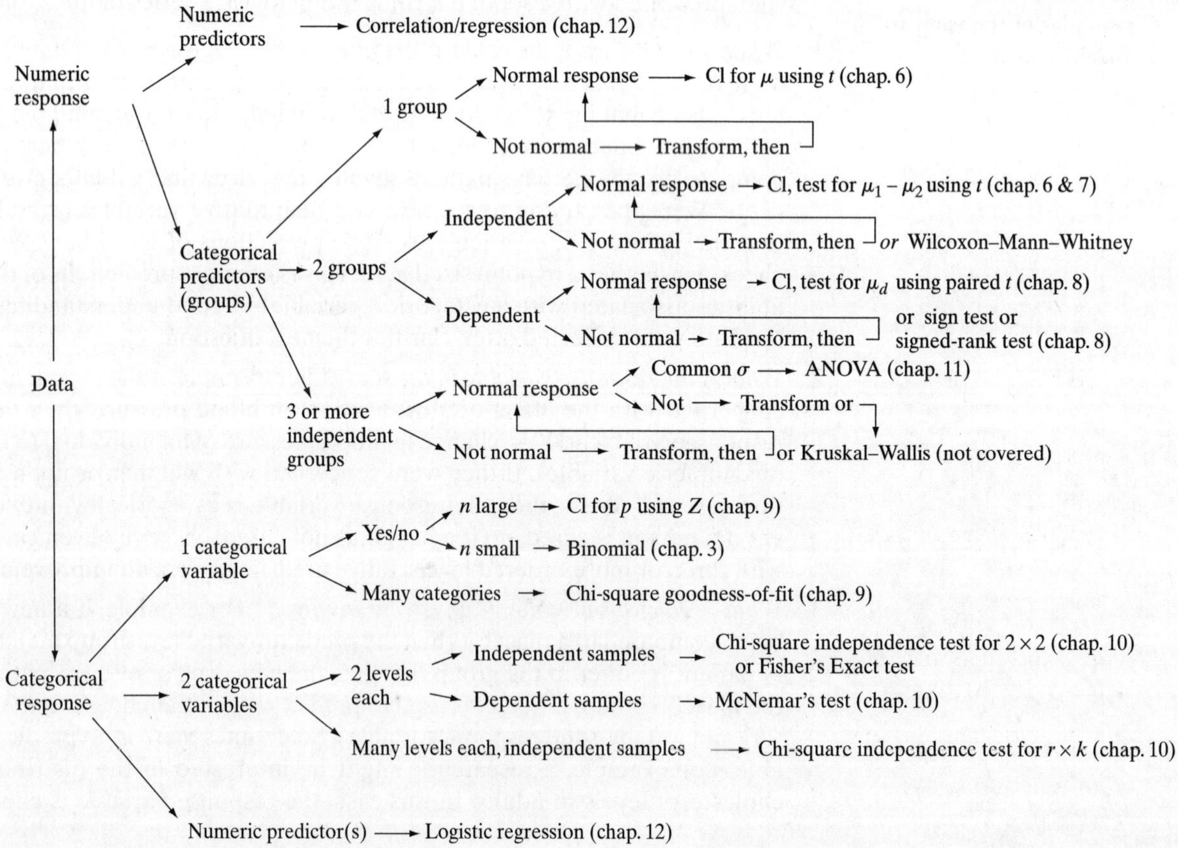

Figure 13.1.1 A flowchart of inference methods

than it is for small samples, due to the Central Limit Theorem.) Nonnormal data can often be transformed to approximate normality and normal-based methods can then be applied. If such transformation fails to achieve approximate normality, then nonparametric methods, such as the Wilcoxon-Mann-Whitney test or the Wilcoxon Signed-Rank test, can be used. Randomization tests may always be used regardless of normality, though for many study designs their implementation requires simulations that require some computer programming.

Note that the flowchart only directs attention to the collection of inference methods presented in the previous chapters; this is not an exhaustive list. Beware of the Mark Twain fallacy: "When your only tool is a hammer, every problem looks like a nail." Not every statistical inference problem can be addressed with the methods presented here. In particular, these methods center on consideration of parameters, such as a population mean, μ, or proportion, p. Sometimes researchers are interested in other aspects of distributions, such as the 75th percentile. When in doubt about how to proceed in an analysis, consult a statistician.

EXPLORATORY DATA ANALYSIS

No matter what type of analysis is being considered, it is always a good idea to start by making one or more graphs of the data. The choice of graphics depends on the type of data being analyzed. For example, when comparing two samples of quantitative data, side-by-side dotplots or boxplots are informative—both as a visual comparison of the two samples and for assessing whether or not the data satisfy the normality condition. When analyzing categorical data, bar charts are useful. When dealing with two quantitative variables, scatterplots are helpful.

Bear in mind that a statistical analysis is intended to help us understand the scientific problem at hand. Thus, conclusions should be stated in the context of the scientific study. In Section 13.2 we present some examples of data sets and the kinds of analyses that might be performed on them.

13.2 Data Analysis Examples

In this section we consider several data sets and the kinds of analyses that are appropriate for each. The three questions stated in Section 13.1 and the flowchart given in Figure 13.1.1 provide a framework for the discussion of the following examples.

Example 13.2.1

Gibberellic Acid Gibberellic acid (GA) is thought to elongate the stems of plants. Researchers conducted an experiment to investigate the effect of GA on a mutant strain of the genus *Brassica* called *ros.* They applied GA to 17 plants and applied water to 15 control plants. After 14 days they measured the growth of each of the 32 plants. For the 15 control plants the average growth was 26.7 mm, with an SD of 37.5 mm. For the 17 plants treated with GA the average growth was 92.6 mm, with an SD of 41.7 mm. The data are given in Table 13.2.1 and are graphed in Figure 13.2.1.[1]

Let us turn to the three questions stated in Section 13.1. (1) In this experiment, the researchers were trying to establish whether GA affects the growth rate of *ros*; (2) the response variable is 14-day growth of *ros*, which is numeric; (3) the predictor variable is group membership (GA group or control group) and is categorical; the two groups are independent of one another.

Table 13.2.1 Growth of *ros* plants (mm) after 14 days

	Control	GA
	3	71
	2	87
	34	117
	13	80
	6	112
	118	66
	14	128
	107	153
	30	131
	9	45
	3	38
	3	137
	49	57
	4	163
	6	47
		108
		35
Mean	26.7	92.6
SD	37.5	41.7

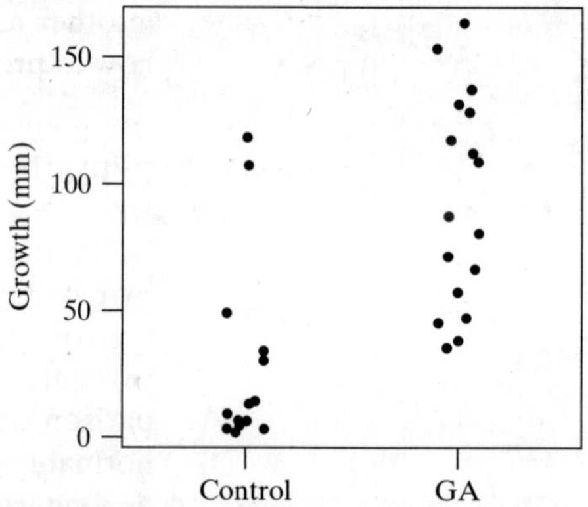

Figure 13.2.1 Dotplots of growth of *ros* plants (mm) after 14 days

The flowchart in Figure 13.1.1 directs us to consider a two-sample t test, if the data are normal or can be transformed to normality, or a Wilcoxon-Mann-Whitney test. Figure 13.2.2 shows that the distribution of the control sample of data is markedly nonnormal; thus, a transformation is called for.

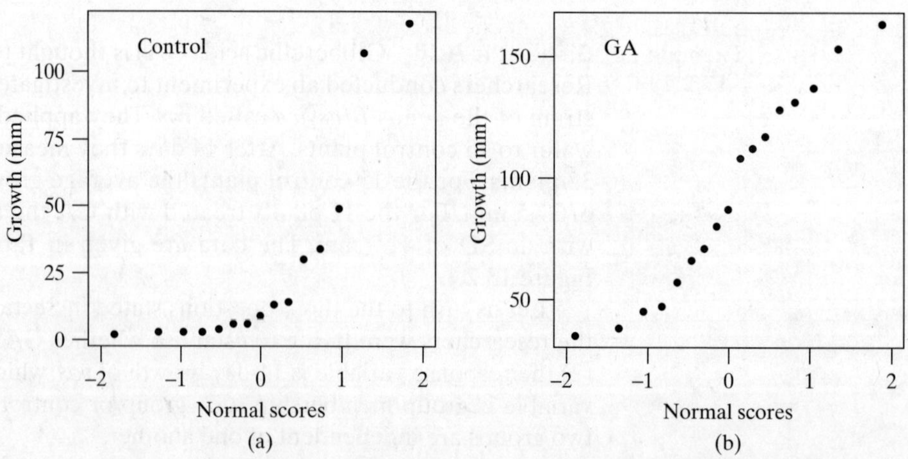

Figure 13.2.2 Normal quantile plots of (a) control data and (b) GA data

Taking logarithms of each of the observations produces the dotplots and normal quantile plots in Figures 13.2.3 and 13.2.4.

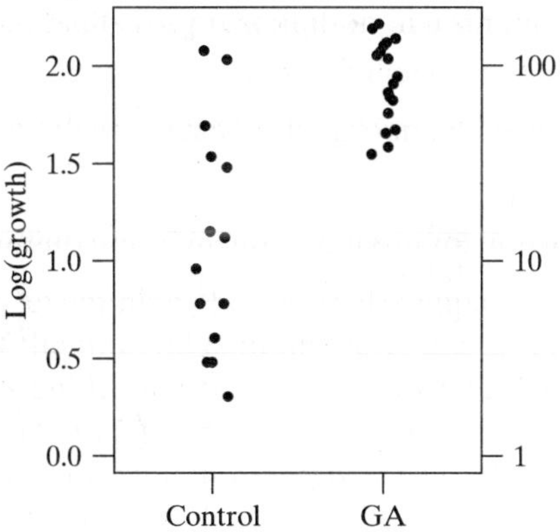

Figure 13.2.3 Dotplots of log(growth) of *ros* plants (mm) after 14 days

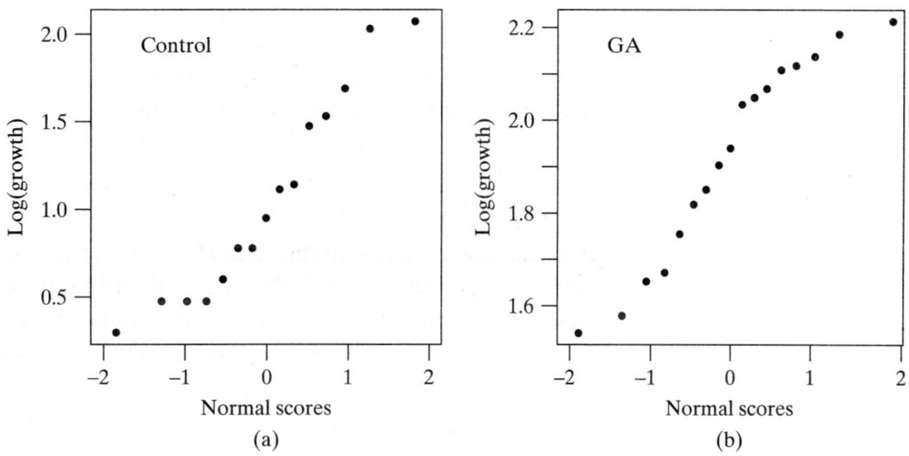

Figure 13.2.4 Normal quantile plots of (a) control data and (b) GA data in log scale

In log scale the data do not show marked evidence of abnormality (Shapiro–Wilk *P*-values for Control and GA are 0.2083 and 0.2296, respectively), so we can proceed with a two-sample t test. The standard deviations of the two samples are clearly quite different, as can be seen from Figure 13.2.3. However, an unpooled t test is still appropriate. The following computer output shows that $t_s = -5.392$ and the *P*-value is very small. Thus, we have strong evidence that GA increases growth of *ros*. ■

```
Two Sample t-test

data: log10(Growth)

t = -5.3917, df = 17.445, p-value < 0.0001

alt. hypothesis: true difference in means is not equal to 0

95 percent confidence interval:

-1.1943596, -0.5234687
```

Example 13.2.2 **Whale Swimming Speed** A biologist was interested in the relationship between the velocity at which a beluga whale swims and the tail-beat frequency of the whale. A sample of 19 whales was studied, and measurements were made on swimming velocity, measured in units of body lengths of the whale per second (so that a value of 1.0 means that the whale is moving forward by one body length, L, per second) and tail-beat frequency, measured in units of hertz (so that a value of 1.0 means one tail-beat cycle per second).[2] Here are the data:

Whale	Velocity (L/sec)	Frequency (Hz)	Whale	Velocity (L/sec)	Frequency (Hz)
1	0.37	0.62	11	0.68	1.20
2	0.50	0.675	12	0.86	1.38
3	0.35	0.68	13	0.68	1.41
4	0.34	0.71	14	0.73	1.44
5	0.46	0.80	15	0.95	1.49
6	0.44	0.88	16	0.79	1.50
7	0.51	0.88	17	0.84	1.50
8	0.68	0.92	18	1.06	1.56
9	0.51	1.08	19	1.04	1.67
10	0.67	1.14			

It would be natural to ask, "When tails beat faster, do whales travel faster?" but the biologist conducting the study focused on the related question, "Does tail-beat frequency depend on velocity?" For the biologist's question, the response variable, frequency, is numeric, and the predictor is velocity, which is also numeric. Thus, we can consider using regression analysis to study the relationship between velocity and frequency. Figure 13.2.5 is a scatterplot of the data, which shows an increasing trend in frequency as velocity increases.

Figure 13.2.5 Scatterplot of frequency versus velocity

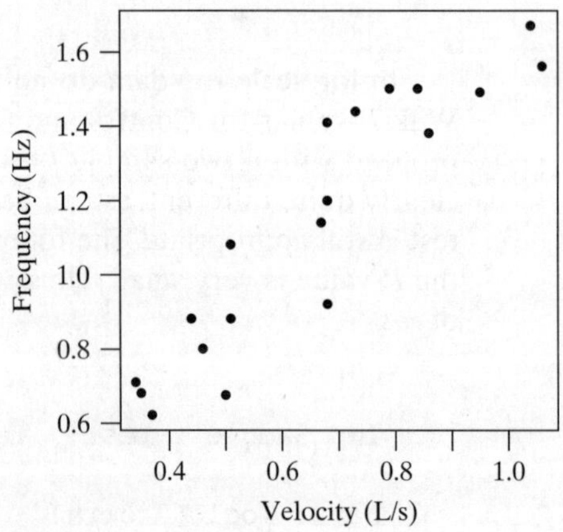

A regression model for these data is $Y = \beta_0 + \beta_1 X + \varepsilon$. Fitting the model to the data gives the equation $\hat{y} = 0.19 + 1.439x$, or Frequency $= 0.19 + 1.439 \times$ Velocity, as shown in the following computer output. Figure 13.2.6 shows the residual plot for this fit. The fact that this plot does not have any clear patterns in it supports the use of the regression model.

```
Coefficients:
              Estimate  Std. Error  t value  Pr(> |t|)
(Intercept)   0.1895    0.1004      1.887    0.0763
Velocity      1.4393    0.1451      9.917    1.75e-08
Residual standard error: 0.1396 on 17 degrees of freedom
R-squared: 0.8526
```

Figure 13.2.6 Residual plot for frequency regression fit

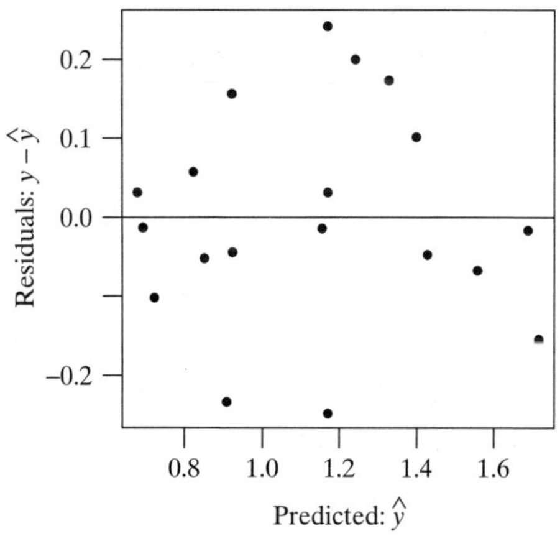

The null hypothesis

$$H_0: \beta_1 = 0$$

is tested with a t test, as shown in the regression output. A normal quantile plot of the residuals, given in Figure 13.2.7, supports the use of the t test here, since it indicates that the distribution of the 19 residuals is consistent with what we would expect to see if the random errors came from a normal distribution. The t statistic has 17 degrees of freedom and a P-value of less than 0.0001. Thus, the evidence that frequency is related to velocity is quite strong; we reject the claim that the linear trend in the data arose by chance.

Continuing the analysis, the computer output shows that r^2 is 85.3%. Thus, in the sample 85.3% of the variability in frequency is accounted for by variability in velocity. (This is significantly different from zero, as indicated with the t test for $H_0: \beta_1 = 0$.) ■

Figure 13.2.7 Normal quantile plot of residuals for frequency regression fit

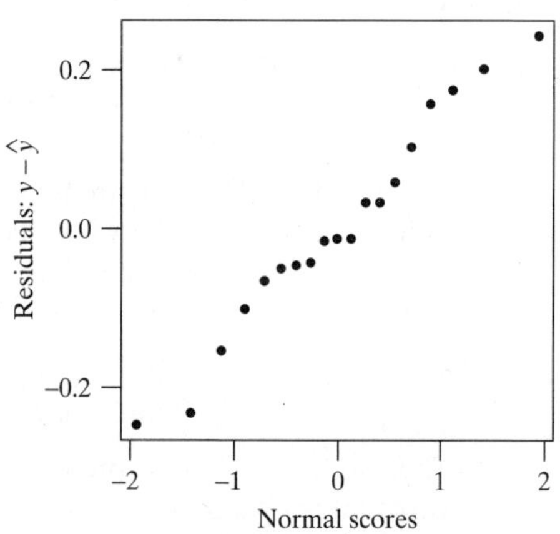

Example 13.2.3 **Progesterone Gel and Preemies** A group of 458 women who were at high risk for preterm birth were randomly assigned to receive either vaginal progesterone gel (n = 235) or a placebo gel (n = 223). The primary outcome of the experiment was preterm birth before 33 weeks of gestation. There were 21 cases of preterm birth in the progesterone gel group, compared to 36 cases in the placebo group.[3]

The purpose of this experiment (which was run double-bind) was to determine whether progesterone gel is effective in preventing preterm birth. Note that because this was an experiment, and not an observational study, we can talk in terms of a cause–effect relationship. The response variable is whether or not a woman gave birth before 33 weeks of gestation. The predictor variable is group membership (i.e., whether or not a woman was given progesterone gel). Figure 13.2.8 is a bar chart of the data, showing that preterm birth was much more common in the placebo group.

These data can be organized into a 2 × 2 contingency table, such as Table 13.2.2. A chi-square test of independence yields $\chi^2_s = 5.45$. With 1 degree of freedom, the P-value for this test is approximately 0.02. There is strong evidence that progesterone gel reduces the probability of preterm birth.

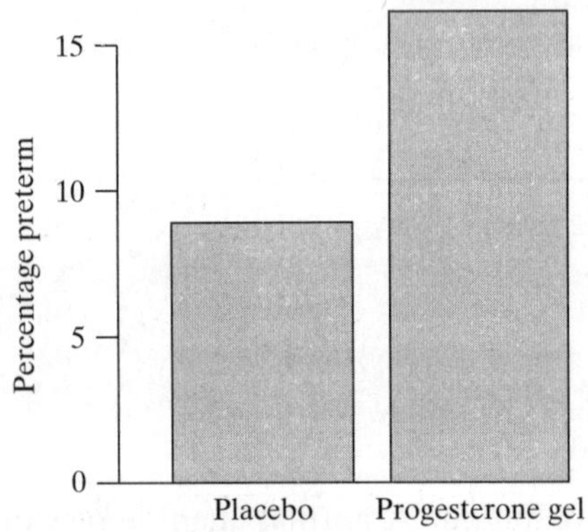

Figure 13.2.8 Bar chart of progesterone gel data

Table 13.2.2 Progesterone gel data

	Treatment		
	Placebo	Progesterone gel	
Preterm	36	21	57
Not preterm	187	214	401
Total	223	235	458

We can also construct a confidence interval with these data. Of placebo patients, $\frac{36}{223}$ or 16.1% had preterm births so that $\tilde{p}_1 = \frac{36+1}{223+2} = 0.164$. Of progesterone gel patients, $\frac{21}{235}$ or 8.9% had preterm births, so $\tilde{p}_2 = \frac{21+1}{235+2} = 0.093$. The standard error of the difference is

$$\mathrm{SE}_{(\tilde{P}_1 - \tilde{P}_2)} = \sqrt{\frac{(0.164)(1-0.164)}{223+2} + \frac{0.093(1-0.093)}{235+2}}$$
$$= 0.031$$

A 95% confidence interval for $p_1 - p_2$ is $(0.164 - 0.093) \pm 1.96(0.031)$ or $(0.010, 0.132)$. Thus, we are 95% confident that progesterone gel reduces the probability of preterm birth by between 1.0 and 13.2 percentage points.

We can also calculate the relative risk of preterm birth. The estimated relative risk is

$$\frac{\Pr\{\text{Preterm birth} \mid \text{Placebo}\}}{\Pr\{\text{Preterm birth} \mid \text{Progesterone gel}\}} = \frac{0.161}{0.089} = 1.81$$

Thus, we estimate that preterm birth is 1.81 times as likely when taking placebo as when taking progesterone gel. ■

Example 13.2.4

Cystic Fibrosis People who have cystic fibrosis tend to have high levels of chloride in their sweat. In fact, measuring sweat chloride is used to screen for the presence of cystic fibrosis. A group of adults with cystic fibrosis were randomly assigned to receive either a low dose, a medium dose, or a high dose of a drug that was designed to improve lung function.[4]

One outcome that was measured in each patient was sweat chloride concentration (mmol/L) 14 days after beginning to take the drug, which is a numeric variable that we will use as the response variable. The predictor variable, group membership (low, medium, or high dose), is categorical. Dotplots of the data are given in Figure 13.2.9; the data are summarized in Table 13.2.3.

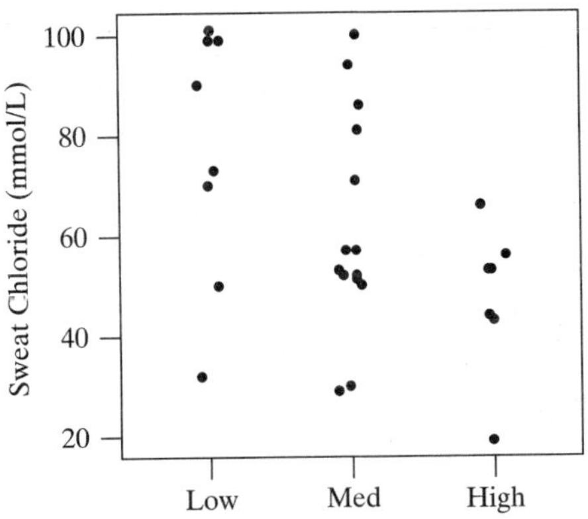

Figure 13.2.9 Dotplots of sweat chloride levels for cystic fibrosis experiment

Table 13.2.3 Sweat chloride levels for cystic fibrosis experiment

Dose	n	Mean	SD
Low	8	76.8	25.4
Medium	14	61.6	21.8
High	7	47.7	14.8

The dotplots suggest an effect due to dose (and we can speak of an effect, not just an association, because this was an experiment). This visual impression can be confirmed with an analysis of variance. The plots show that the distributions are reasonably symmetric. Moreover, the SDs are more or less similar among the groups. (The largest SD is no more than twice the smallest.) Thus, we can have confidence in the ANOVA P-value. The following ANOVA computer output confirms that there is evidence against H_0: $\mu_1 = \mu_2 = \mu_3$, although the evidence is not terribly strong. At the $\alpha = 0.05$ level, we conclude that chloride sweat concentration decreases as the dose of the drug increases.

	Df	Sum Sq	Mean Sq	F value	(Pr > F)
Group	2	3165	1583	3.42	0.048
Residuals	26	12044	463		
Total	28	15209			

Having found evidence that dose matters, we could compare doses with confidence intervals. There are three possible pairwise comparisons, the most interesting of which is the low versus high dose difference. The standard error of the difference in means is $\sqrt{463}\sqrt{\frac{1}{8} + \frac{1}{7}} = 11.1$. A 95% Bonferroni-adjusted confidence interval uses 2.566 as the t multiplier (from Table 11 at the end of the book). The confidence interval is $(76.8 - 47.7) \pm 2.566(11.1)$ or (1, 58). We are 95% confident that the high dose decreases sweat chloride on average by between 1 and 58 mmol/L more than does the low dose. The fact that this interval is entirely above zero confirms that there is a difference between the two dose levels.

Example 13.2.5 **Therapeutic Touch** Therapeutic touch (TT) is a form of alternative medicine in which a practitioner manipulates the human energy field of the patient. However, many persons have questioned the ability of TT practitioners to detect the human energy field—and whether the human energy field even exists. An experimenter tested the abilities of 28 TT practitioners as follows. A screen was set up between the experimenter and the practitioner, who sat on opposite sides of a table. The practitioner extended his or her hands under the screen and rested them, palms up, on the table. The researcher tossed a coin to choose one of the practitioner's hands. The experimenter then held her right hand, palm down, above the chosen hand of the practitioner. The practitioner was then asked to identify which hand had been chosen, as a test of whether the practitioner could detect a human energy field extending from the hand of the experimenter.

Each of the 28 TT practitioners was tested 10 times. The number of correct detections, "hits," in 10 trials varied from 1 to 8, with an average of 4.4. There were 123 hits in the 280 total trials.[5] Table 13.2.4 shows the distribution of hits among the 28 tested practitioners.

Table 13.2.4 Distribution of hits per 10 trials in therapeutic touch experiment

Number of hits	Number of practitioners
0	0
1	1
2	1
3	8
4	5
5	7
6	2
7	3
8	1
9	0
10	0
Total	28

The goal of this experiment was to determine the ability of TT practitioners to detect the human energy field. The response variable is a yes/no (categorical) variable: yes for a hit and no for a miss. There is no predictor variable here; there is just a single group of 28 TT practitioners who were tested.

Let p denote the probability of a hit in one of the trials of the experiment. The natural null hypothesis is H_0: $p = 0.5$. One way to analyze the data would be to conduct a chi-square goodness-of-fit test of H_0 using the 280 total trials, with a directional alternative H_A: $p > 0.5$. The P-value for this test is greater than 0.50, since the data do not deviate from H_0 in the direction specified by H_A.

One might argue that p might be greater than 0.5 for some TT practitioners, but perhaps not for all of them. If p is not the same for each TT practitioner (whether or not p is 0.5 for anyone), then the chi-square goodness-of-fit test using all 280 trials is not appropriate, since the 280 trials are not independent of one another. However, the data for each of the 28 practitioners could be analyzed separately. A binomial model could be used in these analyses, since the sample size of $n = 10$ is rather small. The binomial probabilities are given in Table 13.2.5. The probability of 8 or more hits in 10 trials, for a binomial with $p = 0.5$, is $0.04395 + 0.00977 + 0.00098 = 0.0547$. Thus, if the data from each of the 28 practitioners were analyzed separately, testing $H_0: p = 0.5$ versus $H_A: p > 0.5$, the smallest of the 28 P-values would be 0.0547, and again provides no significant evidence in support of H_A.*

Table 13.2.5 Observed and expected numbers (if $p = 0.5$) of hits per 10 trials in the therapeutic touch experiment

Number of hits	Binomial probability	Observed number, O	Expected number, E
0	0.00098	0	0.027
1	0.00977	1	0.273
2	0.04395	1	1.231
3	0.11719	8	3.281
4	0.20508	5	5.742
5	0.24609	7	6.891
6	0.20508	2	5.742
7	0.11719	3	3.281
8	0.04395	1	1.231
9	0.00977	0	0.273
10	0.00098	0	0.027
Total	1.00000	28	27.999

A different way to conduct the analysis is to investigate whether the 280 observations presented in Table 13.2.4 are consistent with a binomial model. In particular, we can check the model that states that Y has a binomial distribution, with $n = 10$ and $p = 0.5$, where Y is the number of hits in 10 trials. (This is similar to the analysis presented in Section 3.7.) A goodness-of-fit test can be used here. Table 13.2.5 shows the observed numbers (from Table 13.2.4) and expected numbers for each of the 11 possible outcomes.† (The expected numbers don't sum to 28 due to round-off error.)

The chi-square statistic is $\chi_s^2 = \Sigma \frac{(o_i - e_i)^2}{e_i} = 11.7$. The test statistic has 10 degrees of freedom, since there are 11 categories in the model. The P-value for this test is 0.306, which is quite large. Thus, the data are consistent with a binomial distribution for which $p = 0.5$ (i.e., the TT practitioners might as well have tossed coins to choose a hand, rather than trying to detect the human energy field of the experimenter). (*Note*: These data do not disprove the existence of the human energy field; they only fail to provide evidence for its existence.)

BRIEF EXAMPLES

We will now consider some examples for which we will identify the type of analysis that is appropriate, but we won't conduct the analysis.

*Considering the material in optional Section 11.9 on multiple comparisons, note that if we were to consider all 28 of these tests, we ought to require a great deal of evidence before rejecting H_0. Using the Bonferroni correction, we would require that an individual P-value be less than $\alpha_{cw} = 0.05/28 = 0.0018$ before rejecting H_0.

†This chi-square test as shown is for demonstration purposes. The reliability of these results is somewhat questionable as shown since many of the expected counts are small. Instead, we might consider forming number of hit categories (e.g., 0 through 3, 4, 5, 6, 7 through 10) so that the expected counts for each of these six categories is greater than 5.

Example 13.2.6 **Seastars** Researchers measured the length of the longest ray on each of over 200 members of the species *Phataria unifascialis* (a seastar found in the waters of the Gulf of California, Mexico). For a sample of 184 individuals found near Loreto, the average length was 6.78 cm, with an SD of 1.21 cm. For a sample of 77 individuals found near Bahia de Los Angeles, the average length was 8.13 cm, with an SD of 1.33 cm.[6]

The response variable is numeric, and there are two independent groups. Thus, a two-sample t test is appropriate, along with a confidence interval for the difference in population means. (*Note*: The normality condition is not essential, since the sample sizes are quite large.) ■

Example 13.2.7 **Twins** Researchers in Finland studied the physical activity levels of hundreds of sets of same-sex twins. In 1975 they classified subjects into the physical activity categories "exerciser" and "sedentary." They kept track of the health of the subjects through 1994, by which time there were several pairs of twins for whom one twin was alive, but the other had died. In this group there were 49 "sedentary" twins who were living, but whose "exerciser" twin pair was dead. There were 76 "exerciser" twins who were living, but whose "sedentary" twin pair was dead.[7]

The response variable in this observational study is whether or not a subject is alive, a categorical variable. The predictor is also categorical: whether the person is "sedentary" or is an "exerciser." Since the data are paired, McNemar's test is appropriate. ■

Example 13.2.8 **Soil Samples** Researchers took eight soil samples at each of six locations in Mediterranean pastures. They divided the samples into four pairs and put the soil in pots. One pot from each pair was watered continuously, while the other pot was watered for 13 days, then not watered for 18 days, and then watered again for 30 days. The researchers recorded the number of germinations in each pot during the experiment.[8]

This example is similar to Example 13.2.6, in that there are two samples to be compared and the response variable is numeric. However, the samples here are paired, so a paired analysis (Chapter 8) is called for. If the 24 sample differences show a normal distribution, then a paired t test or confidence interval could be used; if not, a transformation could be tried, or a Wilcoxon signed-rank or sign test could be used. ■

Example 13.2.9 **Vaccinations** In 1996 there was an outbreak of the disease varicella in a child care center in Georgia. Some of the children had been vaccinated against varicella, but others had not. Varicella occurred in 9 out of 66 vaccinated children and in 72 out of 82 unvaccinated children.[9]

The response and predictor variables in this experiment are both categorical. The data could be arranged into a 2×2 contingency table and analyzed with a chi-square test of independence. The difference in sample proportions is obviously quite large. However, this is an observational study and not an experiment. Thus, we cannot conclude that the difference in proportions is entirely due to the effect of the vaccine, since the effects of other variables, such as economic status, are confounded with the effect of the vaccine. ■

Example 13.2.10 **Estrogen and Steroids** Plasma estrone plus estradoil (Plasma E_{1+2}) steroid levels were measured in women given estrogen (Premarin) and in a control group of women. The women given estrogen were divided into three treatment groups. One group was given a daily dose of 0.625 mg, one group was given 1.25 mg, and the third group was given 2.5 mg. The researchers noted that the plasma steroid levels were not normally distributed, but became so after a logarithm transformation was applied. In log scale, the data are given in Table 13.2.6.[10]

Table 13.2.6 Log ng/100 ml plasma E_{1+2} concentration for estrogen study

Group	n	Mean	SD
Control	30	2.01	0.27
0.625	16	2.10	0.31
1.25	24	2.34	0.39
2.5	21	2.20	0.24

The response variable in this experiment, log(plasma E_{1+2} concentration), is numeric. It has already been transformed to normality. There are four independent groups to be compared, so an analysis of variance is appropriate. A contrast that compares the control to the average of the three treatment groups would also be useful. ■

Example 13.2.11

Damselflies A researcher captured male damselflies and randomly assigned them to one of three groups. For those in the first group, the sizes of red spots on the wing were artificially enlarged with red ink. For those in the second group, the wing spots were enlarged with clear ink. The third group served as a control. The damselflies were then released into a contained area. The numbers surviving in each of the three groups 22 days later were determined. There were 312 damselflies in each of the three groups. After 22 days there were 41 survivors in the "artificially enlarged with red ink" group, 49 survivors in the "enlarged with clear ink" group, and 57 survivors in the control group.[11]

The response variable in this experiment, survival, is categorical, as is the predictor variable, ink status/type. These data could be arranged into a 2 × 3 contingency table and analyzed with a chi-square test of independence. ■

Example 13.2.12

Tobacco Use Prevention In the Hutchinson Smoking Prevention Project 40 school districts in the state of Washington were formed into 20 pairs on the basis of size, location, and prevalence of high school tobacco use as of the beginning of the study. In each pair, one district was randomly assigned to be in an intervention group and the other was assigned to the control group. If a school district was in the intervention group, then the third-grade students in the district were given a curriculum on preventing tobacco use and the teachers in the district were given special training to help students refrain from smoking. This was repeated one year later with the next new cohort of third-grade students. All the students were then followed for several years. A primary outcome measurement of the study was whether or not students were smoking two years after graduating from high school.

The experimental unit here is an entire school district, so it is natural to use as the response variable the percentage of students from a district who smoke, a numeric variable. The predictor is categorical: intervention group or control group. There are two groups, which are paired together by the design of the experiment. Out of the 20 pairs, there were 13 pairs in which the smoking rate was higher in the control district and 7 pairs in which the smoking rate was higher in the intervention district.[12] A sign test could be used to analyze these data. ■

Example 13.2.13 **Reaction Times** Subjects in a reaction time study were asked to press a button as fast as possible after being exposed to either a visual stimulus (a circle flashing on a computer screen) or an auditory stimulus (a burst of white noise). Average reaction times (ms) were recorded for 10–20 trials for each type of stimulus for each subject. Reaction time to the visual stimulus can be used to predict reaction time to the auditory stimulus.[13]

The response variable is reaction time to the auditory stimulus, which is numeric. The predictor is also numeric: reaction time to the visual stimulus. A scatterplot of the data could be used to check for linearity (and in this case, the relationship is, indeed, linear). Then a regression line can be fit and a confidence interval constructed for the slope. ■

Exercises 13.2.1–13.2.24

13.2.1 Researchers conducted a randomized, double-blind, clinical trial in which some patients with schizophrenia were given the drug clozapine and others were given haloperidol. After one year 61 of 163 patients in the clozapine group showed clinically important improvement in symptoms, compared with 51 out of 159 in the haloperidol group.[14] Identify the type of statistical method that is appropriate for these data, but do not actually conduct the analysis.

13.2.2 Consider the data of Exercise 13.2.1. Conduct an appropriate complete analysis of the data that also includes a graphical display and discussion of how the data do or do not meet the necessary conditions for validity.

13.2.3 A biologist collected data on the height (in inches) and peak expiratory flow (PEF—a measure of how much air a person can expire, measured in l/min) for 10 women.[15] Here are the data:

Subject	Height	PEF
1	63	410
2	63	440
3	66	450
4	65	510
5	64	340
6	62	360
7	67	380
8	64	380
9	65	360
10	67	570

Is PEF related to height? Identify the type of statistical method that is appropriate for these data and this question, but do not actually conduct the analysis.

13.2.4 Consider the data of Exercise 13.2.3. Maria is 1 inch taller than Anika. Using the information from Exercise 13.2.3, how much greater would you predict Maria's PEF to be than Anika's?

13.2.5 A geneticist self-pollinated pink-flowered snapdragon plants and produced 97 progeny with the following colors: 22 red plants, 52 pink plants, and 23 white plants.[16] The purpose of this experiment was to investigate a genetic model that states that the probabilities of red, pink, and white are 0.25, 0.50, and 0.25. Identify the type of statistical method that is appropriate for these data, but do not actually conduct the analysis.

13.2.6 Consider the data of Exercise 13.2.5. Conduct an appropriate complete analysis of the data that also includes a graphical display and discussion of how the data do or do not meet the necessary conditions for validity.

13.2.7 The effect of diet on heart disease has been widely studied. As part of this general area of investigation, researchers were interested in the short-term effect of diet on endothelial function, such as the effect on triglyceride level. To study this, they designed an experiment in which 20 healthy subjects were given, in random order, a high-fat breakfast and a low-fat breakfast at 8 A.M., following a 12-hour fast, on days one week apart from each other. Serum triglyceride levels were measured on each subject before each breakfast and again 4 hours after each breakfast.[17] If you had access to all of the measurements collected in this experiment, how would you analyze the data?

13.2.8 Biologists were interested in the distribution of trees in a wooded area. They intended to use the number of trees per 100-square meter plot as their unit of measurement. However, they were concerned that the shapes of the plots might affect the data collection. To investigate the possibility, they counted the numbers of trees in square plots, round plots, and rectangular plots. The data are shown in the following table.[18] What type of analysis is appropriate for these data?

	Plot shape		
	Square	Round	Rectangular
	5	5	10
	5	7	2
	5	5	3
	8	2	12
	8	4	9
	7	4	5
	4	4	3
	9	7	6
	9	7	5
	7	10	3
	5	9	8
	2	2	9
	8	7	3
Mean	6.3	5.6	6.0
SD	2.14	2.47	3.27

13.2.9 Consider the data of Exercise 13.2.8. Conduct an appropriate complete analysis of the data that also includes a graphical display and discussion of how the data do or do not meet the necessary conditions for validity.

13.2.10 A sample of 15 patients was randomly split into two groups as part of a double-blind experiment to compare two pain relievers.[19] The 7 patients in the first group were given Demerol and reported the following numbers of hours of pain relief:

$$2, 6, 4, 13, 5, 8, 4$$

The 8 patients in the second group were given an experimental drug and reported the following numbers of hours of pain relief.

$$0, 8, 1, 4, 2, 2, 1, 3$$

How might these data be analyzed?

13.2.11 Consider the data of Exercise 13.2.10. Conduct an appropriate complete analysis of the data that also includes a graphical display and discussion of how the data do or do not meet the necessary conditions for validity.

13.2.12 A researcher was interested in the relationship between forearm length and height. He measured the forearm lengths and heights of a sample of 16 women and obtained the following data.[20] How might these data be (i) visualized and (ii) analyzed?

Forearm		Forearm	
Height (cm)	Length (cm)	Height (cm)	Length (cm)
163	25.5	157	26
161	26	178	27
151	25	163	24.5
163	25	161	26
166	27.2	173	28
168	26	160	24.5
170	26	158	25
163	26	170	26

13.2.13 A randomized, double-blind, clinical trial was conducted on patients who had coronary angioplasty to compare the drug lovastatin to a placebo. The percentage of stenosis (narrowing of the blood vessels) following angioplasty was measured on 160 patients given lovastatin and on 161 patients given the placebo. For the lovastatin group the average was 46%, with an SD of 20%. For the placebo group the average was 44%, with an SD of 21%.[21] What type of analysis is appropriate for these data?

13.2.14 Consider the data of Exercise 13.2.13.

(a) Conduct an appropriate analysis of the data.

(b) Describe a graphical procedure to visualize these data.

(c) Discuss how the data likely meet the necessary conditions for validity even though you do not have access to the raw data.

13.2.15 Researchers studied persons who had received intravenous immune globulin (IGIV) to see if they had developed infections of hepatitis C virus (HCV). In part of their analysis, they considered doses of Gammagard (an IGIV product) received by 201 patients. They divided the patients into 4 groups according to the number of doses of "Gammagard made from unscreened or first-generation anti-HCV-screened plasma." Among 48 persons who received 0 to 3 doses, there were 4 cases of HCV infection. There were 2 cases of HCV infection among 45 persons who received 4 to 20 doses, there were 7 cases of HCV infection in the 57 persons who received between 21 and 65 doses, and there were 10 cases of HCV infection among the 51 persons who received more than 65 doses.[22] What type of analysis is appropriate for these data?

13.2.16 Consider the data of Exercise 13.2.15. Conduct an appropriate analysis of the data.

13.2.17 An experiment was conducted to study the effect of tamoxifen on patients with cervical cancer. One of the measurements made, both before and again after tamoxifen was given, was microvessel density (MVD). MVD, which is measured as number of vessels per mm^2, is a measurement that relates to the formation of blood vessels that feed a tumor and allow it to grow and spread. Thus, small values of MVD are better than are large values. Data for 18 patients are shown.[23] How might these data be analyzed?

Patient	MVD before	MVD after	Patient	MVD before	MVD after
1	98	75	10	70	60
2	100	60	11	60	65
3	82	25	12	88	45
4	100	55	13	45	36
5	93	78	14	159	144
6	119	102	15	65	27
7	70	58	16	98	90
8	78	70	17	66	16
9	104	90	18	67	53

13.2.18 Consider the data of Exercise 13.2.17. Conduct an appropriate complete analysis of the data that also includes a graphical display and discussion of how the data do or do not meet the necessary conditions for validity.

13.2.19 As part of a large experiment, researchers planted 2,400 sweetgum, 2,400 sycamore, and 1,200 green ash seedlings. After 18 years, the survival rates were 93% for the sweetgum trees, 88% for the sycamore trees, and 95% for the green ash trees.[24] What type of analysis is appropriate for these data?

13.2.20 Consider the data of Exercise 13.2.19. Conduct an appropriate complete analysis of the data that also includes a graphical display and discussion of how the data do or do not meet the necessary conditions for validity.

13.2.21 A group of female college students were divided into three groups according to upper body strength. Their leg strength was tested by measuring how many consecutive times they could leg press 246 pounds before exhaustion. (The subjects were allowed only one second of rest between consecutive lifts.) The data are shown in the following table.[25] What type of analysis is appropriate for these data?

	Upper body strength group		
	Low	Middle	High
	55	40	181
	70	200	85
	45	250	416
	246	192	228
	240	117	257
	96	215	316
	225		134
Mean	140	169	231
SD	93	77	112

13.2.22 Consider the data of Exercise 13.2.21. Conduct an appropriate complete analysis of the data that also includes a graphical display and discussion of how the data do or do not meet the necessary conditions for validity.

13.2.23 In a study to better understand what sort of mammals are most vulnerable to being hit by cars on a particular highway in upstate New York, researchers sampled stretches of road in the investigation area over a period of 1 year and found 287 dead mammals killed by traffic. Of the 287 hit, 9 were carnivores, 195 were herbivores, and 83 were omnivores. Other research published by the United States Geological Survey (USGS) indicated that among mammals in this habitat, 22% are carnivores, 53% herbivores, and 25% are omnivores.[26] What type of analysis is appropriate for these data?

13.2.24 Consider the data of Exercise 13.2.23. Conduct an appropriate analysis of the data to investigate whether or not some mammals (based on classification by diet) are more vulnerable to being hit than others. In addition, discuss how the data do or do not meet the necessary conditions for validity.

CHAPTER APPENDICES

Appendix

Appendix 3.1
More on the Binomial Distribution Formula

In this appendix we explain more about the reasoning behind the binomial distribution formula.

THE BINOMIAL DISTRIBUTION FORMULA

We begin by deriving the binomial distribution formula for $n = 3$. Suppose that we conduct three independent trials and that each trial results in success (S) or failure (F). Further, suppose that on each trial the probabilities of success and failure are

$$\Pr\{S\} = p$$

$$\Pr\{F\} = 1 - p$$

There are eight possible outcomes of the three trials. Reasoning as in Example 3.6.3 shows that the probabilities of these outcomes are as follows:

Outcome	Number of Successes	Number of Failures	Probability
FFF	0	3	$(1-p)^3$
FFS	1	2	$p(1-p)^2$
FSF	1	2	$p(1-p)^2$
SFF	1	2	$p(1-p)^2$
FSS	2	1	$p^2(1-p)$
SFS	2	1	$p^2(1-p)$
SSF	2	1	$p^2(1-p)$
SSS	3	0	p^3

Again by reasoning parallel to Example 3.6.3, these probabilities can be combined to obtain the binomial distribution formula for $n = 3$ as shown in the table:

Number of Successes, j	Failures, $n - j$	Probability
0	3	$1p^0(1 - p)^3$
1	2	$3p^1(1 - p)^2$
2	1	$3p^2(1 - p)^1$
3	0	$1p^3(1 - p)^0$

This distribution illustrates the origin of the binomial coefficients. The coefficient ${}_3C_1(= 3)$ is the number of ways in which 2 S's and 1 F can be arranged; the coefficient ${}_3C_2(= 3)$ is the number of ways in which 1 S and 2 F's can be arranged.

An argument similar to this shows that the general formula (for any n) is

$$\Pr\{j \text{ successes and } n - j \text{ failures}\} = {}_nC_j p^j(1 - p)^{n-j}$$

where

$${}_nC_j = \text{the number of ways in which } j \text{ S's and } (n - j) \text{ F's can be arranged.}$$

Combinations The binomial coefficient ${}_nC_j$ is also known as the number of combinations of n items taken j at a time; it is equal to the number of different subsets of size j that can be formed from a set of n items.

THE BINOMIAL COEFFICIENTS: A FORMULA

Binomial coefficients can be calculated from the formula

$${}_nC_j = \frac{n!}{j!(n - j)!}$$

where $x!$ ("x-factorial") is defined for any positive integer x by

$$x! = x(x - 1)(x - 2) \cdots (2)(1)$$

and $0! = 1$.

For example, for $n = 7$ and $j = 4$ the formula gives

$$\begin{aligned}{}_7C_4 &= \frac{7!}{4!3!} = \frac{7 \times 6 \times 5 \times 4 \times 3 \times 2 \times 1}{(4 \times 3 \times 2 \times 1)(3 \times 2 \times 1)} \\ &= 35\end{aligned}$$

To see why this is correct, let us consider in detail why the number of ways of rearranging 4 S's and 3 F's should be equal to

$$\frac{7!}{4!3!}$$

Suppose 4 S's and 3 F's were written on cards, like this:

$$\boxed{S_1}\ \boxed{S_2}\ \boxed{S_3}\ \boxed{S_4}\ \boxed{F_1}\ \boxed{F_2}\ \boxed{F_3}$$

Temporarily we put subscripts on the S's and F's to distinguish them. First let us see how many ways there are to arrange the 7 cards in a row:

There are 7 choices for which card goes first;
for each of these, there are 6 choices for which card goes second;
for each of these, there are 5 choices for which card goes third;
for each of these, there are 4 choices for which card goes fourth;
for each of these, there are 3 choices for which card goes fifth;
for each of these, there are 2 choices for which card goes sixth;
for each of these, there is 1 choice for which card goes last.

It follows that there are 7! ways of arranging the 7 cards. Now consider the locations of the 4 S's. There are 4! ways in which the S's can be rearranged among themselves. Likewise, there are 3! ways in which the F's can be rearranged among themselves. If we were to ignore the subscripts on the S's and F's, then some of the 7! ways of arranging the 7 cards would be indistinguishable. Indeed, any rearrangement of the S's *among themselves* leaves the 7 card arrangement looking the same. Similarly, any rearrangement of the F's *among themselves* leaves the 7 card arrangement looking the same. Thus, the number of *distinguishable* arrangements is

$$\frac{7!}{4!3!}$$

Appendix 3.2 Mean and Standard Deviation of the Binomial Distribution

Suppose that Y is a binomial random variable with n trials and p as the probability of success on each trial. Then we can think of Y as the sum of n variables X_1, $X_2, \ldots, X_n$, where each X_i is equal to either 0 or 1 (0 for a failure or 1 for a success). That is, $Y = \Sigma X_i$, with $\Pr\{X_i = 0\} = 1 - p$ and $\Pr\{X_i = 1\} = p$. The n X_i's are a random sample from a hypothetical population of X's that has average $\mu_X = p$ (since $0 \times (1 - p) + 1 \times p = p$).

Now consider the population standard deviation, σ_X, for the population of X's. Recall, from Section 2.8 that for a variable X the definition of σ is

$$\sigma = \sqrt{\text{Population average value of } (X - \mu)^2}$$

For the population of X's, the mean is $\mu_X = p$. Thus, for this population,

$$\sigma_X = \sqrt{\text{Population average value of } (X - p)^2}.$$

In the population of X's, the quantity $(X - p)^2$ takes on only two possible values:

$$(X - p)^2 = \begin{cases} (0 - p)^2 & \text{if } X = 0 \\ (1 - p)^2 & \text{if } X = 1 \end{cases}$$

Furthermore, these values occur in the proportions $(1 - p)$ and p, respectively, so that the population average value of $(X - p)^2$ is equal to

$$(0 - p)^2 \times (1 - p) + (1 - p)^2 \times p$$

This can be simplified to

$$\begin{aligned} p^2 \times (1 - p) + (1 - p)^2 \times p &= p\{p(1 - p) + (1 - p)^2\} \\ &= p\{p - p^2 + 1 - 2p + p^2\} \\ &= p(1 - p) \end{aligned}$$

Hence, the population average value of $(X - p)^2$ is $p(1 - p)$, so $\sigma_X = \sqrt{p(1 - p)}$.

The binomial random variable Y is ΣX_i. To find the mean and standard deviation of Y, we need two facts:

Fact 1: For any collection of random variables $X_1, X_2, \ldots, X_n$ the mean of $\Sigma X_i = \Sigma(\text{mean of } X_i)$.

Fact 2: For a collection of independent random variables $X_1, X_2, \ldots, X_n$ the variance of $\Sigma X_i = \Sigma(\text{variance of } X_i)$.

(Recall that the variance, σ^2, is the square of the standard deviation, σ.)

Using Fact 1, we see that the mean of Y is the mean of ΣX_i, which is Σp. Thus, the mean of Y is $\mu_Y = np$.

Using Fact 2, the variance of Y is the variance of ΣX_i, which equals $\Sigma(\text{Variance of } X_i)$ or $np(1 - p)$. Thus, the standard deviation of Y is $\sigma_Y = \sqrt{np(1 - p)}$.

Appendix 5.1
Relationship between Central Limit Theorem and Normal Approximation to Binomial Distribution

Consider sampling from a dichotomous population. Theorem 5.4.1 states that the sampling distribution of $\hat{P}$, and the equivalent binomial distribution, can be approximated by normal distributions. In this appendix we show how these approximations are related to Theorem 5.2.1 and the Central Limit Theorem.

As shown in Appendix 3.2, if Y is a binomial random variable with n trials and p as the probability of success on each trial, then we can think of Y as the sum of n variables $X_1, X_2, \ldots, X_n$, where each X_i is equal to either 0 or 1 $-$ 0 for a failure or 1 for a success. For a population of 0's and 1's, where the proportion of 1's is given by p, the mean is p and the standard deviation is $\sigma = \sqrt{p(1-p)}$. The sample mean of $X_1, X_2, \ldots, X_n$ is $\bar{X}$, which is the same as the proportion of 1's in the sample (that is, $\hat{P}$). Thus, the sample proportion $\hat{P}$ can be regarded as a sample mean, and so its sampling distribution is described by Theorem 5.2.1.

From part 3 of Theorem 5.2.1 (the Central Limit Theorem), the sampling distribution of $\hat{P}$ is approximately normal if n is large. From part 1 of Theorem 5.2.1, the mean of the sampling distribution of $\hat{P}$ is equal to the population mean—that is, p; this value is given in Theorem 5.4.1(b). From part 2 of Theorem 5.2.1, the standard deviation of the sampling distribution of $\hat{P}$ is equal to

$$\frac{\sigma}{\sqrt{n}}$$

where σ represents the standard deviation of the population of 0's and 1's, which is $\sqrt{p(1-p)}$. Thus, the standard deviation of the sampling distribution of $\hat{P}$ is equal to

$$\frac{\sqrt{p(1-p)}}{\sqrt{n}} = \sqrt{\frac{p(1-p)}{n}}$$

which is the value given in Theorem 5.4.1(b).

Note that the binomial distribution is just a rescaled version of the sampling distribution of $\hat{P}$: $\hat{P} = \frac{Y}{n}$, so $Y = n\hat{P}$. It follows that the binomial distribution also can be approximated by a normal curve with suitably rescaled mean and standard deviation. The mean of $\hat{P}$ is p and the SD of $\hat{P}$ is

$$\sqrt{\frac{p(1-p)}{n}}$$

The rescaled mean is np and the rescaled standard deviation is

$$n\sqrt{\frac{p(1-p)}{n}} = \sqrt{np(1-p)}$$

which are as given in Theorem 5.4.1(a).

Appendix 7.1
How Power Is Calculated

The required sample sizes given in Table 5 were determined by calculating the power of the t test. For large samples, an appropriate power calculation can be based on the normal curve (Table 3). In this appendix we indicate how such an approximate calculation is done.

Recall that the power is the probability of rejecting H_0 when H_A is true. In order to calculate power, therefore, we need to know the sampling distribution of t_s when H_A is true. For large samples, the sampling distribution can be approximated by a normal curve, as shown by the following theorem.

Theorem A.1 Suppose we choose independent random samples, each of size n, from normal populations with means μ_1 and μ_2 and a common standard deviation σ. If n is large, the sampling distribution of t_s can be approximated by a normal distribution with

$$\text{Mean} = \frac{\mu_1 - \mu_2}{\sqrt{\dfrac{\sigma^2}{n} + \dfrac{\sigma^2}{n}}} = \sqrt{\frac{n}{2}}\left(\frac{\mu_1 - \mu_2}{\sigma}\right)$$

and

$$\text{Standard deviation} = 1$$

To illustrate the use of Theorem A.1 for power calculations, suppose we are considering a one-tailed t test with $\alpha = 0.025$. The hypotheses are

$$H_0\colon \mu_1 = \mu_2$$
$$H_A\colon \mu_1 > \mu_2$$

If we want a power of 0.80 for an effect size of 0.4, then Table 5 recommends samples of size of $n = 100$. Let us confirm this recommendation using Theorem A.1.

If H_0 is true, so that $\mu_1 = \mu_2$, then the sampling distribution of t_s is approximately a normal distribution with mean equal to 0 and SD equal to 1. This is the null distribution of t_s; it is shown as the dashed curve in Figure A.7.1.

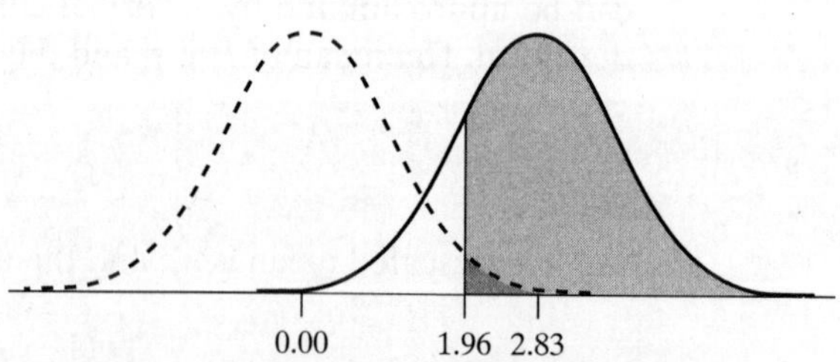

Figure A.7.1

Suppose that in fact H_A is true, that the effect size is

$$\frac{\mu_1 - \mu_2}{\sigma} = 0.4$$

and that we are using samples of size $n = 100$. Then, according to Theorem A.1, the sampling distribution of t_s will be approximately a normal distribution with SD equal to 1 and mean equal to

$$\sqrt{\frac{n}{2}}\left(\frac{\mu_1 - \mu_2}{\sigma}\right) = \sqrt{\frac{100}{2}}(0.4) = 2.83$$

This distribution is the solid curve in the figure.

For $n_1 = n_2 = 100$, we have df $\approx \infty$, so from Table 4 the critical value is equal to 1.96. Thus, the P-value would be less than 0.025 and we would reject H_0 if

$$t_s > 1.96$$

Using the dotted curve, the probability of this event is equal to 0.025; this is shown in the figure as the dark gray area. Using the solid curve, the probability that $t_s > 1.96$ includes all the shaded area in the figure. The shaded area can be determined from Table 3 using

$$Z = \frac{1.96 - 2.38}{1} = -0.87$$

From Table 3, the area is $0.8078 \approx 0.81$. Thus, we have shown that, for $n_1 = n_2 = 100$,

$$\text{if } \frac{\mu_1 - \mu_2}{\sigma} = 0.4, \text{ then } \Pr\{\text{reject } H_0\} \approx 0.81$$

We have found that the power against the specified alternative is approximately equal to 0.81; this agrees well with Table 5, which claims that the power is equal to 0.80.

If we were concerned with a two-tailed test at $\alpha = 0.05$, the critical value would again be 1.96, and so the power would again be approximately equal to 0.81, because the area under the solid curve corresponding to the left-hand tail of the dashed curve is negligible.

Of course, in constructing Table 5, one begins with the specified power (0.80) and determines n, rather than the other way around. This "inverse" problem can be solved using an approach similar to the foregoing. In the figure, the shaded area (0.80) would be given; this would determine the Z value and in turn determine n, once the effect size is specified.

Chapter Notes

Chapter 1

8. The data are approximate, having been reconstructed from the dotplots and summary information given by Allen and Gorski. Regarding the first concern mentioned in Example 1.2.2, the authors were mindful of the effect that the two largest observations could have on their conclusions and calculated the average for the homosexual men a second time, after deleting these two values. As for the second concern, the authors calculated the averages for those who had AIDS and those who did not in each group of men. They found that AIDS is associated with smaller, not larger, AC areas, so that when only persons without AIDS are compared, the difference between homosexual and heterosexual men is even larger than the difference found in the full data set.

16. Several variables were measured on each subject at the start of the experiment. Adjusting for the effects of these covariates within the placebo group only slightly reduces the difference in mortality rates between adherers and nonadherers. Thus, differences in overall health explain only a small part of the "adherer versus nonadherer" mortality rate difference.

Chapter 3

16. This is one of the crosses performed by Gregor Mendel in his classic studies of heredity; heterozygous plants (which are yellow seeded because yellow is dominant) are crossed with each other.

Chapter 4

5. The fish are young of the year, observed in October; they are quite small. (The distribution of lengths in older populations is not approximately normal.)

7. The standard deviation given in this problem is realistic for an idealized "uniform" field, in which yield differences between plots are due to local random variation rather than large-scale and perhaps systematic variation.

15. The distribution is actually a discrete distribution called a Poisson distribution; however, a Poisson distribution with large mean is approximately normal.

Chapter 5

11. The distribution included additional peaks, because sometimes the subject fumbled the button more than once on a single trial.

Chapter 6

35. All subjects were men, age 18–21, with heights between 175 and 183 cm. Because vital capacity is related to height, the raw data were adjusted slightly, using linear regression, to control for the effect of height.

Chapter 7

6. Some of the animals *lost* weight during the 78 days, so that the mean weight gains are based on both positive and negative values.

60. There were two control groups in this study. The control group included in this analysis is "patients undergoing reverse augmentation mammaplasty" (the "scar" group discussed in the article). Also, the authors neglected to transform the data before conducting a t test. Thus, they got a large P-value, although they noted that the two groups looked quite different.

65. It is sometimes stated that the validity of the Mann–Whitney test requires that the two population distributions have the same shape and differ only by a shift. This is not correct. The computations underlying Table 6 require only that the common population distribution (under the null hypothesis) be continuous. A further property, technically called *consistency* of the test, requires that the two distributions be *stochastically ordered*, which is the technical way of saying that one of the variables has a consistent tendency to be larger than the other. In fact, the title of Mann and Whitney's original paper is "On a test of whether one of two random variables is stochastically larger than the other" (*Annals of Mathematical Statistics* **18**, 1947). In Section 7.12 we discuss the requirement of stochastic ordering, calling it an "implicit assumption." (The confidence interval procedure mentioned at the end of Section 7.10 does require the stronger assumption that the distributions have the same shape.)

66. Simulations have been used to show that the Wilcoxon-Mann-Whitney test is more powerful than the t test in the presence of outliers, but that in the absence of outliers, the t test is slightly preferable for a variety of population distributions.

Chapter 8

10. In a study in which there is no natural pairing (for example, if identical twins are not available), one may wish to take two equal size groups and create pairs by using covariates such as age and weight. If an experiment is conducted in which members of a pair are randomly assigned to opposite treatment groups, then a paired data analysis has good properties. However, if an observational study is conducted (so that there is no random assignment within pairs), then a paired analysis, such as a paired t test, will tend to understate the true variability of the difference being studied and the true Type I error rate of a t test will be greater than the nominal level of the test. For discussion, see David, H. A., and Gunnink, J. L. (1997). The paired t test under artificial pairing. *The American Statistician* **51**, 9–12.

Chapter 9

3. Agresti, A., and Coull, B. A. (1998). Approximate is better than "exact" for interval estimation of binomial proportions. *The American Statistician* **52**, 119–126. The authors show that 95% confidence intervals based on $\tilde{p}$ are superior to other commonly used confidence intervals. They also note that if one uses $\tilde{p}$, then it is not necessary to construct tables or rules for how large the sample size needs to be in order for the confidence interval to have good coverage properties.

Chapter 10

47. Agresti, A., and Caffo, B. (2000). Simple and effective confidence intervals for proportions and differences of proportions result from adding two successes and two failures. *The American Statistician* **54**, 280–288. Agresti and Caffo conduct a series of simulations which show that adding 1 to each cell results in good coverage properties when the sample sizes, n_1 and n_2, are as small as 10. Unpublished calculations done by J. Witmer show that these good properties are also obtained when n_1 and n_2 are as small as 5, provided p_1 and p_2 are not both close to 0 or both close to 1, in which case the interval becomes quite conservative (i.e., the coverage rate approaches 100% for a nominal 95% confidence interval).

Chapter 11

15. In order to demonstrate the variability of plot yields, the experimenters planted the *same* variety of barley in all 16 plots.

Answers to Selected Exercises

Chapter 1

1.2.3 The acupuncturist expects acupuncture to work better than aspirin, so she or he is apt to "see" more improvement in someone given acupuncture than in someone given aspirin—even if the two groups are truly equivalent to each other in their response to treatment.

1.3.1 **(a)** Cluster sampling. The three clinics are the three clusters.

1.3.2 **(a)** The sample is non-random and likely non-representative of the general population because it consists of (1) volunteers from (2) nightclubs. (i) The social anxiety level of people who attend nightclubs is likely lower than the social anxiety level of the general public. (ii) A better sampling strategy would be to recruit subjects from across the population.

Chapter 2

2.1.2 **(a)** (i) Height and weight, (ii) Continuous variables, (iii) A child, (iv) 37 **(b)** (i) Blood type and cholesterol level, (ii) Blood type is categorical, cholesterol level is continuous, (iii) A person, (iv) 129

2.2.1 **(a)** There is no single correct answer. One possibility is:

Molar width	Frequency (no. specimens)
[5.4, 5.6)	1
[5.6, 5.8)	5
[5.8, 6.0)	7
[6.0, 6.2)	12
[6.2, 6.4)	8
[6.4, 6.6)	2
[6.6, 6.8)	1
Total	36

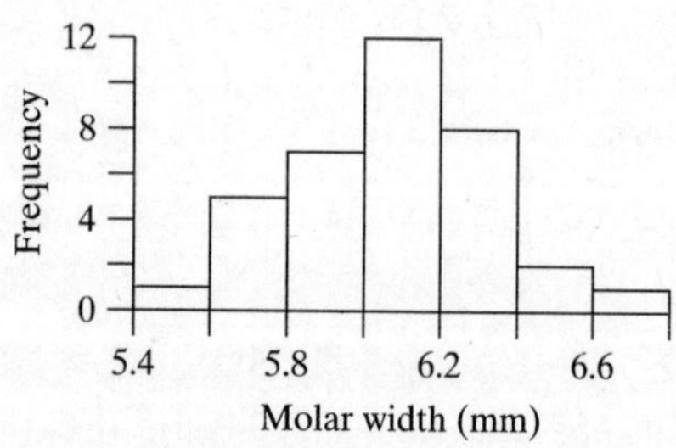

(b) The distribution is fairly symmetric

2.2.7 There is no single correct answer. One possibility is:

Glucose (%)	Frequency (no. of dogs)
[70, 75)	3
[75, 80)	5
[80, 85)	10
[85, 90)	5
[90, 95)	2
[95, 100)	2
[100, 105)	1
[105, 110)	1
[110, 115)	0
[115, 120)	1
[120, 125)	0
[125, 130)	0
[130, 135)	1
Total	31

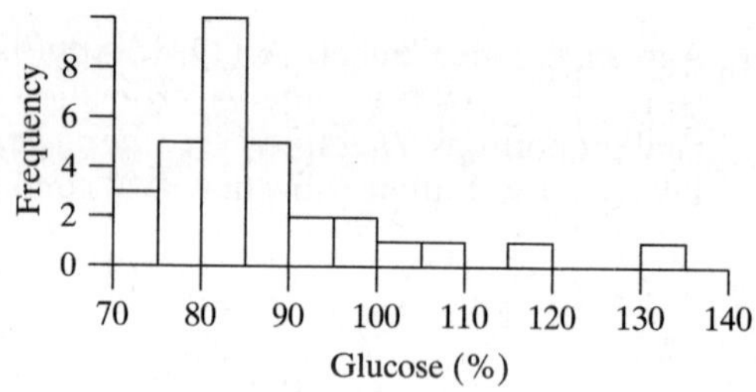

2.3.1 Any sample with $\Sigma y_i = 100$ would be a correct answer. For example: 18, 19, 20, 21, 22.

2.3.5 $\bar{y} = 293.8$ mg/dl; median $= 283$ mg/dl.

2.3.6 $\bar{y} = 309$ mg/dl; median $= 292$ mg/dl.

2.3.11 Median $= 10.5$ piglets

2.3.13 Mean $\approx$ median ≈ 50

2.4.2 **(a)** Median $= 9.2$, $Q_1 = 7.4$, $Q_3 = 11.9$ **(b)** IQR $= 4.5$ **(c)** Upper fence $= 18.65$ **(d)**

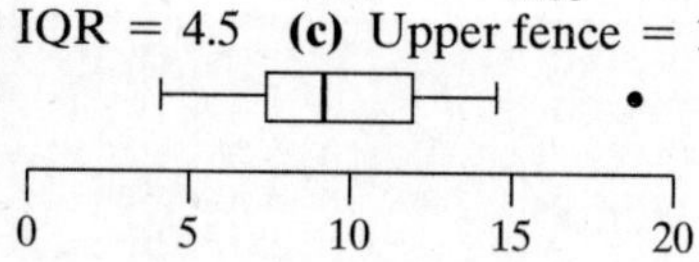

2.6.1 **(a)** $s = 2.45$ **(b)** $s = 3.32$

2.6.4 $\bar{y} = 33.10$ lb; $s = 3.44$ lb.

2.6.9 **(a)** 32.23 ± 8.07 contains 10/15 or 67% of the observations **(b)** 16.09 to 48.37 contains 15/15 or 100% of the observations

2.6.14 4%.

2.6.15 $\bar{y} = 45; s = 12.$

2.7.1 Mean = 37.3; SD = 12.9

2.S.8 30%

2.S.15 **(a)** Mean = 38.45 **(b)** SD = 3.20 **(c)** 100% **(d)** CV = 8.3%

Chapter 3

3.2.1 **(a)** 0.51 **(b)** 0.94 **(c)** 0.46 **(d)** 0.54

3.2.6 **(a)** 0.107 **(b)** 0.0585

3.2.7 **(a)** 0.916 **(b)** 0.838

3.3.1 **(a)** 0.185 **(b)** 0.117 **(c)** No; Pr{Smoke}≠Pr{Smoke | High income}

3.4.3 **(a)** 0.62 **(b)** 0.65 **(c)** 0.35

3.5.5 0.9

3.5.6 0.794

3.6.8 **(a)** 0.3746 **(b)** 0.0688 **(c)** 0.1254

3.6.11 **(a)** $0.75^6 = 0.1780$ **(b)** $1 - 0.1780 = 0.8220$

3.7.1 expected frequencies: 939.5; 5,982.5; 15,873.1; 22,461.8; 17,879.3; 7590.2; 1,342.6.

3.S.3 0.3675

3.S.7 **(a)** $1 - 0.95^{100} = 0.9941$
(b) $1 - 0.95^n \geq 0.94$, so $n \geq \log(.06)/\log(0.95)$, so $n \leq 55$

3.S.10 **(a)** 0.0076 **(b)** 0.0019

3.S.12 **(a)** 9 **(b)** 20862.

Chapter 4

4.3.3 **(a)** 84.13% **(b)** 61.47% **(c)** 77.34% **(d)** 22.66% **(e)** 20.38% **(f)** 20.38%

4.3.4 **(a)** 22.66% **(b)** 20.38%

4.3.8 **(a)** 90.7 lb; **(b)** 85.3 lb

4.3.12 **(a)** 98.76% **(b)** 98.76% **(c)** 1.24%

4.4.3

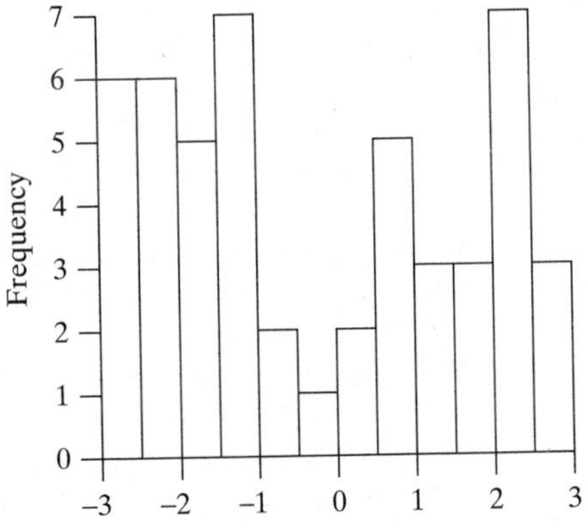

(a)

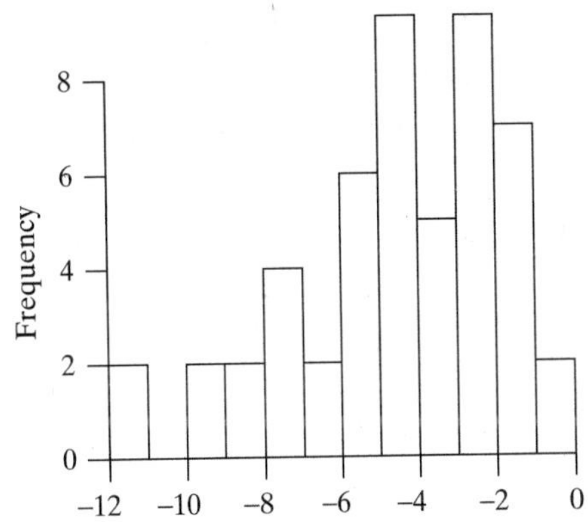

(b)

4.4.8 **(a)** No; the small *P*-value provides evidence of non-normality.

4.S.4 **(a)** 97.98% **(b)** 12.71% **(c)** 46.39% **(d)** 10.69% **(e)** 35.51% **(f)** 5.59% **(g)** 59.10%

4.S.5 0.0559

4.S.6 173.2 cm

4.S.8 **(a)** 1.3 **(b)** 1.65

4.S.15 0.1056

4.S.18 200

Chapter 5

5.1.2 23/64

5.2.4 **(a)** 28.86% **(b)** 73.30% **(c)** 0.7330

5.2.6 **(a)** 0.6680

5.2.10 **(a)** 0.1861 **(b)** 0.9044

5.2.13 **(a)** 0.1056 **(b)** 0.0150

5.2.15 **(a)** 41.5 **(b)** 2.35

5.3.1 **(a)** 0.66 **(b)** 0.29

5.4.2 **(a)** 0.1762 **(b)** 0.1742

5.4.5 **(a)** 0.7198

5.4.7 0.9708

5.S.1 0.9947

5.S.9 71.23%

Unit I

I.1 **(a)** Graph I **(b)** The median is less than the mean

I.4 The sampling distribution of the sample percentage is the distribution of $\hat{P}$, the sample percentage of mice weighing more than 26 gm, as it varies from one sample of 20 mice to another in repeated samples.

I.5 **(a)** 0.514 **(b)** 0.873

I.10 0.674

I.15 **(a)** 68.48 mg/L **(b)** 85th percentile **(c)** 0.0010

Chapter 6

6.2.1 **(a)** 51.3 ng/gm **(b)** 26.5 ng/gm

6.3.3 **(a)** SE = 3.9 mg **(b)** (23.4,40.0)

6.3.11 **(a)** $4.1 < \mu < 21.9$ pg/ml.

(b) We are 95% confident that the average drop in HBE levels from January to May in the population of all participants in physical fitness programs like the one in the study is between 4.1 and 21.9 pg/ml.

6.3.15 $1.17 < \mu < 1.23$ mm

6.3.20 2.81

6.4.2 178 men

6.4.6 36 plants

6.5.1 The SD is larger than the mean, but negative values are not possible. Thus, the distribution must be skewed to the right.

6.5.6 The observational units (aliquots) are nested within the units (flasks) that were randomly allocated to treatments. Therefore, the six observations may not be independent of one another, because of the hierarchical structure in the data, so that the experimenter's method of calculating the SE may not be valid.

6.6.1 2.41

6.6.7 0.44

6.7.1 $1.46 < \mu_1 - \mu_2 < 3.34$

6.7.6 **(a)** $-5 < \mu_1 - \mu_2 < 9$ sec.

(b) We are 90% confident that the population mean prothrombin time for rats treated with an antibiotic (μ_1) is smaller than that for control rats (μ_2) by an amount that might be as much as 5 seconds or is larger than that for control rats (μ_2) by an amount that might be as large as 9 seconds.

6.S.2 **(a)** $\bar{y} = 2.275$; $s = 0.238$; SE = 0.084 **(b)** (2.08,2.47)

(c) μ = population mean stem diameter of plants of Tetrastichon wheat three weeks after flowering.

6.S.4 23 plants

6.S.9 **(a)** We must be able to view the data as a random sample of independent observations from a large population that is approximately normal.

(b) Normality of the population

(c) Independence of the observations would be questionable, because birthweights of the members of a twin pair might be dependent.

Chapter 7

Remark concerning tests of hypotheses The answer to a hypothesis testing exercise includes verbal statements of the hypotheses and a verbal statement of the conclusion from the test in the context of the problem. In phrasing these statements, we have tried to capture the essence of the biological question being addressed; nevertheless the statements are necessarily oversimplified and they gloss over many issues that in reality might be quite important. For instance, the hypotheses and conclusion may refer to a causal connection between treatment and response; in reality the validity of such a causal interpretation usually depends on a number of factors related to the design of the investigation (such as unbiased allocation of animals to treatment groups) and to the specific experimental procedures (such as the accuracy of assays or measurement techniques). In short, the student should be aware that the verbal statements are intended to clarify the *statistical* concepts; their *biological* content may be open to question.

7.1.2 **(b)** 9

7.2.1 **(a)** $t_s = -3.13$ so $0.02 < P\text{-value} < 0.04$

(b) $t_s = 1.25$ so $0.20 < P < 0.40$

(c) $t_s = 4.62$ so $P < 0.001$

7.2.3 **(a)** yes **(b)** no **(c)** yes **(d)** no

7.2.7 **(a)** H_0: Mean serotonin concentration is the same in heart patients and in controls ($\mu_1 = \mu_2$); H_A: Mean serotonin concentration is not the same in heart patients and in controls ($\mu_1 \neq \mu_2$). $t_s = -1.38$. H_0 is not rejected. There is insufficient evidence ($0.10 < P < 0.20$) to conclude that serotonin levels are different in heart patients than in controls.

7.2.11 H_0: Flooding has no effect on ATP ($\mu_1 = \mu_2$); H_A: Flooding has some effect on ATP ($\mu_1 \neq \mu_2$). $t_s = -3.92$. H_0 is rejected.

7.3.4 Type II.

7.3.6 Yes; because zero is outside of the confidence interval, we know that the P-value is less than .05, so we reject the hypothesis that $\mu_1 - \mu_2 = 0$.

7.4.1 People with respiratory problems move to Arizona (because the dry air is good for them).

7.4.4 **(a)** Coffee consumption rate

(b) Coronary heart disease (present or absent)

(c) Subjects (i.e., the 1,040 persons)

7.5.1 **(a)** $0.10 < P < 0.20$ **(b)** $0.03 < P < 0.04$

7.5.3 **(a)** yes **(b)** yes **(c)** yes **(d)** no

7.5.9 H_0: Wounding the plant has no effect on larval growth ($\mu_1 = \mu_2$); H_A: Wounding the plant tends to diminish larval growth ($\mu_1 < \mu_2$), where 1 denotes wounded and 2 denotes control. $t_s = -2.69$. H_0 is rejected. There is sufficient evidence ($0.005 < P < 0.01$) to conclude that wounding the plant tends to diminish larval growth.

7.5.10 **(a)** H_0: The drug has no effect on pain ($\mu_1 = \mu_2$); H_A: The drug increases pain relief ($\mu_1 > \mu_2$). $t_s = 1.81$. H_0 is rejected. There is sufficient evidence ($0.03 < P < 0.04$) to conclude that the drug is effective.

(b) The P-value would be between 0.06 and 0.08. At $\alpha = 0.05$ we would not reject H_0.

7.6.4 No, according to the confidence interval the data do not indicate whether the true difference is "important."

7.6.6 0.33

7.7.1 **(a)** 23 **(b)** 11

7.7.4 **(a)** 71 **(b)** 101 **(c)** 58

7.7.6 0.5.

7.10.1 **(a)** $P > 0.20$ **(b)** $0.02 < P < 0.05$ **(c)** $P < 0.01$

7.10.3 **(a)** H_0: Toluene has no effect on dopamine in rat striatum; H_A: Toluene has some effect on dopamine in rat striatum. $U_s = 32$. H_0 is rejected. There is sufficient evidence ($0.02 < P < 0.05$) to conclude that toluene increases dopamine in rat striatum.

7.S.2 H_0: Mean platelet calcium is the same in people with high blood pressure as in people with normal blood pressure ($\mu_1 = \mu_2$); H_A: Mean platelet calcium is different in people with high blood pressure than in people with normal blood pressure ($\mu_1 \neq \mu_2$).
$t_s = 11.2$. H_0 is rejected. There is sufficient evidence ($P < 0.0001$) to conclude that platelet calcium is higher in people with high blood pressure.

7.S.4 No; the t test is valid because the sample sizes are rather large.

7.S.8 H_0: Stress has no effect on growth; H_A: Stress tends to retard growth. $U_s = 148.5$. H_0 is rejected. There is sufficient evidence ($P < 0.0005$) to conclude that stress tends to retard growth.

Chapter 8

8.2.1 **(a)** 0.34.

8.2.3 H_0: Progesterone has no effect on cAMP($\mu_1 = \mu_2$); H_A: Progesterone has some effect on cAMP ($\mu_1 \neq \mu_2$). $t_s = 3.4$. H_0 is rejected. There is sufficient evidence ($0.04 < P\text{-value} < 0.05$) to conclude that progesterone decreases cAMP under these conditions.

8.2.6 **(a)** $-0.50 < \mu_1 - \mu_2 < 0.74$°C, where 1 denotes treated and 2 denotes control.

8.4.1 **(a)** $P > 0.20$ **(b)** $P = 0.180$ **(c)** $P = 0.039$ **(d)** $P = 0.004$

8.4.4 H_0: Weight of the cerebral cortex is not affected by environment ($p = 0.5$); H_A: Environmental enrichment increases cortex weight ($p > 0.5$). $B_s = 10$. H_0 is rejected. There is sufficient evidence $P = 0.0195$) to conclude that environmental enrichment increases cortex weight.

8.4.8 0.000061

8.4.11 $n = 6$; P-value $= 0.03125$

8.5.1 **(a)** $P > 0.20$ **(b)** $P = 0.078$ **(c)** $P = 0.047$ **(d)** $P = 0.016$

8.5.3 H_0: Hunger rating is not affected by treatment (mCPP vs. placebo); H_A: Treatment does affect hunger rating. $W_s = 27$ and $n_d = 8$. H_0 is not rejected. There is insufficient evidence ($P > 0.20$) to conclude that treatment has an effect.

8.6.4 No. "Accurate" prediction would mean that the individual differences (d's) are small. To judge whether this is the case, one would need the individual values of the d's; using these, one could see whether most of the magnitudes ($|d|$'s) are small.

8.S.8 H_0: The average number of species is the same in pools as in riffles ($\mu_1 = \mu_2$); H_A: The average numbers of species is less in pools than in riffles ($\mu_1 < \mu_2$). $t_s = 4.58$. H_0 is rejected. There is sufficient evidence ($P < 0.001$) to conclude that the average number of species in pools is lesser than in riffles.

8.S.12 H_0: Caffeine has no effect on RER ($\mu_1 = \mu_2$); H_A: Caffeine has some effect on RER ($\mu_1 \neq \mu_2$). $t_s = 3.94$. H_0 is rejected. There is sufficient evidence ($0.001 < P < 0.01$) to conclude that caffeine tends to decrease RER under these conditions.

Unit II

II.2 **(a)** $0.001 < P < 0.01$ so reject H_0
(b) There is strong evidence of a difference between average ecological footprint of women and men.
(c) (1.00, 4.46) hectares so the difference is "ecologically important."

II.4 The confidence interval excludes zero, so we would reject H_0; this means that the P-value is less than 0.05. Thus the P-value is less than 0.10, so we would reject H_0.

II.8 **(a)** Power goes up as n goes up. Thus, if there is a true difference we are more likely to detect it when $n = 18$ than when $n = 12$.
(b) The effect size is 1.5.

II.9 **(a)** $P = 5/28 \approx 0.1563$
(b) No; the P-value is greater than 0.10
(c) $P = 3/28 \approx 0.107$

II.20 SD

Chapter 9

9.1.2 **(a)** 0.227 **(b)** 0.435
(c) No; the fewest mutants possible is zero, in which case $\tilde{P}$ is 2/7

9.1.4 **(a)** 0.2501 **(b)** 0.0352

9.1.5 **(a)** (i) 0.3164, (ii) 0.4219, (iii) 0.2109, (iv) 0.0469, (v) 0.0039.

9.1.9 0.5053.

9.2.2 **(a)** 0.040 **(b)** 0.020

9.2.3 **(a)** (0.134, 0.290) **(b)** (0.164, 0.242)

9.2.5 **(a)** (0.164, 0.250);
(b) We are 95% confident that the probability of adverse reaction in infants who receive their first injection of vaccine is between 0.164 and 0.250.

9.2.7 $n \geq 146$.

9.3.4 (0.646, 0.838)

9.4.1 H_0: The population ratio is 12:3:1 (Pr{white} = 0.75, Pr{yellow} = 0.1875, Pr{green} = 0.0625); H_A: The ratio is not 12:3:1. $\chi^2_s = 0.69$. H_0 is not rejected. There is little or no evidence ($P > 0.20$) that the model is not correct; the data are consistent with the model.

9.4.2 H_0 and H_A as in Exercise 10.1. $\chi_s^2 = 6.9$. H_0 is rejected. There is sufficient evidence ($0.02 < P < 0.05$) to conclude that the model is incorrect; the data are not consistent with the model.

9.4.8 H_0: The drug does not cause tumors $\left(\Pr\{T\} = \frac{1}{3}\right)$; H_A: The drug causes tumors $\left(\Pr\{T\} > \frac{1}{3}\right)$, where T denotes the event that a tumor occurs first in the treated rat. $\chi_s^2 = 6.4$. H_0 is rejected. There is sufficient evidence ($0.005 < P < 0.01$) to conclude that the drug does cause tumors.

9.S.2 **(a)** 0.2182 **(b)** 0.5981

9.S.3 (0.707, 0.853).

9.S.14 **(a)** H_0: Directional choice is random (Pr{toward} = 0.25, Pr{away} = 0.25, Pr{right} = 0.25, Pr{left} = 0.25); H_A: Directional choice is not random. $\chi_s^2 = 4.88$. H_0 is not rejected. There is insufficient evidence ($0.10 < P < 0.20$) to conclude that the directional choice is not random.

9.S.16 H_0: The probability of an egg being on a particular type of bean is 0.25 for all four types of beans; H_A: H_0 is false. $\chi_s^2 = 2.23$. H_0 is not rejected. There is insufficient evidence ($P > 0.20$) to conclude that cowpea weevils prefer one type of bean over the others.

Chapter 10

10.2.3 **(a)**

5	20
10	40

(b) $\hat{p}_1 = 5/15 = 1/3$ and $\hat{p}_2 = 20/60 = 1/3$; yes

10.2.5 H_0: Mites do not induce resistance to wilt ($p_1 = p_2$); H_A: Mites do induce resistance to wilt ($p_1 < p_2$), where p denotes the probability of wilt and 1 denotes mites and 2 denotes no mites. $\chi_S^2 = 7.21$. H_0 is rejected. There is sufficient evidence ($0.0005 < P\text{-value} < 0.005$) to conclude that mites do induce resistance to wilt.

10.2.10 H_0: The two timings are equally effective ($p_1 = p_2$); H_A: The two timings are not equally effective ($p_1 \neq p_2$). $\chi_S^2 = 4.48$. H_0 is rejected. There is sufficient evidence ($0.02 < P\text{-value} < 0.05$) to conclude that the simultaneous timing is superior to the sequential timing.

10.2.13 H_0: Ancrod and placebo are equally effective ($p_1 = p_2$); H_A: Ancrod and placebo are not equally effective ($p_1 \neq p_2$). $\chi_S^2 = 3.82$. We do not reject H_0; there is insufficient evidence ($0.05 < P\text{-value} < 0.10$) to conclude that the treatments differ.

10.3.3 **(a)** $\widehat{\Pr}\{D|S\} = 0.239$, $\widehat{\Pr}\{D|WW\} = 0.305$, $\widehat{\Pr}\{S|D\} = 0.439$, $\widehat{\Pr}\{S|A\} - 0.522$.
(b) H_0: There is no association between antibody and survival (Pr{D|S} = Pr{D|WW}); H_A: There is some association between treatment method (surgery vs. watchful waiting) and survival (Pr{D|S} ≠ Pr{D|WW}). H_0 is rejected. There is insufficient evidence ($0.05 < P\text{-value} < 0.10$) to conclude that the survival rates differ for the two treatments.

10.3.4 **(a)** $\widehat{\Pr}\{RF|RH\} = 0.934$ **(b)** $\widehat{\Pr}\{RF|LH\} = 0.511$ **(c)** $\chi_S^2 = 398$ **(d)** $\chi_S^2 = 1{,}623$

10.4.1

5	1
9	15

6	0
8	16

10.5.3 **(a)** H_0: The blood type distributions are the same for ulcer patients and controls (Pr{O|UP} = Pr{O|C}, Pr{A|UP} = Pr{A|C}, Pr{B|UP} = Pr{B|C}, Pr{AB|UP} = Pr{AB|C}); H_A: The blood type distributions are not the same. H_0 is rejected. There is sufficient evidence ($P\text{-value} < 0.0001$, df = 3) to conclude that the blood type distribution of ulcer patients is different from that of controls.

10.5.5 **(a)** H_0: Change in ADAS-Cog score is independent of treatment; H_A: Change in ADAS-Cog score is related to treatment. $\chi_S^2 = 10.26$, df = 4. H_0 is rejected. There is sufficient evidence ($0.02 < P\text{-value} < 0.05$) to conclude that EGb and placebo are not equally effective.

10.6.2 This analysis is not appropriate because the observational units (mice) are nested within the units (litters) that were randomly allocated to treatments. This hierarchical structure casts doubt on the condition that the observations on the 224 mice are independent, especially in light of the investigator's comment that the response varied considerably from litter to litter.

10.7.3 $0.001 < p_1 - p_2 < 0.230$. No; the confidence interval suggests that bed rest may actually be harmful.

10.7.5 **(a)** $0.067 < p_1 - p_2 < 0.118$.
(b) We are 95% confident that the proportion of persons with type O blood among ulcer patients is higher than the proportion of persons with type O blood among healthy individuals by between 0.067 and 0.118. That is, we are 95% confident that p_1 exceeds p_2 by between 0.067 and 0.118.

10.8.1 H_0: There is no association between oral contraceptive use and stroke ($p = 0.5$); H_A: There is an association between oral contraceptive use and stroke ($p \neq 0.5$), where p denotes the probability that a discordant pair will be Yes(case)/No(control). $\chi_S^2 = 6.72$. H_0 is rejected. There is sufficient evidence ($0.001 < P\text{-value} < 0.01$) to conclude that stroke victims are more likely to be oral contraceptive users. ($p > 0.5$).

10.9.1 **(a)** (i) 1.339 (ii) 1.356 **(b)** (i) 1.314 (ii) 1.355

10.9.7 **(a)** 1.241 **(b)** (1.036,1.488)
(c) We are 95% confident that taking heparin increases the odds of a negative response by a factor of between 1.036 and 1.488 when compared to taking enoxaparin.

10.S.3 (a) H_0: Sex ratio is 1:1 in warm environment ($p_1 = 0.5$); H_A: Sex ratio is not 1:1 in warm environment ($p_1 \neq 0.5$), where p_1 denotes the probability of a female in the warm environment. $\chi_S^2 = 0.18$. H_0 is not rejected. There is insufficient evidence (P-value > 0.20) to conclude that the sex ratio is not 1:1 in warm environment.

(c) H_0: Sex ratio is the same in the two environments ($p_1 = p_2$); H_A: Sex ratio is not the same in the two environments ($p_1 \neq p_2$), where p denotes the probability of a female and 1 and 2 denote the warm and cold environments. $\chi_S^2 = 4.20$. H_0 is rejected. There is sufficient evidence ($0.02 < P$-value < 0.05) to conclude that the probability of a female is higher in the cold than the warm environment.

10.S.12 H_0: Site of capture and site of recapture are independent (Pr{RI|CI} = Pr{RI|CII}); H_A: Flies preferentially return to their site of capture (Pr{RI|CI} > Pr{RI|CII}), where *C* and *R* denote capture and recapture and I and II denote the sites. H_0 is rejected. There is sufficient evidence ($0.0005 < P$-value < 0.005) to conclude that flies preferentially return to their site of capture.

10.S.14 (a) 1.709 **(b)** $1.55 < \theta < 1.89$

(c) The odds ratio gives the (estimated) odds of survival for men compared to women. This ratio (of 1.709) is a good approximation to the relative risk of death for women compared to men (which is 1.658), because death is fairly rare.

Unit III

III.2 **(a)** $0.025 < P < 0.05$ so reject H_0
(b) There are 6 tables to consider.

III.3 **(a)** $20x32/57 = 11.23$
(b) $\chi^2 = 1.84$, $df = 2$. $P > 0.20 > 0.05$, so we retain H_0.

III.5 **(a)** $n = 49$ **(b)** $n = 66$

III.10 **(a)** True **(b)** False **(c)** False **(d)** True

Chapter 11

11.2.1 (a) SS(between) = 228, SS(within) = 120
(b) SS(total) = 348
(c) MS(between) = 114, MS(within) = 15, $s_{pooled} = 3.87$

11.2.4 (a)

Source	df	SS	MS
Between groups	3	135	45
Within groups	12	337	28.08
Total	15	472	

(b) 4 **(c)** 16

11.4.2 (a) H_0: The stress conditions all produce the same mean lymphocyte concentration ($\mu_1 = \mu_2 = \mu_3 = \mu_4$); H_A: Some of the stress conditions produce different mean lymphocyte concentrations (the μ's are not all equal). $F_s = 3.84$. H_0 is rejected. There is sufficient evidence ($0.01 < P$-value < 0.02) to conclude that some of the stress conditions produce different mean lymphocyte concentrations.
(b) $s_{pooled} = 2.78$ cells/ml $\times 10^{-6}$

11.4.3 (a) H_0: Mean HBE is the same in all three populations
(b) H_0: ($\mu_1 = \mu_2 = \mu_3$)
(c) $F_s = 0.58$. H_0 is not rejected. There is insufficient evidence (P-value > 0.20) to conclude that mean HBE is not the same in all three populations.
(d) $s_{pooled} = 14.4$ pg/ml

11.6.2 There is no single correct answer. One possibility is as follows:

	Piglet				
Treatment	**Litter 1**	**Litter 2**	**Litter 3**	**Litter 4**	**Litter 5**
1	2	5	2	4	5
2	1	4	1	1	2
3	4	2	5	2	4
4	5	3	3	3	3
5	3	1	4	5	1

11.6.5 Plan II is better. We want units within a block to be similar to each other; plan II achieves this. Under plan I the effect of rain could be confounded with the effect of a variety.

11.7.2 (a)

Source	df	SS	MS
Between species	1	2.19781	2.19781
Between flooding levels	1	2.25751	2.25751
Interaction	1	0.097656	0.097656
Within groups	12	0.47438	0.03953
Total	15	5.027356	

(b) There are 1 numerator and 12 denominator degrees of freedom.
(c) We do not reject H_0. There is insufficient evidence ($P = 0.142$) to conclude that there is an interaction present.
(d) $F_s = 2.19781/0.03953 = 55.60$.
(e) H_0 is rejected. There is strong evidence ($P = 0.000008$) to conclude that species affects ATP concentration.
(f) $s_{pooled} = \sqrt{0.03953} = 0.199$.

11.7.4 $F_s = \dfrac{31.33/1}{30648.81/(223 - 4)} = 31.33/139.95 = 0.22.$

With df = 1 and 140, Table 10 gives $F_{0.20} = 1.66$. Thus, P-value > 0.20 and we do not reject H_0. There is insufficient evidence (P-value > 0.20) to conclude that there is an interaction present.

11.8.2 **(a)** 123 mm Hg **(b)** 123.2 mm Hg
(d) 0.851 mm Hg

11.8.7 $0.60 < \mu_E - \mu_S < 1.55$ gm, where $\mu_E = \frac{1}{2}(\mu_{E,\text{Low}} + \mu_{E,\text{High}})$ and $\mu_S = \frac{1}{2}(\mu_{S,\text{Low}} + \mu_{S,\text{High}})$.

11.8.8 **(b)** $L = 3.685$ nmol/10^8 platelets/hour; $SE_L = 1.048$ nmol/10^8 platelets/hour

11.9.1 The following hypotheses are rejected: H_0: $\mu_C = \mu_D$; H_0: $\mu_A = \mu_D$; H_0: $\mu_B = \mu_D$; H_0: $\mu_C = \mu_E$; H_0: $\mu_A = \mu_E$; H_0: $\mu_B = \mu_E$; H_0: $\mu_B = \mu_C$; H_0: $\mu_A = \mu_C$. The following hypotheses are not rejected: H_0: $\mu_A = \mu_B$; H_0: $\mu_D = \mu_E$. Summary:

C A B E D

There is sufficient evidence to conclude that treatments D and E give the largest means, treatments A and B the next largest, and treatment C the smallest. There is insufficient evidence to conclude that treatments A and B give different means or that treatments D and E give different means.

11.9.2 The following hypotheses are not rejected: H_0: $\mu_A = \mu_B$; H_0: $\mu_B = \mu_D$; H_0: $\mu_B = \mu_E$; H_0: $\mu_D = \mu_E$. Summary:

C A B E D

11.9.4 **(a)** Yes, each of diets B, C, and D differs from A, as none of the intervals includes zero.

11.S.1 **(a)** H_0: The three classes produce the same mean change in fat-free mass ($\mu_1 = \mu_2 = \mu_3$); H_A: At least one class produces a different mean (the μ's are not all equal).
(b) $F_s = 0.64$. We do not reject H_0. There is insufficient evidence (P-value > 0.20) to conclude that the population means differ.

11.S.3 H_0: The mean refractive error is the same in the four populations ($\mu_1 = \mu_2 = \mu_3 = \mu_4$); H_A: Some of the populations have different mean refractive errors (the μ's are not all equal). $F_s = 3.56$. H_0 is rejected. There is sufficient evidence ($0.01 < P$-value < 0.02) to conclude that some of the populations have different mean refractive errors.

11.S.13 Let 1, 2, 3, and 4 denote placebo; probucol; multivitamins; and probucol and multivitamins.
(a) $\bar{y}_2 - \bar{y}_1 = 1.79 - 1.43 = 0.36$.
(b) $\bar{y}_4 - \bar{y}_3 = 1.54 - 1.40 = 0.14$.
(c) The contrast that measures the interaction between probucol and multivitamins is "the difference in differences" from parts (a) and (b):

$$(\bar{y}_2 - \bar{y}_1) - (\bar{y}_4 - \bar{y}_3) = 0.36 - 0.14 = 0.22.$$

[Note: This is not the only correct answer; reversing the signs in (a) and (b), or in (c), is also correct.]

Chapter 12

12.2.1 **(a)** (iv), (i), (ii), (iii), (v). (The correlations are −0.97, −0.63, 0.10, 0.58, and 0.93.)
(b) (iv), (iii), (ii)

12.2.2 **(b)** $r = 0.44$

12.2.3 H_0: There is no correlation between blood urea and uric acid concentration ($\rho = 0$); H_A: Blood urea and uric acid concentration are positively correlated ($\rho > 0$). $t_s = 3.952$. H_0 is rejected. There is strong evidence (P-value < 0.0005) to conclude that blood urea and uric acid concentration are positively correlated.

12.2.6 **(a)** H_0: There is no correlation between plant density and mean cob weight ($\rho = 0$); H_A: Plant density and mean cobb weight are correlated ($\rho \neq 0$). $t_s = -11.9$. H_0 is rejected. There is strong evidence (P-value < 0.001) to conclude that plant density and mean cobb weight are negatively correlated.
(b) Observational study
(c) No; this is an observational study in which plant density was observed but not manipulated. The study suggests that density manipulation is worth exploring in a follow-up experiment.

12.3.1 **(b)** $\widehat{Leucine} = -0.05 + 0.02928 \times Time$; the slope is 0.02928 ng/min.
(d) $s_e = 0.0839$.

12.3.3 **(c)** $\hat{y} = -0.592 + 7.641x$; $s_e = 0.881$°C

12.3.6 **(a)** $\hat{y} = 607.7 + 25.01x$
(b)

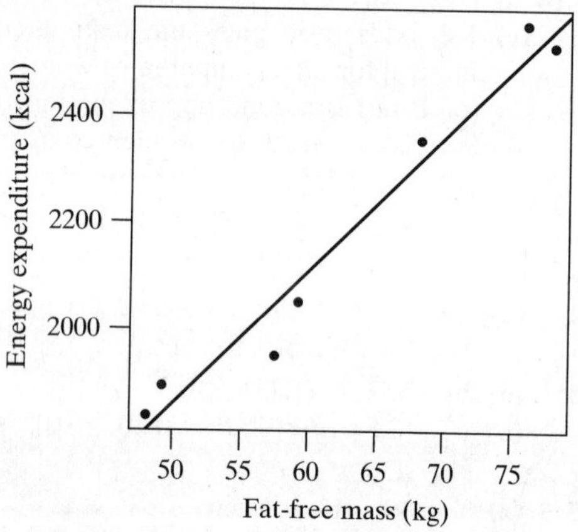

(c) As fat-free mass goes up by 1 kg, energy expenditure goes up by 25.01 kcal, on average.
(d) $s_e = 64.85$ kcal

12.3.10 **(b)** $r^2 = 0.107 = 10.7\%$ **(f)** $12/17 = 71\%$

12.4.5 Estimated mean = 21.1 mm; estimated SD = 1.3 mm

12.4.9 Estimated mean = 658.2 L/min; estimated SD = 115.16 L/min

12.5.1 **(a)** $0.0252 < \beta_1 < 0.0333$ ng/min
(b) We are 95% confident that the rate at which leucine is incorporated into protein in the population of all *Xenopus* oocytes is between 0.0252 ng/min and 0.0333 ng/min.

12.5.5 **(a)** $19.4 < \beta_1 < 30.6$ kcal/kg
(b) $20.6 < \beta_1 < 29.4$ kcal/kg

12.5.7 H_0: There is no linear relationship between respiration rate and altitude of origin ($\beta_1 = 0$); H_A: Trees from higher altitudes tend to have higher respiration rates ($\beta_1 > 0$). $t_s = 6.06$. H_0 is rejected. There is sufficient evidence (P-value < 0.0005) to conclude that trees from higher altitudes tend to have higher respiration rates.

12.6.6 **(a)** – (iii), **(b)** – (i), **(c)** – (ii)

12.7.1 **(a)** The dashed lines, which tell us where the true (population) regression line lies.
(b) The wide band is the prediction band.
(c) With less data both bands would be wider.

12.S.1 0.24 gm

12.S.3 **(a)** Slope = 1.39

12.S.6 **(a)** $s_e = 0.137$ cm
(b) $H_0: \rho = 0$ or $H_0: \beta_1 = 0$. $t_s = 3.01$. H_0 is rejected. There is sufficient evidence ($0.02 < P$-value $<$ 0.04) to conclude that a positive correlation between diameter of forage branch and wing length.

Unit IV

IV.4 0.143.

IV.5 **(a)** $r^2 = 0.23^2 = 0.0529$. We can account for 5.3% of the variability in price by using fiber in a regression model.
(b) $\hat{y} = 17.42 + 0.62 \times 2.63 = 17.42 + 1.63 = 19.05$, so the residual is $17.3 - 19.05 = -1.75$.
(c) The typical size of an error in predicting price is around 3.1 cents/ounce.

IV.7 **(a)** The F statistic is $7.445/7.74 = 0.96$.
(b) There is essentially no evidence that the mean number of tree species per 0.1 hectare plot differs along the three rivers.

(c) We need random samples of independent observations from normally distributed populations that have a common standard deviation.

IV.12 **(a)** $b_1 = -0.9061 \times \left(\frac{5.4176}{11.0090}\right) = -0.4459$,
$b_0 = 41.6858 - (-0.4459 \times 39.4034) = 59.2557$.
$\hat{y} = 59.256 - 0.446x$
(b) $\hat{y} = 59.256 - 0.446 \times 30 = 45.876$
(c) $\mu_{Y|X=30} = 59.256 - 0.446 \times 30 = 45.876$
(d) Since 30mm falls within the range of observed wing lengths, the prediction is an interpolation.
(e) We would expect hummingbird B's wings to beat about 0.446 Hz less (slower) than hummingbird A's.

Chapter 13

13.2.1 A chi-square test of independence would be appropriate. The null hypothesis of interest is $H_0: p_1 = p_2$, where p_1 = Pr{clinically important improvement if given clozapine} and p_2 = Pr{clinically important improvement if given haloperidol}. A confidence interval for $p_1 - p_2$ would also be relevant.

13.2.10 A two-sample comparison is called for here, but the data do not support the condition of normality. Thus, the Wilcoxon-Mann-Whitney test, or a randomization test, is appropriate.

13.2.12 It would be natural to consider correlation and regression with these data. For example, we could regress Y = forearm length on X = height; we could also find the correlation between forearm length and height and test the null hypothesis that the population correlation is zero.

Credits

Chapter 1

[1]Excerpt from LOUIS PASTEUR: THE STORY OF HIS MAJOR DISCOVERIES by Jacques Nicolle. Published by Basic Books, © 1961, **p. 12**; [4]Excerpt from MAO and Schizophrenia by Steven G. Potkin, H. Eleanor Cannon, Dennis L. Murphy, and Richard Jed Wyatt from Are Paranoid Schizophrenics Biologically Different From Other Schizophrenics? from THE NEW ENGLAND JOURNAL OF MEDICINE. Published by Massachusetts Medical Society, © 1978, **p. 14**.

Chapter 2

[33]"Sorbus Aucuparia" by A. M. Barclay and R. M. M. Crawford from Seedling Emergence in the Rowan (Sorbus Aucuparia) From an Altitudinal Gradient. JOURNAL OF ECOLOGY. Published by Blackwell Publishing Ltd, © 1984, **p. 69**.

Chapter 6

[41]"Question on Ferulic Acid" by Ann E. Hagerman, Ralph L. Nicholson from High-Performance Liquid Chromatographic Determination of Hydroxycinnamic Acids in the Maize Mesocotyl. JOURNAL OF AGRICULTURAL AND FOOD CHEMISTRY. Published by American Chemical Society, © 1982, **p. 225**; [58]"Average Human Blood Pressure Calculation" by Paul Erne, Peter Bolli, Ernst Bürgisser, Fritz R. Bühler from Correlation Of Platelet Calcium With Blood Pressure. NEW ENGLAND JOURNAL OF MEDICINE. Copyright © 1984 by Massachusetts Medical Society. Used by permission of Massachusetts Medical Society, **p. 231**.

Chapter 7

[32]"Early Study Of The Relationship Between Diet" by Yerushalmy J, Hilleboe H. E from Fat in the Diet And Mortality From Heart Disease. NEW YORK STATE JOURNAL OF MEDICINE. Copyright © 1957 by The Medical Society of the State of New York. Used by permission of The Medical Society of the State of New York, **p. 266**;[46]Excerpt from Result Of Serum LD In Healthy Young People by Williams G. Z, Widdowson G. M, Penton. J from Individual Character of Variation in Time-Series Studies of Healthy People: II. Differences in Values for Clinical Chemical Analytes in Serum Among Demographic Groups, by Age And Sex from CLINICAL CHEMISTRY. Published by American Association for Clinical Chemistry, Inc, © 1978, **p. 279**; [49]"Gender Determination of Fetus from Heart Rate" by Bruce Petrie, Sidney J. Segalowitz from Use of Fetal Heart Rate, Other Perinatal and Maternal Factors as Predictors of sex. Copyright © 1980 by Ammons Scientific, Ltd. Used by permission of Ammons Scientific, Ltd, **p. 284**; [50]"Measurement of The Concentration of Coumaric Acid In Corn Seedlings" by Ann E. Hagerman, Ralph L. Nicholson from High-Performance Liquid Chromatographic Determination of Hydroxycinnamic Acids in the Maize Mesocotyl. JOURNAL OF AGRICULTURAL AND FOOD CHEMISTRY. Published by American Chemical Society, © 1982, **p. 284**; [71]Excerpt from Investigation on Relationship Between Intracellu-Lar Calcium And Blood Pressure by Paul Erne, Peter Bolli, Ernst Bürgisser, Fritz R. Bühler from Correlation Of Platelet Calcium With Blood Pressure from NEW ENGLAND JOURNAL OF MEDICINE. Published by Massachusetts Medical Society, © 1984, **p. 310**.

Unit III

[4]Excerpt From Brain-To-Brain Interface Allows Transmission of Tactile And Motor Information Between Rats by Duke Medicine News and Communications, Scientific Reports 3, 1319. Copyright © Feb 28, 2013 by Duke University Health System. Used by Permission of Duke University Health System, **p. 447**.

INDEX

D

E

F

INDEX OF EXAMPLES

Table 1 Areas Under the Normal Curve

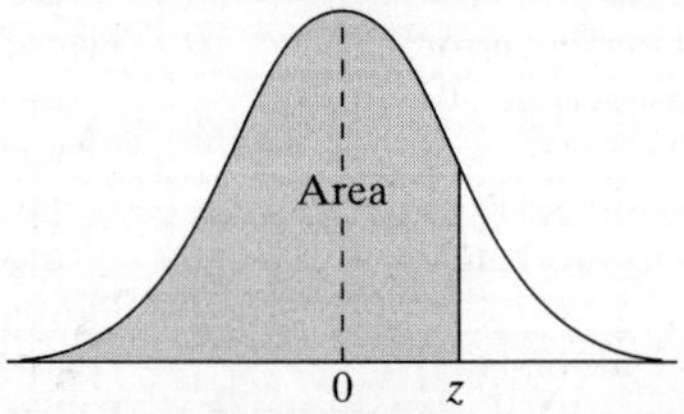

z	.00	.01	.02	.03	.04	.05	.06	.07	.08	.09
–3.4	0.0003	0.0003	0.0003	0.0003	0.0003	0.0003	0.0003	0.0003	0.0003	0.0002
–3.3	0.0005	0.0005	0.0005	0.0004	0.0004	0.0004	0.0004	0.0004	0.0004	0.0003
–3.2	0.0007	0.0007	0.0006	0.0006	0.0006	0.0006	0.0006	0.0005	0.0005	0.0005
–3.1	0.0010	0.0009	0.0009	0.0009	0.0008	0.0008	0.0008	0.0008	0.0007	0.0007
–3.0	0.0013	0.0013	0.0013	0.0012	0.0012	0.0011	0.0011	0.0011	0.0010	0.0010
–2.9	0.0019	0.0018	0.0017	0.0017	0.0016	0.0016	0.0015	0.0015	0.0014	0.0014
–2.8	0.0026	0.0025	0.0024	0.0023	0.0023	0.0022	0.0021	0.0021	0.0020	0.0019
–2.7	0.0035	0.0034	0.0033	0.0032	0.0031	0.0030	0.0029	0.0028	0.0027	0.0026
–2.6	0.0047	0.0045	0.0044	0.0043	0.0041	0.0040	0.0039	0.0038	0.0037	0.0036
–2.5	0.0062	0.0060	0.0059	0.0057	0.0055	0.0054	0.0052	0.0051	0.0049	0.0048
–2.4	0.0082	0.0080	0.0078	0.0075	0.0073	0.0071	0.0069	0.0068	0.0066	0.0064
–2.3	0.0107	0.0104	0.0102	0.0099	0.0096	0.0094	0.0091	0.0089	0.0087	0.0084
–2.2	0.0139	0.0136	0.0132	0.0129	0.0125	0.0122	0.0119	0.0116	0.0113	0.0110
–2.1	0.0179	0.0174	0.0170	0.0166	0.0162	0.0158	0.0154	0.0150	0.0146	0.0143
–2.0	0.0228	0.0222	0.0217	0.0212	0.0207	0.0202	0.0197	0.0192	0.0188	0.0183
–1.9	0.0287	0.0281	0.0274	0.0268	0.0262	0.0256	0.0250	0.0244	0.0239	0.0233
–1.8	0.0359	0.0352	0.0344	0.0336	0.0329	0.0322	0.0314	0.0307	0.0301	0.0294
–1.7	0.0446	0.0436	0.0427	0.0418	0.0409	0.0401	0.0392	0.0384	0.0375	0.0367
–1.6	0.0548	0.0537	0.0526	0.0516	0.0505	0.0495	0.0485	0.0475	0.0465	0.0455
–1.5	0.0668	0.0655	0.0643	0.0630	0.0618	0.0606	0.0594	0.0582	0.0571	0.0559
–1.4	0.0808	0.0793	0.0778	0.0764	0.0749	0.0735	0.0722	0.0708	0.0694	0.0681
–1.3	0.0968	0.0951	0.0934	0.0918	0.0901	0.0885	0.0869	0.0853	0.0838	0.0823
–1.2	0.1151	0.1131	0.1112	0.1093	0.1075	0.1056	0.1038	0.1020	0.1003	0.0985
–1.1	0.1357	0.1335	0.1314	0.1292	0.1271	0.1251	0.1230	0.1210	0.1190	0.1170
–1.0	0.1587	0.1562	0.1539	0.1515	0.1492	0.1469	0.1446	0.1423	0.1401	0.1379
–0.9	0.1841	0.1814	0.1788	0.1762	0.1736	0.1711	0.1685	0.1660	0.1635	0.1611
–0.8	0.2119	0.2090	0.2061	0.2033	0.2005	0.1977	0.1949	0.1922	0.1894	0.1867
–0.7	0.2420	0.2389	0.2358	0.2327	0.2296	0.2266	0.2236	0.2206	0.2177	0.2148
–0.6	0.2743	0.2709	0.2676	0.2643	0.2611	0.2578	0.2546	0.2514	0.2483	0.2451
–0.5	0.3085	0.3050	0.3015	0.2981	0.2946	0.2912	0.2877	0.2843	0.2810	0.2776
–0.4	0.3446	0.3409	0.3372	0.3336	0.3300	0.3264	0.3228	0.3192	0.3156	0.3121
–0.3	0.3821	0.3783	0.3745	0.3707	0.3669	0.3632	0.3594	0.3557	0.3520	0.3483
–0.2	0.4207	0.4168	0.4129	0.4090	0.4052	0.4013	0.3974	0.3936	0.3897	0.3859
–0.1	0.4602	0.4562	0.4522	0.4483	0.4443	0.4404	0.4364	0.4325	0.4286	0.4247
–0.0	0.5000	0.4960	0.4920	0.4880	0.4840	0.4801	0.4761	0.4721	0.4681	0.4641

Table I Areas Under the Normal Curve *(continued)*

z	.00	.01	.02	.03	.04	.05	.06	.07	.08	.09
0.0	0.5000	0.5040	0.5080	0.5120	0.5160	0.5199	0.5239	0.5279	0.5319	0.5359
0.1	0.5398	0.5438	0.5478	0.5517	0.5557	0.5596	0.5636	0.5675	0.5714	0.5753
0.2	0.5793	0.5832	0.5871	0.5910	0.5948	0.5987	0.6026	0.6064	0.6103	0.6141
0.3	0.6179	0.6217	0.6255	0.6293	0.6331	0.6368	0.6406	0.6443	0.6480	0.6517
0.4	0.6554	0.6591	0.6628	0.6664	0.6700	0.6736	0.6772	0.6808	0.6844	0.6879
0.5	0.6915	0.6950	0.6985	0.7019	0.7054	0.7088	0.7123	0.7157	0.7190	0.7224
0.6	0.7257	0.7291	0.7324	0.7357	0.7389	0.7422	0.7454	0.7486	0.7517	0.7549
0.7	0.7580	0.7611	0.7642	0.7673	0.7704	0.7734	0.7764	0.7794	0.7823	0.7852
0.8	0.7881	0.7910	0.7939	0.7967	0.7995	0.8023	0.8051	0.8078	0.8106	0.8133
0.9	0.8159	0.8186	0.8212	0.8238	0.8264	0.8289	0.8315	0.8340	0.8365	0.8389
1.0	0.8413	0.8438	0.8461	0.8485	0.8508	0.8531	0.8554	0.8577	0.8599	0.8621
1.1	0.8643	0.8665	0.8686	0.8708	0.8729	0.8749	0.8770	0.8790	0.8810	0.8830
1.2	0.8849	0.8869	0.8888	0.8907	0.8925	0.8944	0.8962	0.8980	0.8997	0.9015
1.3	0.9032	0.9049	0.9066	0.9082	0.9099	0.9115	0.9131	0.9147	0.9162	0.9177
1.4	0.9192	0.9207	0.9222	0.9236	0.9251	0.9265	0.9278	0.9292	0.9306	0.9319
1.5	0.9332	0.9345	0.9357	0.9370	0.9382	0.9394	0.9406	0.9418	0.9429	0.9441
1.6	0.9452	0.9463	0.9474	0.9484	0.9495	0.9505	0.9515	0.9525	0.9535	0.9545
1.7	0.9554	0.9564	0.9573	0.9582	0.9591	0.9599	0.9608	0.9616	0.9625	0.9633
1.8	0.9641	0.9649	0.9656	0.9664	0.9671	0.9678	0.9686	0.9693	0.9699	0.9706
1.9	0.9713	0.9719	0.9726	0.9732	0.9738	0.9744	0.9750	0.9756	0.9761	0.9767
2.0	0.9772	0.9778	0.9783	0.9788	0.9793	0.9798	0.9803	0.9808	0.9812	0.9817
2.1	0.9821	0.9826	0.9830	0.9834	0.9838	0.9842	0.9846	0.9850	0.9854	0.9857
2.2	0.9861	0.9864	0.9868	0.9871	0.9875	0.9878	0.9881	0.9884	0.9887	0.9890
2.3	0.9893	0.9896	0.9898	0.9901	0.9904	0.9906	0.9909	0.9911	0.9913	0.9916
2.4	0.9918	0.9920	0.9922	0.9925	0.9927	0.9929	0.9931	0.9932	0.9934	0.9936
2.5	0.9938	0.9940	0.9941	0.9943	0.9945	0.9946	0.9948	0.9949	0.9951	0.9952
2.6	0.9953	0.9955	0.9956	0.9957	0.9959	0.9960	0.9961	0.9962	0.9963	0.9964
2.7	0.9965	0.9966	0.9967	0.9968	0.9969	0.9970	0.9971	0.9972	0.9973	0.9974
2.8	0.9974	0.9975	0.9976	0.9977	0.9977	0.9978	0.9979	0.9979	0.9980	0.9981
2.9	0.9981	0.9982	0.9982	0.9983	0.9984	0.9984	0.9985	0.9985	0.9986	0.9986
3.0	0.9987	0.9987	0.9987	0.9988	0.9988	0.9989	0.9989	0.9989	0.9990	0.9990
3.1	0.9990	0.9991	0.9991	0.9991	0.9992	0.9992	0.9992	0.9992	0.9993	0.9993
3.2	0.9993	0.9993	0.9994	0.9994	0.9994	0.9994	0.9994	0.9995	0.9995	0.9995
3.3	0.9995	0.9995	0.9995	0.9996	0.9996	0.9996	0.9996	0.9996	0.9996	0.9997
3.4	0.9997	0.9997	0.9997	0.9997	0.9997	0.9997	0.9997	0.9997	0.9997	0.9998

Table 2 Critical Values of Student's *t* distribution

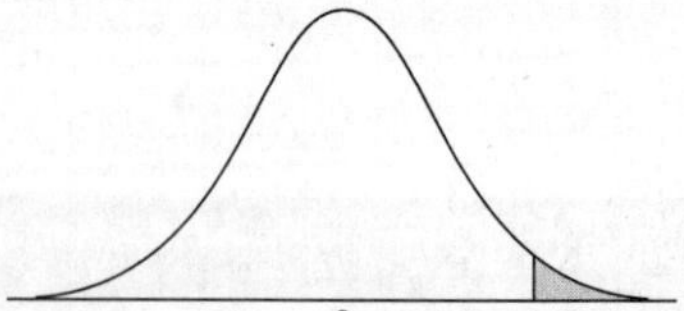

df	UPPER TAIL PROBABILITY									
	0.20	0.10	0.05	0.04	0.03	0.025	0.02	0.01	0.005	0.0005
1	1.376	3.078	6.314	7.916	10.579	12.706	15.895	31.821	63.657	636.619
2	1.061	1.886	2.920	3.320	3.896	4.303	4.849	6.965	9.925	31.599
3	0.978	1.638	2.353	2.605	2.951	3.182	3.482	4.541	5.841	12.924
4	0.941	1.533	2.132	2.333	2.601	2.776	2.999	3.747	4.604	8.610
5	0.920	1.476	2.015	2.191	2.422	2.571	2.757	3.365	4.032	6.869
6	0.906	1.440	1.943	2.104	2.313	2.447	2.612	3.143	3.707	5.959
7	0.896	1.415	1.895	2.046	2.241	2.365	2.517	2.998	3.499	5.408
8	0.889	1.397	1.860	2.004	2.189	2.306	2.449	2.896	3.355	5.041
9	0.883	1.383	1.833	1.973	2.150	2.262	2.398	2.821	3.250	4.781
10	0.879	1.372	1.812	1.948	2.120	2.228	2.359	2.764	3.169	4.587
11	0.876	1.363	1.796	1.928	2.096	2.201	2.328	2.718	3.106	4.437
12	0.873	1.356	1.782	1.912	2.076	2.179	2.303	2.681	3.055	4.318
13	0.870	1.350	1.771	1.899	2.060	2.160	2.282	2.650	3.012	4.221
14	0.868	1.345	1.761	1.888	2.046	2.145	2.264	2.624	2.977	4.140
15	0.866	1.341	1.753	1.878	2.034	2.131	2.249	2.602	2.947	4.073
16	0.865	1.337	1.746	1.869	2.024	2.120	2.235	2.583	2.921	4.015
17	0.863	1.333	1.740	1.862	2.015	2.110	2.224	2.567	2.898	3.965
18	0.862	1.330	1.734	1.855	2.007	2.101	2.214	2.552	2.878	3.922
19	0.861	1.328	1.729	1.850	2.000	2.093	2.205	2.539	2.861	3.883
20	0.860	1.325	1.725	1.844	1.994	2.086	2.197	2.528	2.845	3.850
21	0.859	1.323	1.721	1.840	1.988	2.080	2.189	2.518	2.831	3.819
22	0.858	1.321	1.717	1.835	1.983	2.074	2.183	2.508	2.819	3.792
23	0.858	1.319	1.714	1.832	1.978	2.069	2.177	2.500	2.807	3.768
24	0.857	1.318	1.711	1.828	1.974	2.064	2.172	2.492	2.797	3.745
25	0.856	1.316	1.708	1.825	1.970	2.060	2.167	2.485	2.787	3.725
26	0.856	1.315	1.706	1.822	1.967	2.056	2.162	2.479	2.779	3.707
27	0.855	1.314	1.703	1.819	1.963	2.052	2.158	2.473	2.771	3.690
28	0.855	1.313	1.701	1.817	1.960	2.048	2.154	2.467	2.763	3.674
29	0.854	1.311	1.699	1.814	1.957	2.045	2.150	2.462	2.756	3.659
30	0.854	1.310	1.697	1.812	1.955	2.042	2.147	2.457	2.750	3.646
40	0.851	1.303	1.684	1.796	1.936	2.021	2.123	2.423	2.704	3.551
50	0.849	1.299	1.676	1.787	1.924	2.009	2.109	2.403	2.678	3.496
60	0.848	1.296	1.671	1.781	1.917	2.000	2.099	2.390	2.660	3.460
70	0.847	1.294	1.667	1.776	1.912	1.994	2.093	2.381	2.648	3.435
80	0.846	1.292	1.664	1.773	1.908	1.990	2.088	2.374	2.639	3.416
100	0.845	1.290	1.660	1.769	1.902	1.984	2.081	2.364	2.626	3.390
140	0.844	1.288	1.656	1.763	1.896	1.977	2.073	2.353	2.611	3.361
1000	0.842	1.282	1.646	1.752	1.883	1.962	2.056	2.330	2.581	3.300
∞	0.842	1.282	1.645	1.751	1.881	1.960	2.054	2.326	2.576	3.291
	60%	80%	90%	92%	94%	95%	96%	98%	99%	99.9%
	CRITICAL VALUE FOR CONFIDENCE LEVEL									

Table 3 Critical Values of the Chi-Square Distribution

Note: Column headings are non-directional (omni-directional) *P*-values. If H_A is directional (which is only possible when df = 1), the directional *P*-values are found by dividing the column headings in half.

	TAIL PROBABILITY						
df	0.20	0.10	0.05	0.02	0.01	0.001	0.0001
1	1.64	2.71	3.84	5.41	6.63	10.83	15.14
2	3.22	4.61	5.99	7.82	9.21	13.82	18.42
3	4.64	6.25	7.81	9.84	11.34	16.27	21.11
4	5.99	7.78	9.49	11.67	13.28	18.47	23.51
5	7.29	9.24	11.07	13.39	15.09	20.51	25.74
6	8.56	10.64	12.59	15.03	16.81	22.46	27.86
7	9.80	12.02	14.07	16.62	18.48	24.32	29.88
8	11.03	13.36	15.51	18.17	20.09	26.12	31.83
9	12.24	14.68	16.92	19.68	21.67	27.88	33.72
10	13.44	15.99	18.31	21.16	23.21	29.59	35.56
11	14.63	17.28	19.68	22.62	24.72	31.26	37.37
12	15.81	18.55	21.03	24.05	26.22	32.91	39.13
13	16.98	19.81	22.36	25.47	27.69	34.53	40.87
14	18.15	21.06	23.68	26.87	29.14	36.12	42.58
15	19.31	22.31	25.00	28.26	30.58	37.70	44.26
16	20.47	23.54	26.30	29.63	32.00	39.25	45.92
17	21.61	24.77	27.59	31.00	33.41	40.79	47.57
18	22.76	25.99	28.87	32.35	34.81	42.31	49.19
19	23.90	27.20	30.14	33.69	36.19	43.82	50.80
20	25.04	28.41	31.41	35.02	37.57	45.31	52.39
21	26.17	29.62	32.67	36.34	38.93	46.80	53.96
22	27.30	30.81	33.92	37.66	40.29	48.27	55.52
23	28.43	32.01	35.17	38.97	41.64	49.73	57.08
24	29.55	33.20	36.42	40.27	42.98	51.18	58.61
25	30.68	34.38	37.65	41.57	44.31	52.62	60.14
26	31.79	35.56	38.89	42.86	45.64	54.05	61.66
27	32.91	36.74	40.11	44.14	46.96	55.48	63.16
28	34.03	37.92	41.34	45.42	48.28	56.89	64.66
29	35.14	39.09	42.56	46.69	49.59	58.30	66.15
30	36.25	40.26	43.77	47.96	50.89	59.70	67.63